Muscles of the Shoulder and Arm 1

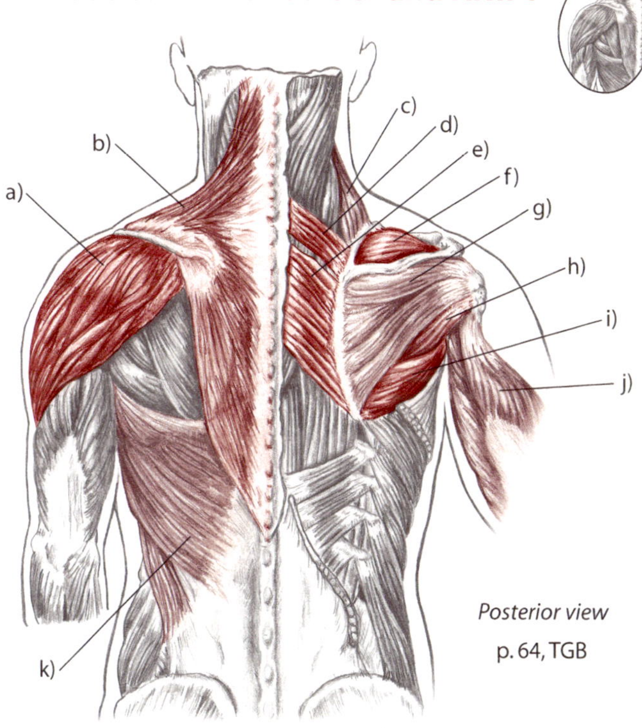

Posterior view

p. 64, TGB

a) Deltoid
b) Trapezius
c) Levator scapula
d) Rhomboid minor
e) Rhomboid major
f) Supraspinatus
g) Infraspinatus
h) Teres minor
i) Teres major
j) Triceps brachii
k) Latissimus dorsi

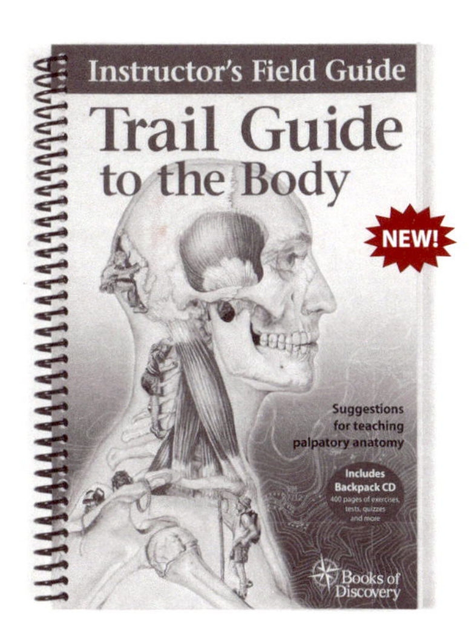

■ ***NEW! Instructor's Field Guide and Backpack (Quiz and Test Bank) CD:***
In response to instructor requests, we created this 120-page Field Guide to assist teachers who are using *Trail Guide* in the classroom. Whether you are a new instructor, an experienced instructor looking for fresh ideas, or are just becoming familiar with *Trail Guide to the Body*, this will be an invaluable resource for you. This Field Guide follows the chapters and structures in *Trail Guide*, providing you with an easy-to-follow template and optional teaching elements.

The Instructor's Backpack CD (included with the Instructor's Field Guide) is stuffed with quizzes, fill-in illustrations, take-home assignments, word finds, crossword puzzles and more! 400+ pages, 900 illustrations. These items are FREE to institutions that require the *Trail Guide to the Body* textbook!

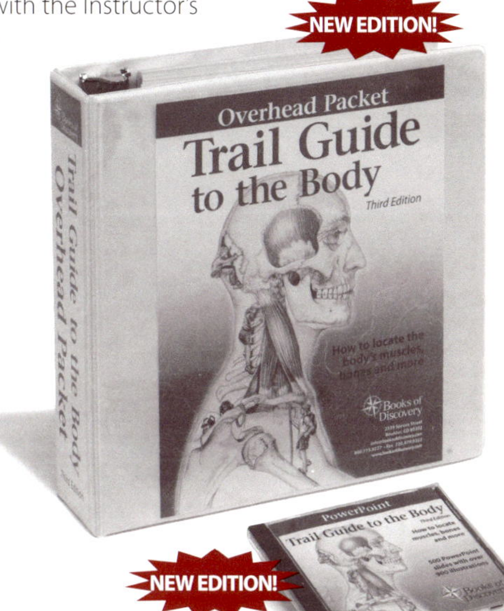

■ ***Visual Aids:*** Available in PowerPoint or Overhead format, these products feature illustrations and text to assist instructors in the hands-on, kinesthetic part of class. The Overhead Packet contains 750 illustrations on 234 transparencies, and the PowerPoint features more than 900 illustrations on 500 slides.

■ ***Textbook/Student Handbook/ Flashcard Set Combo:***
Purchase the 3rd Edition of the *Trail Guide to the Body* and get 15% off the Student Handbook and the Volume I and Volume II flashcards.

■ ***Textbook/Flashcard Set Combo:***
Purchase the 3rd Edition of the *Trail Guide to the Body* and get 15% off the Volume I and Volume II flashcards.

■ ***Textbook/Handbook Combo:***
Purchase the 3rd Edition of the *Trail Guide to the Body* textbook and get 15% off the Student Handbook.

Combos available for U.S. retail customers only.

To order call 800-775-9227 or order online at www.booksofdiscovery.com

Wholesale and student discounts are available. See our website or call for pricing.

International Distributors

United Kingdom/Europe
Ultimate Massage Solutions
www.ultimatemassagesolutions.com
44-28-9059-0594 • jearls@eim.dnet.co.uk

Australia
Akasha (Australia) Distributors
0409-669-531 • info@akasha.net.au

New Zealand
Akasha Books Limited - New Zealand
www.akasha.co.nz
64-4-296-1551 • info@akasha.co.nz

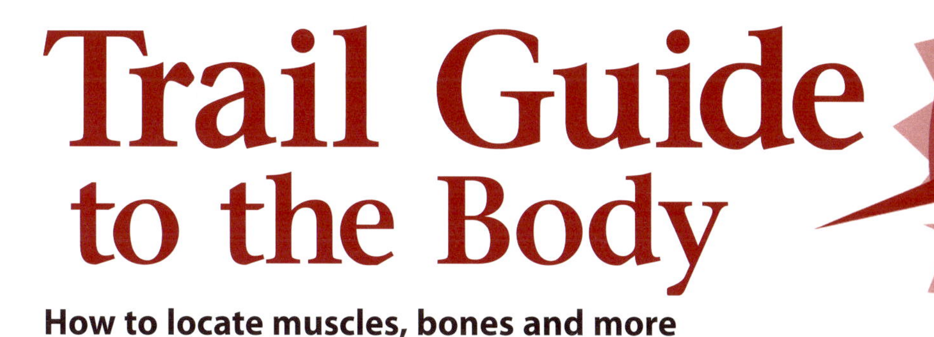

Trail Guide to the Body

How to locate muscles, bones and more

Third Edition

Andrew Biel, LMP
Licensed Massage Practitioner

Illustrations by Robin Dorn, LMP
Licensed Massage Practitioner

Third Edition

Published by Books of Discovery
2539 Spruce St., Boulder, CO 80302 USA
www.booksofdiscovery.com
info@booksofdiscovery.com
800.775.9227

Associate Editors:
Marty Ryan, LMP
Clint Chandler, LMP
Kate Bromley, MA, LMP
Lauriann Greene, LMP

Graphic coloring by Rupert Grange, Esq.
Printed in Canada by Printcrafters, Winnipeg

Library of Congress Cataloging-in-Publication Data

Biel, Andrew R.
Trail Guide to the Body: How to locate muscles, bones and more
Third Edition

Includes bibliographical references.
Includes index.

ISBN: 0-9658534-5-4
Library of Congress Control Number: 2005902119

15 14 13 12 11 10 9 8 7 6 5 4

Grateful acknowledgment is
made to reprint an excerpt from:

The Magic Mountain by Thomas Mann
Copyright © 1927. Used by permission
of Random House, a division of Alfred Knopf, Inc.

Four Quartets by T.S. Eliot
Copyright © 1943. Used by permission
of Harcourt Brace & Company

Disclaimer
The purpose of Books of Discovery's products is to provide information for hands-on therapists on the subject of palpatory anatomy. This book does not offer medical advice to the reader and is not intended as a replacement for appropriate healthcare and treatment. For such advice, readers should consult a licensed physician.

Table of Contents

Table of Contents

Preface

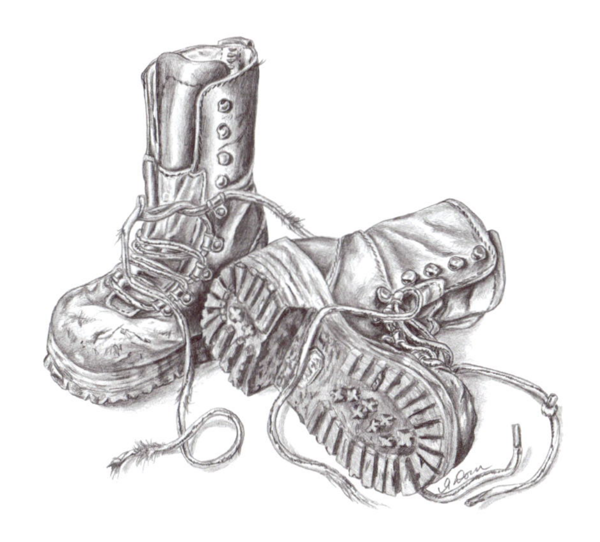

We shall not cease from exploration.
And the end of all our exploring
Will be to arrive where we started
And know the place for the first time.

T.S. Eliot, *Four Quartets*

Many years ago, as a skinny ten-year old, I remember pinching the flesh under my armpit only to accidentally locate a muscle. When I moved my arm in a certain way, the flesh would harden and slip into my fingers. "Wow," I thought, "I didn't think I *had* any muscles!"

I told my parents about my discovery, and they suggested that I check the encyclopedia to see which muscle I had found. The Latin names I encountered only confused me, but for months I showed everyone I met my one and only muscle.

I continued to be fascinated with the parts and pieces of the body and with how these all seemed to work together to produce movement, breath, even life itself. During my training as a bodyworker, I learned that the mysterious muscle of my armpit was the *latissimus dorsi*. Soon I learned how to palpate other muscles as well as the various tendons, bones and tissues located throughout the body. I also realized the importance of palpation for tissue assessment and for performing safe and effective manual therapy techniques.

Later, as an instructor of bodywork and palpatory anatomy, I became familiar with many books describing and illustrating the anatomy of the body. I found few, however, that demonstrated how to locate and explore the body's structures manually. *Trail Guide to the Body* is designed to do just that: to teach you to map, navigate and "gain your bearings" on the human body.

In preparation for any journey, it helps if you know the lay of the land you will be traveling. For every healthcare provider, a thorough understanding of the location and interrelationship of the body's structures is essential. The "hands-on" practitioner, however, cannot merely take a guided bus tour of the body, viewing it from afar and only hearing of its amazing qualities. She must undertake instead the actual/physical exploration through a geography that is never exactly the same on any two individuals. Rolling up her sleeves, she must rely on her hands and her senses to learn about the most challenging and fascinating of all terrains - the human body.

So welcome! You are about to embark on the journey of a lifetime with this book as your trusty guide.

Acknowledgments

The long and winding path of creativity is often strewn with boulders, lacking in signposts and intersected by dead end trails. Luckily my path was cleared by the sharp machetes and skilled help of many expert field guides and hiking partners.

It is always a pleasure to work with an artist as committed and talented as Robin Dorn. A heartfelt thanks to Lyn Gregory for her encouragement, patience and suggestions, and Marty Ryan for his editing, numerous ideas and voice of confidence.

I was blessed to have a wonderful support team for the third edition: Many thanks to Jessica Xavier for her design concepts and continual patience, Dana Ecklund for his persistance and sharp eye, Melinda Helmick for holding down the fort and the dedicated staff at Books of Discovery - Rhoni Hirst, Jeni Breezley, BJ Conway, Linda Giandinoto, Sean Griffin, Teal Meiling, Linda Lee, Christopher Westfall and Audra King.

Thank you to Joan E. Ryan, LMT, MD, Aaron Adams, Ashley Bechel, Miranda Legge, Christine Malles, Gene Martinez and Mindy Morton for their proofreading and editorial suggestions.

Many thanks for the patience of Jennifer Spinelli, Jason Glunt, Alex Gregory, Johanna Kasten, Shane Nicholsen, Steve Snyder, Nathan Musselman, Holadia, David Mason and Matt Samet for their help with the modeling and photography.

Thanks also to Chris Grauch, Christina Goehrig, Jessica Basamanowicz, Kendra Busby, Kathryn Dean, Kathy Eike, Jean Marie Fay, Joanna Gardner, Nicholas Hammersley, Anne Hartshorn, Meghan Heath, Carrie Henderson, Mary Lynn Jackson, Kimberly Kiriaki, Elizabeth Milliken, Rama Newton, Thea Satrom, Sare Selko, Penelope Thompson, Jaime Tousignant and Ashley Wilson.

Special thanks to Roger Williams and Martha Austen for their tremendous support of Robin.

Continued thanks to Jennifer 'JJ' Booksh, Kate Bromley, Clint Chandler, Claire Gipson, Lauriann Greene, Robert Karman, Chris Maisto, Jackie Phillips, Anthony Sayre, Diana Thompson, Summer Westfall and the entire staff at Printcrafters.

I am very grateful to the following people for their expertise, research and encouragement: Leon Chaitow, Sandy Fritz, Darlene Hertling, John White, Sharon Babcock, Cynthia Christy, Ann Ekes, Barb Frye, Daniel Gebo, Jim Holland, George C. Kent, Don Kelley, Lee Haines, Mary Marzke, Susan Parke, Annie Thoe, Jeannie Waschow and John Zurhourek.

Thank you also to Jamie Alagna, Adam Bailey, Nancy Benerofe, Alexis Brereton, Deb Brockman, Mary Bryan, Patrick Bufi, Sylvia Burns, Kirk Butler, Sean Castor, Thomas Crown, Jessica Elliott, Vicky Fosie, Dawn Fosse, Joanne Fowler, Gaye Franklin, Steve Goldstein, Laura Goularte, Alyce Green-Davis, Leslie Grounds, Joanne Guidici, Petra Guyer, Debra Harrison, Chad Herrin, Llysa Holland, Ian Hubner, Melissa Iverson, Leslie Jowett, Diana Kincaid, Alison Kim, Erica King, Elinore Knutson, Beth Langston, Dave Lawrence, Andrew Litzky, Kate McConnell, Sean McDaniel, Becky Masters, Micheal Max, Audra Meador, Chris Meier, Sandy Merrell, Steve Miller, Debra Nelli, Eric Newberg, Sally Nurney, Dave Oder, Jillian Orton, Vicky Panzeri, Paula Pelletier, Anita Quinton, Dee Reeder, Coleen Renee, Obie Roe, Penny Rosen, Dawn Schmidt, Janice Schwartz, Gerald Sexton, Joy Shaw, Danny Tseng, Zdenka Vargas, Brian Weyand, Damon Williams, Cynthia Wold, Tonya Yuricich and Pantelis Zafiriou.

Special thanks to my family for their support and encouragement. The third edition of *Trail Guide to the Body* is dedicated to the students of bodywork and manual therapy around the world - past, present and future.

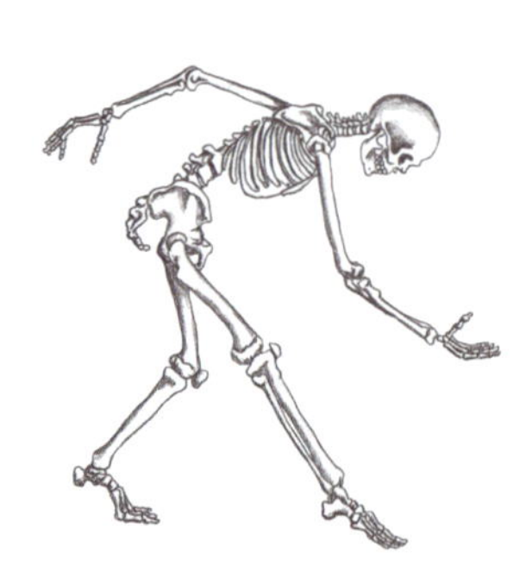

Introduction

Tour Guide Tips

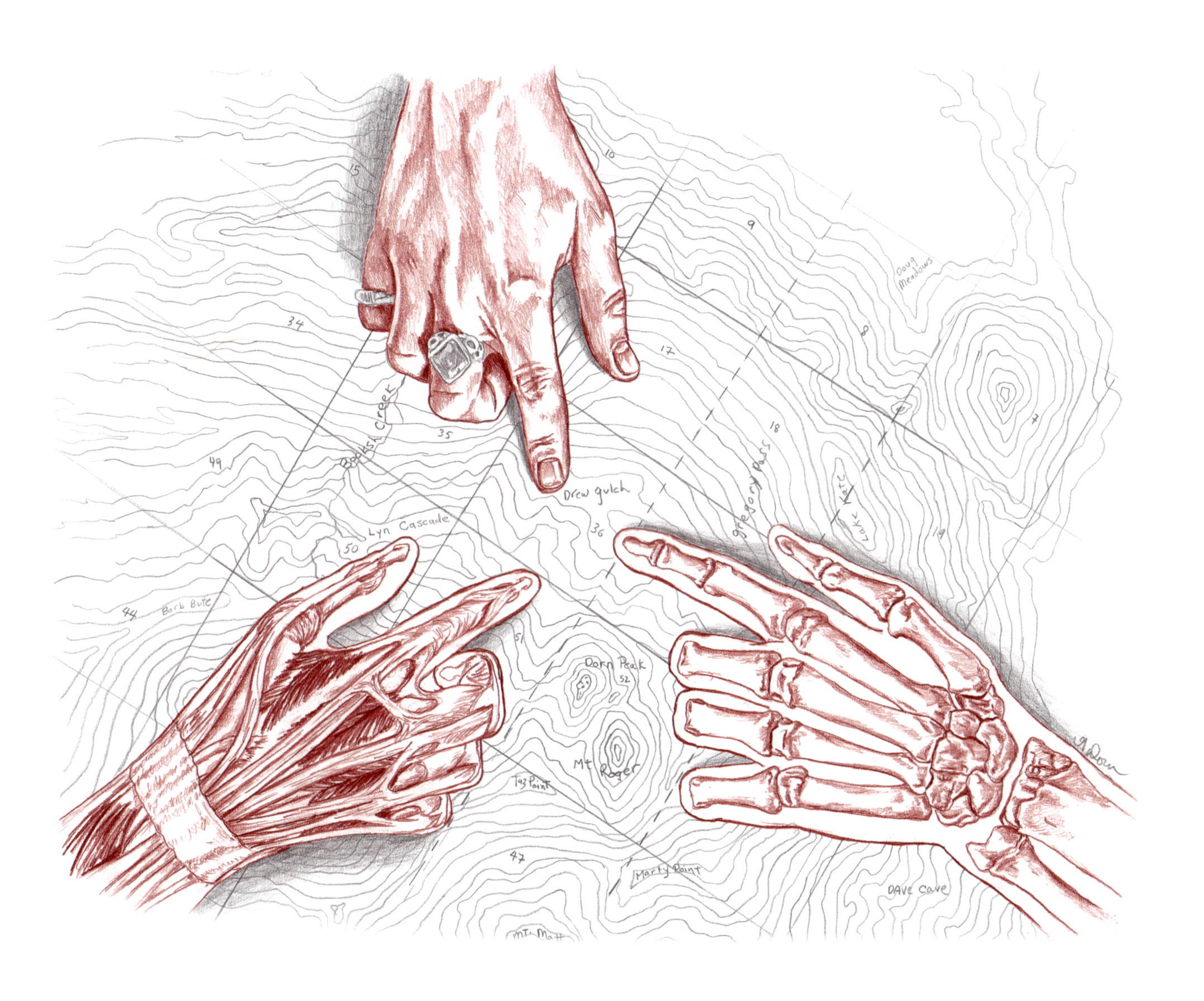

How To Use This Book

Trail Guide to the Body has seven chapters, six of which focus on a different region of the body. The topographical contours that can be seen on the surface of the skin and exercises to explore the skin and fascia are outlined first. These are followed by the bones and bony landmarks (the bone's hills, dips and ridges). The bony landmarks can be thought of as "trail markers." They are used as stepping off points to locate muscles and tendons. Finally, other structures, such as ligaments, nerves, arteries and lymph nodes, are accessed.

Wherever possible, a region's bony landmarks have been strung together to form a trail (0.1). These trails are designed to help you understand the connections between structures. Without a path to follow, you, the traveler, would be lost in a jungle of flesh and bones with no idea of your trail's location. You and your travel partner will find the journey more enjoyable and valuable if you have a trail to lead you to your destination point.

Since bodies come in a variety of sizes and shapes, it may seem unrealistic that one trail guide could apply to all of them. If the terrain is never the same, what is the use of a map? Even though the topography, shape and proportion of each person are unique, the body's composition and structures are virtually identical on all individuals. The differences are simply qualitative: It is easy to find many structures on a person with a slender build and more challenging on a physique with bulky muscles or a large amount of adipose (fatty) tissue (0.2).

Trail Guide to the Body is designed around the following scenario: You follow along with the text and palpate on a partner (friend or classmate) who is on a bodywork table or seated in a chair. If you are a student, you are advised to proceed step-by-step, repeat certain methods when necessary, and explore the body along the way. If you are a more experienced practitioner, you may want to pick and choose your destinations.

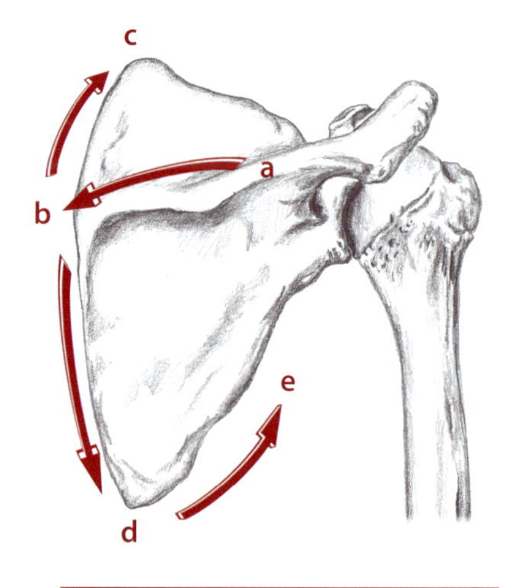

(0.1) A bony landmark trail of the shoulder

- **a** Spine of the scapula
- **b** Medial border
- **c** Superior angle
- **d** Inferior angle
- **e** Lateral border

The procedures outlined in *Trail Guide to the Body* are gentle and rarely uncomfortable, yet it is best to practice on an individual with no serious health conditions. Your partner may either wear loose, thin clothing or be undressed and draped under a sheet to enable you to palpate more easily.

Sometimes your partner will be asked to lie or sit passively on the table. At other times, she may be asked to move a limb, bend a joint or contract a group of muscles. These movements should be done smoothly and according to the specific instructions of the text to enable you to explore the region thoroughly.

Talk to your partner before palpating so she will understand her role. Also, clarify beforehand which areas of the body you would like to palpate and explore so she will know what to expect.

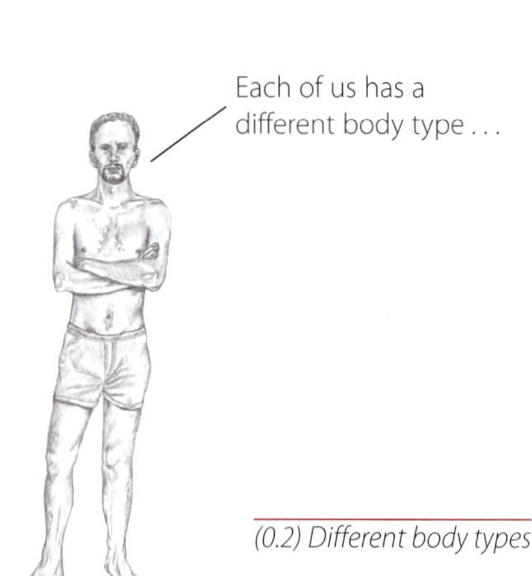

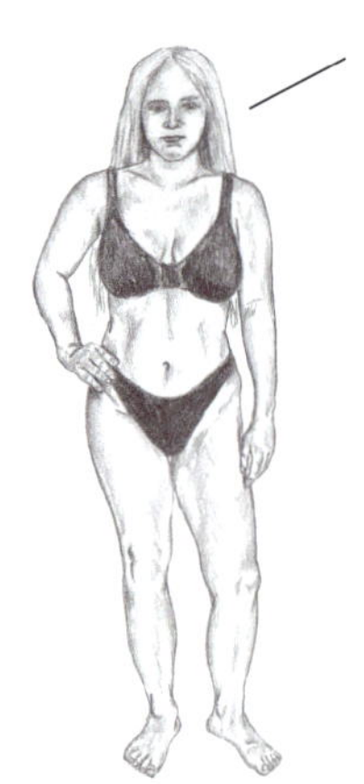

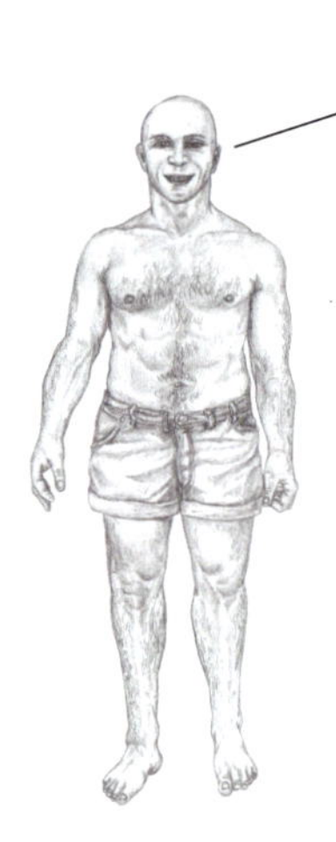

(0.2) Different body types

Trail Guide to the Body

Order and simplification are the first steps toward the mastery of a subject - the actual enemy is the unknown.

Thomas Mann, *The Magic Mountain*

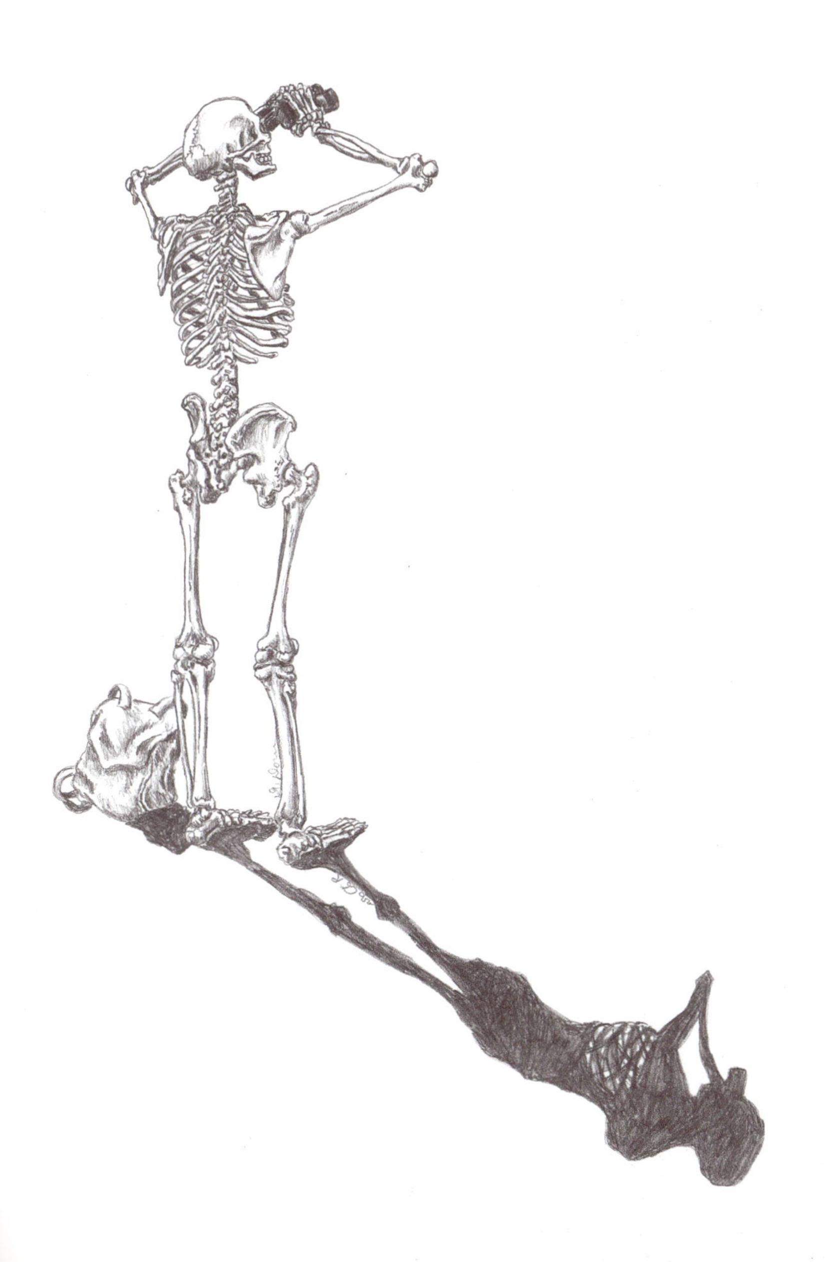

Books of Discovery brings healthcare providers around the world the very best in palpatory and anatomical information. We are committed to providing quality educational products for students, instructors and practitioners.

Over the past seven years I have often remarked that our products offer more than "how to find muscles, bones and more." Our books, study aids and flashcards help a practitioner locate structures so they can apply their skills accurately.

It will always make us feel proud knowing that a mother in Portland, Maine or an athlete in Sarasota, Florida or even a bricklayer in Liverpool, England is receiving quality bodywork from a practitioner who has benefited from our products.

Andrew Biel

Trail Guide to the Body is produced with vegetable-based ink and paper composed of 40% recycled stock and 10% post-consumer waste. Books of Discovery donates a portion of its profits to reforestation projects. Please visit www.yourtruenature.com for more information.

Key

Name of structure

Introduction describing a structure's function, depth and relationship to other structures

A list of the **A**ction, **O**rigin, **I**nsertion and **N**erve innervation of the muscle

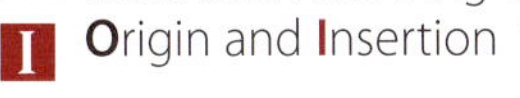

O **I** Illustration showing the **O**rigin and **I**nsertion

Step-by-step instructions on how to **palpate** a structure

"Check It" questions will confirm your location. They may ask you about your location in relation to a nearby structure or ask you or your partner to create a movement. Unless otherwise indicated, the answers to the questions should be "Yes!"

Alternative palpatory routes

Sternocleidomastoid

The sternocleidomastoid (SCM) is located on the lateral and anterior aspects of the neck. It has a large belly composed of two heads: a flat, clavicular head and a slender, sternal head (5.30). Both heads merge to attach behind the ear at the mastoid process. The carotid artery passes deep and medial to the SCM; the external jugular lies superficial to it.

(5.30)

A *Unilateral:*
Laterally flex the head and neck to the same side
Rotate the head and neck to the opposite side
Bilateral:
Extend the neck
Flex the neck
Assist in inhalation

O *Sternal head:* Top of manubrium
Clavicular head: Medial 1/3 of clavicle

I Mastoid process of temporal bone, lateral superior nuchal line of occiput

N Spinal accessory

1) Supine with practitioner at head of table. Locate the mastoid process of the temporal bone, the medial clavicle and the top of the sternum.
2) Draw a line between these landmarks to delineate the location of the SCM. Note how the two SCMs form a "V" on the front of the neck.
3) Ask your partner to raise her head very slightly off the table as you palpate the SCM. It will usually protrude visibly (5.31).

(5.31) Partner supine

With your partner relaxed, can you grasp the SCM between your fingers and outline its shape?

Hey!

sternocleidomastoid **ster**-no-**kli**-do-**mas**-toyd

Pronunciation and etymology of anatomical terms

Look for **Mr. Bones** sharing cautionary advice or other helpful hints

Check out the boxes for palpation tips, comparative anatomy and other curiosities

The techniques described in *Trail Guide to the Body* should be viewed as helpful tour guides. When first palpating, it is best to follow the specific instructions. After you have located a structure, it is recommended that you adapt and explore other methods to find the approach that works best for you. Wherever possible, an optional method for locating a structure has been included. As with any worthwhile journey, veering off course to explore other areas often leads to wonderful discoveries. Please feel free to veer.

etymology **et**-i-**mol**-o-gee the science of the origin and development of a word

Palpation Hints

Palpation means "to examine or explore by touching (an organ or area of the body), usually as a diagnostic aid." It is an art and a skill which involves **1)** locating a structure, **2)** becoming aware of its characteristics and **3)** assessing its quality or condition so you can determine how to treat it.

The first two aspects of palpation - locating and being attentive to the body's structures - require a thorough knowledge of functional anatomy and experience through mindful, hands-on practice. This is the focus of *Trail Guide to the Body*. Assessment - the third aspect of palpation - is a vast subject requiring a book of its own.

As an experience involving all the senses, palpation requires receptive hands and fingers, open eyes, listening ears, calm breath and a quiet mind. As you explore the terrain and texture of the body, be sure to bring along all of your sensing tools.

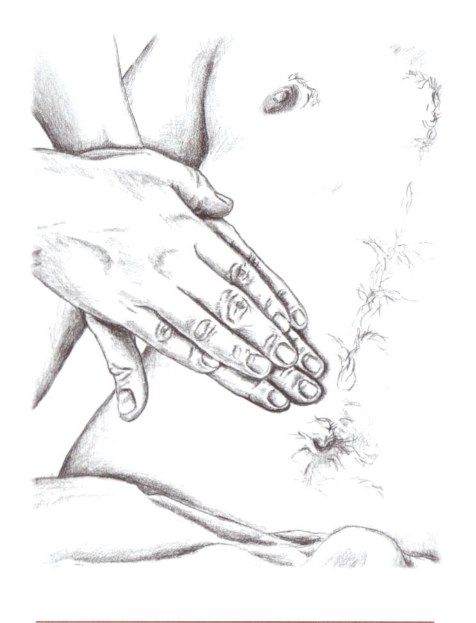

(0.3) A firm top hand and a soft bottom hand

Making Contact

Let your hands and fingers be responsive and sensitive. Relaxed, patient hands will allow the body's contours, temperature and structures to come more easily into your awareness.

When palpating, you may want to close your eyes periodically to enhance your awareness. For greater sensitivity and stability, try laying one hand upon the other, using the top hand to create the necessary pressure, while the bottom hand remains relaxed (0.3). This will allow the bottom hand to stay receptive as the top hand directs movement and depth.

Smaller structures can be located by using one or two fingertips (0.4). Larger structures are best palpated with your whole hand. By sculpting out all of the sides and edges, full hand contact helps to define the complete shape of a region or structure and also allows for a greater understanding of the interrelationships of structures (0.5).

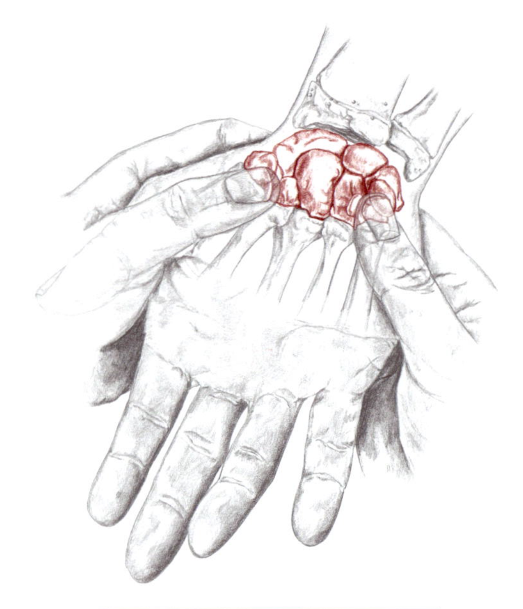

(0.4) Using your thumbpads to explore the small carpal bones in the wrist

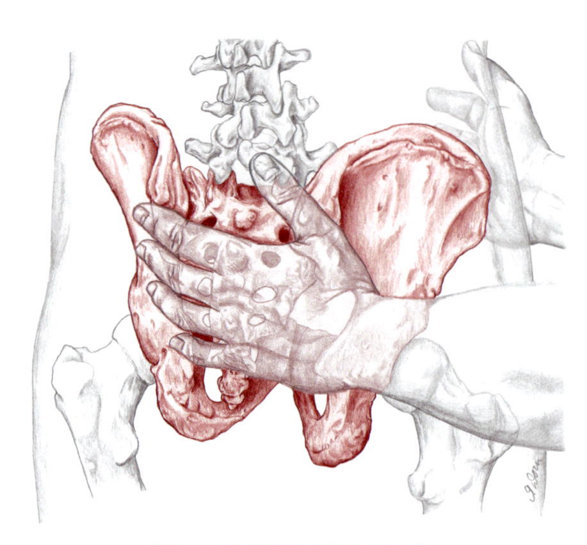

(0.5) Using your entire hand to palpate the pelvis and sacrum

palpate **pal**-pate L. *palpare*, to touch

Working Hard vs. Working Smart

Often in the excitement of trying to locate something (whether it be a muscle or a set of car keys), you search so earnestly that your mental and physical awareness begins to diminish. Frustration arises, your breath stalls and your hands ultimately become insensitive. You begin to "work hard." Instead of working hard, you can "work smart" by reading the information about the structure before you palpate. Also, as you palpate, visualize what you are trying to access and verbalize to your partner what you are feeling.

Work smart by first locating the structure you wish to palpate on your own body before palpating it on your partner's. Self-palpation will improve your kinesthetic understanding of what you are looking for on your partner. Also, read the information aloud. Hearing the language as you are reading the text will improve your understanding and retention of the information.

Lastly, be patient with your learning process. Allow yourself to "make a wrong turn and get lost" on the body. Chances are you are close to what you are seeking. By letting your senses recognize the body's trail signs, you will get to where you want to be.

Dogs, cats, horses and other animals offer a wonderful opportunity to compare musculoskeletal anatomy through palpation. For example, the next time you are petting your neighbor's cat, take a moment to locate its scapula. Compare the scapula's shape, location and surrounding tissues to those of a human's or dog's. The anatomical differences may surprise you, but the similarities will amaze you.

Less Is More

As you begin exploring the body, you may not be able to access things as readily as you might wish. A common reaction is to press harder and deeper with your hands and fingers; however, instead of pushing into the muscles and other tissues, try to invite the tissues into your hands. Gentle contact will allow your hands to be sensitive, while excessive pushing only numbs the fingers, making for an uncomfortable experience for your partner (0.6).

Even deep structures are best accessed with mild pressure. Paradoxically, the deeper you move into the body, the slower and softer your touch needs to be. Ultimately, palpation at different levels of the body is not a question of pressure, but of intention. Having a clear intention as you seek out various structures will make for an easier, smoother journey.

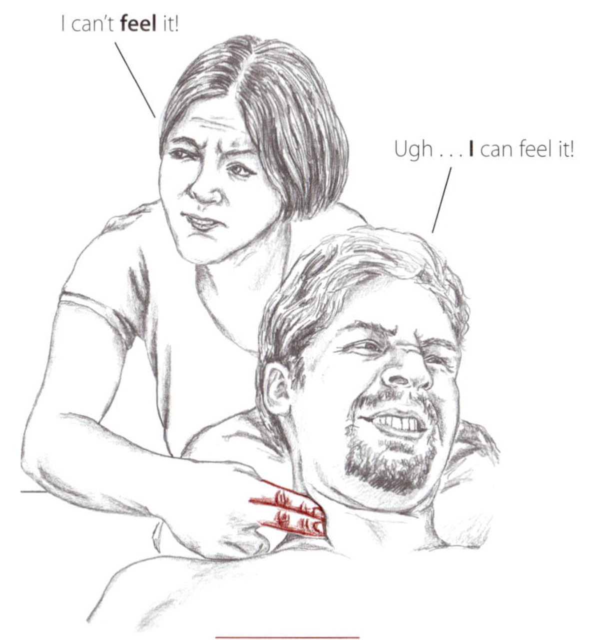

(0.6) Less is more

Rolling and Strumming

When outlining the shape or edge of a bone, try rolling your fingers or thumb *across*, rather than *along*, its surface. This is similar to checking the sharpness of a knife by sliding your finger across the blade. Do the same with the ropy fibers of muscle tissue. Like strumming the strings of a guitar, this method will help you ascertain the muscle's fiber direction and tensile state (0.7).

(0.7) Strumming across the fibers of the brachioradialis

Here is a simple exercise to increase your tactile sensitivity and palpatory skills. You will need a phone book and a human hair. Lay the hair beneath a single page of the phone book. Close your eyes, palpate

through the page and try to locate the hair. When you find it, reposition the hair and add another page. Continue to add pages until you can no longer locate the hair. How many pages can you palpate through? 5? 10? 15?!

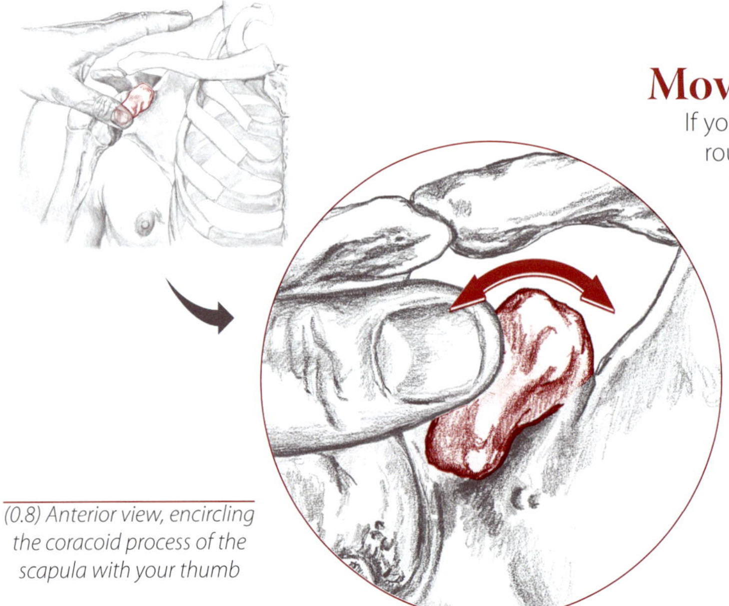

(0.8) Anterior view, encircling the coracoid process of the scapula with your thumb

Movement and Stillness

If you were to compare the texture of newspaper with rough sandpaper, you would naturally want to rub your fingers across their surfaces. In contrast, when you lay your hand on an expectant mother's abdomen, hoping to feel the fetus move, you naturally keep your hand still and quiet. Similarly, when you want to determine the fiber direction of a muscle or sculpt the shape of a bone, move your hands along its surface (0.8). However, when you want to feel a muscle contract or a bone *move*, keep your hands still and follow the movement. In other words, if the structure you are palpating is stationary, move your hands across it. If it is moving, stay still.

Movement as a Palpation Tool

Throughout the text, you will be asked to create specific movement on a partner's body with or without that person's help. These movements will help to verify the location of structures as well as any changes occurring in the tissues as a result.

Active movement is performed by your partner. She actively moves her body while you palpate or observe the movement. For example, the text may say, "Ask your partner to slowly flex her elbow while you palpate her biceps brachii muscle." All active movements performed by your partner should be slow and smooth - as changes in tissue are difficult to follow during fast, jerky motions (0.9).

Sometimes your partner will be asked to contract and relax a muscle. For example, "To feel the forearm flexors, lay your hand on your partner's forearm and ask her to alternately flex and relax her wrist." The on-and-off aspect of this technique will not only help you locate muscles and tendons, but will also give you the opportunity to feel the difference between contracted and relaxed tissue.

Passive movement is the opposite of active movement: Your partner relaxes while you move her body. For example, when the text says, "Passively flex and extend the elbow," you will move the forearm while your partner remains passive and allows the action to occur (0.10).

Resisted movement requires both of you to act: Your partner attempts to perform an action against your gentle resistance. For example, "To feel the elbow flexors contract, ask your partner to flex her elbow against your resistance" (0.11). As she meets the gentle resistance of your hand, no movement will occur at your partner's elbow. In this text, resisted movements are used to distinguish and compare the lengths, shapes and edges of different muscle bellies and tendons.

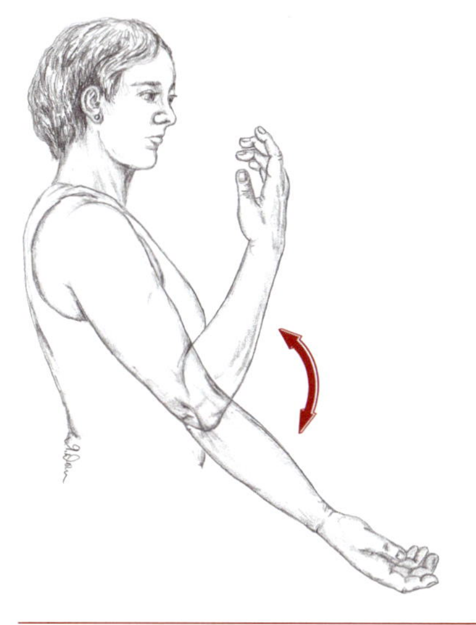

(0.9) Active flexion and extension of the elbow

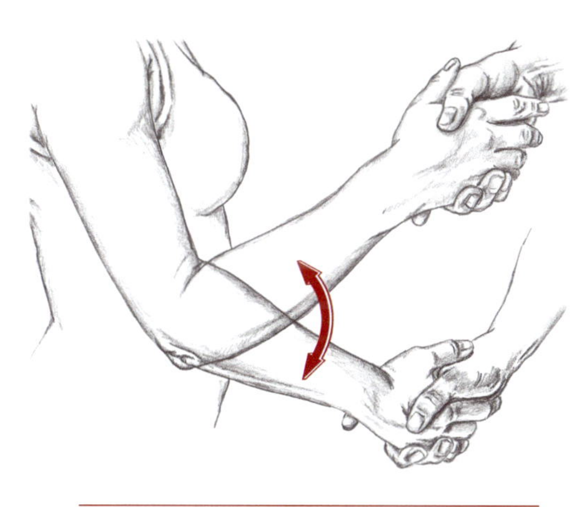

(0.10) Passive flexion and extension of the elbow

An adult has over 600,000 sensory receptors in the skin - more nerve endings than any other part of the body. The fingertips are one of the most sensitive areas, with up to 50,000 nerve endings every square inch. The fingertips are so sensitive that a single touch sensor can respond to a pressure of less than 1/1400 of an ounce - the weight of an average house fly.

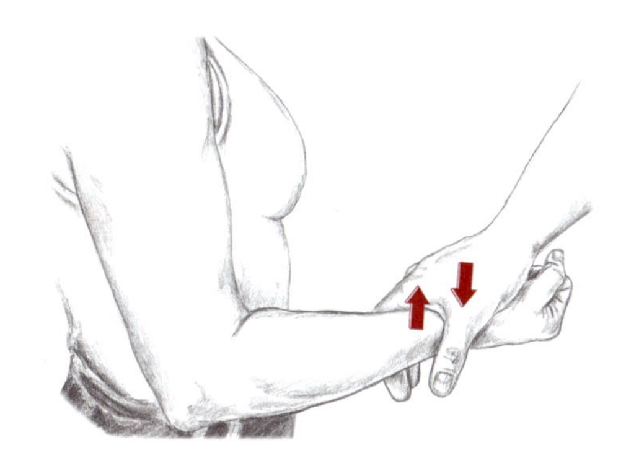

(0.11) Resisted flexion of the elbow

Leonardo da Vinci (1452-1519), who dissected bodies secretly at night, was the first to depict his anatomical findings. His anatomical illustration, laid out in over 750 drawings, is not only detailed and accurate, but also reveals many of the structural variations that can be seen when comparing bodies.

The anomalies shown in the drawings were not a case of Leonardo the artist dominating Leonardo the scientist; as a true Renaissance man there can be little question that he drew exactly what he saw in the cadavers.

The structures of the human body do not always conform to the standard anatomical model. Structural differences have been recorded in almost every muscle, bone, major blood vessel and organ in the body. Recognizing that the guidebook may not always coincide exactly with the geography of a particular body will help to prevent confusion and possible frustration.

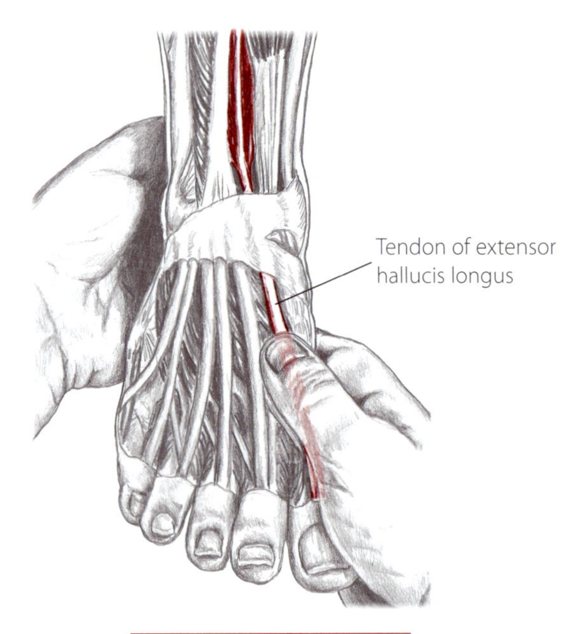

(0.12) Exploring a skinny tendon on the dorsal surface of the foot

When in Doubt, Ask the Body

While palpating, you may be confused or have questions about the body's structures and their whereabouts. When in doubt, ask the body you are palpating. For example, you may wonder, "What skinny tendon is this I see running along the top of the foot?" (0.12) The best advice would be to follow it in both directions and see where it leads you. If it runs from the big toe to the ankle and becomes taut when the toe is extended, it is the tendon of extensor hallucis longus (p. 373). Always remember, you are never alone; the body is waiting to help you.

All of the structures outlined in *Trail Guide to the Body* with their Latin or Greek names, unique shapes and buried positions, are inside you, your partner and your patients. These structures have been there for years waiting to be discovered by you. Have faith and you will be able to locate them.

Three Principles of Palpation

1) Move slowly. Haste only interferes with sensation. **2)** Avoid using excessive pressure. Less is truly more. **3)** Focus your awareness on what it is you are feeling. In other words, be present.

Also, you can practice your palpation skills on yourself at anytime. Yes, you may get a few curious glances, but daily routines such as waiting in line or riding the bus are wonderful opportunities to explore the malleable skin, tiny bones and sinewy muscles of your forearms and hands.

 Creating Your Palpatory Journal

Do you remember the first movie you ever saw? How about that initial bite of (what would soon become) your favorite food? Chances are that these encounters created lasting impressions. You might recall details of later films or subsequent helpings of that scrumptious dish, but over time your senses and memory of those secondary encounters probably diminished.

Learning to palpate is no different. Our initial hands-on experiences can cast long shadows over future encounters. For example, exploring the shape, density and fibers of the deltoid muscle for the first time can be formative. But as you become more familiar and less surprised by the muscle, later encounters will leave less of an impact.

The repetitive practice involved with learning a new skill, such as the martial arts, dance or palpation, requires constant presence of the mind and body. It's a difficult journey, but an invaluable one that can be enhanced by creating a palpatory journal. Like a personal diary, your journal is a chronicle of your hands-on experiences. You could store your palpatory stories in your head, but it's certainly more effective to record them in a small notebook or on your computer.

Initially your journal remarks may be broad and undefined. "The deltoid was tight." "The hamstrings felt ropy." As your palpatory instinct develops greater awareness of the body's nuances, so will your ability to articulate your findings. "I was able to shift the fascia of the upper chest caudally, but not laterally." "Left iliotibial tract was inseparable from vastus lateralis. Hypertonicity was an eight, on a scale from from one to ten."

Your notebook can also include impressions, ideas, questions and correlations. For example, "This week I palpated several different gastrocnemius muscles and noticed that four were particularly tender and had limited range-of-motion. Is this common or just coincidence?" Or "67-year-old male: the superficial fascia surrounding his hamstrings felt like bubble wrap. I've noticed this with two other seniors."

Of course, journaling is a "head-based" activity and palpation is strongest when it is connected to the hands, heart and gut. You may want to abandon words altogether and, instead, use colored pens to draw your experiences, or speak your findings into a small tape recorder. The best part is that there are no right or wrong answers.

Over time, whether you have explored the tissues of twenty or two hundred individuals, your journal will begin to fill with your thoughts and findings. Your palpatory journal will have evolved into something else - a memoir where you can read through and reflect on all of your adventures.

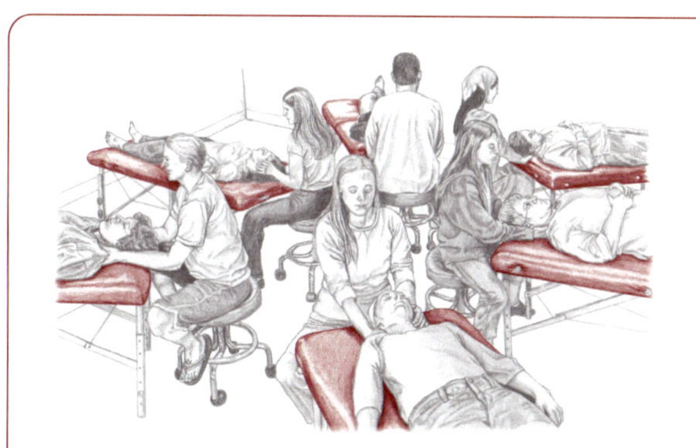

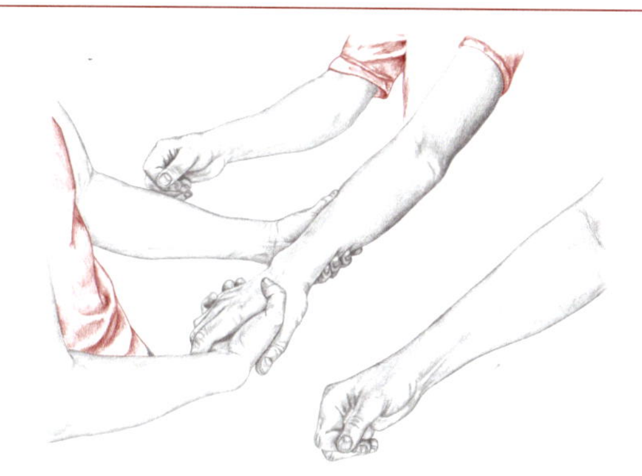

Palpating a variety of bodies in succession can create an unparalleled hands-on experience. This can be easily accomplished with a "round robin," where you rotate with others to palpate a series of people. Classroom settings (above), study sessions with friends or even social gatherings offer opportunities for a round robin. The key to a productive round robin is maintaining awareness of the similarities and differences you are feeling from one person to the next.

Exploring the Textural Differences of Structures

This section is designed to help you identify and compare the physical characteristics of the various structures and tissues in the body. Understanding the textural differences between structures will help you to determine which techniques to apply on a particular body part in your hands-on practice.

Following are descriptions of various structures in their "normal," healthy condition. The tissue's basic structural design will be identical on everyone, but, of course, the particular quality or feel of a tissue will be as unique as the individual you are palpating. For example, a long-distance runner may have lean, sinewy bands of muscle tissue while an individual leading a sedentary lifestyle may have a very different quality to his muscles. Although the feel of the muscle tissue is different, its design and composition are the same.

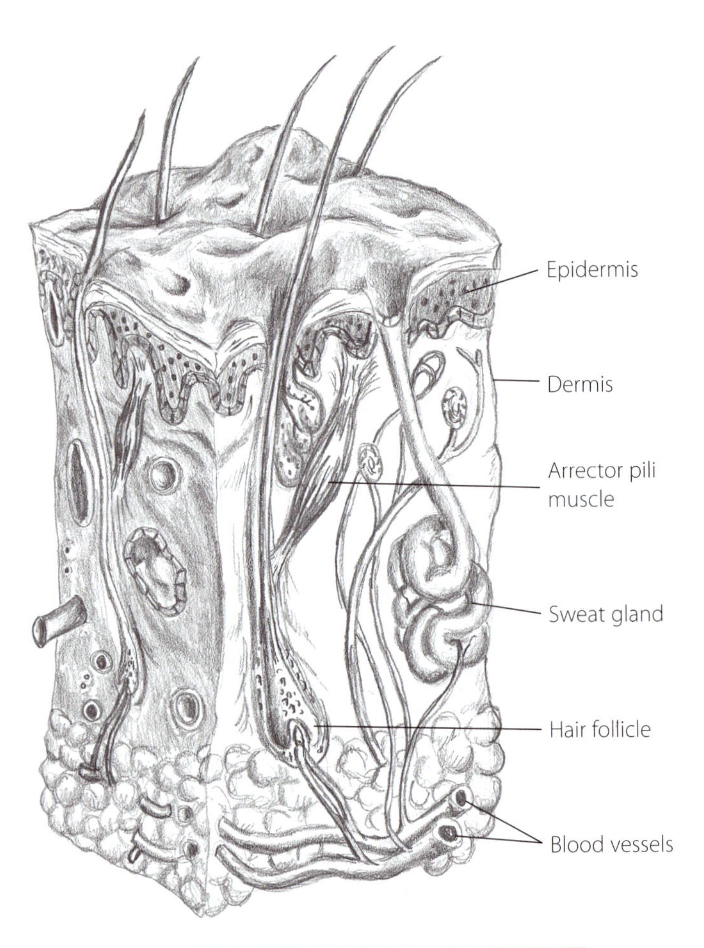

(0.13) Cross section of the skin. If you do not like the skin you have, just wait a month. An average adult sheds about 600,000 particles of skin every hour, amounting to one and a half pounds of skin each year. Altogether, the outer skin changes about every twenty-seven days. Add it up and that is nearly 1000 new skins in a lifetime.

Skin

Although often regarded as merely the body's covering, the skin is, in fact, the largest organ of the body (0.13). On an adult male, the skin can cover a surface area of nineteen square feet and weigh nearly ten percent of the total body weight. The skin averages about 1/20 of an inch in thickness, with the eyelids having the thinnest skin - less than 1/500 of an inch. The skin is intimately connected with the superficial fascia and deeper tissues, and its texture, thickness and flexibility vary throughout the body.

For example, palpate the skin on the back of your hand. Note its thin, delicate and pliable quality. Then turn your hand over and explore the palmar surface. Here the skin has a thicker, tougher layering.

Bone

Bones and bony landmarks (the hills, valleys and bumps on the surface of bones) are easy to distinguish from other tissues because they have a solid feel. Of course, the bones shift along with their surrounding structures during movement.

Sometimes other structures can feel like bone; for example, when a muscle contracts against resistance, its belly and tendons become very hard. Ligaments also can have a particularly solid quality. The shape and rigidity of bones and bony landmarks are constant, unlike muscles which can transform from a soft to a hard state and back again.

Muscle

Skeletal muscle, the voluntary contractile tissue that moves the skeleton, is composed of muscle cells (fibers), layers of connective tissue (fascia) and numerous nerves and blood vessels.

A muscle's infrastructure is similar to that of an orange: A broad sheet of fascia encases the whole fruit, deeper layers of fascia separate the orange into "wedges" (the portions you eat after peeling) and, finally, a thin coating of tissue surrounds each individual, tiny "bud" of fruit (0.14).

If we then apply this analogy to a muscle, a layer of fascia (epimysium) encases the muscle "belly," a deeper layer (perimysium) wraps the long muscle fibers into bundles and, finally, each microscopic muscle fiber is bound in fascia (endomysium) (0.15). Unlike an orange, however, a muscle's layers of connective tissue merge at either end of the muscle to form a strong tendon. The tendon attaches the muscle to a bone.

Muscle tissue has three specific physical characteristics which help to distinguish it from other tissues. First, **muscle tissue has a striated texture** - similar to a plank of unsanded wood. This is different from tendons, which have a smoother feel. The fibrous quality of a muscle belly is caused by bundles of muscle fibers running in a particular direction.

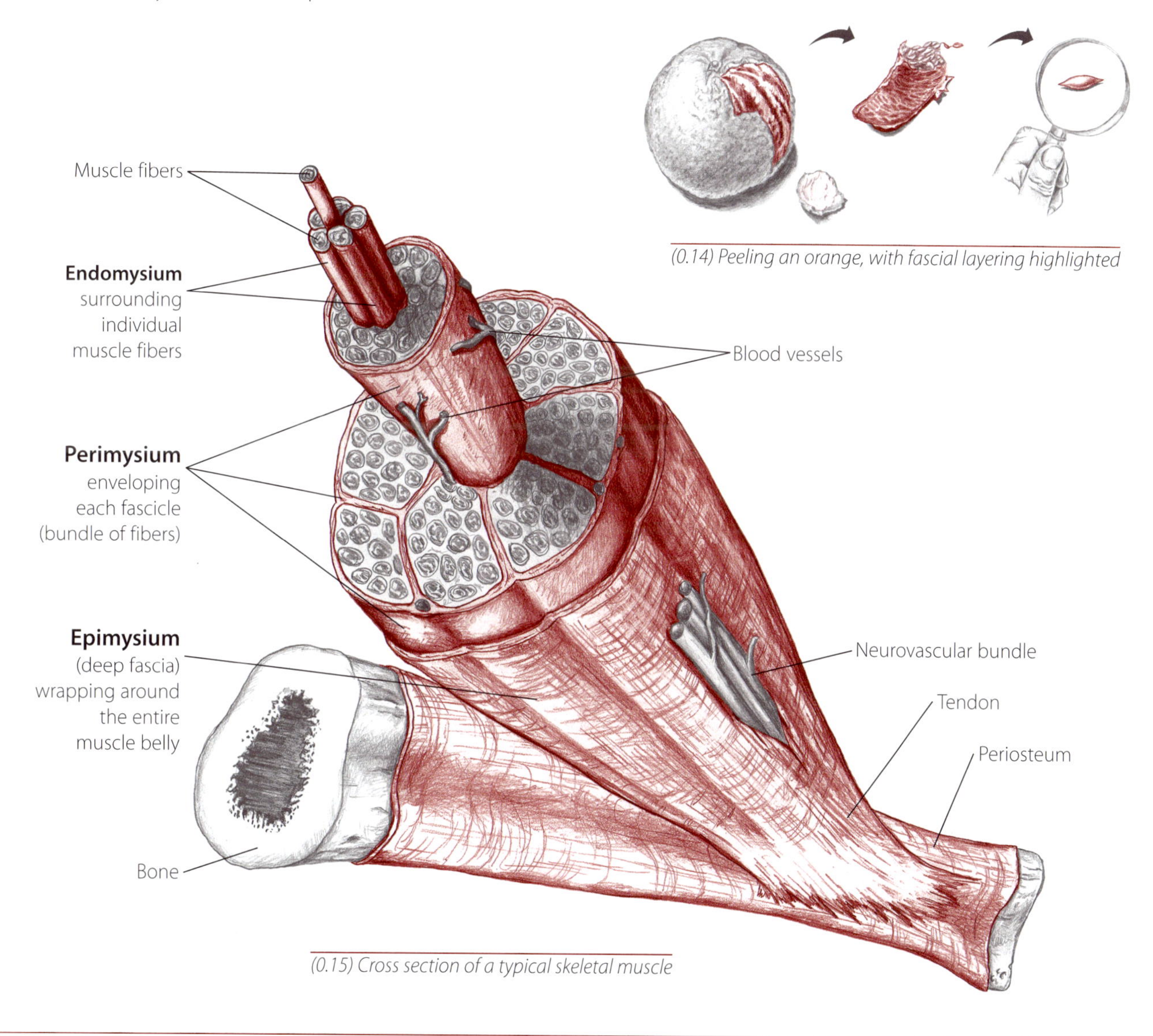

(0.14) Peeling an orange, with fascial layering highlighted

(0.15) Cross section of a typical skeletal muscle

In order for a specific movement to occur, muscles have to play particular roles. The muscle that carries out an action is called the **prime mover**, while muscles that support the prime mover are called **synergists**. Muscles which have an opposite action of the prime mover are called **antagonists**.

So when you *dorsiflex* your ankle (p. 39), the prime mover is the tibialis anterior. It is assisted in this movement by two synergists, the extensor digitorum longus and extensor hallucis longus (p. 373). Playing the role of antagonists to the tibialis anterior are the gastrocnemius, soleus and plantar flexors of the ankle.

When you *plantar flex* your ankle, the roles reverse: Now the prime mover is the soleus (p. 364), the synergists are the gastrocnemius and plantar flexors, and the antagonists are the tibialis anterior, extensor digitorum longus and extensor hallucis longus.

Second, **the direction of the muscle fibers** can be used to determine the specific muscle you are palpating. Depending on the shape and design of a muscle (0.17), the direction of its fibers may be parallel, convergent or diagonal. For example, the erector spinae muscles (p. 202) have vertical fibers that run parallel to the spine. Identifying their fiber direction can help you distinguish the erector spinae from the oblique and horizontal fibers of the other back muscles.

Lastly, **muscle tissue is unique because it can be in a contracted or relaxed state**. When a muscle is relaxed, it often has a soft, malleable feel; when contracted, it has a firm, solid quality. As the tension in muscle tissue changes, surrounding tissues like tendons and fascia also change, becoming taut or loose.

How can you palpate a muscle that is deep to a superficial, overlying muscle? In some areas, the overlying muscle can be shifted to the side. At other times, you can slowly compress your fingerpads beyond the superficial muscle into the deeper tissues, using the different textures and fiber directions as guides. This is similar to palpating through your sweater, shirt and skin to access a muscle in your arm.

Discover the three distinguishing features of muscle tissue by palpating your biceps brachii - the muscle on the front of the arm (0.16). Keep your arm relaxed and feel for the biceps' ropy fibers. Note how its fiber direction runs distally (down the arm). Then contract and relax the biceps and sense how it tightens into a solid mass and relaxes into a soft wad.

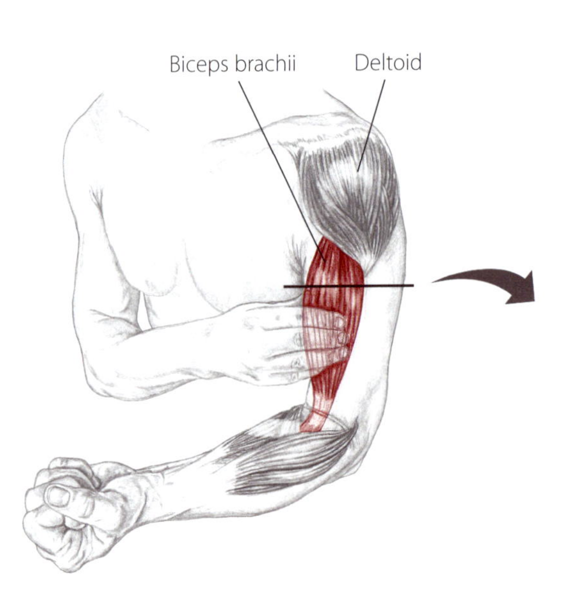

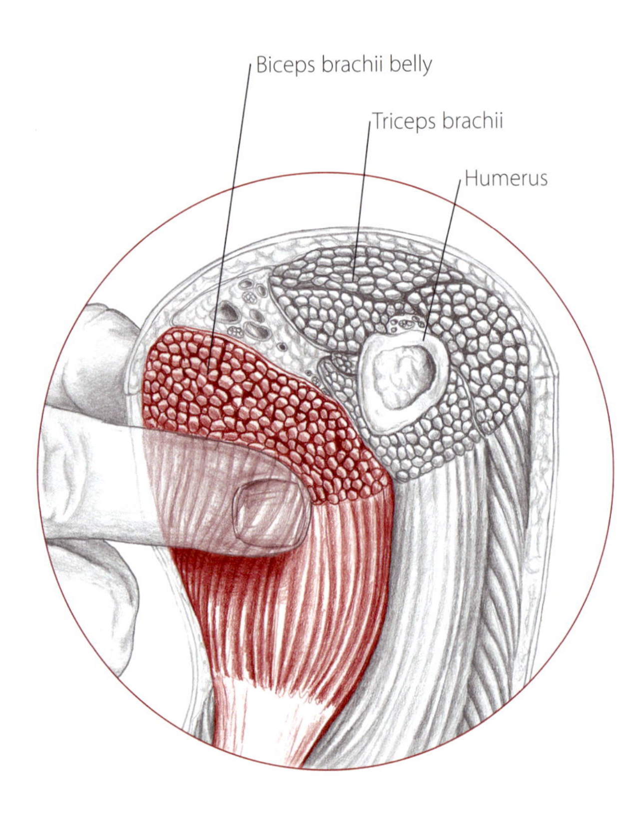

(0.16) Palpating the belly of your biceps brachii muscle with a cross section close-up (right)

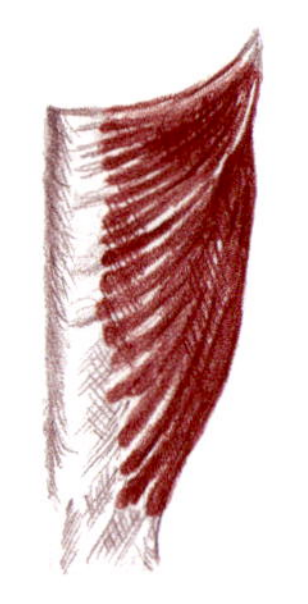

convergent
(latissimus dorsi, p. 79)

biceps
(biceps brachii, p. 103)

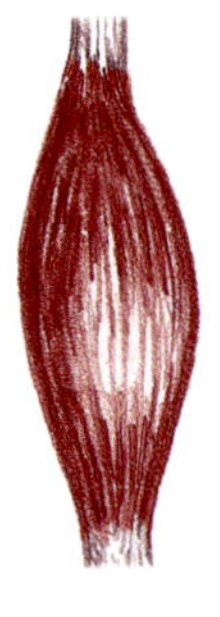

fusiform
(brachialis, p. 140)

unipennate
(extensor digitorum longus, p. 371)

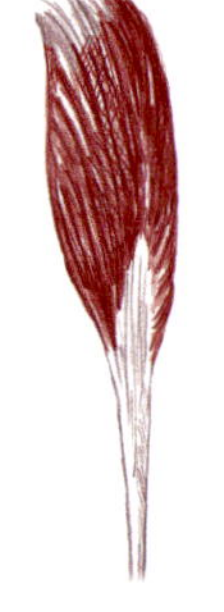

bipennate
(rectus femoris, p. 300)

multibelly
(rectus abdominis, p. 215)

(0.17) Different shapes of muscle bellies

aponeurosis	**ap**-o-nu-**ro**-sis	Grk. *apo*, from + *neuron*, nerve or tendon
ligament	**lig**-a-ment	L. a band

Tendon

Tendons attach muscle to bone. More accurately, they connect muscles to the periosteum - the connective tissue which surrounds the bone (p. 21). Tendons are composed of dense connective tissue shaped into bundles of parallel collagen fibers. Each end of a muscle has one or more tendons.

Tendons come in a variety of shapes and sizes. Some are short and wide like those of the gluteus maximus at the buttocks. Others are long and thin such as the tendinous cables of your wrist. A broad, flat tendon is called an aponeurosis. An example is the galea aponeurotica (p. 263) that extends across the top of your cranium. All tendons have a smooth, tough, almost resilient feel to them, regardless of their shapes.

Locate the distal tendon of the biceps brachii by holding your elbow in a flexed position (0.18). First, locate the biceps' muscle belly and follow it distally toward your inner elbow. As you progress, the muscle belly will become more slender and, at the crease of the inner elbow, it will become a smooth, thin tendon. It may feel like a taut strand of cable. Explore around either side of this tendon.

Ligament

Ligaments connect bones together at a joint. Their task is to strengthen and stabilize joints. Like tendons, ligaments are made of dense connective tissue. But unlike a tendon's parallel fiber arrangement, a ligament's fibers have a more uneven configuration.

The design and length of ligaments vary. Many simply cross a joint and blend in with the deeper joint capsule, like the ankle's deltoid ligament (0.19). Others span a distance between several bones, like the supraspinous ligament of the back (p. 225).

Ligaments often have a dense, taut feel and sometimes their fiber directions are palpable. If you want to distinguish a tendon from a ligament, explore its attachments and variable tension. A *tendon* connects a muscle belly to a bone, while a *ligament* attaches a bone to another bone. A *tendon* will become taut or slack depending on whether it is shortened or lengthened or if its muscle belly is contracted. A *ligament* will remain taut throughout all movements or states of contraction.

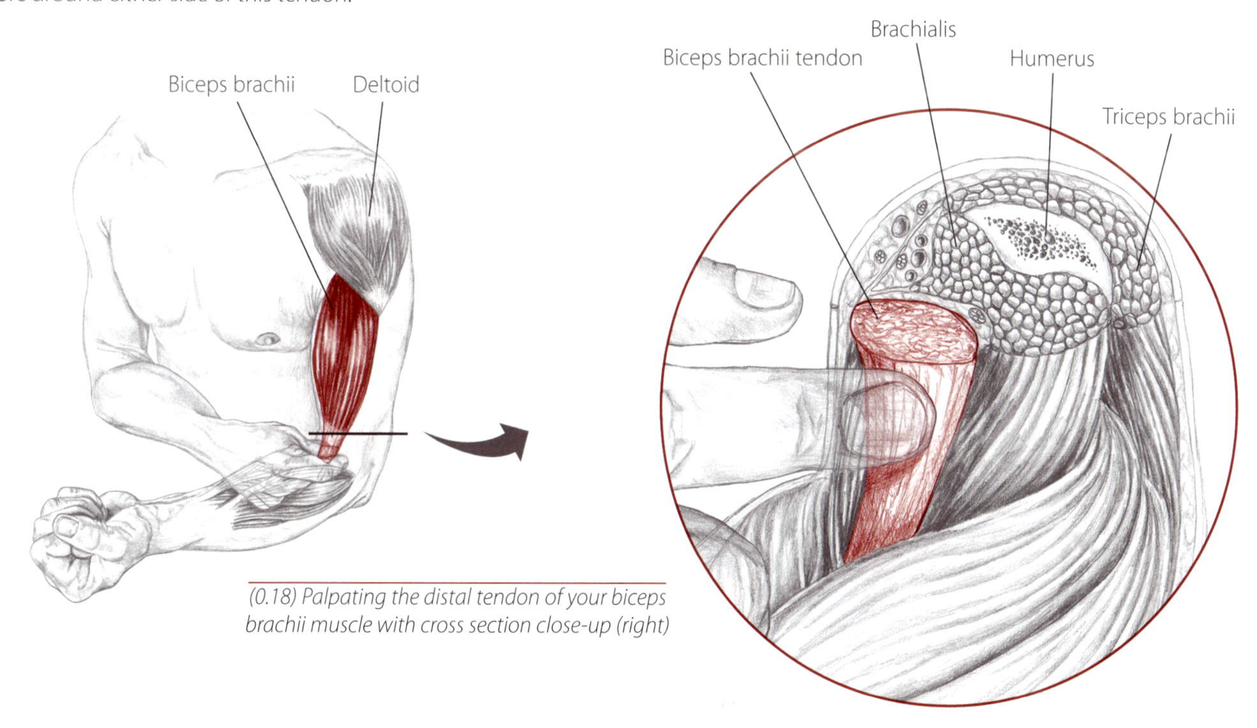

(0.18) Palpating the distal tendon of your biceps brachii muscle with cross section close-up (right)

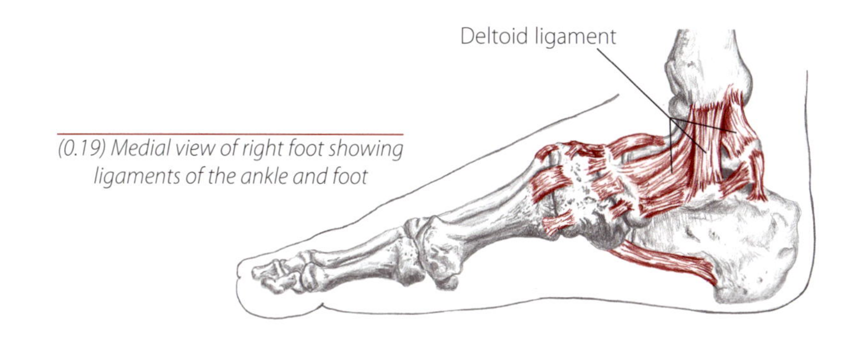

(0.19) Medial view of right foot showing ligaments of the ankle and foot

Fascia

Like tendons and ligaments, fascia is a form of dense connective tissue. It is a continuous sheet of fibrous membrane located beneath the skin and around muscles and organs. This fascial system forms a three-dimensional matrix of connective tissue extending throughout the body from head to toe.

There are two types of fascia: superficial and deep. **Superficial fascia** is located immediately deep to the skin and covers the entire body. Often perceived as a thin sheet, superficial fascia is actually a spacial layering filled with adipose tissue, nerves, blood and lymph vessels, and connective tissue (0.20). The density of the superficial fascia varies from very thin (on the back of the hand) to quite thick (the sole of the foot).

Deep fascia has a more complex design. It surrounds muscle bellies, holding them together and separating them into functional groups. It also fills in the spaces between muscles and, like superficial fascia, carries blood vessels and nerves. Portions of the deep fascia penetrate into the muscle belly and encase each tiny muscle fiber.

Because of its ubiquitous quality, precise palpation of the fascial system requires an experienced, sensitive touch. On the next page are three simple exercises that can help you get a basic feel of the fascia and its relationship to other structures.

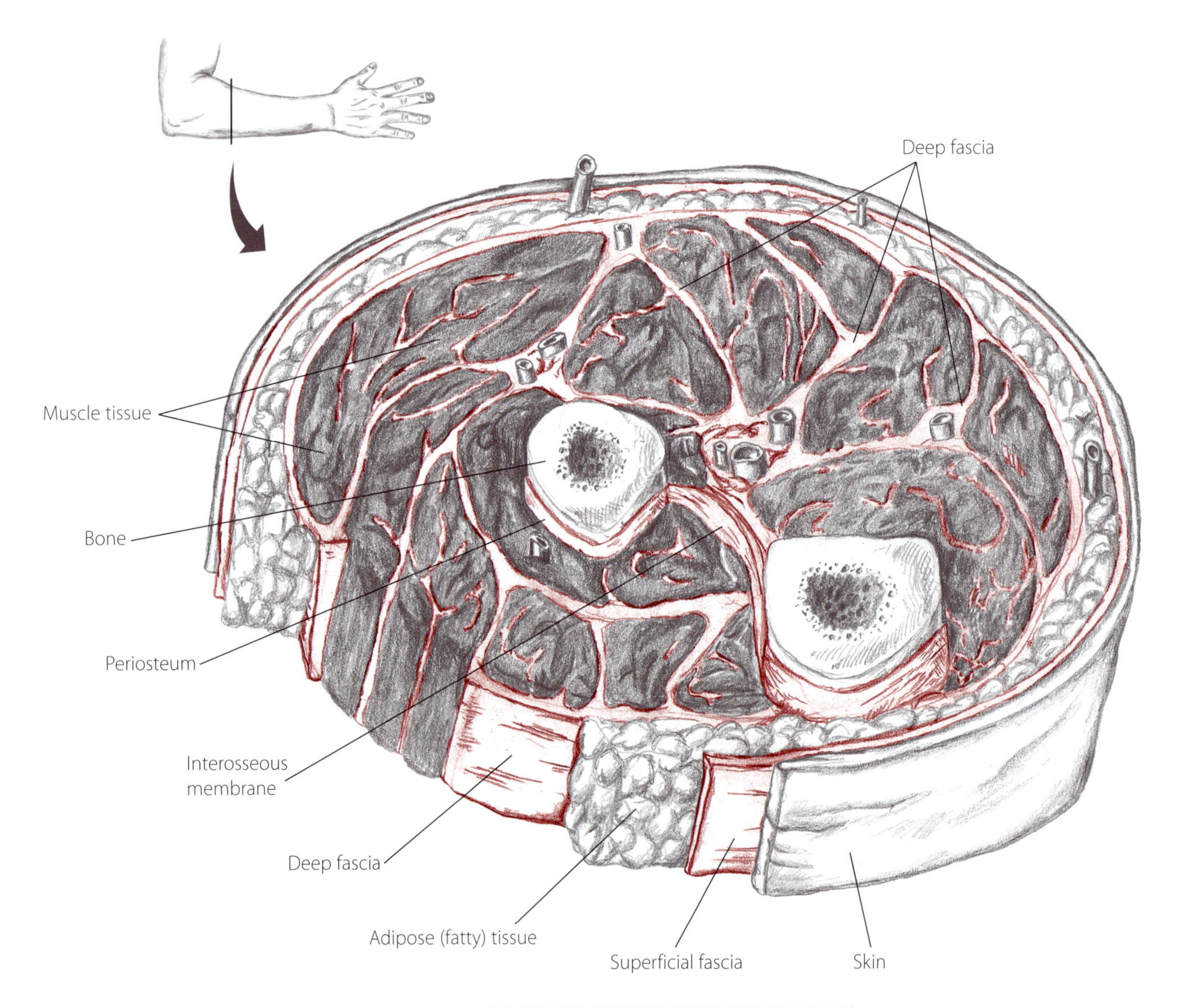

(0.20) Cross section of the forearm showing the arrangement of bone, muscle and fascia

septum	**sep**-tum	L. enclosure
septa	**sep**-ta	plural for septum

Explore Your Fascia

Pull up the skin on the back of your hand (0.21). Notice how the skin does not pull up entirely (as when you pull a baggy shirt away from your body). This is because the fascia is holding the skin down. Try this on your knee and various other parts of your body and notice how it is easier to lift the skin and fascia in some areas and more difficult in others (0.22).

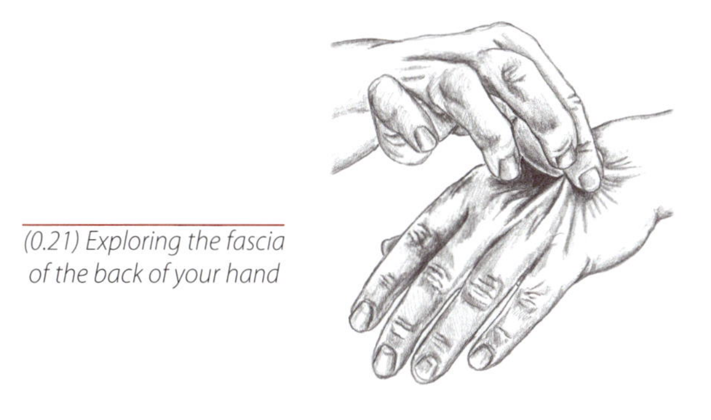

(0.21) Exploring the fascia of the back of your hand

This exercise is designed to give you a sense of the continuity of the fascial sheet throughout the body and of how pulling on one portion of this sheet can affect another.

Draw a small "X" on your forearm. Place your fingerpads approximately two inches away from the "X." Using the gentle pressure of your fingerpads, slowly move the skin of your arm in various directions away from the mark (0.23).

Notice how the "X" stretches and responds more easily when you pull in a certain direction, yet may not move as easily when pulled in another direction. As you continue, reposition your fingers farther away from the "X," so eventually you are pulling across the skin of the hand.

(0.22) Exploring the fascia of your knee

Here is an exercise to demonstrate the omnipresent, yet phantomlike, nature of fascia. Put a latex glove on your partner's hand followed by a thick winter glove. If you explore your partner's hand, you will immediately detect the texture and thickness of the winter glove and the general shape of the hands and fingers. The latex glove (representing the fascia), however, may be more challenging to detect.

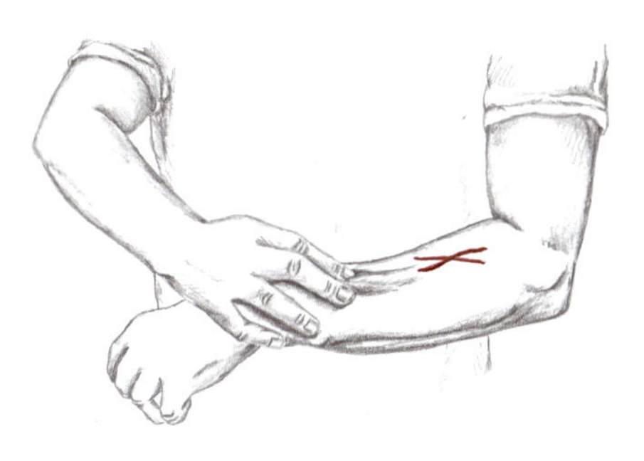

(0.23) Exploring the superficial fascia with an "X" drawn on the forearm

Retinaculum

A retinaculum is a structure that holds an organ or tissue in place. In relation to muscular connective tissue, a retinaculum is a transverse thickening of the deep fascia which straps tendons down in a particular location or position. For example, the retinacula of the ankle stabilize the tendons which traverse the sharp curve of the ankle (0.24).

Most retinacula are superficial and accessible. A retinaculum can be distinguished from its deeper tendons by its different fiber direction. A retinaculum will have transverse fibers that run perpendicular to the deeper tendons.

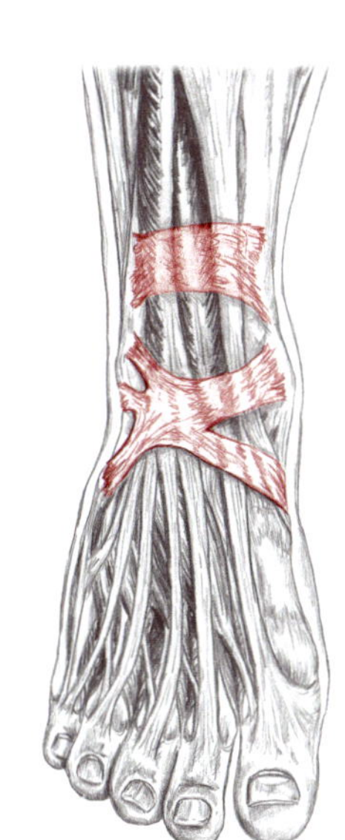

(0.24) Retinacula of the ankle

retinaculum	**ret**-i-**nak**-u-lum	L. halter, band, rope
retinacula	**ret**-i-**nak**-u-la	plural for retinaculum

Artery and Vein

Arteries and veins have distinct features that you can palpate. For example, the pulse of the heart can be felt when pressing on an artery but not on a vein. Arteries are often situated on the protected side of an appendage and buried deep to the musculature. Some veins can be palpated superficially and are easily seen on the dorsal surfaces of the hands and feet.

Locating an artery is not only necessary for determining the pulse, but also important when palpating other structures. For example, when palpating the sternocleidomastoid muscle in the neck, it is crucial for you to be aware of the location of the carotid artery (p. 268), the chief blood vessel supplying the head and neck, so you avoid pressing on it. If an artery is occluded for a sustained period of time during palpation, the distal portion of the appendage will begin to tingle or become numb.

Let your arm hang at your side for a minute, allowing the blood to fill the superficial veins of your hand and forearm. The veins will swell with the increased pressure and become visible (0.25). For more dramatic results, gently squeeze your forearm with your opposite hand or apply a tourniquet.

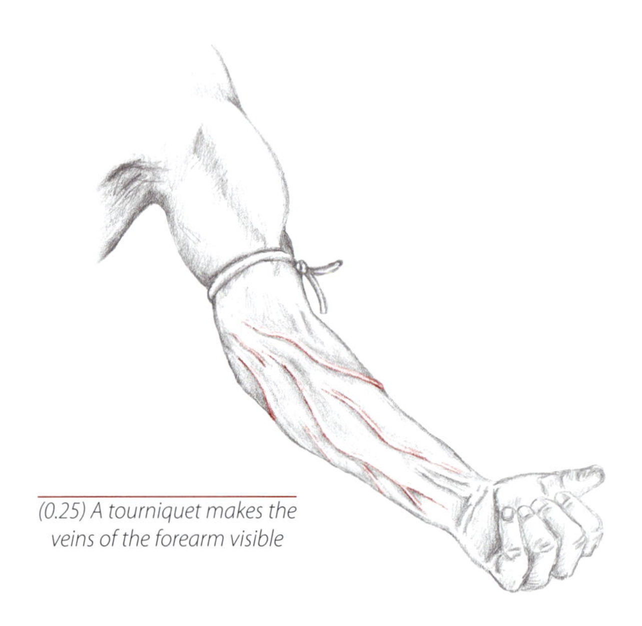

(0.25) A tourniquet makes the veins of the forearm visible

Bursa

A bursa is a small, fluid-filled sack that reduces friction between two structures (0.26). Situated primarily around joints, most of the body's six hundred bursae cushion skin, tendons, ligaments, muscle or organs from the hard surfaces of bones. They are also located between two muscles, two tendons, a tendon and ligament, or a muscle and ligament.

Bursitis, inflammation of a bursa, is a common disorder accompanied by tenderness in the area and crepitation (cracking and clicking sounds) of the joint. When inflamed, superficial bursae are easily palpable and sometimes visible. In their normal state, however, bursae are generally not palpable.

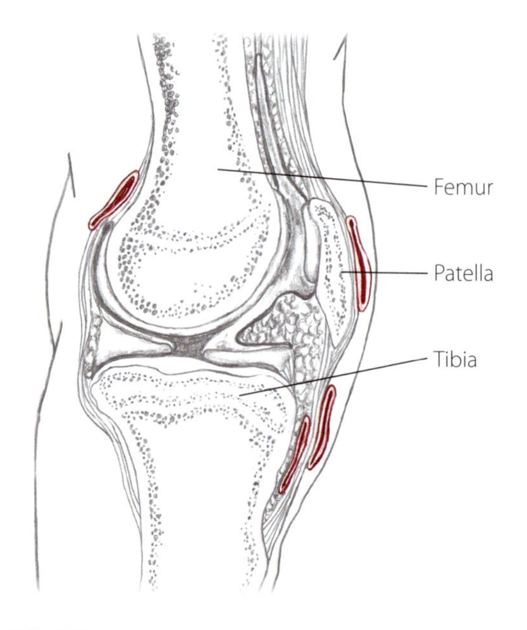

(0.26) Cross section view of knee highlighting some of the bursae of the knee joint

William Harvey (1578-1657), often regarded as the first experimental scientist, discovered that blood circulates throughout the body. Along with his descriptions of the cardiovascular system, he explained how veins are equipped with valves that prevent blood from flowing backwards between heartbeats. To prove his theory, Harvey tied a tourniquet around an

assistant's arm and allowed the blood to pool in the distal veins. He observed small swellings along the paths of veins which he thought were valves. Harvey pressed on a valve and pushed the blood out of the vein to the next valve. As he held his finger on the distal valve, the proximal valve prevented blood from flowing backwards and the vein remained empty.

adipose	**a**-di-**pos**	L. fat, copious
bursa	**ber**-sah	L. purse
plexus	**plek**-sus	L. interwoven

Nerve

Nerve vessels are tube-shaped, mobile and tender when compressed (0.27). Although sections of nerves and plexuses (bundles of nerves) can be accessed throughout the body, they are best avoided. Compression or impingement of a nerve may create a sharp, shooting sensation locally or down the corresponding appendage.

Lymph Node

Lymph nodes collect lymphatic fluid from lymphatic vessels. They are bean-shaped and may range in size from a tiny pea to an almond. Lymph nodes are located throughout the body with palpable groups of nodes found in the body's creases such as the groin, axilla and neck (0.28). Healthy lymph nodes are roundish, slightly movable and nontender. They differ from other glands which are usually larger and have irregular, lumpy surfaces.

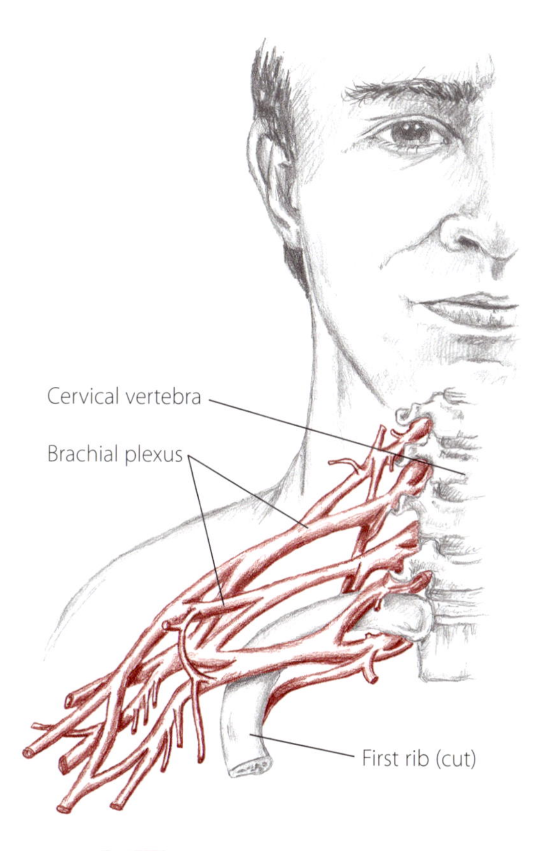

(0.27) Anterior view of the brachial plexus. Nerve impulses travel along nerve fibers at 210 miles an hour (or 320 feet a second).

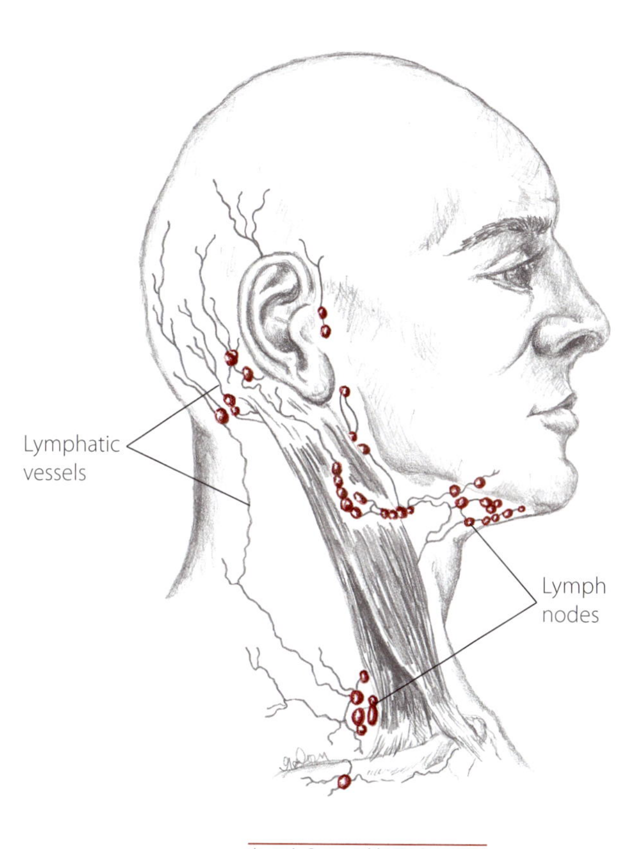

(0.28) Cervical lymph nodes

Adipose Tissue

Adipose (fatty) tissue is a form of loose connective tissue. It is deposited at many levels throughout the body including the marrow of long bones, around the kidneys, the padding around joints and behind the eyeballs. Needless to say, some of these areas are outside the reach of this text.

The most palpable location for adipose tissue is in the subcutaneous layer of tissue between the skin and superficial fascia. This layer of adipose varies in thickness throughout the body and may have different consistencies. Adipose usually has a gelatinous (jellylike) consistency, making it easy to sink the fingers into and detect deeper structures.

Stand up and squeeze the flesh of your own buttocks to feel adipose tissue. Yes, you might feel silly, but note the superficial layer of adipose. Then tighten the muscles of your buttocks and feel the textural difference between the adipose and the deeper muscles.

NOTES

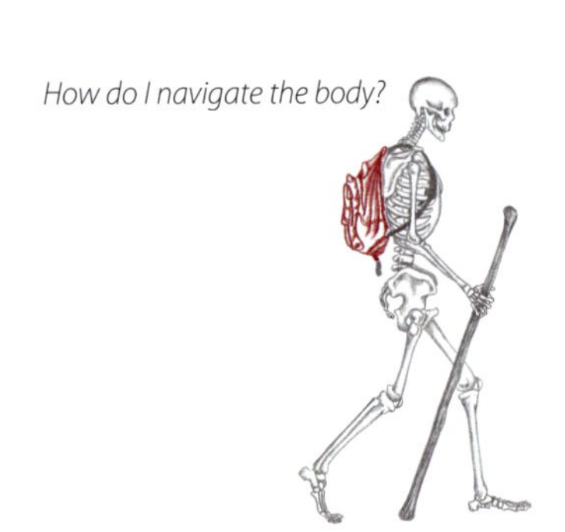

1

Navigating the Body

The nature of this book demands that we explore specific, individual structures and regions on our journey. However, before we set out into the hills and valleys of the body, some preparation is in order. This chapter will familiarize you with important mapping and navigational terms. It will also show you the "big picture" of the body's systems highlighted in the text. This way, when the trail guide leads you in a certain direction, you will know which way to go!

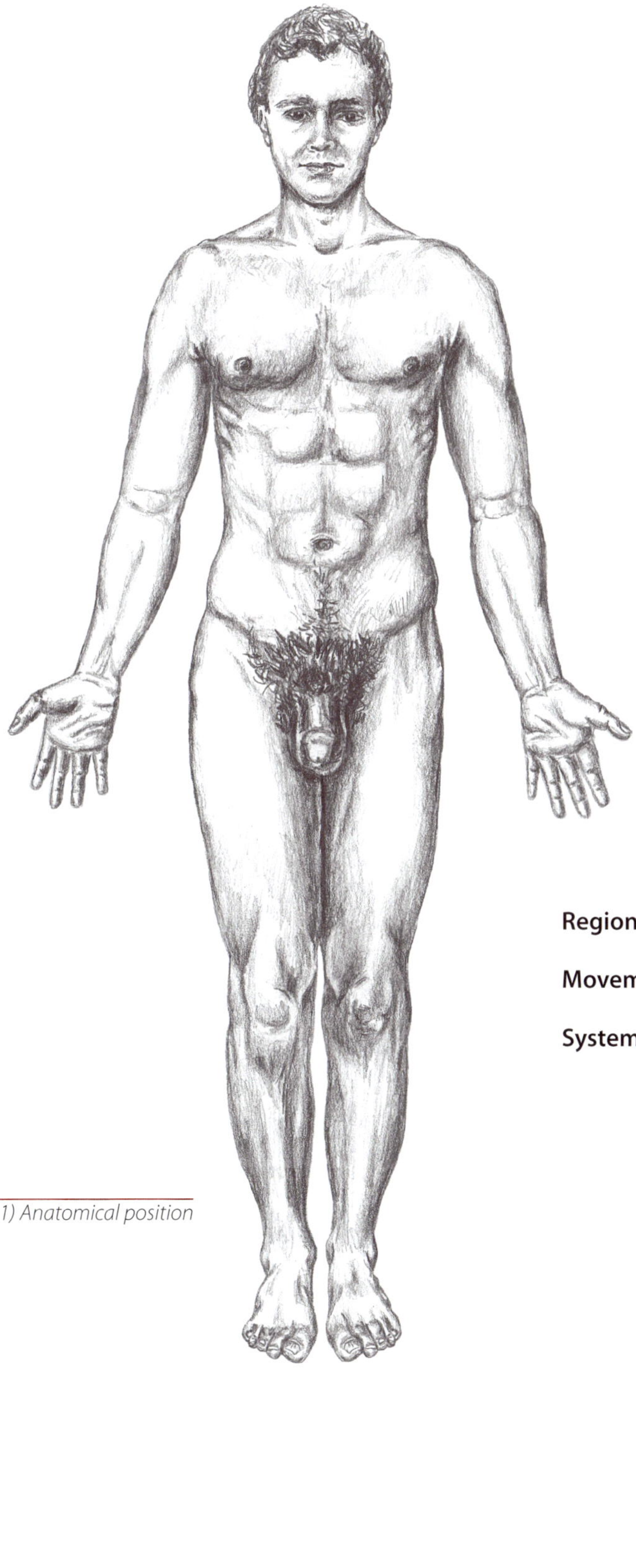

(1.1) Anatomical position

Regions of the Body

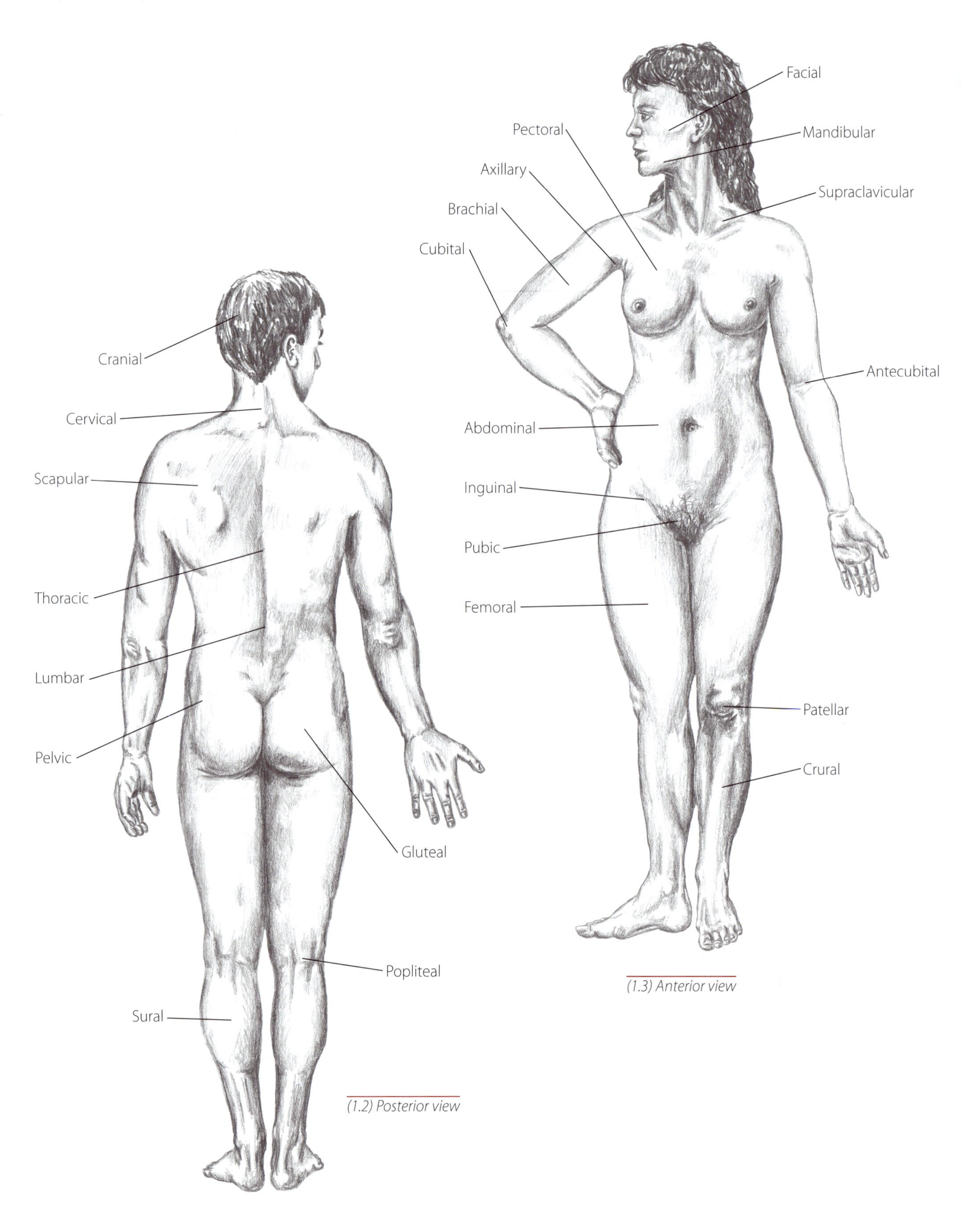

(1.2) Posterior view

(1.3) Anterior view

Planes of Movement

When the body is in the standard anatomical position, standing erect with the palms facing forward (p. 29), it can be divided into three imaginary planes (1.4). These planes help clarify and specify movements.

The **sagittal plane** divides the body into left and right halves. The descriptive terms medial and lateral correlate to the sagittal plane; the actions of flexion and extension occur along this plane. The midline (or midsagittal plane) runs down the center of the body, dividing the sagittal plane in two symmetrical halves.

The **frontal (or coronal) plane** divides the body into front and back portions. The terms anterior and posterior relate to the frontal plane; the actions of adduction and abduction happen along this plane.

Dividing the body into upper and lower parts is the **transverse plane**. The terms superior and inferior refer to the transverse plane; rotation happens within this plane.

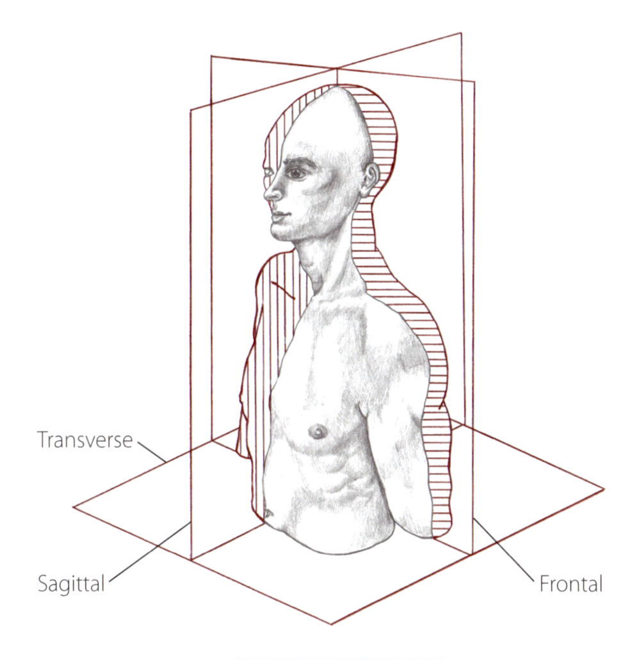

(1.4) Planes of the body

Directions and Positions

Specific terms are used to help communicate location, direction and position of body structures. These terms replace more general references like "up there" or "north of here," which are less precise and can be confusing. Each direction is paired up with its complementary direction.

Superior refers to a structure closer to the head. **Inferior** means closer to the feet. "The nose is superior to the navel." "The navel is inferior to the nose." (1.5) The terms **cranial** (closer to the head) and **caudal** (closer to the buttocks) are used when referring to structures on the trunk.

Posterior concerns a structure further toward the back of the body than another structure. **Anterior** refers to a structure further in front. "The sternum is anterior to the spine." (1.5) These directions are also referred to as dorsal (posterior) and ventral (anterior).

Medial pertains to a structure closer to the midline (or center) of the body. **Lateral** refers to a structure further away from the midline. "The last (pinkie) toe is lateral to the big toe." (1.6)

Distal means a structure further away from the trunk or the body's midline. **Proximal** designates a structure closer to the trunk. These directions are used only when referring to the arms and legs. "The foot is distal to the thigh." (1.6) "The forearm is proximal to the hand."

Superficial describes a structure closer to the body's surface. **Deep** refers to a structure deeper in the body. "The abdominal muscles are superficial to the intestines." "The intestines are deep to the abdominal muscles." (1.7)

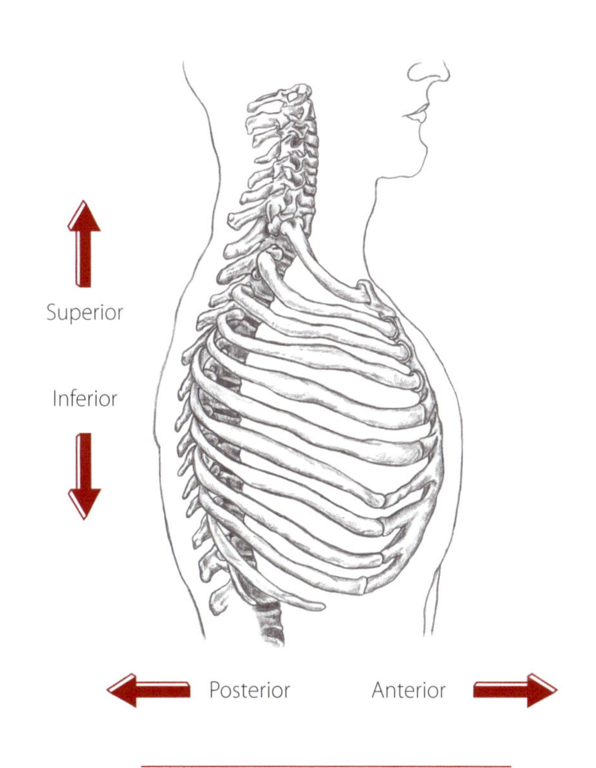

(1.5) Lateral view of rib cage and vertebrae

sagittal	**saj**-i-tal	L. arrowlike
coronal	ko-**ro**-nal	L. crownlike
transverse	**trans**-verse	L. across, turned across

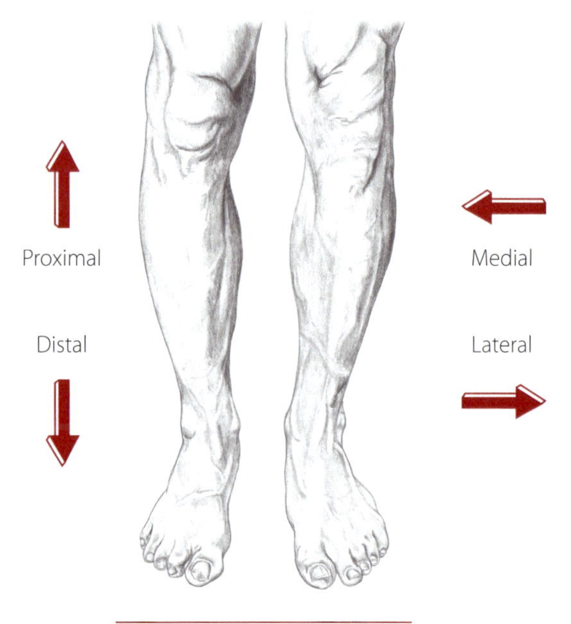

(1.6) Anterior view of legs and feet

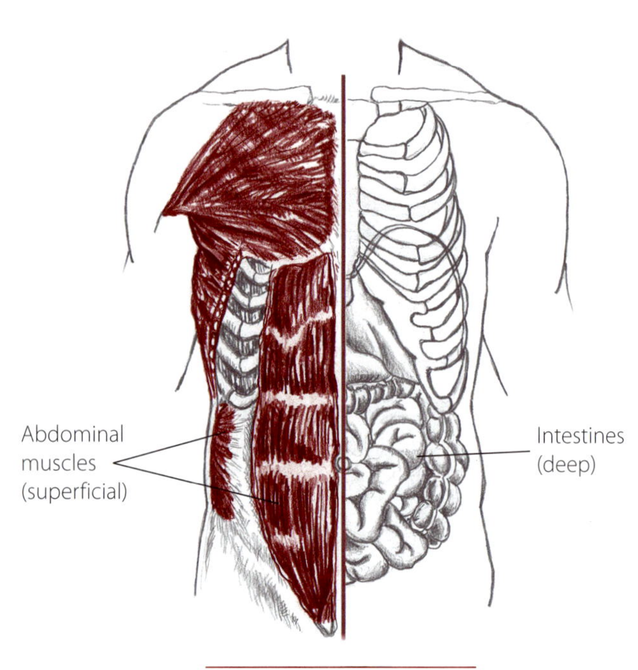

(1.7) Anterior view of abdomen

Movements of the Body

Movement of the body occurs at the joints, where bones articulate (or connect). Although movement affects the placement of bones, the terminology of movement always refers to joints. Bending your knee is called "flexion of the knee." "Flexion of the leg" would require an ambulance. See pages 34-39 for a description of movement at specific joints.

Extension is movement that straightens or opens a joint. In anatomical position, most joints are extended. When a joint can extend beyond its normal range of motion it is called hyperextension. **Flexion** is movement that bends a joint or brings the bones closer together. In a fetal position most joints are in a flexed position (1.8). Both flexion and extension take place along the sagittal plane.

Adduction of a joint brings a limb medially toward the body's midline ("adding to the body"). **Abduction** moves a limb laterally away from the midline ("abduct or carry away"). These actions happen along the frontal plane and pertain only to the appendages. To adduct the fingers or toes is to bring them together; to abduct is to spread them apart.

Medial rotation and **lateral rotation** (sometimes referred to as internal and external rotation) occur at the shoulder and hip joints. When the joint medially rotates, the limb turns in toward the midline. Lateral rotation swings the limb away from the midline.

Rotation pertains only to the axial skeleton (p. 40), specifically the head and vertebral column. Rotation of the head and neck occurs as a driver turns to check whether a car is coming from behind in the next lane. These movements happen along the transverse plane.

Circumduction is possible only at the shoulder and hip joints. It involves a combination of flexion, extension, adduction and abduction; together these actions create a cone-shaped movement (1.9). Swimming the backstroke requires circumduction at the shoulder joint.

Prone is the position of lying on the table face down. **Supine** ("on your spine") is to lie face up. **Sidelying** is just that - lying on your side.

ab- (as in *abduct*) L. away from
ad- (as in *adduct*) L. toward

Lateral flexion occurs only at the axial skeleton. For example, when the head or vertebral column bend laterally to the side.

Supination and **pronation** describe the pivoting action of the forearm. Supination ("carrying a bowl of soup") occurs when the radius and ulna lie parallel to one another. Pronation ("prone to spill it") takes place when the radius crosses over the ulna, turning the palm down. Supination and pronation also occur at the feet.

Opposition happens only at the carpometacarpal joint of the thumb. It occurs when the thumbpad crosses the palm toward the last (pinkie) finger.

Inversion and **eversion** occur as a combination of movements of several joints of the feet. Inversion ("turn in") elevates the foot's medial side and brings the sole of the foot medially. Eversion ("turn out") elevates the foot's lateral side and moves the sole laterally.

Plantar flexion and **dorsiflexion** only refer to movement at the ankle. Plantar flexion is performed by moving the ankle to point your foot into the earth or stepping on a car's gas pedal. Dorsiflexion is the opposite movement, such as moving the ankle to let off the gas pedal.

Protraction and **retraction** pertain to the scapula, clavicle, head and jaw. Protraction ("protrude") occurs when one of these structures moves anteriorly. Retraction ("retreat") is movement posteriorly.

Elevation and **depression** refer to the movement of the scapula and jaw. Elevation is movement superiorly. Depression is movement inferiorly.

Deviation means to wander from the usual course. Lateral deviation occurs at the mandible during talking or chewing.

(1.8) In the fetal position most joints are flexed

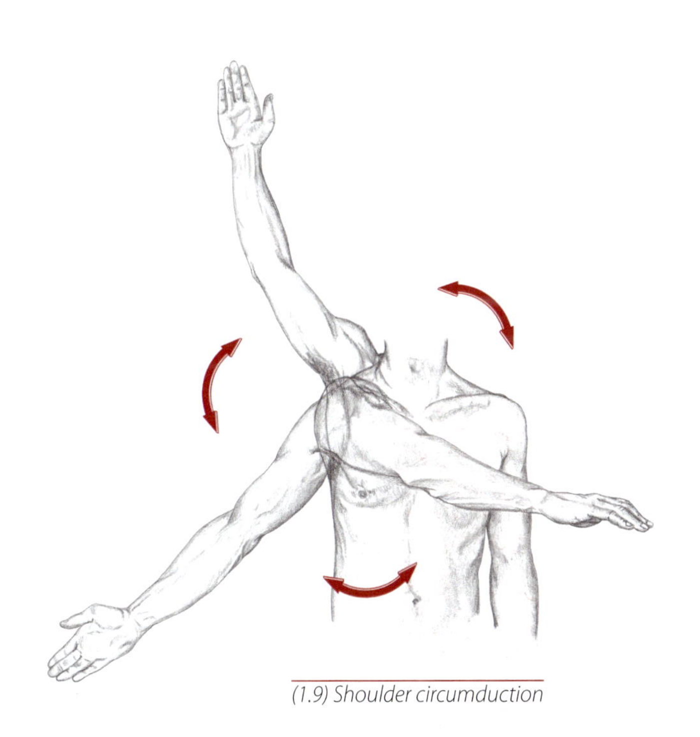

(1.9) Shoulder circumduction

The names of many bones, bony landmarks and muscles may initially look and sound foreign. They are - most anatomical terms are Latin or Greek. However, the source or story behind the terms can help to clarify their meaning. Take the phrase "infraspinous fossa of the scapula." The **scapula** is a flat bone of the shoulder. In Latin, scapula means "shoulder blade" - its common name. **Fossa** translates as "shallow depression." **Infraspinous** is a directional term (like north or south-west). It means inferior (infra-) to the spine of the scapula

Spine of the scapula

Infraspinous fossa

(-spinous). Put this all together and the "infra-spinous fossa of the scapula" translates as "the shallow depression located below the spine of the shoulder blade." Keep an eye peeled for translations and phonetic descriptions at the bottom of pages.

dorsi	**dor**-si	L. of the back
plantar	**plan**-tar	L. the sole of the foot

Movements of the Body

Spine and Thorax
(vertebral column)

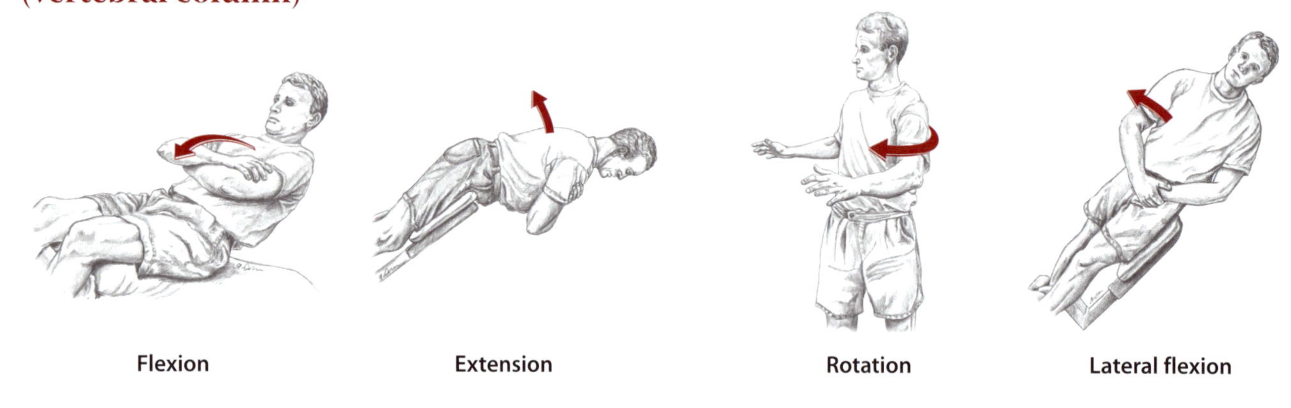

Flexion | Extension | Rotation | Lateral flexion

Neck
(cervical spine)

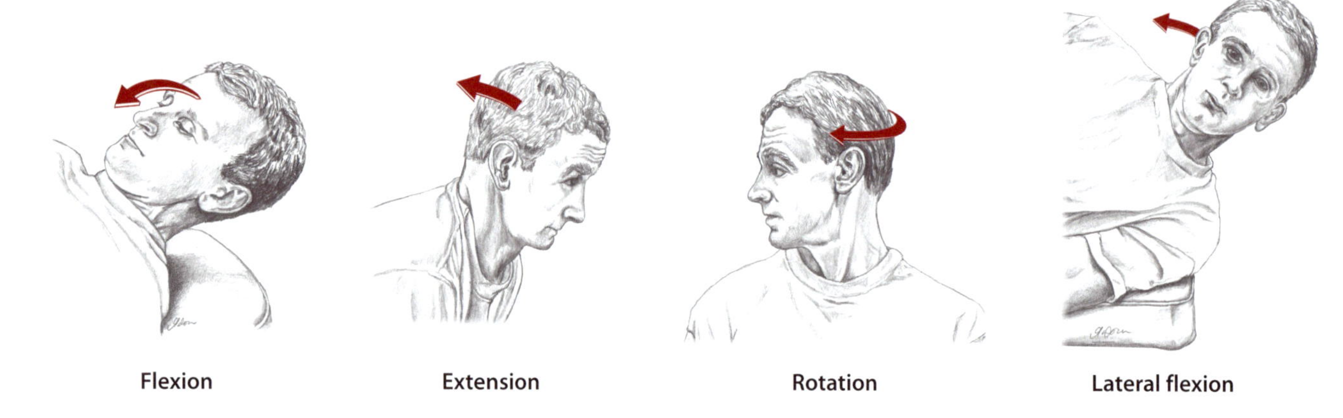

Flexion | Extension | Rotation | Lateral flexion

Ribs/Thorax

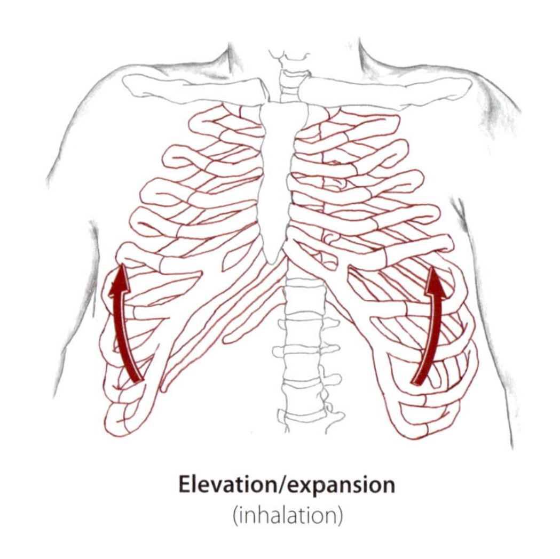

Elevation/expansion
(inhalation)

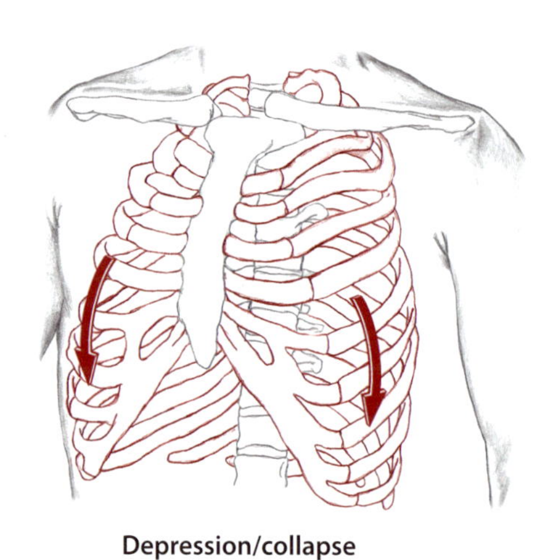

Depression/collapse
(exhalation)

Scapula

(scapulothoracic joint)

Shoulder

(glenohumeral joint)

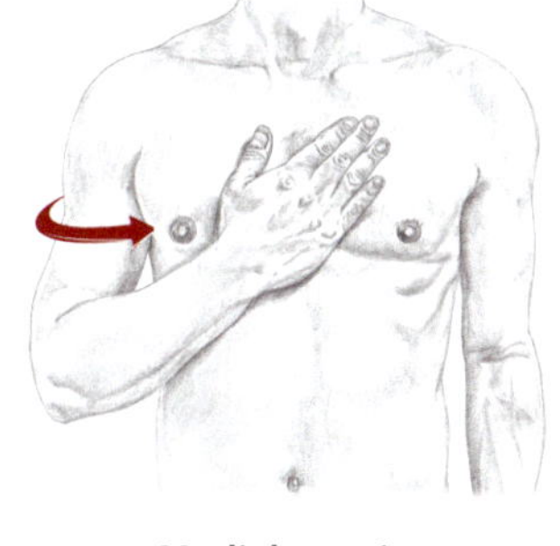

Medial rotation
(internal rotation)

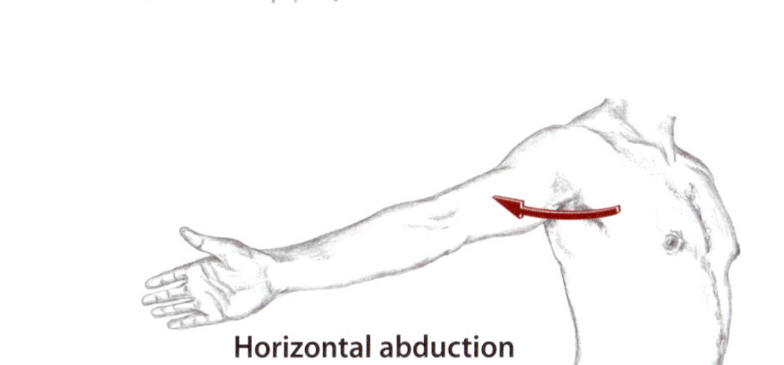

Horizontal abduction

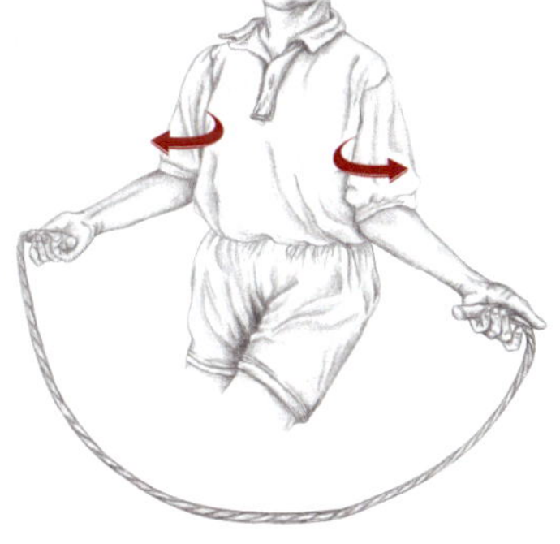

Lateral rotation
(external rotation)

Elbow and Forearm

(humeroulnar and humeroradial joints - elbow, proximal and distal radioulnar joints - forearm)

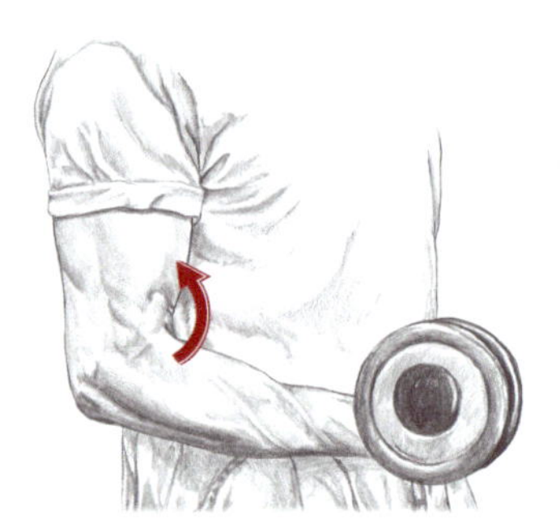

Flexion of the elbow

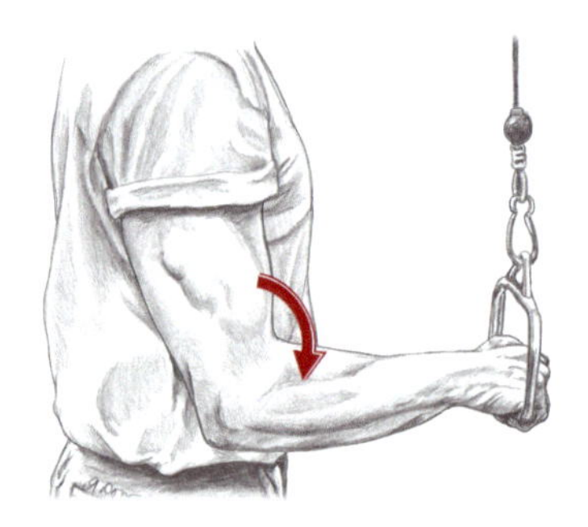

Extension of the elbow

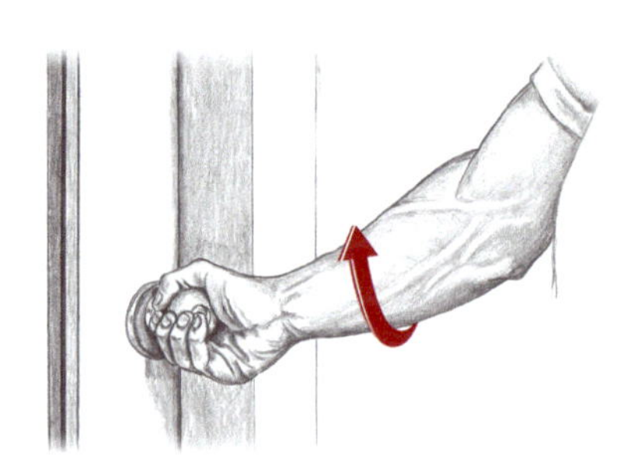

Supination of the forearm

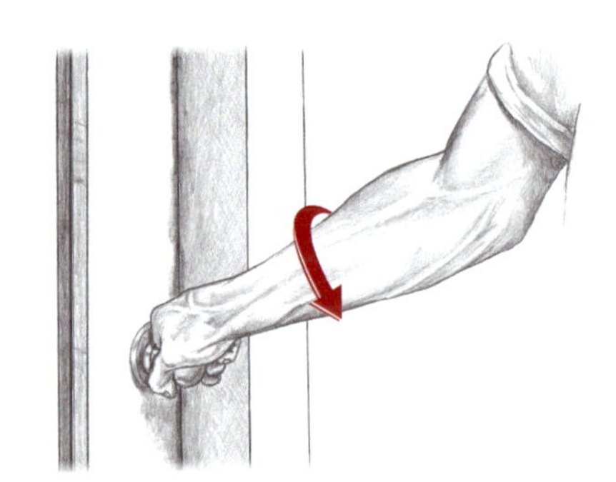

Pronation of the forearm

Wrist

(radiocarpal joint)

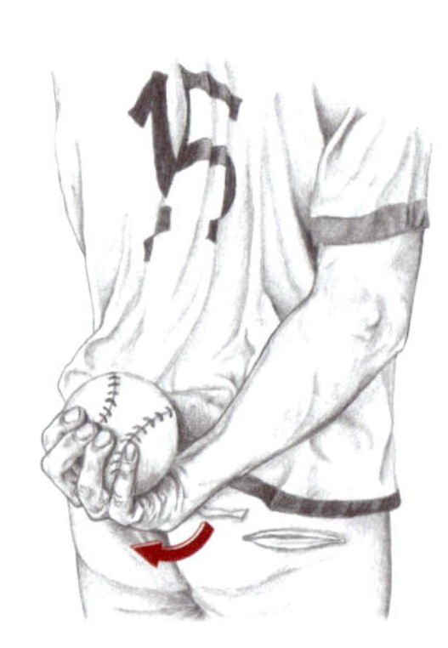

Flexion

Extension

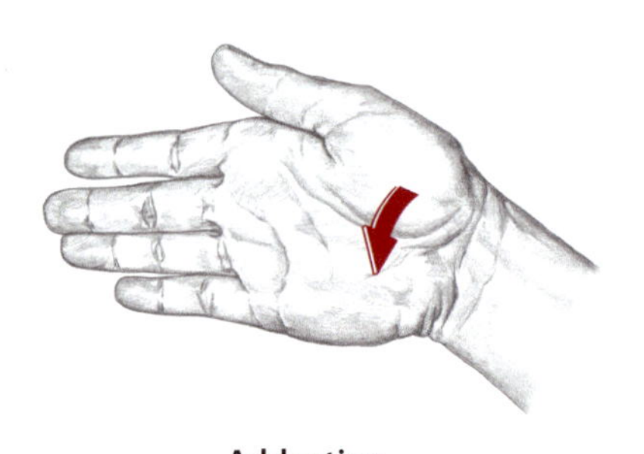

Adduction
(ulnar deviation)

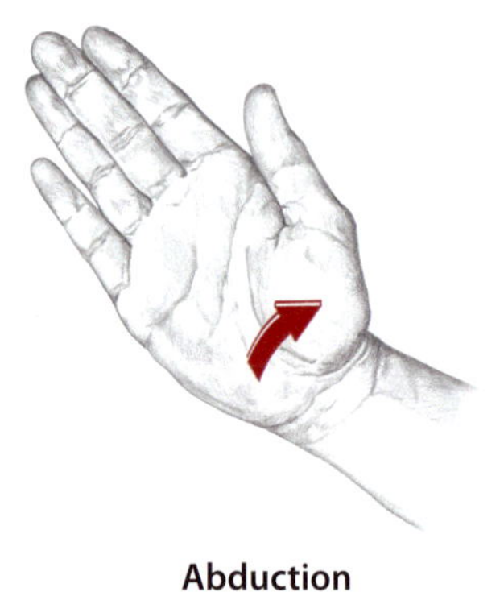

Abduction
(radial deviation)

Thumb

(first carpometacarpal and metacarpophalangeal joints)

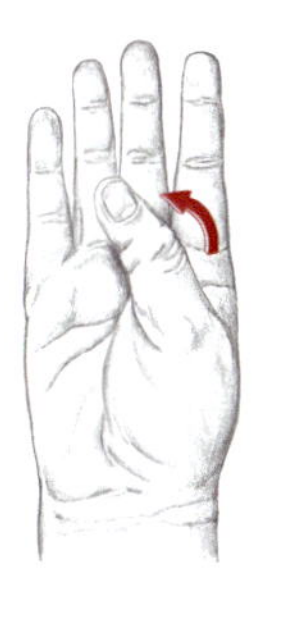

Flexion

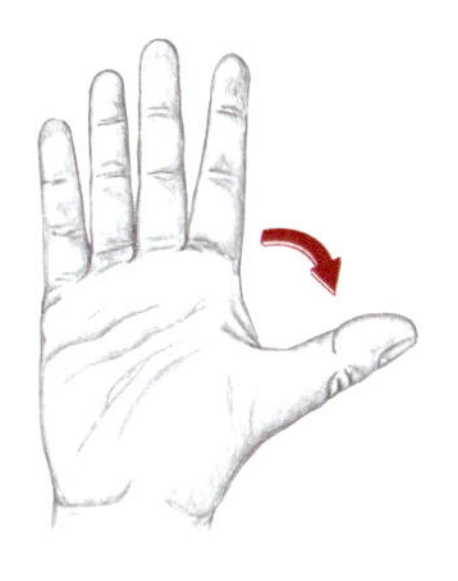

Extension

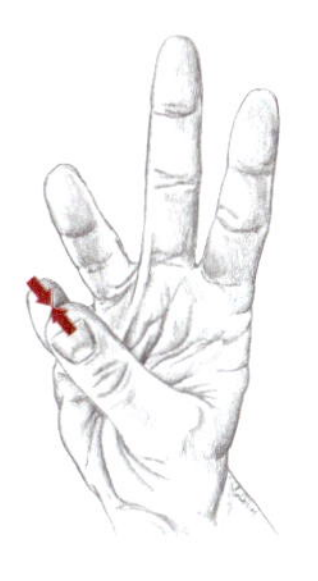

Opposition

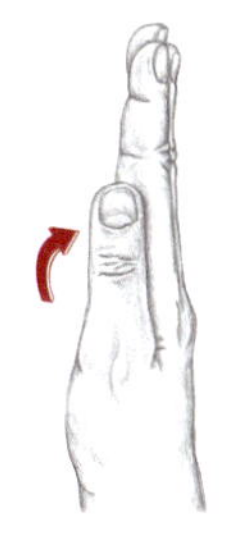

Adduction

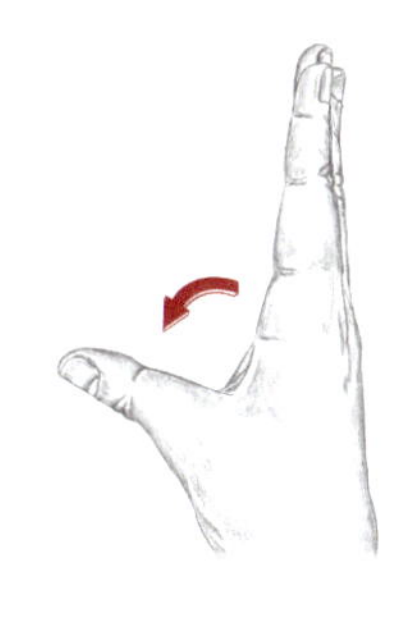

Abduction

Fingers

(metacarpophalangeal, proximal and distal interphalangeal joints)

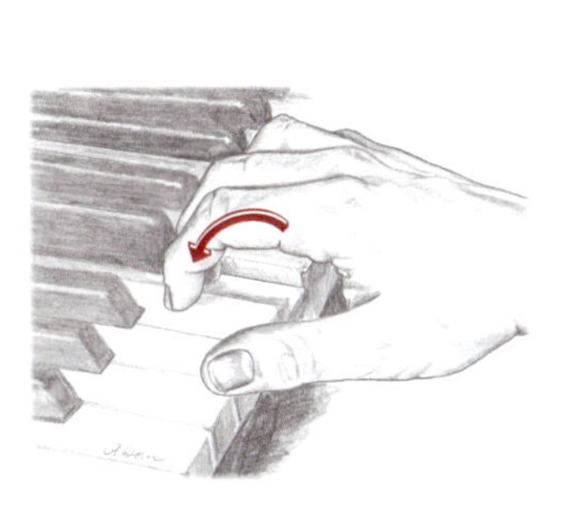

Flexion

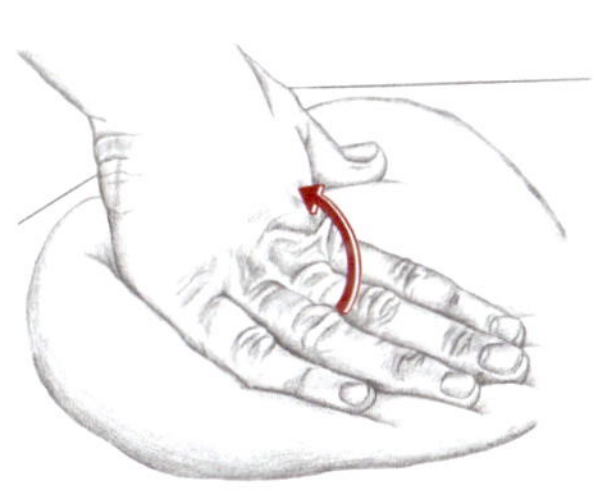

Extension

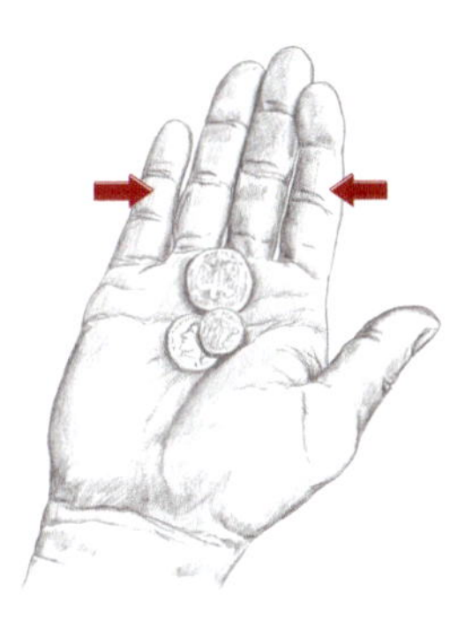

Adduction

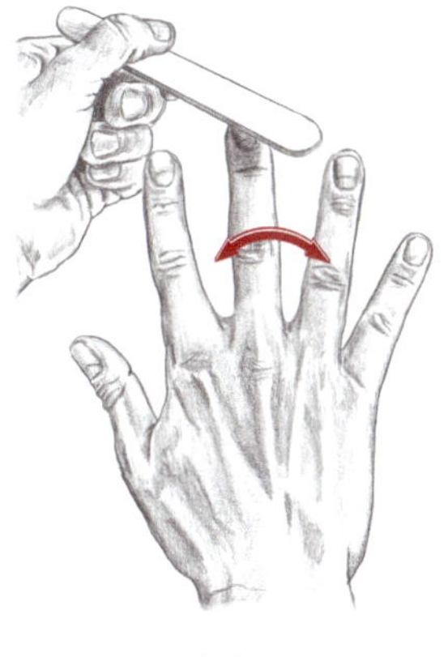

Abduction

Mandible

(temporomandibular joint)

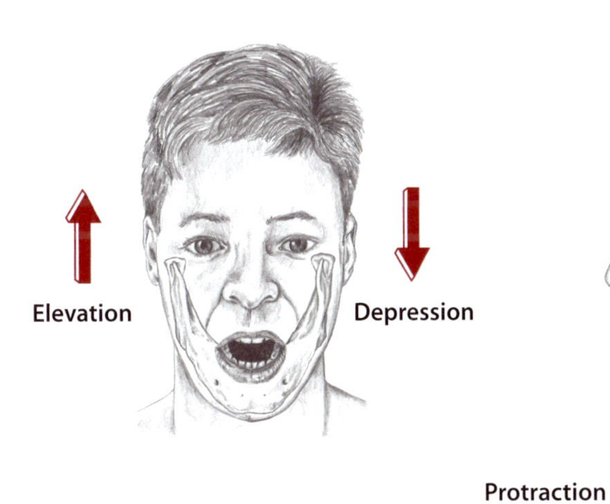

Elevation

Depression

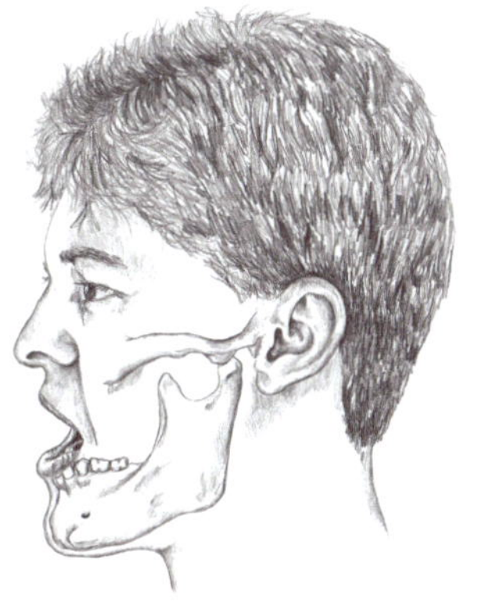

Protraction

Retraction

Lateral deviation

Pelvis

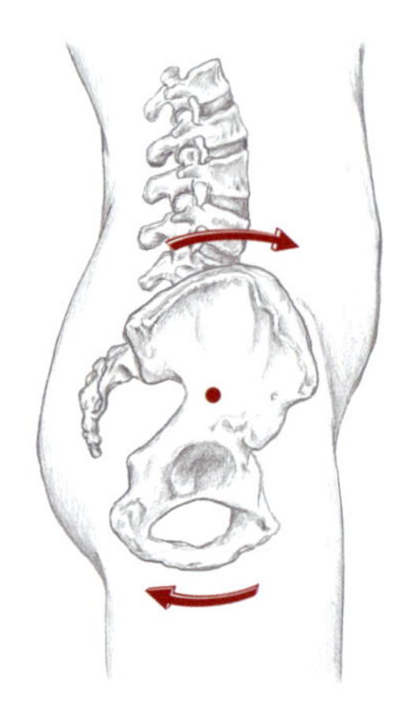

Anterior tilt
(downward rotation)

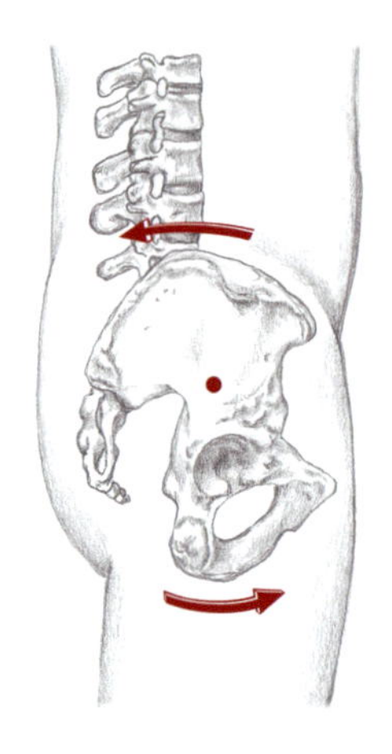

Posterior tilt
(upward rotation)

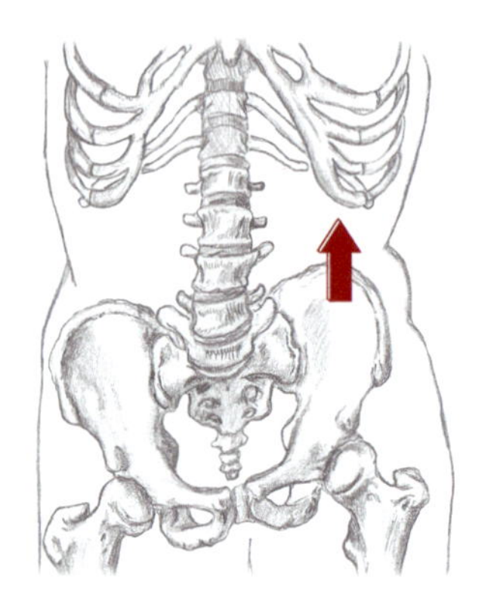

Lateral tilt
(elevation)

Hip

(coxal joint)

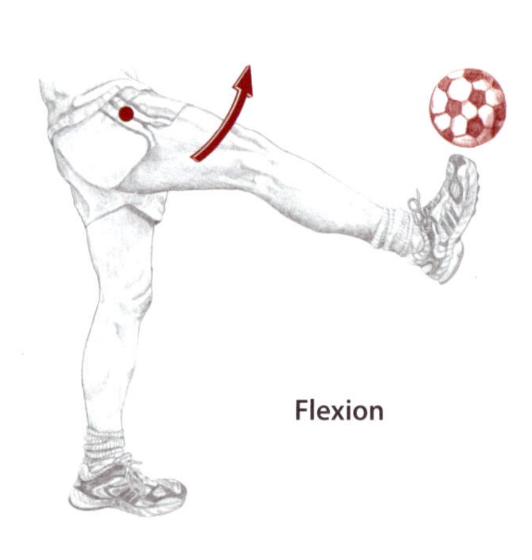

Flexion

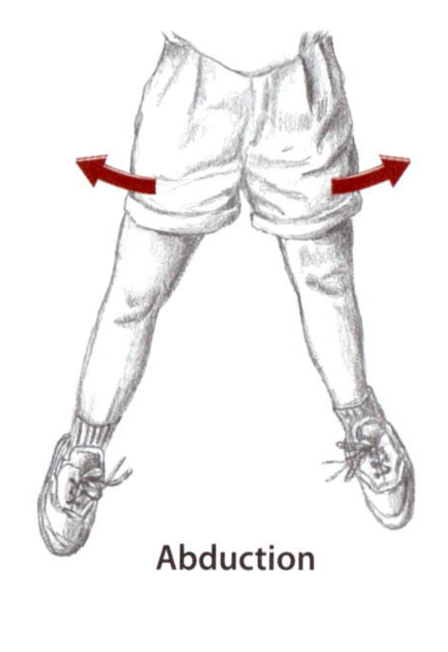

Abduction

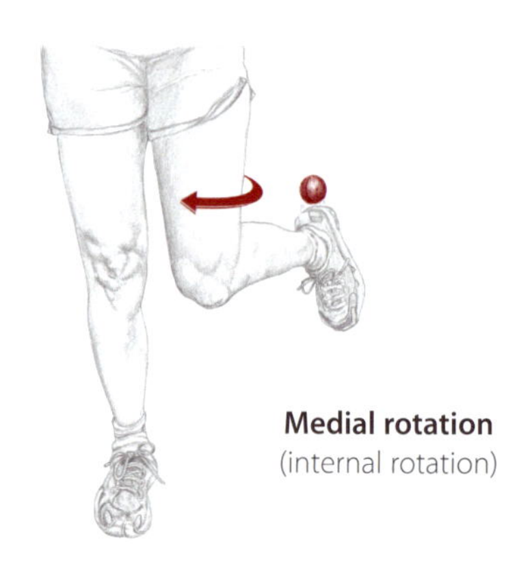

Medial rotation
(internal rotation)

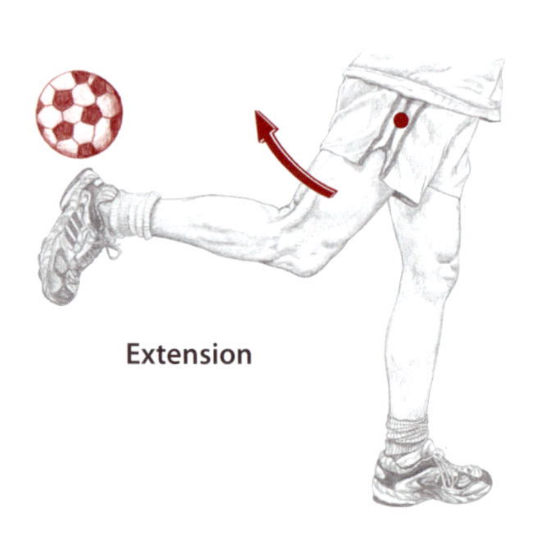

Extension

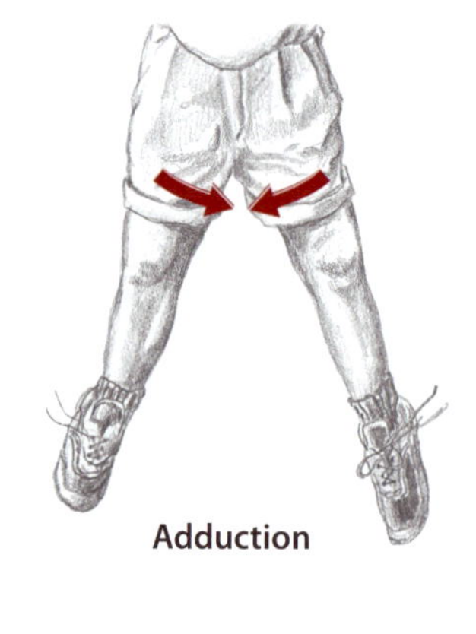

Adduction

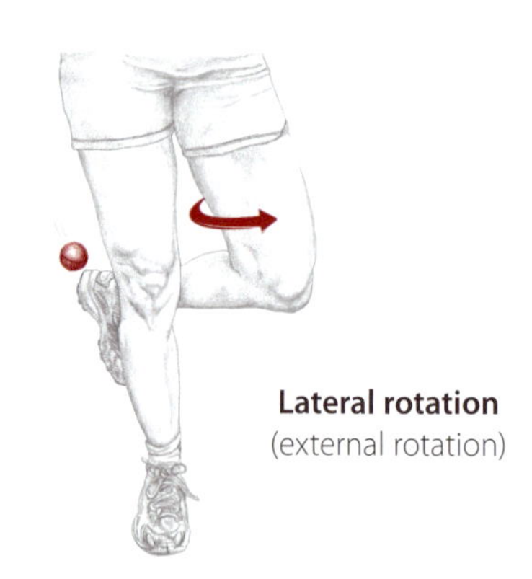

Lateral rotation
(external rotation)

Knee

(tibiofemoral joint)

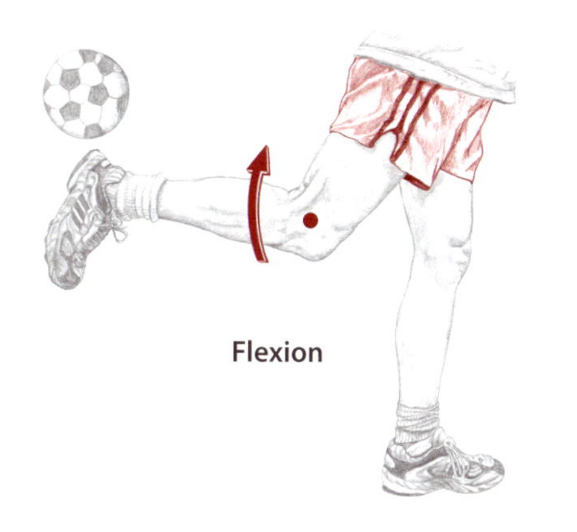

Flexion

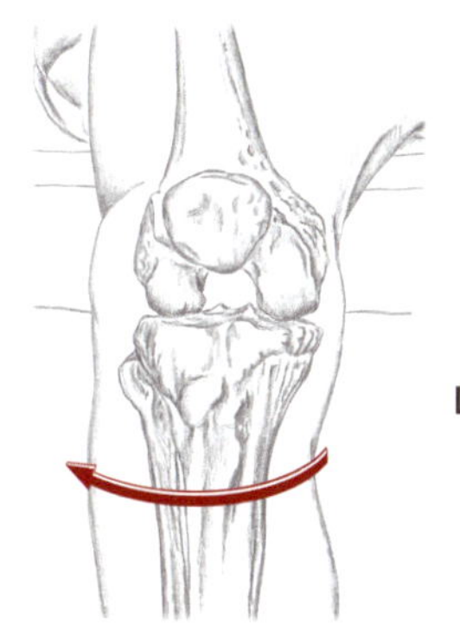

Lateral rotation of flexed knee
(right knee)

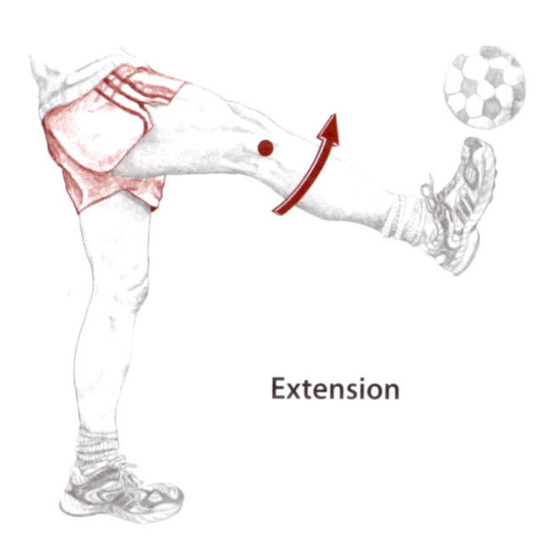

Extension

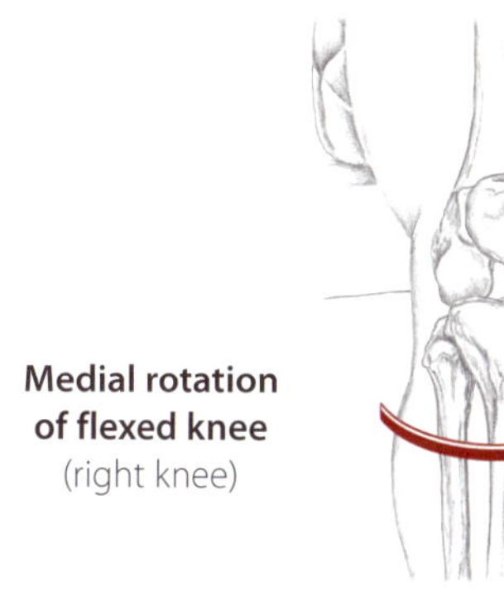

Medial rotation of flexed knee
(right knee)

Ankle, Foot and Toes

(talocrural, talotarsal, midtarsal, tarsometatarsal, metatarsophalangeal and interphalangeal joints)

Dorsiflexion of ankle

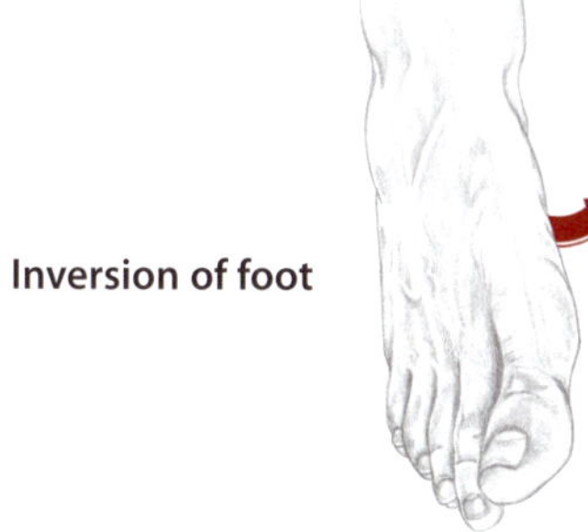

Inversion of foot

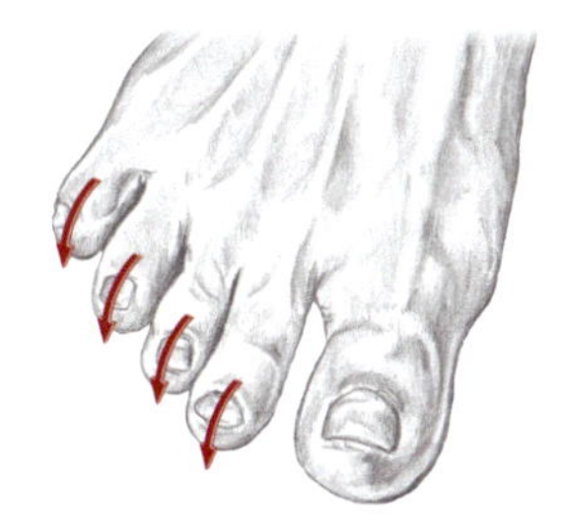

Flexion of toes
"curling the toes"

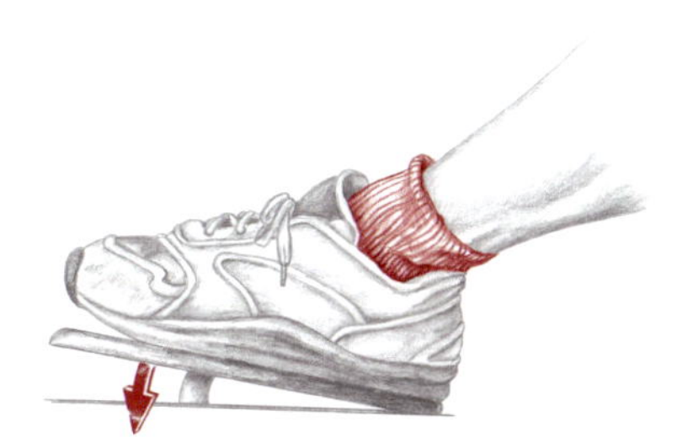

Plantar flexion of ankle

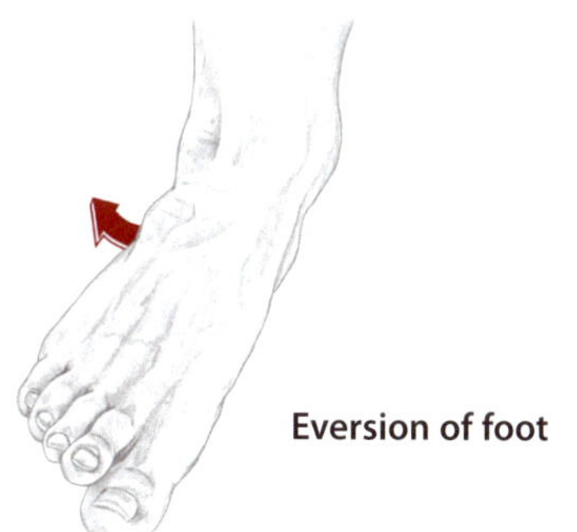

Eversion of foot

Extension of toes
"straighten the toes"

Systems of the Body

The Skeletal System

The bones are linked together to form the skeleton. The skeleton is divided into two sections: the axial and the appendicular skeletons. The **axial** skeleton is the skeleton's center. It includes the cranium, vertebral column, ribs, sternum and hyoid bone. The **appendicular** ("appendages") skeleton is composed of the arms and legs, including the pectoral girdle (scapula and clavicle) and pelvic girdle (hips).

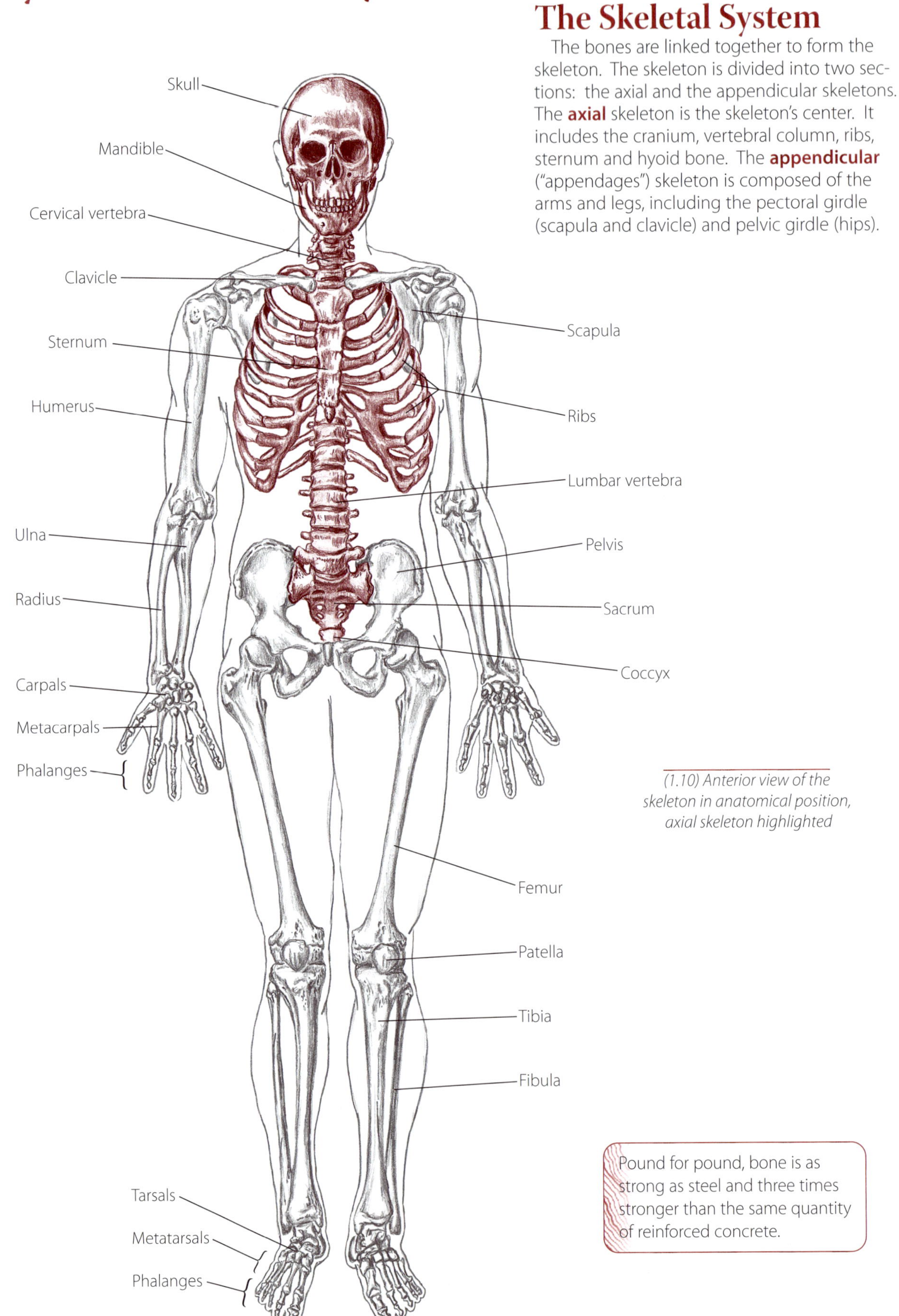

(1.10) Anterior view of the skeleton in anatomical position, axial skeleton highlighted

Pound for pound, bone is as strong as steel and three times stronger than the same quantity of reinforced concrete.

appendicular	**ap**-en-**dik**-u-lar	L. to hang to
axial	**ak**-see-al	L. axle
skeleton	**skel**-et-on	Grk. dried up

The Skeletal System

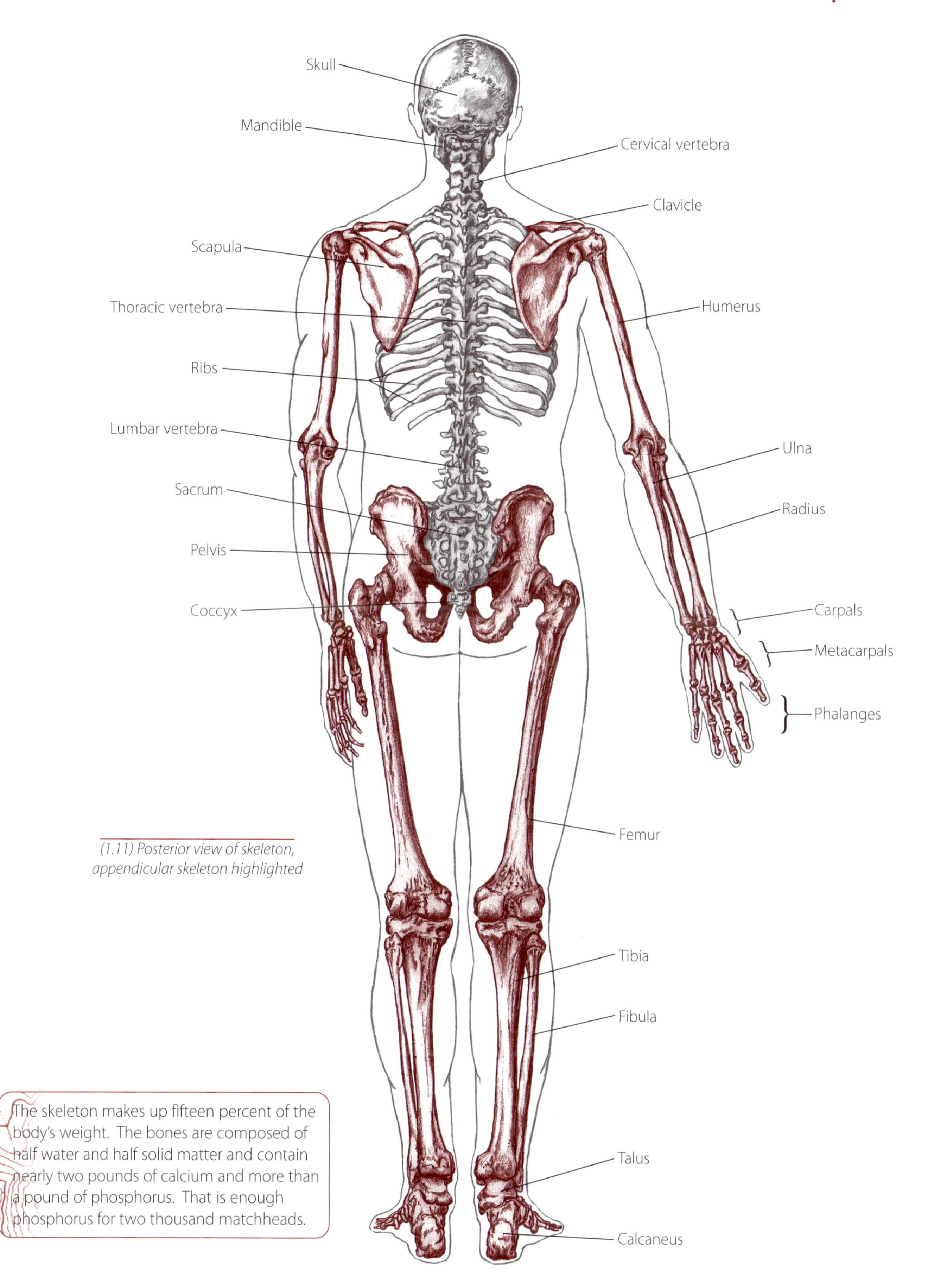

(1.11) Posterior view of skeleton, appendicular skeleton highlighted

The skeleton makes up fifteen percent of the body's weight. The bones are composed of half water and half solid matter and contain nearly two pounds of calcium and more than a pound of phosphorus. That is enough phosphorus for two thousand matchheads.

Types of Joints

A joint or articulation is the point of contact between bones. A joint's structure determines its function. All articulations have a fibrous, cartilaginous or synovial structure. Because of their design, fibrous and cartilaginous joints have little or no movement capability. Synovial joints, however, contain a joint cavity (absent in fibrous and cartilaginous joints). This space allows for movement at the synovial joint. Although synovial joints all have the same basic structural components, they have different movement capabilities. There are six types of synovial joints: ball-and-socket, ellipsoid, hinge, saddle, gliding and pivot.

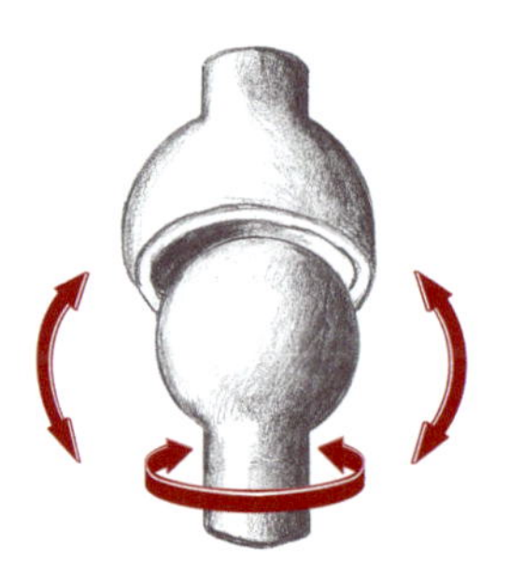

A **ball-and-socket joint** is self-explanatory: A spherical surface of one bone fits into the dish-shaped depression of another bone. Such a joint is capable of movement in every plane. The shoulder (or glenohumeral joint) is an example of a joint capable of circumduction.

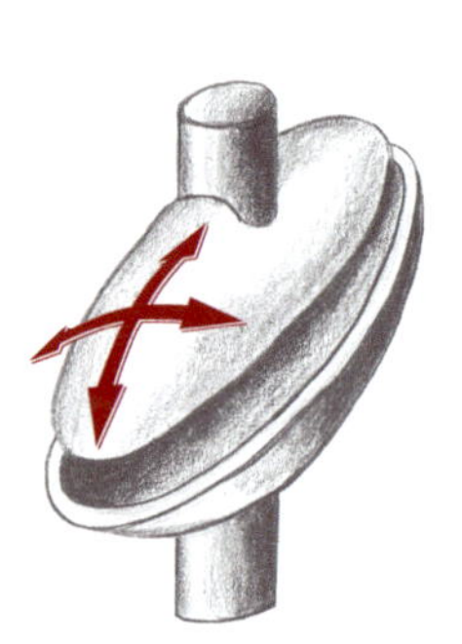

An **ellipsoid joint** consists of the oval-shaped end of one bone articulating with the elliptical basin of another bone. It permits flexion/extension and abduction/adduction as seen at the wrist (radiocarpal) joint.

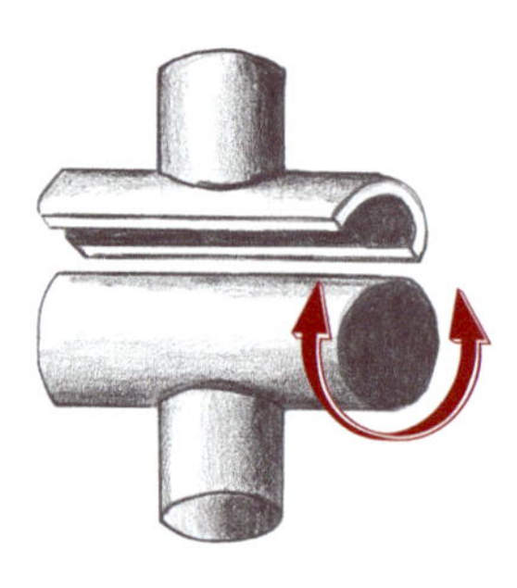

A **hinge** joint allows only flexion and extension, similar to the movements of a door hinge. An example of a hinge joint is the elbow (humeroulnar) joint.

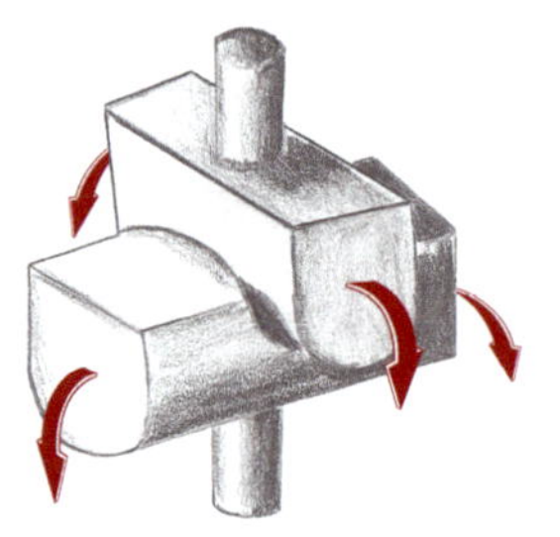

A **saddle joint** is a modified ellipsoid joint composed of convex and concave articulating surfaces - like two saddles. The joint between the trapezium (one of the small carpal bones in the wrist) and the first metacarpal bones is an example of a saddle joint.

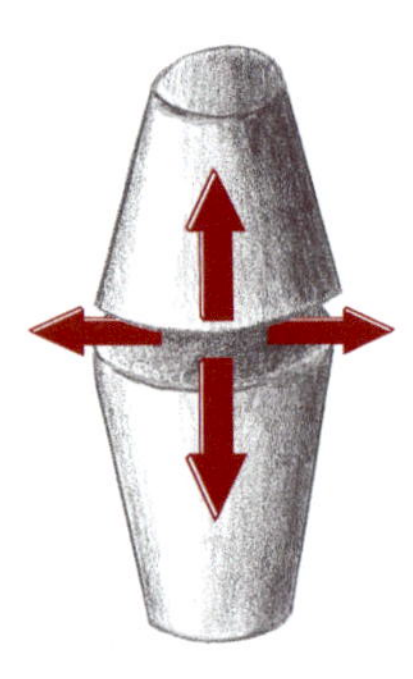

A **gliding joint** is usually between two flat surfaces and allows the least amount of movement of all synovial joints. Only small shifting movements are possible at these articulations, such as those between the carpal bones in the wrist or tarsal bones in the foot.

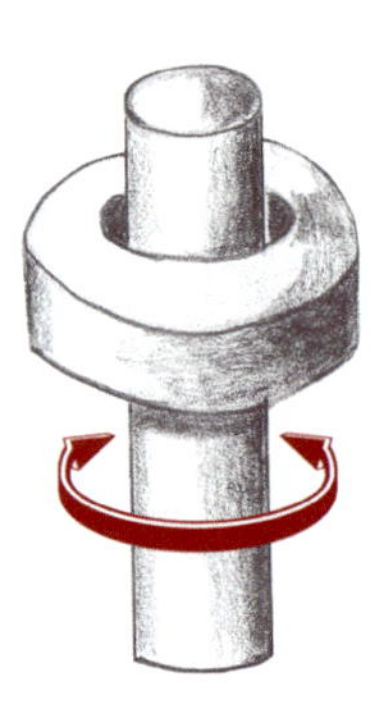

A **pivot joint** is designed to allow one bone to rotate around the surface of another bone. For example, rotation of the head occurs because of the pivot joint between the first and second cervical vertebrae (the atlantoaxial joint).

synovial sin-**o**-ve-al L. *synovia*, joint fluid

The Muscular System

A muscle's name can give you clues to its specific features. The name reflects either a muscle's shape (rhomboid), location (temporalis), fiber direction (external oblique), action (adductors) or attachment sites (coracobrachialis).

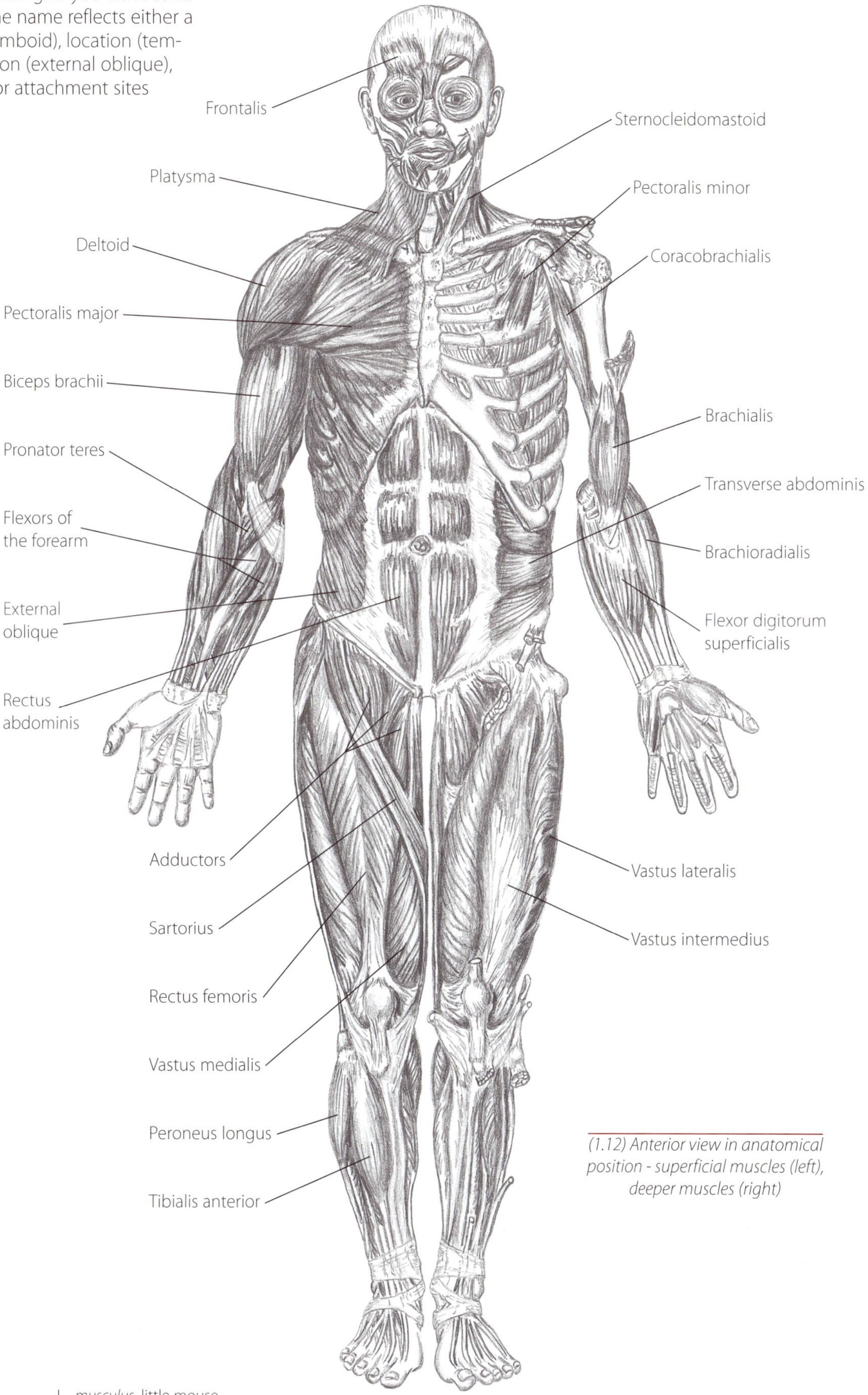

(1.12) Anterior view in anatomical position - superficial muscles (left), deeper muscles (right)

muscle	**mus**-el	L. *musculus*, little mouse
myo-		Grk. muscle
tendon	**ten**-dun	L. to stretch

The Muscular System

At either end of the muscle is a tendon which attaches the muscle to a bone. Each muscle has an origin and an insertion. The origin is the attachment to the more stationary bone while the insertion is the connection to the more mobile bone.

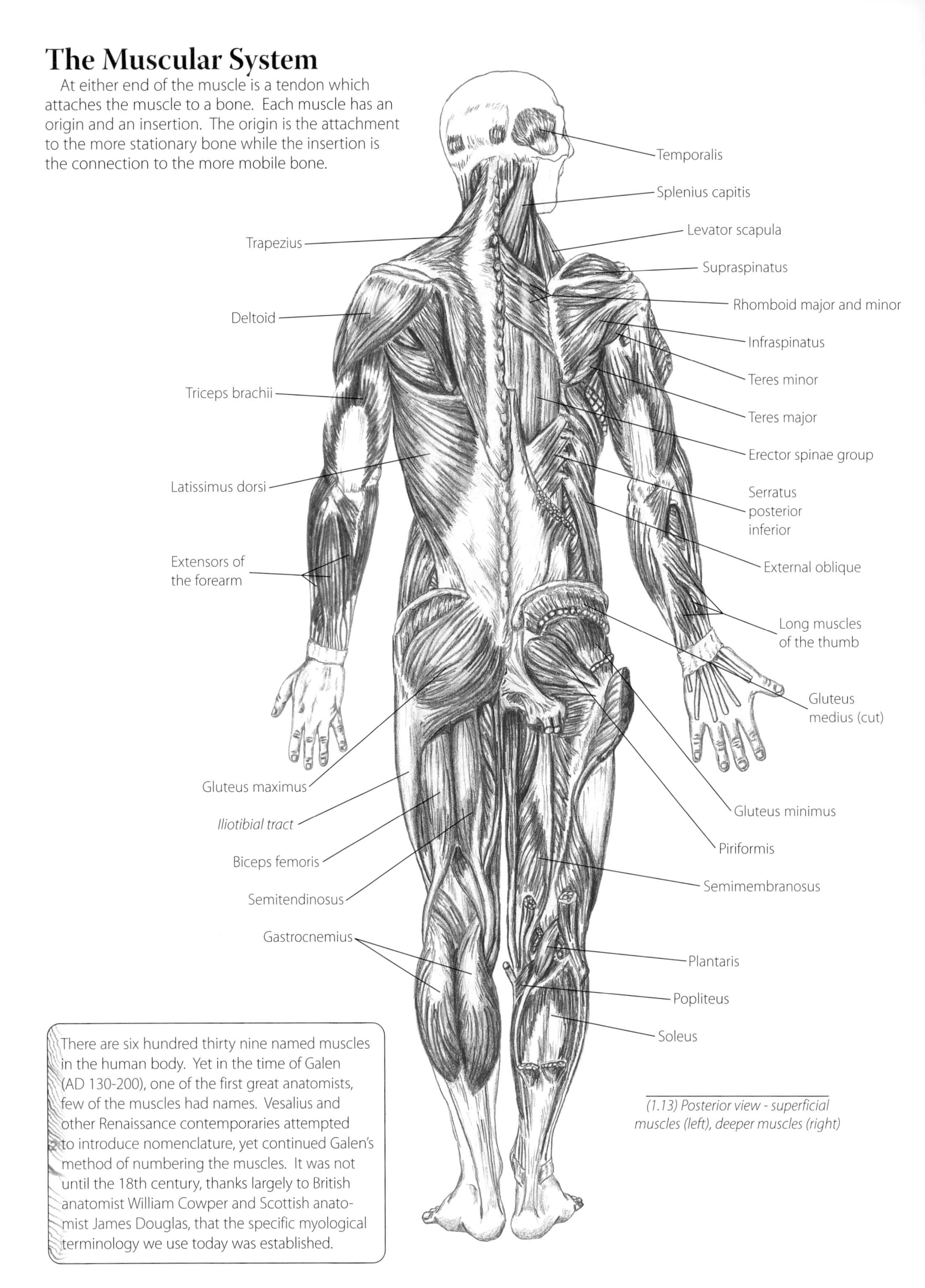

(1.13) Posterior view - superficial muscles (left), deeper muscles (right)

There are six hundred thirty nine named muscles in the human body. Yet in the time of Galen (AD 130-200), one of the first great anatomists, few of the muscles had names. Vesalius and other Renaissance contemporaries attempted to introduce nomenclature, yet continued Galen's method of numbering the muscles. It was not until the 18th century, thanks largely to British anatomist William Cowper and Scottish anatomist James Douglas, that the specific myological terminology we use today was established.

The Muscular System

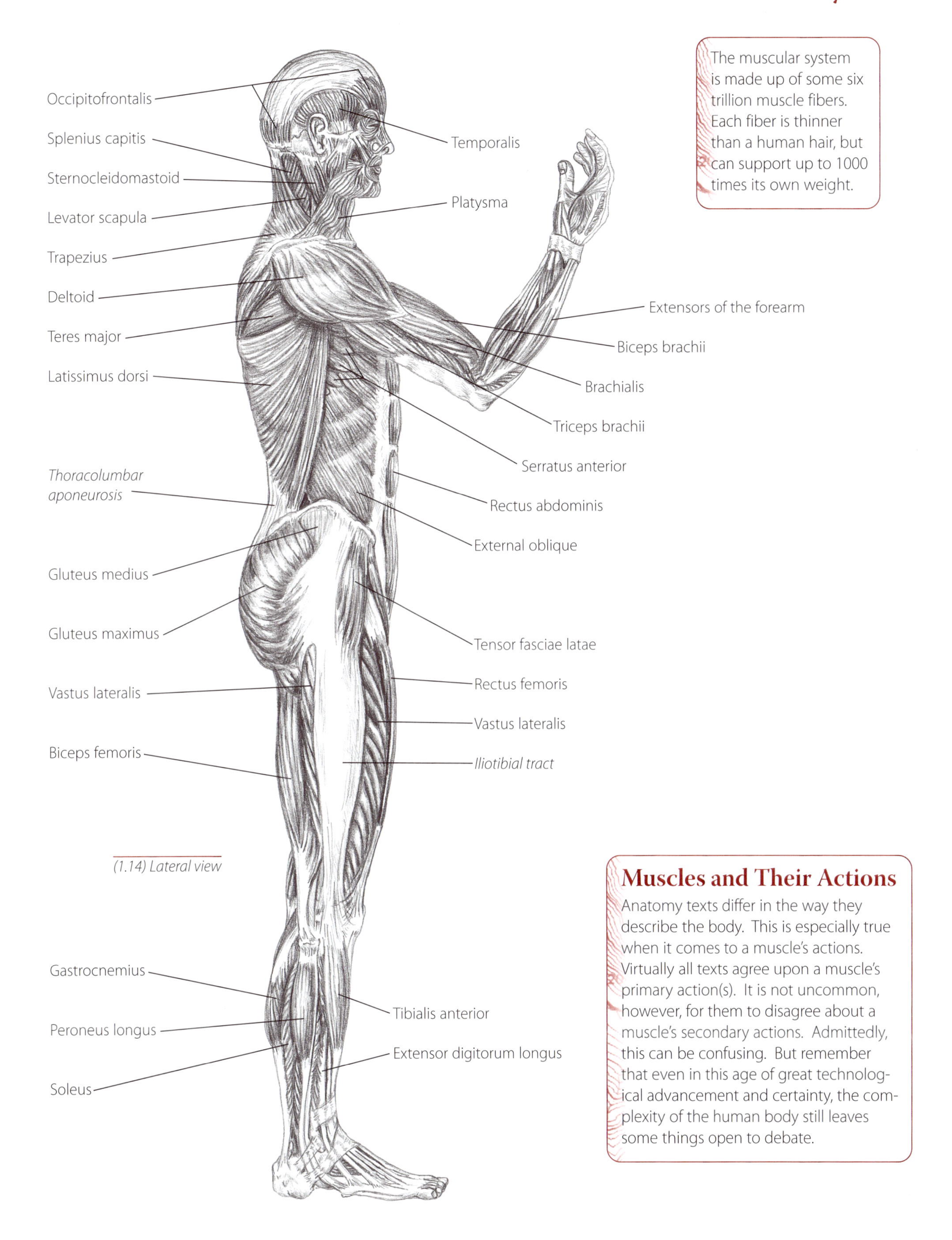

(1.14) Lateral view

The muscular system is made up of some six trillion muscle fibers. Each fiber is thinner than a human hair, but can support up to 1000 times its own weight.

Muscles and Their Actions

Anatomy texts differ in the way they describe the body. This is especially true when it comes to a muscle's actions. Virtually all texts agree upon a muscle's primary action(s). It is not uncommon, however, for them to disagree about a muscle's secondary actions. Admittedly, this can be confusing. But remember that even in this age of great technological advancement and certainty, the complexity of the human body still leaves some things open to debate.

The Fascial System

The following illustrations show aspects of the fascia from both topographical and cross section viewpoints.

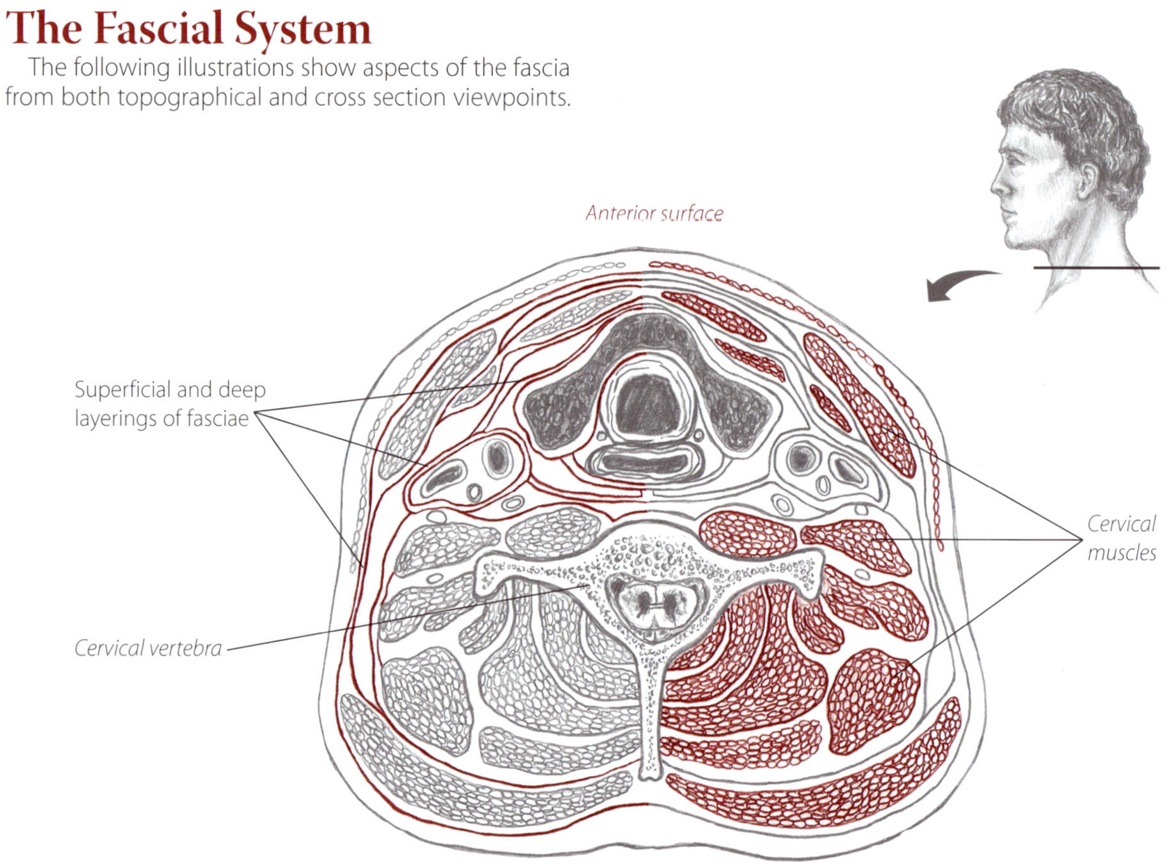

(1.15) Cross section of the neck highlighting layers of fasciae (left) and cervical muscles (right)

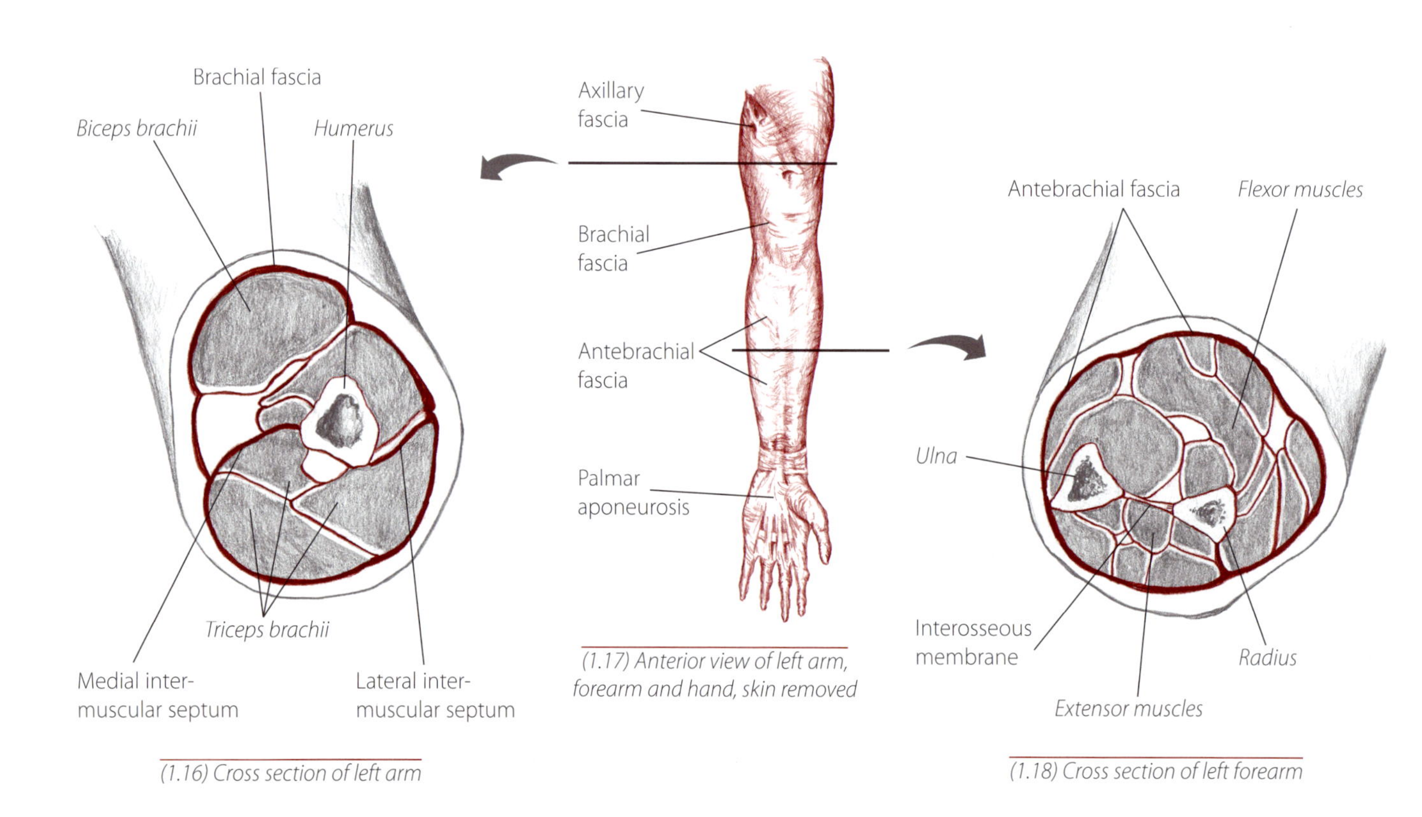

(1.16) Cross section of left arm

(1.17) Anterior view of left arm, forearm and hand, skin removed

(1.18) Cross section of left forearm

fascia	**fash**-ah	L. a band, bandage
retinaculum	**ret**-i-**nak**-u-lum	L. halter, band, rope
septum	**sep**-tum	L. enclosure

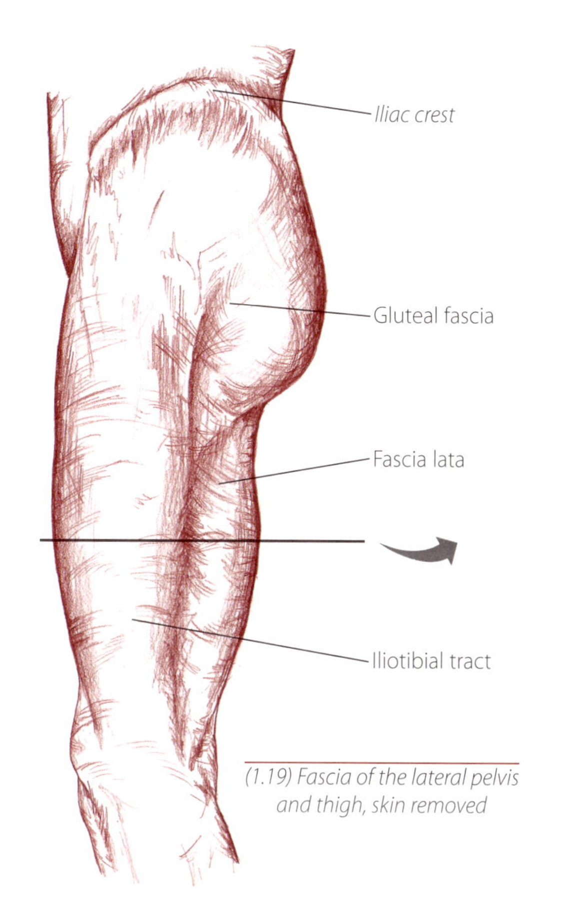

(1.19) Fascia of the lateral pelvis and thigh, skin removed

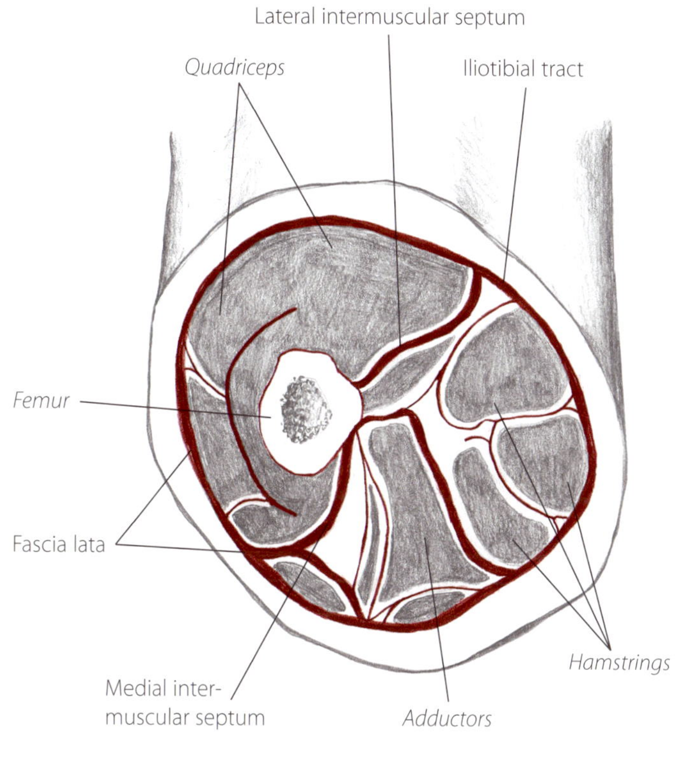

(1.20) Cross section of left thigh

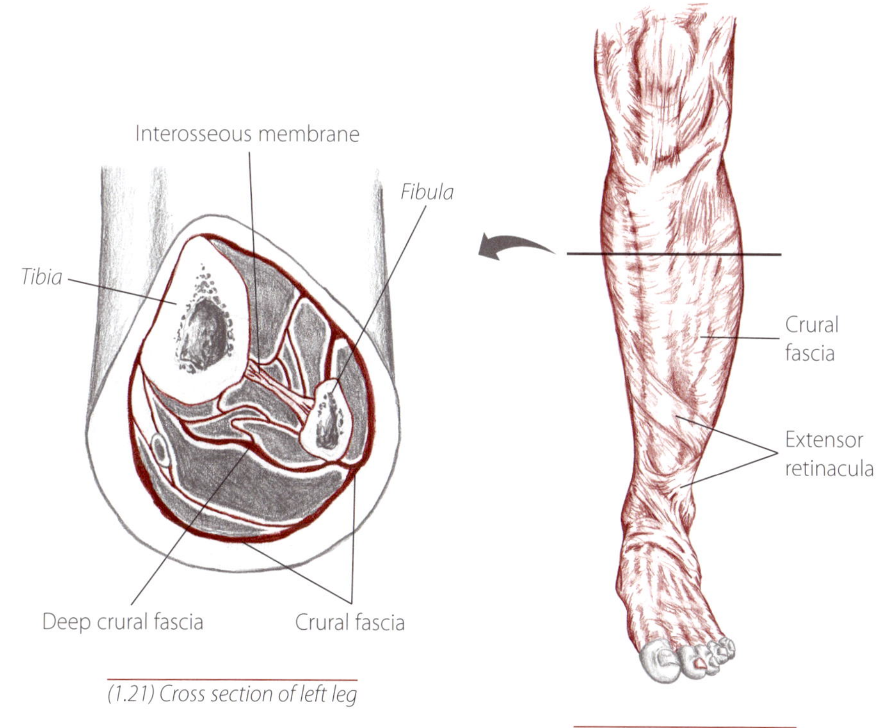

(1.21) Cross section of left leg

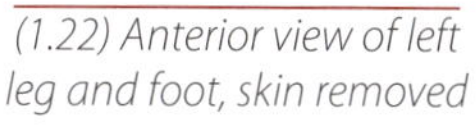

(1.22) Anterior view of left leg and foot, skin removed

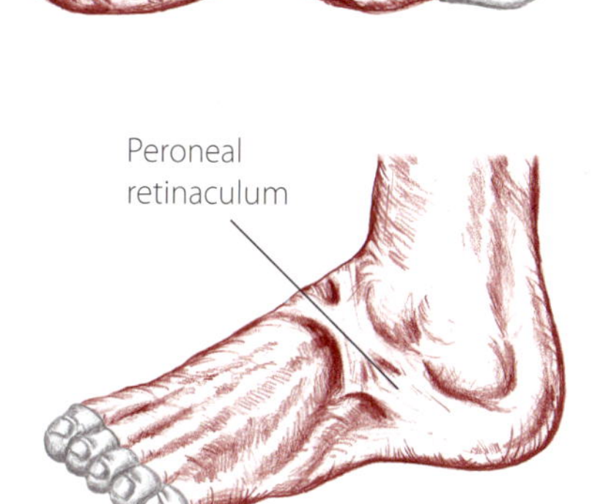

(1.23) Medial and lateral views of left foot, skin removed

aponeurosis **ap**-o-**nu**-ro-sis Grk. *apo*, from + *neuron*, nerve or tendon

The Cardiovascular System

Arteries and veins are the blood vessels of the cardiovascular system. They form an amazing network that transports blood from the heart, brings it to the body's tissues and then carries it back to the heart.

Arteries carry blood away from the heart. As an artery moves away from the heart, it divides into smaller branches. Arterioles, the smallest branches, divide into millions of microscopic vessels called capillaries. The walls of the capillaries serve as nutrient and waste exchange sites between the body's tissues and the blood. The capillaries then merge back together, creating small veins or venules which then unite to form the larger veins that carry the blood back to the heart.

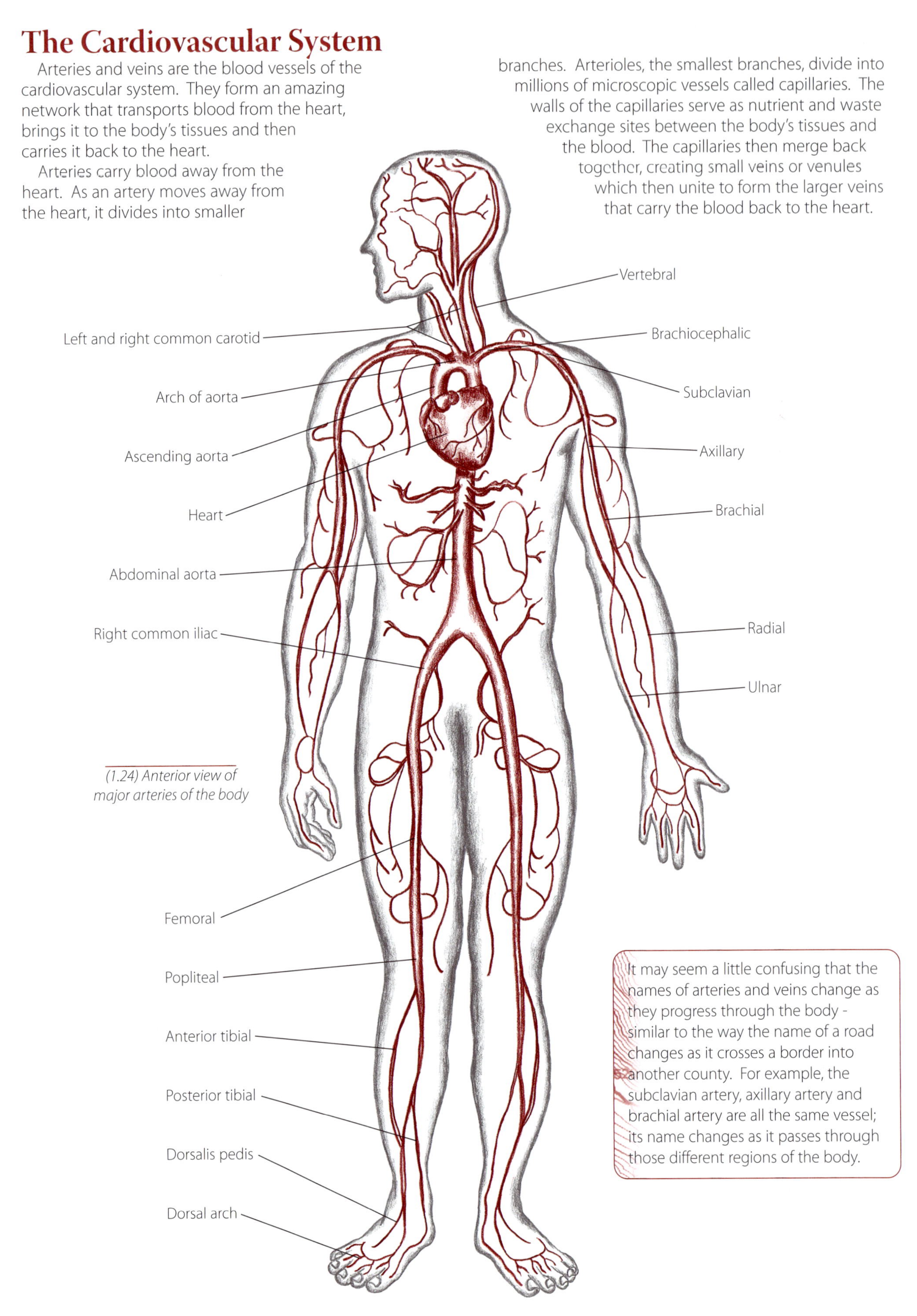

(1.24) Anterior view of major arteries of the body

It may seem a little confusing that the names of arteries and veins change as they progress through the body - similar to the way the name of a road changes as it crosses a border into another county. For example, the subclavian artery, axillary artery and brachial artery are all the same vessel; its name changes as it passes through those different regions of the body.

artery		Grk. windpipe
capillary	**kap**-i-**lar**-ee	L. hairlike
vein		L. vessel

The Cardiovascular System

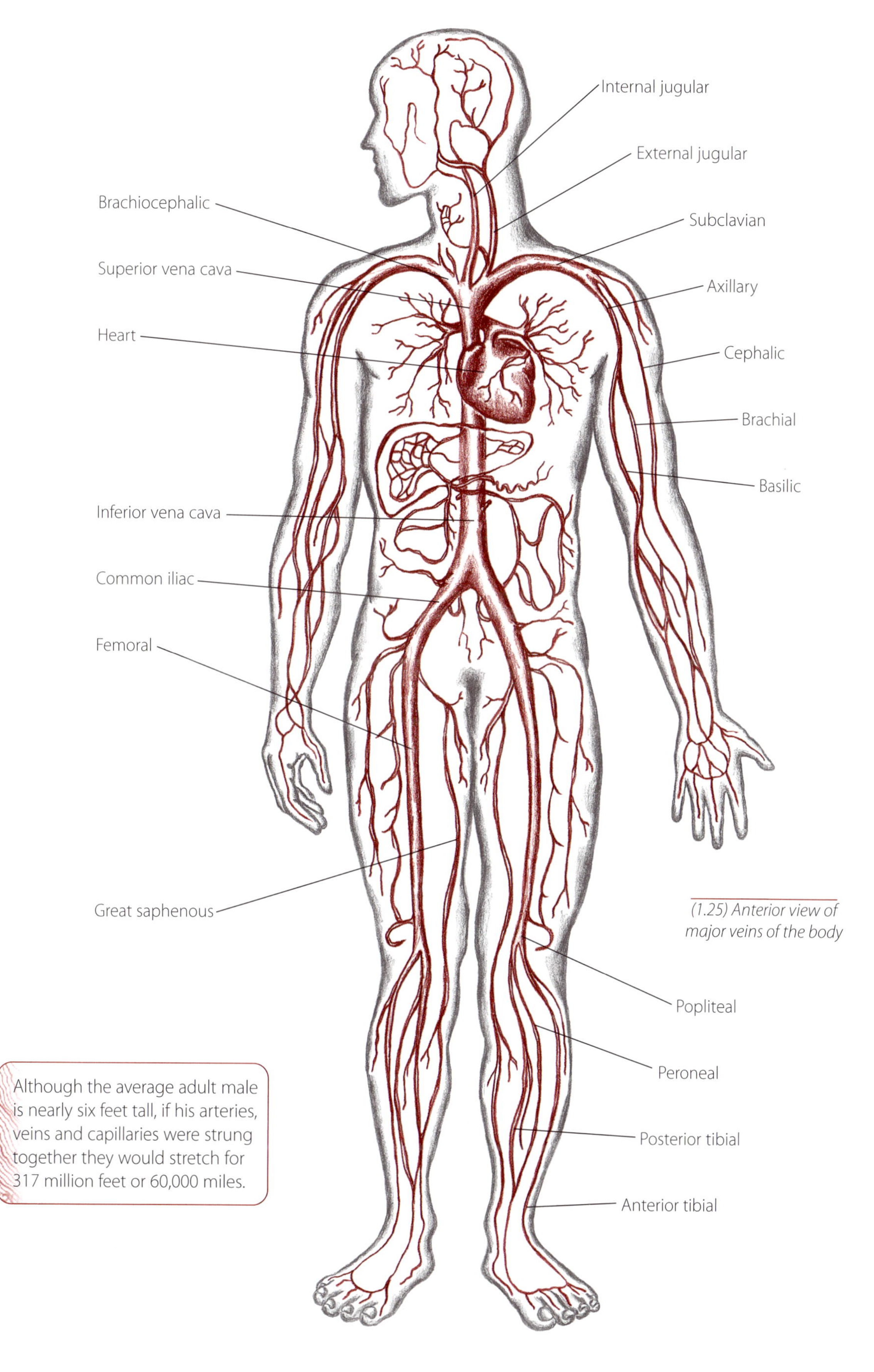

(1.25) Anterior view of major veins of the body

Although the average adult male is nearly six feet tall, if his arteries, veins and capillaries were strung together they would stretch for 317 million feet or 60,000 miles.

The Nervous System

The nervous system is the body's functional headquarters. It senses, interprets and responds to the body's needs in order to maintain homeostasis or equilibrium. The brain and spinal cord make up the central nervous system while the remaining aspects form the peripheral nervous system.

Many nerves branch off the spinal cord and exit through the sides of the vertebrae. Some of these nerves regroup to form a plexus. The main plexuses are the cervical, brachial, lumbar and sacral. The individual branches of a nerve plexus split off and have names that correspond to the regions they innervate.

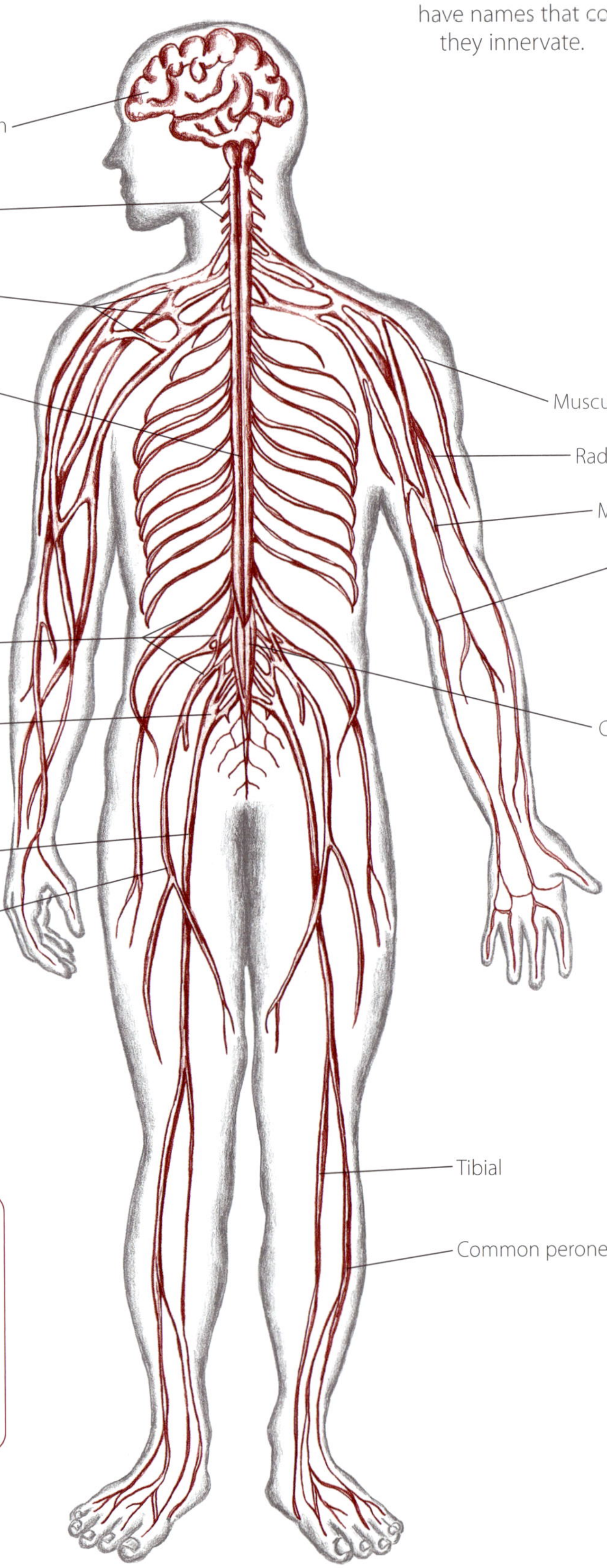

Brain
Cervical plexus
Brachial plexus
Spinal cord
Musculocutaneous
Radial
Median
Ulnar
Lumbar plexus
Sacral plexus
Cauda equina
Sciatic
Femoral
Tibial
Common peroneal

(1.26) Anterior view of major nerves of the body

Ants and bees, both revered for their intelligence and diligence, have roughly 250 and 900 nerve cells, respectively, in their entire bodies. Humans, who do not always demonstrate such qualities, have an estimated 10,000,000,000 nerve cells in the brain alone.

cauda equina **kaw**-da eh-**kwy**-na L. horse's tail

The Lymphatic System

The lymphatic system is composed of several organs, yellow fluid called lymph, small microscopic vessels called lymphatics and lymph nodes. These structures perform many functions throughout the body such as draining the interstitial fluid which escapes from capillaries and transporting it back to the heart. Lymphatic vessels carry fats from the intestines to the blood. Lymphatic tissue also helps the body's immune system defend against foreign cells, microbes and cancer cells.

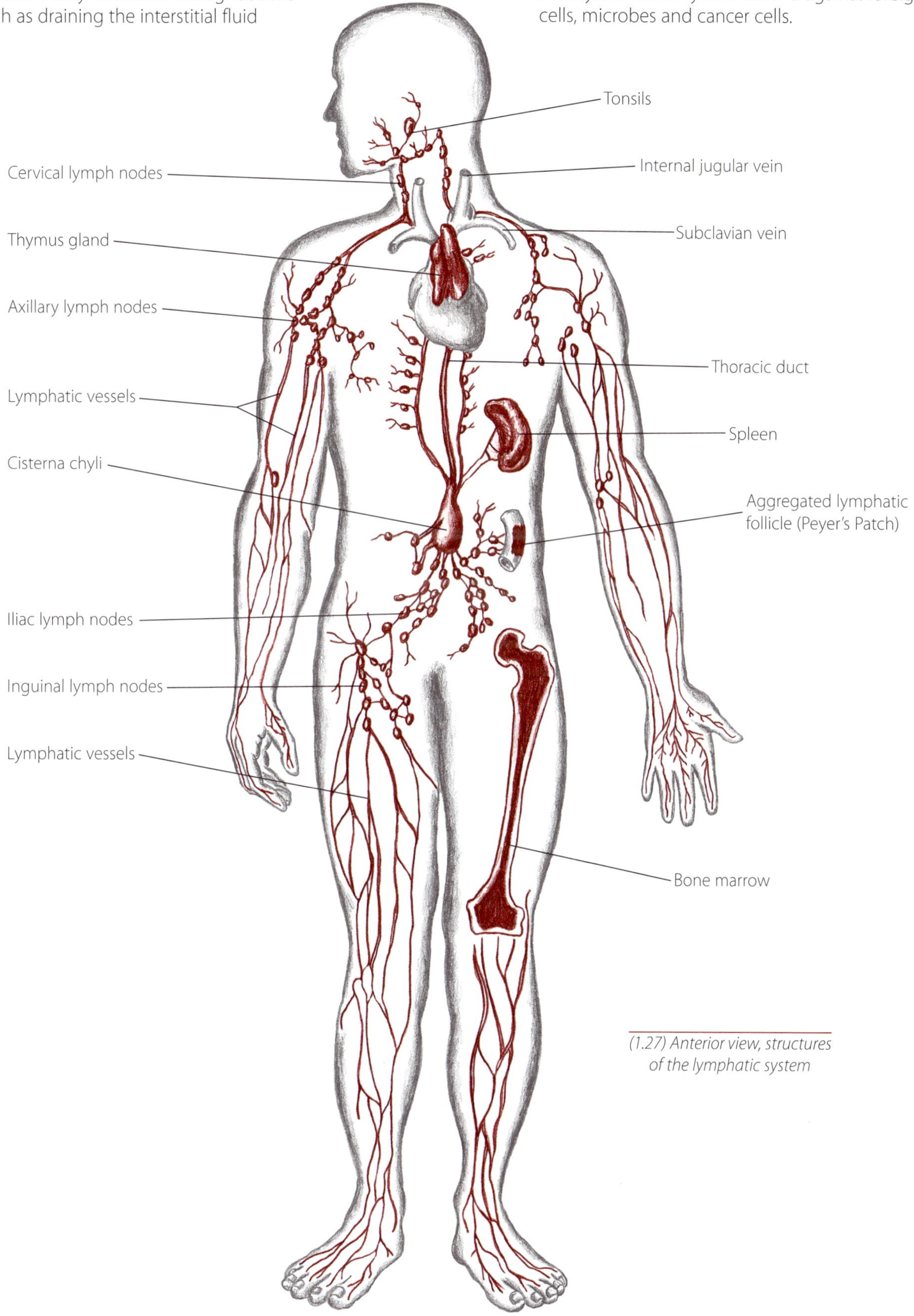

Tonsils
Cervical lymph nodes
Internal jugular vein
Thymus gland
Subclavian vein
Axillary lymph nodes
Thoracic duct
Lymphatic vessels
Spleen
Cisterna chyli
Aggregated lymphatic follicle (Peyer's Patch)
Iliac lymph nodes
Inguinal lymph nodes
Lymphatic vessels
Bone marrow

(1.27) Anterior view, structures of the lymphatic system

interstitial	in-ter-**stish**-al	L. placed between
lymph	limf	L. pure spring water
cisterna chyli	sis-**turn**a **ki**-lee	

NOTES

To the shoulder and arm...

2

Shoulder & Arm

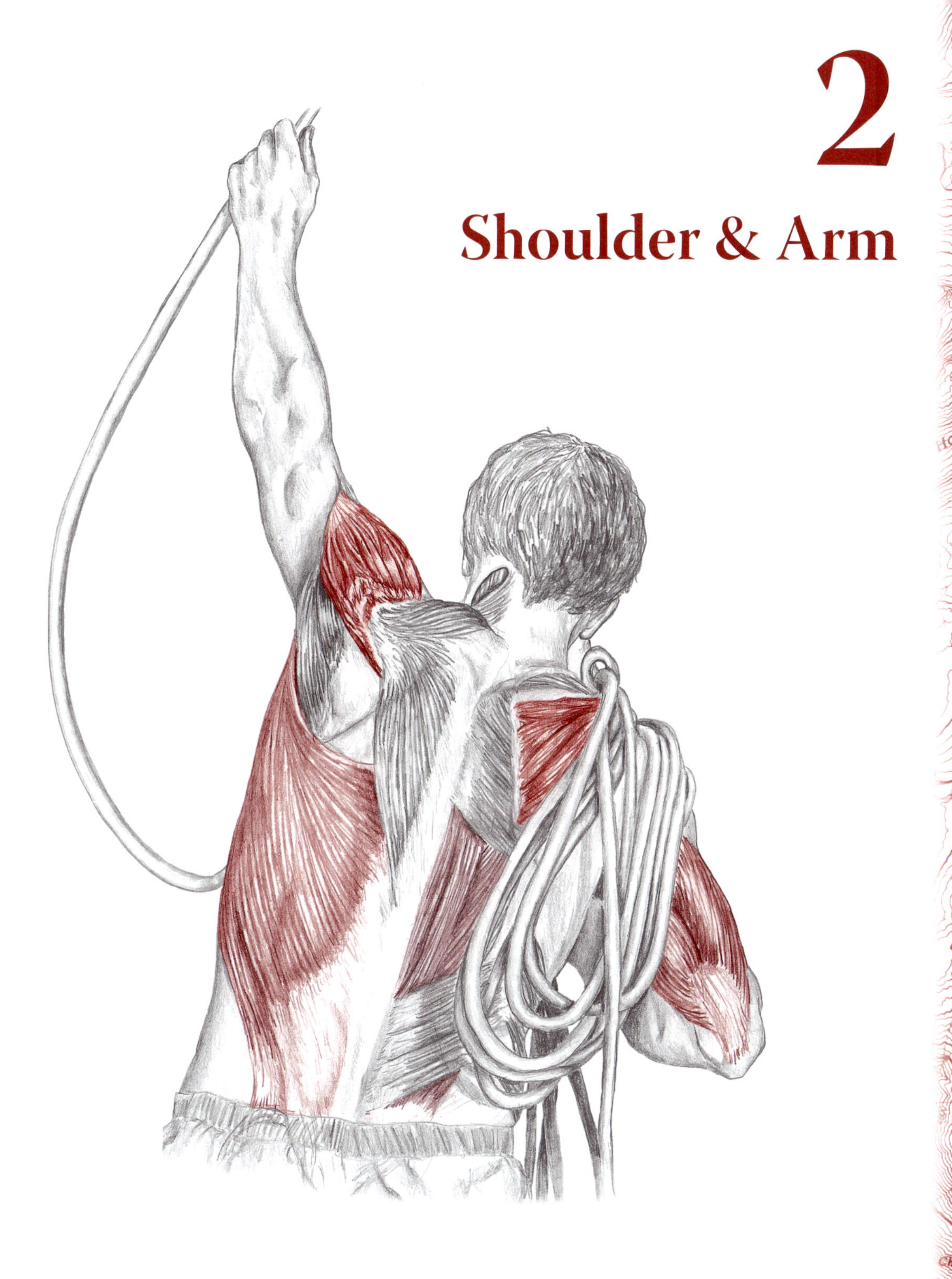

Topographical Views

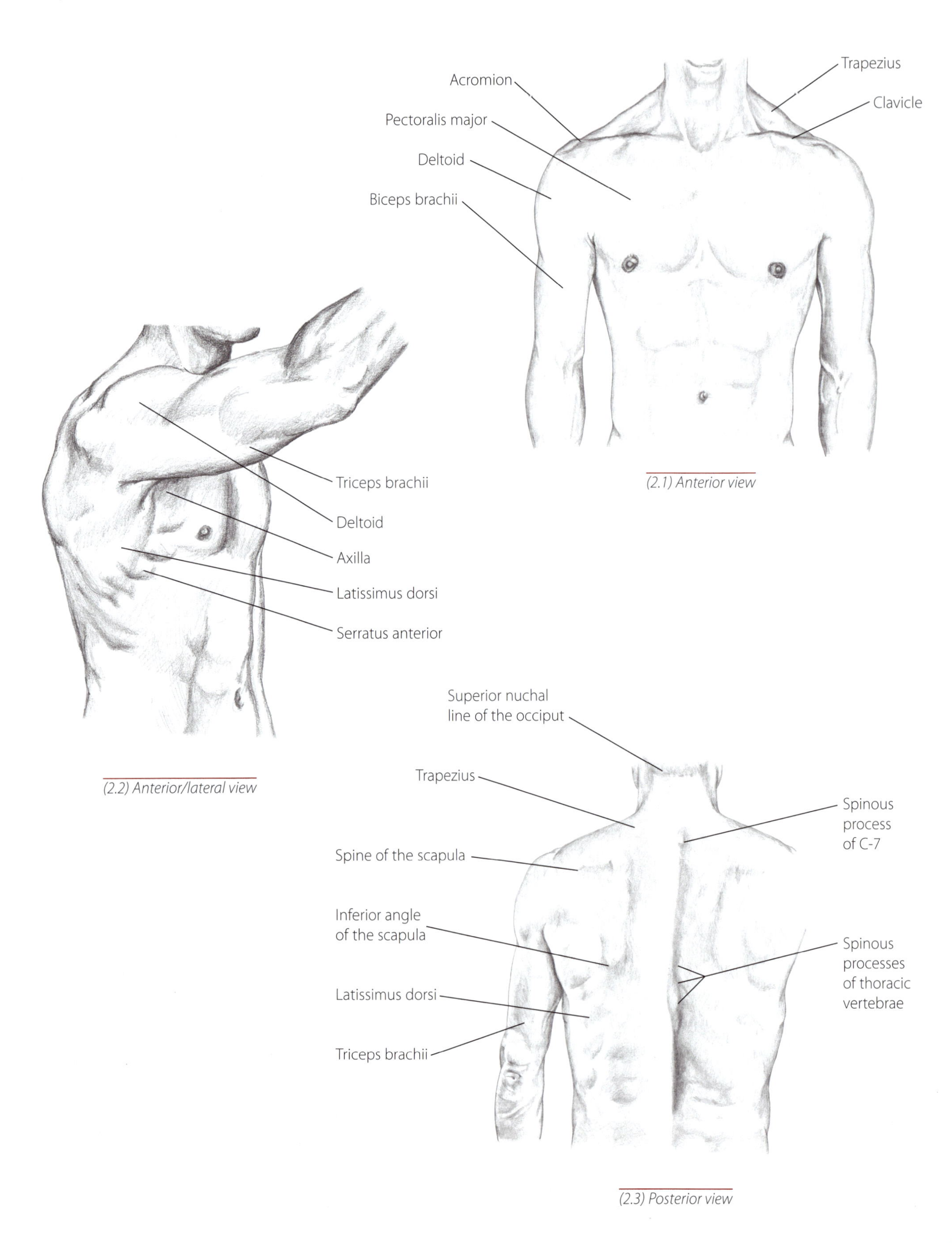

(2.1) Anterior view

(2.2) Anterior/lateral view

(2.3) Posterior view

Exploring the Skin and Fascia

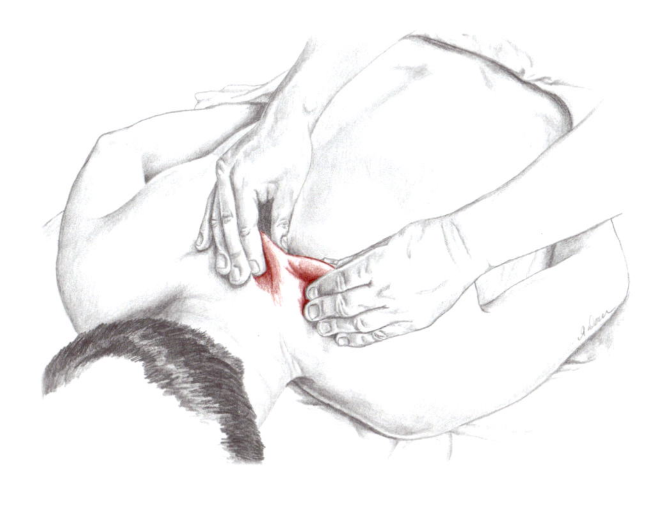

(2.4) Partner prone

1) Partner prone. Begin by gently lifting the skin and fascia of the upper back. As you raise it away from the thicker, deeper musculature, twist the tissue from side to side (2.4). Compare the changes in tissue as you explore the top of the shoulders, arms and upper chest.
2) Take particular note of the tissue's changes in thickness and elasticity. For example, the skin and fascia superficial to the spine of the scapula may be dense and matted, while the tissue at the top of the shoulder, only a few inches away, may be thin and mobile.

1) Partner supine. Slowly sink your fingers into the skin of the upper chest. Then gently shift the tissue from side to side (2.5). Try moving it in all directions, sensing its mobility, resistance and temperature.
2) Compare this tissue with other areas of the shoulder and arm, including the axilla (armpit) and the area near the clavicle.

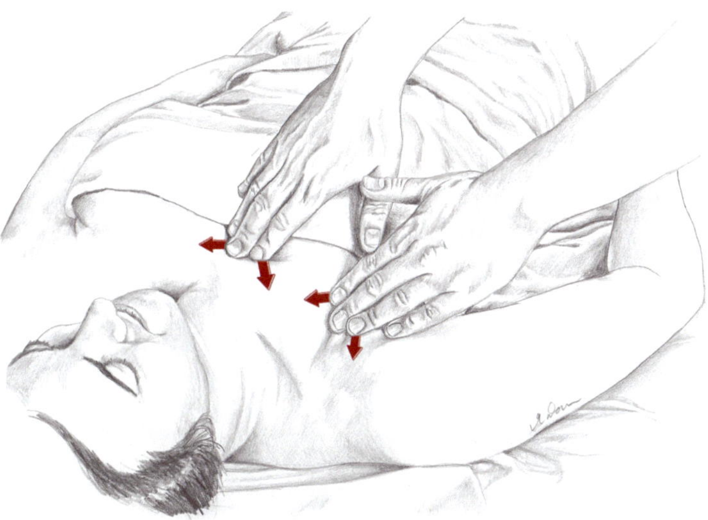

(2.5) Partner supine

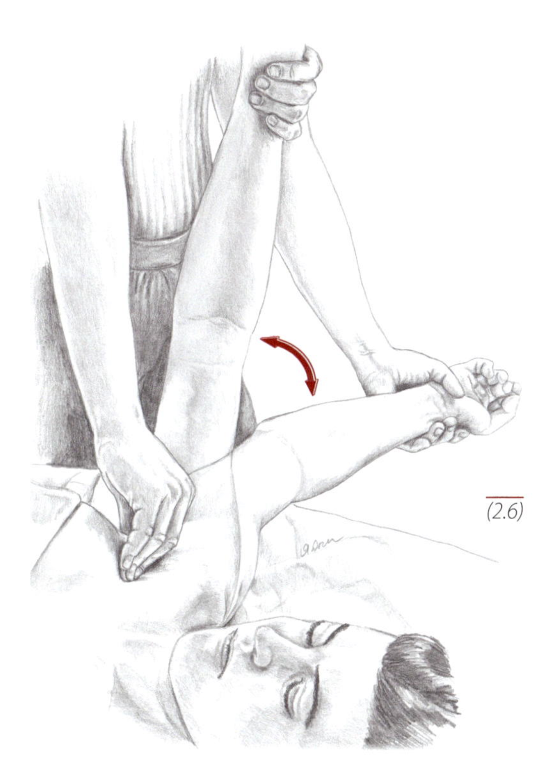

(2.6)

1) Partner supine. Here is an opportunity to feel the skin and fascia shorten or stretch. Holding your partner's arm at the wrist, gently grasp the tissue of the upper chest.
2) Encourage your partner to relax her arm as you passively move it up and down (horizontal abduction and adduction). Note the changes you feel in the tissues.
3) Try this same action while grasping the tissue near the clavicle, sternum or latissimus dorsi. Explore different movements at the shoulder, feeling how virtually all the skin of the upper chest, shoulder and arm shifts to accommodate even a simple action (2.6).

Bones of the Shoulder and Arm

The shoulder complex is made up of three bones: the clavicle, scapula and humerus (2.7). The **clavicle** or collarbone is superficial and runs horizontally along the top of the chest at the base of the neck. It articulates laterally with the acromion of the scapula (acromioclavicular joint) and medially with the sternum (sternoclavicular joint). Both joints are synovial joints. The sternoclavicular joint is the single attachment site between the upper appendicular and axial skeletons.

The **scapula** is the triangular-shaped bone of the upper back. Along with the clavicle, the scapula plays a vital role in stabilization and movement of the arm. The scapula has several fossae, corners and ridges which serve as attachment sites for sixteen muscles. The scapula glides across the posterior surface of the thorax to form the scapulothoracic joint. However, because this articulation does not have any of the usual joint components, it is considered a false joint.

The **humerus** is the bone of the arm. The proximal humerus articulates with the glenoid fossa of the scapula to form the glenohumeral joint. The glenohumeral joint is a synovial, ball-and-socket joint with a wide range of movement. The deltoid muscle and numerous tendons surround the proximal humerus and the glenohumeral joint.

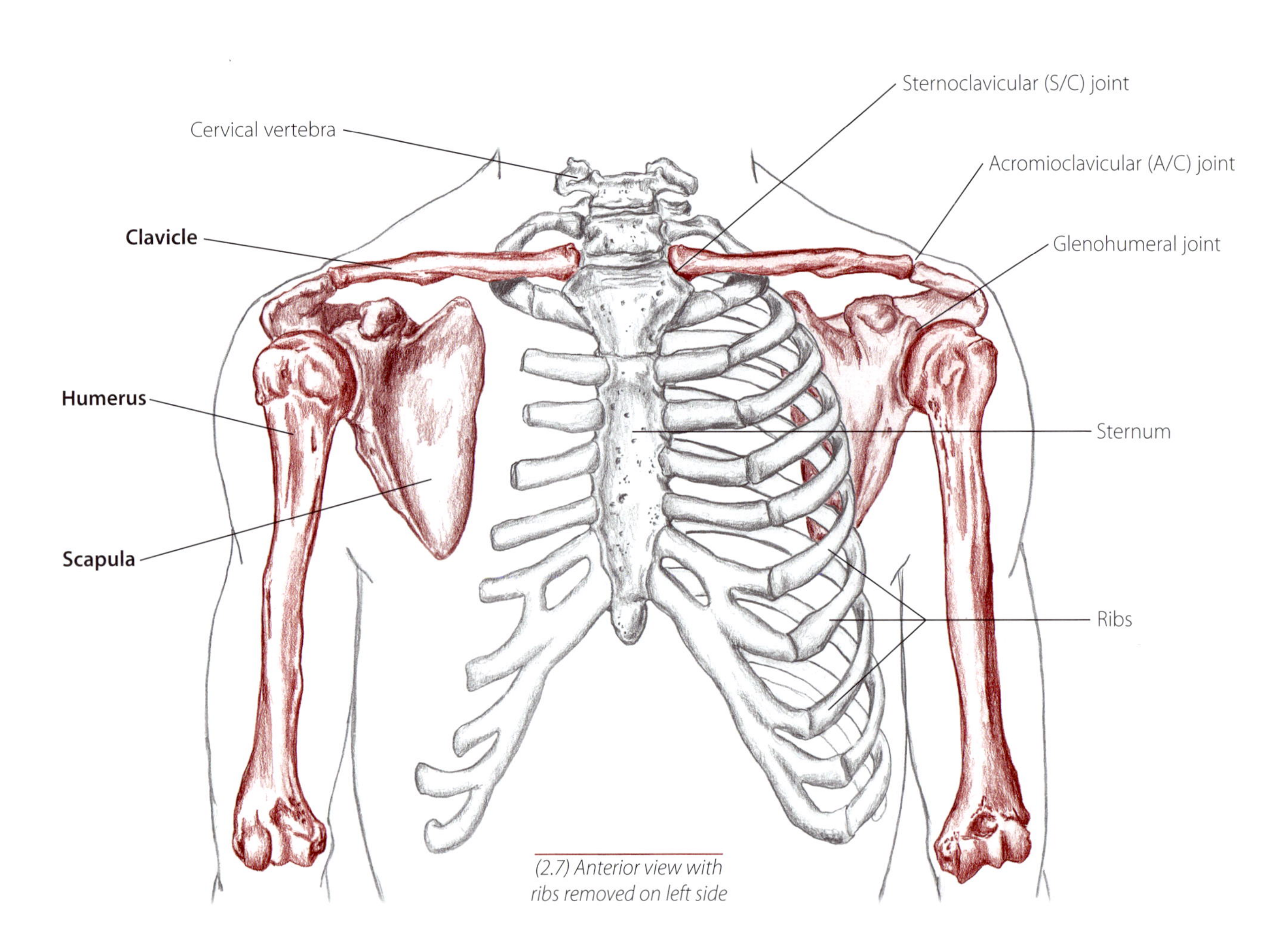

(2.7) Anterior view with ribs removed on left side

The clavicle is the first bone to start ossifying (hardening) in a human fetus, yet paradoxically it is the last to completely develop - often not until the late teens or early twenties. This fact, along with its superficial location, may explain why the clavicle is one of the most frequently broken bones in the body.

A quadruped, such as a dog or cat, however, is not concerned with breaking its clavicle. Since a quadruped's scapula is positioned on the lateral side of the trunk (as opposed to a human's, which lies on the posterior side of the trunk), its clavicle is not as essential to the movement of the shoulder complex. Actually, cats have a thin sliver for a clavicle and dogs have just a small piece of cartilage.

A bird's clavicles are joined to form a furcula. The single unit of the furcula acts as a strut, offering greater stability to the large pectoral muscles during flight. The furcula is what we split apart when vying for the long end of the "wishbone."

clavicle	**klav**-i-k'l	L. little key
furcula	**fur**-ku-la	L. a little fork
humerus	**hu**-mer-us	L. upper arm

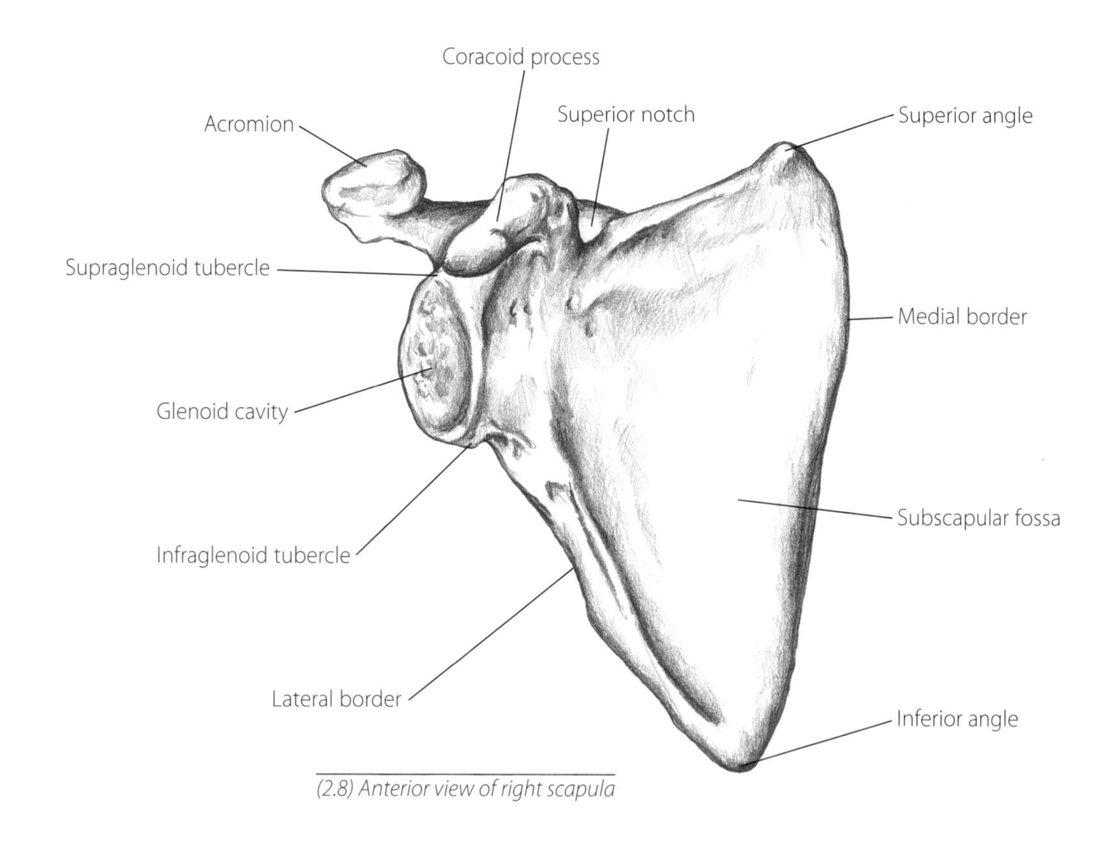

(2.8) Anterior view of right scapula

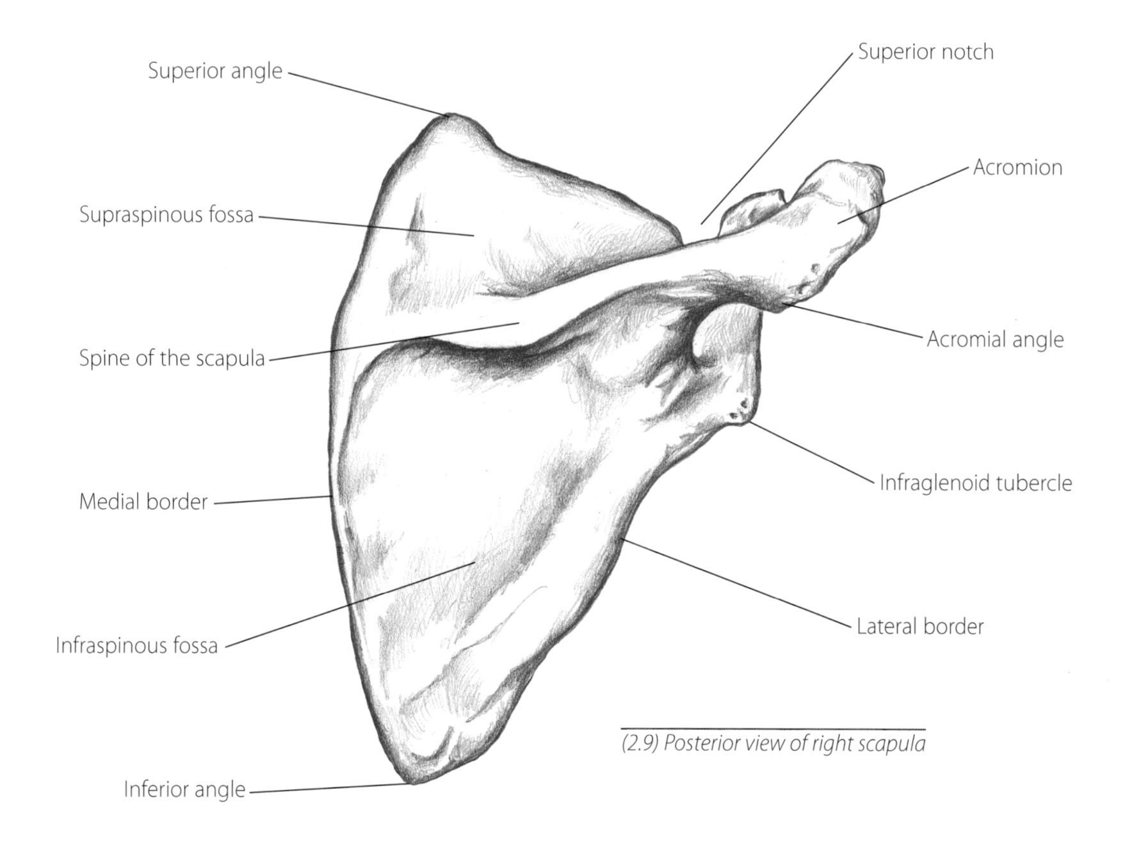

(2.9) Posterior view of right scapula

process	**pros**-es	L. going forth
scapula	**skap**-u-la	L. shoulder, blade
scapulae	**skap**-u-lay	plural for scapula

Bony Landmarks

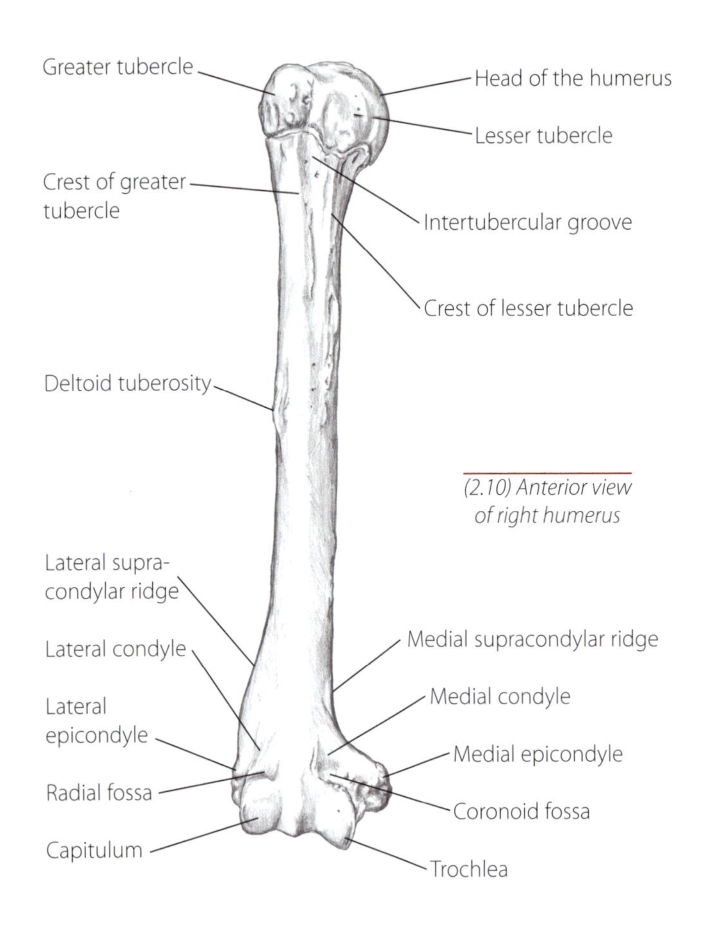

(2.10) Anterior view of right humerus

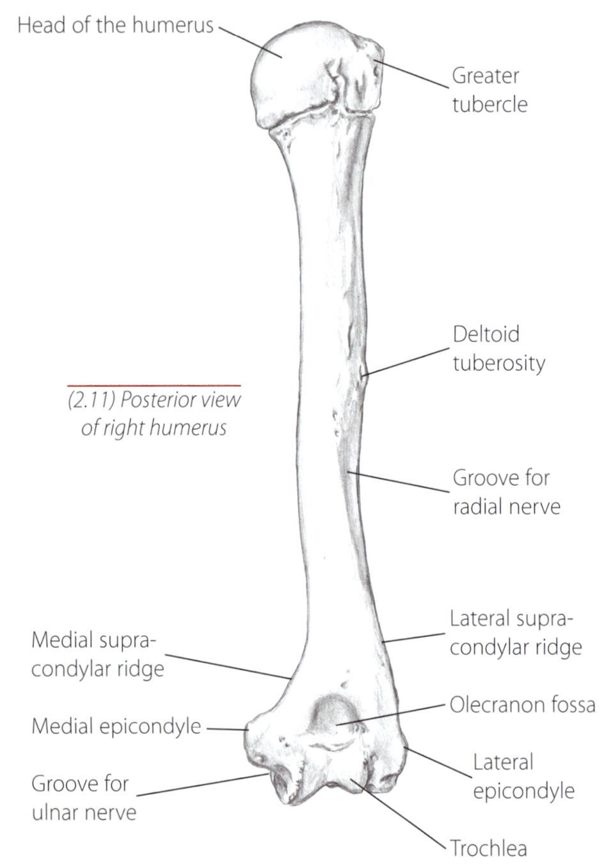

(2.11) Posterior view of right humerus

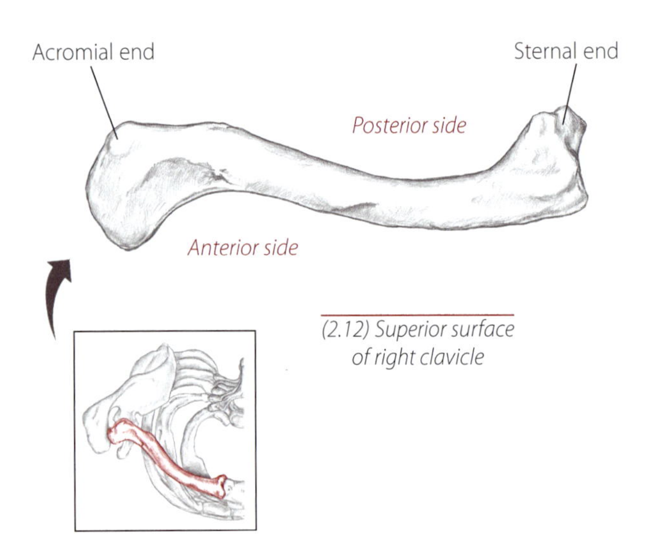

(2.12) Superior surface of right clavicle

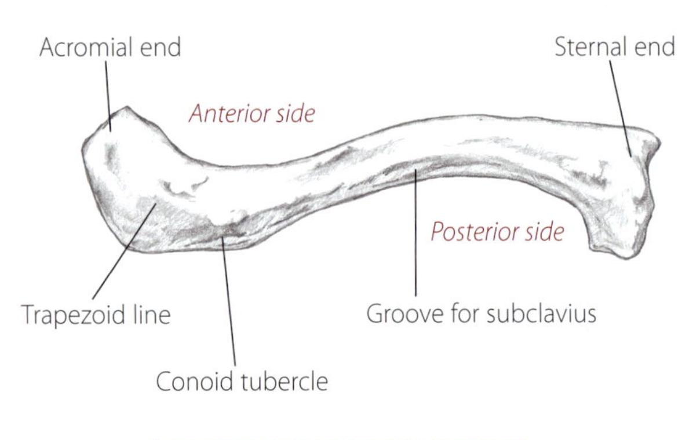

(2.13) Inferior surface of right clavicle

Bony Landmark Trails

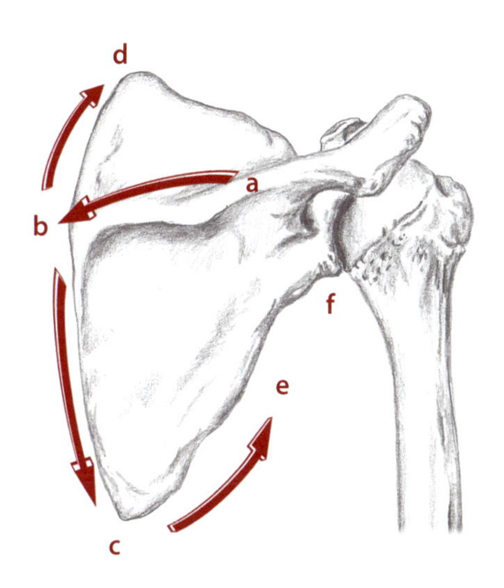

Trail 1 "Along the Edges" explores the sides and corners of the posterior scapula.

- **a** Spine of the scapula
- **b** Medial border
- **c** Inferior angle
- **d** Superior angle
- **e** Lateral border
- **f** Infraglenoid tubercle

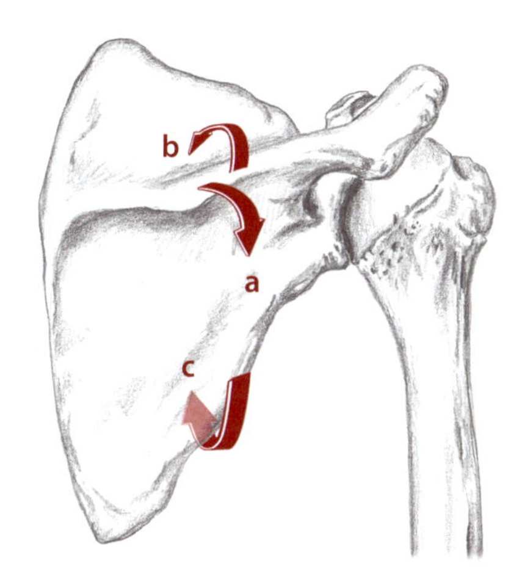

Trail 2 "In the Trenches" leaps off the spine of the scapula and sinks into the three basins of the scapula.

- **a** Infraspinous fossa
- **b** Supraspinous fossa
- **c** Subscapular fossa

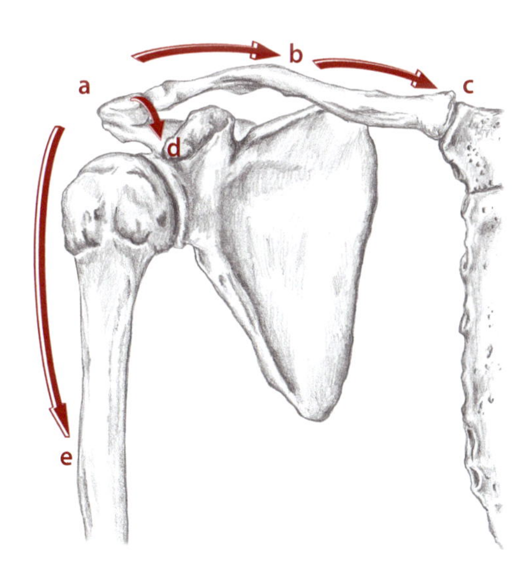

Trail 3 "Springboard Ledge" leads around to the anterior shoulder, using the scapula's acromion as a jumping-off point.

- **a** Acromion
- **b** Clavicle
- **c** Acromioclavicular and sternoclavicular joints
- **d** Coracoid process
- **e** Deltoid tuberosity

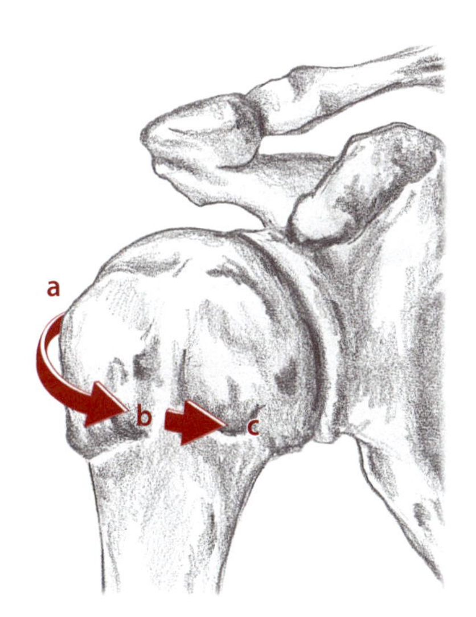

Trail 4 "Two Hills and a Valley" focuses on the three landmarks located along the anterior, proximal humerus.

- **a** Greater tubercle
- **b** Intertubercular groove
- **c** Lesser tubercle

Trail 1 "Along the Edges"

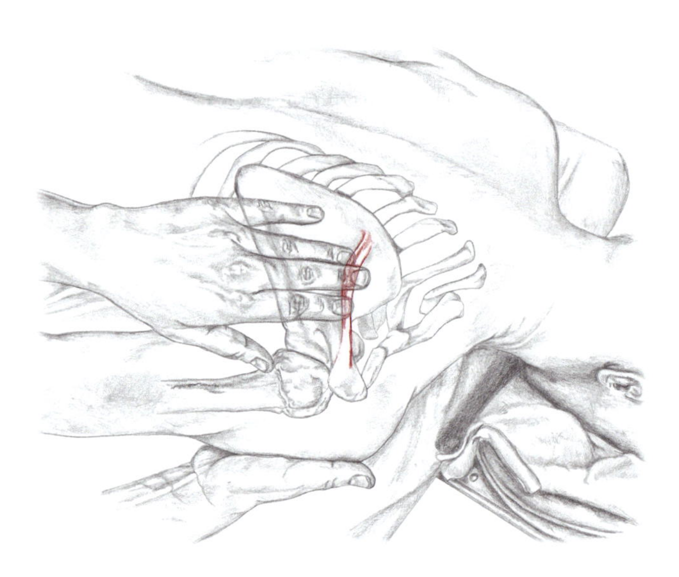

(2.14) Partner prone with spine of the scapula highlighted

Spine of the Scapula

The spine of the scapula is a superficial ridge located just off the top of the shoulder. It runs at an oblique angle to the body, spanning from the acromion to the medial border. It is an attachment site for the posterior deltoid (p. 75) and middle and lower fibers of the trapezius (p. 77).

1) Partner prone. Lay your hand across the upper back and slide your fingertips inferiorly until they roll over the superficial spine (2.14).
2) Strum your fingers vertically, palpating its width and edges. Also explore its entire length by palpating laterally toward the acromion and medially toward the vertebral column.

As you strum your fingers over the spine, do you feel a ditch of soft tissue above and below it? If your partner slowly elevates his scapula, does the spine elevate as well?

Because of its central location, the spine of the scapula makes a great base camp for locating other landmarks. If you become lost or confused while palpating the scapula, return to its spine.

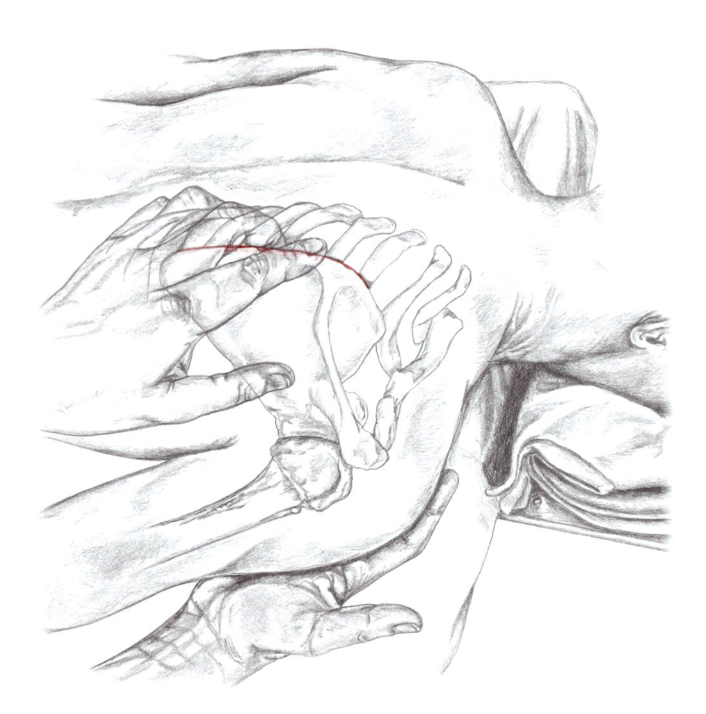

(2.15) Sculpting out the medial border

Medial Border

The medial border is the long edge of the scapula that runs parallel to the vertebral column. It can measure five to seven inches in length, depending on body type. The medial border is an attachment site for the rhomboids (p. 90) and serratus anterior (p. 94) and is deep to the trapezius.

1) Partner prone. Place your partner's hand in the small of his back to raise the medial border off the ribs. For more exposure, scoop and raise the shoulder with one hand.
2) Locate the spine of the scapula and glide your fingertips medially until they slide off the spine onto the medial border (2.15).
3) Follow the medial border inferiorly and superiorly; note that it extends further inferiorly from the spine of the scapula than superiorly.

Does the edge you feel run vertically?

Inferior Angle

There are two angles of the scapula, one on either end of the medial border. The inferior angle is superficial and located at the medial border's lower end.

1) Prone. Place your partner's hand in the small of his back. Glide your fingers inferiorly along the medial border.
2) At the end of the medial border, the edge of the scapula will turn a corner and start to rise superiorly and laterally. This corner is the inferior angle (2.16).

Can you sculpt around the inferior angle and pinch it with your fingertip and thumb?

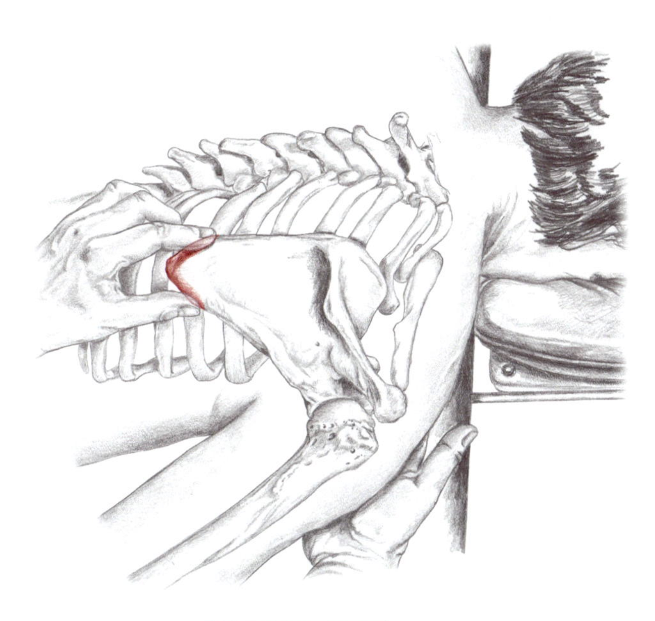

(2.16) Partner prone, pinching the inferior angle

Superior Angle

The superior angle is located at the superior end of the medial border. It serves as the inferior attachment site for the levator scapula muscle. Because the angle is located deep to the trapezius muscles (p. 76), it may not be as easy to isolate as the inferior angle.

1) Prone. Scoop the shoulder with your hand to raise it off the table. This will soften the overlying muscles.
2) Locate the medial border. Slide your fingertips superiorly along the border to find the superior angle (2.17).
3) You may need to move an inch superior to the spine of the scapula to reach the superior angle.

Sculpt out the superior angle and note if it is continuous with the medial border. Locate both the inferior angle and the superior angle. Note the distance between them and gently slide the scapula superiorly and inferiorly.

With your partner sidelying, elevate the scapula toward the ear. As the scapula falls away from the rib cage, the superior angle will be quite palpable.

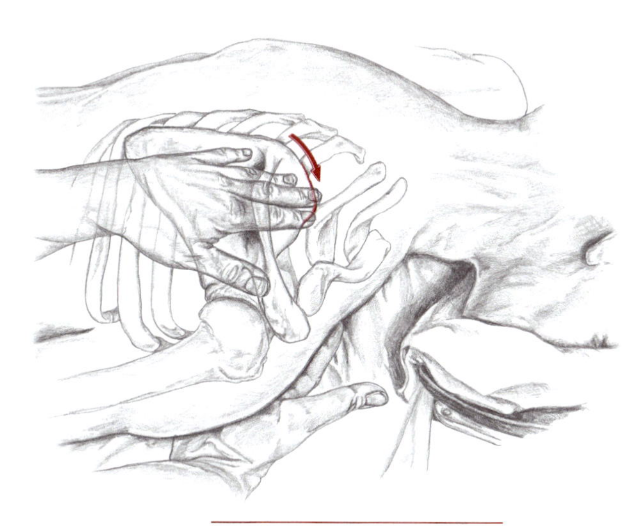

(2.17) Isolating the superior angle

The term "winged scapula" refers to a postural condition in which the medial border falls away from the rib cage and visibly protrudes posteriorly. Often indicating a weak serratus anterior muscle, a winged scapula may also involve the muscles which pull the shoulder girdle anteriorly such as the pectoralis major and minor.

Actually, a degree of scapular winging normally occurs with scapular abduction. For example, when a boxer throws a punch (and fully abducts the scapula) the glenoid fossa is facing anteriorly. In order for this to happen, the medial border must shift posteriorly away from the rib cage. If this winging of the scapula did not occur, the glenoid fossa would not move anteriorly and the boxer would be flat on the canvas in the first round.

(2.18) Partner prone, palpating the lateral border

(2.19) Partner prone, accessing infraglenoid tubercle

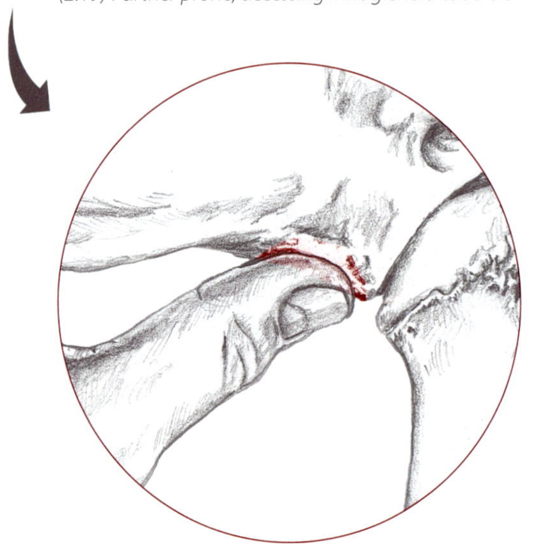

Lateral Border

The lateral border extends superiorly and laterally from the inferior angle toward the axilla or "armpit." It is an attachment site for the teres major and teres minor muscles (p. 79, 82) and, due to the thickness of these tissues, may not be as clearly defined as the medial border.

1) Prone. Drape the arm off the side of the table. Slide your thumb from the inferior angle superiorly along the lateral border.
2) Follow the border in the direction of the axilla. If the musculature is too thick to palpate through, try curling your thumb underneath the tissue (2.18). This is most effective when locating the infraglenoid tubercle (see below).

Is the edge of bone you are palpating continuous with the inferior angle? As you follow it superiorly, does it lead you in the direction of the axilla?

Try the above method with your partner's hand in the small of his back.

Infraglenoid Tubercle

The infraglenoid tubercle is located at the most superior aspect of the lateral border. The tubercle is not a distinguishable point, but a small spot which serves as an attachment site for the long head of the triceps brachii (p. 105). It lies deep to the teres minor and deltoid muscles.

Exploring the infraglenoid tubercle often elicits tenderness in the surrounding tissues. By using your broad thumbpad, you will be able to palpate more precisely without causing pain.

1) Prone. Locate the lateral border.
2) Slide along the lateral border to its most superior portion (2.19). To access the landmark directly, you can either compress through the overlying muscles or curl underneath them.

Are you along the edge of the lateral border? Are you on the posterior side of the axilla?

tubercle **tu**-ber-kl L. a little swelling

Trail 2 "In the Trenches"

Infraspinous Fossa

The scapula contains three fossae or depressions: the infraspinous, supraspinous and subscapular. Each fossa is designed to accommodate a muscle belly and its tendinous attachments. The infraspinous fossa is the triangular area inferior to the spine of the scapula; it is filled with the infraspinatus muscle (p. 82).

1) Prone. Palpate the spine of the scapula, its medial border and its lateral border to isolate the infraspinous fossa.
2) Cradle the inferior angle in the webbing between your index finger and thumb. Your index finger will rest along the medial border, your thumb along the lateral border (2.20).
3) Place a finger of the opposite hand along the length of the spine of the scapula. The triangular-shaped area you isolate is the infraspinous fossa.

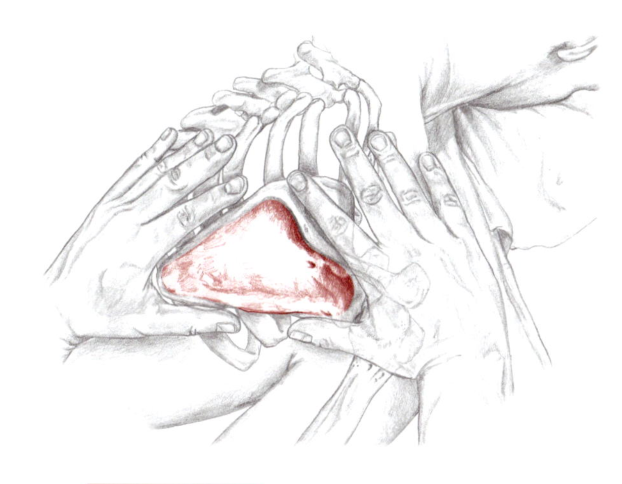

(2.20) Partner prone

Supraspinous Fossa

The supraspinous fossa is located superior to the spine of the scapula. It is small in size, yet quite deep. Because the supraspinatus muscle (p. 82) attaches to and lies in this basin, the supraspinous fossa is difficult to access directly.

1) Prone. Drop your thumbpad inferiorly and laterally from the superior angle into the fossa, or lay your thumb along the spine of the scapula and raise it superiorly into the fossa.
2) Although the fossa is covered by the trapezius and supraspinatus muscles, explore as much as you can of its size and shape (2.21).
3) Slide your fingers laterally, noting how the fossa becomes thinner and finally ends at the junction of the acromion and clavicle. Actually, the fossa continues underneath the acromion although it is inaccessible.

Are you superior to the spine of the scapula? If you strum your fingers vertically, can you palpate the supraspinatus fibers running horizontally toward the acromion?

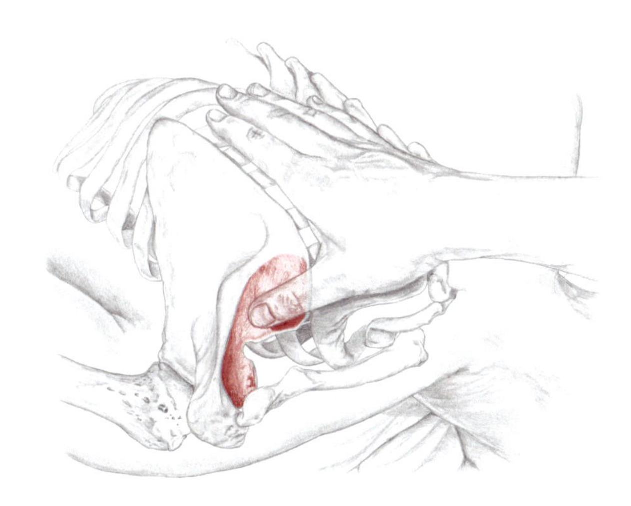

(2.21) Partner prone

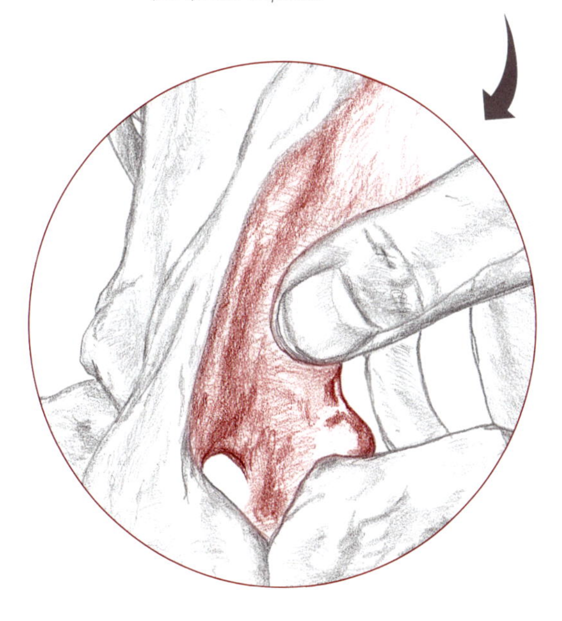

fossa **fos**-a L. a shallow depression

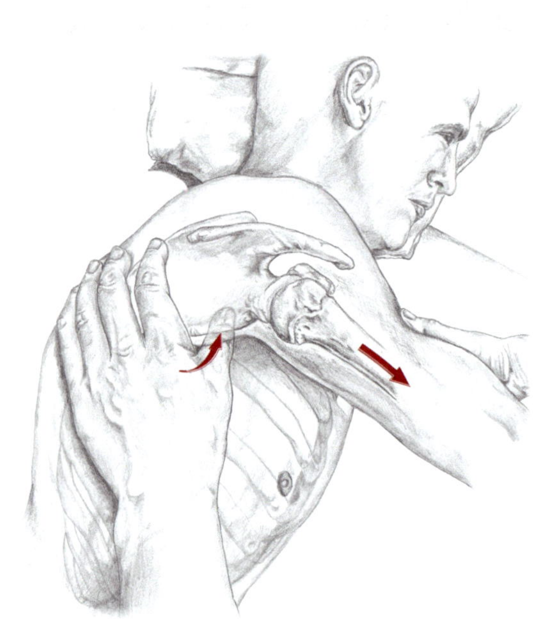

(2.22) Partner sidelying, accessing the lateral side of subscapular fossa. Palpating this fossa can be tender so move slowly and check in with your partner.

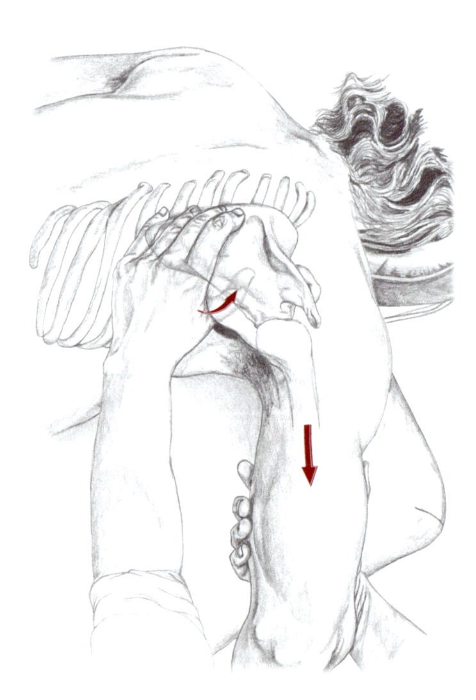

(2.23) Partner prone, palpating the lateral side of the subscapular fossa

Subscapular Fossa

The subscapular fossa is located on the scapula's anterior (or underside) surface, next to the rib cage. It is the attachment site for the subscapularis and the location of the serratus anterior muscle (p. 94).

The fossa can be challenging to access, due to the scapula's close proximity to the rib cage and the numerous muscle bellies surrounding it.

1) Sidelying. This position allows the scapula to slide away from the rib cage for easier access (2.22).
2) Place your thumb at the middle of the lateral border. Be sure to position your thumb anterior to the large mass of muscles along the lateral border.
3) Slowly sink and curl your thumbpad onto the surface of the fossa. Use your other hand to maneuver the arm and scapula for a position that best allows your thumb to sink into the tissue. You may only be able to sink an inch into the fossa.

Can you feel the rib cage and anterior surface of the scapula on either side of your thumb? Try this same approach with your partner prone (2.23).

Here is a method for palpating the medial portion of the subscapular fossa. The fossa may or may not be accessible, depending on the tissue's flexibility.

1) Sidelying. Flex your partner's shoulder and lay your fingertips along the medial border. With the other hand, move the scapula posteriorly (bringing the medial border off the ribs) (2.24).
2) Slowly curl your fingers through the rhomboid and trapezius muscles, under the scapula and onto the fossa.

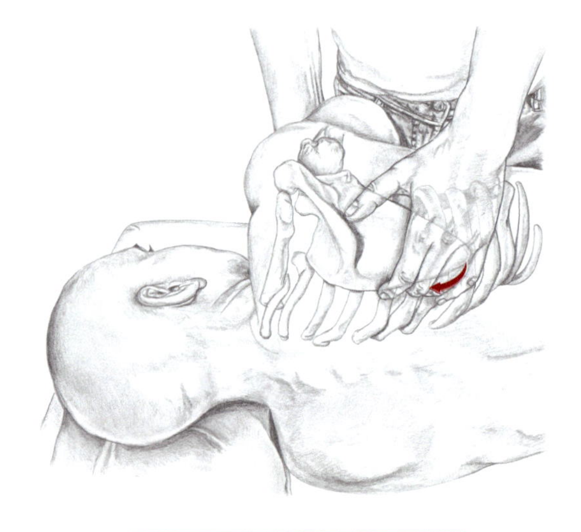

(2.24) Partner sidelying, accessing the medial side of the subscapular fossa

acromion a-**cro**-me-on Grk. *akron*, top + *omos*, shoulder

Trail 3 "Springboard Ledge"

Acromion

The acromion is the lateral aspect of the spine of the scapula and is located at the top of the shoulder. It has a flat surface and articulates with the clavicle's lateral end. The acromion serves as an attachment site for the trapezius and deltoid muscles.

The acromial angle is the small corner that can be felt along the acromion's lateral/posterior aspect (see p. 57).

1) Seated or supine. Locate the spine of the scapula.
2) Follow the spine as it rises superiorly and laterally to the top of the shoulder. Use your fingerpads to explore the acromion's flat surface (2.25).
3) Explore and sculpt around all sides of the acromion and its attachment to the clavicle.

Is the bone you are palpating superficial and directly on the top of the shoulder? Can you feel the small point of the acromial angle on the posterior edge of the acromion?

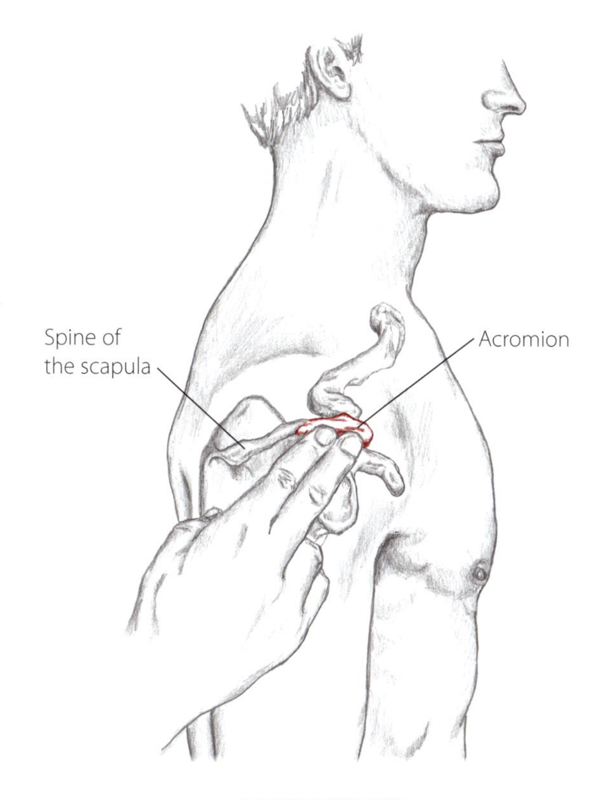

(2.25) Lateral view

Clavicle

The superficial clavicle lies horizontally across the upper chest and has a gentle "S" shape. It is an attachment site for a number of muscles. Both ends of the clavicle are superficial and accessible. The lateral end is relatively flat and often rises slightly higher than the acromion. The medial end is round and articulates with the sternum.

1) Seated. Locate the acromion and walk your fingers medially onto the shaft of the clavicle.
2) Grasp the clavicle's cylindrical body between your finger and thumb and explore its length from the acromion to the sternum. Observe how its acromial end rises superiorly while its sternal end curves inferiorly (2.26).

Have your partner move his shoulder anteriorly and the shaft of the clavicle will protrude visibly. Can you locate the medial and lateral ends of the clavicle simultaneously?

With your fingers at either end of the clavicle, ask your partner to elevate and depress, then adduct and abduct his scapula. As the scapula moves, notice how the ends of the clavicle shift their positions.

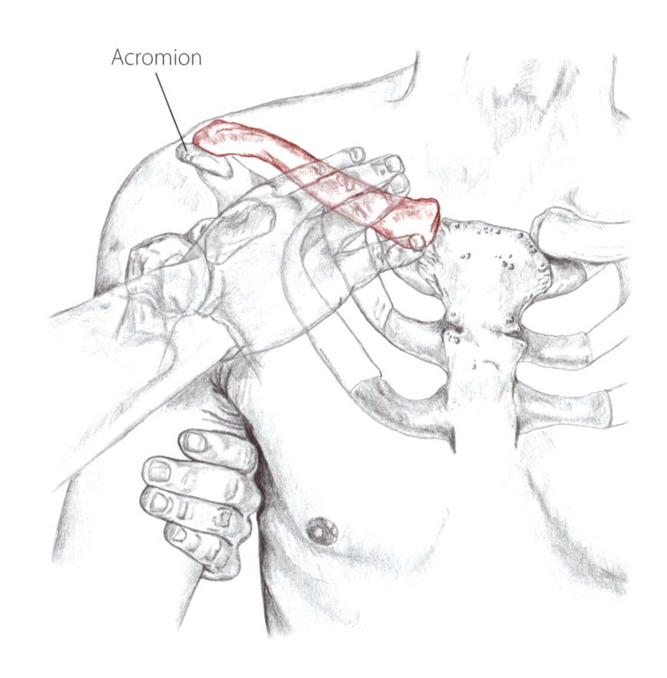

(2.26) Anterior view, clavicle highlighted

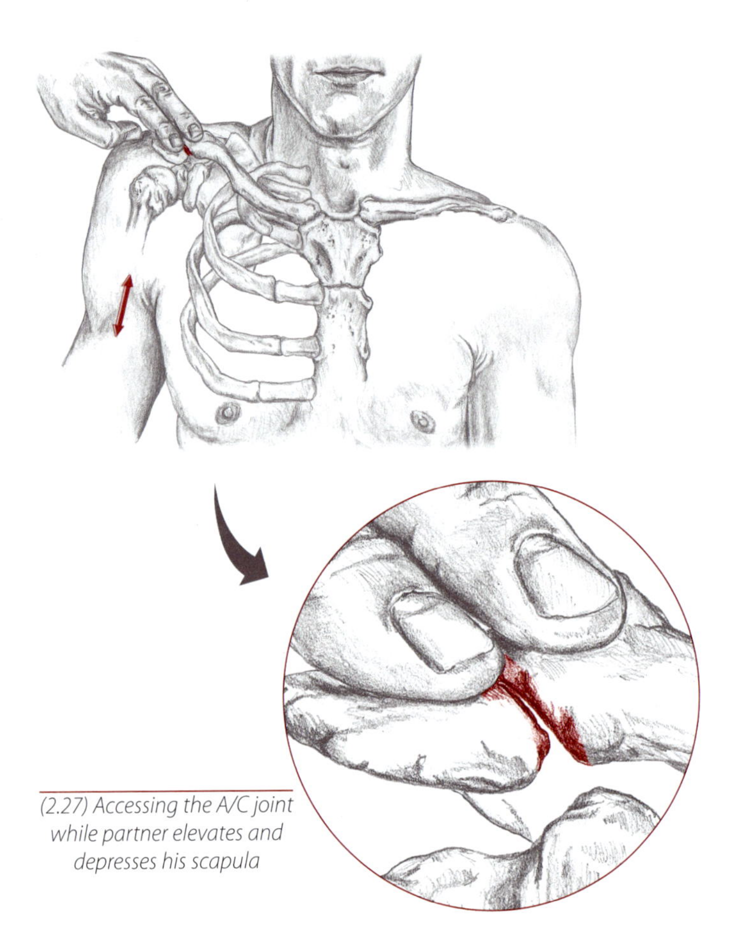

(2.27) Accessing the A/C joint while partner elevates and depresses his scapula

Acromioclavicular and Sternoclavicular Joints

The **acromioclavicular (A/C) joint** is the small articulation between the acromion of the scapula and the acromial end of the clavicle. The anterior and superior surfaces of this thin crevice can be palpated directly.

The **sternoclavicular (S/C) joint** is the articulation between the sternal end of the clavicle and the sternum. Unlike the slender, smooth A/C joint, the S/C joint is wedge-shaped and contains a small, impalpable fibrous disk. At rest, only the inferior portion of the sternal end makes contact with the sternum. When the clavicle is elevated, the sternal end pivots on the sternum.

A/C Joint

1) Seated or supine. Locate the acromion.
2) Glide medially toward the clavicle. Your finger will feel a small "step" as you rise up onto the surface of the clavicle.
3) Backtrack slightly. Just lateral to the step will be the A/C joint's slender ditch.

Does the acromial end of the clavicle lie slightly higher than the acromion? Place a finger where you believe the A/C joint to be and ask your partner to slowly elevate and depress his scapula (2.27). As the scapula rises, do you feel the joint space widen slightly? As it depresses, does the joint space diminish?

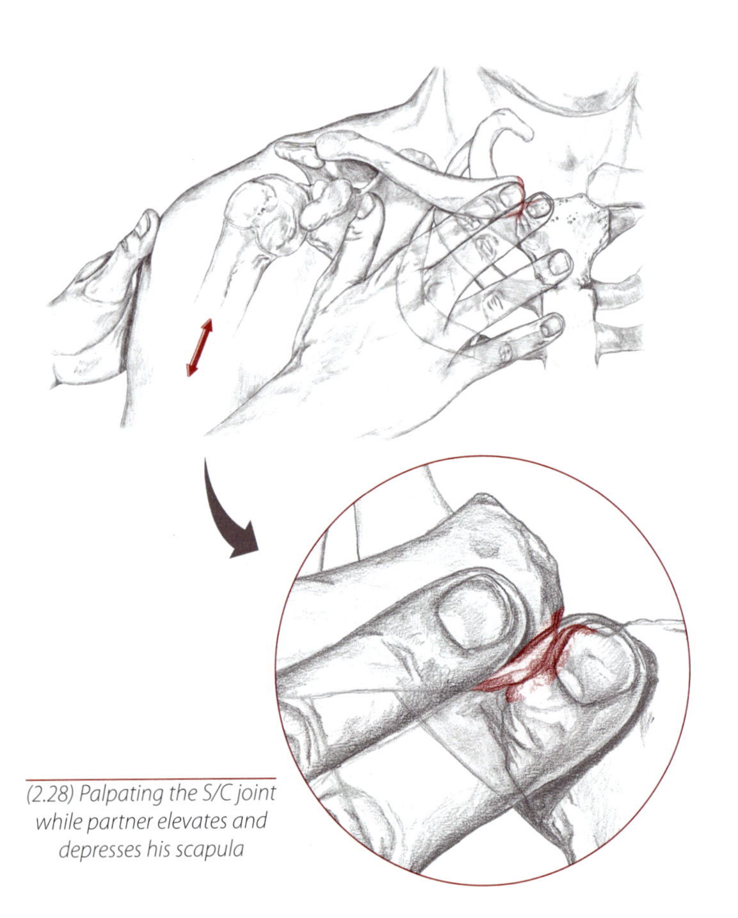

(2.28) Palpating the S/C joint while partner elevates and depresses his scapula

S/C Joint

1) Seated or supine. Slide your fingers medially along the shaft of the clavicle.
2) Just lateral to the body's centerline, the shaft will broaden to become the bulbous sternal end.
3) Locate the S/C joint by sliding your finger medially off the sternal end. Passively elevate, depress and abduct the scapula. Then explore the changes occurring at the S/C joint.

Place a finger where you believe the S/C joint to be and ask your partner to slowly elevate and depress his scapula (2.28). Can you feel the joint space widen and diminish?

acromioclavicular a-**kro**-me-o-kla-**vik**-u-lar
sternoclavicular **ster**-no-kla-**vik**-u-lar

Coracoid Process

The coracoid process of the scapula is the beak-like projection found inferior to the shaft of the clavicle. Depending on the position of the scapula, it is often found in the deltopectoral groove (p. 70) between the deltoid and pectoralis major fibers. The coracoid process can be tender when palpated, so proceed carefully.

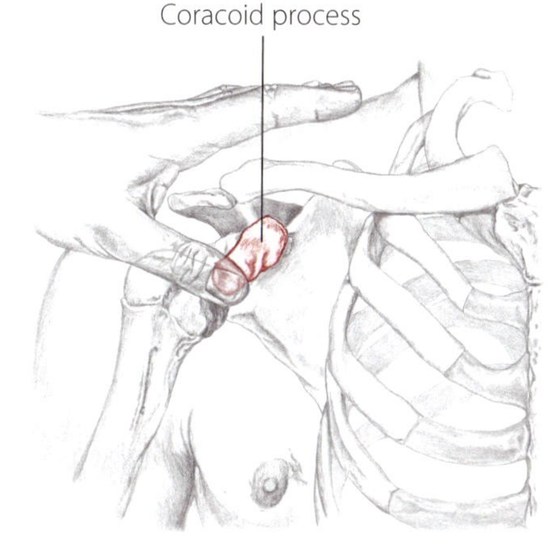

(2.29) Anterior view of right shoulder

1) Seated or supine. Lay your thumb along the lateral shaft of the clavicle.
2) Slide inferiorly off the clavicle no more than an inch and a half. Locate the tip of the coracoid process by compressing your fingerpads into the tissue (2.29).
3) As the coracoid becomes more apparent, get a better understanding of its shape and size by sculpting a circle around its edges.

Are you inferior to the shaft of the clavicle? Passively move the scapula with your other hand and feel the coracoid follow your movements.

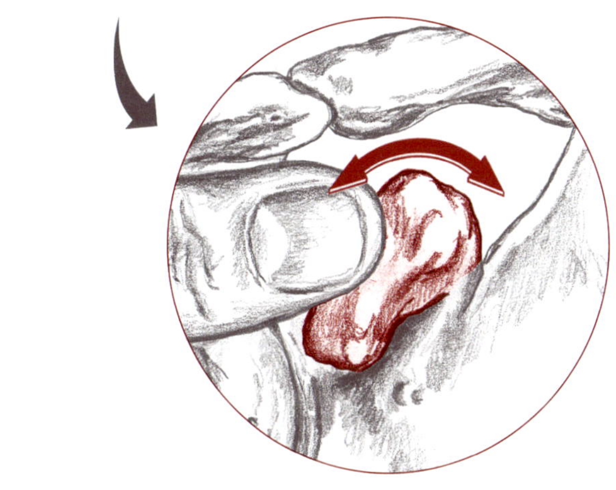

Deltoid Tuberosity

The deltoid tuberosity is located on the lateral side of the mid-humeral shaft. It is a small, low bump that serves as an attachment site for the converging fibers of the deltoid muscle (p. 75).

1) Seated or supine. Locate the acromion.
2) Slide off the acromion and down the lateral aspect of the arm (2.30).
3) When you reach the halfway point between the shoulder and elbow, there will be a small mound on the lateral side of the arm.

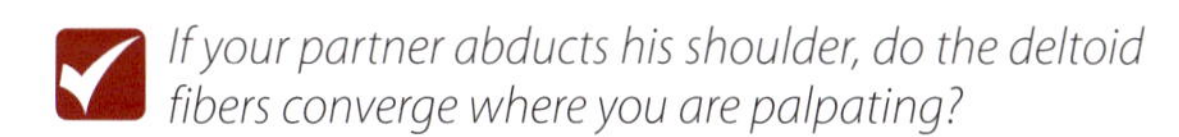

If your partner abducts his shoulder, do the deltoid fibers converge where you are palpating?

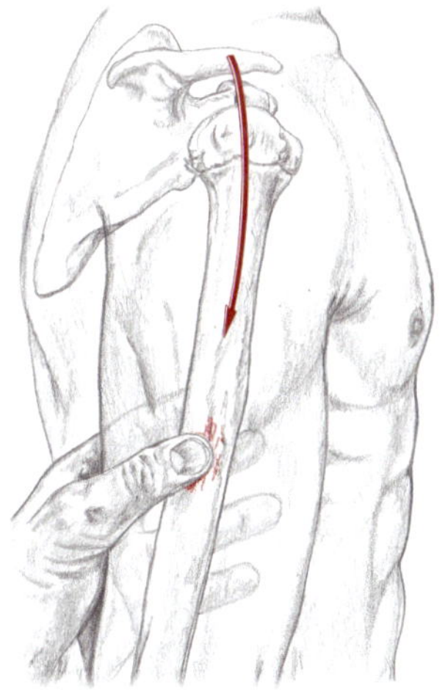

(2.30) Lateral view accessing the deltoid tuberosity

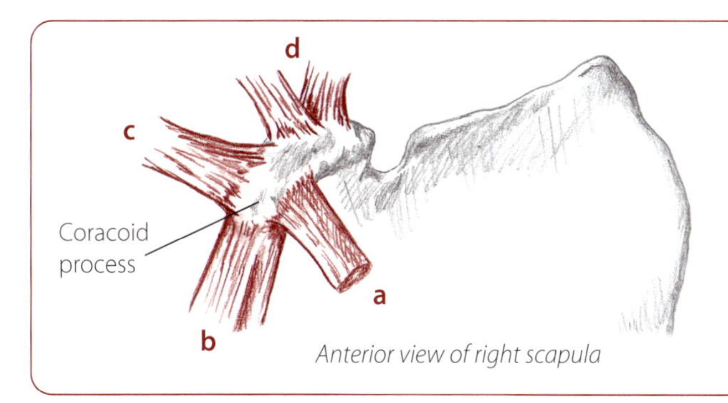

Anterior view of right scapula

The coracoid process is an attachment site for several tendons and ligaments. The arrangement of these structures can be illustrated in a clockwise fashion. On the right scapula, the pectoralis minor tendon **(a)** connects at four o'clock, while the coracobrachialis and biceps brachii tendons **(b)** lie at seven o'clock. The coracoacromial ligament **(c)** attaches at ten o'clock and the coracoclavicular ligaments **(d)** attach further posteriorly on the coracoid process at eleven o'clock and midnight.

coracoid	**kor**-a-koyd	Grk. raven's beak
tuberosity	tu-ber-**os**-i-tee	L. a swelling

Trail 4 "Two Hills and a Valley"

Greater and Lesser Tubercles, Intertubercular Groove

These three landmarks are located on the proximal humerus deep to the deltoid muscle. The **greater tubercle** is located inferior and lateral to the acromion. It is shaped more like a low mound than a pointy hill. The greater tubercle is an attachment site for three of the four rotator cuff muscles - supraspinatus, infraspinatus and teres minor (p. 82).

The **lesser tubercle** is smaller than the greater tubercle and is an attachment site for the fourth rotator cuff muscle - subscapularis. The **intertubercular groove** is situated between the greater and lesser tubercles, and is roughly a pencil's width in diameter. Within the groove lies the tendon of the long head of the biceps brachii, which can be tender, so you should palpate gently in this region.

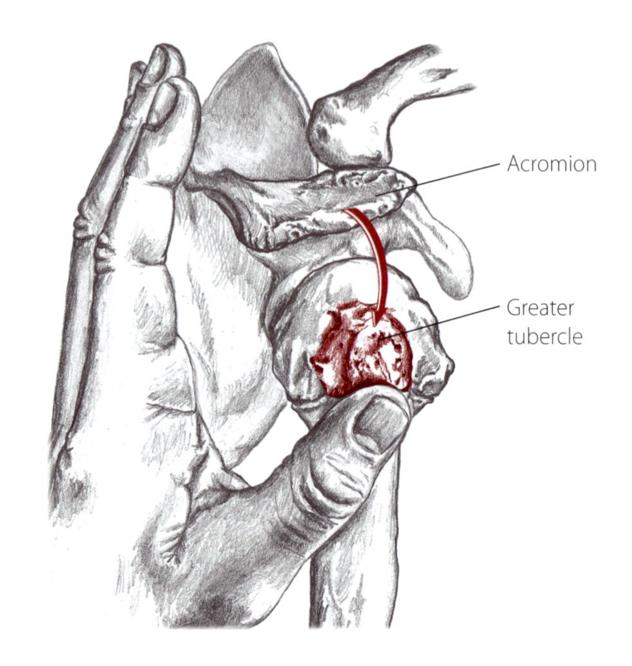

(2.31) Lateral view of right shoulder, sliding off the acromion to the greater tubercle

Greater tubercle

1) Seated or supine. Shaking hands with your partner, locate the acromion.
2) Slide off the acromion inferiorly and laterally approximately one inch (2.31).
3) The solid surface located deep to the deltoid fibers will be the greater tubercle. You may feel a small dip between the acromion and the tubercle.

Intertubercular groove and lesser tubercle

1) Place your thumb on the greater tubercle (2.32, a).
2) Begin to rotate the arm laterally. As the humerus rotates, the greater tubercle **(a)** will move out from under your thumb and be replaced by the slender ditch of the intertubercular groove **(b)**.
3) As you continue to laterally rotate the arm, your thumb will rise out of the groove onto the lesser tubercle **(c)**.

✓ *Place your thumb at the greater tubercle and passively rotate the arm medially and laterally. Do you feel the "bump-ditch-bump" sequence as the three landmarks pass beneath your thumb? Are you horizontal to the level of the coracoid process?*

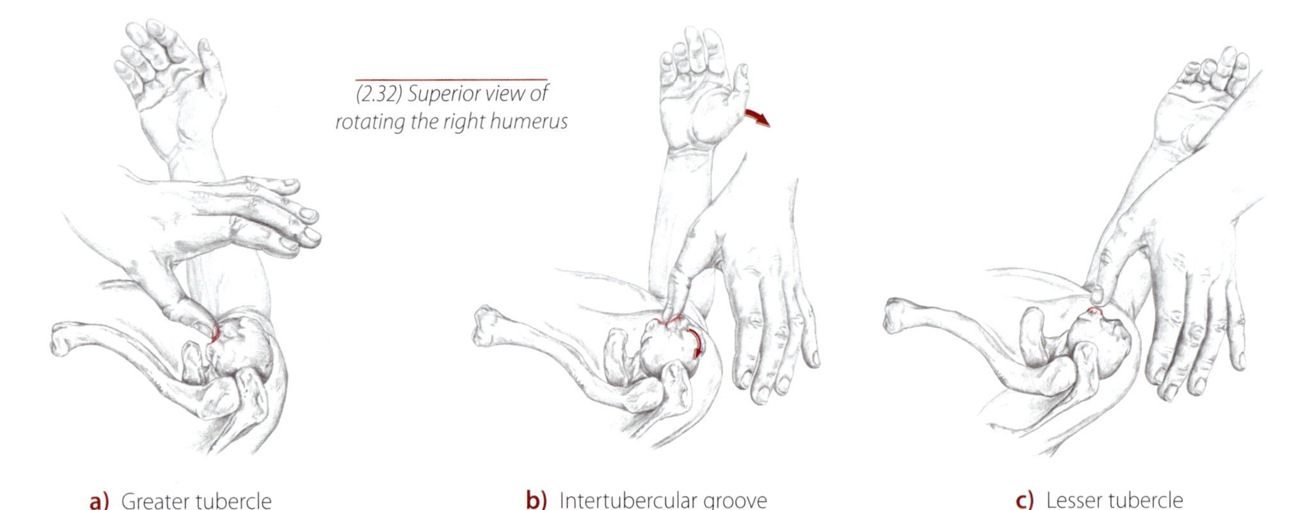

(2.32) Superior view of rotating the right humerus

a) Greater tubercle **b)** Intertubercular groove **c)** Lesser tubercle

intertubercular **in**-tur-tu-**ber**-ku-lar

Muscles of the Shoulder and Arm

The muscles of the shoulder and arm are an amazingly diverse group. Some of them span across the back and rib cage, some attach at the cranium while others extend down to the elbow. All of the muscles create movement at the shoulder complex (formed by the scapula, clavicle and humerus). Some also elevate the ribs, extend the head and cervical vertebrae or bend the elbow (2.33 - 2.35).

The superficial muscles of the shoulder and back are presented first, followed by the deeper muscles of the back, and lastly, the muscles of the arm. Some muscles are presented together to better understand how they function as a group.

Although the instructions for each muscle or muscle group specify the position in which to place your partner (prone, supine or seated), exploration in all positions is encouraged for a better understanding of the muscle(s) and the surrounding structures.

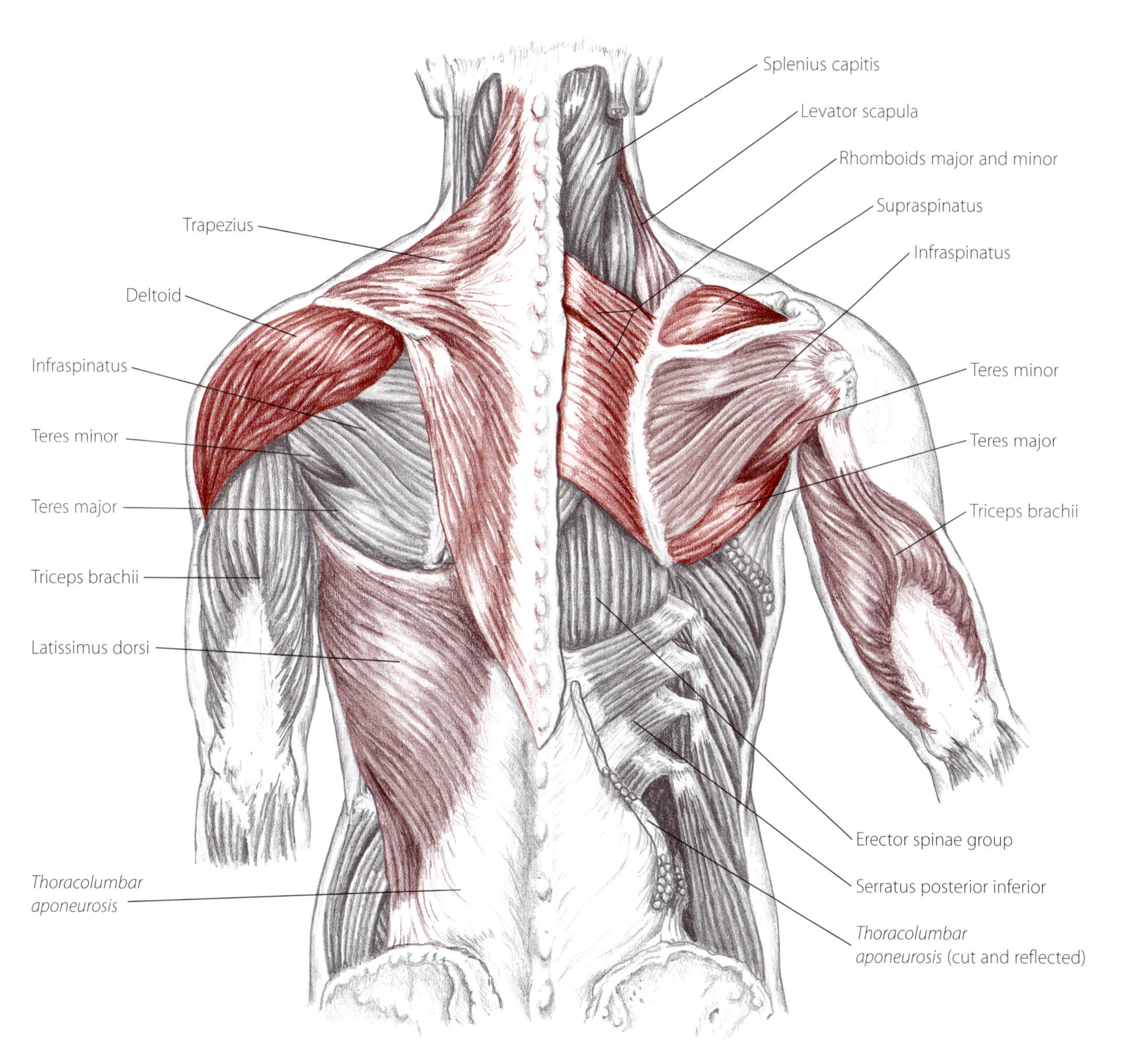

(2.33) Posterior view of shoulder and back. Latissimus dorsi, trapezius and deltoid are removed on right side.

The trapezius received its present name from the British anatomist William Cowper (c. 1700). Previously, it was called the *musculus cucullaris* (L. muscle hood), since the two trapezius muscles together resemble a monk's hood.

Muscles of the Shoulder and Arm

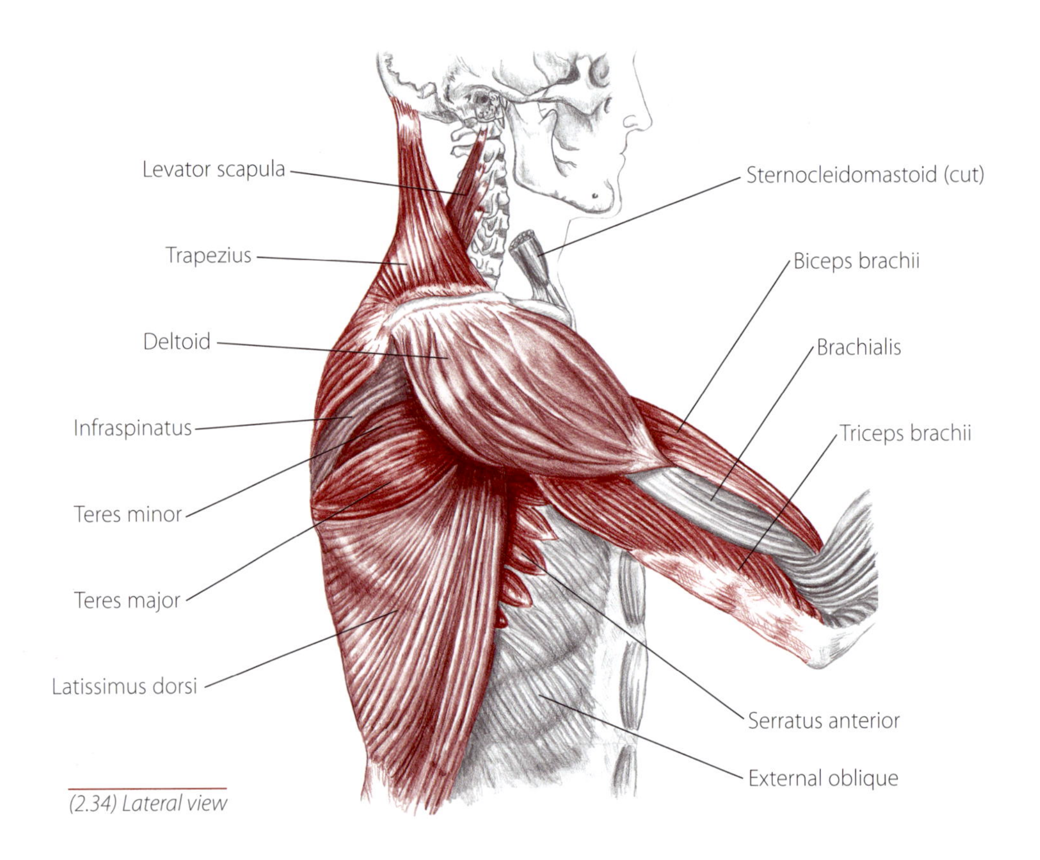

(2.34) Lateral view

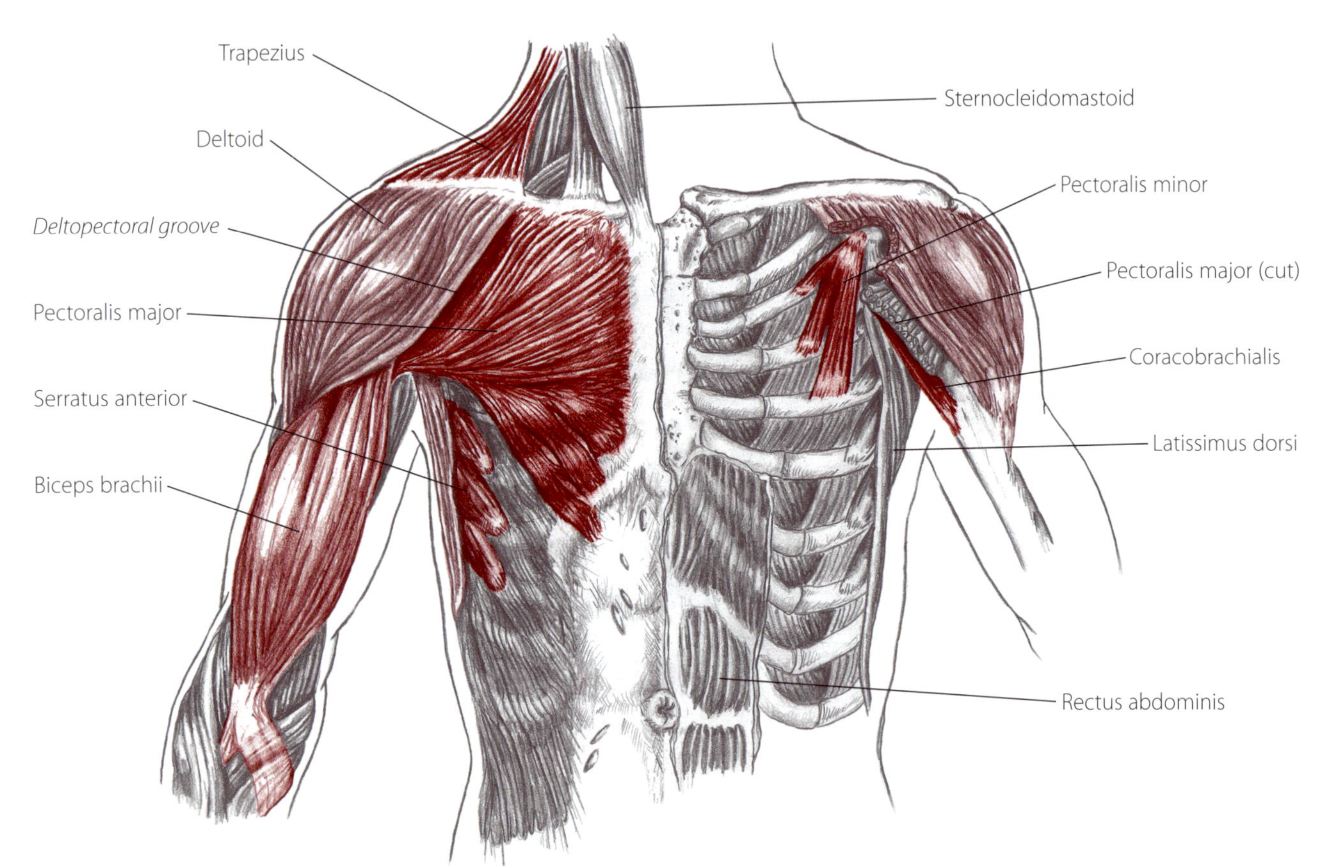

(2.35) Anterior view, pectoralis major and biceps brachii removed on right side

Synergists - Muscles Working Together

*muscles not shown

Shoulder

(glenohumeral joint)

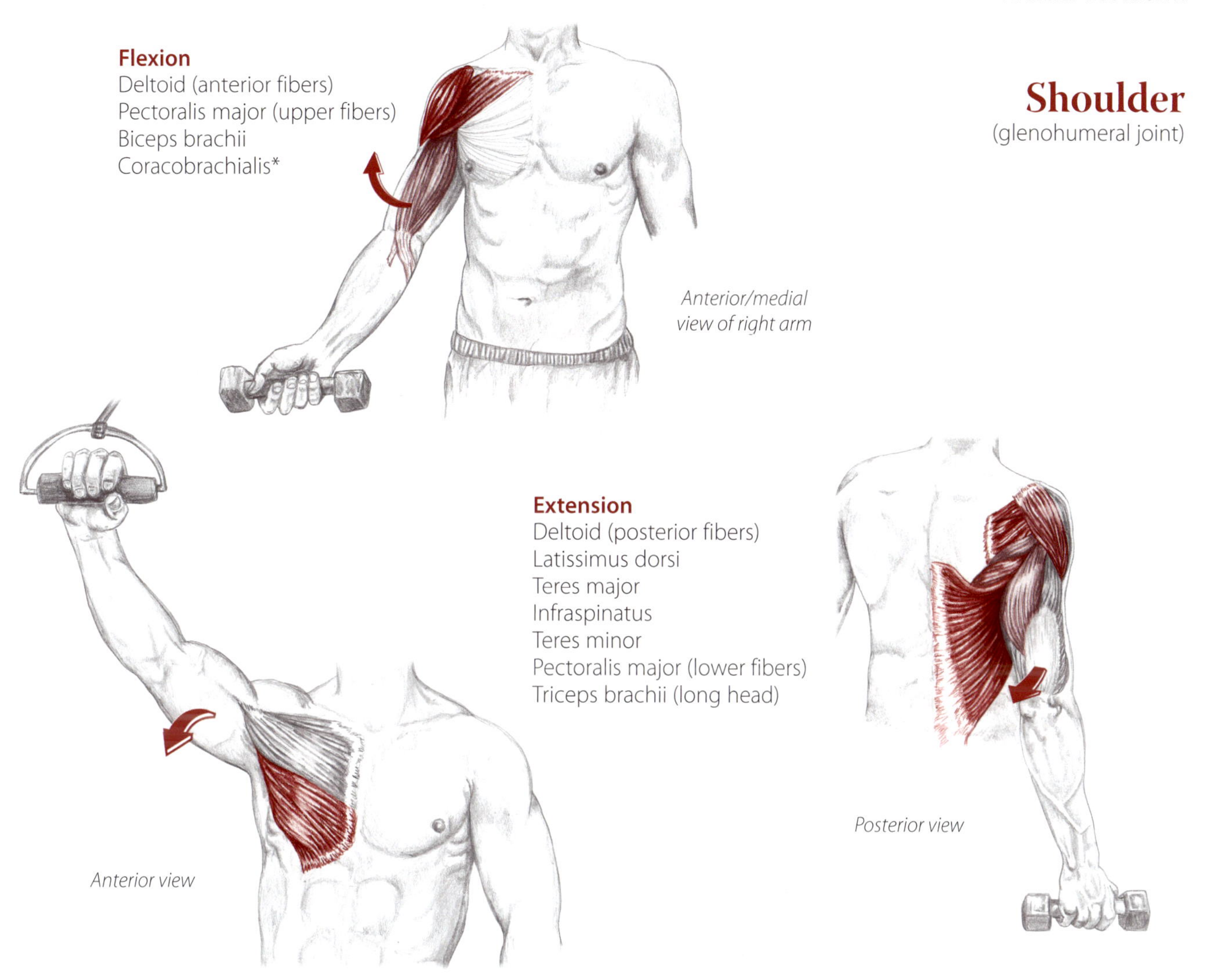

Flexion
Deltoid (anterior fibers)
Pectoralis major (upper fibers)
Biceps brachii
Coracobrachialis*

Anterior/medial view of right arm

Extension
Deltoid (posterior fibers)
Latissimus dorsi
Teres major
Infraspinatus
Teres minor
Pectoralis major (lower fibers)
Triceps brachii (long head)

Anterior view

Posterior view

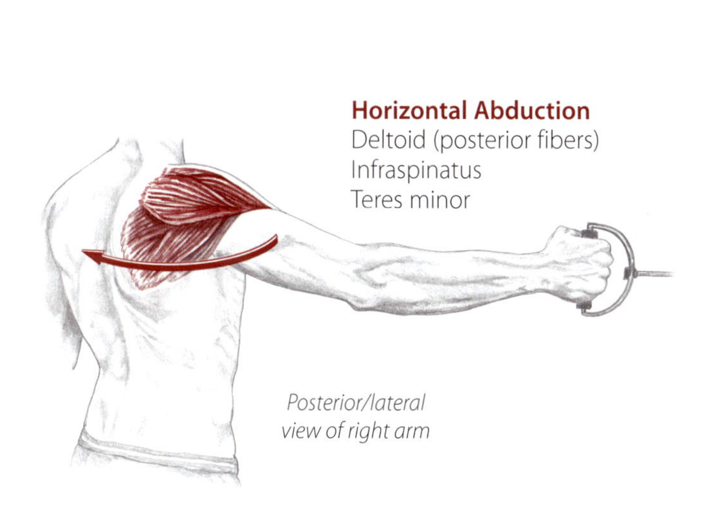

Horizontal Abduction
Deltoid (posterior fibers)
Infraspinatus
Teres minor

Posterior/lateral view of right arm

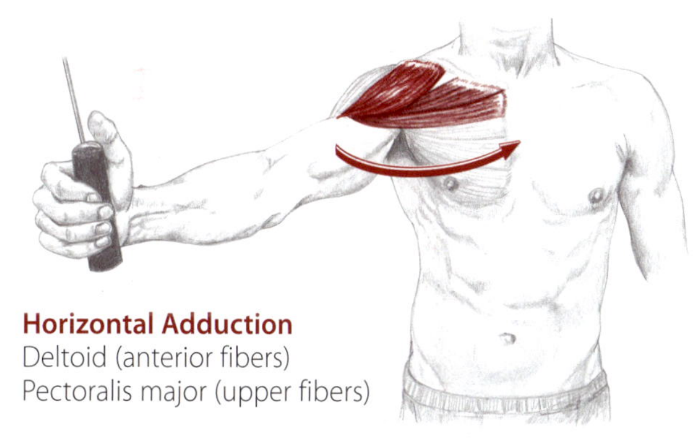

Horizontal Adduction
Deltoid (anterior fibers)
Pectoralis major (upper fibers)

Anterior view

Shoulder
(glenohumeral joint)

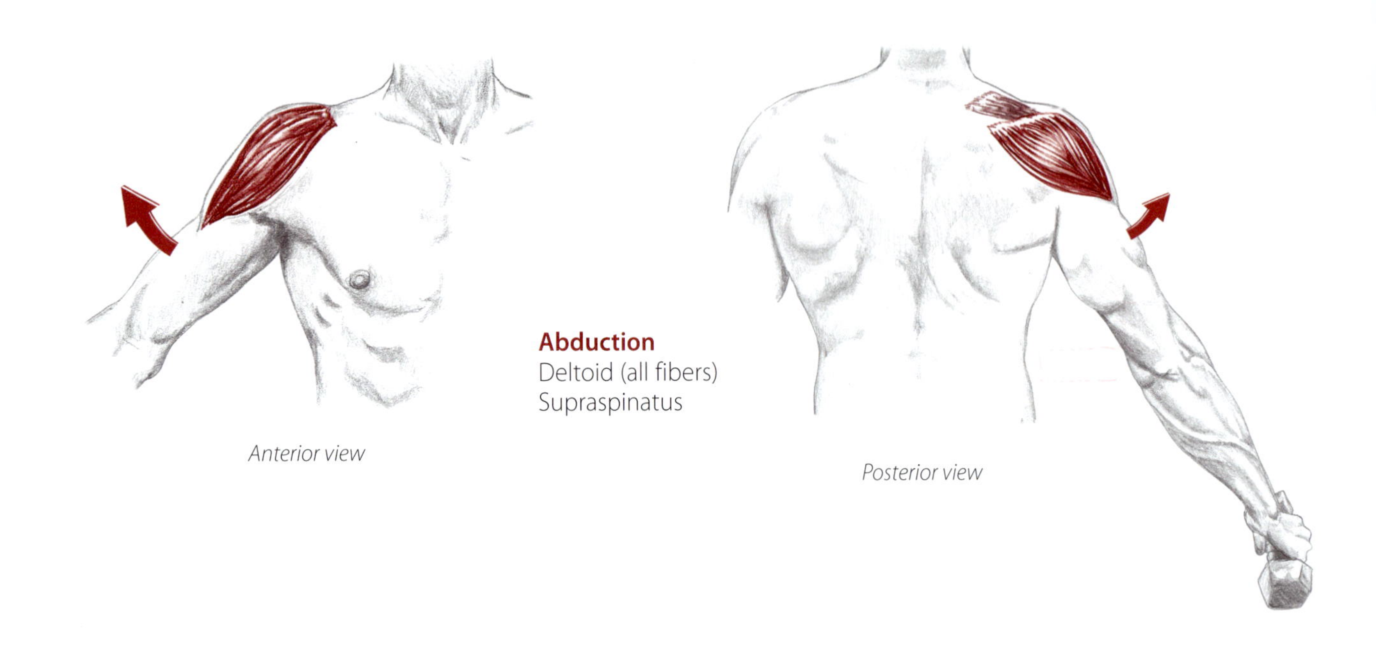

Abduction
Deltoid (all fibers)
Supraspinatus

Adduction
Latissimus dorsi
Teres major
Infraspinatus
Teres minor
Pectoralis major (all fibers)
Triceps brachii (long head)
Coracobrachialis

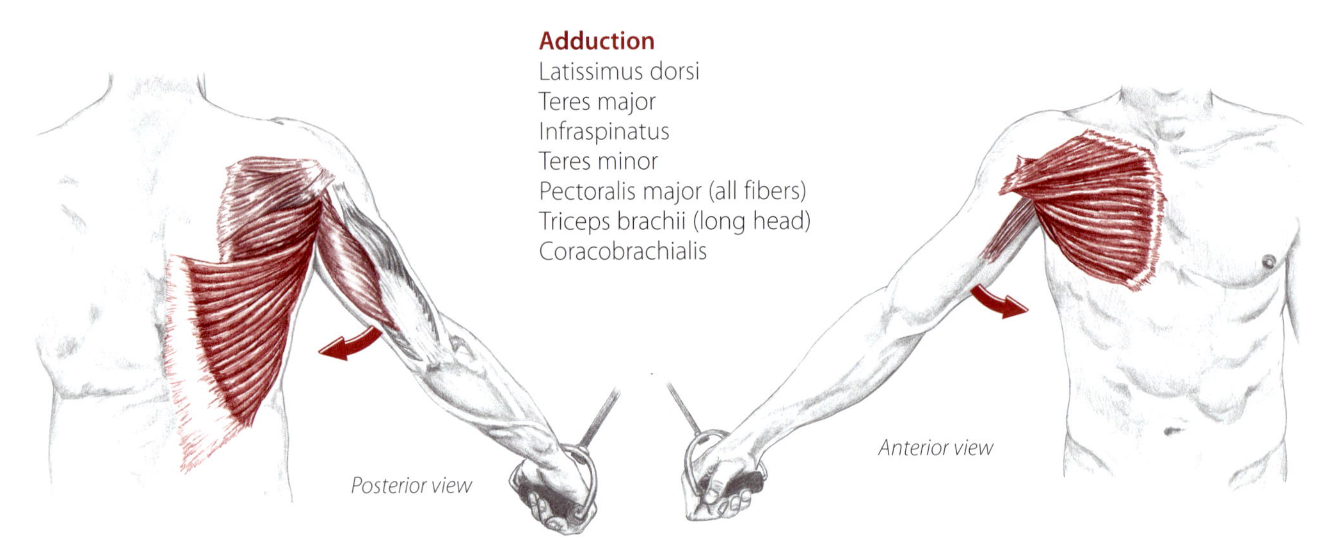

Lateral Rotation (external rotation)
Deltoid (posterior fibers)
Infraspinatus
Teres minor

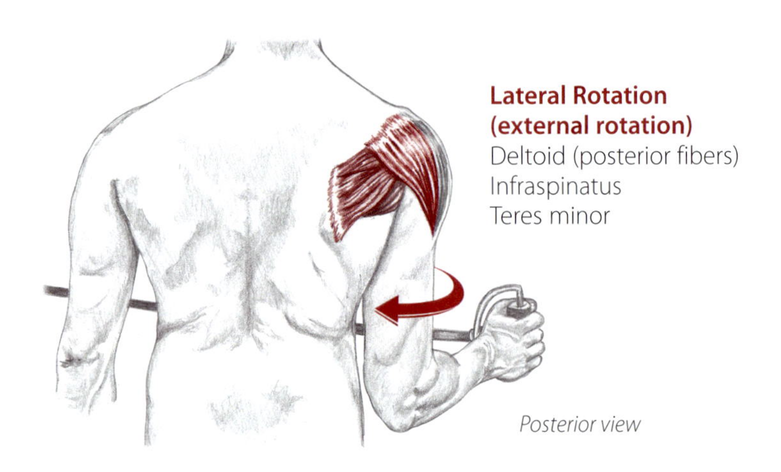

Shoulder

(glenohumeral joint)

Medial Rotation (internal rotation)

Deltoid (anterior fibers)
Latissimus dorsi
Teres major
Subscapularis
Pectoralis major (all fibers)

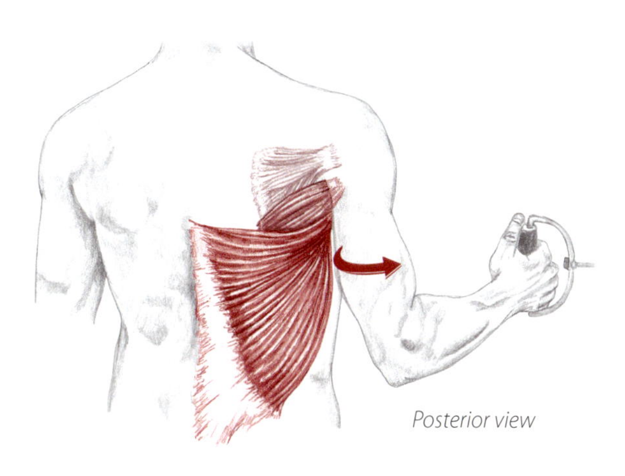

Posterior view

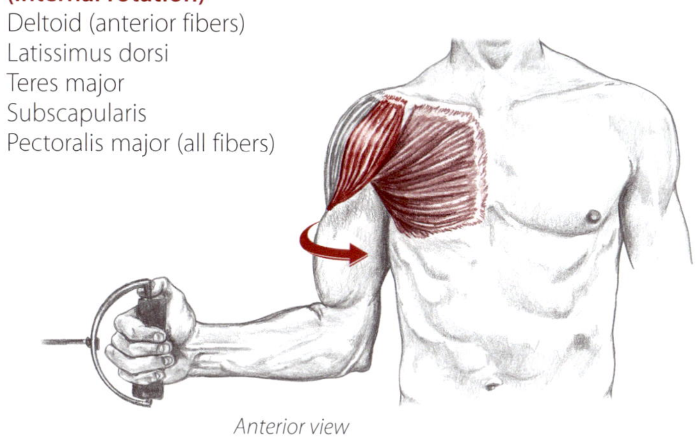

Anterior view

Scapula

(scapulothoracic joint)

Elevation

Trapezius (upper fibers)
Rhomboid major
Rhomboid minor
Levator scapula

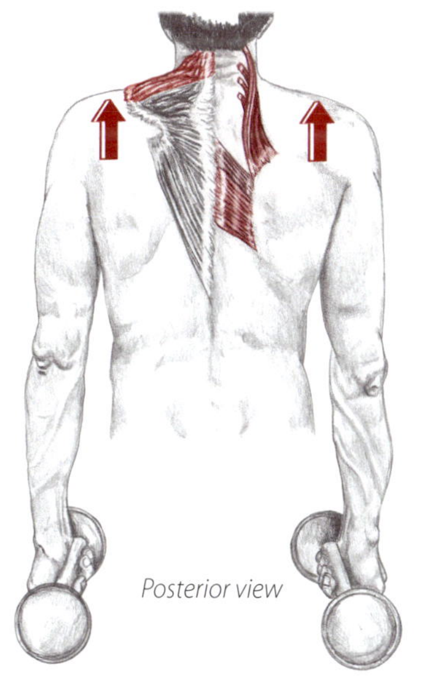

Posterior view

Depression

Trapezius (lower fibers)
Serratus anterior (with the origin fixed)
Pectoralis minor

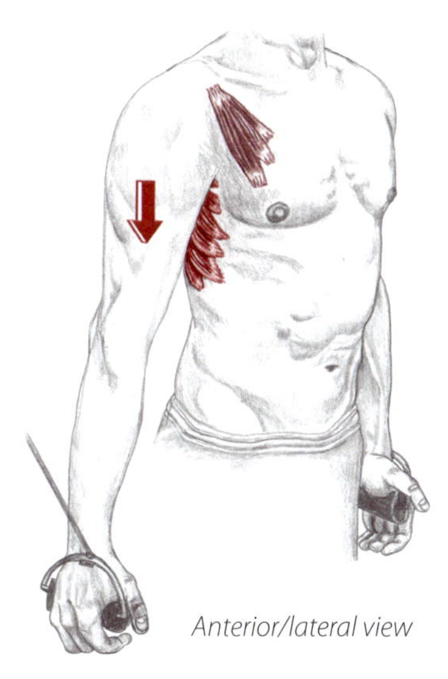

Anterior/lateral view

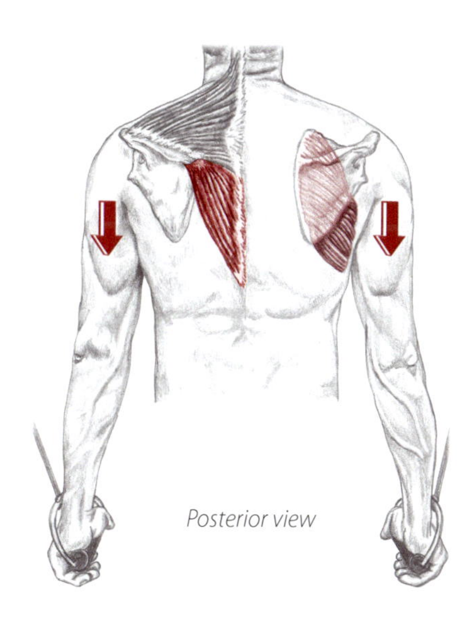

Posterior view

Scapula

(scapulothoracic joint)

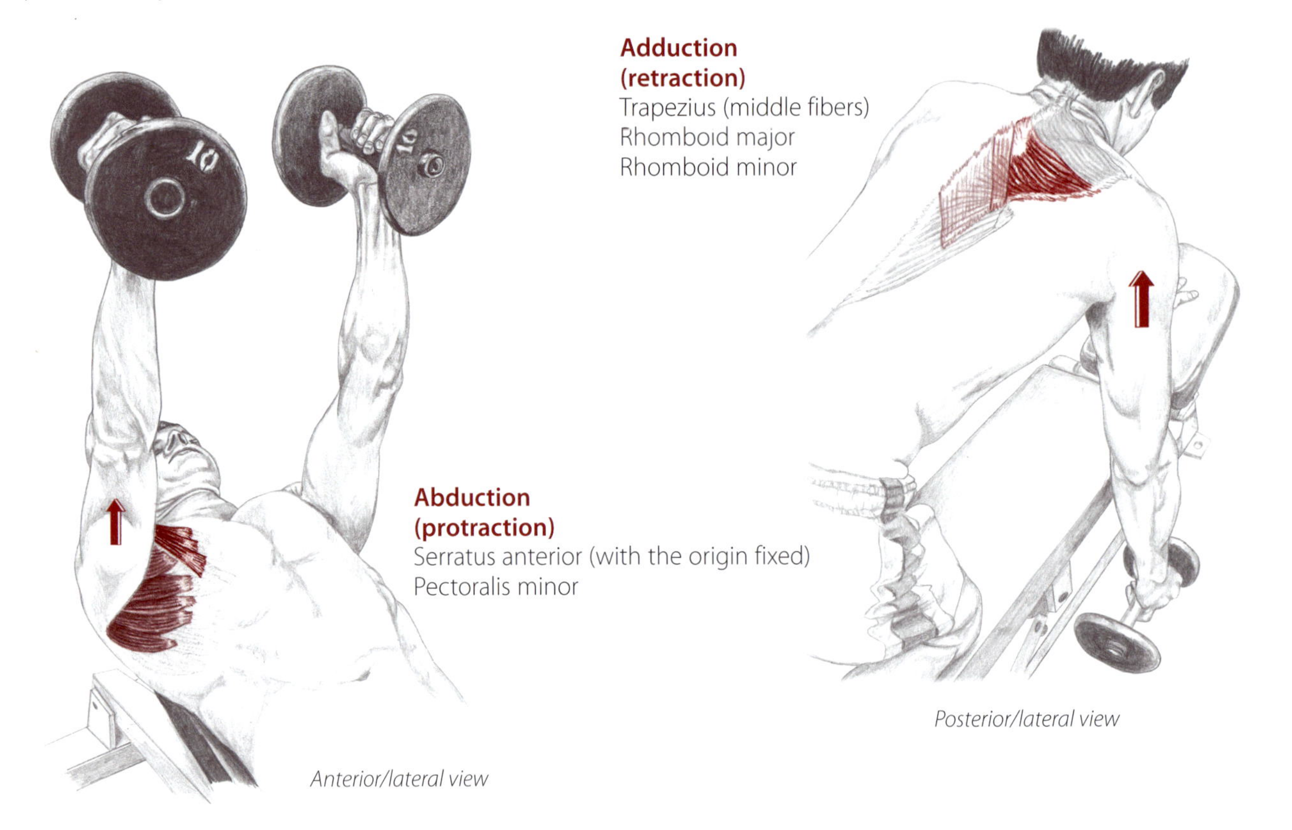

Adduction (retraction)
Trapezius (middle fibers)
Rhomboid major
Rhomboid minor

Abduction (protraction)
Serratus anterior (with the origin fixed)
Pectoralis minor

Anterior/lateral view

Posterior/lateral view

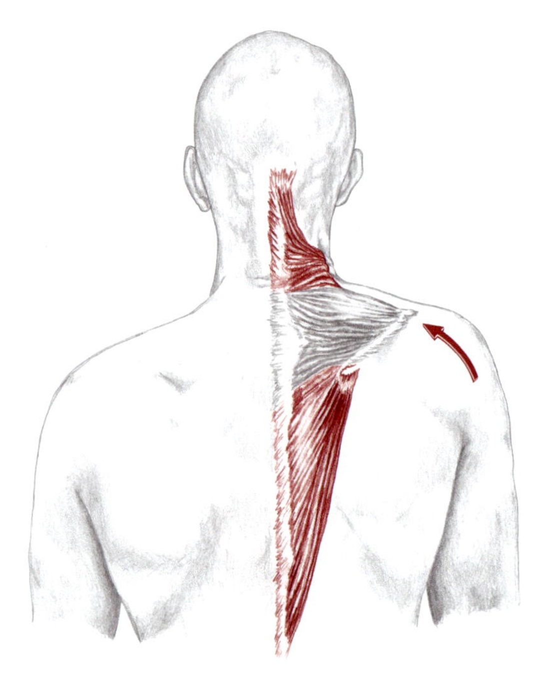

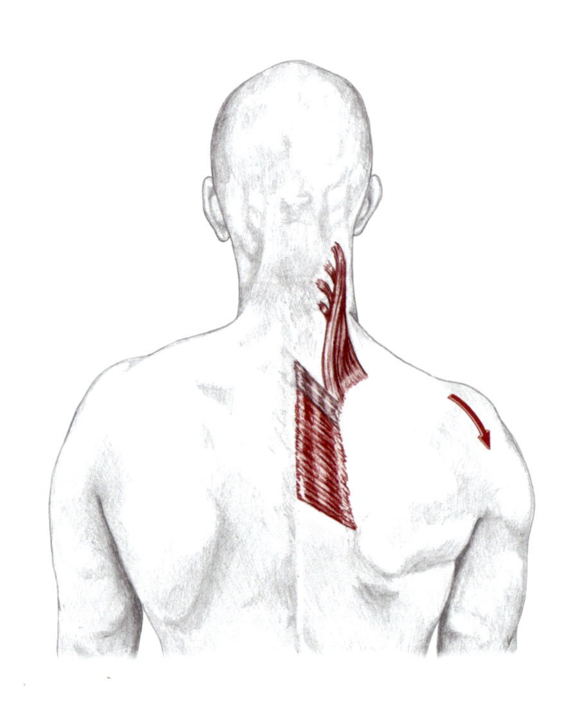

Upward Rotation
Trapezius (upper and lower fibers)

Posterior views

Downward Rotation
Rhomboid major
Rhomboid minor
Levator scapula

Deltoid

The triangle-shaped deltoid is located on the cap of the shoulder. The origin of the deltoid (which is interestingly enough identical to the insertion of the trapezius) curves around the spine of the scapula and clavicle forming a "V" shape. From this broad origin, the fibers converge down the arm to attach at the deltoid tuberosity (2.36).

The deltoid fibers can be divided into three segments: the anterior, middle and posterior fibers. All three groups abduct the humerus, but the anterior and posterior fibers are antagonists in both flexion/extension and medial/lateral rotation.

A *All fibers:*

Abduct the shoulder (glenohumeral joint)

Anterior fibers:

Flex the shoulder (g/h joint)

Medially rotate the shoulder (g/h joint)

Horizontally adduct the shoulder (g/h joint)

Posterior fibers:

Extend the shoulder (g/h joint)

Laterally rotate the shoulder (g/h joint)

Horizontally abduct the shoulder (g/h joint)

O Lateral one-third of clavicle, acromion and spine of scapula

I Deltoid tuberosity

N Axillary from brachial plexus

Belly of the deltoid

1) Seated. Locate the spine of the scapula, the acromion and the lateral one-third of the clavicle. Note the "V" shape these landmarks form.
2) Locate the deltoid tuberosity.
3) Palpate between these landmarks to isolate the superficial, convergent fibers of the deltoid. Be sure to explore the deltoid's most anterior and posterior aspects.

Are the fibers you feel superficial and do they converge toward the deltoid tuberosity? If your partner alternately abducts and releases, do you feel the fibers contract and relax (2.38)?

A: All: abd. GH
Ant: Flx + IR
Mid: Abd.
Post: Ext. + ER.

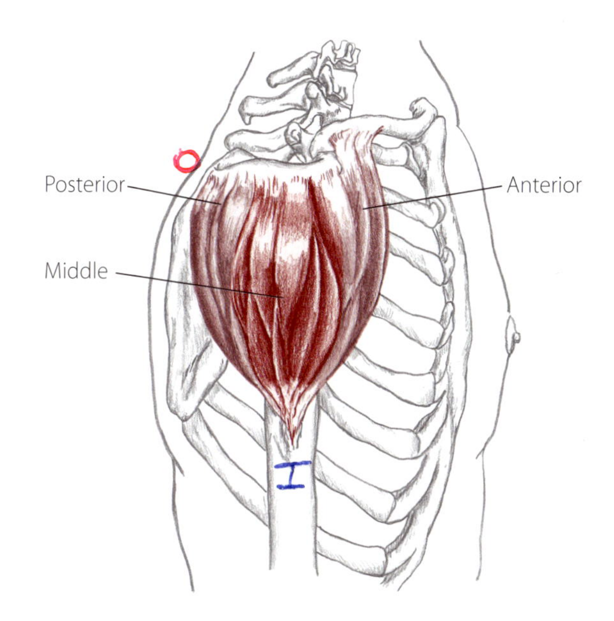

(2.36) Lateral view of deltoid showing the three segments

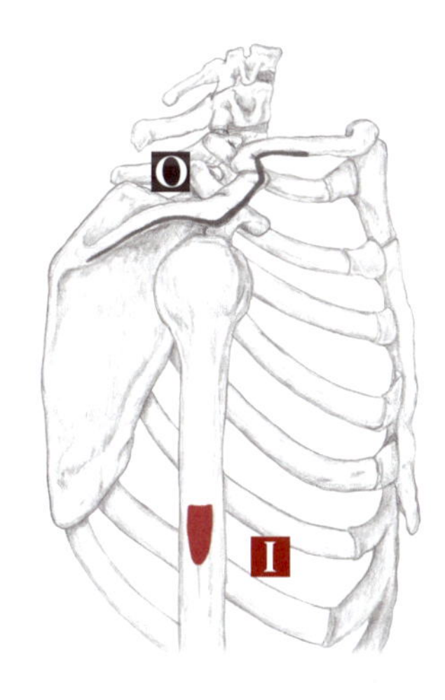

(2.37) Origin and insertion of deltoid

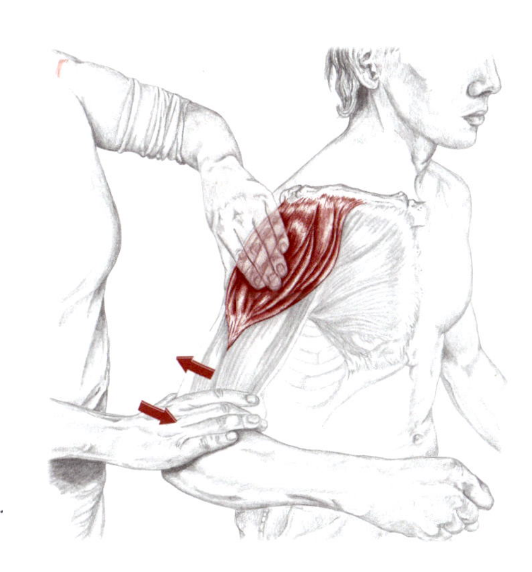

(2.38) Anterior/lateral view

deltoid **del**-toid Grk. *delta*, capital letter D (Δ) in the Greek alphabet

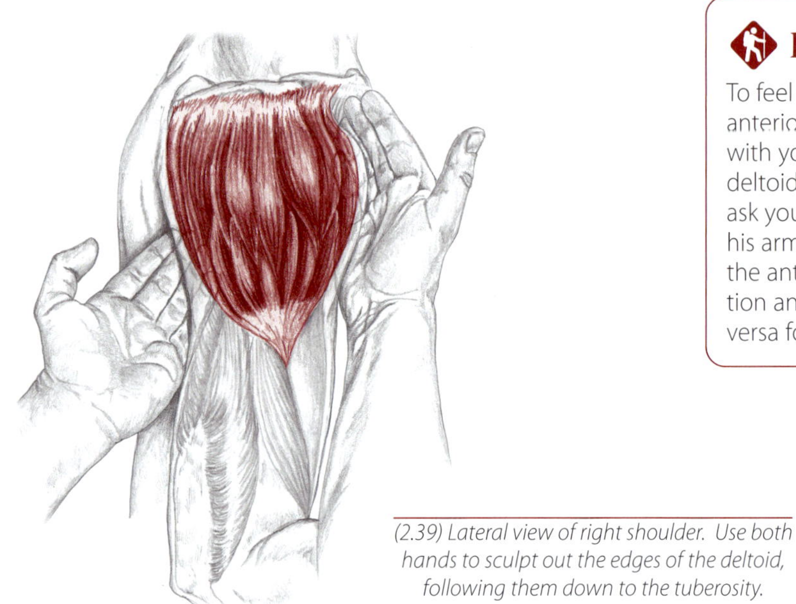

(2.39) Lateral view of right shoulder. Use both hands to sculpt out the edges of the deltoid, following them down to the tuberosity.

Deltoid as antagonist to itself

To feel the antagonistic abilities of the deltoid's anterior and posterior fibers: **1)** Shaking hands with your partner, place your other hand on the deltoid. **2)** Keeping his elbow next to his side, ask your partner to medially and laterally rotate his arm against your resistance. Can you sense the anterior fibers contract upon medial rotation and relax upon lateral rotation and vice versa for the posterior fibers?

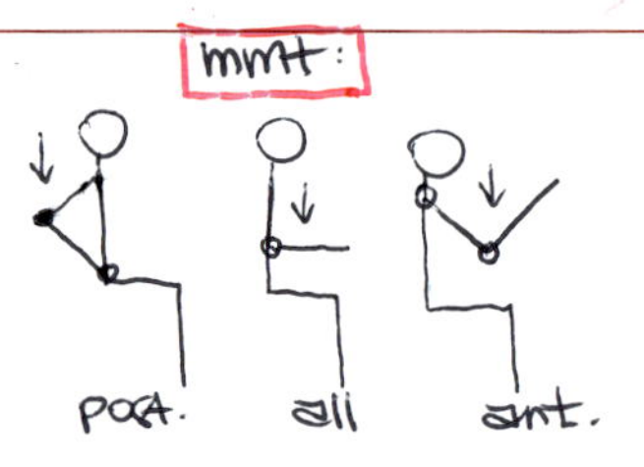

Trapezius

The trapezius lies superficially along the upper back and neck. Its broad, thin fibers blanket the shoulders, attaching to the occiput (the bone at the base of the head, p. 237), lateral clavicle, scapula and spinous processes of the thoracic vertebrae (2.40, 2.42).

The trapezius fibers can be divided into three groups: upper (descending) fibers, middle fibers and lower (ascending) fibers. The upper and lower fibers are antagonists in elevation and depression of the scapula, respectively. All fibers of the trapezius are easy to palpate.

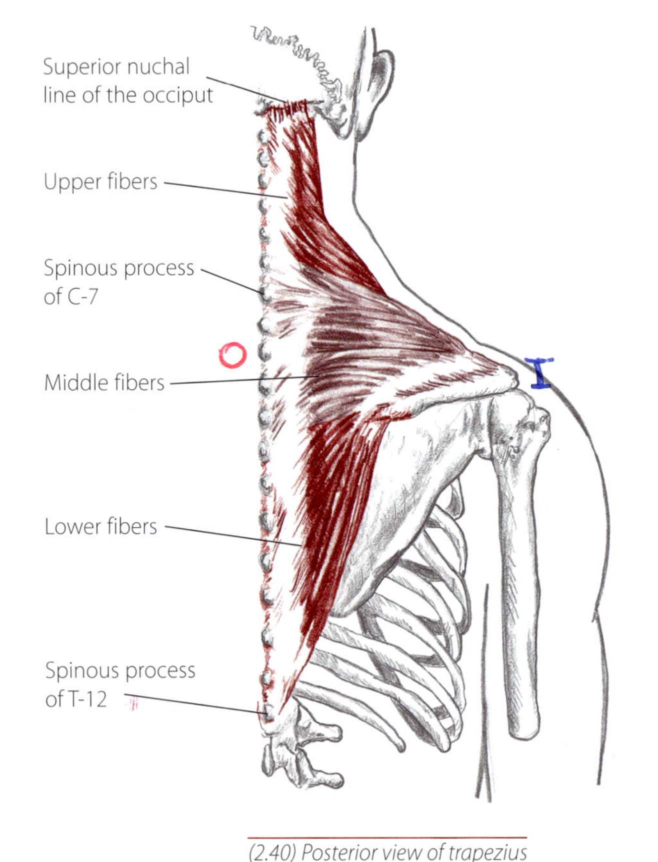

(2.40) Posterior view of trapezius

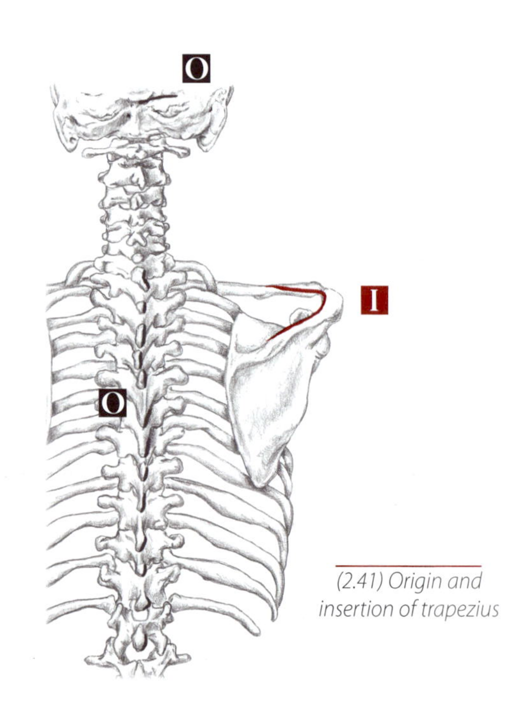

(2.41) Origin and insertion of trapezius

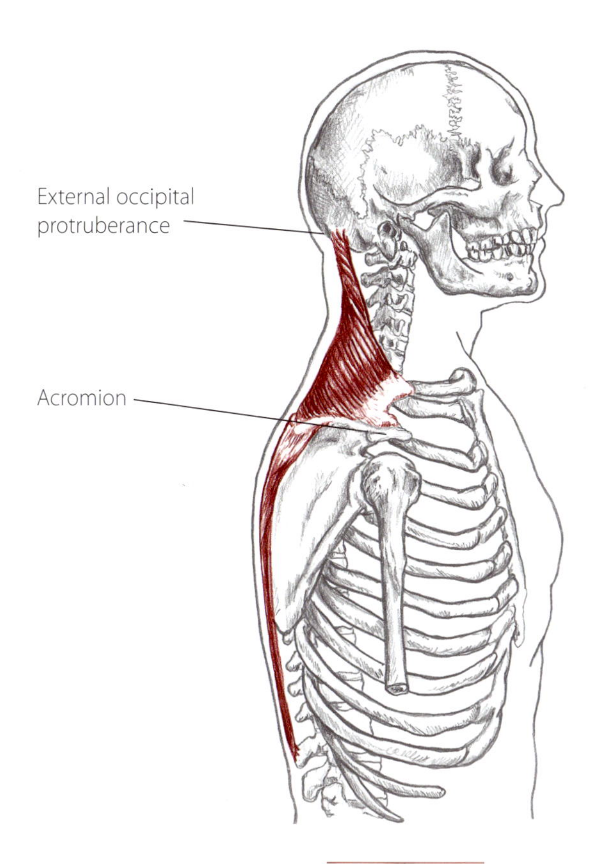

(2.42) Lateral view of trapezius

A *Upper fibers:*
Bilaterally
Extend the head and neck

Unilaterally
Laterally flex the head and neck to the same side
Rotate the head and neck to the opposite side
Elevate the scapula (scapulothoracic joint)
Upwardly rotate the scapula (s/t joint)

Middle fibers:
Adduct the scapula (s/t joint)
Stabilize the scapula (s/t joint)

Lower fibers:
Depress the scapula (s/t joint)
Upwardly rotate the scapula (s/t joint)

O External occipital protuberance, medial portion of superior nuchal line of the occiput, ligamentum nuchae and spinous processes of C-7 through T-12

I Lateral one-third of clavicle, acromion and spine of the scapula

N Spinal accessory and cervical plexus

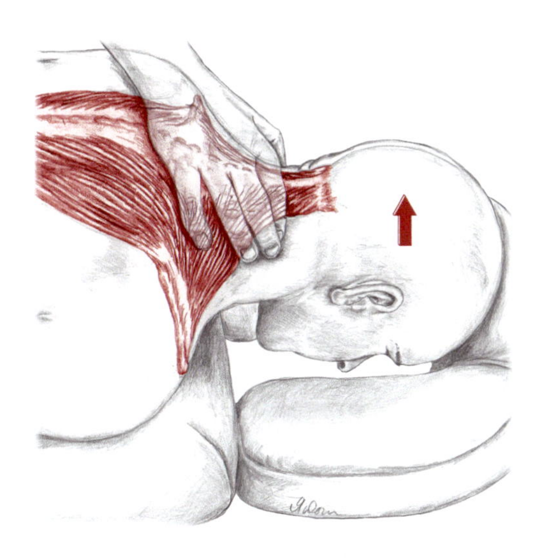

(2.43) Partner prone

As your partner extends his head, you will likely see two parallel "speed bumps" running along the posterior neck. These bulges are formed primarily by the deeper semispinalis capitis muscle (p. 206), with the trapezius muscles draped on top.

Upper fibers of the trapezius

1) Prone. These fibers form the easily accessible flap of muscle lying across the top of the shoulder. Along the posterior neck they are surprisingly skinny, each being only an inch wide.
2) Grasp the superficial tissue on the top of the shoulder and feel the upper trapezius fibers. Take note of their slender quality (2.43).
3) Follow the fibers superiorly toward the base of the head at the occiput. To feel the fibers along the posterior neck contract, stand at the head of the table and ask your partner to extend his head "a quarter inch off the face cradle." Then follow the fibers inferiorly to the lateral clavicle.

Is the muscle you are grasping thin and superficial? Grasp the fibers along the top of the shoulder and have your partner elevate his scapula gently toward his ear. Do the muscle fibers become taut?

MMT: (hd. rot. away, elevate sh.)

A: -Bilat: Ext's hd.
-Ipsilat. Lat flx.
-Contralat. rot.
-Sup. rot. of scap.

trapezius	tra-**pee**-ze-us	Grk. a little table or trapezoid shape
nuchae	**nu**-kay	L. nape of neck
occiput	**ok**-si-put	L. the back of the skull

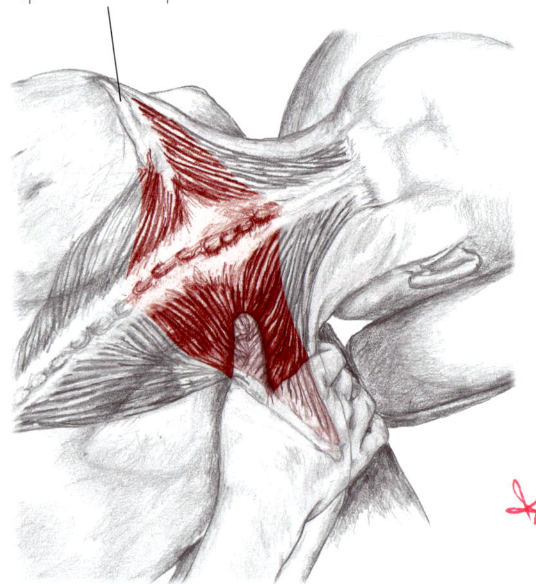

(2.44) Partner prone, accessing the middle fibers of the trapezius

Middle fibers of the trapezius

1) Locate the spine of the scapula.
2) Slide medially from the spine of the scapula onto the trapezius and move your fingers across its fibers (2.44). The trapezius fibers are superficial and thin, so explore at a superficial level and not deeper into the rhomboids or erector spinae muscles.

✔ *Palpate the middle fibers and ask your partner to adduct his scapula. "Bring your shoulder up off the table." Can you feel any contraction in the fibers?*

MMT: SH @ 90°, thumb up
-resist Horiz. abd.
A: Retracts scap.

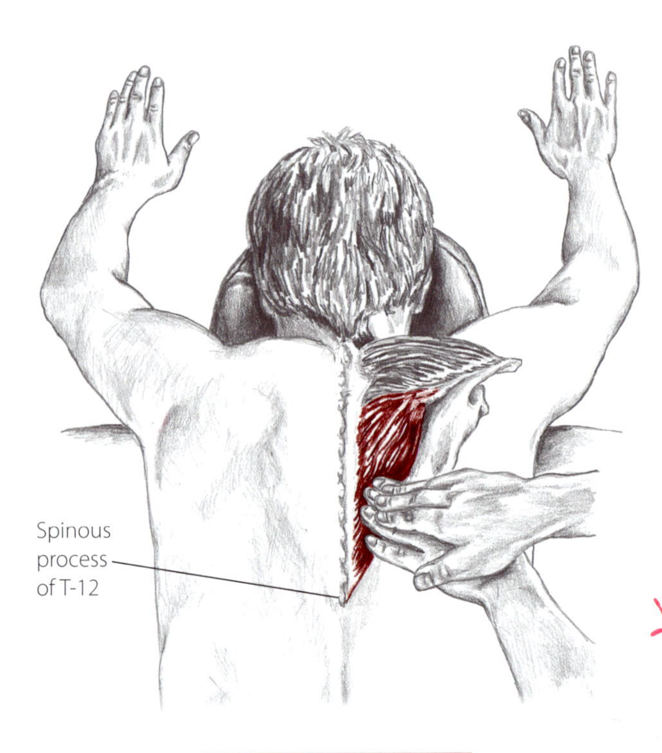

(2.45) Partner prone, with arms reaching out in front of him

Lower fibers of the trapezius

1) Locate the edge of the lower fibers by drawing a line from the spine of the scapula to the spinous process of T-12 (p. 183).
2) Palpate along this line and push your fingers into the edge of the lower fibers. Ask your partner to hold his arms out in front of him (like Superman) and feel for the superficial fibers of the trapezius (2.45).
3) Attempt to lift the lower fibers between your fingers, raising them off the underlying musculature.

✔ *Another action to feel the lower fibers contract is to ask your partner to depress his shoulder. Do the lower fibers run at a gentle angle toward the scapula (rather than parallel with the vertebral column like the erector spinae muscles)?*

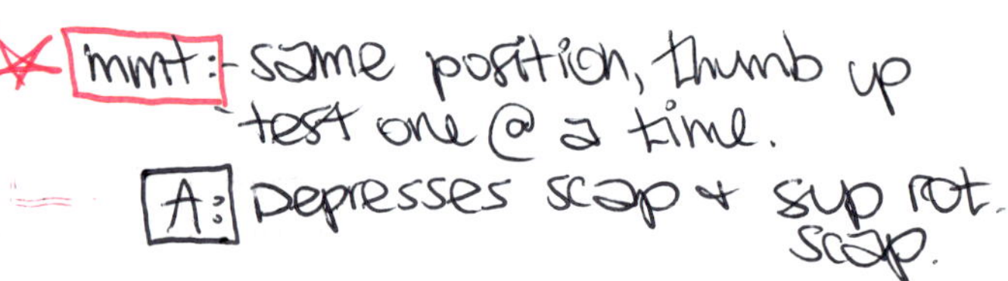

Latissimus Dorsi and Teres Major

The **latissimus dorsi** is the broadest muscle of the back. Its thin, superficial fibers originate at the low back, ascend the side of the trunk and merge into a thick bundle at the axilla (2.46). Both ends of the latissimus dorsi are difficult to isolate; however, its middle portion next to the lateral border of the scapula is easy to grasp.

The **teres major** is called "lat's little helper" because it is a complete synergist with the latissimus dorsi (2.47). It is superficial and located along the scapula's lateral border between the latissimus dorsi and teres minor. Although they share names, the teres major and teres minor rotate the arm in opposite directions - the major medially, the minor laterally.

The latissimus dorsi and teres major are sometimes called the "handcuff muscles," since their actions collectively bring the arms into the "arresting" position!

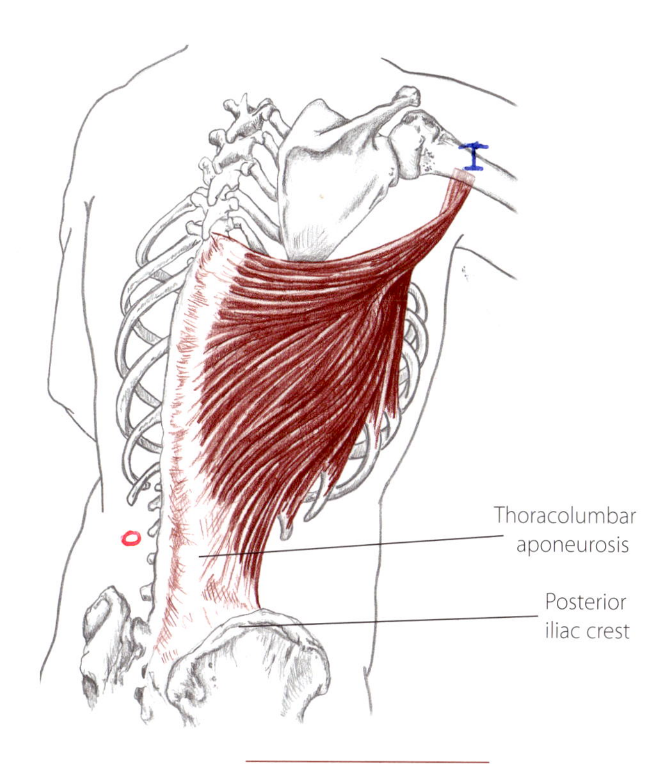

(2.46) Lateral/posterior view of latissimus dorsi

Latissimus Dorsi

A Extend the shoulder (glenohumeral joint)

Adduct the shoulder (g/h joint)

Medially rotate the shoulder (g/h joint)

* MMT: Prone
- Resist add. + ext. w/ palm facing med.

O Spinous processes of last six thoracic vertebrae, last three or four ribs, thoracolumbar aponeurosis and posterior iliac crest

I Crest of the lesser tubercle of the humerus

N Thoracodorsal

Teres Major

A Extend the shoulder (glenohumeral joint)

Adduct the shoulder (g/h joint)

Medially rotate the shoulder (g/h joint)

O Lateral side of inferior angle and lower half of lateral border of the scapula

I Crest of the lesser tubercle of the humerus

N Lower subscapular

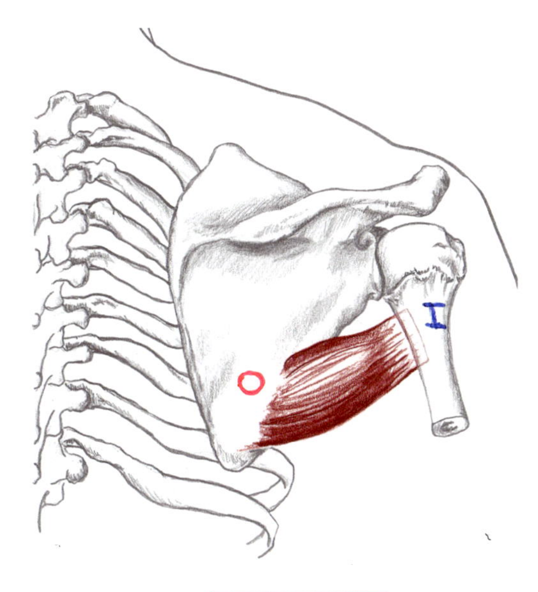

(2.47) Posterior view of teres major

* MMT: Prone:
- Resist add + ext. w/ SH IR'd + back of hand resting on LB.

> The latissimus dorsi not only moves the arm, but because of its broad origin can also affect the trunk and spine. Contraction of the left latissimus dorsi assists in lateral flexion of the trunk to the left. If the arm is fixed, as when hanging from a bar, the latissimus will assist in extension of the spine and tilting of the pelvis anteriorly and laterally.

latissimus dorsi	la-**tis**-i-mus **dor**-si	L. widest of the back
teres	**teh**-reez	L. rounded, finely shaped

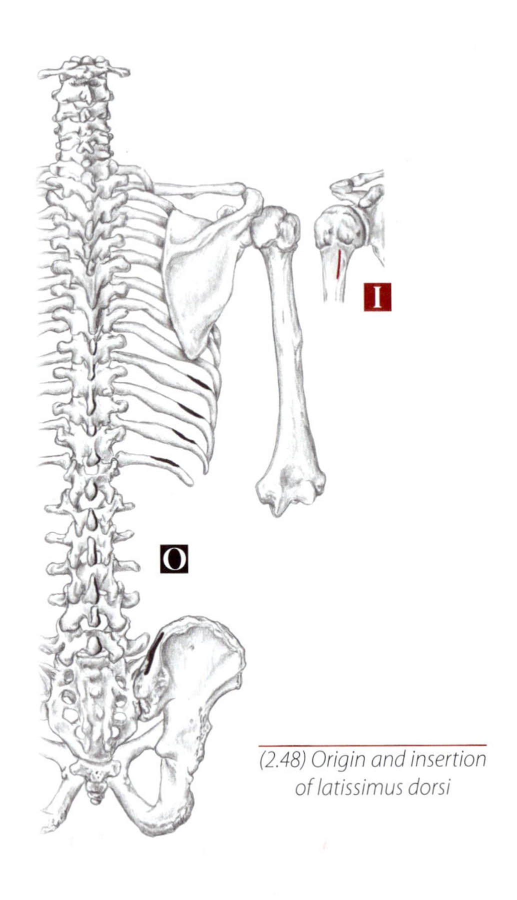

(2.48) Origin and insertion of latissimus dorsi

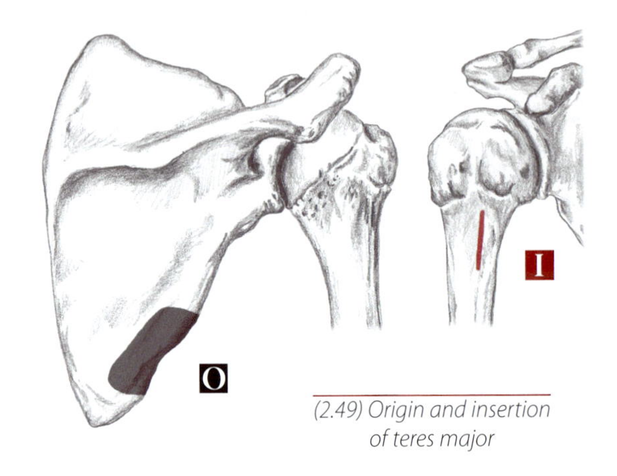

(2.49) Origin and insertion of teres major

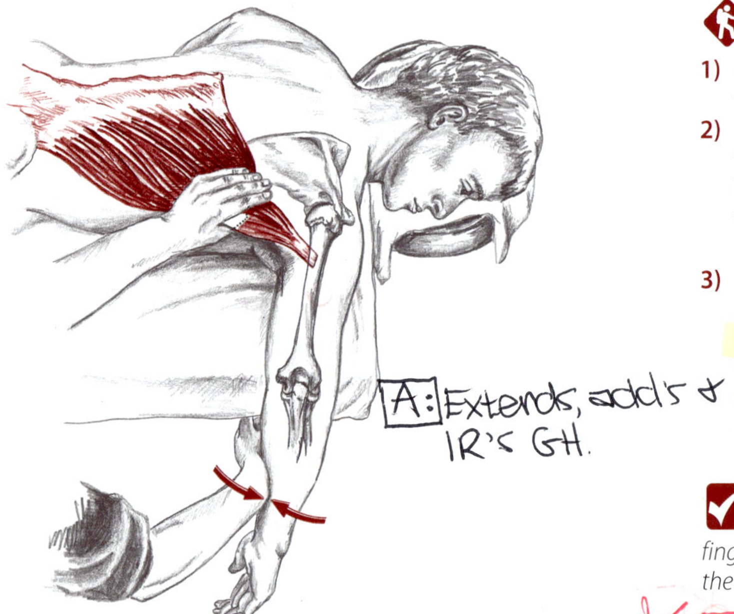
(2.50) Partner prone, medially rotating at the shoulder

A: Extends, add's & IR's GH.

Latissimus dorsi

1) Prone with the arm off side of the table. Locate the scapula's lateral border.
2) Using your fingers and thumb, grasp the thick wad of muscle tissue lateral to the lateral border. This is the latissimus dorsi (and perhaps some of teres major). Note how this muscle tissue flares off the side of the trunk.
3) Feel the latissimus fibers contract by asking your partner to medially rotate his shoulder against your resistance. "Swing your hand up toward your hip." (2.50) As this occurs, follow the latissimus fibers superiorly into the axilla and inferiorly on the ribs.

To ensure you are not just lifting the skin, grasp the tissue and let it slowly slip out between your fingers. Do you feel the muscle's fibrous texture or just the skin's jellylike quality?

MMT: Prone, arm ext'd, palm facing med.
Resist Ext. + add.

Latissimus dorsi

1) With your partner supine, cradle the arm in a flexed position. Then grasp the tissue of the latissimus located beside the lateral border.
2) Ask your partner to extend his shoulder against your resistance. "Press your elbow toward your hip." This will force the latissimus to contract (2.51).

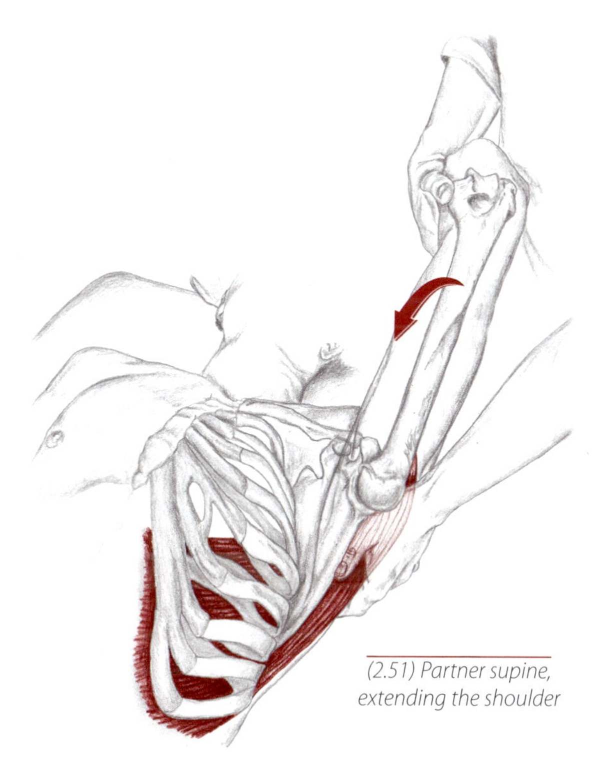

(2.51) Partner supine, extending the shoulder

 Teres major

1) Prone with the arm off the side of the table. Locate and grasp the latissimus dorsi fibers between your fingers and thumb.
2) Move your fingers and thumb medially to where you feel the scapula's lateral border. The muscle fibers that lie medial to the latissimus and attach to the lateral border will be the teres major.
3) Follow these fibers toward the axilla where they blend with the latissimus dorsi.

Lay your thumb on the inferior aspect of the lateral border and have your partner medially rotate the shoulder joint to distinguish the teres major from the latissimus dorsi (2.52). The fibers of both muscles will contract. Those that attach directly to the lateral border belong to teres major; the more lateral fibers belong to latissimus dorsi.

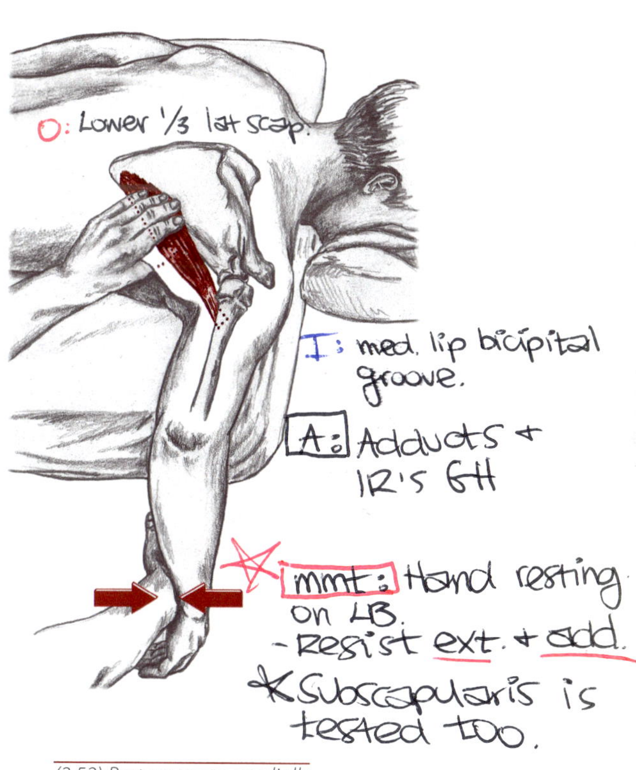

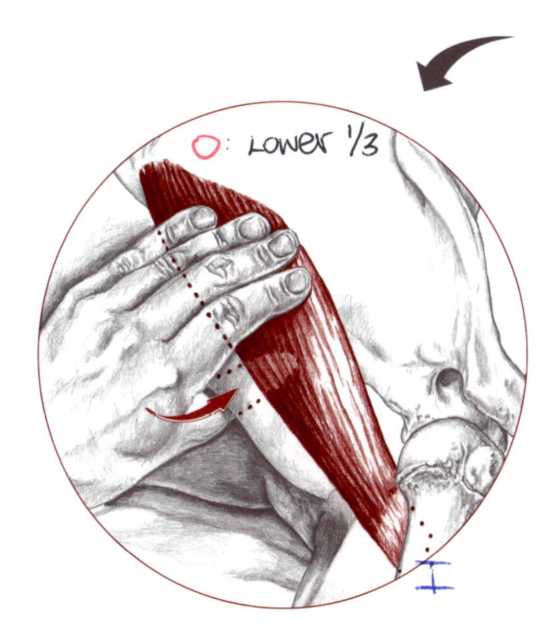

(2.52) Partner prone, medially rotating at the shoulder

Rotator Cuff Muscles

Supraspinatus
Infraspinatus
Teres Minor
Subscapularis

Supraspinatus, infraspinatus, teres minor and subscapularis are known as the rotator cuff muscles. Together they encompass, and therefore stabilize, the glenohumeral joint. All of the rotator cuff muscles are accessible, including their tendons, which attach to the head of the humerus.

The chunky **supraspinatus** is located in the supraspinous fossa, deep to the trapezius' upper fibers. Its belly runs underneath the acromion and attaches to the humerus' greater tubercle (2.53). The supraspinatus assists the deltoid with abduction of the shoulder and is the only muscle of the group not involved in shoulder rotation.

The flat, convergent belly of the **infraspinatus** is located in the infraspinous fossa. Most of its belly is superficial with a medial portion deep to the trapezius and a lateral portion beneath the deltoid (p. 75). The infraspinatus attaches immediately posterior to the supraspinatus on the greater tubercle (2.54) and is a synergist with the teres minor in lateral rotation of the shoulder. The unique, dense quality of the infraspinatus muscle is due to its thick, layered fascia.

The **teres minor** is a small muscle squeezed between the infraspinatus and teres major. It is located high in the axilla and can be challenging to grasp (2.55). The teres minor and teres major are antagonists in rotation of the humerus.

The deep **subscapularis** (2.56), located on the scapula's anterior surface, is sandwiched between the subscapular fossa and serratus anterior muscle (p. 94). With only a small portion of its muscle belly accessible, the subscapularis is the only rotator cuff muscle that attaches to the humerus' lesser tubercle. It rotates the shoulder medially.

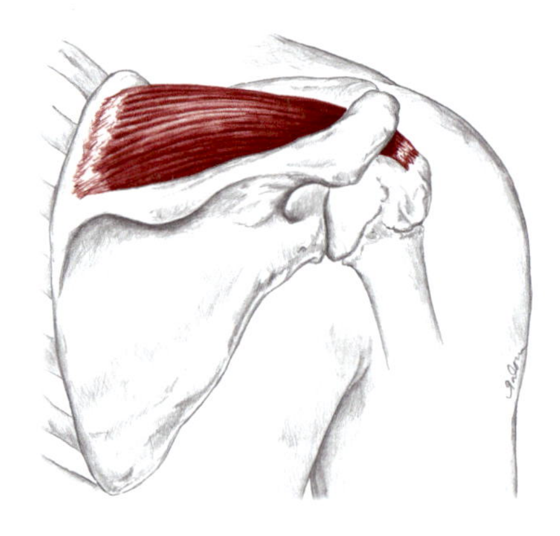

Supraspinatus

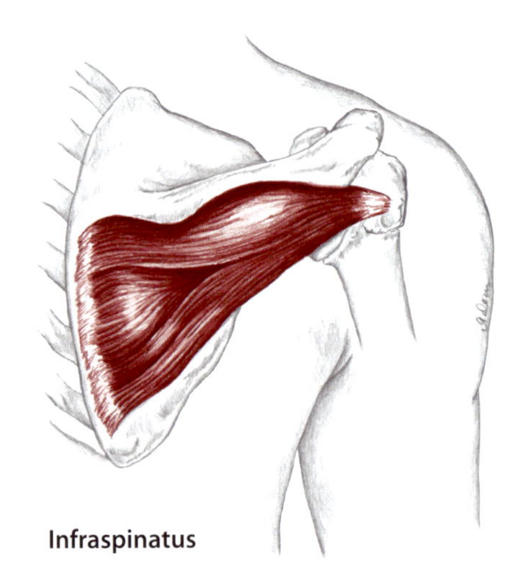

Infraspinatus

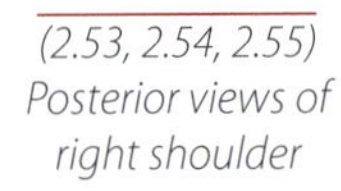

(2.53, 2.54, 2.55) Posterior views of right shoulder

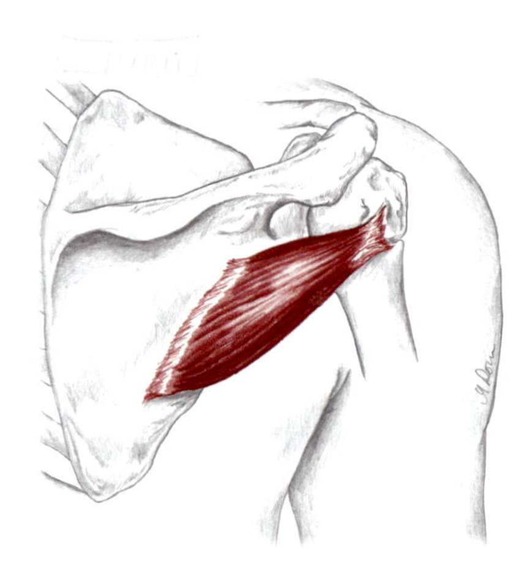

Teres minor

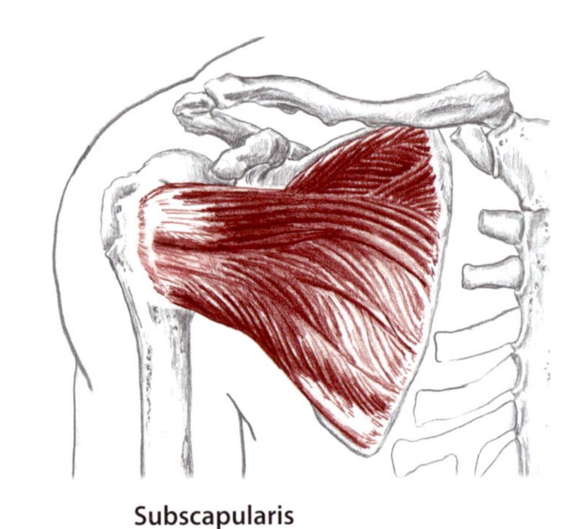

Subscapularis

(2.56) Anterior view of right shoulder with ribs removed

infraspinatus	**in**-fra-spi-**na**-tus
subscapularis	sub-**skap**-u-**lar**-is
supraspinatus	**soo**-pra-spi-**na**-tus

Supraspinatus

A Abduct the shoulder (glenohumeral joint)
Stabilize head of humerus in glenoid cavity

O Supraspinous fossa of the scapula

I Greater tubercle of the humerus

N Suprascapular

Infraspinatus

A Laterally rotate the shoulder (glenohumeral joint)
Adduct the shoulder (g/h joint)
Extend the shoulder (g/h joint)
Horizontally abduct the shoulder (g/h joint)
Stabilize head of humerus in glenoid cavity

O Infraspinous fossa of the scapula

I Greater tubercle of the humerus

N Suprascapular

Teres Minor

A Laterally rotate the shoulder (glenohumeral joint)
Adduct the shoulder (g/h joint)
Extend the shoulder (g/h joint)
Horizontally abduct the shoulder (g/h joint)
Stabilize head of humerus in glenoid cavity

O Superior half of lateral border of the scapula

I Greater tubercle of the humerus

N Axillary

Subscapularis

A Medially rotate the shoulder (glenohumeral joint)
Stabilize head of humerus in glenoid cavity

O Subscapular fossa of the scapula

I Lesser tubercle of the humerus

N Upper and lower subscapular

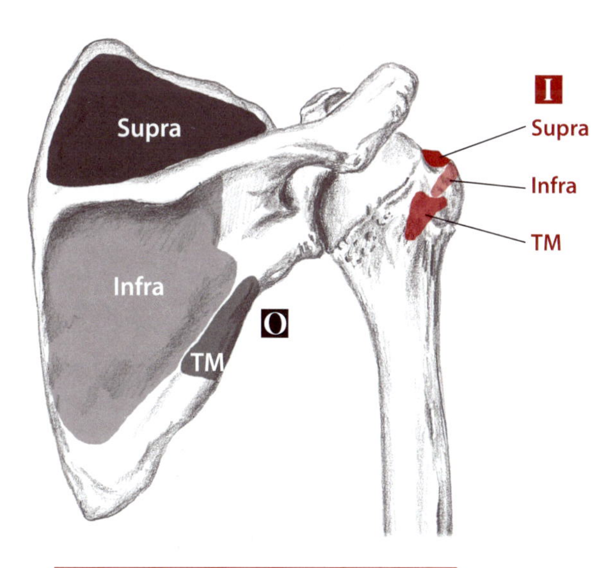

(2.57) Posterior view of right shoulder showing origins and insertions of supraspinatus, infraspinatus and teres minor

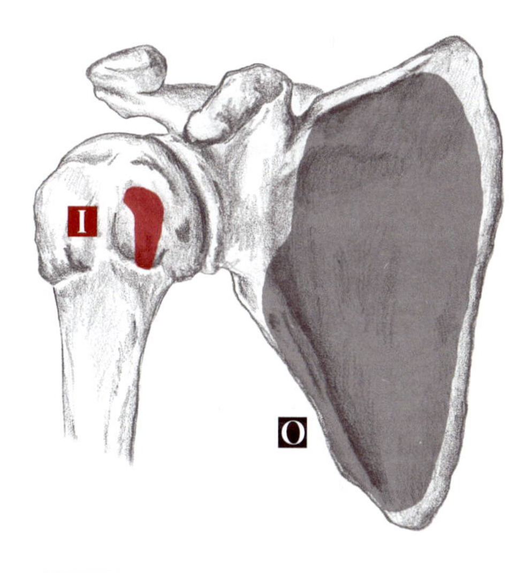

(2.58) Anterior view of right shoulder showing origin and insertion of subscapularis

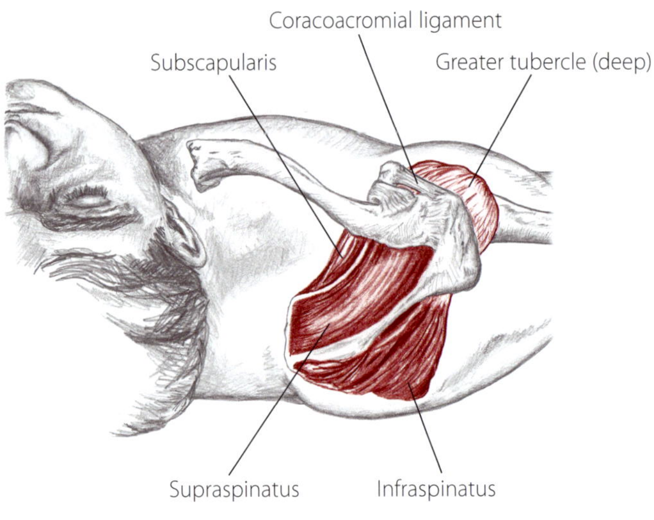

(2.59) Superior view of right shoulder

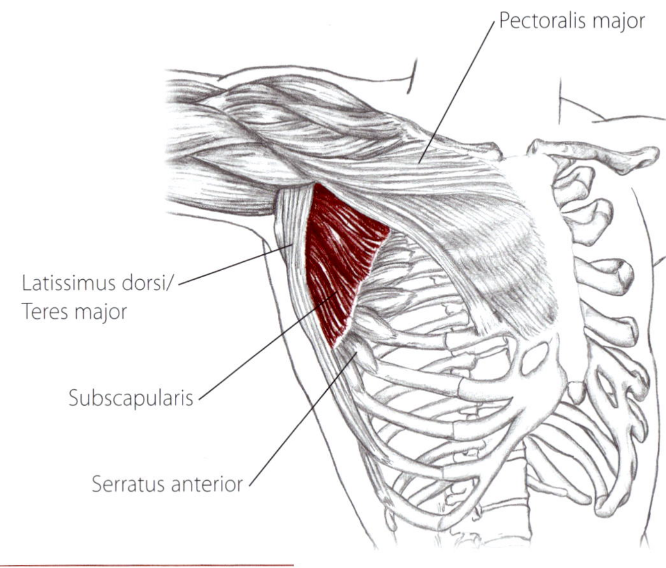

(2.60) Lateral/inferior view of axilla

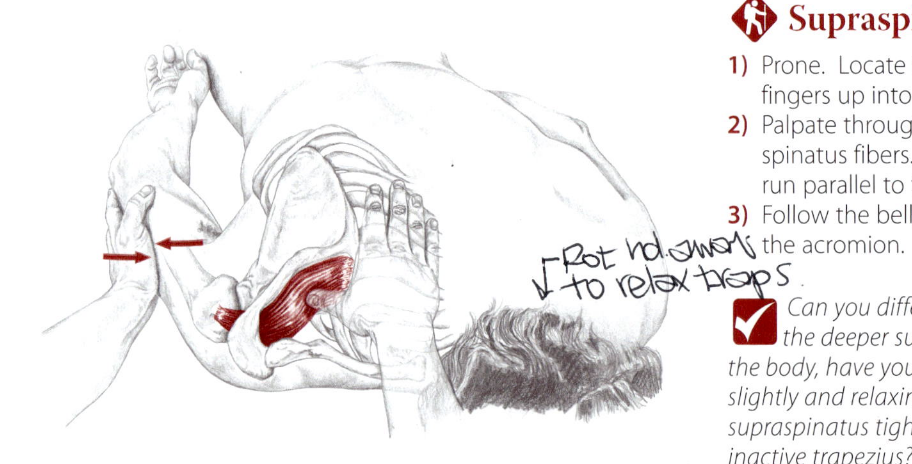

(2.61) Partner prone, alternately abducting and relaxing his shoulder to feel supraspinatus contract

Supraspinatus

1) Prone. Locate the spine of the scapula. Slide your fingers up into the supraspinous fossa.
2) Palpate through the trapezius and onto the supraspinatus fibers. As you palpate, note how the fibers run parallel to the spine.
3) Follow the belly laterally until it tucks under the acromion.

Can you differentiate the fibers of the trapezius and the deeper supraspinatus? With the arm alongside the body, have your partner alternate between abducting slightly and relaxing the shoulder (2.61). Can you feel the supraspinatus tighten and soften underneath the inactive trapezius?

Infraspinatus

1) Prone, with the forearm off the side of the table. Locate the spine, medial border and lateral border of the scapula.
2) Form a triangle around the infraspinatus by laying a finger along each of these landmarks.
3) Palpate in the triangle and strum across the infraspinatus fibers. Follow them laterally as they converge underneath the deltoid to attach to the humerus.

With the forearm off the side of the table, ask your partner to alternately raise his elbow one inch toward the ceiling and relax (2.62). Do you feel the infraspinatus contract and tighten?

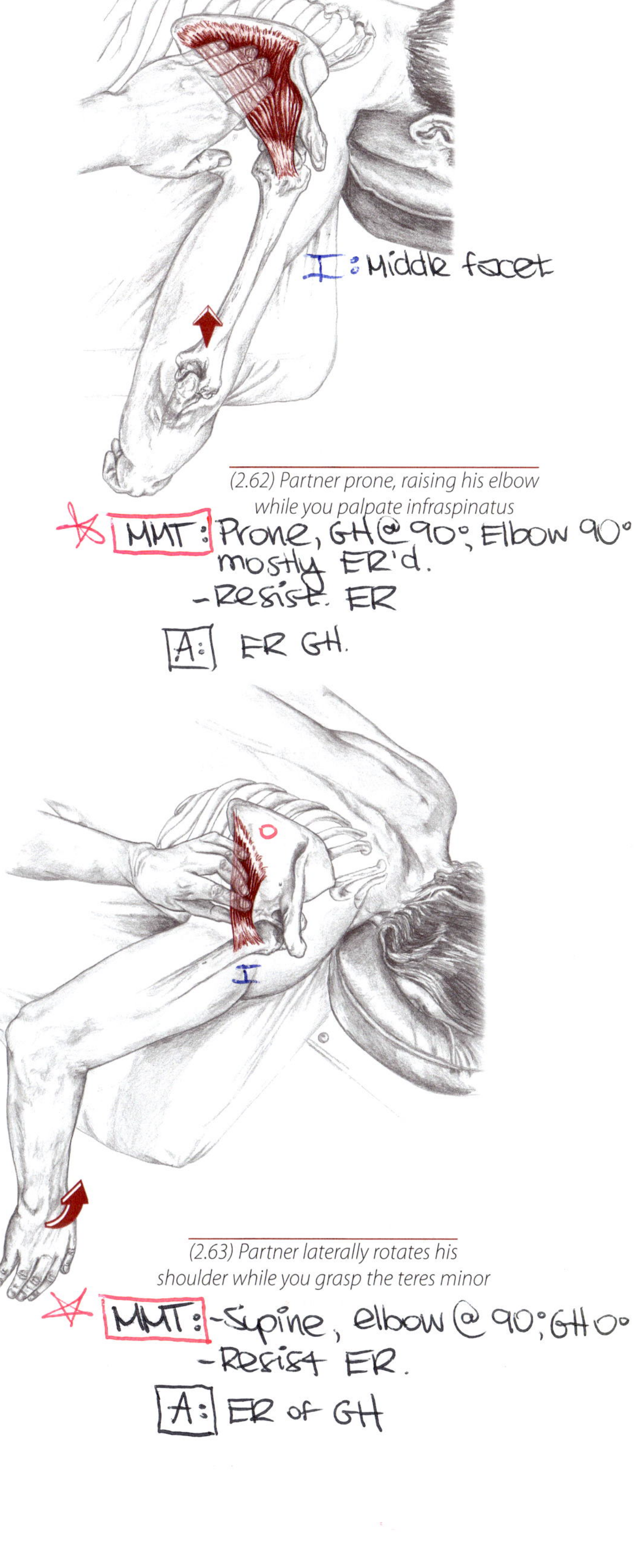

(2.62) Partner prone, raising his elbow while you palpate infraspinatus

(2.63) Partner laterally rotates his shoulder while you grasp the teres minor

Teres minor

1) Prone, with the arm off the side of the table. Locate the lateral border of the scapula; specifically, its superior half. Slide laterally off the lateral border onto the surface of the teres minor.
2) Compress into and across its tube-shaped belly. Move inferiorly and compare it in size to the teres major. Also, reach your thumb up into the axilla and grasp the belly of the teres minor as you would a hamburger (2.63).
3) Ask your partner to laterally rotate his shoulder. "Swing your hand up toward your head." Bringing the elbow toward the ceiling also forces the teres minor to contract.

Does the muscle you are palpating attach along the superior half of the scapula's lateral border?

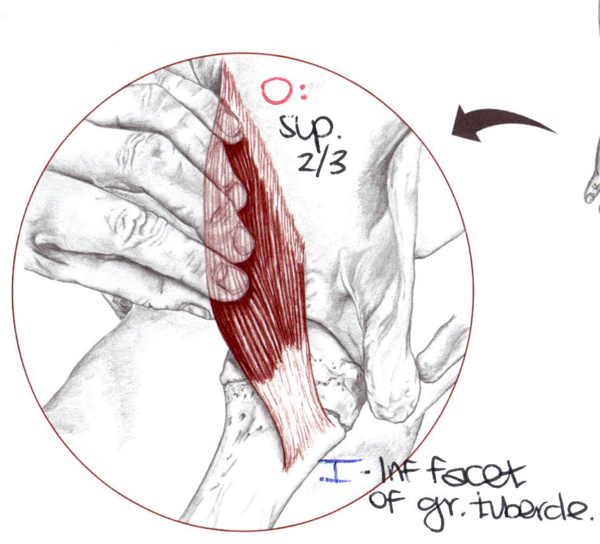

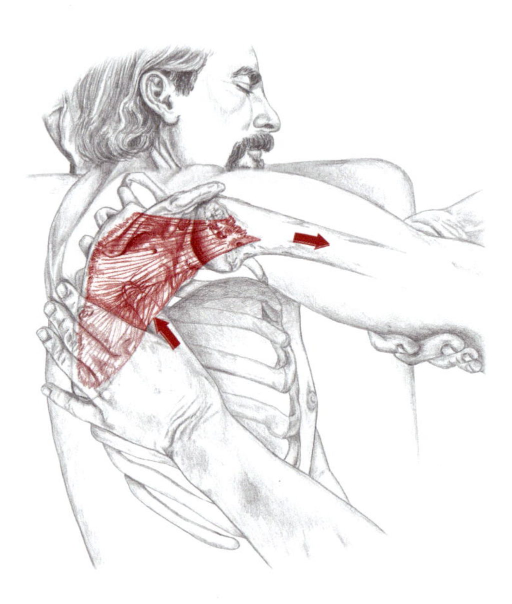

(2.64) Partner sidelying, accessing subscapularis

Subscapularis

1) Sidelying. Flex the shoulder and pull the arm anteriorly as much as possible. This will allow easier access to the scapula's anterior surface.
2) Hold the arm with one hand while the thumb of the other locates the lateral border. Hint: Slide your thumb underneath the latissimus dorsi and teres major fibers instead of going through them (2.64).
3) Slowly and gently curl your thumb onto the subscapular fossa. You may not feel the subscapularis fibers immediately, but if your thumb is on the anterior surface of the scapula, you will be accessing a portion of the fibers.

Ask your partner to gently rotate his shoulder medially. Can you feel the subscapularis fibers contract beneath your thumb? Explore the subscapularis by moving your thumb more superiorly or inferiorly.

Supine. Cradle the arm in a flexed position and locate the lateral border. Slowly sink your thumbpad onto the subscapular fossa, adjusting the arm and scapula as you progress (2.65).

MMT: Prone, hand resting on LB
→ Resist ext. + add.
A: IR's GH

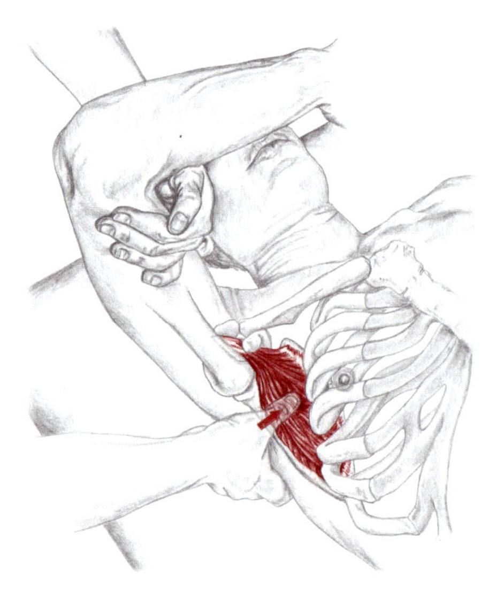

(2.65) Partner supine, accessing subscapularis

Here is a fun way to experience the opposite rotational capabilities of the teres major and teres minor. With your partner prone, lay your hand on the surfaces of the teres major and minor. Ask your partner to alternately medially and laterally rotate his arm. (Be sure he does not raise his elbow, because then they both contract.)

Can you feel the teres major contract while the teres minor softens upon medial rotation? Vice versa for lateral rotation?

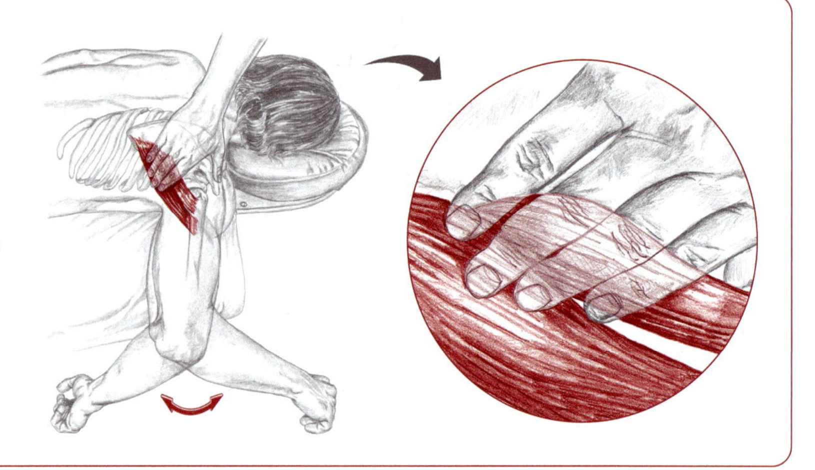

Rotator Cuff Tendons

The tendons of the rotator cuff muscles can be difficult to access in anatomical position (2.66). The supraspinatus and infraspinatus tendons are situated deep to the acromion, while the tendons of the subscapularis and teres minor lie deep to the thick belly of the deltoid.

This dilemma can be overcome, however, and the individual tendons isolated by placing the humerus in the positions outlined below. Since the rotator cuff tendons lie against the surface of the greater or lesser tubercles of the humerus, they cannot be separated from the underlying bone.

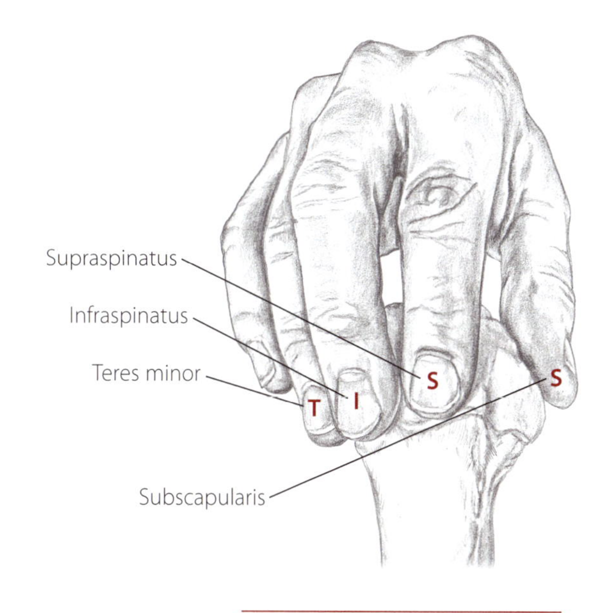

(2.66) *Anterior/lateral view of rotator cuff tendon attachment sites*

 Supraspinatus tendon

1) The attachment of the tendon will be located just distal to the acromion on the greater tubercle.
2) Supine or seated, with the arm at the side of the body. Locate the acromion and slide inferiorly onto the surface of the greater tubercle (2.67). Between these two landmarks will be a palpable portion of the tendon.
3) Sink your thumb tip through the deltoid fibers. Using firm pressure, roll your thumb across the small mound of the supraspinatus tendon.

Are you palpating on the surface of the greater tubercle or on the superficial deltoid fibers?

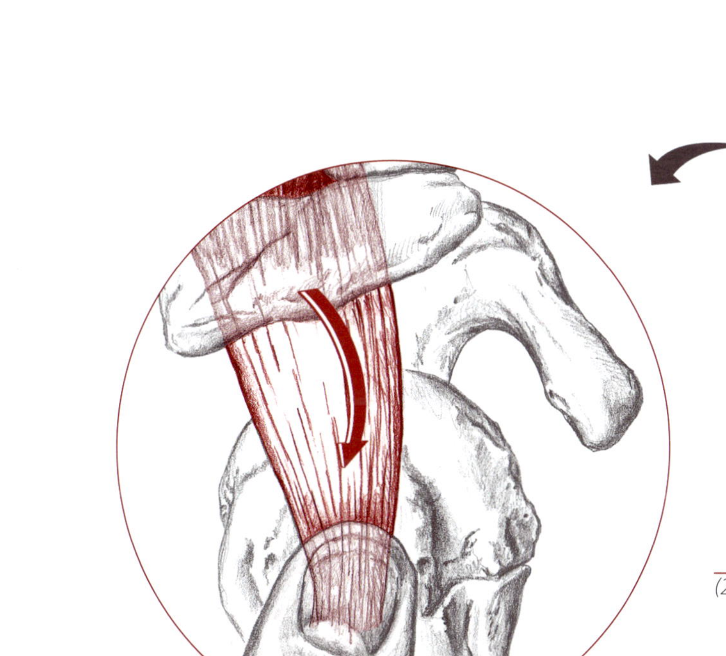

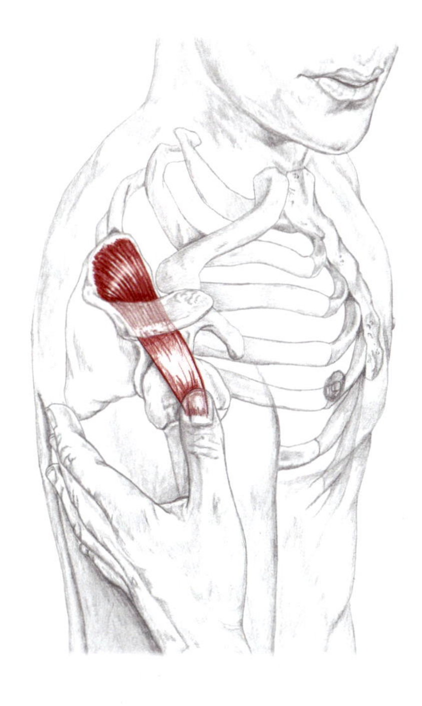

(2.67) *Anterior/lateral view accessing the supraspinatus tendon*

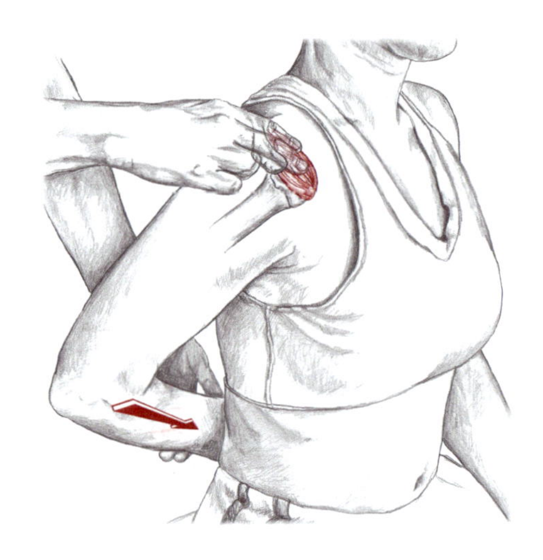

(2.68) Partner seated, forearm behind her back palpating the supraspinatus tendon

Supraspinatus tendon

1) Seated. Place your partner's arm behind her back. This position will medially rotate and extend the humerus.
2) Passively extend the arm as far as is comfortable for your partner (2.68). This position brings the supraspinatus tendon out from under the acromion, just anterior and inferior to the acromioclavicular (A/C) joint.

Is the arm medially rotated and extended as far as comfortably possible? Are you palpating inferior to the A/C joint?

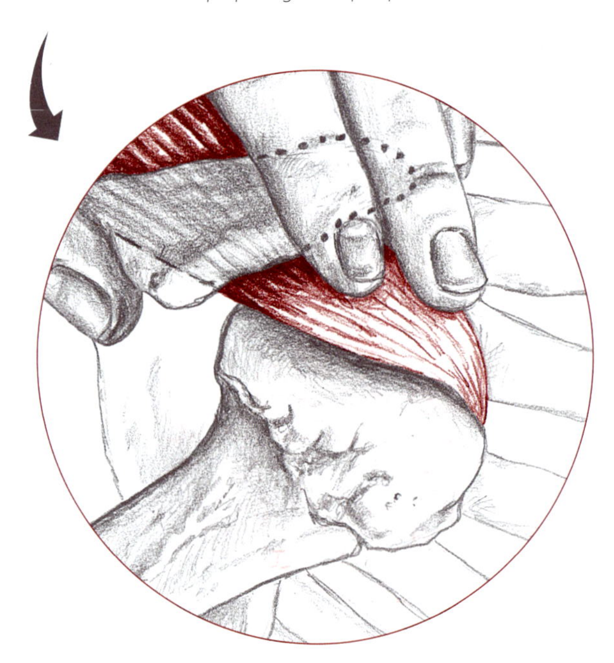

Infraspinatus and teres minor tendons

1) Prone, with arm off the side of the table. Locate the bellies of these muscles.
2) Strumming across their fibers, follow their bellies laterally as they pass inferior the acromion. Palpating through the deltoid, roll across their slender tendinous attachments at the greater tubercle (2.69).
3) Turn your partner supine. Locate the tendinous attachment of supraspinatus. Move posteriorly along the greater tubercle and feel for the small tendinous attachments of the infraspinatus and teres minor.

Are you palpating deep to the deltoid fibers? Do you feel the solid surface of the greater tubercle beneath your fingers?

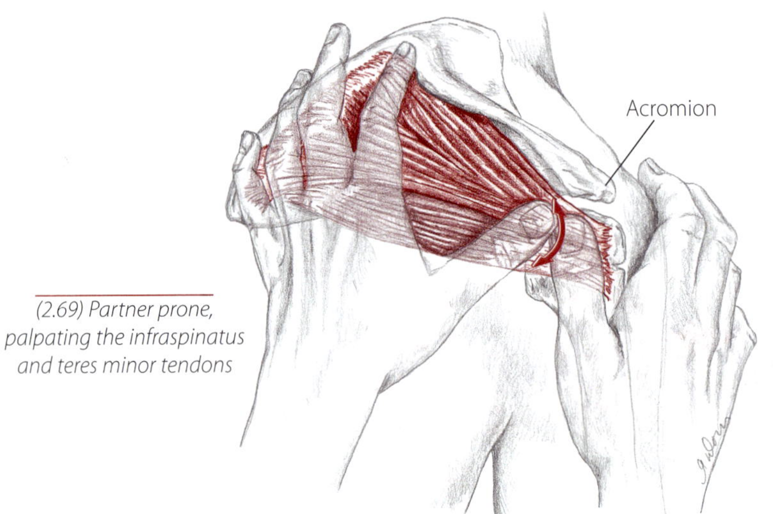

(2.69) Partner prone, palpating the infraspinatus and teres minor tendons

The supraspinatus was long thought to be a mere "spark plug" for shoulder abduction, initiating the movement before the stronger deltoid took over. Research shows, however, that the supraspinatus contracts through the full movement of abduction and can single-handedly bring the arm to 90°.

Infraspinatus and teres minor tendons

1) Supine or seated. Flex the shoulder to 90°. Then horizontally adduct and laterally rotate slightly (10 to 20°).
2) Although deep to the posterior deltoid, this position causes the infraspinatus tendon to move below the acromion and be accessible (2.70).
3) Locate the acromial angle. Drop inferiorly off the angle and explore this region.

Is the shoulder flexed, adducted and laterally rotated? Do you feel the solid surface of the greater tubercle beneath your fingers? Return the arm to a neutral position and notice how the posterior humerus slides back under the acromion.

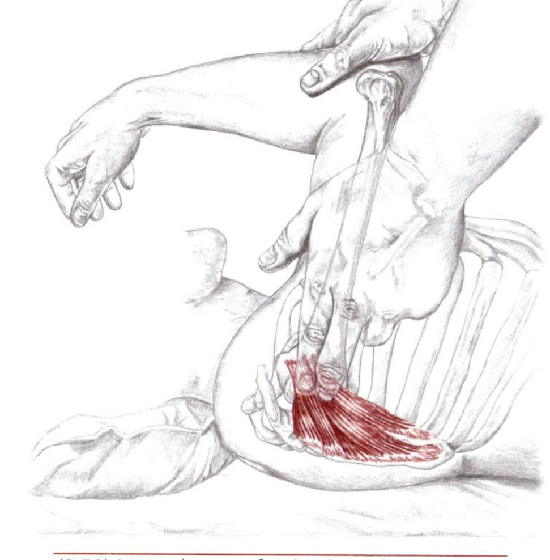

(2.70) Lateral view of right shoulder, accessing the infraspinatus and teres minor tendons

Subscapularis tendon

1) Seated or supine. Place the arm next to the trunk in anatomical position.
2) Locate the coracoid process of the scapula. Slide one inch inferiorly and laterally from the coracoid. You will be between the two tendons of the biceps brachii.
3) Palpate through the deltoid fibers, exploring the deeper tissue which lies along the lesser tubercle of the humerus (2.71). This is the location of the subscapularis tendon. Explore for more of the tendon by moving medially off the lesser tubercle.

Is the arm positioned next to the body? Are you palpating deep to the deltoid fibers? Can you feel the solid surface of the lesser tubercle?

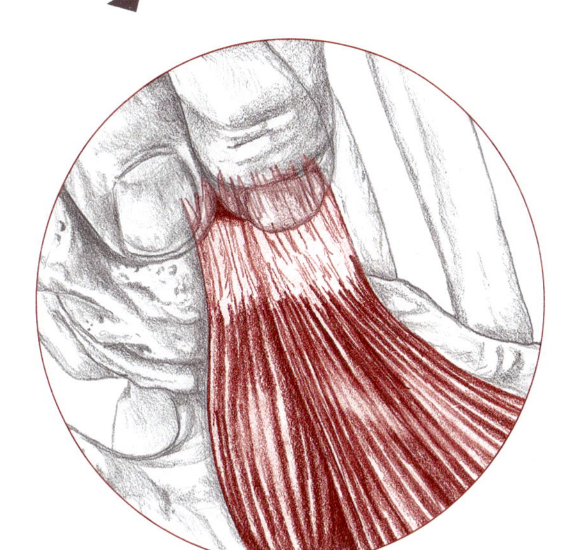

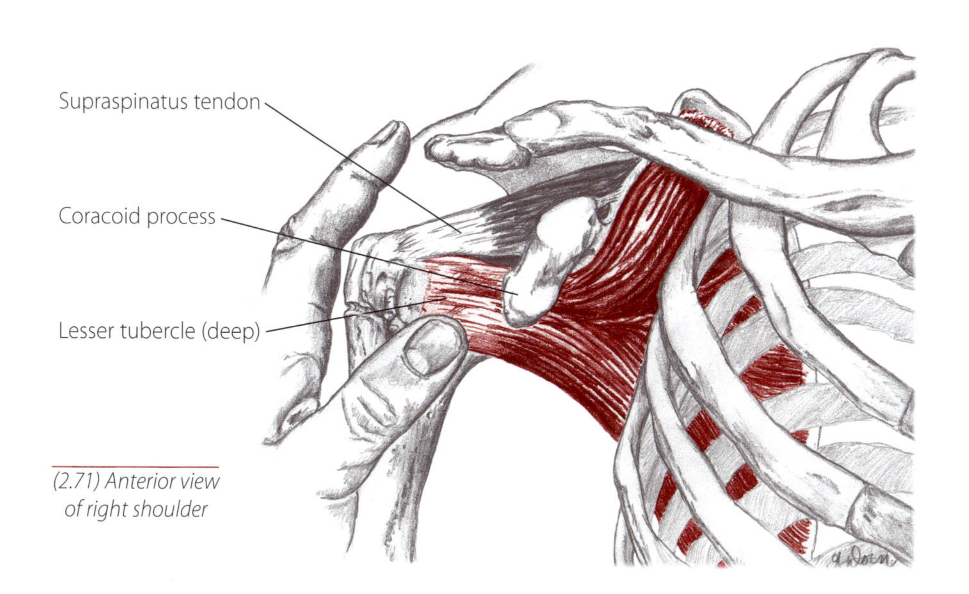

(2.71) Anterior view of right shoulder

Rhomboid Major and Minor

The rhomboid muscles are located between the scapula and vertebral column. Named for their geometric shape, the major (2.72) is larger than the minor (2.73). The muscles are difficult to distinguish individually. They have thin fibers that lie deep to the trapezius and superficial to the erector spinae muscles (p. 202).

A Adduct the scapula (scapulothoracic joint)

Elevate the scapula (s/t joint)

Downwardly rotate the scapula (s/t joint)

O *Major:*
Spinous processes of T-2 to T-5
Minor:
Spinous processes of C-7 and T-1

I *Major:*
Medial border of the scapula between the spine of the scapula and inferior angle

Minor:
Upper portion of medial border of the scapula, across from spine of the scapula

N Dorsal scapular from brachial plexus

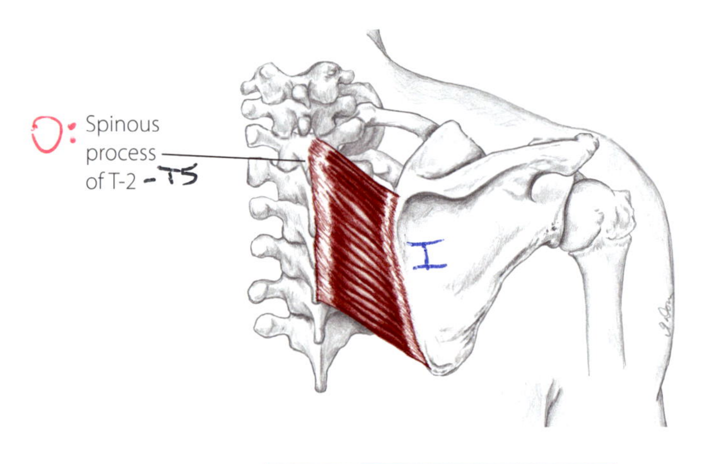

(2.72) Posterior view of rhomboid major

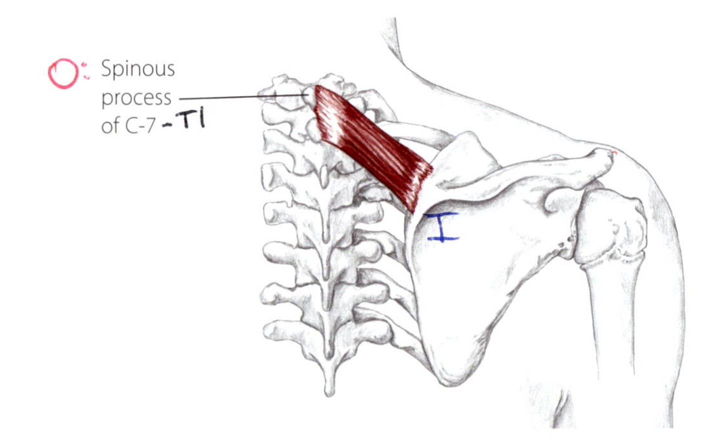

(2.73) Posterior view of rhomboid minor

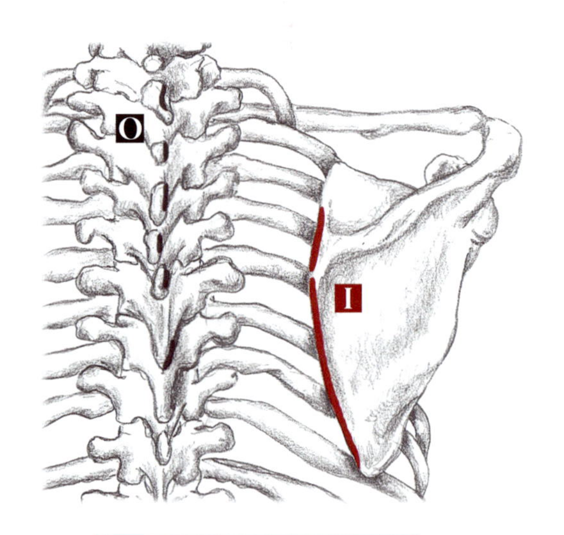

(2.74) Posterior view of right shoulder showing origins and insertions of rhomboids

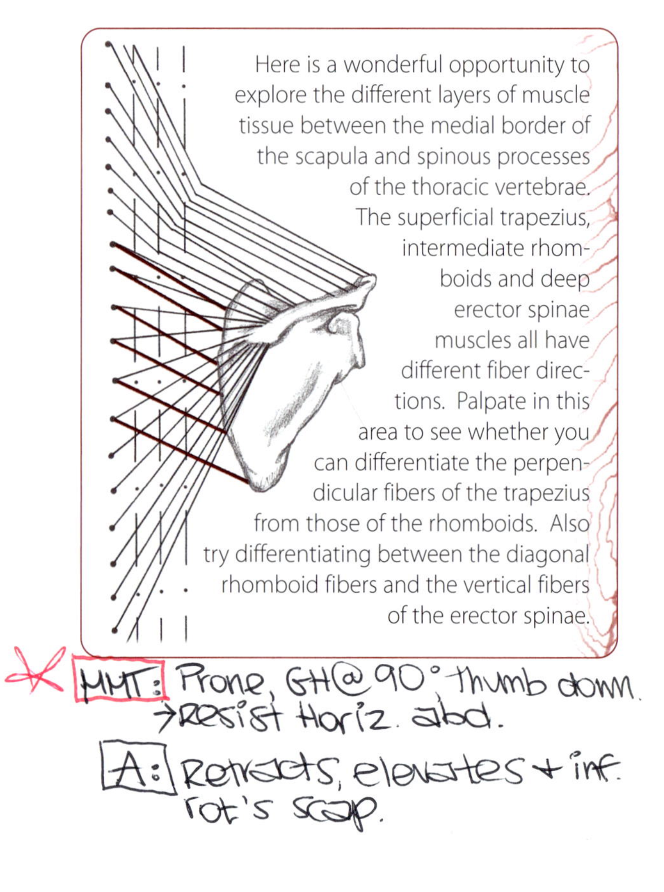

Here is a wonderful opportunity to explore the different layers of muscle tissue between the medial border of the scapula and spinous processes of the thoracic vertebrae. The superficial trapezius, intermediate rhomboids and deep erector spinae muscles all have different fiber directions. Palpate in this area to see whether you can differentiate the perpendicular fibers of the trapezius from those of the rhomboids. Also try differentiating between the diagonal rhomboid fibers and the vertical fibers of the erector spinae.

rhomboid **rom**-boyd Grk. in geometry, a parallelogram with oblique angles and only the opposite sides equal

Rhomboids

1) Prone. Locate the scapula's medial border and the spinous processes of C-7 through T-5 (p. 182).
2) Palpating through the thin trapezius, explore the area you have identified and strum vertically across the fibers of the rhomboids. Palpate all sides of the rhomboids. On some individuals you can press your fingers into the lower border of the rhomboid major and locate its edge.

Are you deep to the trapezius fibers? Do the fibers you are palpating run at an oblique angle? Place your partner's hand in the small of his back and ask him to slightly raise his elbow toward the ceiling (2.75). Although this action will engage the superficial trapezius, can you feel the deeper rhomboids contract?

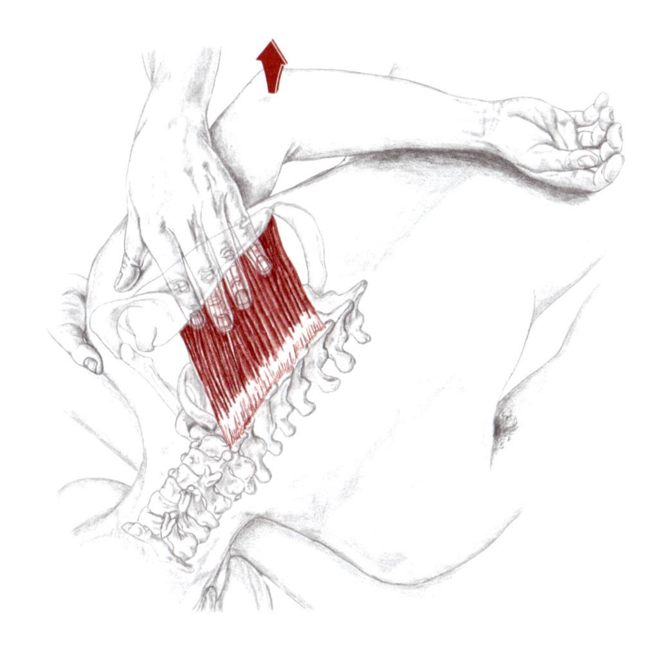

(2.75) Partner prone, raising his elbow toward the ceiling

MMT: Arm @ 90°, thumb down.

Levator Scapula

The levator scapula is located along the lateral and posterior sides of the neck. Its inferior portion is deep to the upper trapezius; however, as the levator ascends the lateral side of the neck, its fibers come out from under the trapezius and become superficial (2.77). Its belly is approximately two fingers wide with fibers that naturally twist around themselves (2.76).

The levator scapula attaches to the transverse processes of the cervical vertebrae (p. 186). Located on the lateral side of the neck, all of these small protuberances extend laterally at approximately the same width, except for the processes of C-1 which are broader.

The brachial plexus, a large group of nerves which innervates the arm, exits from the transverse processes of the cervical vertebrae. When accessing the processes to locate the origin of the levator scapula, begin by using your soft fingerpads to avoid compressing a nerve.

The levator is completely accessible by palpating either through the upper fibers of the trapezius or directly from the side of the neck.

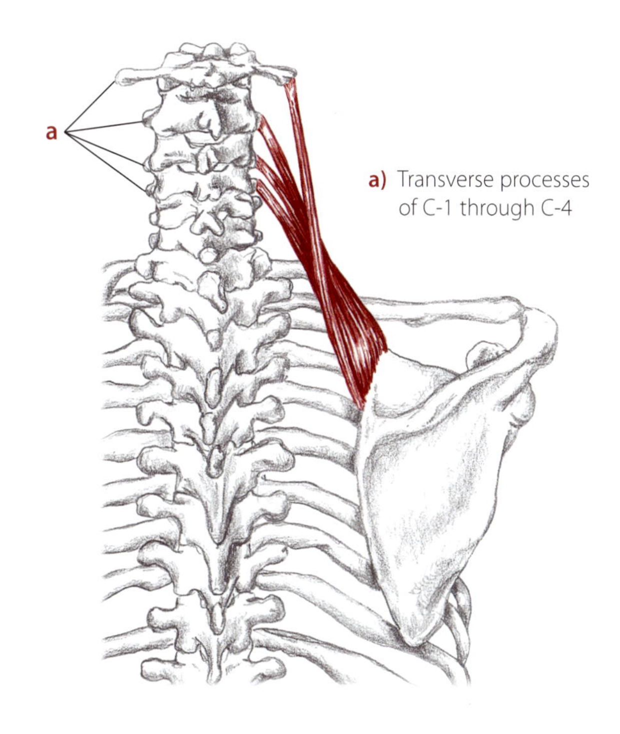

(2.76) Posterior view of levator scapula

levator leh-**va**-tor L. lifter

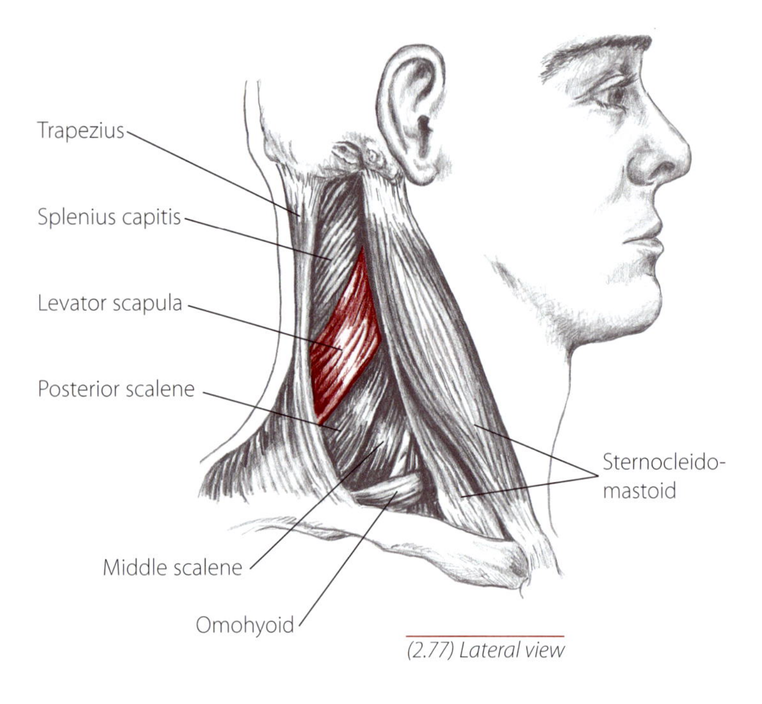

(2.77) Lateral view

A *Unilaterally:*
Elevate the scapula (scapulothoracic joint)
Downwardly rotate the scapula (s/t joint)
Laterally flex the head and neck
Rotate the head and neck to the same side

Bilaterally:
Extend the head and neck

O Transverse processes of first through fourth cervical vertebrae

I Upper region of medial border and superior angle of scapula

N Dorsal scapular and cervical nerves

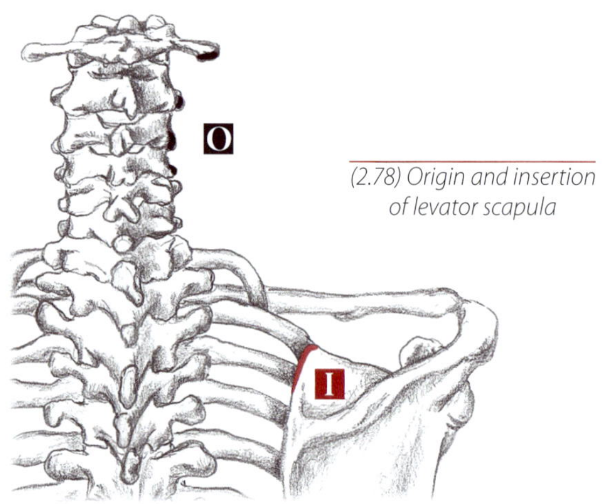

(2.78) Origin and insertion of levator scapula

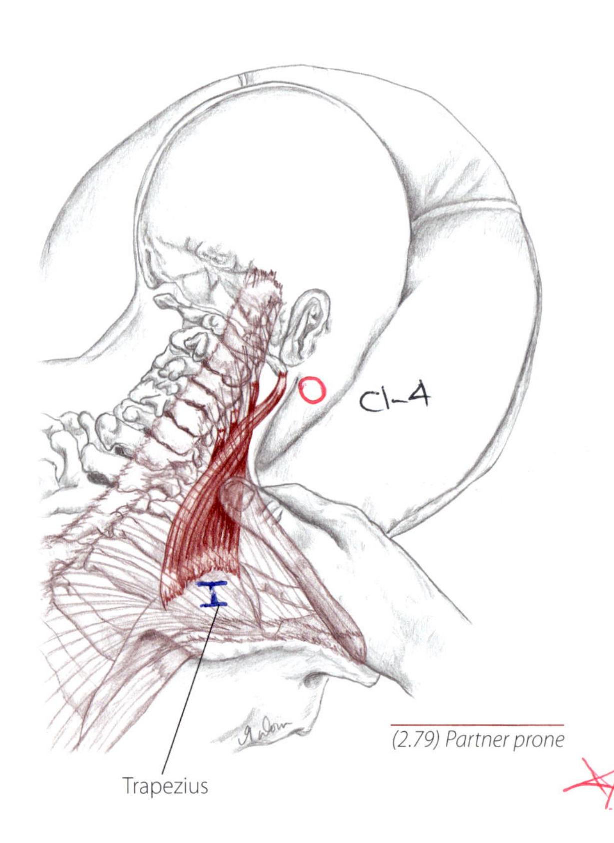

(2.79) Partner prone

Levator scapula

1) Prone, supine or sidelying. Palpating through the trapezius, locate the superior angle of the scapula (p. 61) and the upper region of the medial border.
2) Place your fingers just off the superior angle and firmly strum across the belly of the levator. The fibers will likely have a ropy texture (2.79).
3) Follow these fibers superiorly as they extend to the lateral side of the neck to the transverse processes of the cervical vertebrae (p. 186).

Can you differentiate the levator fibers from the trapezius fibers? Do the fibers you are palpating lead toward the lateral side of the neck?

MMT: Prone, arm add'd + slightly ext'd, + E elbow flx'd
→ Resist add. + SH elevation
A: Elevates scap., rot's inf.
Bilat. ext. NK
Ipsilat: lat flx., + rot.

Here is an alternate route for palpating the levator's superficial fibers on the lateral side of the neck.

1) Prone, supine or sidelying. Locate the upper fibers of the trapezius.
2) Roll two fingers anteriorly off the trapezius and press into the tissue of the neck.
3) Gently strum your fingers anteriorly and posteriorly across the levator fibers (2.80). Often you will feel a distinct band of tissue that leads superiorly toward the lateral neck and inferiorly under the trapezius.
4) Place your fingertips on the levator and ask your partner to alternately elevate and relax his scapula. Do you feel the levator scapula contract and relax beneath your fingertips?

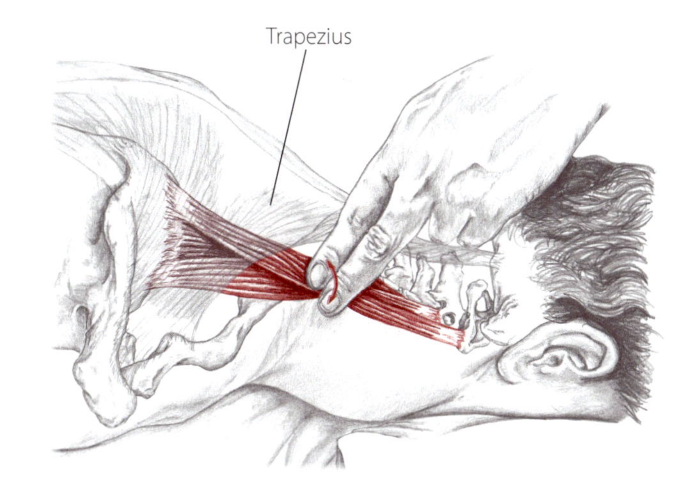

(2.80) Partner prone, lateral view of neck. Strumming across the superficial fibers of levator scapula.

The levator scapula is situated between the splenius capitis and posterior scalene muscles on the lateral side of the neck (2.77). The levator can be distinguished from these neighboring muscles during palpation because it moves the scapula. No other muscle deep to the upper trapezius or attaching to the lateral cervical vertebrae is capable of this action.

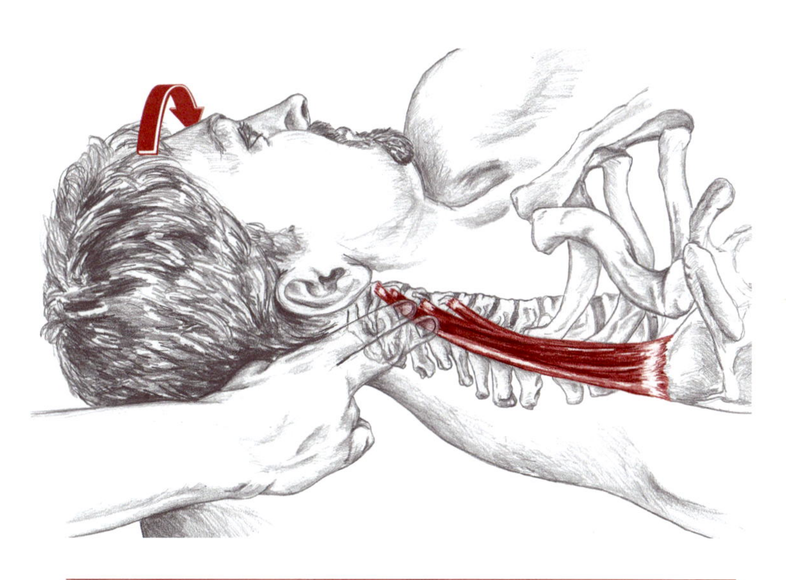

(2.81) Partner supine. Passively rotating the head 45° away from the side you are palpating shifts the cervical transverse processes further anteriorly. Also, it gives the levator scapula more palpable tension. Conversely, this position shortens and softens the overlying trapezius fibers.

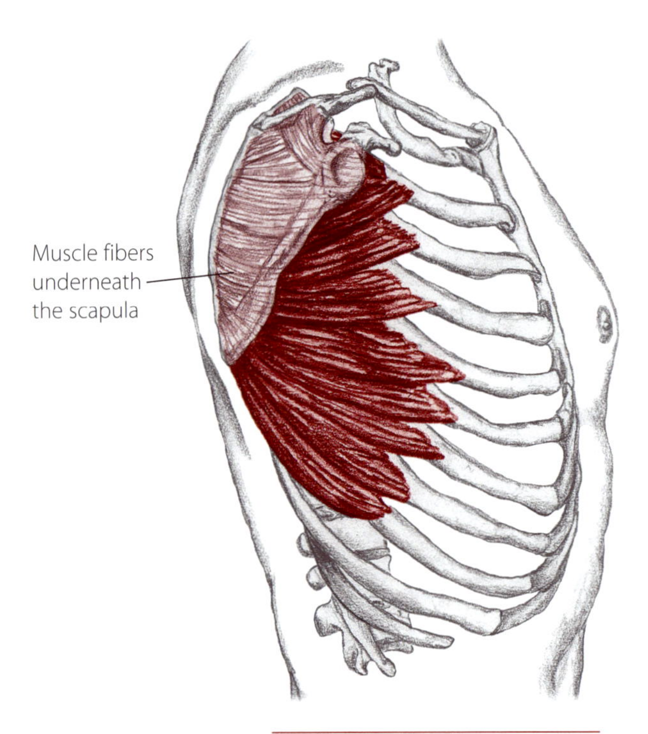

(2.82) *Lateral view of serratus anterior*

Serratus Anterior

Always well-developed on superheroes, the serratus anterior lies along the posterior and lateral rib cage. Its oblique fibers extend from the ribs underneath the scapula and attach to its medial border (2.82). Most of the serratus anterior is deep to the scapula, latissimus dorsi or pectoralis major; however, the portion of the serratus below the axilla (armpit) is superficial and easily accessible (2.84). This muscle is unique in its ability to abduct the scapula, making it an antagonist to the rhomboids.

Palpating along the sides of the ribs can tickle, so use slow, firm pressure. Also, if you are accessing the right serratus, it may be easier to stand on the left side of the table.

A *With the origin fixed:*
Abduct the scapula (scapulothoracic joint)

Depress the scapula (s/t joint)

Hold the medial border of the scapula against the rib cage

If scapula is fixed:
May act in forced inhalation

O Surfaces of upper eight or nine ribs

I Anterior surface of medial border of the scapula

N Long thoracic

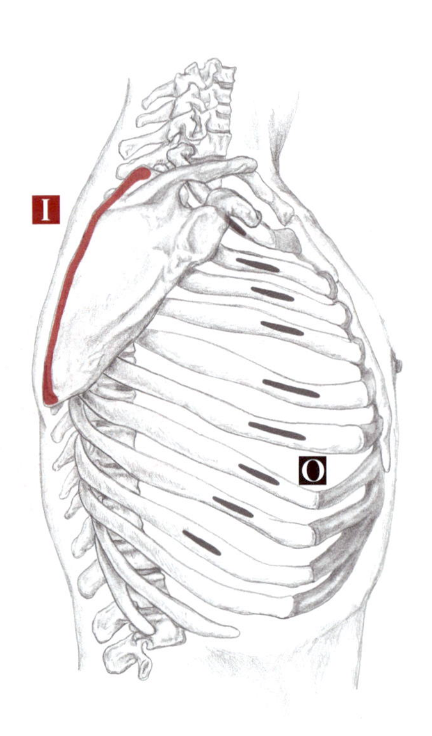

(2.83) Origin and insertion of serratus anterior

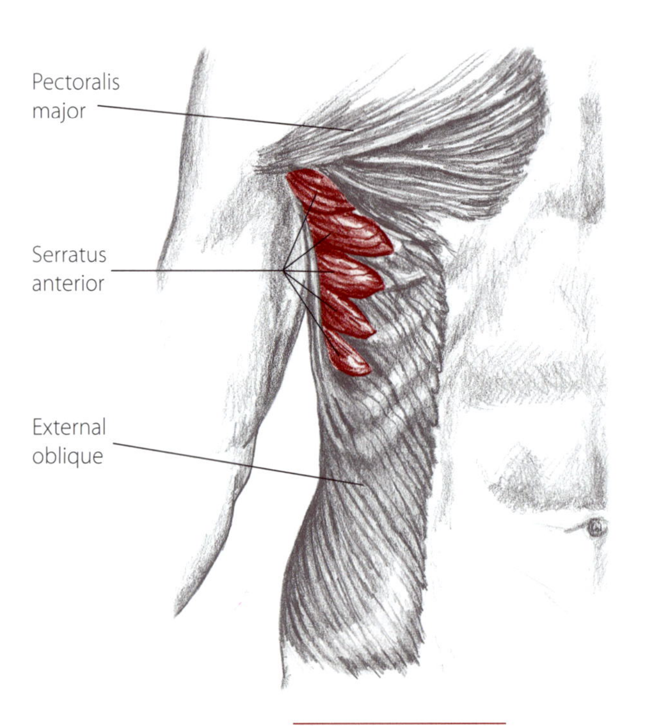

(2.84) Anterior view with serratus highlighted

serratus ser-**a**-tus L. notched

Serratus anterior

1) Supine. Isolate the location of the serratus by abducting the arm slightly and locating the lower edge of the pectoralis major (p. 97). Then locate the anterior border of the latissimus dorsi.
2) Place your fingerpads along the side of the ribs between the pectoralis major and latissimus dorsi.
3) Strum your fingers across the ribs and palpate for the serratus anterior fibers. To differentiate between the ribs and the serratus fibers (both have a similar "speed bump" shape), remember that the ribs are deep and have a solid texture while the serratus fibers are superficial and malleable.

To feel the serratus anterior contract (2.85): Ask your partner to flex his shoulder so his fist is raised toward the ceiling. Place one hand upon the serratus fibers and your other hand on top of his raised fist. Ask him to alternately abduct his scapula and relax: "Reach toward the ceiling and then relax." Do you feel the serratus fibers contract and soften? Can you follow the fibers along the ribs to where they tuck underneath the latissimus dorsi?

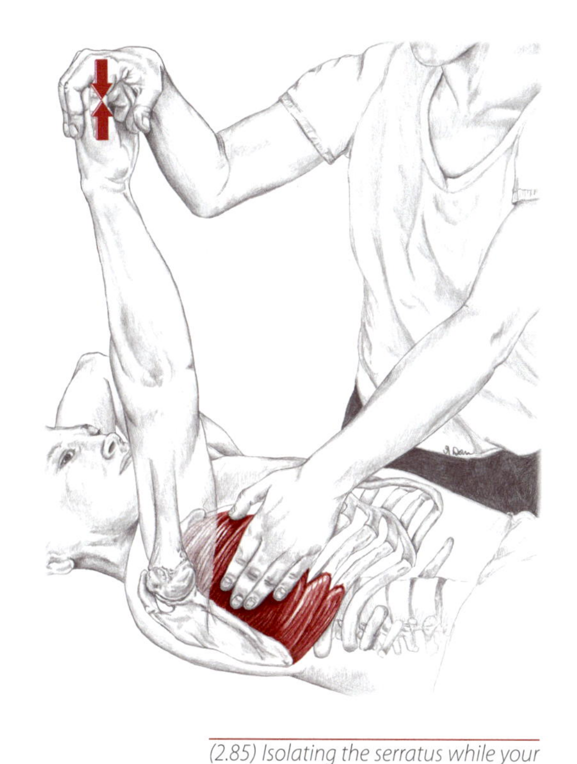

(2.85) Isolating the serratus while your partner reaches his hand toward the ceiling

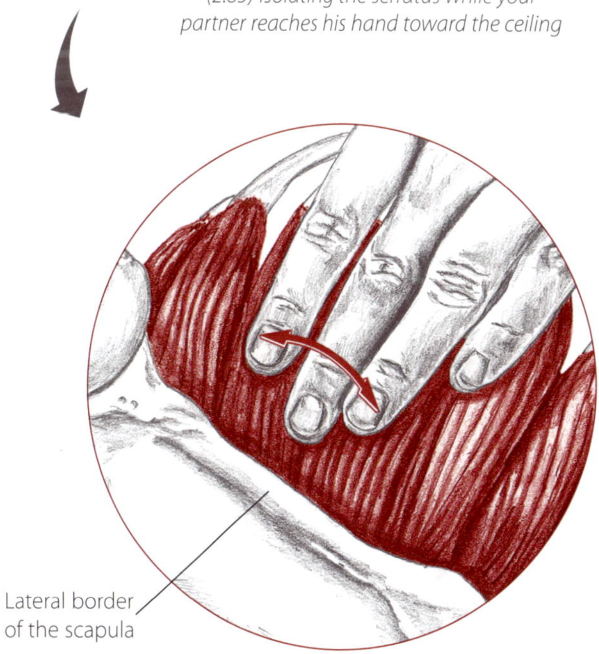

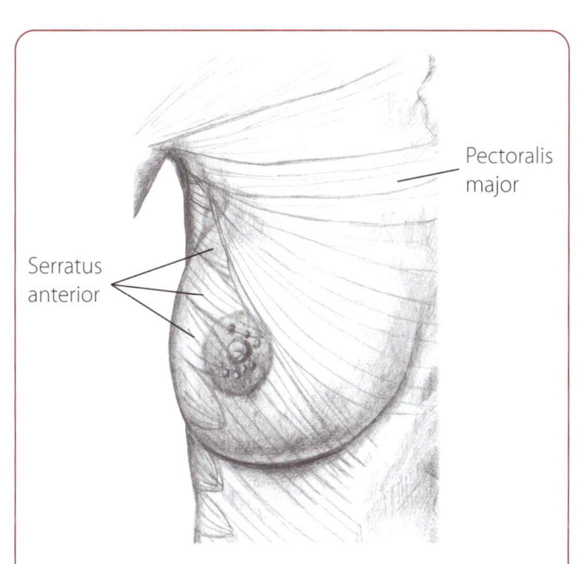

The breast is essentially made up of subcutaneous fat and is primarily supported by suspensory ligaments connecting the skin with the deep fascia over the pectoralis major. Although it varies widely in shape and size, the breast generally extends inferiorly from the second to the sixth ribs, medially to the sternum and laterally to the anterior axilla. Nearly two-thirds of the breast covers the pectoralis major while the inferior/lateral aspect covers the serratus anterior.

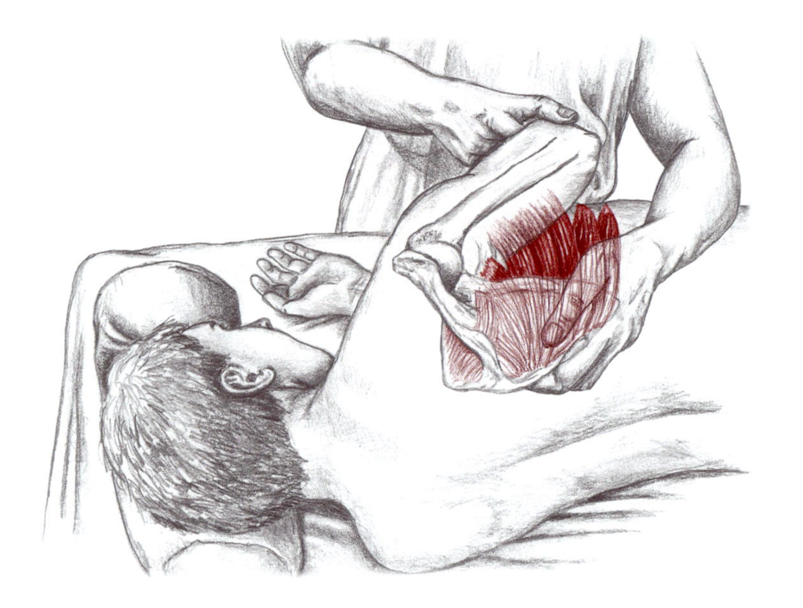

(2.86) Partner sidelying, curling your fingers under the medial border of the scapula

Turn your partner sidelying with his arm at his side. Locate the medial border of the scapula to access the insertion of the serratus anterior. Curl your fingers beneath the medial border (and through the trapezius and rhomboid fibers) onto the beginnings of the subscapular fossa and explore the area where the serratus attaches (2.86).

The function of a dog's or quadruped's serratus anterior is different from that of a human's. Unlike a human, a dog (right, anterior view) carries part of his body weight on his front legs. Together, a dog's serratus muscles form a sling from either scapula to the thorax that cradles and supports the weight of the trunk and stabilizes the pectoral girdle against the thorax.

On humans, the serratus anterior is primarily responsible for abducting the scapula or resisting a push against the shoulder. If you get down on your hands to do a push-up, you will see (and feel) how this position forces your serratus muscles to function as a dog's would.

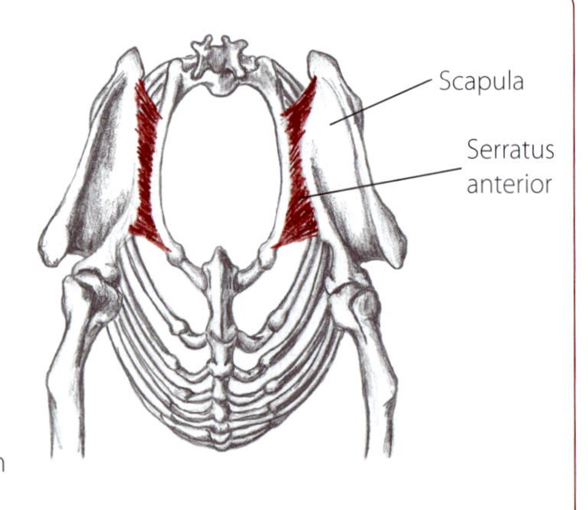

Pectoralis Major

The pectoralis major is a broad, powerful muscle located on the chest. Except for the part beneath breast tissue, its convergent, superficial fibers are accessible. Pectoralis major is divided into three segments - the clavicular, sternal and costal fibers (2.87). The upper and lower fibers perform opposing actions at the shoulder joint - flexion and extension, respectively - making this muscle an antagonist to itself.

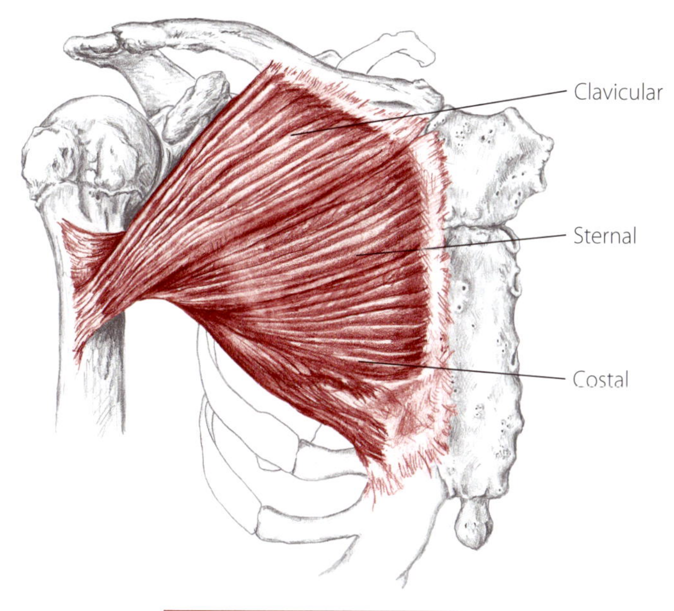

(2.87) Anterior view identifying the three segments of pectoralis major

A *All fibers:*

Adduct the shoulder (glenohumeral joint)

Medially rotate the shoulder (g/h joint)

Assist in elevating the thorax in forced inhalation (if arm is fixed)

Upper fibers:

Flex the shoulder (g/h joint)

Horizontally adduct the shoulder (g/h joint)

Lower fibers:

Extend the shoulder (g/h joint)

O Medial half of clavicle, sternum and cartilage of first through sixth ribs

I Crest of greater tubercle of humerus

N Medial and lateral pectoral

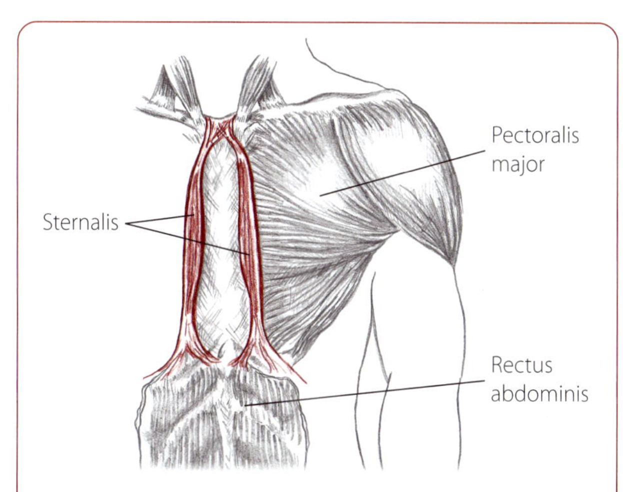

Present in roughly 5% of the population, the sternalis is a thin, superficial muscle lying on the sternum. Its vertical fibers run from the manubrium down to the level of the seventh costal cartilage. The function of the sternalis is unknown. Palpate the surface of your partner's sternum and explore for a sternalis.

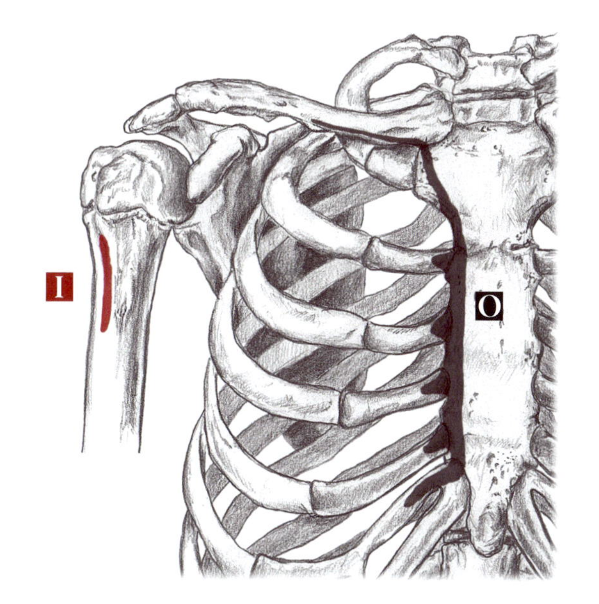

(2.88) Origin and insertion of pectoralis major

pectoralis **pek**-to-**ra**-lis L. chest

When exploring the pectoralis major and minor, it is advisable to palpate around breast tissue and not directly into it. This raises the question, "When palpating on a female, how do you access these and other chest muscles without contacting breast tissue?"

The most important aspect when palpating near breast tissue is communicating your intentions to your partner. Also, encourage her to let you know if at any time she wishes to stop.

Assuming your partner is draped under a sheet or wearing a sports bra, the key to safe and comfortable palpation around breast tissue is positioning your client so the breast tissue will naturally shift away from where you are accessing. For instance, in a supine position, the breast will shift laterally, allowing easier access to the sternal and upper pectoral regions. In this position, however, larger breasts may crowd the axillary region. In such situations you could either ask your partner to shift and hold her breast medially, allowing you to access the axilla, or use the back of your own hand to push the tissue medially (above).

In a sidelying position, the breast will fall medially, opening up the axillary region. The axilla can be opened up further by passively shifting the shoulder anteriorly (left).

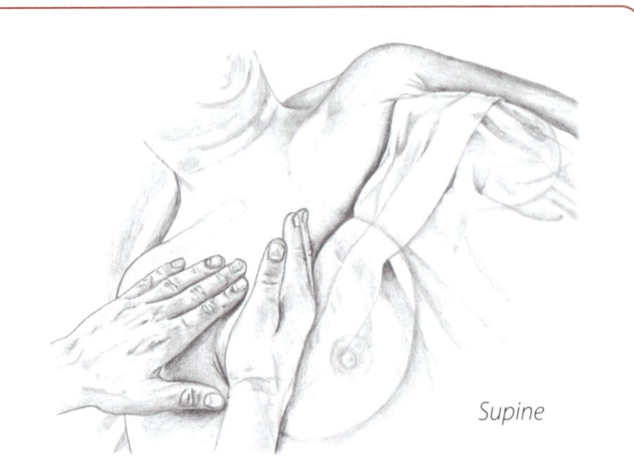
Supine

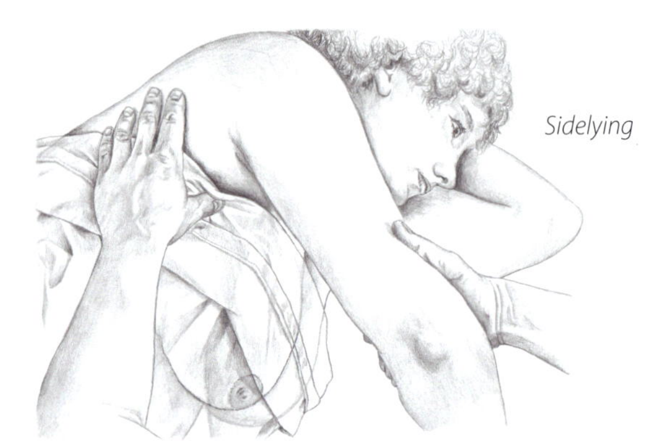
Sidelying

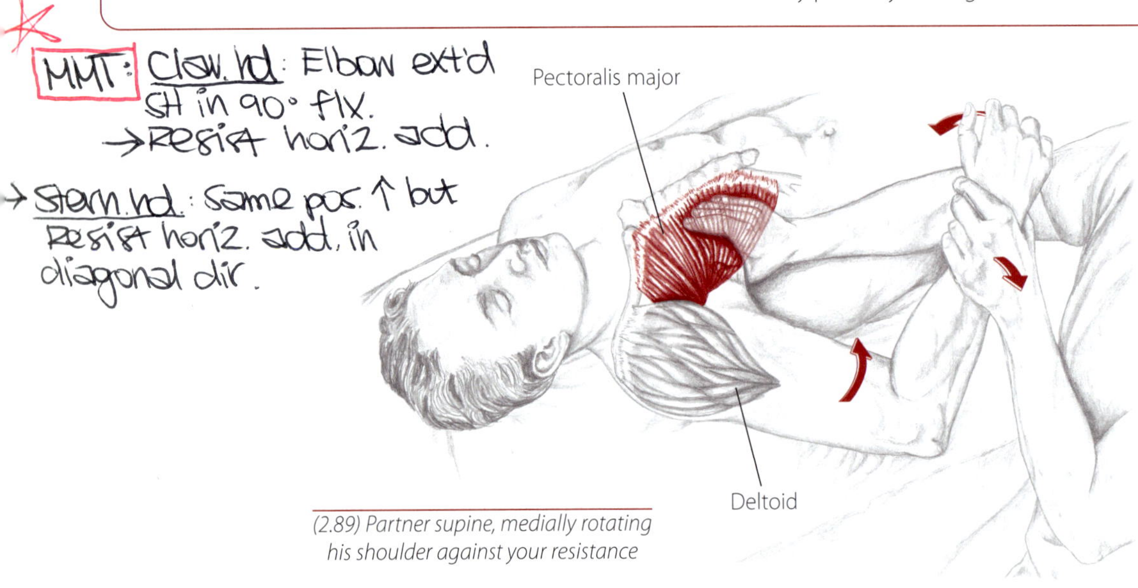

(2.89) Partner supine, medially rotating his shoulder against your resistance

Supine

1) With your partner's shoulder slightly abducted, sit or stand facing him.
2) Locate the medial shaft of the clavicle and move inferiorly onto the clavicular fibers.
3) Explore the surface of the pectoralis major. Follow the fibers laterally as they blend with the deltoid and attach at the greater tubercle.
4) Grasp the belly of the pectoralis by sinking your thumb into the axilla. Ask your partner to medially rotate his shoulder against your resistance. "Press your hand toward your belly." (2.89) Note the contraction of the pectoralis.

✔ *Do the clavicular fibers run parallel with the anterior deltoid? As you grasp the belly, do you sense its thickness and how it lies across the rib cage?*

Sidelying

1) Supporting your partner's arm, flex the shoulder and pull it anteriorly toward you. This position not only brings the pectoralis major off the chest wall, but also allows the breast tissue to fall away from the area you are palpating.
2) Grasping the pectoralis major, explore its mass from the ribs to the humerus (2.90). Passively flex and extend the shoulder, perceiving the changes in the tension of the tissues.

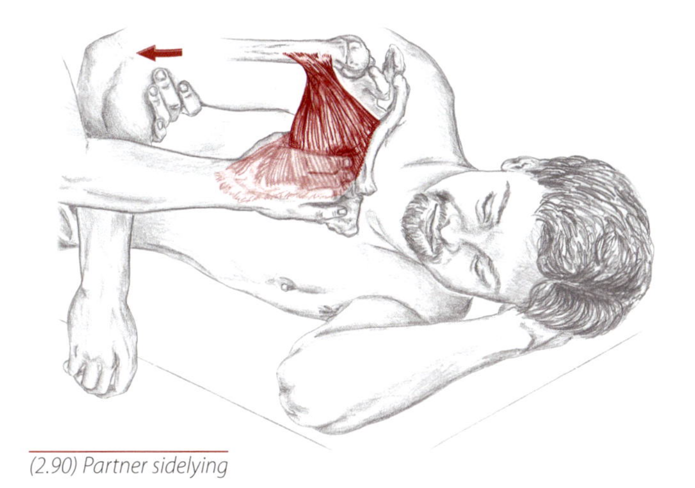

(2.90) Partner sidelying

Here is a way to get a sense of the antagonistic movements of the pectoralis major's upper and lower fibers.

1) Supine. Begin with your partner's hand raised up toward the ceiling. As you create resistance, ask your partner to flex his shoulder. "Meeting my resistance, try to bring your hand over your head." The upper fibers will contract while the lower fibers remain lax.
2) Ask him to extend against your resistance. "Now try to bring your hand toward your hips." Here the lower fibers will contract while the upper fibers relax (2.91).

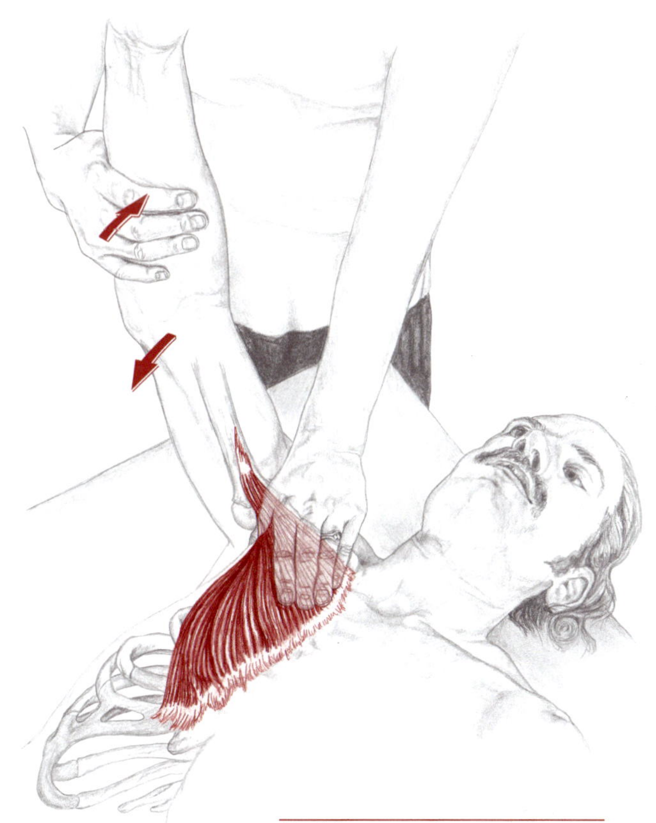

(2.91) Partner supine with arm raised, feeling the lower fibers contract

When choosing between the "white or dark meat" of a cooked bird, be sure to thank its different intramuscular connective tissues. Dark and white meat are present in all mammals, but are more distinct in birds. The reason is that light-colored musculature is rich in muscle fibers and poor in sacroplasm - the tissue that surrounds the muscle fiber - while dark meat has the exact opposite composition. And if you are fond of the "breast," chew on this fact: a bird's pectoralis majors make up 20-35% of its body weight.

Pectoralis Minor

The pectoralis minor lies next to the rib cage deep to the pectoralis major (2.92). Its fibers run perpendicular to the pectoralis major fibers from the scapula's coracoid process to the upper ribs. During aerobic activity the pectoralis minor helps to elevate the rib cage for inhalation. The major vessels serving the arm - the brachial plexus, axillary artery and vein - cross underneath the pectoralis minor, creating the potential for neurovascular compression by this muscle (2.94).

Access to the minor can be achieved by either pressing through or sliding underneath the thick pectoralis major. The second method is more specific and will be outlined here. The pectoralis minor can be sensitive, so palpate slowly, allowing your fingers or thumb to sink into the tissue.

(2.92) Anterior view of pectoralis minor

A Depress the scapula (scapulothoracic joint)

Abduct the scapula (s/t joint)

Tilt the scapula anteriorly (s/t joint)

If scapula is fixed:
Assist in forced inhalation

O Third, fourth and fifth ribs

I Coracoid process of the scapula

N Medial pectoral

(2.93) Origin and insertion of pectoralis minor

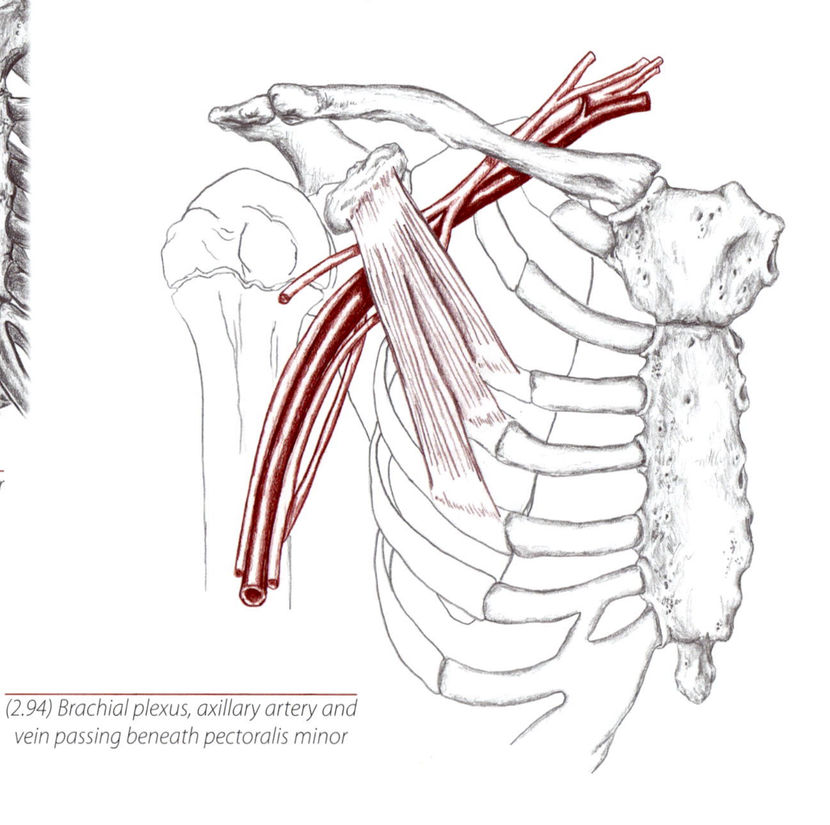

(2.94) Brachial plexus, axillary artery and vein passing beneath pectoralis minor

Supine

1) Abduct the arm and place your fingerpads at the lateral edge of the pectoralis major.
2) Slowly and gently slide under the pectoralis major, following along the surface of the ribs.
3) Eventually your fingerpads will come in contact with the small wall of muscle lying next to the ribs (2.95). This is the side of the pectoralis minor. If you do not feel its tissue, visualize its location next to the ribs.

Ask your partner to depress his scapula. "Ever-so-slightly press your shoulder down toward your hip." When he depresses, do you feel the pectoralis minor contract? Do the fibers you feel run toward the coracoid process?

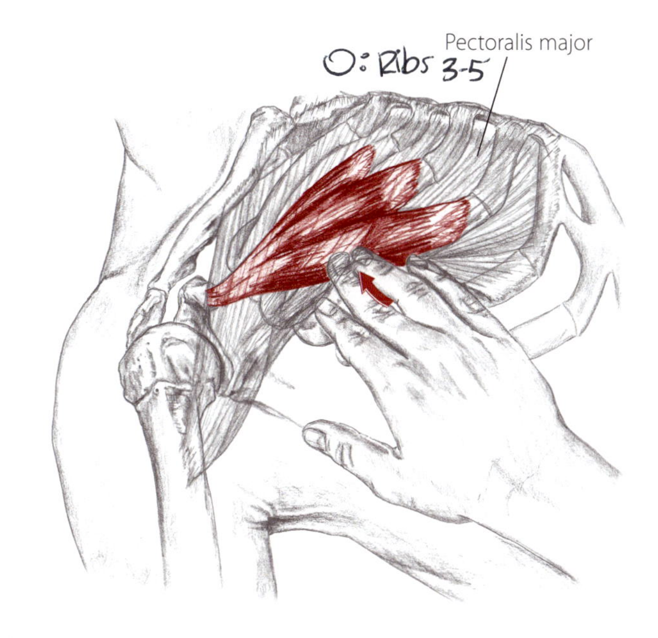

(2.95) Partner supine, sliding under the pectoralis major to access the pectoralis minor

MMT: Forward thrust of SH.

A: Upward tilt, depresses SH
- Elevates ribs on inspiration.

Sidelying

1) Support the arm in a flexed position and pull it anteriorly. This brings the pectoralis major off the chest wall and allows the breast tissue to fall away from the area you are palpating.
2) Slowly slide your thumb under the pectoralis major, following along the surface of the ribs (2.96). Your thumb will press into the side and onto the surface of the pectoralis minor. Then ask your partner to gently depress his scapula while you feel for the minor's contraction.

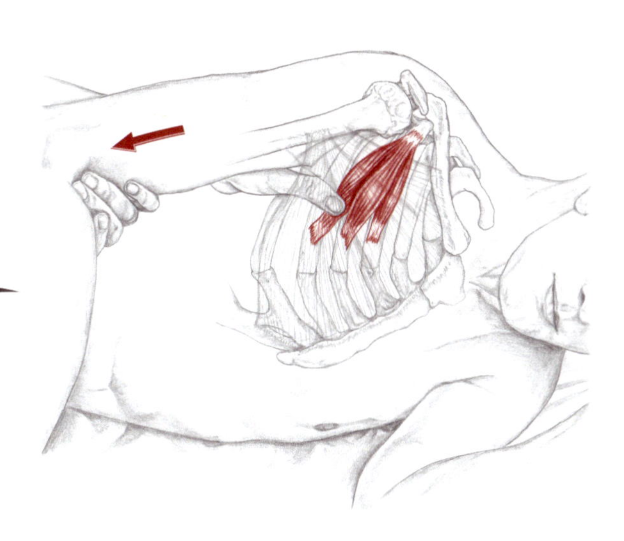

(2.96) Partner sidelying with arm in flexed position

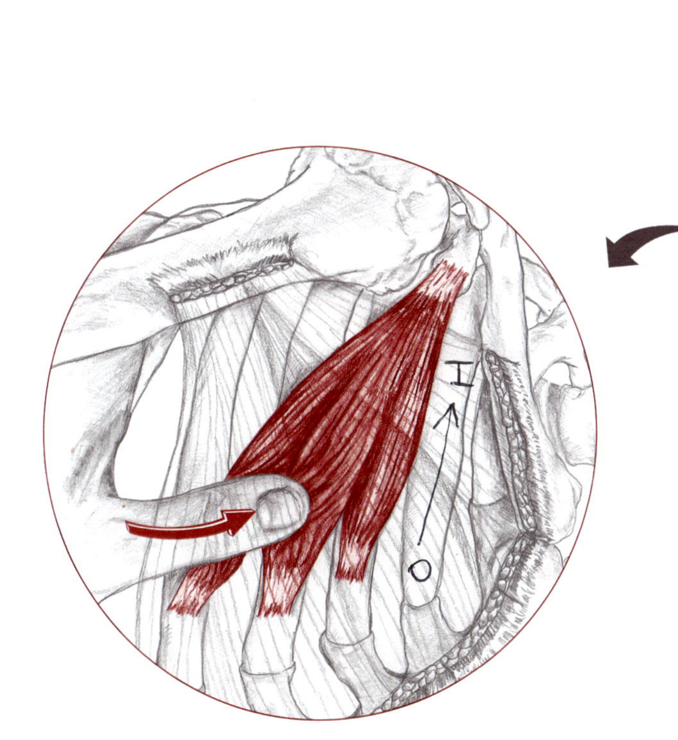

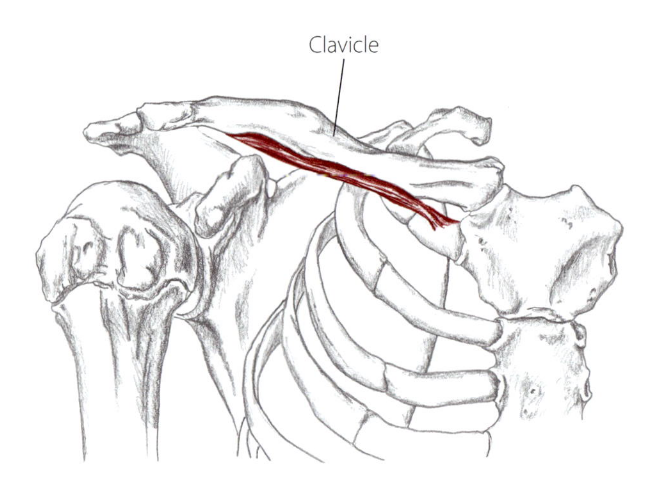

(2.97) Anterior view of subclavius

Subclavius

As its name suggests, the subclavius is located underneath the clavicle. Its fibers run parallel to the clavicle, deep to the pectoralis major, and can be challenging to truly isolate (2.97).

On quadrupeds (four-legged animals), the subclavius is quite large and plays an important role in stabilizing the clavicle and shoulder girdle during locomotion. A human's subclavius, on the other hand, is a small, secondary muscle.

A Draw clavicle inferiorly and anteriorly

Elevate first rib (to assist in inhalation)

Stabilize the sternoclavicular joint

O First rib and cartilage

I Inferior, lateral aspect of the clavicle (subclav. groove

N Subclavian

1) Sidelying. Support the arm in a flexed position and pull it anteriorly. This position brings the clavicle and pectoralis major off the rib cage and allows your thumb to curl even further around the clavicle.
2) Place your thumb and fingers at the center of the clavicle. Slowly curl your thumb around the clavicle's underside, trying to access the subclavius (2.98). You may not access a muscle belly, but may feel instead some slightly dense tissue tucked under the clavicle.
3) Try this method with your partner in a supine position.

Can you detect a slender strip of tissue deep to the clavicle? Can you distinguish between the superficial pectoralis major fibers (heading toward the axilla) and the subclavius fibers (parallel to the clavicle)?

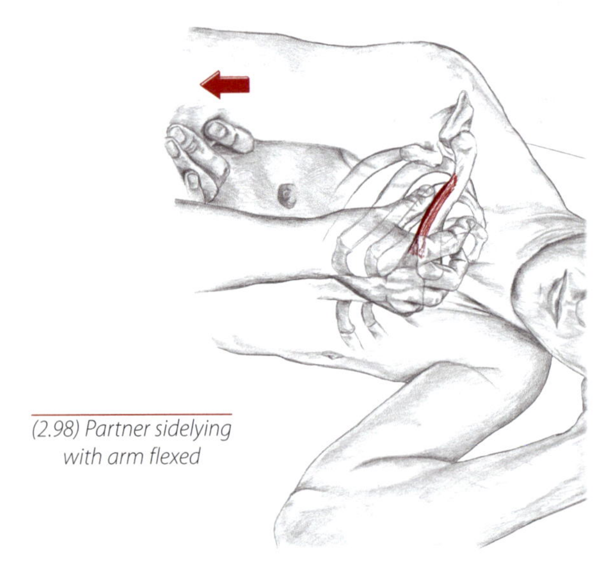

(2.98) Partner sidelying with arm flexed

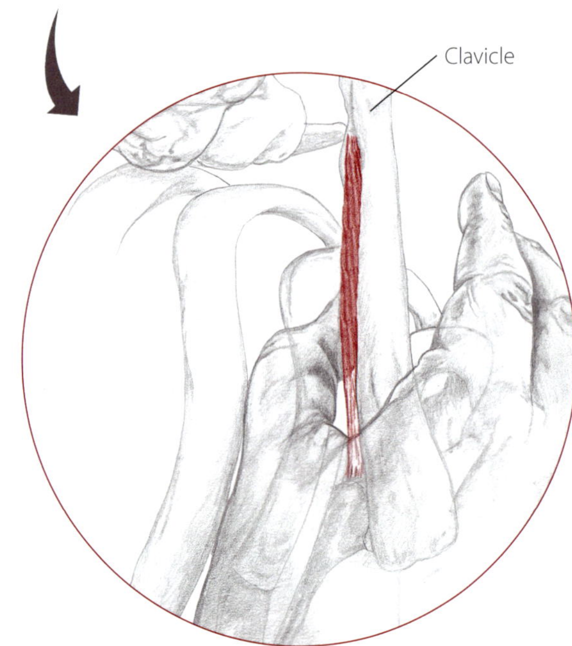

subclavius sub-**klay**-vee-us

Biceps Brachii

The biceps brachii lies superficially on the anterior arm. It has a long head and a short head which merge to form a long, oval belly. The tendon of the long head passes through the intertubercular groove of the humerus (p. 68). This groove helps to stabilize the tendon as it rises over the top of the shoulder (2.99).

The distal tendon of the biceps dives into the antecubital space (inner elbow) to attach at the radius, allowing this muscle to be the primary muscle of forearm supination. The majority of the biceps brachii is easily palpable.

A Flex the elbow (humeroulnar joint)

Supinate the forearm (radioulnar joints)

Flex the shoulder (glenohumeral joint)

O *Short head:*
Coracoid process of scapula
Long head:
Supraglenoid tubercle of scapula

I Tuberosity of the radius and aponeurosis of the biceps brachii

N Musculocutaneous

1) Supine or seated. Bend the elbow and shake hands with your partner.
2) Ask your partner to flex his elbow against your resistance. Palpate the anterior surface of the arm and locate the hard, round belly of the biceps (2.101).
3) Follow the belly distally to the inner elbow. Note how the muscle belly thins, becoming a solid, distinct tendon. Then follow the biceps proximally to where it tucks beneath the anterior fibers of the deltoid.

Ask your partner to flex his elbow and see if you can sculpt out the biceps' distal tendon and distinguish it from the deeper brachialis muscle (p. 140). Also, shake hands with your partner and ask him to alternately pronate and supinate his forearm against your resistance. Do you feel the muscle belly and tendon contract upon supination?

Supraglenoid tubercle
Coracoid process
Intertubercular groove (deep)
Short head
Long head
Bicipital aponeurosis

(2.99) Anterior view of biceps brachii

O
I

(2.100) Origin and insertion of biceps brachii

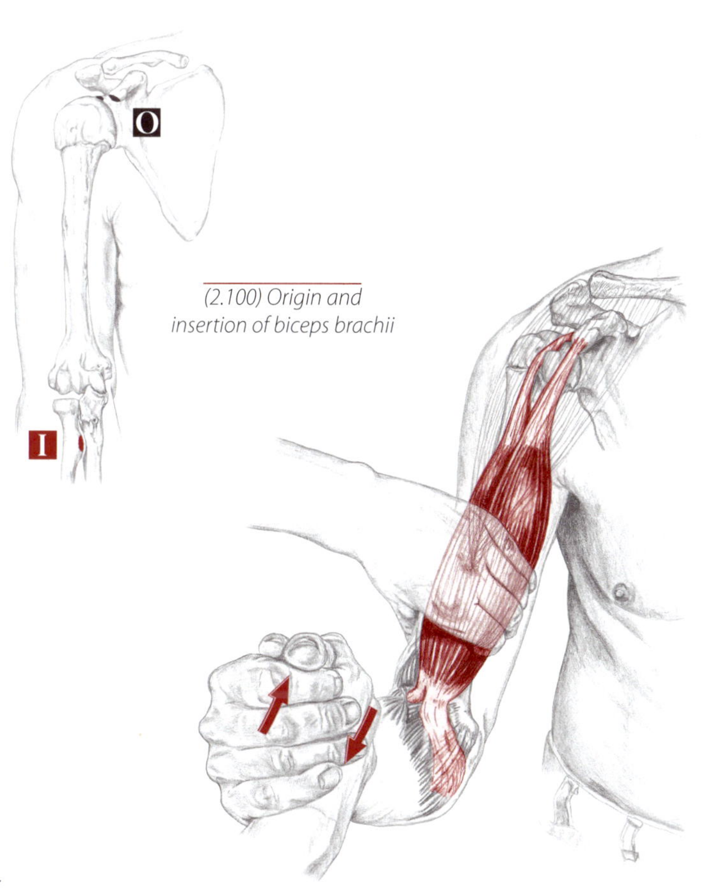

(2.101) Feeling biceps contract as your partner tries to flex his elbow

MMT: Resist elbow flx. w/ forearm supinated
A: Flx. + supinates elbow
Long hd: SH flx.

biceps brachii **bi**-seps **bray**-key-i L. two-headed muscle of the arm

In exploring the distal tendon of the biceps, you may notice a smaller tendonlike band that expands off it medially. This is the bicipital aponeurosis, a thin sheet of fascia that curves around the forearm flexors and blends into the antebrachial fascia. It stabilizes the ulna during flexion and supination, and - similar to a "tennis elbow" armband - supports the forearm flexors.

1) With the elbow flexed, shake hands with your partner. As you locate the biceps' distal tendon, ask your partner to flex her elbow against your resistance, making the tendon more discernible.
2) Slide over to the tendon's medial aspect and explore for the aponeurosis. When the biceps contracts, it is sometimes visible. Follow this fascial strip as far as you can around the medial forearm.

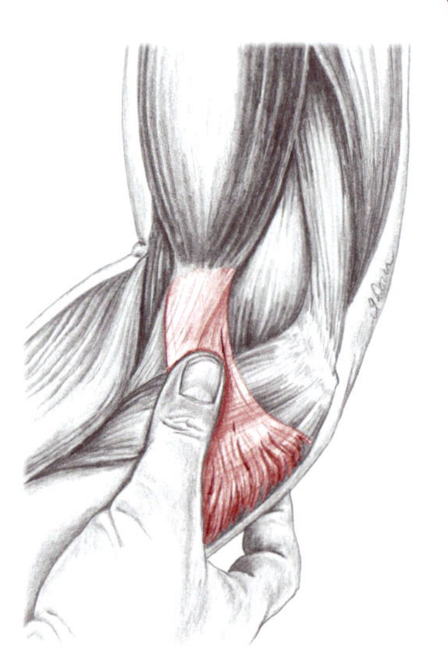

Medial view of right elbow

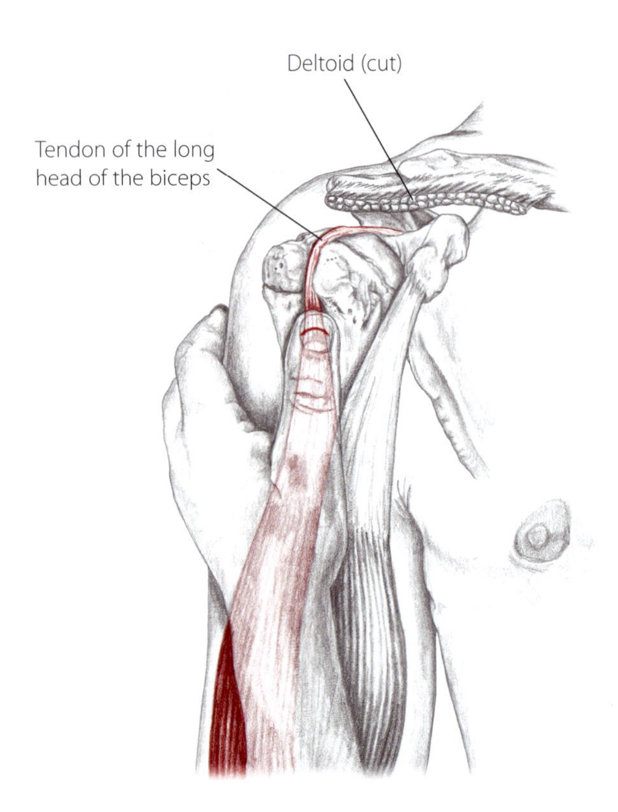

(2.102) Anterior view of right shoulder

The Tendon of the Long Head of the Biceps Brachii

Because the biceps tendon is situated in the intertubercular groove of the humerus and runs parallel to the superficial deltoid fibers, it can be difficult to truly isolate.

1) Locate the intertubercular groove (p. 68). Laterally rotating the arm may make it easier to pinpoint the tendon (2.102).
2) Ask your partner to gently flex his elbow against your resistance in order to feel the biceps tendon become taut in the intertubercular groove. Be aware that the deltoid's anterior fibers will also contract upon flexion of the shoulder.

In addition to a long head and a short head, the biceps may have a head which attaches to the humerus. Reported in less than 10% of the population, this extra head originates along the medial humerus next to the coracobrachialis before joining the short head.

Triceps Brachii

The triceps brachii is the only muscle located on the posterior arm. Creating extension at the elbow and shoulder, it is an antagonist at both these joints to the biceps brachii.

The triceps has three heads: long, lateral and medial (2.103, 2.104). The long head extends off the infraglenoid tubercle of the scapula (p. 62), weaving between the teres major and minor. The lateral head lies superficially beside the deltoid while the medial head lies mostly underneath the long head. All three heads converge into a thick, distal tendon proximal to the elbow.

Aside from its proximal portion, which is deep to the deltoid, the triceps is superficial and easily accessible.

A *All heads:*
Extend the elbow (humeroulnar joint)

Long head:
Extend the shoulder (glenohumeral joint)

Adduct the shoulder (g/h joint)

O *Long head:*
Infraglenoid tubercle of the scapula
Lateral head:
Posterior surface of proximal half of the humerus
Medial head:
Posterior surface of distal half of the humerus

I Olecranon process of the ulna

N Radial

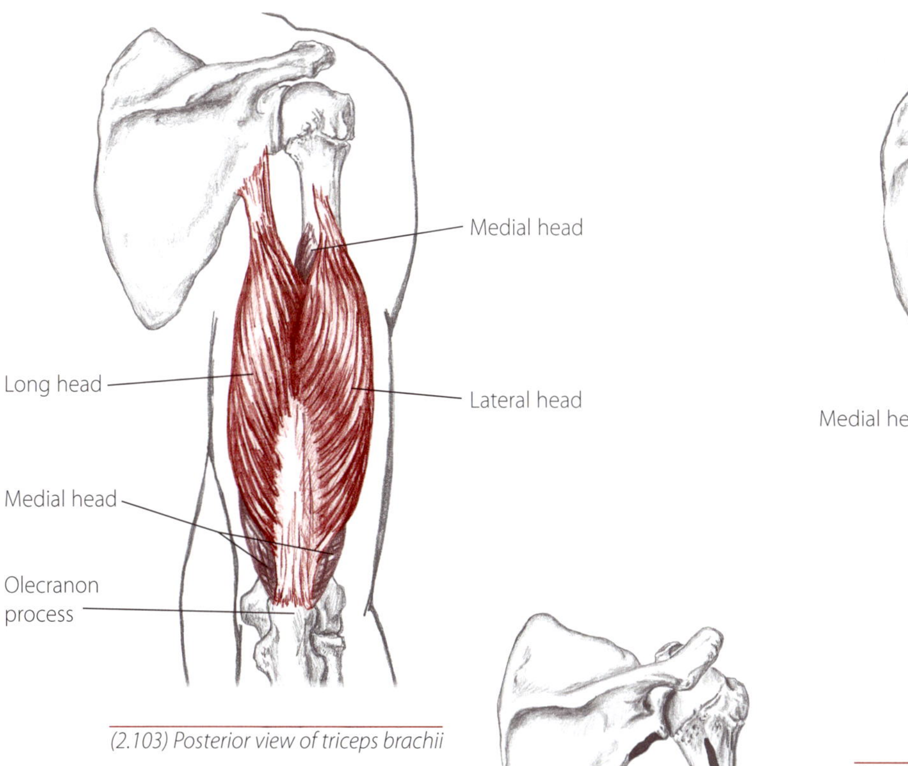

(2.103) Posterior view of triceps brachii

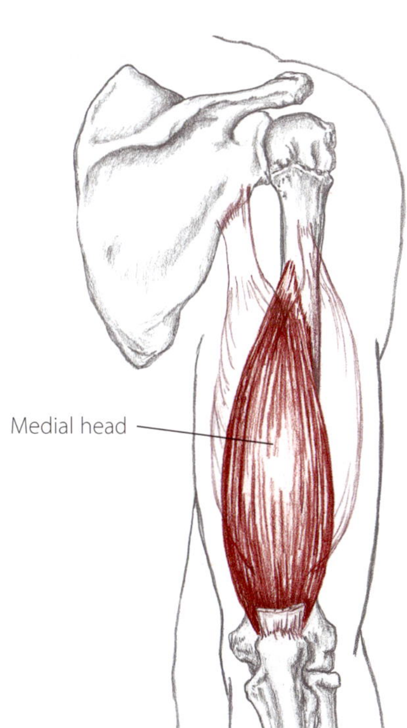

(2.104) Posterior view of the medial head of the triceps brachii, deep to the lateral and long heads

O

I

(2.105) Origins and insertion of triceps brachii

triceps brachii **tri**-seps **bray**-key-i L. three-headed muscle of the arm

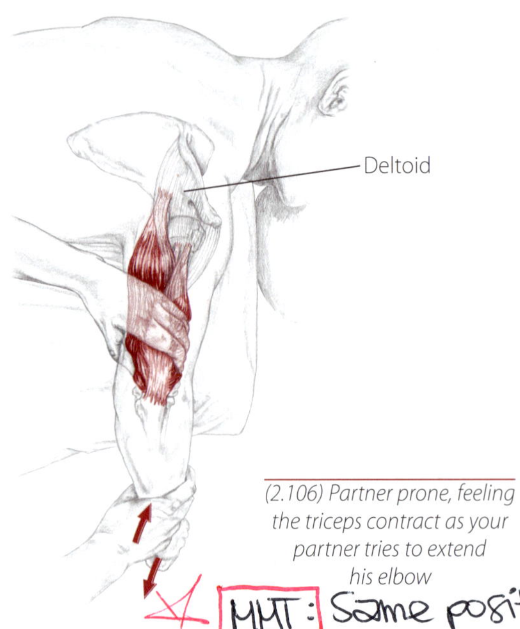

(2.106) Partner prone, feeling the triceps contract as your partner tries to extend his elbow

1) Prone. Bring the arm off the side of the table and palpate the posterior aspect of the arm. Outline the edge of the posterior deltoid and then explore the size and shape of the triceps.
2) Locate the olecranon process to outline the distal tendon of the triceps. Then ask your partner to extend his elbow as you apply resistance at his forearm (2.106). Slide your other hand off the olecranon process proximally and onto the broad triceps tendon.
3) With your partner still contracting, widen your fingers and palpate the medial and lateral heads on either side of the tendon.

Does the muscle tighten when your partner extends his elbow? Can you feel the medial and lateral triceps heads bulge on either side of the distal tendon?

MMT: Same position, arm mostly extended.
→ Resist elbow ext.
A: Long hd. Exts. GH + elbow
Med & lat.: Exts elbow.

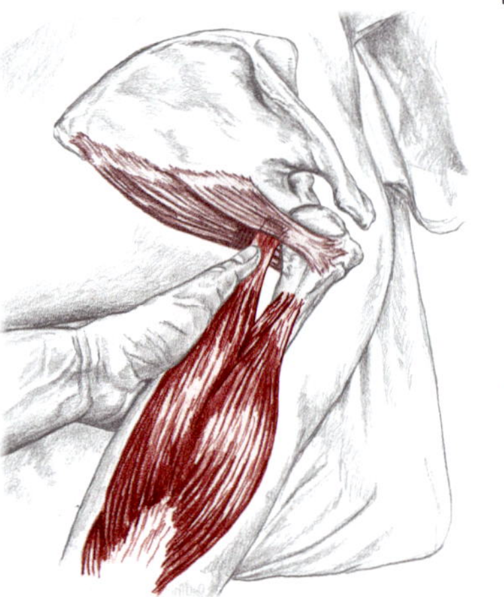

(2.107) Partner prone, isolating the tendon of the long head

The Tendon of the Long Head of the Triceps Brachii

A helpful hint for locating the long head of the triceps is the fact that it is the only band of muscle on the posterior arm that runs superiorly along the proximal and medial aspect of the arm. The deltoid fibers run at a more diagonal direction than the long head of the triceps.

1) Prone. Place one hand on the proximal elbow and ask your partner to bring his elbow toward the ceiling against your resistance. This action will contract the long head of the triceps.
2) Locate its belly along the proximal and medial aspect of the arm. Follow the muscle proximally by strumming across the belly. Note how it disappears underneath the posterior deltoid toward the infraglenoid tubercle.
3) With the arm relaxed, press through the posterior deltoid and strum across its skinny tendon as it attaches to the infraglenoid tubercle.

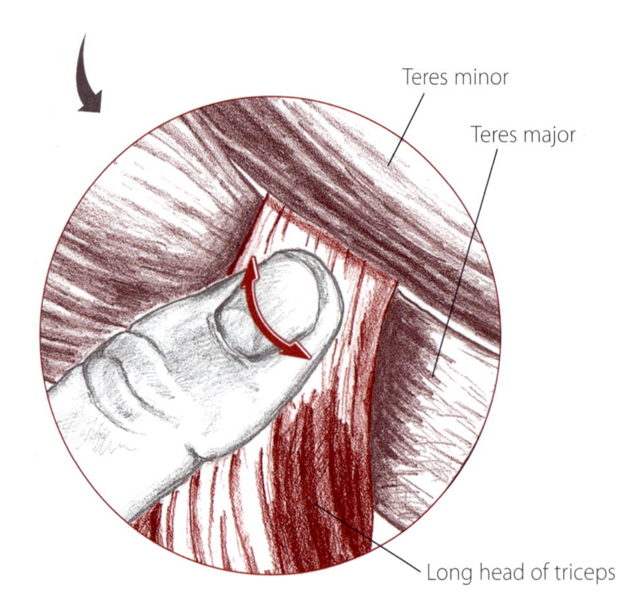

The long head of the triceps crosses over the teres major and under the teres minor (2.107). Can you follow the long head up to the division of the teres muscles? Have your partner medially and laterally rotate his shoulder to differentiate the teres muscles (p. 79, 82).

Coracobrachialis

The coracobrachialis is a small, tubular muscle located in the axilla (2.108). Sometimes known as the "armpit" muscle, it is a secondary flexor and adductor of the shoulder. In anatomical position, the coracobrachialis is deep to the pectoralis major and anterior deltoid and lies anterior to the axillary artery and brachial plexus. Abducting the shoulder (opening up the axilla) brings the belly of the coracobrachialis to a superficial and palpable position.

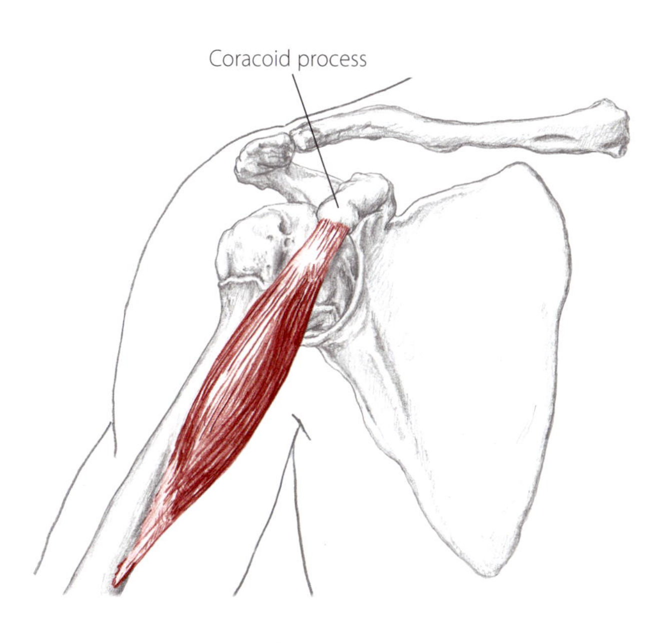

(2.108) Anterior view of coracobrachialis

A Flex the shoulder (glenohumeral joint)

Adduct the shoulder (g/h joint)

O Coracoid process of the scapula

I Medial surface of mid-humeral shaft

N Musculocutaneous

1) Supine. Laterally rotate and abduct the shoulder to 45˚. Locate the fibers of the pectoralis major. This tissue forms the axilla's anterior wall and will be a reference point for locating coracobrachialis.
2) Lay one hand along the medial side of the arm and move your fingerpads into the armpit.
3) Have your partner horizontally adduct gently against your resistance (2.110). Isolate the solid edge of the pectoralis major then slide off the pectoralis major fibers posteriorly (into the axilla) and explore for the slender, contracting belly of the coracobrachialis. Its belly may be visible upon adduction.

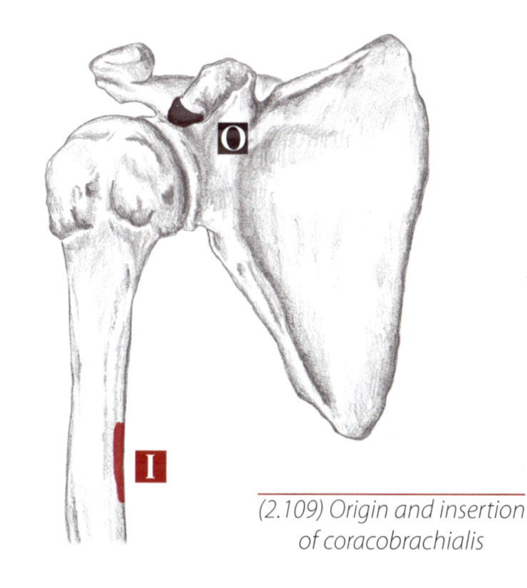

(2.109) Origin and insertion of coracobrachialis

Is the muscle you are palpating on the medial side of the upper arm? Does its belly lie posterior to the overlying flap of the pectoralis major? Can you strum along its cylindrical belly?

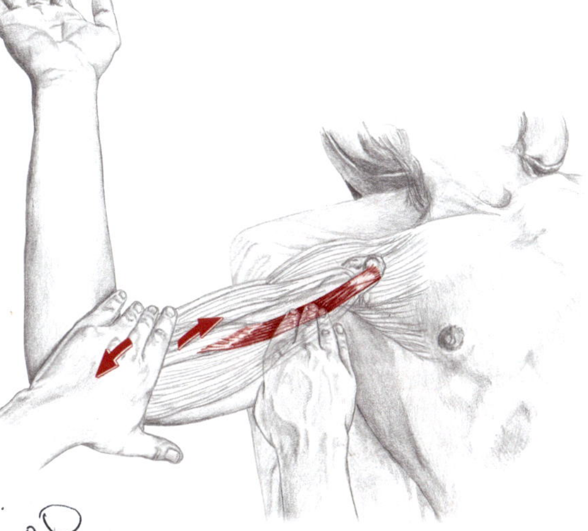

(2.110) Partner supine, palpating coracobrachialis as your partner horizontally adducts against your resistance

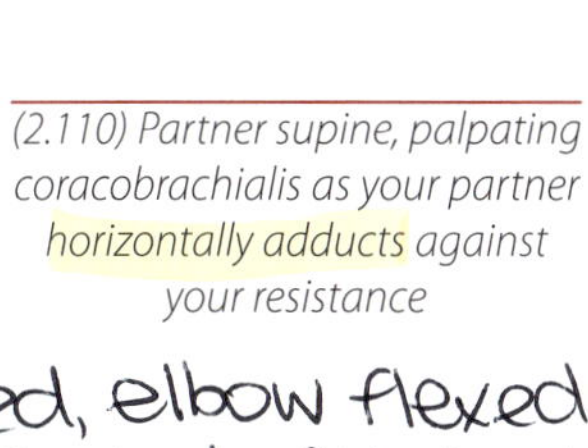

MMT: Seated, elbow flexed, forearm supinated, GH fixed
→ Resist SH flx. + slight add.
A: flx. + add. of GH.

Other Structures of the Shoulder and Arm

Axilla

The axilla is the cone-shaped area commonly called the armpit (2.111). It is formed by four walls: **(a)** the lateral wall (biceps brachii and coracobrachialis), **(b)** the posterior wall (subscapularis and latissimus dorsi), **(c)** anterior wall (pectoralis major), and **(d)** the medial wall (rib cage and serratus anterior). There are several important vessels which pass through the axillary region (2.112) including the brachial artery and the brachial plexus (nerves).

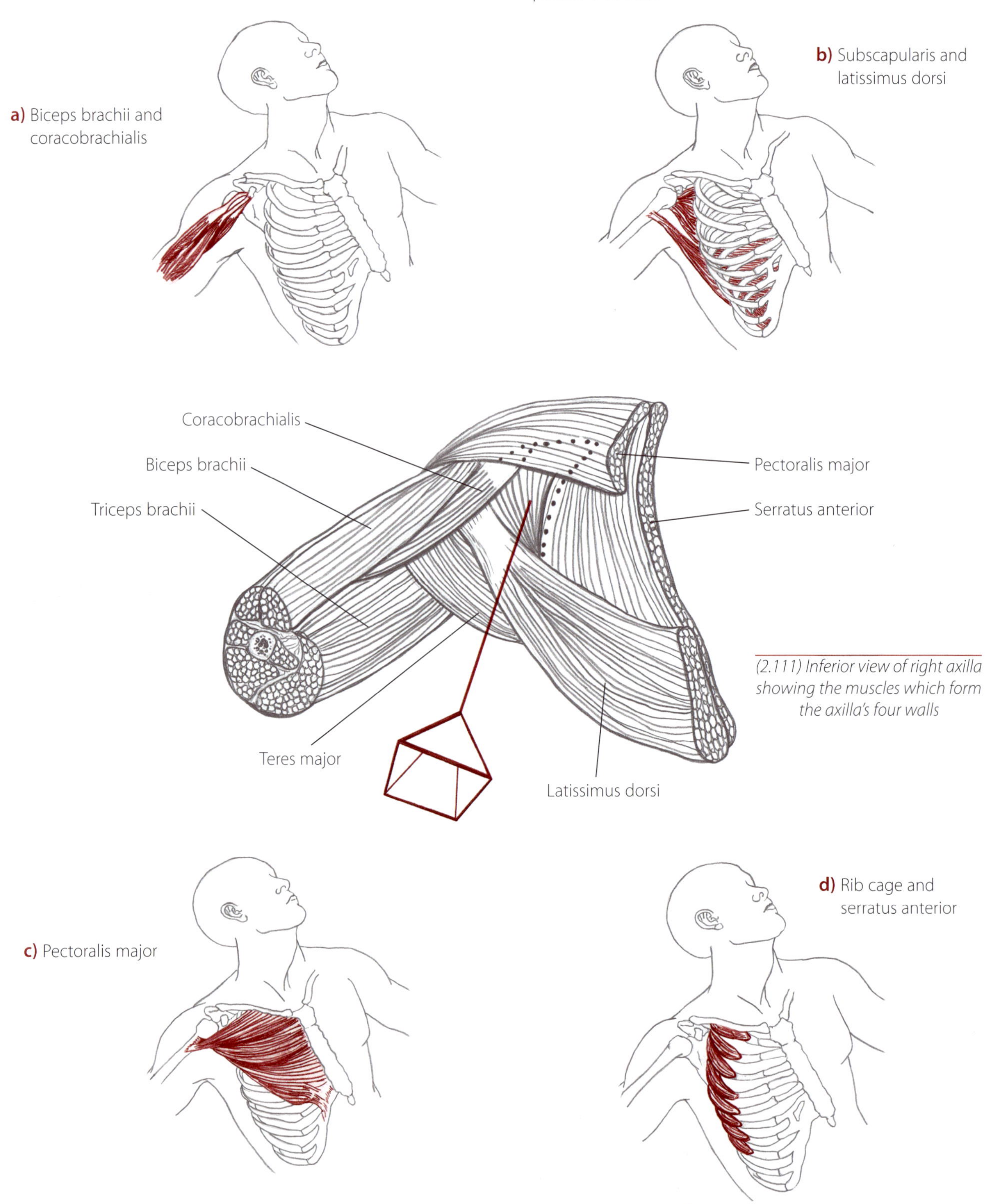

(2.111) Inferior view of right axilla showing the muscles which form the axilla's four walls

axilla	**ak**-sil-a	L. armpit
axillary	**ak**-si-**lar**-ee	

Compression or impingement of the brachial plexus or one of its nerves can create a sharp, shooting sensation down the arm. If this occurs, immediately release and adjust your position posteriorly. Also, ask your partner for feedback.

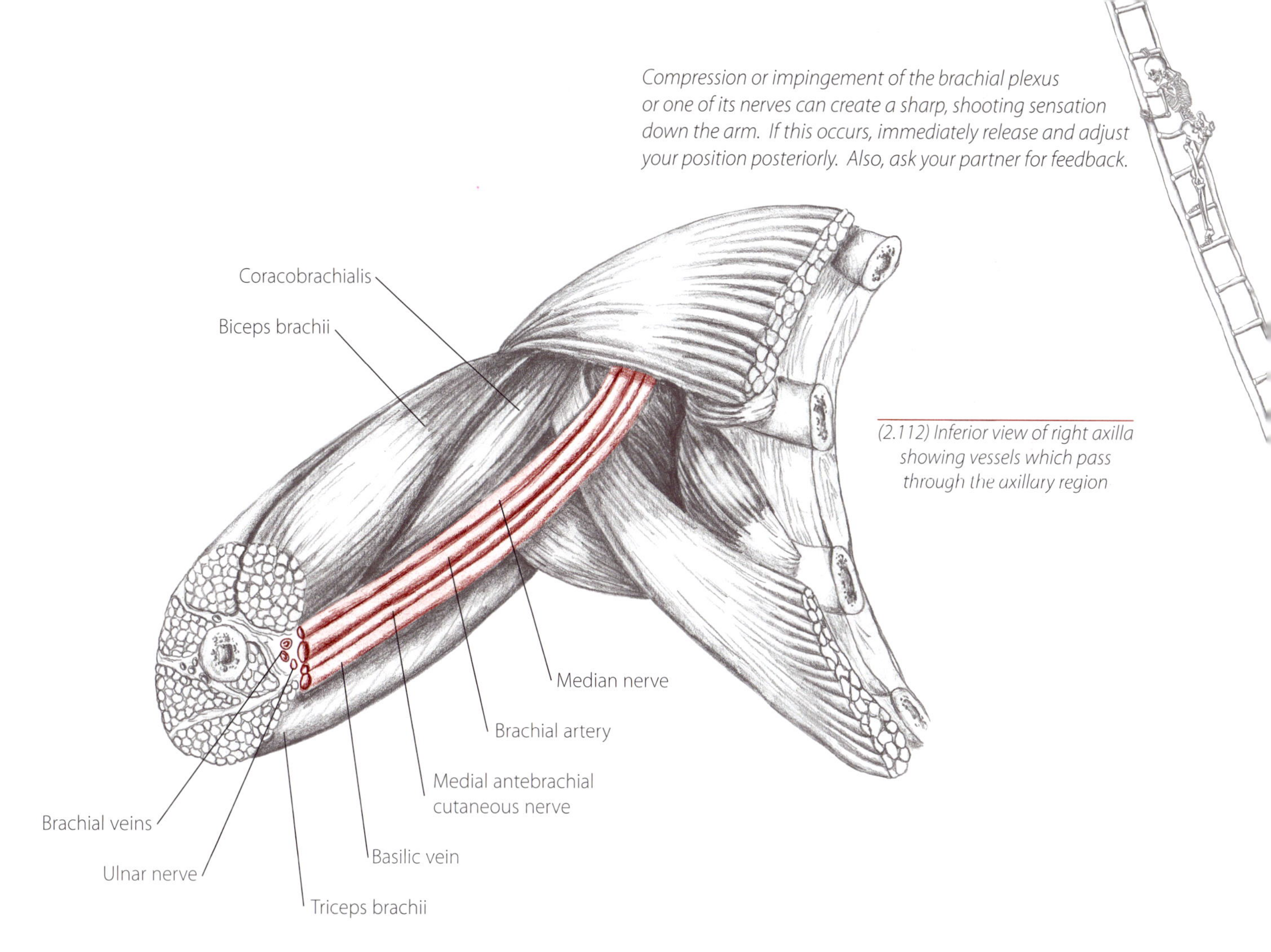

(2.112) Inferior view of right axilla showing vessels which pass through the axillary region

Sternoclavicular Joint

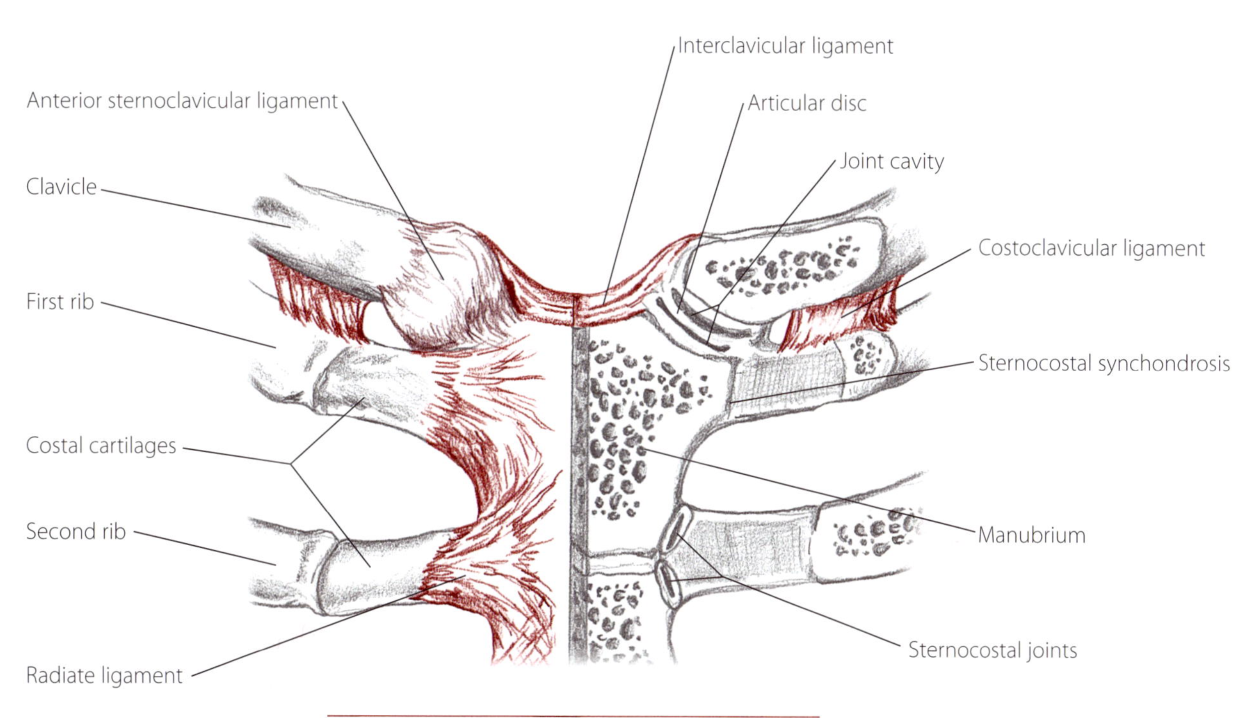

(2.113) Anterior view, right side shown in coronal section

brachial	**bray**-key-al	L. relating to the arm
gland		L. acorn
synchondrosis	**sin**-con-**dro**-sis	

Ligaments of the Shoulder and Glenohumeral Joint

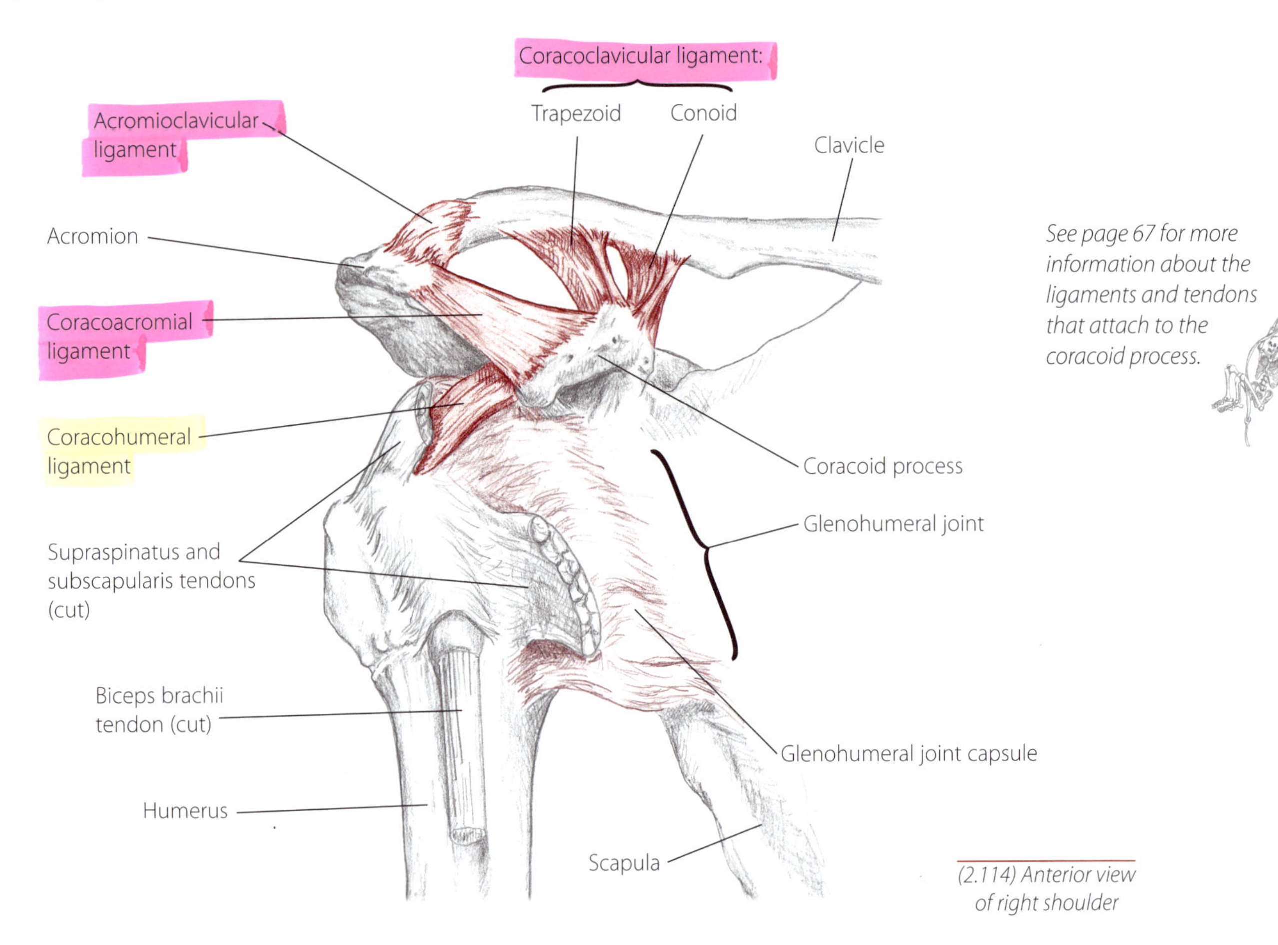

(2.114) Anterior view of right shoulder

See page 67 for more information about the ligaments and tendons that attach to the coracoid process.

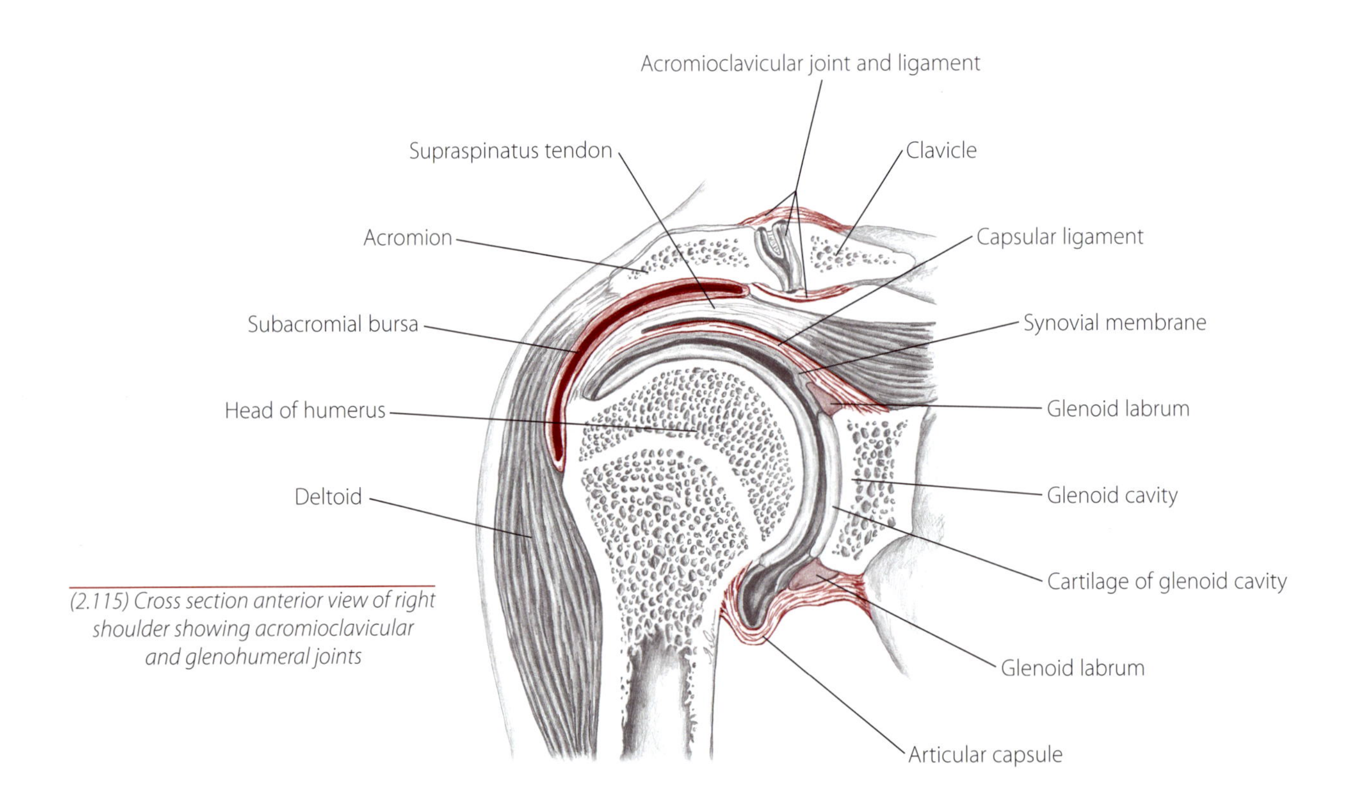

(2.115) Cross section anterior view of right shoulder showing acromioclavicular and glenohumeral joints

coracoacromial **cor**-a-**ko**-a-**kro**-mi-ul
coracoclavicular **cor**-a-**ko**-cla-**vic**-u-lar
ligament **lig**-a-ment L. a band

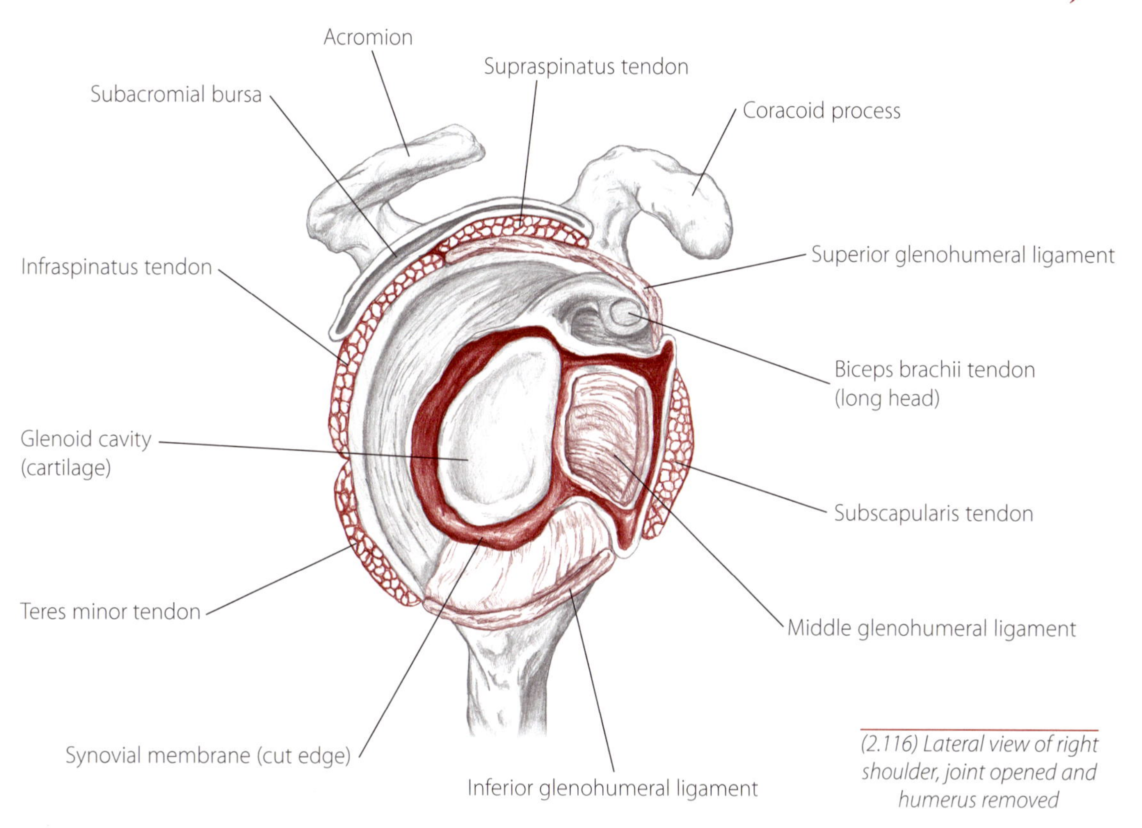

(2.116) Lateral view of right shoulder, joint opened and humerus removed

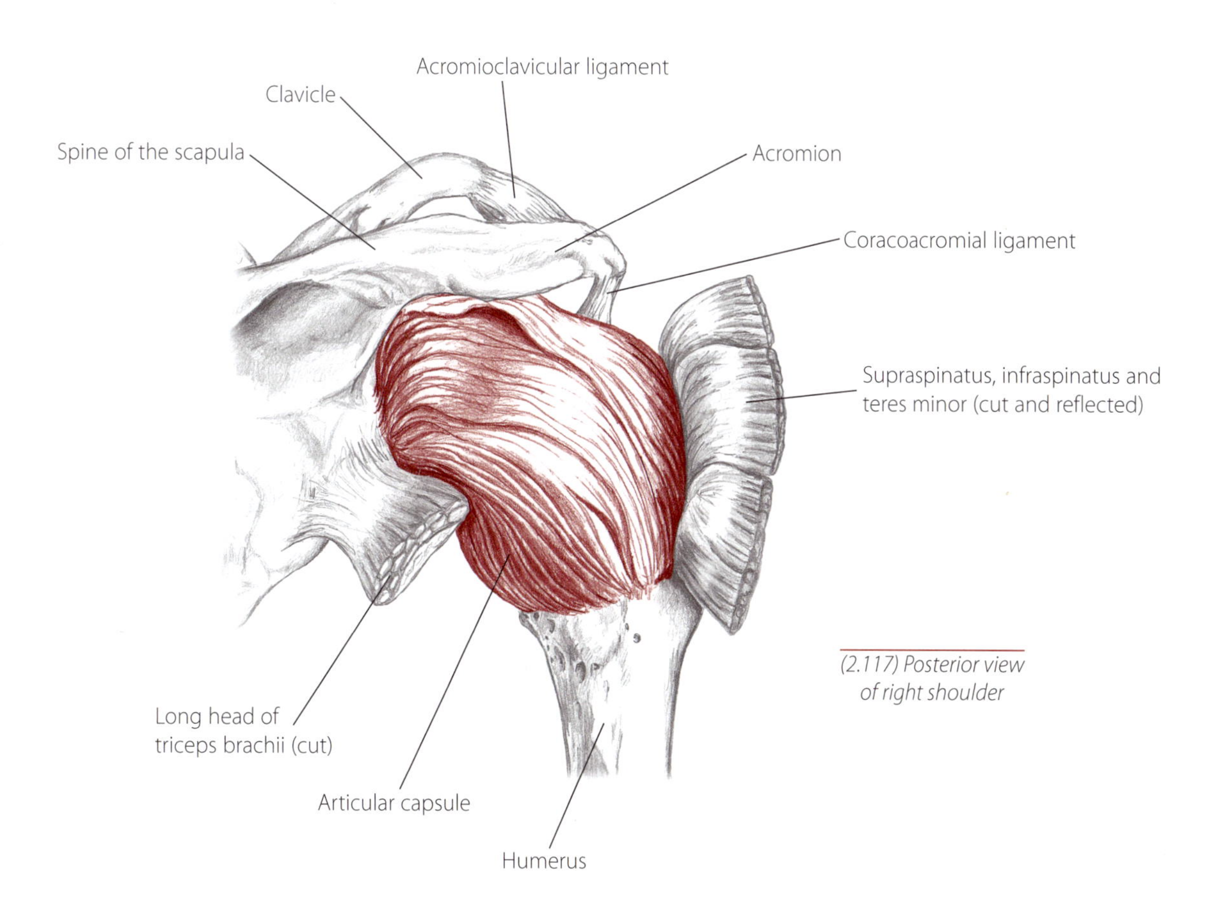

(2.117) Posterior view of right shoulder

labrum **lay**-brum L. lip

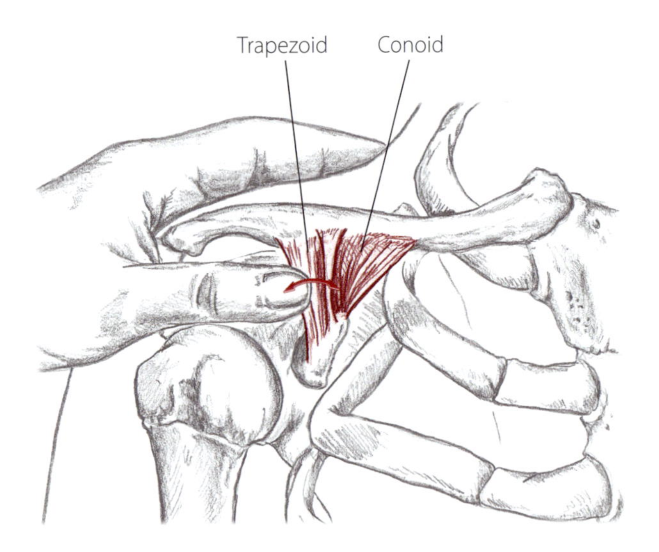

(2.118) Anterior view of right shoulder palpating coracoclavicular ligament

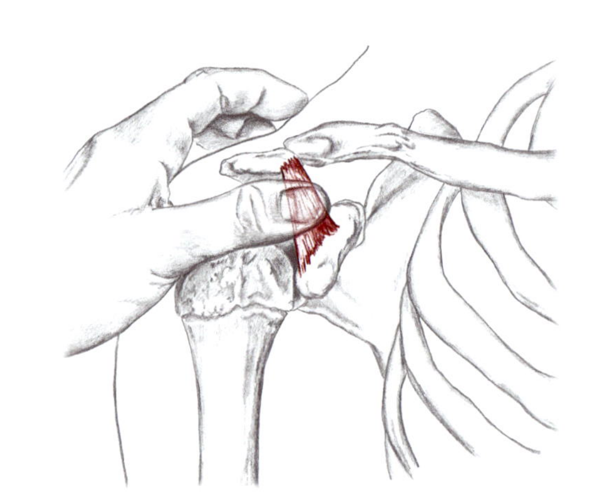

(2.119) Anterior view, palpating the coracoacromial ligament

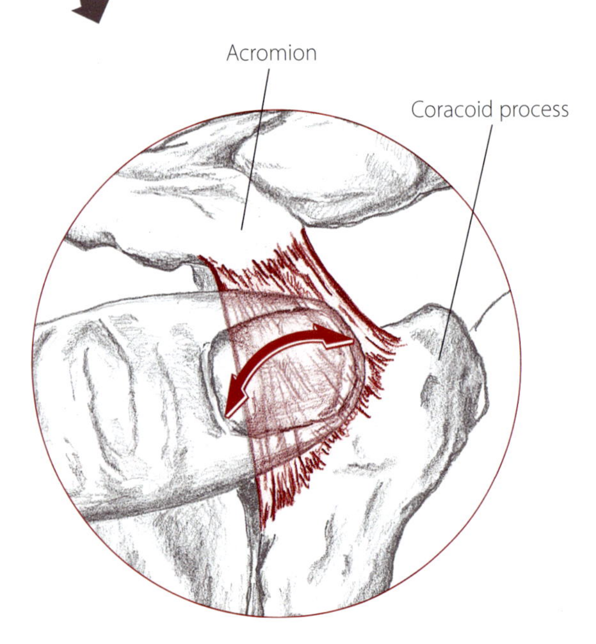

Coracoclavicular Ligament

The coracoclavicular ligament is composed of two smaller ligaments: the trapezoid and conoid. Both ligaments stretch from the coracoid process of the scapula to the inferior surface of the clavicle (2.114). Together they provide stability for the acromioclavicular joint and form a strong bridge between the scapula and clavicle.

The coracoclavicular ligament can be accessed by palpating between the clavicle and coracoid process or curling under the anterior aspect of the clavicle.

1) Seated or supine. Abduct and medially rotate the shoulder. This position brings the ligaments more to the surface.

2) Locate the coracoid process of the scapula and the shaft of the clavicle.

3) Palpate in the space between these landmarks. Roll your thumbpad across its fibers (2.118). Unlike the superficial pectoralis major fibers, the ligaments will feel like solid, taut bands.

Passively move the shoulder girdle in several directions and see if a particular position allows you greater access to the ligaments.

Coracoacromial Ligament

Unlike most ligaments which hold two bones together, the coracoacromial ligament attaches the scapula's coracoid process to its acromion (2.119). Along with the acromion, this ligament forms the coracoacromial arch across the top of the shoulder. This arch helps to protect the rotator cuff tendons and subacromial bursa from direct trauma by the acromion. The wide band of the coracoacromial ligament lies deep to the deltoid but is still accessible.

1) Supine or seated. Locate the coracoid process. Then locate the anterior edge of the acromion.

2) Palpating deep to the deltoid fibers, explore between these landmarks for the wide band of the coracoacromial ligament. Strum your finger across its fibers (2.119).

3) To bring the ligament closer to the surface, try extending the arm. This position will roll the humeral head anteriorly and press the ligament forward.

Are you between the acromion and the coracoid process? Place one finger on the ligament and passively move the shoulder girdle in various positions. Can you feel how the ligament's relationship to the surrounding tissues changes as the position of the shoulder changes?

Subacromial Bursa

Also known as the subdeltoid bursa, this sizable fluid sac has two major sections (2.120). The lateral portion creates a smooth surface for the acromion and deltoid to glide over the head of the humerus and rotator cuff tendons. The medial part cushions the coracoacromial ligament from the supraspinatus tendon.

With the arm at the side, most of the bursa is underneath the acromion and inaccessible. Extending the shoulder joint, however, will bring the bursa forward. Since abduction of the shoulder compresses the bursa, this action (when accompanied by pain and tenderness) can be used as an indicator of subacromial bursitis.

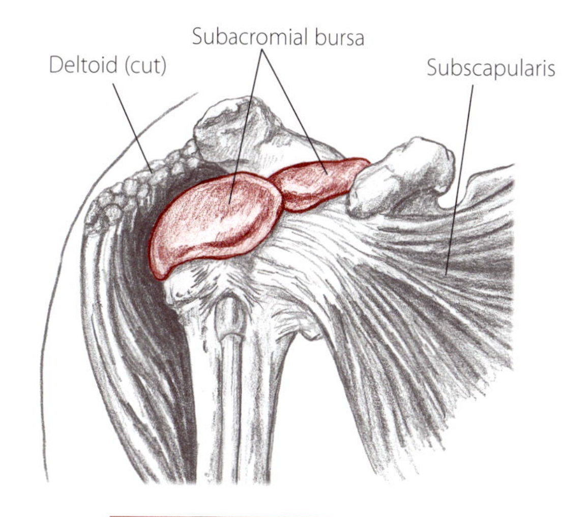

(2.120) Anterior view of right shoulder

1) With your partner seated, stand behind him and locate the acromion.
2) Drop your fingers off the anterior edge of the acromion. Then, with your other hand, slowly extend the shoulder by pulling the elbow posteriorly; this will bring the bursa out from under the acromion. You will be palpating at the depth between the deltoid and rotator cuff tendons (2.121).
3) A little hint - palpate gently. Bursae are delicate structures and best accessed with a soft touch. If it is inflamed it will be acutely tender.

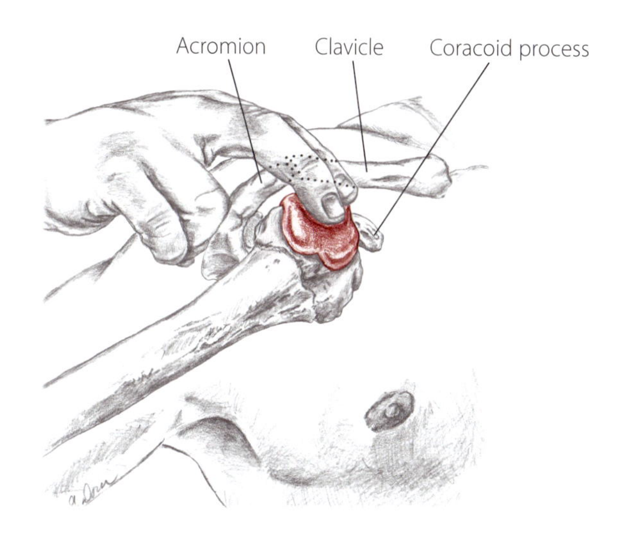

(2.121) Anterior/lateral view with right shoulder extended, palpating the subacromial bursa

Axillary Lymph Nodes

The axillary lymph nodes are located in the axilla. When palpating in the axillary region, use a deliberate yet gentle touch to avoid tickling your partner. Also, move slowly, using gentle pressure to avoid impinging the artery and nerves.

1) Supine or seated. Abduct the arm and slowly sink two fingers up into the axilla. Then bring the arm back to the side of the body to soften the axillary tissue further.
2) Slide your fingers up to the top of the axilla and then medially toward the rib cage. Often there will be a few lymph nodes located against the ribs (2.122).
3) Move to the lateral side of the axilla and use light pressure against the humerus to locate the strong pulse of the brachial artery. The vessel will be positioned between the stringy coracobrachialis and long head of the triceps brachii muscle.

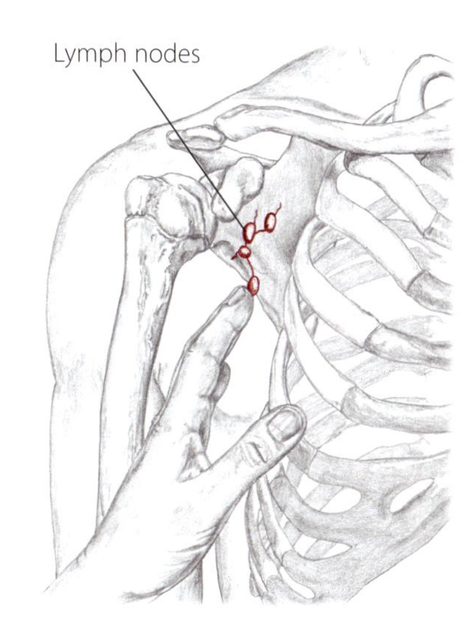

(2.122) Anterior view of right shoulder

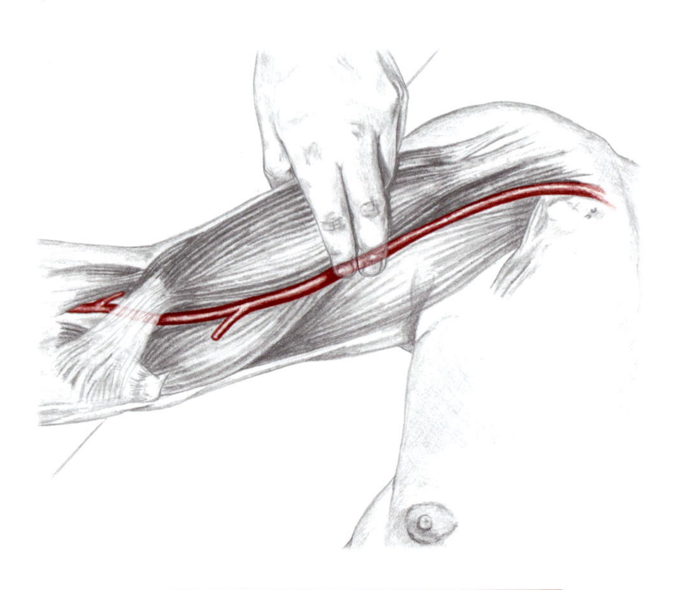
(2.123) Anterior/medial view, palpating the brachial pulse between the biceps and triceps brachii

Brachial Artery

The brachial artery is a continuation of the axillary artery and runs between the biceps and triceps brachii. Its pulse can be felt between these muscles on the medial side of the arm (p. 109). Before the brachial artery branches off to the radial and ulnar arteries, its pulse can be felt at the elbow, just medial to the biceps brachii tendon.

1) Seated or supine. Abduct the arm and place your fingerpads on the medial side of the arm. A helpful guide is the shallow dip which forms between the biceps and triceps (2.123).
2) Gently press your fingers toward the shaft of the humerus to feel the brachial pulse.
3) The brachial pulse can also be detected just medial to the distal tendon of the biceps brachii.

NOTES

To the forearm and hand...

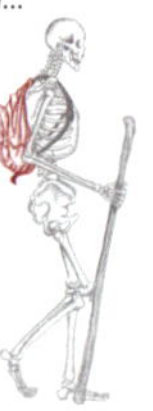

3
Forearm & Hand

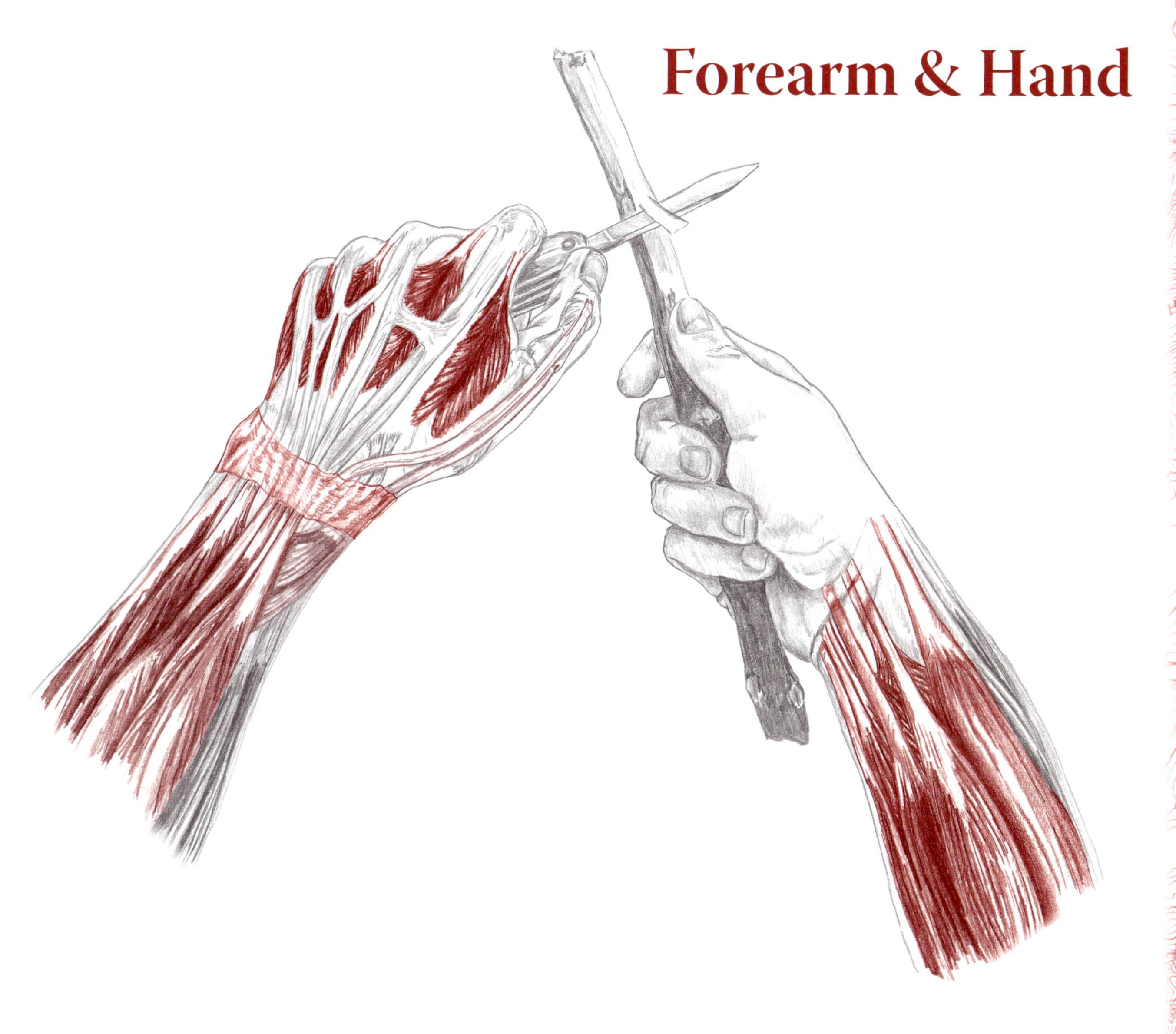

Topographical Views

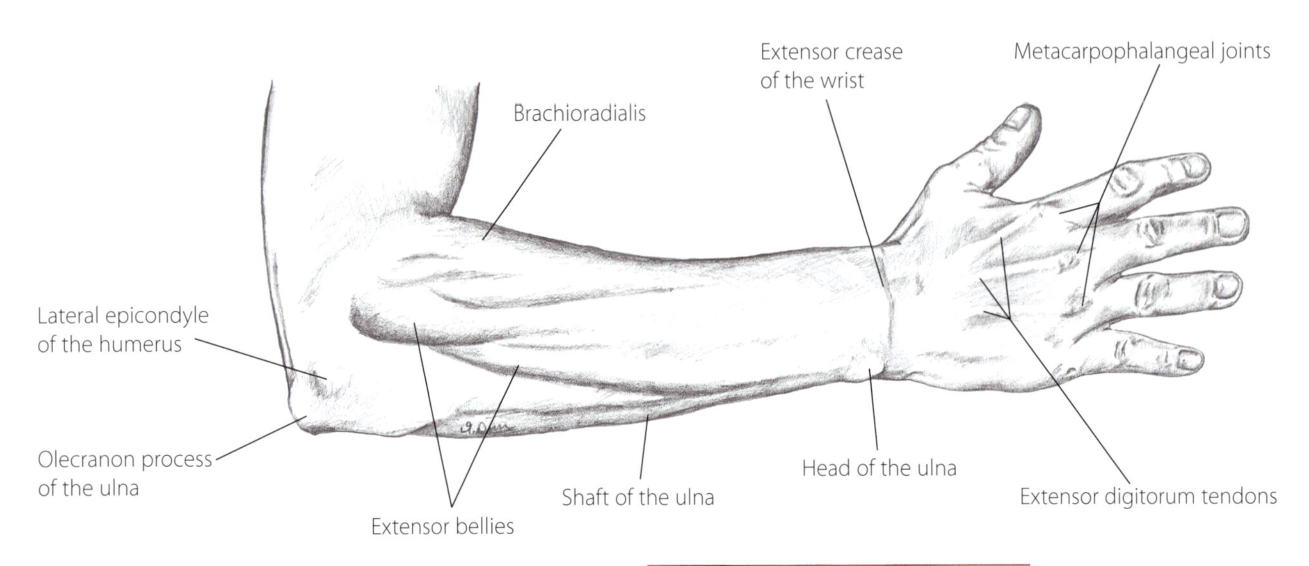

(3.1) Lateral view of right forearm and hand

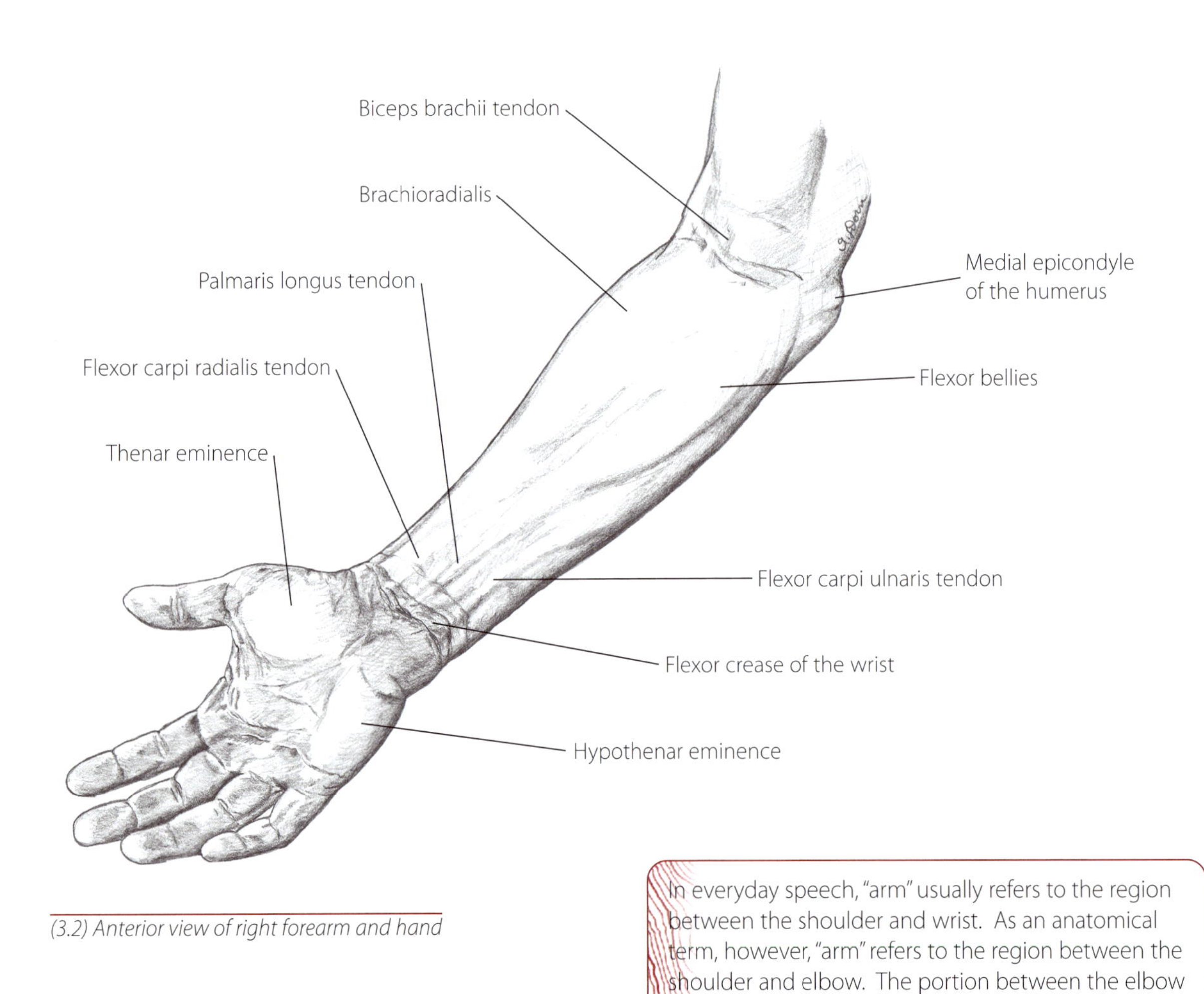

(3.2) Anterior view of right forearm and hand

In everyday speech, "arm" usually refers to the region between the shoulder and wrist. As an anatomical term, however, "arm" refers to the region between the shoulder and elbow. The portion between the elbow and wrist is called the "forearm."

thenar	**thee**-nar	Grk. palm, flat of the hand
hypothenar	**hi**-po-**thee**-nar	Grk. *hypo*, under or below

Exploring the Skin and Fascia

1) Begin by gently lifting the skin and fascia of the forearm. Compare the thickness and elasticity of the posterior (hairy) side to the anterior (hairless) side (3.3).
2) Explore the length of the forearm. Note how the tissue along the shaft of the forearm may be more challenging to grasp than the tissue at the wrist and elbow regions.

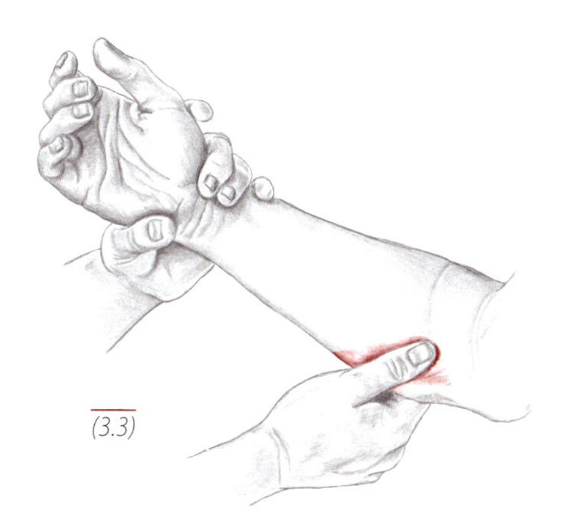

(3.3)

1) Using one hand to stabilize the forearm, use your other hand to gently twist the skin and fascia around the forearm's shaft (3.4).
2) Now try to tug the skin superiorly and inferiorly. Often the tissue will have more elasticity in a horizontal direction (around the forearm's shaft) than in a vertical one.

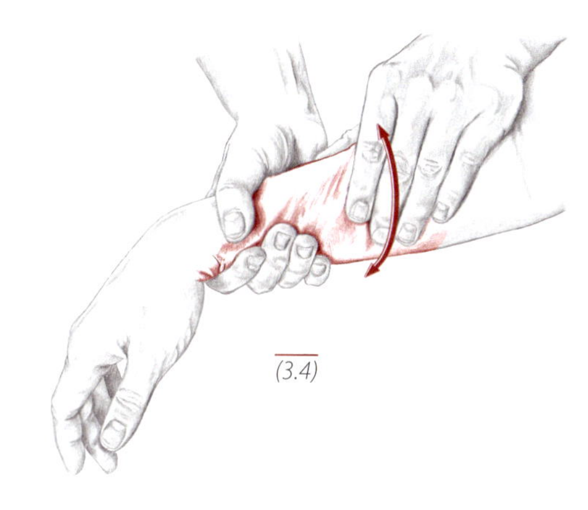

(3.4)

1) Here is an opportunity to feel your partner's skin and fascia stretch during passive movement. Grasp the tissue near the wrist and *passively* flex and extend the wrist joint (3.5). Feel how supple and plentiful the tissue is when the wrist is flexed. As the wrist is extended, however, the skin may be pulled from between your fingers.
2) Continue to move the wrist while grasping the tissue on all sides of the forearm. Supinate and pronate the forearm, feeling how these actions cause the tissue to move differently.
3) Ask your partner to *actively* move his wrist and fingers while you grasp the skin and fascia. Encourage him to move slowly. Play with isolating specific actions - for instance, extension of the wrist as opposed to extension of the fingers - to feel the tissues shift with different actions.

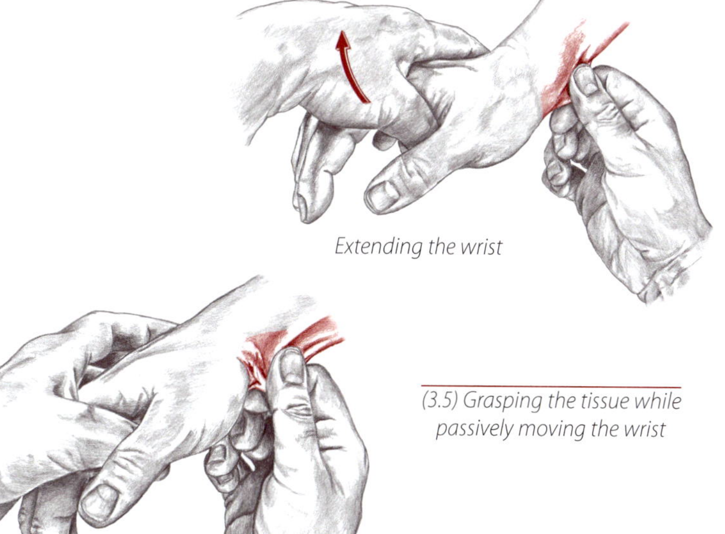

Extending the wrist

Wrist in neutral

(3.5) Grasping the tissue while passively moving the wrist

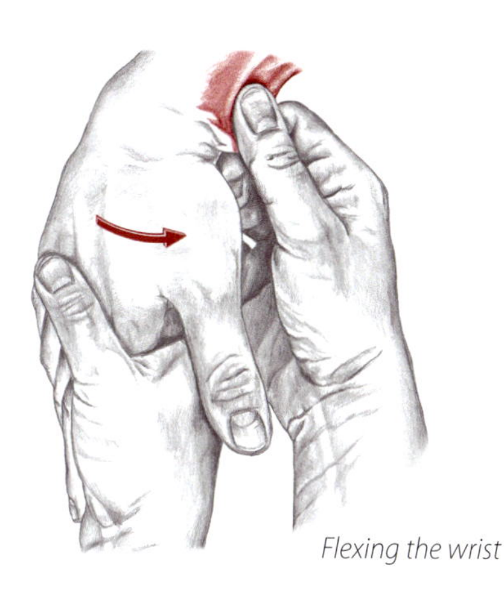

Flexing the wrist

Bones of the Forearm and Hand

The **humerus** is the bone of the arm. Its proximal end articulates with the scapula to form the glenohumeral joint. Its distal end joins with the ulna and radius at the elbow. The elbow has two joints: the humeroulnar and humeroradial.

The **radius** and **ulna** make up the bones of the forearm (3.6). The ulna is superficial and has a palpable edge that extends from the elbow to the wrist. The radius ("on the thumb side") is lateral to the ulna and is partially buried in muscle. Pronation and supination of the forearm are created by the radius pivoting around the ulna at the proximal and distal radioulnar joints.

The three groups of bones in the wrist and hand are the carpals, metacarpals and phalanges. The **carpals** are eight, pebble-sized bones that form two rows (proximal and distal), each containing four carpal bones (3.9). Located distal to the "flexor crease" of the wrist, the carpals are accessible from all sides - the palmar, dorsal, radial and ulnar surfaces of the hand.

The **metacarpals** are five long bones spanning the palm of the hand. The metacarpals' proximal end is the base, the long midsection is the shaft and the distal end is the head (3.7). The metacarpals are easily palpable along the hand's dorsal surface. They are deep to the muscles on the palmar side.

The **phalanges** are the bones of the fingers. The thumb has two phalange bones and the fingers have three. All sides of the phalanges are accessible (3.8).

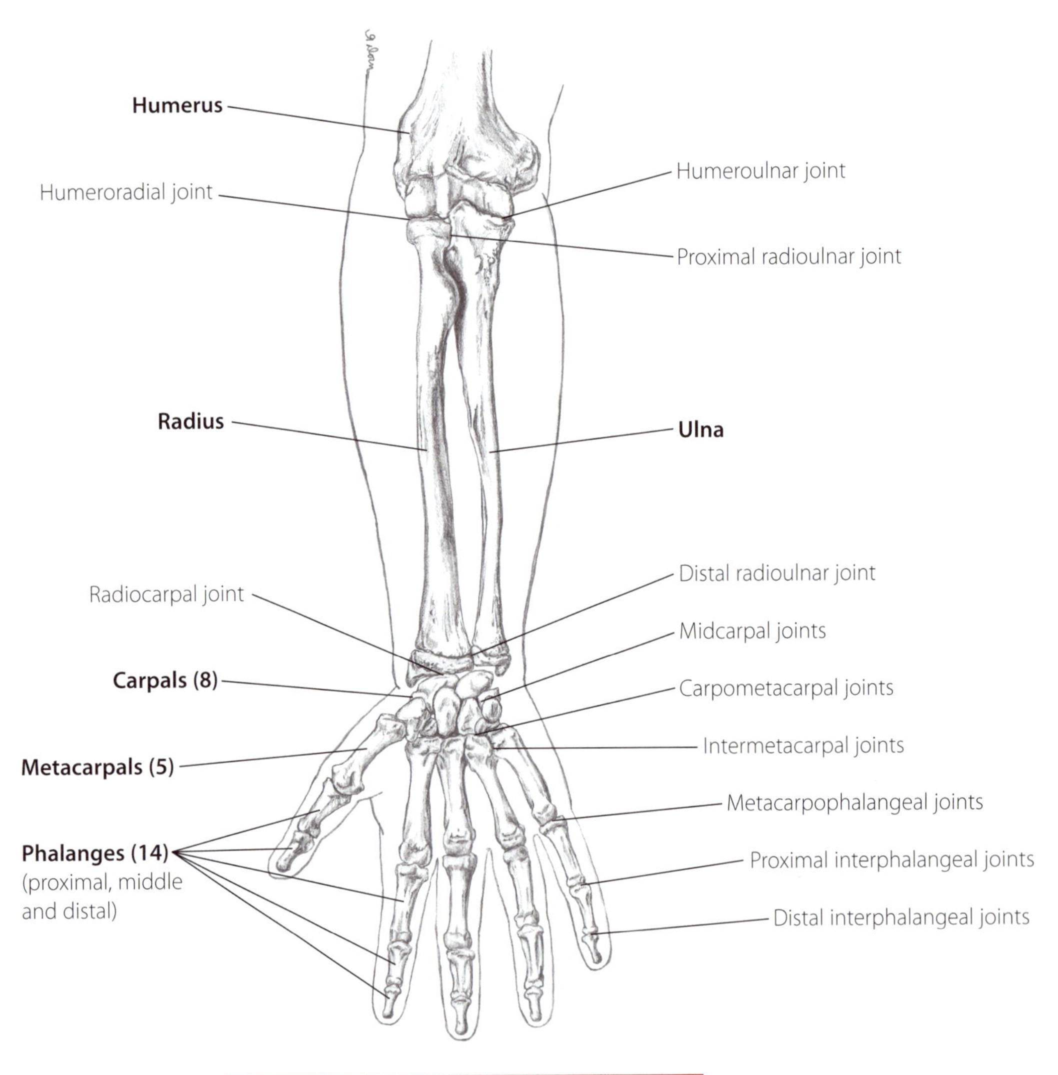

(3.6) Anterior (palmar) view of right forearm and hand

Let's talk joints! The **radiocarpal** joint (wrist), formed by the radius and proximal carpals, is an ellipsoid joint. The gliding joints at the **midcarpal** and second to fifth **carpometacarpal** joints allow for only small, shifting movements.

The first **carpometacarpal** joint of the thumb is an ellipsoid-shaped saddle joint. The **metacarpophalangeal** joints, the large "knuckles" of the hand, are also ellipsoid joints. The **interphalangeal** joints of the fingers are hinge joints.

humerus	**hu**-mer-us	L. upper arm
radius	**ray**-dee-us	L. staff, spoke of a wheel
ulna	**ul**-na	L. elbow, arm

Bony Landmarks

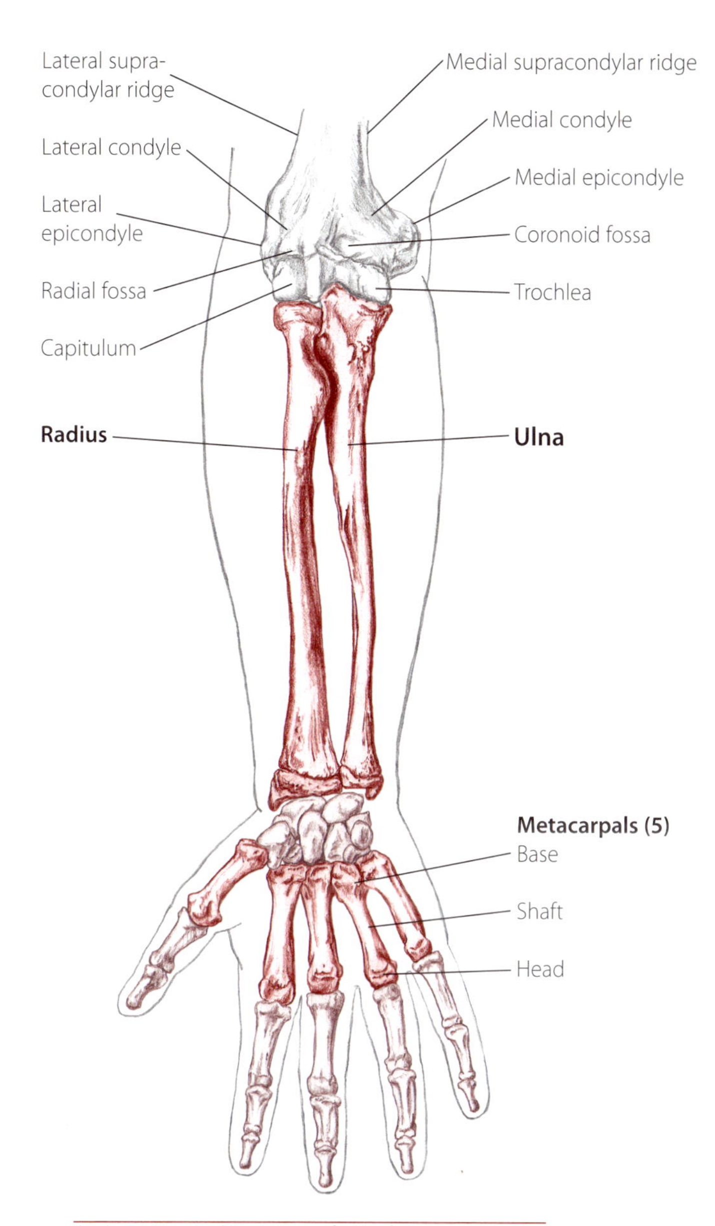

(3.7) Anterior (palmar) view of right forearm and hand

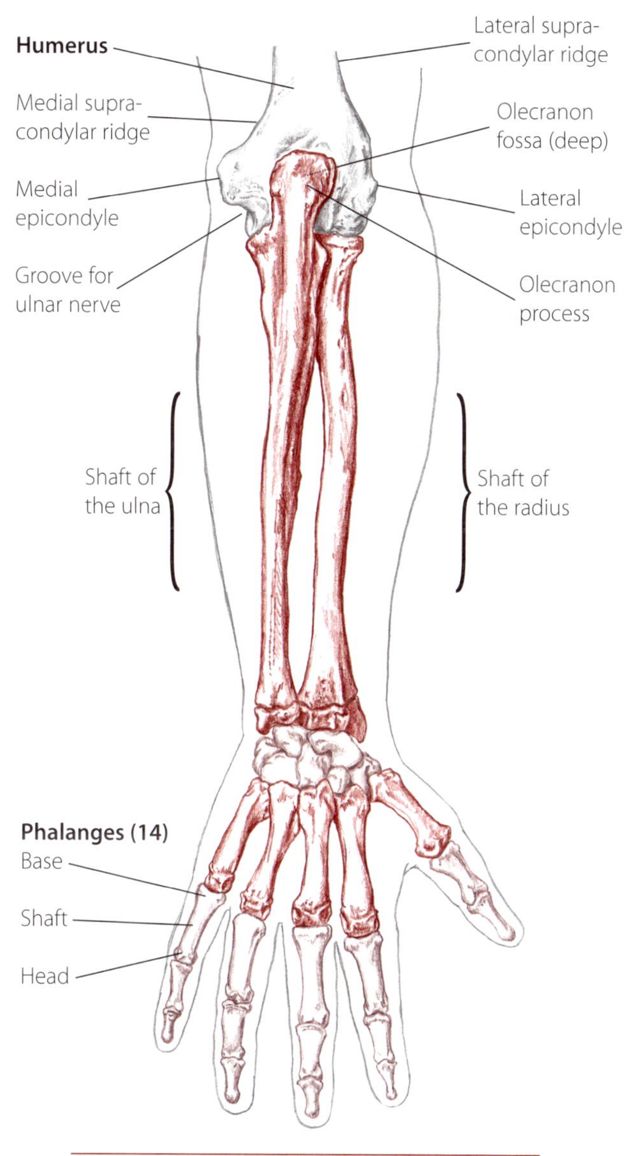

(3.8) Posterior (dorsal) view of right forearm and hand

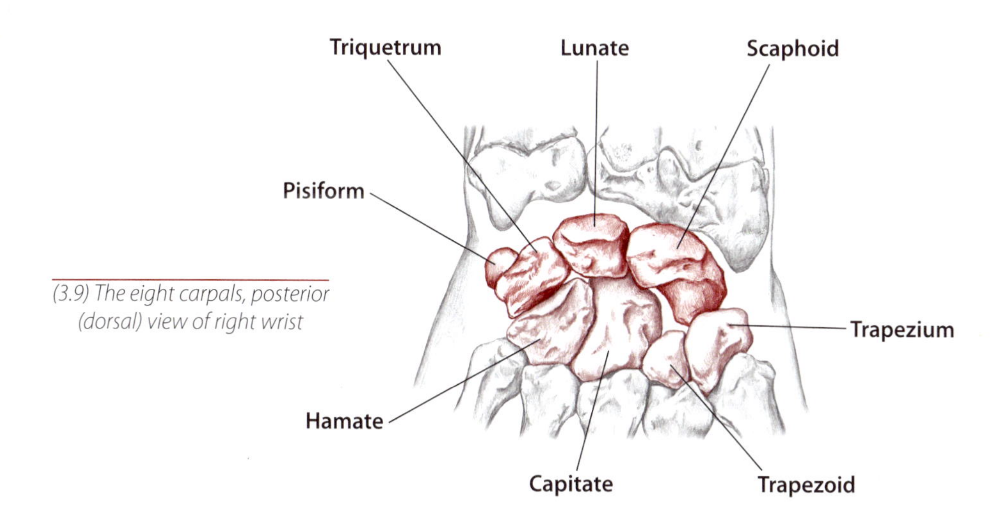

(3.9) The eight carpals, posterior (dorsal) view of right wrist

carpal	**kar**-pul	Grk. pertaining to the wrist
metacarpal	**met**-a-**kar**-pul	Grk. *meta*, after, beyond
phalange	fa-**lan**-jee	Grk. closely knit row, line of battle

Bony Landmarks

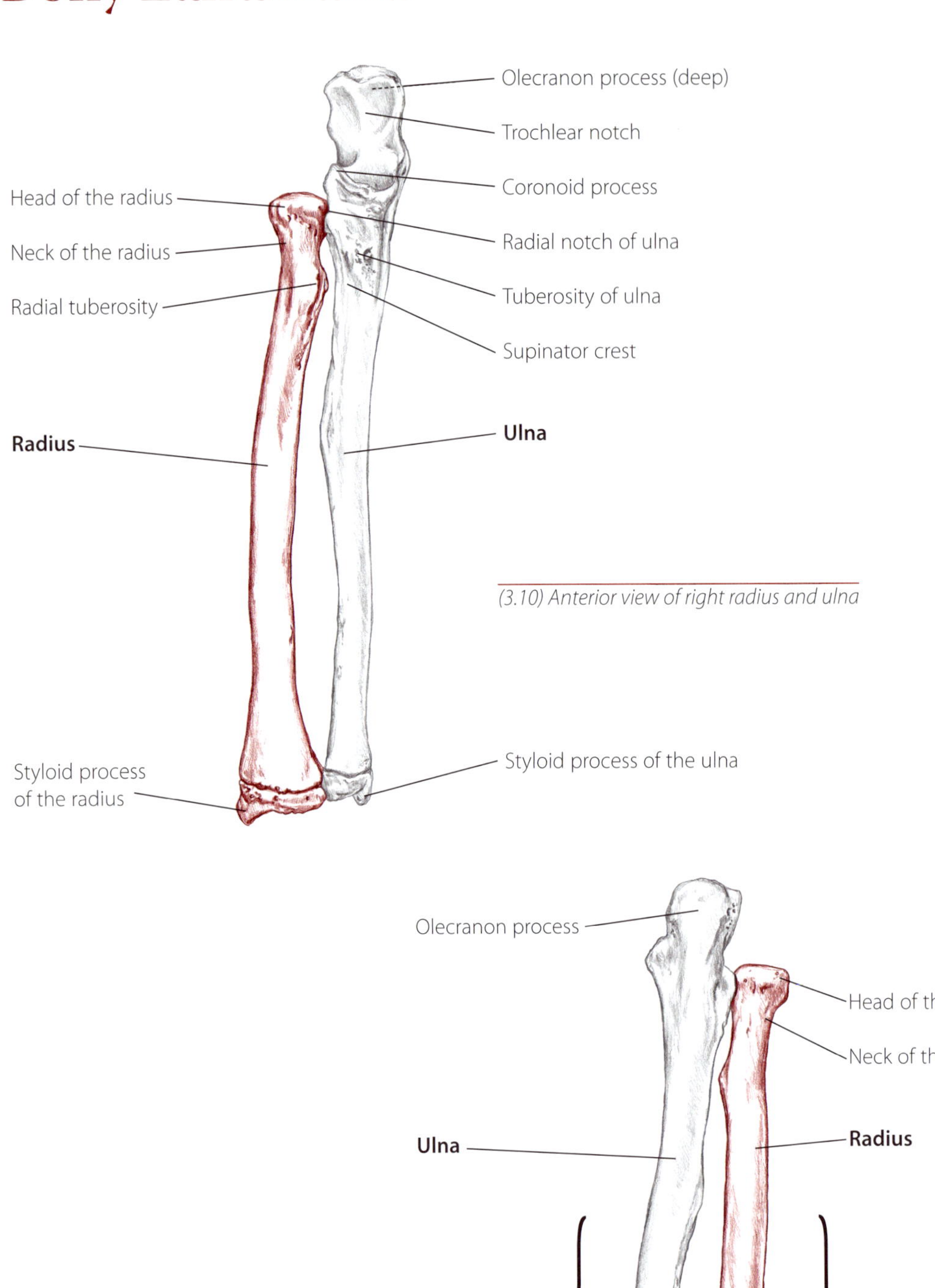

(3.10) Anterior view of right radius and ulna

(3.11) Posterior view of right radius and ulna

Bony Landmark Trails

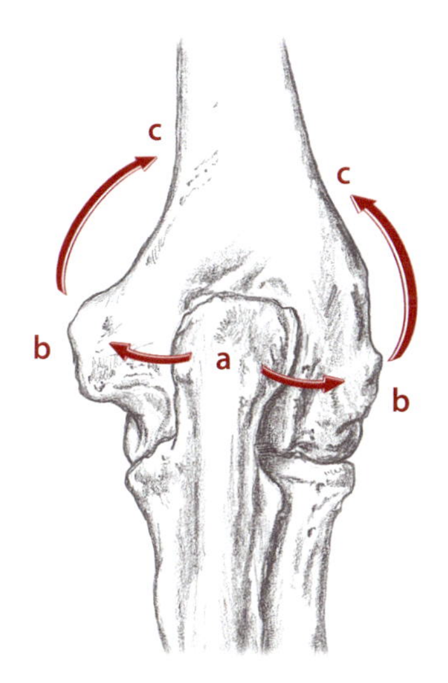

Trail 1 "Knob Hill" explores the elbow and distal humerus.

a Olecranon process and fossa
b Epicondyles of the humerus
c Supracondylar ridges of the humerus

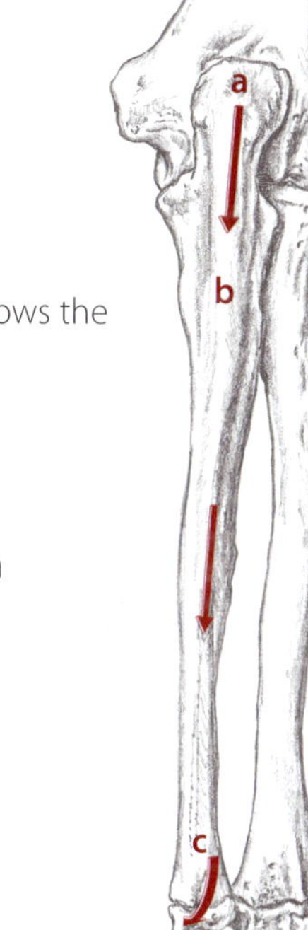

Trail 2 "The Razor's Edge" follows the length of the superficial ulna.

a Olecranon process
b Shaft of the ulna
c Head of the ulna
d Styloid process of the ulna

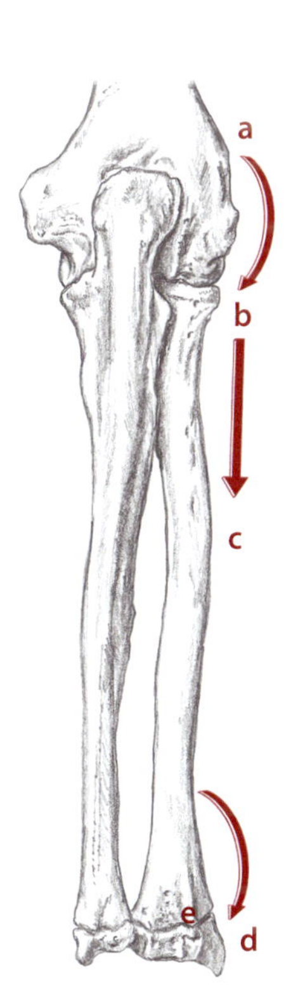

Trail 3 "Pivot Pass" travels the length of the radius, the bone which creates the pivoting action of the forearm.

a Lateral epicondyle of the humerus
b Head of the radius
c Shaft of the radius
d Styloid process of the radius
e Lister's tubercle

Trail 4 "Walking On Your Hands" explores the small carpal bones of the wrist as well as the bones and joints of the hand.

Some translations for the names of bones may cause you to scratch your head and wonder what early anatomists were thinking. The carpals, luckily, cause no such puzzlement.

capitate	L. head-shaped	*scaphoid*	L. boat-shaped
hamate	L. hooked	*trapezium*	Grk. little table
lunate	L. crescent-shaped	*trapezoid*	Grk. table-shaped
pisiform	L. pea-shaped	*triquetrum*	L. three-cornered

Trail 1 "Knob Hill"

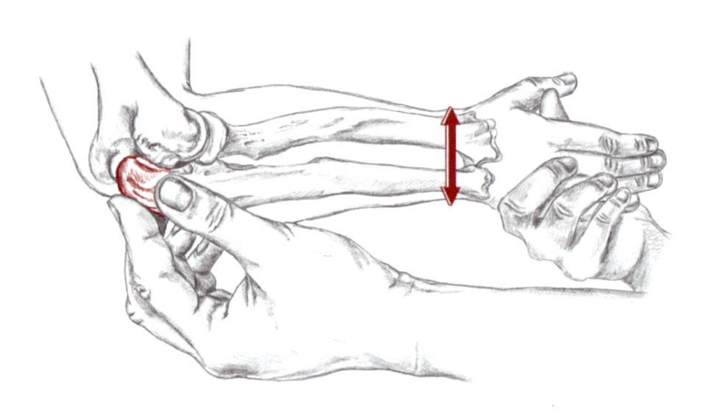

(3.12) Palpating the olecranon process

Olecranon Process and Fossa

The olecranon process (or elbow) is located on the proximal end of the ulna and articulates with the distal humerus. Its large surface is the attachment site for the triceps brachii muscle. It forms the "point" of the elbow and is easily located.

The olecranon fossa is a large cavity on the posterior, distal end of the humerus designed to accommodate the olecranon process when the elbow is extended. Located deep to the triceps brachii tendon, the fossa is only partially accessible.

Olecranon process

1) Partner seated. Shake hands with your partner and explore the large, superficial knob at the elbow. Palpate and explore its angular surface and sides.
2) Passively flex and extend the elbow, noticing how the olecranon process feels in various positions (3.12).

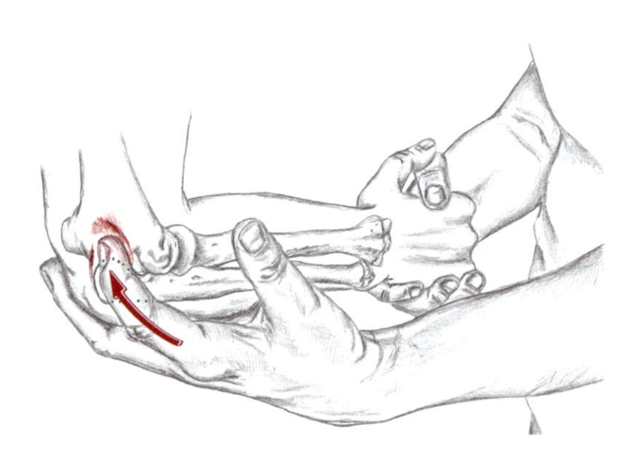

(3.13) Locating the olecranon fossa

Olecranon fossa

1) Flex the elbow and locate the olecranon process.
2) Roll your finger proximally around the top of the process, pressing through the triceps tendon and into the fossa.
3) Because of the presence of the triceps brachii tendon and the proximity of the olecranon process, only a small crescent-shaped ditch will be accessible (3.13).

When locating the fossa, are you proximal to the tip of the olecranon process? If you flex and extend the elbow slightly, do you feel a change in the fossa's shape and size?

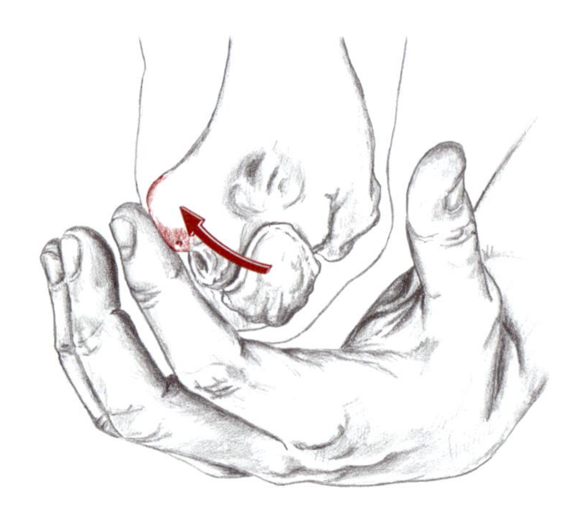

(3.14) Posterior view of right elbow, palpating the medial epicondyle

Epicondyles of the Humerus

As the humerus extends down the arm, its distal end broadens medially and laterally. Directly medial from the olecranon process is the medial epicondyle. It is superficial and has a protruding, spherical shape designed to accommodate the tendons of the wrist and hand flexors.

The lateral epicondyle is smaller than its medial counterpart and is located lateral to the olecranon process. It is an attachment site for the tendons of the wrist and hand extensors.

fossa	**fos**-a	L. a shallow depression
olecranon	o-**lek**-ran-on	Grk. elbow
process	**pros**-es	L. going forth

Epicondyles of the humerus

1) With your partner seated, shake hands and locate the olecranon process.
2) Slide your finger medially off the olecranon. You will encounter a small ditch before rising up onto the large, superficial medial epicondyle. Explore its bulbous shape (3.14).
3) Return to the olecranon. Slide laterally to the lateral epicondyle. Note that it is smaller than the medial epicondyle (3.15).

Set a finger on each epicondyle and slowly flex and extend the elbow. The surrounding muscle tissue may move, but the epicondyles should remain stationary. Do they?

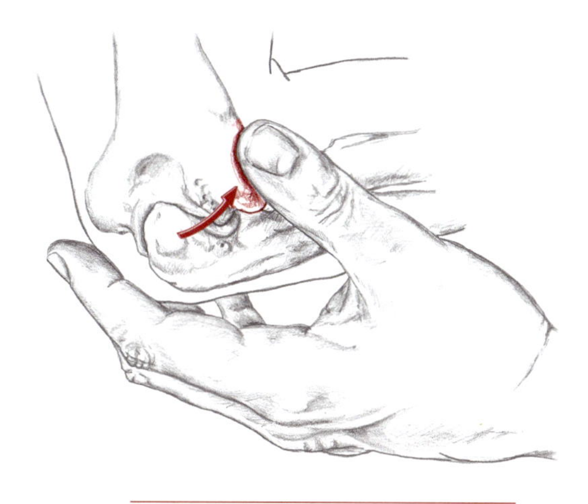

(3.15) Posterior/lateral view of right elbow, palpating the lateral epicondyle

The ulnar nerve (p. 167), which creates the "funny bone" sensation when struck, courses between the medial epicondyle and the olecranon process.

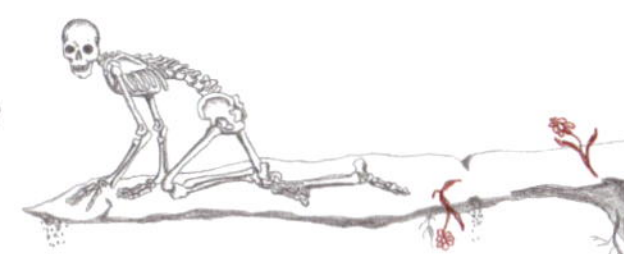

Supracondylar Ridges of the Humerus

These two ridges extend proximally from the respective epicondyles of the humerus. Both serve as attachment sites for the forearm muscles. The lateral supracondylar ridge is located superficially, while the medial ridge sinks into the arm and is situated close to the ulnar nerve.

1) With your partner seated, shake hands and locate the medial epicondyle.
2) Move proximally from the medial epicondyle. The bony ridge which extends from the epicondyle is the medial supracondylar ridge (3.16). Roll your fingers back and forth across the ridge to sense its distinct edge.
3) Explore the lateral supracondylar ridge.

Can you follow the ridges proximally a few inches before they disappear under the muscles of the arm?

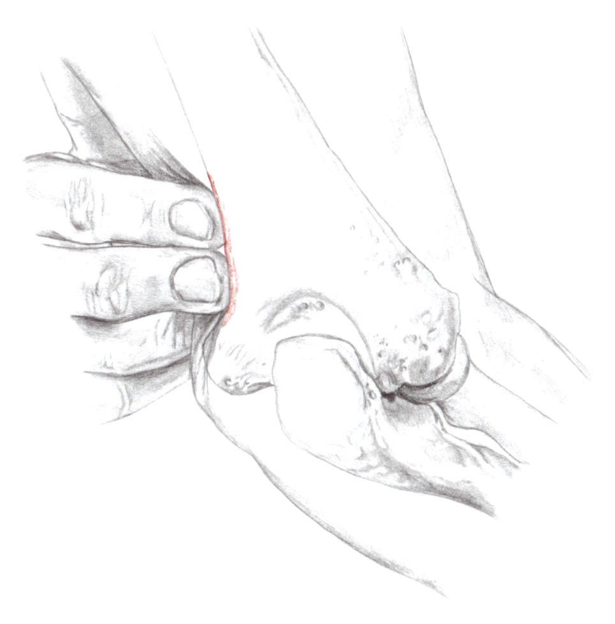

(3.16) Posterior/lateral view of right elbow, exploring the medial supracondylar ridge

condyle	**con**-dial	Grk. knuckle
epi-	**eh**-pee	Grk. above, upon
lateral	**lat**-er-al	L. to the side

Trail 2 "The Razor's Edge"

Shaft of the Ulna

The long, straight shaft of the ulna extends from the olecranon process to the head of the ulna. Although numerous muscles lie beside the shaft, it has a superficial, palpable edge that runs along the forearm's posterior/medial aspect.

(3.17) Lateral view of right forearm, palpating shaft of ulna

1) Shaking hands with your partner, locate the olecranon process. Slide your fingers distally along the shaft.
2) To define its shape and location, roll your fingers across its edge. Follow it down the length of the forearm.

Is the bone you are palpating superficial? Does it stretch the length of the forearm (3.17)?

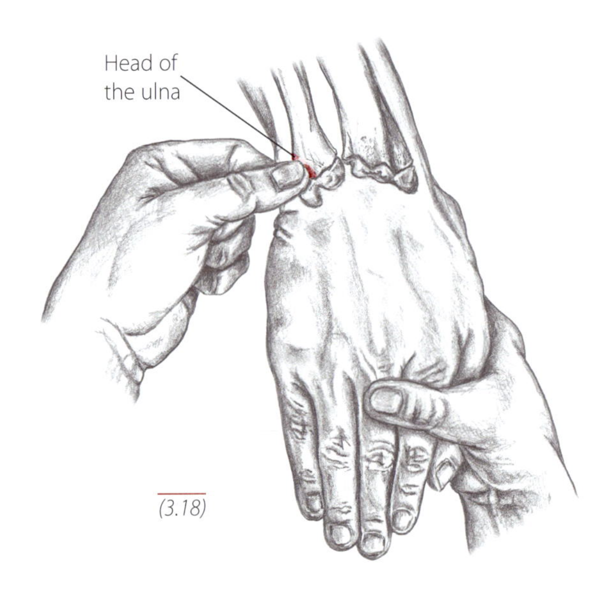

(3.18)

Head of the Ulna

The shaft of the ulna swells to form the head of the ulna. The head is the superficial knob visible along the posterior/medial side of the wrist that can disrupt the placement of a watchband.

1) Slide your fingers distally along the ulnar shaft.
2) Just proximal to the wrist, the shaft will bulge to become the head of the ulna. Palpate all sides of the bulbous head (3.18).

Is the knob you are palpating connected to the shaft of the ulna? In a neutral position, is it on the posterior/medial side of the forearm?

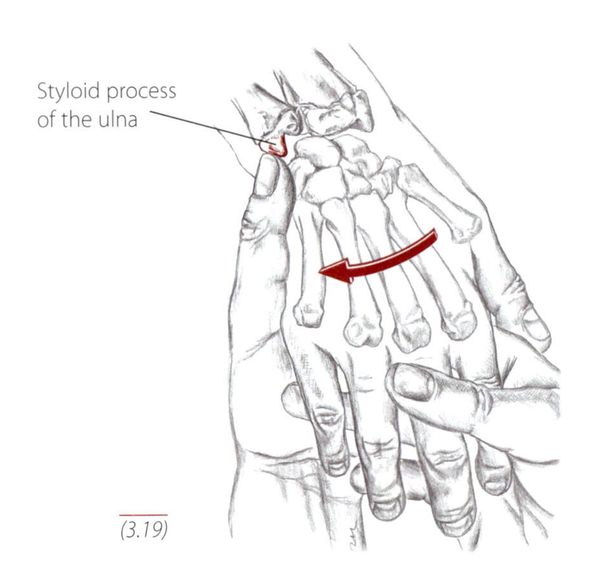

(3.19)

Styloid Process of the Ulna

Both the ulna and the radius have styloid processes at their distal ends. The radius' styloid process (p. 126) is larger and extends farther distally. The ulna's styloid process is sharper and more pronounced. It is a tooth-like projection pointing distally off the head of the ulna. It is located on the posterior/medial side of the wrist. Both styloid processes are superficial, and the tendons of the forearm muscles pass beside them.

1) Shake hands with your partner. Passively adduct the wrist to soften the surrounding tendons.
2) Use your thumb to locate the posterior aspect of the ulnar head. Slide distally off the head to palpate the small tip of the styloid process (3.19).

Is the bone you are palpating connected to the ulnar head (as opposed to a separate carpal bone)? If you slowly flex and extend the wrist, it should remain stationary. Does it?

Head of the Radius

The head of the radius is distal to the humerus' lateral epicondyle. It forms the radius' proximal end and has a circular, bell shape. The head is stabilized by the annular ligament (p. 167) and is a pivoting point for supination and pronation of the forearm. Although it is deep to the supinator and extensor muscles, the head's posterior, lateral aspect can be accessed.

1) Shake hands with your partner and locate the lateral epicondyle.
2) Slide distally off the lateral epicondyle, across the small ditch between the humerus and radius and onto the head of the radius (3.20).
3) The head of the radius is the only bony structure in this vicinity. Explore its ring-shaped, superficial surface.

Are you distal to the lateral epicondyle? Place your thumb on the head and, with your other hand, slowly supinate and pronate the forearm (3.21). Do you feel the head's rotating movement under your thumb?

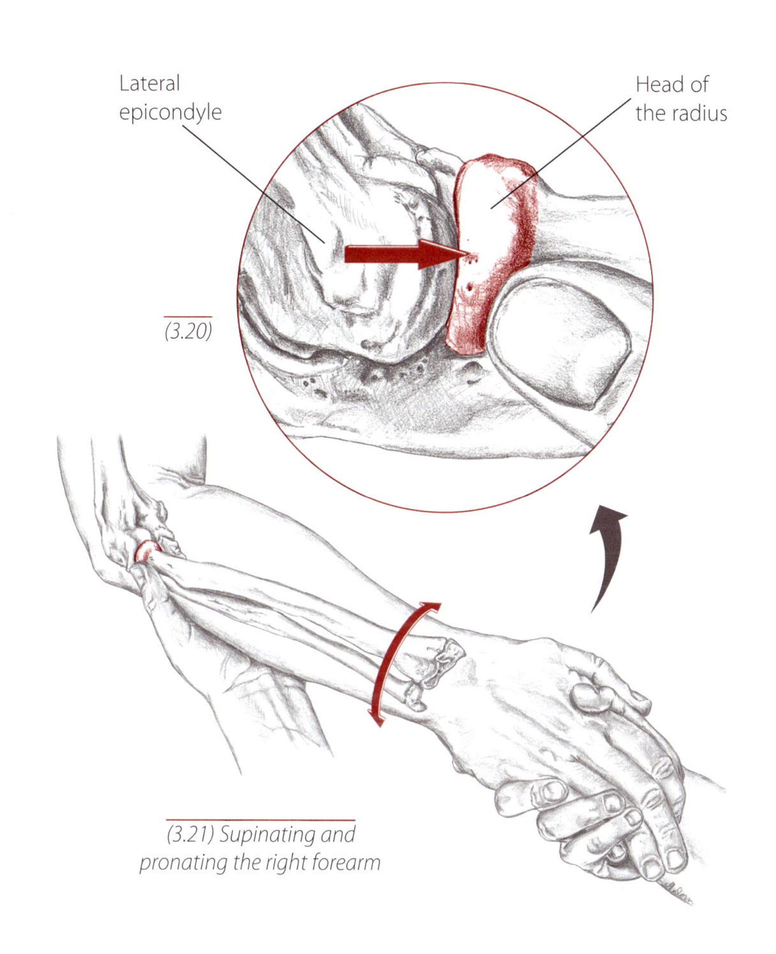

(3.20)

(3.21) Supinating and pronating the right forearm

Shaft of the Radius

The shaft of the radius is located on the lateral side (thumb side) of the forearm. Unlike the superficial edge of the ulnar shaft, most of the shaft of the radius is buried under muscle tissue. Its distal portion, however, is superficial and can be directly accessed.

1) Flex the elbow to 90° and put the forearm in a neutral "handshake" position.
2) Locate the head of the radius. Slide distally off the head, noting how the radius sinks beneath the forearm muscles. Continue down the forearm and feel the radius become superficial near the wrist (3.22).
3) Along the distal forearm, explore all sides of the superficial shaft of the radius.

Is the bone you are palpating along the lateral side of the forearm? Place one hand upon the radial shaft, while the other hand slowly supinates and pronates the forearm. Do you feel the shaft of the radius pivot around the shaft of the ulna?

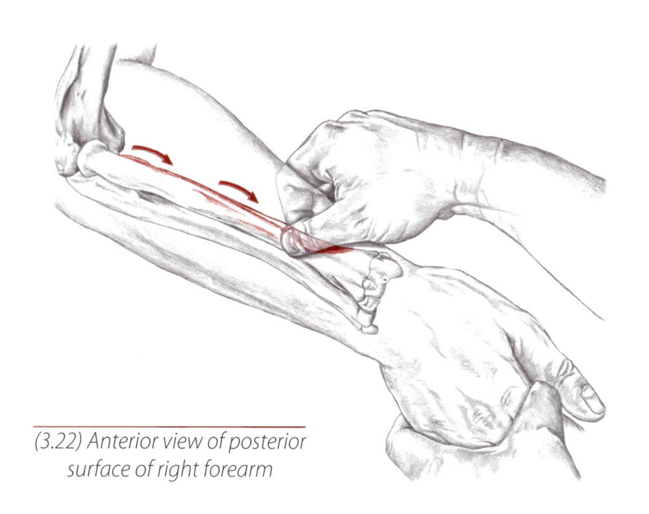

(3.22) Anterior view of posterior surface of right forearm

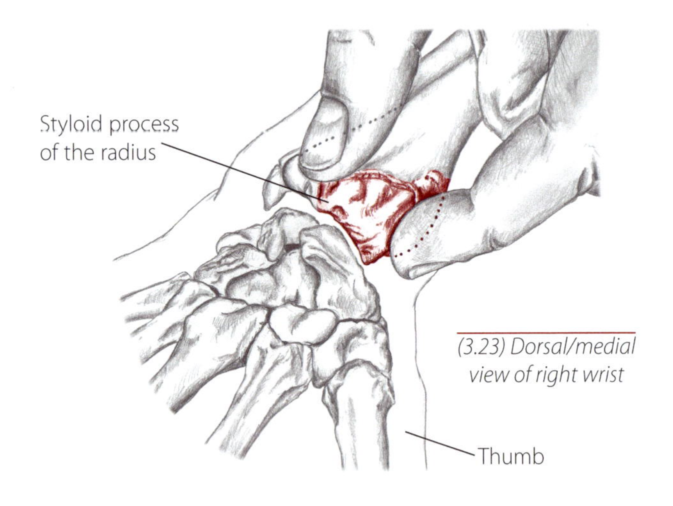

(3.23) Dorsal/medial view of right wrist

Styloid Process of the Radius

The styloid process of the radius, in comparison to the toothlike styloid process of the ulna, is a wider, more substantial mound of bone. Located on the lateral side of the radius, the styloid process is surrounded by the extensor tendons and is the attachment site for the brachioradialis (p. 141).

1) Begin by grasping the distal radial shaft between your thumb and finger. Slide distally, noting how the radius broadens in all directions.
2) Palpate along the lateral side (thumb side) of the radius to the tip of the styloid process (3.23).

Are you proximal to the "flexor crease" of the wrist? Is the portion of bone you are palpating surrounded by several thin tendons? If you passively flex and extend the wrist, the styloid process should remain stationary. Does it?

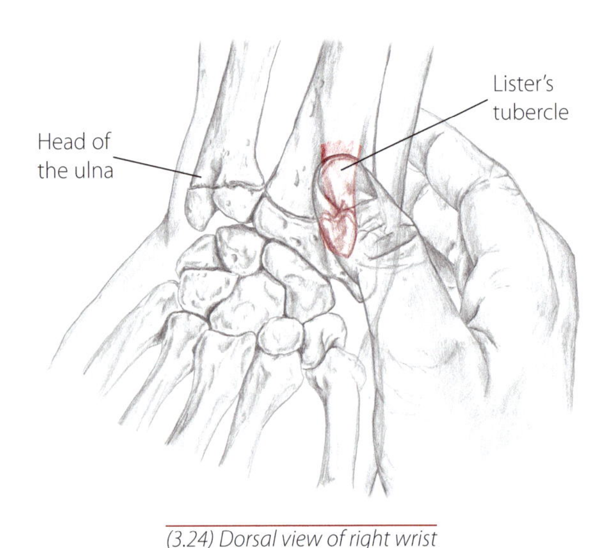

(3.24) Dorsal view of right wrist

Lister's Tubercle

Named in honor of Joseph Lister, father of modern antiseptic surgery, this superficial knob is located on the dorsal surface of the radial styloid process. With its oblong shape, Lister's tubercle (also known as the dorsal tubercle) acts as a hook for the extensor pollicis longus tendon (p. 160). For our purposes, however, it will serve as a bench mark for finding two of the carpals - the lunate and capitate.

1) Using your thumb, locate the dorsal surface of the styloid process of the radius.
2) Slide your thumb in the direction of the head of the ulna and explore for the oblong knob of Lister's tubercle.
3) The tubercle will be felt directly across from the head of the ulna - perhaps an inch away (3.24).

Are you on the dorsal surface of the radius? Is the bump you feel superficial, oval and across from the head of the ulna? If you passively flex and extend the wrist, the tissue over the tubercle should move, yet the tubercle itself should stay stationary. Does it?

The styloid processes of the radius and ulna serve as important jumping-off points for locating the carpals. Locate both processes and determine if the radial styloid process extends farther distally than the ulnar. (It should.) Then explore just distal to the processes, noting how your fingers naturally sink into the tissue of the wrist. This is the general location of the proximal row of carpals.

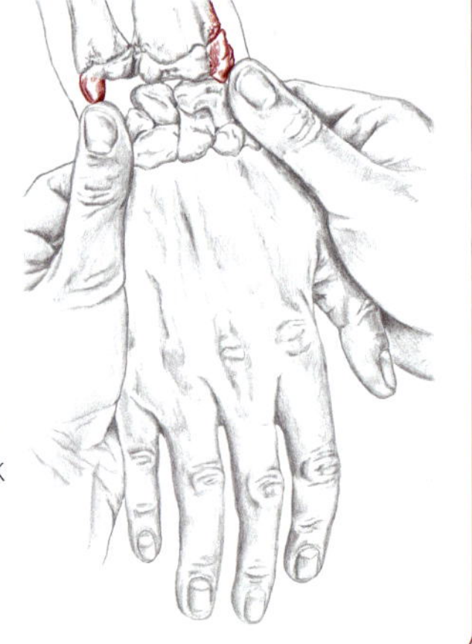

styloid **sti**-loyd Grk. a pillar

Carpals

There are eight carpal bones located at the wrist. Small and uniquely shaped, the carpals are closely wedged together between the distal radius and ulna and the metacarpals. The carpals form two rows, each composed of four bones (3.27).

Located distal to the flexor crease at the wrist, under the heel of the hand, the carpals lie deep to numerous flexor and extensor tendons. These overlying tendons, combined with the carpal's compact arrangement, make isolating individual bones a challenge.

The next few pages present the carpals in pairs. We will begin by exploring the carpals as one large group. We will then access the pisiform, triquetrum and the hamate, as these are possibly the easiest to isolate.

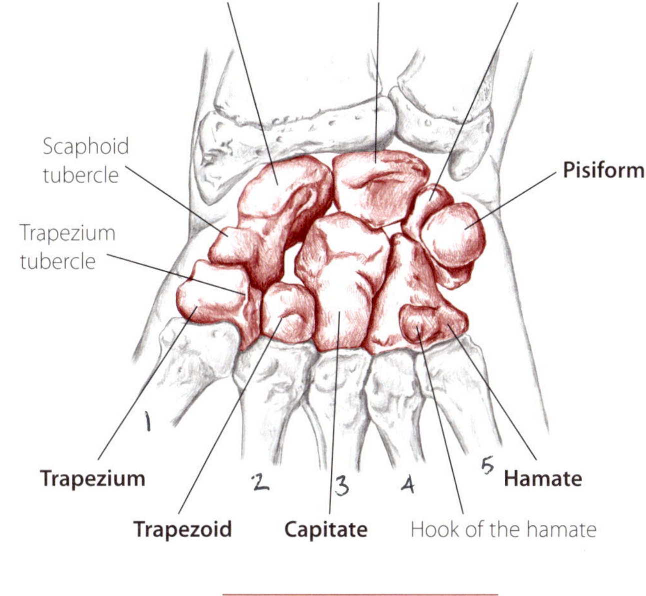

(3.25) Palmar view of right wrist

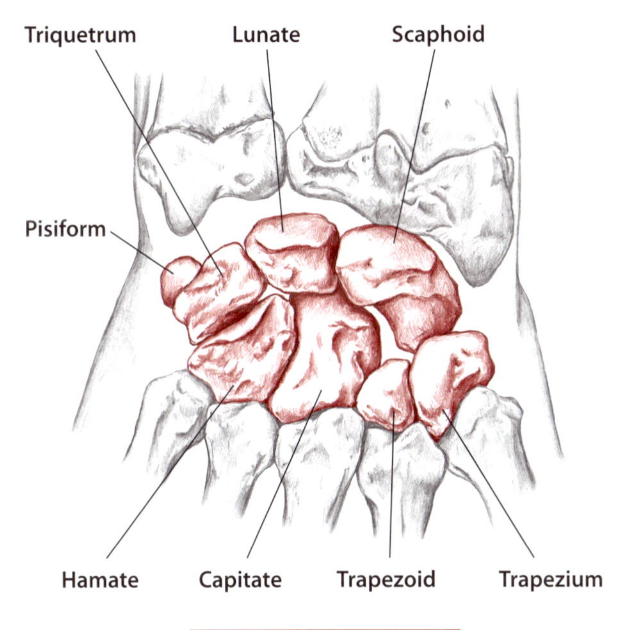

(3.26) Dorsal view of right wrist

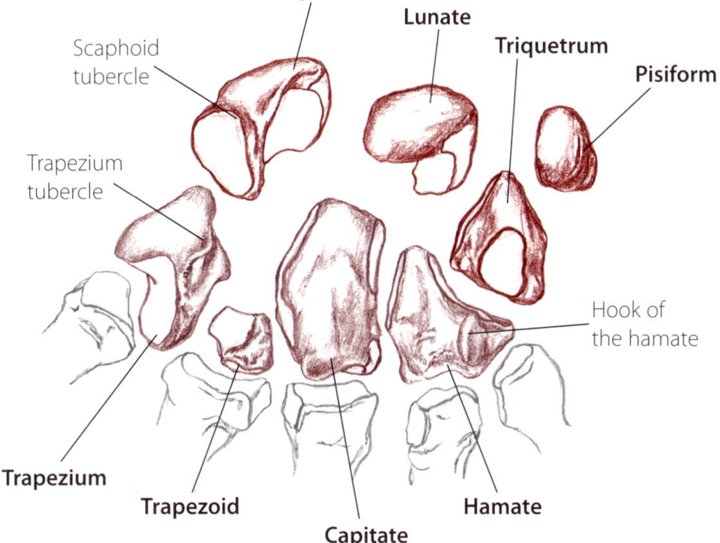

(3.27) The expanded carpals - palmar view of right hand. The scaphoid, lunate, triquetrum and pisiform form the proximal row, while the trapezium, trapezoid, capitate and hamate make up the distal row.

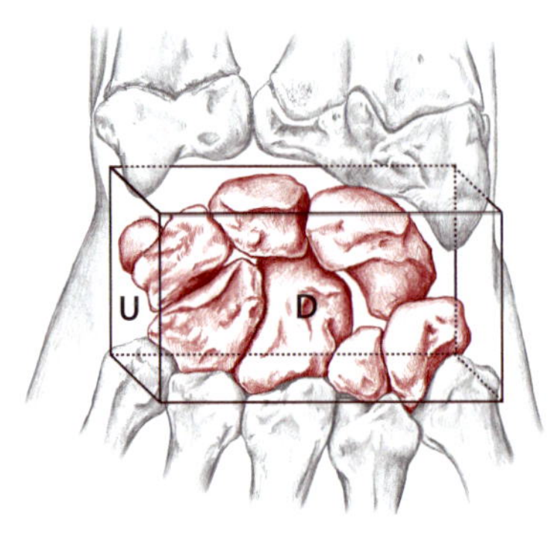

Ulnar/dorsal view
of right wrist

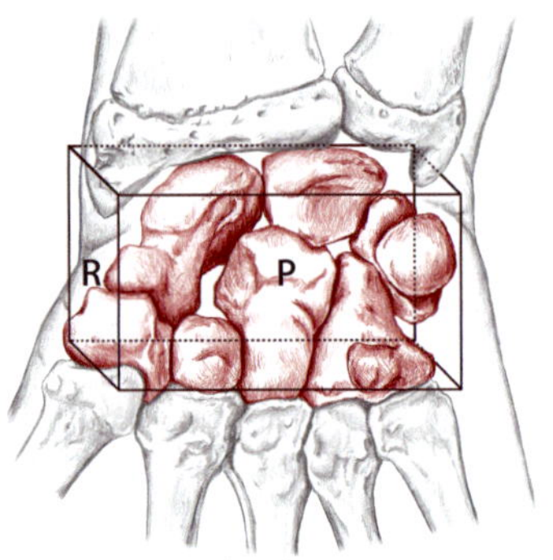

Radial/palmar view
of right wrist

(3.28) Luckily, the wrist has four surfaces from which to explore - the palmar, dorsal, radial and ulnar sides. Some carpals are accessible from one side of the wrist, while others can be explored from a few sides. In any case, exploring all of the wrist's surfaces will allow you to feel at least a portion of each carpal.

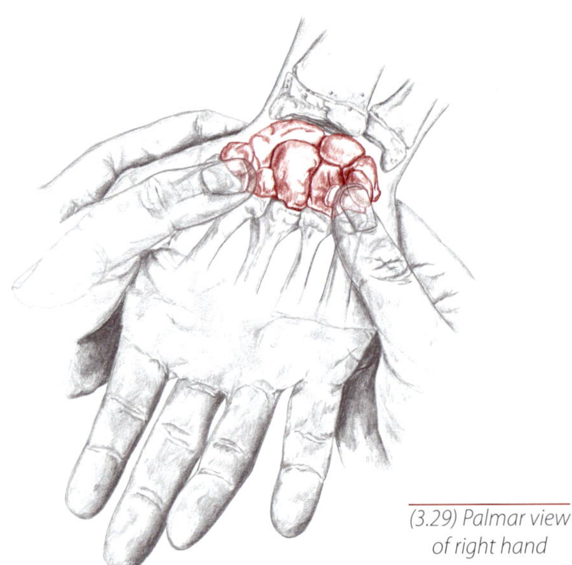

(3.29) Palmar view of right hand

Carpals as a Group

1) Position the hand with the palm up and locate the styloid processes of the ulna and radius.
2) Sliding just distal from the styloid processes, explore the palmar surface for the carpals.
3) Rest your thumbpads on the heel of the hand and then passively move the wrist in all directions (3.29). Note how the carpals shift and undulate slightly like small stones in a pouch. Turn the hand over and explore the dorsal surface (3.30).

Are you distal to the wrist's flexor crease? When the wrist is flexed, can you sense how the carpals press into the palm of the hand? When the wrist is extended, can you feel the carpals shift and become more prominent on the hand's dorsal surface?

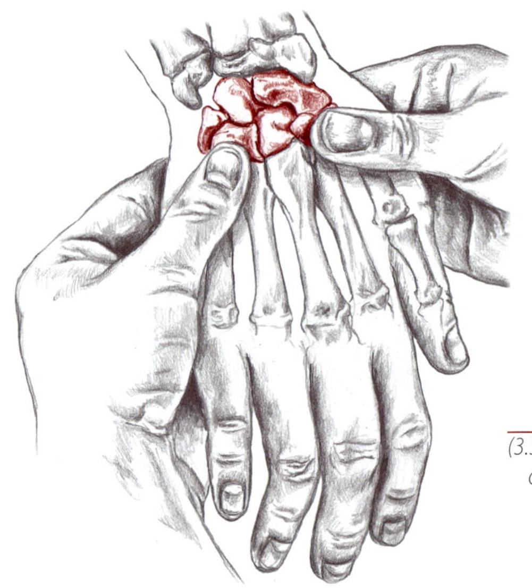

(3.30) Dorsal view of right hand

carpal	**kar**-pul	hamate	**ham**-ate
metacarpal	**met**-a-**kar**-pul	lunate	**lu**-nate
phalange	fa-**lan**-jee	navicular	na-**vik**-u-lar

Pisiform

The knobby pisiform is an attachment site for the flexor carpi ulnaris (p. 148). Protruding along the ulnar/palmar surface of the wrist, the pisiform is just distal to the flexor crease.

1) Locate the flexor crease of your partner's wrist. Then slide over to the "pinky" side of the crease.
2) Move slightly distal to the crease, rolling your thumb-pad in small circles. Explore under the thick tissue of the palm for the nuggetlike pisiform (3.31).

✔ *Passively flex the wrist and notice how the pisiform can be wiggled from side to side (3.32). Extend the wrist and observe how it becomes immobile. (This immobility is due to the tension created by the flexor carpi ulnaris tendon.) Then ask your partner to actively adduct her wrist. Can you feel the tendon of flexor carpi ulnaris as it comes down the medial wrist and attaches to the pisiform?*

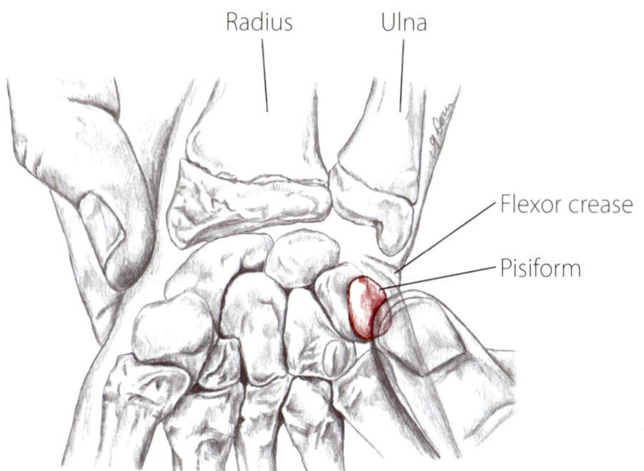

(3.31) Palmar surface of right hand

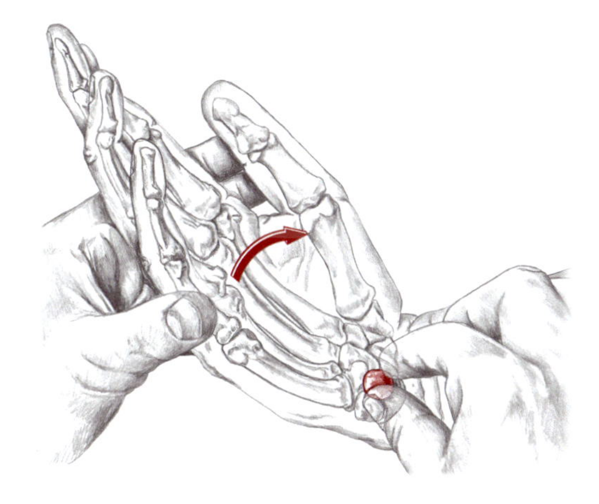

(3.32) Ulnar view of right hand

Triquetrum

This pyramid-shaped bone is located on the dorsal surface of the pisiform, just distal to the styloid process of the ulna. In a neutral position, only the dorsal surface of the triquetrum is palpable; however, abduction shifts the triquetrum so it is accessible on the wrist's ulnar surface.

1) With the palm of your partner's hand facing away from you, locate the styloid process of the ulna. Slide distally, noting a slender ditch, before rising to the surface of the triquetrum (3.33).
2) Keeping your finger stationary, abduct the wrist and note how the triqetrum protrudes to the side (3.34). Adduct and feel the bone disappear back into the wrist.

✔ *During abduction and adduction, do you feel the triquetrum protrude and then disappear? Locate the pisiform on the wrist's palmar surface. Can you locate the triquetrum by beginning at the pisiform and slowly sliding around to the ulnar side of the wrist?*

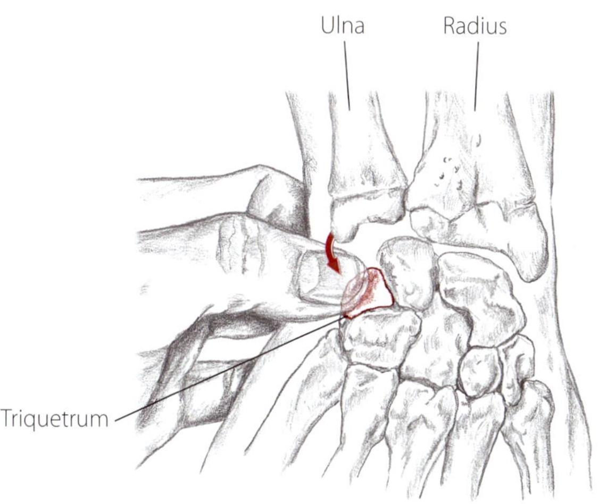

(3.33) Dorsal view of right hand with wrist in neutral, accessing the triquetrum

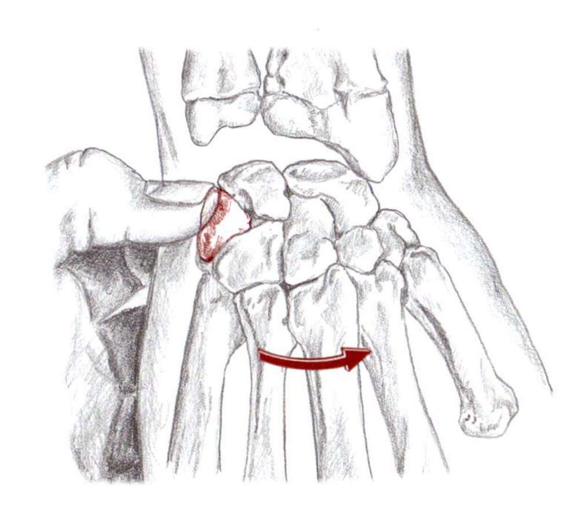

(3.34) Dorsal view of right hand with wrist abducted

pisiform	**pi**-si-form
scaphoid	**skaf**-oyd
styloid	**sti**-loyd
trapezium	tra-**pee**-ze-um
triquetrum	tri-**kwe**-trum

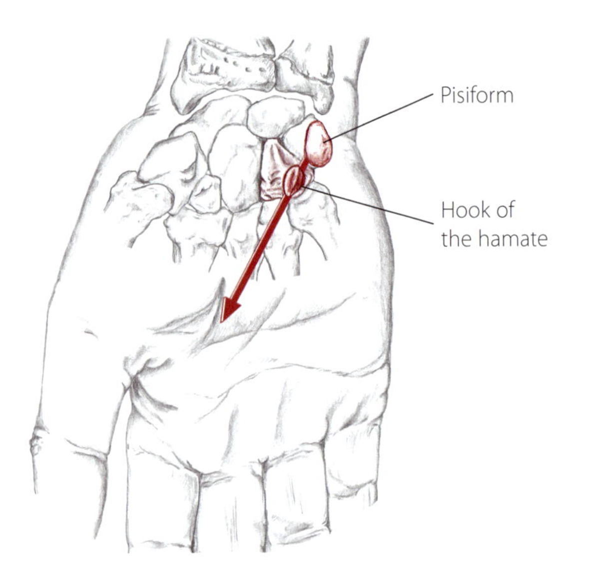

(3.35) Palmar view of right hand

Keeping your thumbpad in place and rolling it gently around the hook will give you the best sense of its shape and locale!

Hamate

Located distal and lateral to the pisiform, the hamate has a small protuberance or "hook" that is palpable on the hand's palmar surface. The pisiform and the hook of the hamate serve as medial attachment sites for the flexor retinaculum, the connective tissue band that forms the "roof" of the carpal tunnel. The flat surface of the hamate's body is accessible on the hand's dorsal surface where the bases of the fourth and fifth metacarpals merge. When palpated, the hook is often tender.

1) Locate your partner's pisiform. Draw an imaginary line from the pisiform to the base of the index finger (3.35).

2) Using your thumbpad, slide off the pisiform along this line. Approximately half of an inch from the pisiform, explore for this subtle mound beneath the padding of the hand (3.36).

Are you between the pisiform and the base of the index finger? Using gentle pressure, can you sense a small ditch between the pisiform and the hook of the hamate?

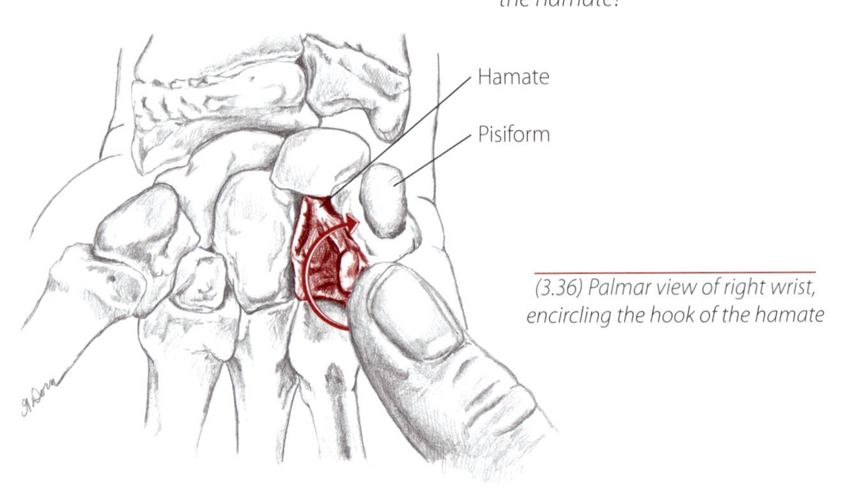

(3.36) Palmar view of right wrist, encircling the hook of the hamate

The pisiform and hook of the hamate form a small channel called the Tunnel of Guyon. The ulnar nerve and artery pass through this canal, under shelter of its roof - the pisohamate ligament (p. 171). The ulnar nerve is particularly vulnerable in the vicinity of the Tunnel of Guyon to compression injuries. Activities such as repeated use of a pneumatic jackhammer or leaning on the handlebars during long-distance bicycling can put chronic pressure on the nerve.

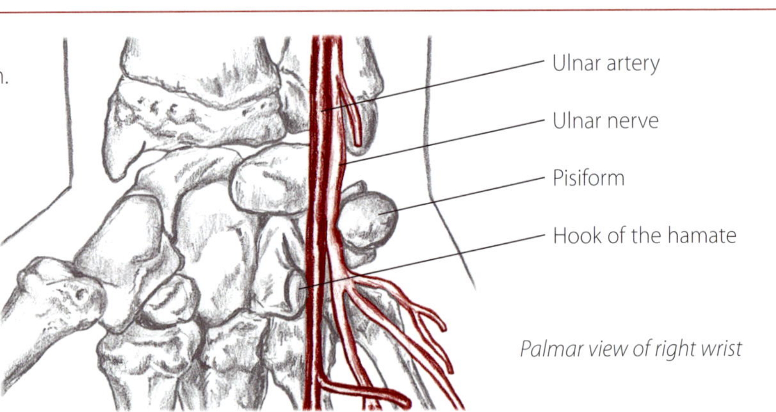

Palmar view of right wrist

Scaphoid

The peanut-shaped scaphoid (or navicular) is the most commonly fractured carpal. It is located on the radial side of the hand, distal to the styloid process of the radius. Although the scaphoid forms the floor of the tendinous "anatomical snuffbox" (p. 162), it is still accessible from the dorsal, palmar and radial sides of the wrist.

1) Beginning on the wrist's radial surface, locate the radius' styloid process. Slide your thumb distally off the process, falling between the superficial tendons and into the natural ditch where the scaphoid will be found (3.37).
2) Maintain your position and passively adduct the wrist. As you do so, feel for the scaphoid to bulge into your thumb (3.38). Now abduct the wrist and feel how the scaphoid disappears back into the wrist.
3) From here, explore the scaphoid's dorsal and palmar surfaces. On the palmar surface, along the flexor crease, is the scaphoid tubercle (p. 132).

Are you distal to the end of the styloid process of the radius? During adduction and abduction, do you feel the scaphoid protrude and then disappear?

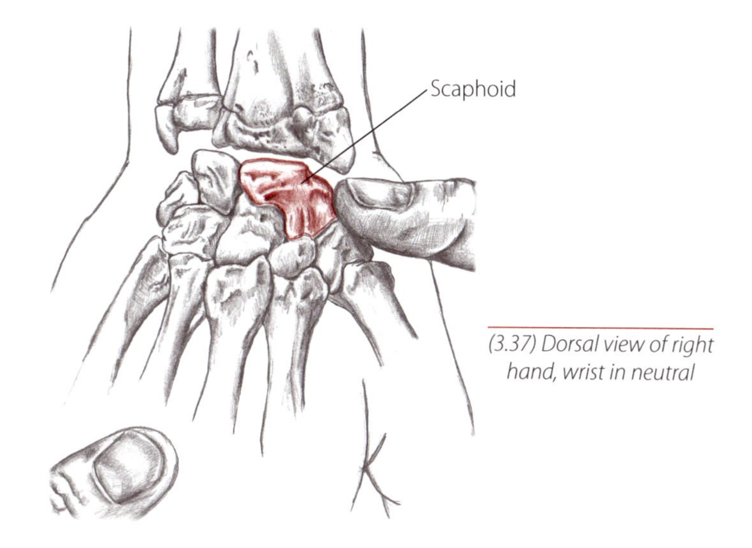

(3.37) Dorsal view of right hand, wrist in neutral

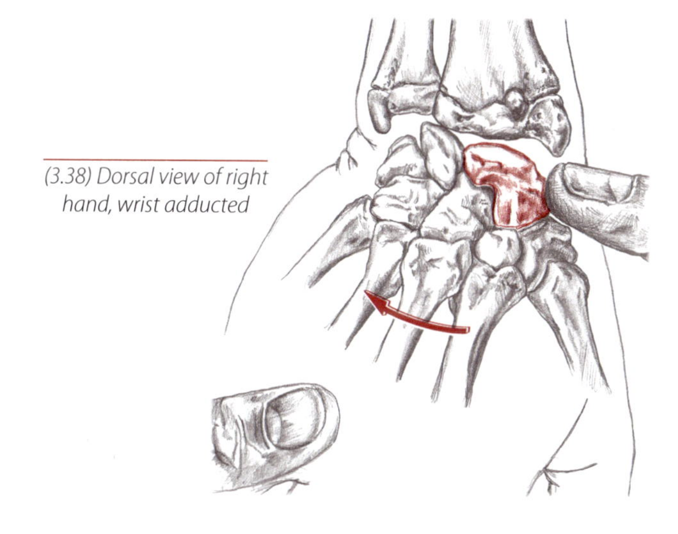

(3.38) Dorsal view of right hand, wrist adducted

Trapezium

Located distal to the scaphoid, the small trapezium articulates with the base of the first metacarpal. This articulation, the first metacarpophalangeal joint, is the source of the thumb's unique movements.

The trapezium is most accessible on its radial or dorsal side, and can be isolated either distally from the scaphoid or proximally from the first metacarpal.

1) Palpating along the hand's radial/dorsal side, locate the scaphoid. Then slide distally (3.39).
2) You may, by accident, pass the trapezium and ride up onto the base of the first metacarpal. If so, simply slide back proximally to the trapezium.

To check if you are indeed feeling the base of the first metacarpal instead of the trapezium, ask your partner to slowly flex and extend his thumb. With this action, the base of the first metacarpal should clearly move. To check for the trapezium: Are you distal to the scaphoid and proximal to the base of the first metacarpal?

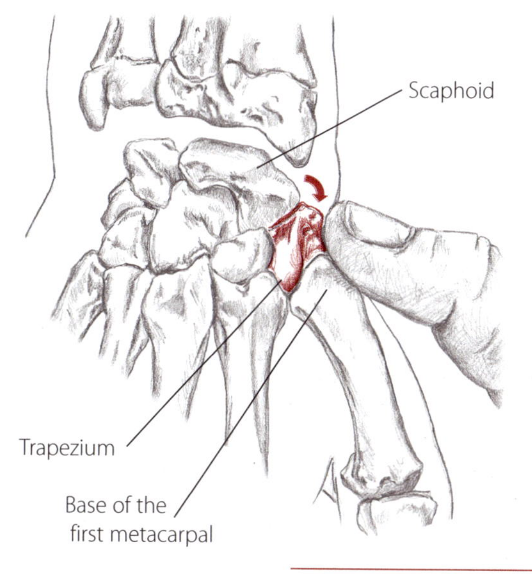

(3.39) Dorsal view of right hand

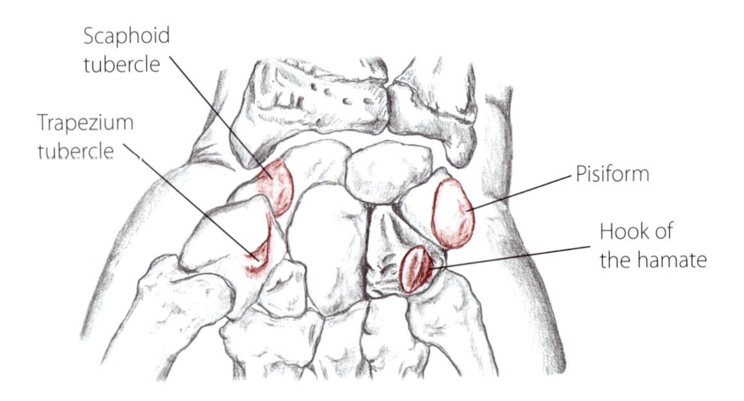

(3.40) Palmar view of right wrist, showing the four attachment sites of the flexor retinaculum

Scaphoid and Trapezium Tubercles

The scaphoid and trapezium tubercles serve as lateral attachment sites for the flexor retinaculum (p. 169), the connective tissue band that forms the "roof" of the carpal tunnel (3.40). Both tubercles are located on the palmar surface of the wrist, near the flexor crease. Oftentimes, the tubercles are situated so closely to each other that they are difficult to distinguish individually. The two, however, are palpable, either separately or together.

1) Locate the radial surface of the scaphoid, along the flexor crease. Then walk your thumb around to the palmar side of the scaphoid.
2) Using your thumbpad, explore just distal to the flexor crease for a prominent, bony knob (3.41).
3) Flex the wrist slightly to soften the surrounding tissue (3.42).

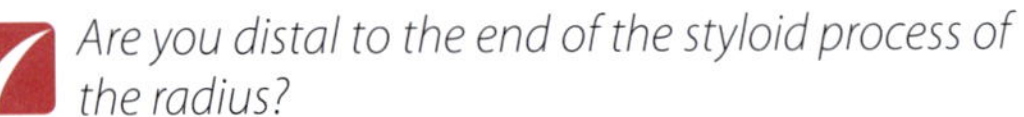

Are you distal to the end of the styloid process of the radius?

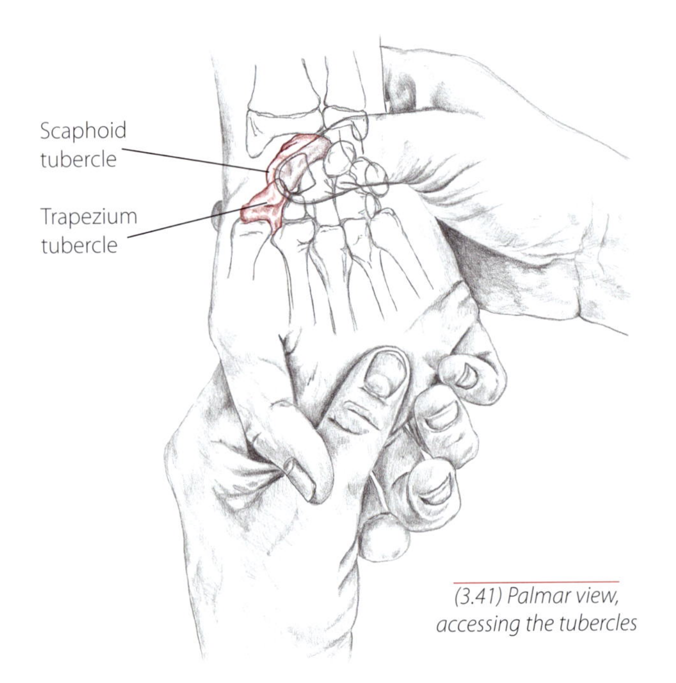

(3.41) Palmar view, accessing the tubercles

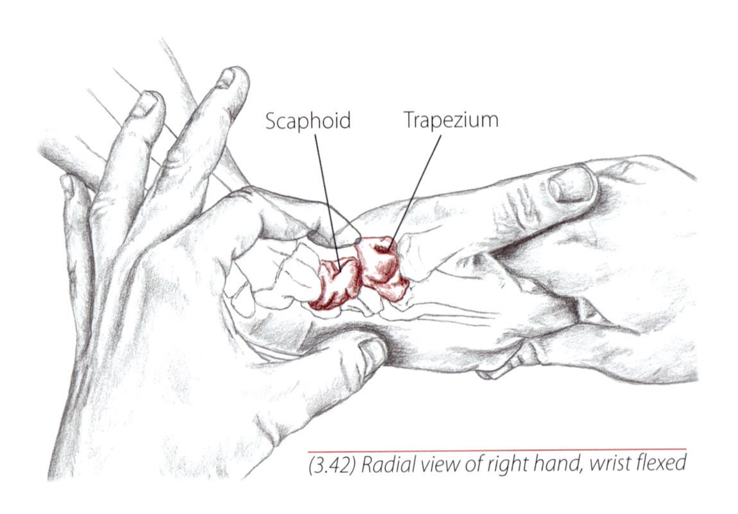

(3.42) Radial view of right hand, wrist flexed

The pisiform is much larger on quadrupeds such as dogs (right), on whom it protrudes posteriorly above the heel of the front paw. This arrangement allows the flexor carpi ulnaris muscle that attaches to the pisiform greater leverage and power to flex the wrist when running on all fours. A human's pisiform is only a pea-sized knob. It is, nevertheless, still useful for kneading bread dough.

Lunate and Capitate

The **lunate** is the most frequently dislocated carpal. Located just distal to Lister's tubercle (p. 126), it is relatively inaccessible when the wrist is in a neutral position; flexing the wrist, however, will slide the lunate to the dorsal surface.

The **capitate** is the largest of the carpals and is located distal to the lunate. It has a shallow ditch on its dorsal surface that can be easily palpated.

Although the lunate and capitate lie deep to the extensor tendons, both carpals are accessible on their dorsal surfaces and can be isolated between Lister's tubercle and the shaft of the third metacarpal (3.43).

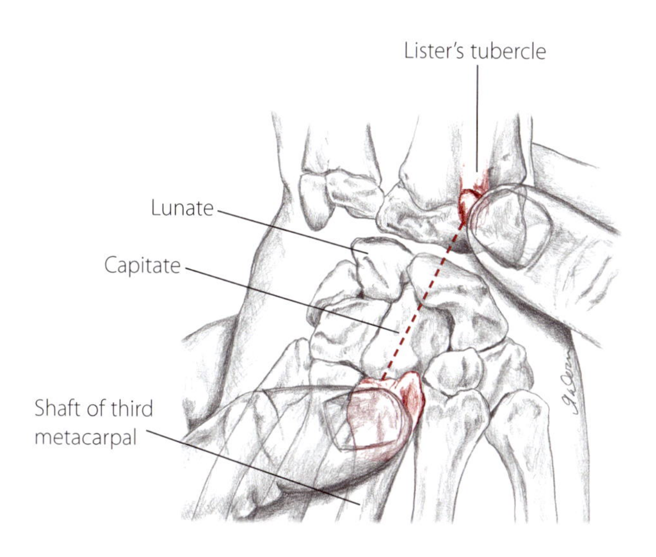

(3.43) Dorsal view of right wrist. With the wrist in neutral, draw an imaginary line between Lister's tubercle and the base of the third metacarpal, marking out the location of the capitate.

1) Locate Lister's tubercle and the base of the third metacarpal. With the wrist slightly extended, lay your thumb between these points and notice how it falls into a small cavity. This is the location of the lunate and capitate (3.44).
2) Set your thumb at the proximal end of this cavity. Then flex the wrist and feel the lunate press into your finger (3.45). Next extend the wrist and feel this carpal disappear back into the wrist.
3) Shift your thumb to the distal end of the cavity and notice how it bumps into the base of the third metacarpal. Passively flex the wrist, noting how the capitate rolls into your finger, "filling" its own cavity.

Are you between Lister's tubercle and the shaft of the third metacarpal? When isolating the lunate are you just distal to the edge of Lister's tubercle? Do you feel a small knob press into your thumb upon flexion?

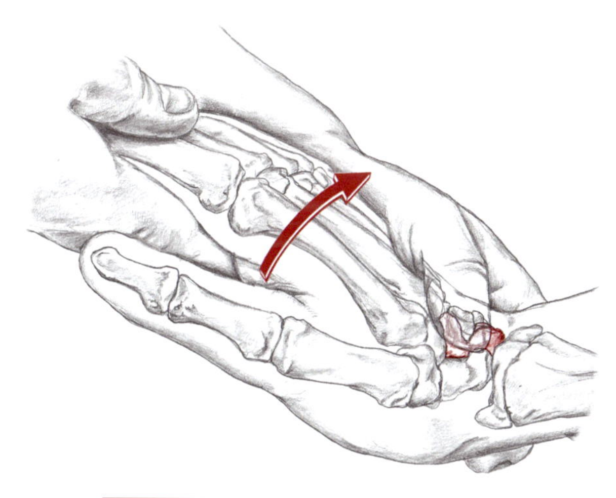

(3.44) Radial view of right hand, wrist extended

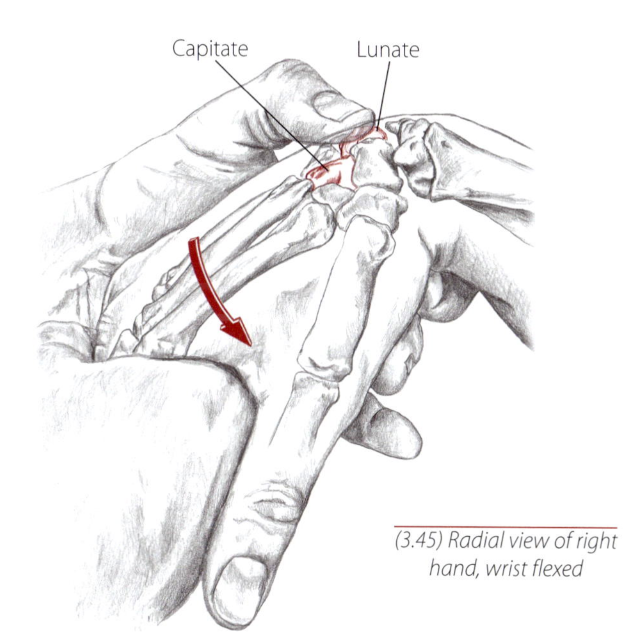

(3.45) Radial view of right hand, wrist flexed

To locate the carpometacarpal joints (right), ask your partner to flex her fingers and wrist. Roughly an inch or two distal to the extensor crease at the wrist will be a series of bumps across the dorsal surface of the hand. These bumps are the bases of the metacarpals which articulate with the carpals to form the carpometacarpal joints.

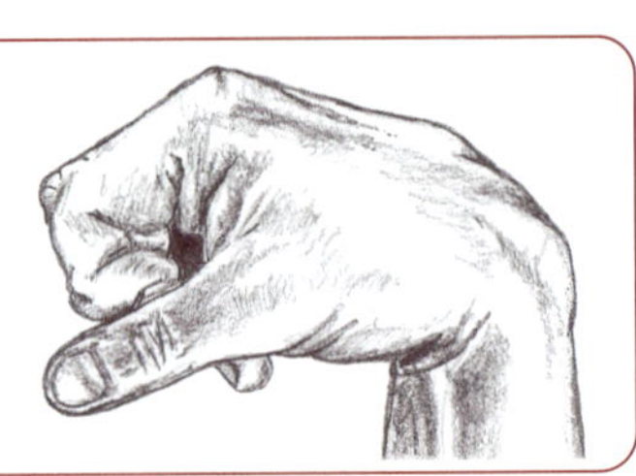

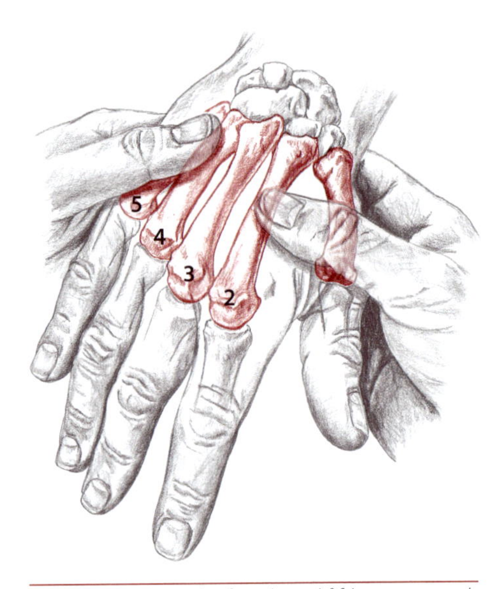

(3.46) Notice how the fourth and fifth metacarpals allow for more movement between them than the second and third metacarpals

Metacarpals and Phalanges

The fingers contain no muscles, only the tendons of the digitorum muscles and strong ligaments which hold together the phalanges of each finger.

1) Palpate the dorsal surface of your partner's hand and feel the superficial metacarpal shafts. Explore the space between the metacarpals for the interossei muscles (p. 163). Then gently undulate the metacarpals up and down (3.46).
2) Turn the hand over, and explore the metacarpals and phalanges from the palmar surface, noting how they lie deep to the tissues of the palm (3.47).
3) Move distally and explore where the heads of the metacarpals join with the phalanges to form the large metacarpophalangeal "knuckle" joints (3.48). Passively flex a metacarpophalangeal joint and distinguish between the head of the metacarpal and the base of the proximal phalange.
4) Move distally to the phalanges and isolate the slender tendons, ligaments and connective tissue of the fingers. Also note the absence of any muscle tissue.

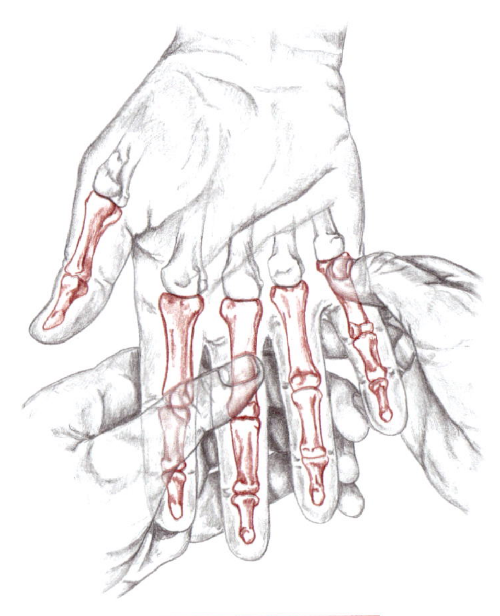

(3.47) Palmar view of right hand

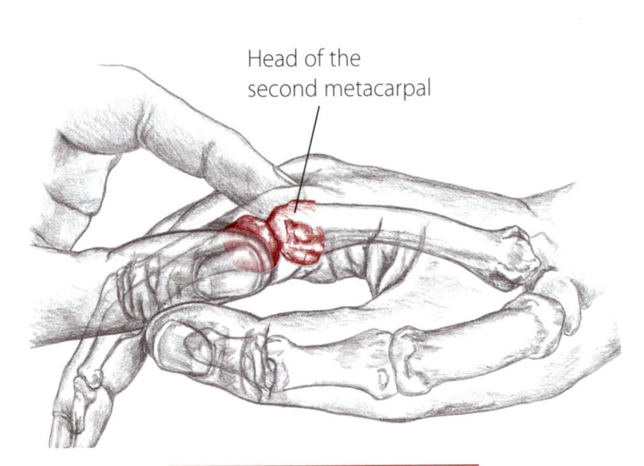

(3.48) Radial view of second finger

metacarpophalangeal **met**-a-**kar**-po-**fa**-lan-**jee**-al

Muscles of the Forearm and Hand

The muscles of the forearm primarily create movement at the wrist and fingers. Many have small, fusiform bellies that connect to space-efficient tendons in the lower forearm. These tendons extend distally into the wrist and hand. The crowded muscle bellies and tendons of the forearm can be challenging to isolate. To simplify matters, the muscles of this chapter have been arranged into four primary groups:

a) Muscles that act primarily at the **elbow**:
Brachialis
Brachioradialis

b) Muscles that move the **wrist and/or fingers** (*carpi*, *digitorum* or *palmaris* muscles). This group can be further subdivided into four smaller groups:
Extensors of the wrist and fingers
Flexors of the wrist and fingers
Adductors of the wrist
Abductors of the wrist

(Some muscles that act upon the wrist can actually move it in two directions. Flexor carpi ulnaris, for example, both flexes and adducts the wrist.)

c) Muscles that create the pivoting action between the **radius and ulna**:
Pronator teres
Pronator quadratus
Supinator

d) Short and long muscles that maneuver the **thumb** (*pollicis* muscles).

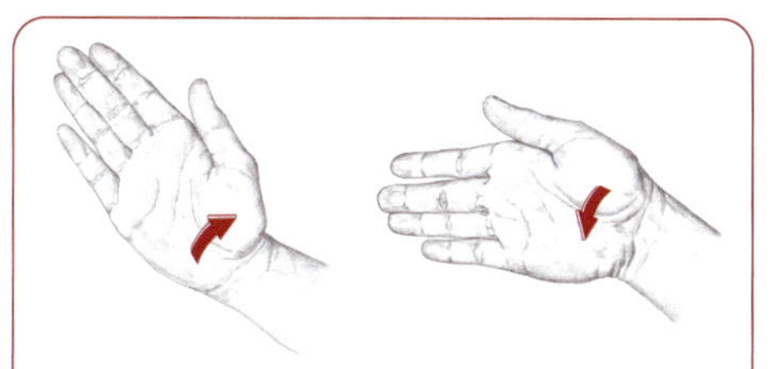

Distinguishing between abduction (left) and adduction (right) of the wrist can be confusing if the forearm is pronated or supinated. For instance, if you pronate your forearm (palm toward the floor) and then adduct at the wrist, your hand will move away from the centerline of your body and it will appear as though you have abducted your wrist. Not so: Both adduction and abduction remain the same regardless of the forearm's position.

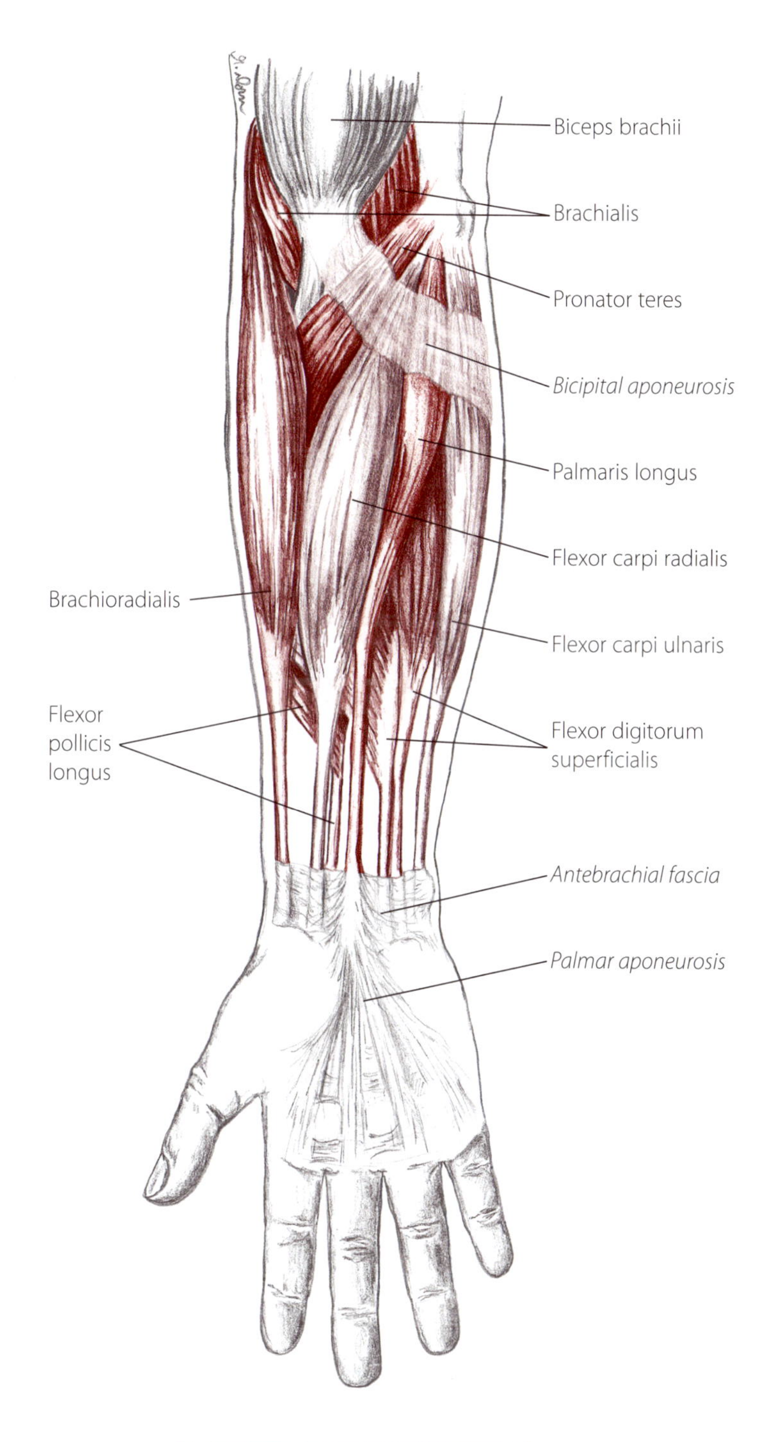

(3.49) Anterior view of right forearm and hand, skin removed from palm of hand

Muscles of the Forearm and Hand

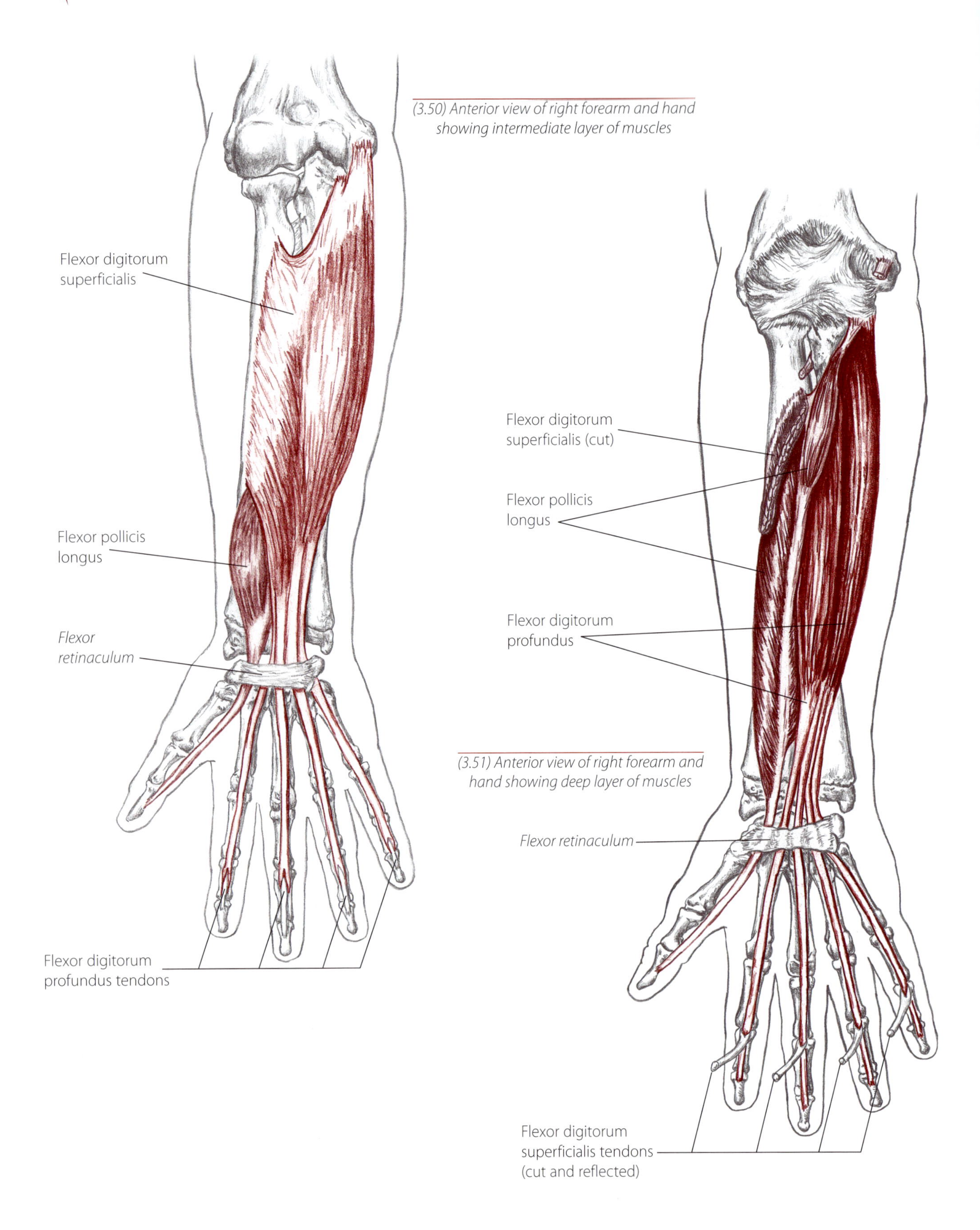

(3.50) Anterior view of right forearm and hand showing intermediate layer of muscles

(3.51) Anterior view of right forearm and hand showing deep layer of muscles

Muscles of the Forearm and Hand

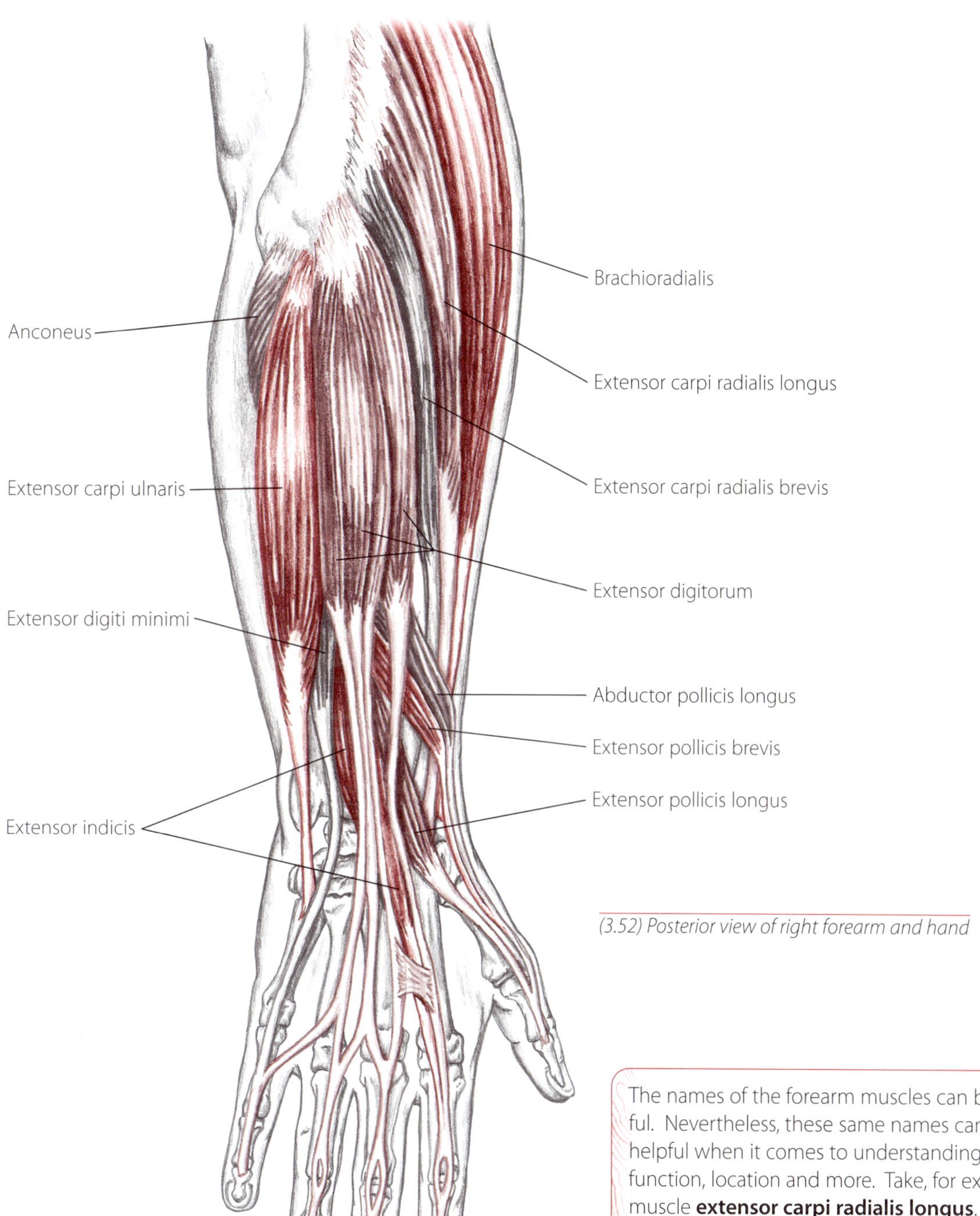

(3.52) Posterior view of right forearm and hand

The names of the forearm muscles can be a mouthful. Nevertheless, these same names can be very helpful when it comes to understanding a muscle's function, location and more. Take, for example, the muscle **extensor carpi radialis longus**. What does its name reveal?

1) It is specified as an **extensor**, so it extends. This also indicates that there is a **flexor** carpi radialis.
2) "Carpi" means it extends the **carpals** (wrist joint). This indicates there is also a different muscle that moves the **digits** - extensor digitorum.
3) It runs along the **radial** side of the forearm. This indicates that there is also an extensor carpi **ulnaris** on the ulnar side.
4) If there is a **longus**, there must also be a **brevis** - extensor carpi radialis brevis.

Synergists - Muscles Working Together

*muscles not shown

Elbow

(humeroulnar and humeroradial joints)

Flexion
Biceps brachii
Brachialis
Brachioradialis
Flexor carpi radialis
Flexor carpi ulnaris (assists)
Palmaris longus
Pronator teres (assists)
Extensor carpi radialis longus (assists)*
Extensor carpi radialis brevis (assists)*

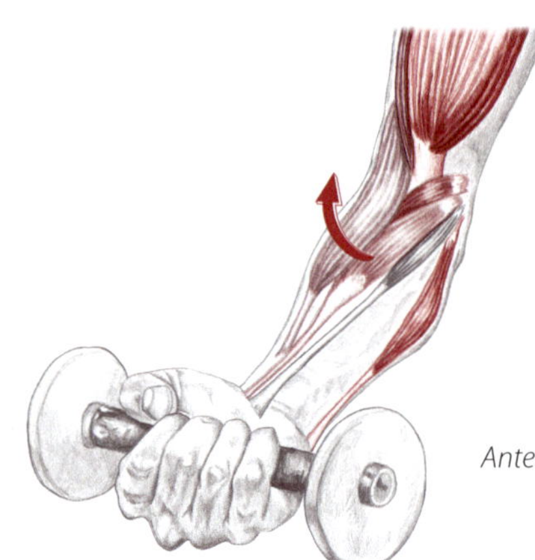

Anterior/medial view

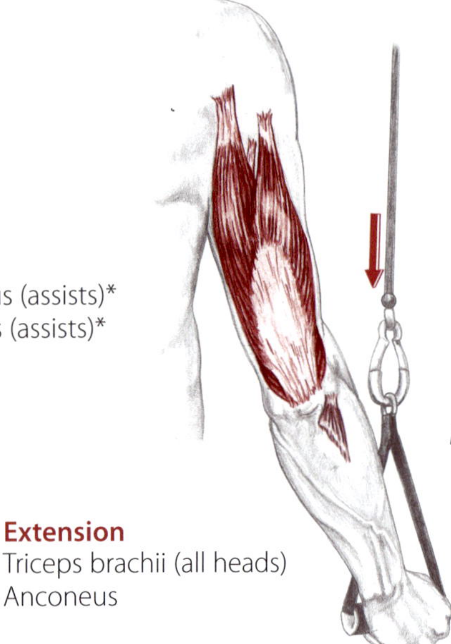

Posterior view

Extension
Triceps brachii (all heads)
Anconeus

Forearm

(proximal and distal radioulnar joints)

Supination
Biceps brachii
Supinator
Brachioradialis (assists)

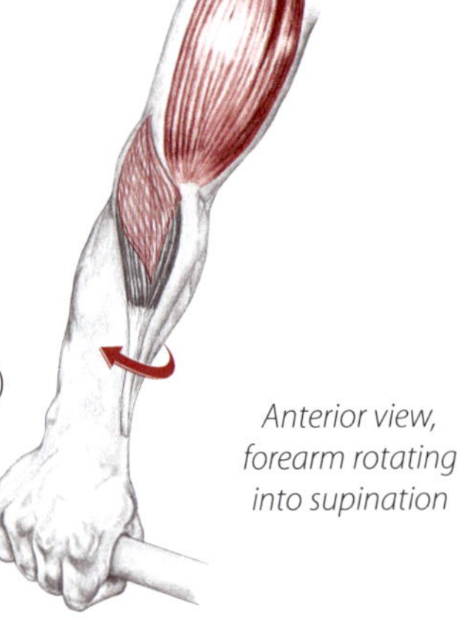

Anterior view, forearm rotating into supination

Pronation
Pronator teres
Pronator quadratus
Brachioradialis (assists)

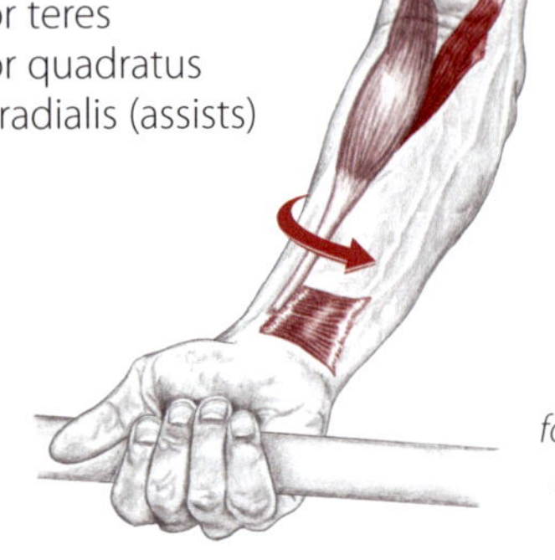

Anterior view, forearm rotating into pronation

Wrist

(radiocarpal joint)

Extension
Extensor carpi radialis longus
Extensor carpi radialis brevis
Extensor carpi ulnaris
Extensor digitorum (assists)

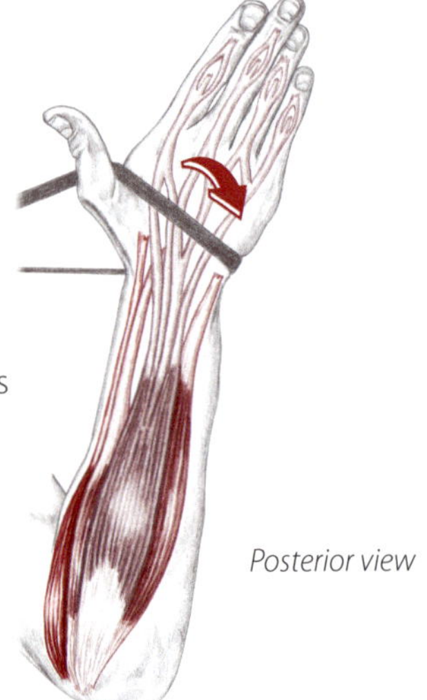

Posterior view

Flexion
Flexor carpi radialis
Flexor carpi ulnaris
Palmaris longus
Flexor digitorum superficialis
Flexor digitorum profundus (assists)*

Anterior/medial view

Wrist

(radiocarpal joint)

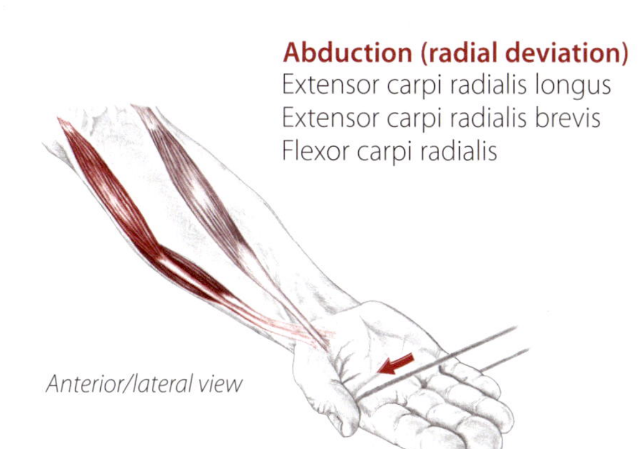

Abduction (radial deviation)
Extensor carpi radialis longus
Extensor carpi radialis brevis
Flexor carpi radialis

Adduction (ulnar deviation)
Extensor carpi ulnaris
Flexor carpi ulnaris

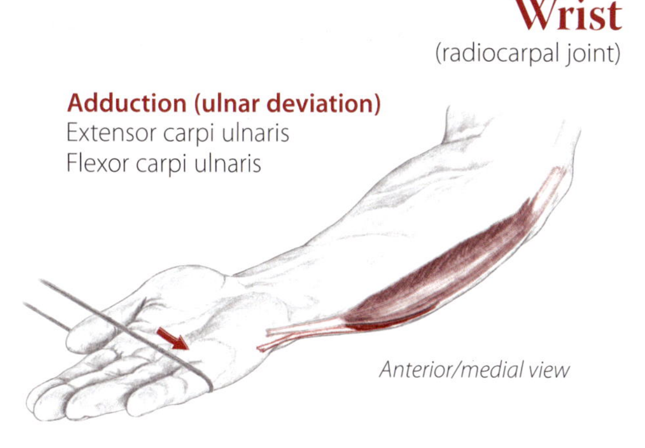

Hand and Fingers

(metacarpophalangeal, proximal and distal interphalangeal joints)

Extension of the second through fifth fingers
Extensor digitorum
Extensor indicis
(2nd)*
Lumbricals
Dorsal interossei
(2nd - 4th) (assists)
Palmar interossei
(2nd, 4th, 5th) (assists)

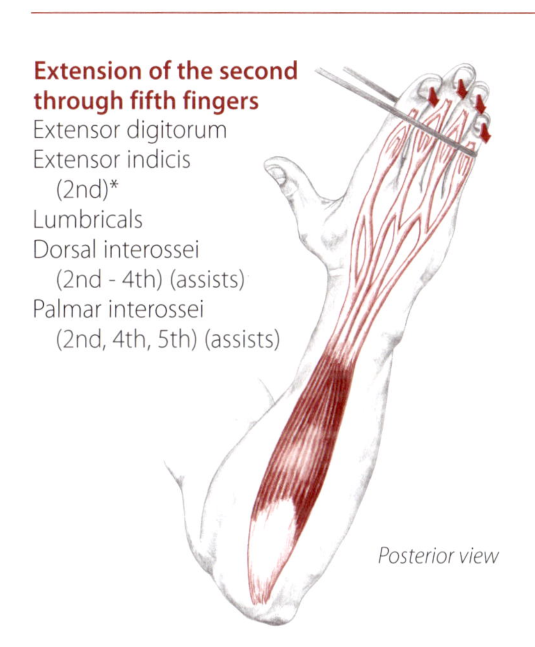

Flexion of the second through fifth fingers
Flexor digitorum superficialis
Flexor digitorum profundus
Flexor digiti minimi brevis (5th)*
Lumbricals
Dorsal interossei (2nd - 4th) (assists)
Palmar interossei (2nd, 4th, 5th) (assists)

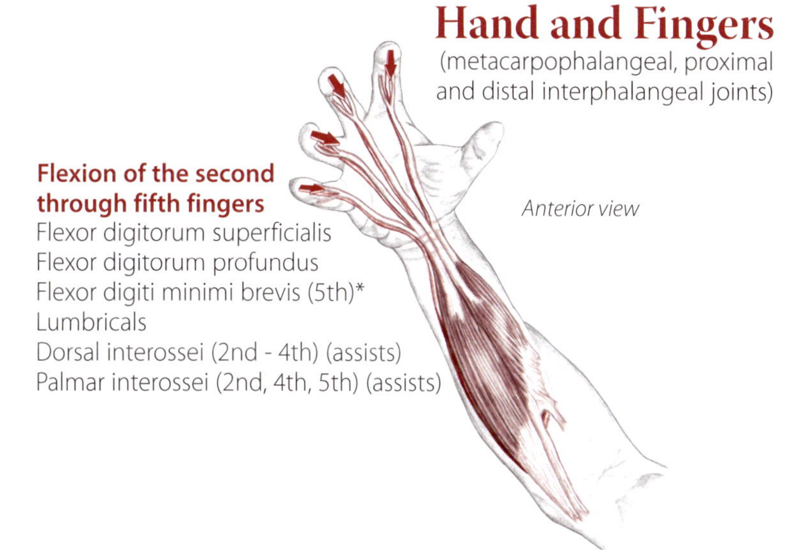

Thumb

(first carpometacarpal and metacarpophalangeal joints)

Opposition
Opponens pollicis
Flexor pollicis brevis (assists)*
Abductor pollicis brevis (assists)*

Flexion
Flexor pollicis longus
Flexor pollicis brevis*
Adductor pollicis (assists)
Palmar interossei (1st) (assists)*

Extension
Extensor pollicis longus
Extensor pollicis brevis
Abductor pollicis longus
Palmar interossei (1st) (assists)

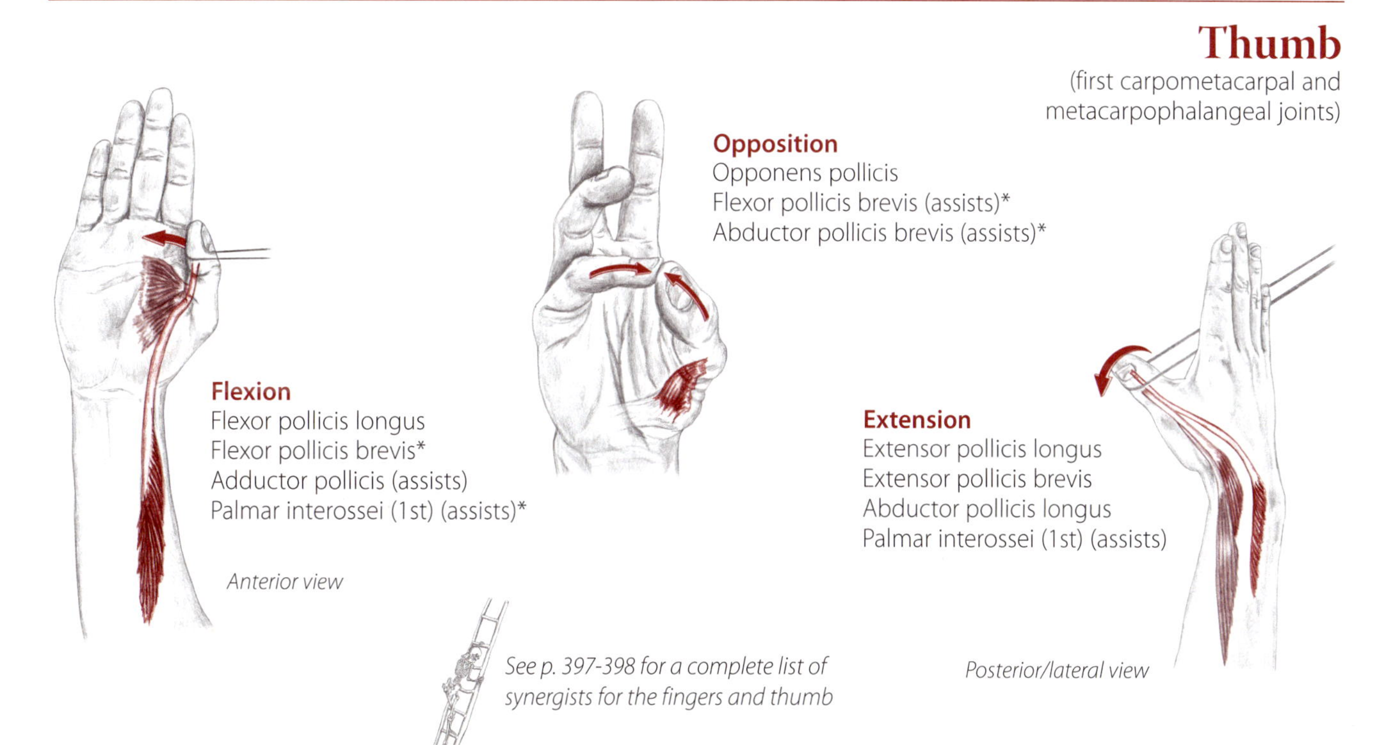

See p. 397-398 for a complete list of synergists for the fingers and thumb

Brachialis

The brachialis is a strong elbow flexor that lies deep to the biceps brachii (p. 103) on the anterior arm. It has a flat yet thick belly (3.53). Ironically, however, the brachialis' girth only helps the biceps to bulge further from the arm, making brachialis the biceps' best friend.

Although it lies underneath the biceps, portions of brachialis are accessible. Its lateral edge, sandwiched between the biceps and triceps brachii, is superficial and palpable. The distal aspect of the brachialis is also accessible as it passes along either side of the biceps tendon.

Deltoid (cut)

Biceps brachii tendon (cut)

(3.53) Anterior view of right arm showing brachialis

O

I

(3.54) Origin and insertion

A Flex the elbow (humeroulnar joint)

O Distal half of anterior surface of humerus

I Tuberosity and coronoid process of ulna

N Musculocutaneous

1) Shake hands with your partner and flex the elbow to 90˚. It is important to distinguish the muscle tissue of the biceps brachii from that of the brachialis. Ask your partner to flex her elbow against your resistance and isolate the edges of the round biceps brachii belly.
2) With the arm relaxed, slide laterally half an inch off the distal biceps. The edge of the brachialis can be detected by rolling your fingers across its surface. As you strum across its solid edge, you will feel a pronounced "thump." (3.55)
3) Continuing to strum across its edge, follow it distally to where it disappears into the elbow.
4) Locate the distal biceps tendon. Palpate along either side of the tendon for portions of the deeper brachialis (3.56).

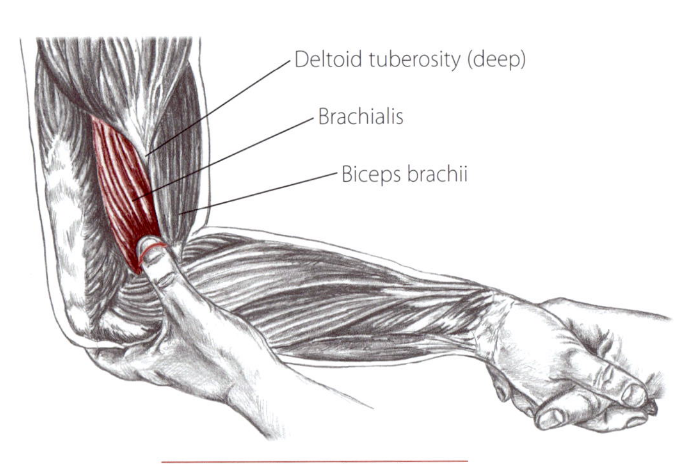

(3.55) Lateral view of right forearm, strumming across the edge of brachialis

Can you roll across a distinct wad of muscle on the lateral side of the arm? Can you follow it distally toward the inner elbow? Locate the triceps and biceps brachii. Are the brachialis fibers between them on the lateral arm?

Locate the deltoid tuberosity. Slide distally straight down the lateral side of the arm and explore for the edge of the brachialis.

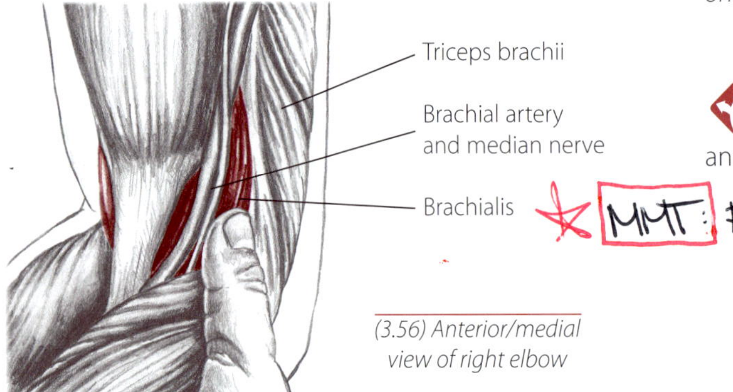

(3.56) Anterior/medial view of right elbow

★ MMT: Elbow flx. in pronation

brachial	**bray**-key-al	L. relating to the arm
brachialis	**bray**-key-**al**-is	

Brachioradialis

The brachioradialis is superficial on the lateral side of the forearm. It has a long, oval belly which forms a helpful dividing line between the flexors and extensors of the wrist and hand. Its muscle belly becomes tendinous halfway down the forearm. It is the only muscle that runs the length of the forearm but does not cross the wrist joint (3.57). Resisted flexion of the elbow causes brachioradialis to visibly protrude on the forearm and become readily palpable.

A Flex the elbow (humeroulnar joint)

Assist to pronate and supinate the forearm when these movements are resisted

O Lateral supracondylar ridge of humerus

I Styloid process of radius

N Radial

1) Shake hands with your partner and flex the elbow to 90°. With the forearm in a neutral position (thumb toward the ceiling), ask your partner to flex her elbow against your resistance.
2) Look for the brachioradialis bulging out on the lateral side of the elbow. If it is not visible, locate the lateral supracondylar ridge of the humerus and slide distally.
3) With your partner still contracting, use your other hand to palpate its superficial, tubular belly (3.60). Try to pinch its belly between your fingers and follow it as far distally as possible. As it becomes more tendinous, strum across its distal tendon toward the styloid process of the radius.

Upon resisted flexion of the elbow, does the belly you are palpating contract and bulge out? Is it superficial? Does it extend off the lateral supracondylar ridge of the humerus?

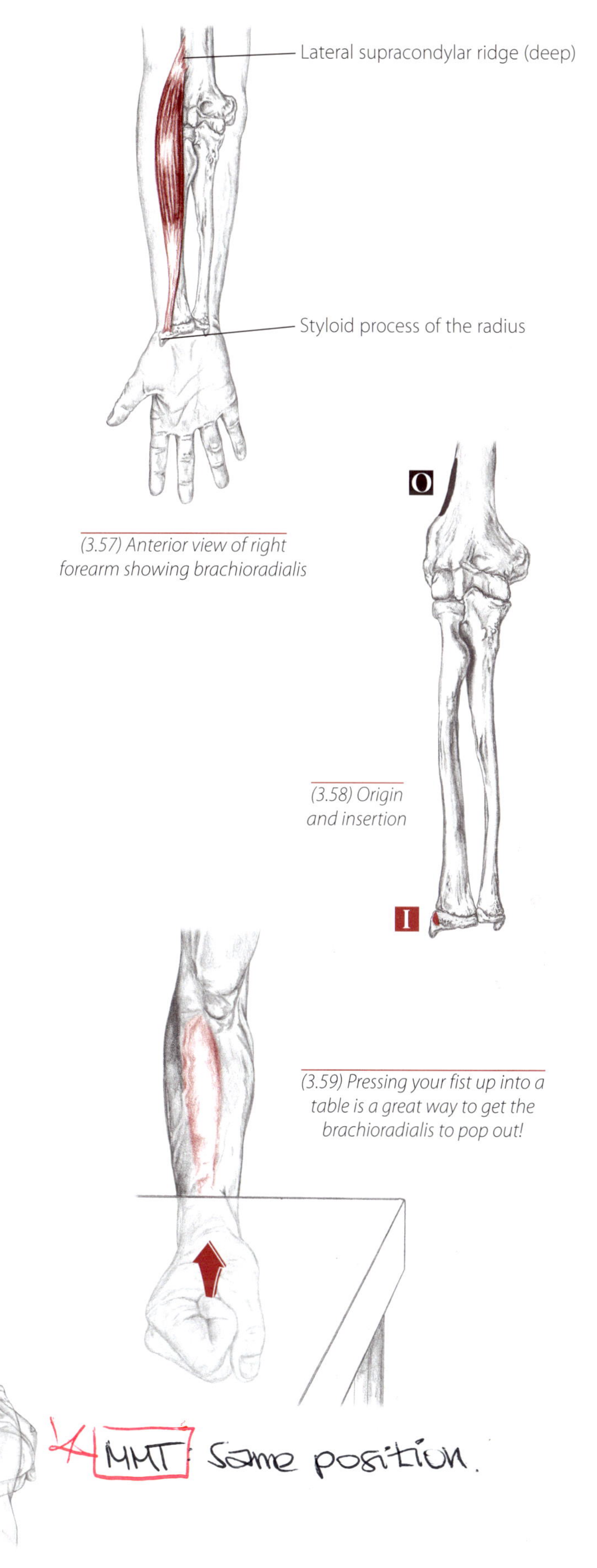

(3.57) Anterior view of right forearm showing brachioradialis

(3.58) Origin and insertion

(3.59) Pressing your fist up into a table is a great way to get the brachioradialis to pop out!

(3.60) Anterior/lateral view of right forearm, partner flexing her elbow against your resistance

(3.61) *Cross section of the right forearm, diagonal line dividing the flexors and extensors*

Distinguishing Between the Flexor and Extensor Groups of the Forearm

Before we isolate specific flexors and extensors, let us first determine the location of these two muscle groups. The flexors and extensors of the hand and wrist are located in the forearm. In anatomical position, the flexors are located on the anterior/medial (hairless) side of the forearm, while the extensors are positioned on the posterior/lateral (hairy) side.

The brachioradialis and shaft of the ulna can be used as clear dividing lines between these muscle groups (3.61). Both of these structures run superficially down the opposite sides of the forearm, separating the flexors and extensors.

1) Shake hands with your partner and flex the elbow to 90°. Locate the brachioradialis and shaft of the ulna (p. 124). Palpate the length of these structures, observing how they divide the forearm into two halves.
2) Move medially from the shaft of the ulna onto the flexors of the forearm. Explore this half of the forearm, noting the girth of these muscles.
3) Ask your partner to slightly flex her wrist against your resistance (3.62). Note the contraction of the flexors.
4) Move to the lateral side of the shaft of the ulna and explore the extensor bellies (3.63). Notice how they are smaller and more sinewy than the flexor bellies. Ask your partner to extend her wrist against your resistance, feeling the extensors contract.

When your partner curls (flexes) her wrist, do the muscles on the hairless side of the forearm contract? Do the extensors contract when the hand moves in the opposite direction (extension)?

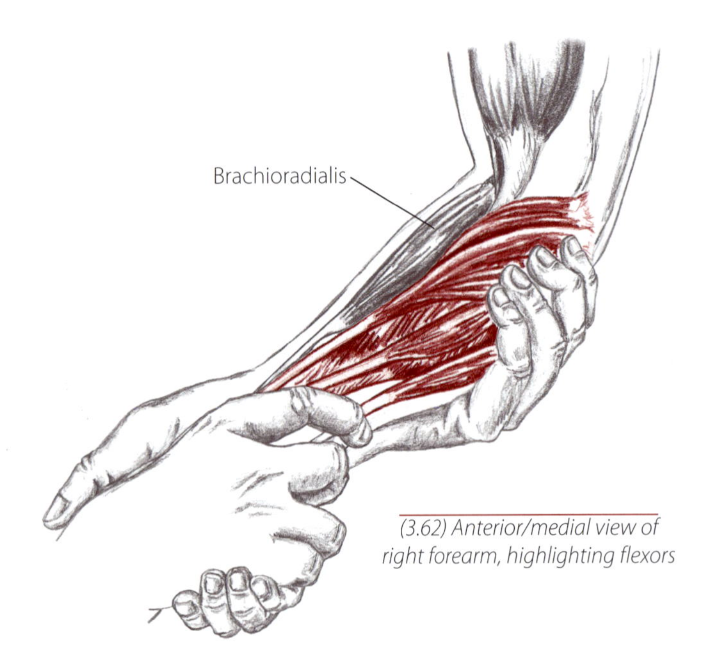

(3.62) *Anterior/medial view of right forearm, highlighting flexors*

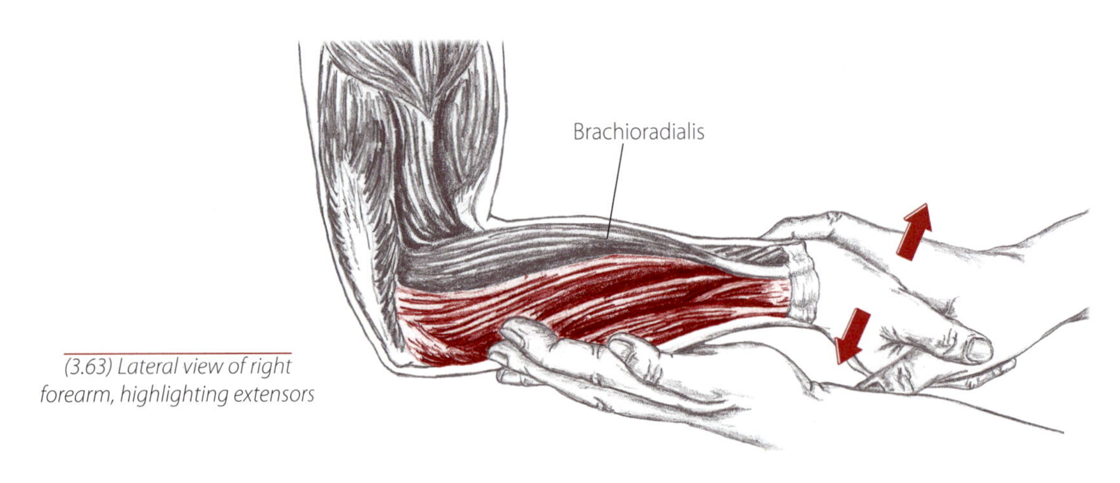

(3.63) *Lateral view of right forearm, highlighting extensors*

Extensors of the Wrist and Hand

Extensor Carpi Radialis Longus
Extensor Carpi Radialis Brevis
Extensor Carpi Ulnaris
Extensor Digitorum

The four extensors create extension primarily at the wrist and fingers. They are situated between the brachioradialis and the shaft of the ulna along the forearm's lateral, posterior surface. All of these muscles are superficial and accessible, though challenging to truly isolate. Originating on the lateral side of the humerus, the bellies of the extensors become tendinous approximately two inches proximal to the wrist joint (3.64). As a group they are smaller and more sinewy than the forearm flexors.

Extensor carpi radialis longus and brevis (discussed here as one muscle) are lateral/posterior to the brachioradialis. **Extensor carpi ulnaris**, as its name suggests, lies beside the ulnar shaft. **Extensor digitorum** is located between these muscles and has four long, superficial tendons stretching along the dorsal surface of the hand and fingers.

(3.64) Posterior view of right forearm

Seated:

MMT: Elbow Ext'd
Resist ext. towards rad. side.
A: Extends + rad. dev. wrist.
O: Dist lat. supracondylar ridge.
I: Base of 2nd MC

MMT: Elbow flxd
Resist wrist ext towards rad. side
A: Extends + radially dev's wrist.
O: Common Ext. tendon.

MMT: Elbow Ext'd
Resist ext. towards ulnar side.
A: Extends + ulnar dev.

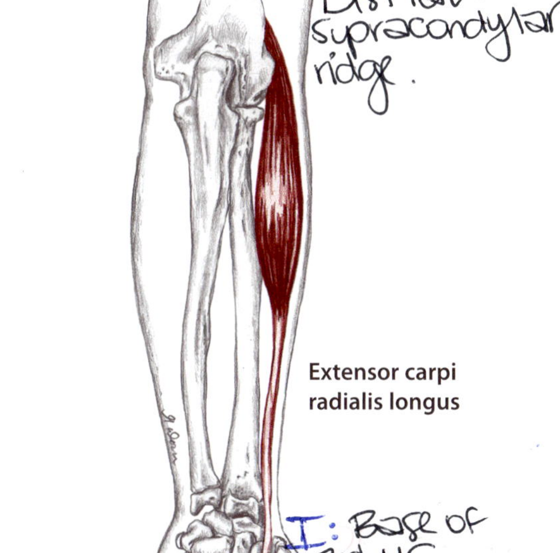

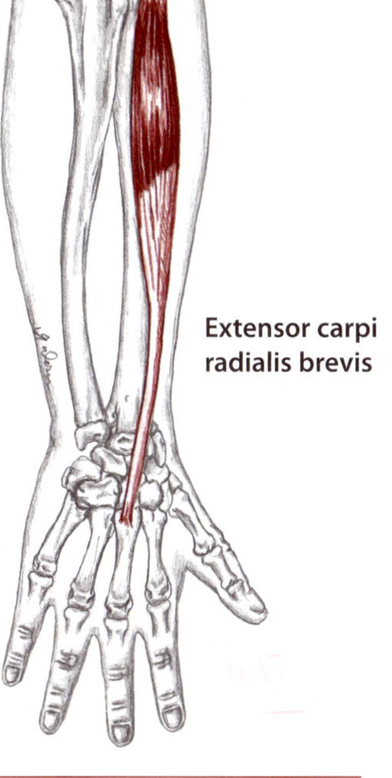

(3.65, 3.66, 3.67) Posterior views of right forearm

brevis	**breh**-vis	L. short
carpi	**kar**-pi	L. of the wrist
digit	**di**-jit	L. finger

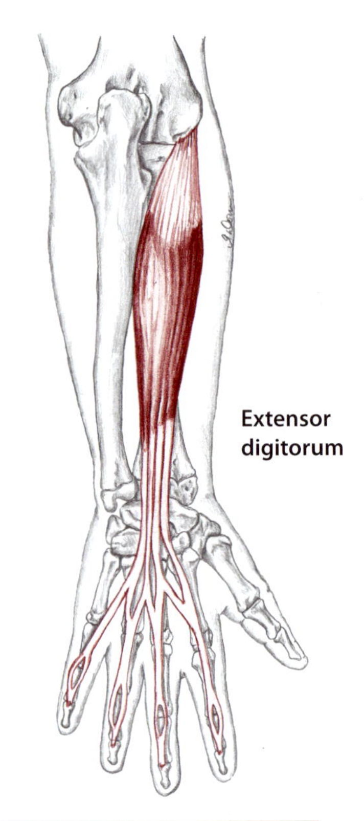

(3.68) Posterior view of right forearm

Extensor Carpi Radialis Longus and Brevis

A Extend the wrist (radiocarpal joint)

Abduct the wrist (radiocarpal joint)

Assist to flex the elbow (humeroulnar joint)

O Lateral supracondylar ridge of humerus

I *Longus:*
Base of second metacarpal
Brevis:
Base of third metacarpal

N Radial

Extensor Carpi Ulnaris

A Extend the wrist (radiocarpal joint)

Adduct the wrist (radiocarpal joint)

O Common extensor tendon from the lateral epicondyle of humerus

I Base of fifth metacarpal

N Radial

Extensor Digitorum

A Extend the second through fifth fingers (metacarpophalangeal and interphalangeal joints)

Assist to extend the wrist (radiocarpal joint)

O Common extensor tendon from the lateral epicondyle of humerus

I Middle and distal phalanges of second through fifth fingers

N Radial

MMT: Forearm pronated. Resist ext. against prox. phalanges.
A: Ext's MCP + IP jnts 2-5 assists wrist ext.

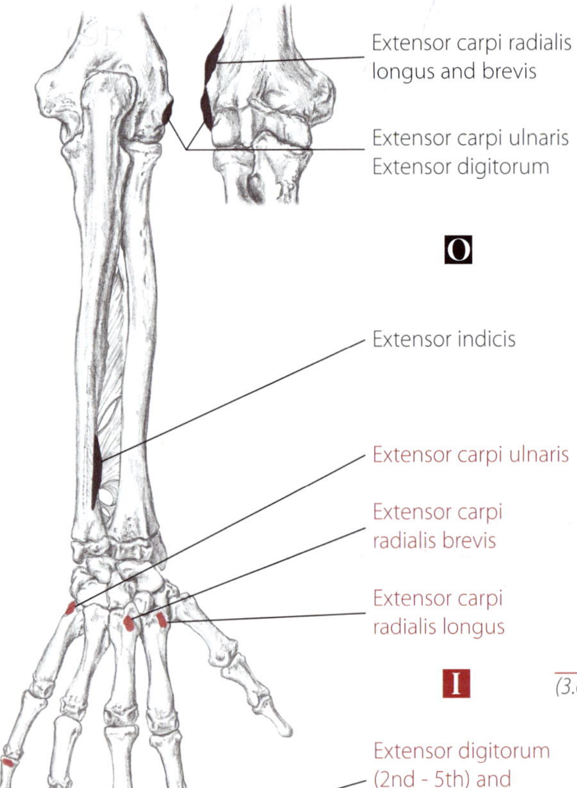

(3.69) Origins and insertions of extensors

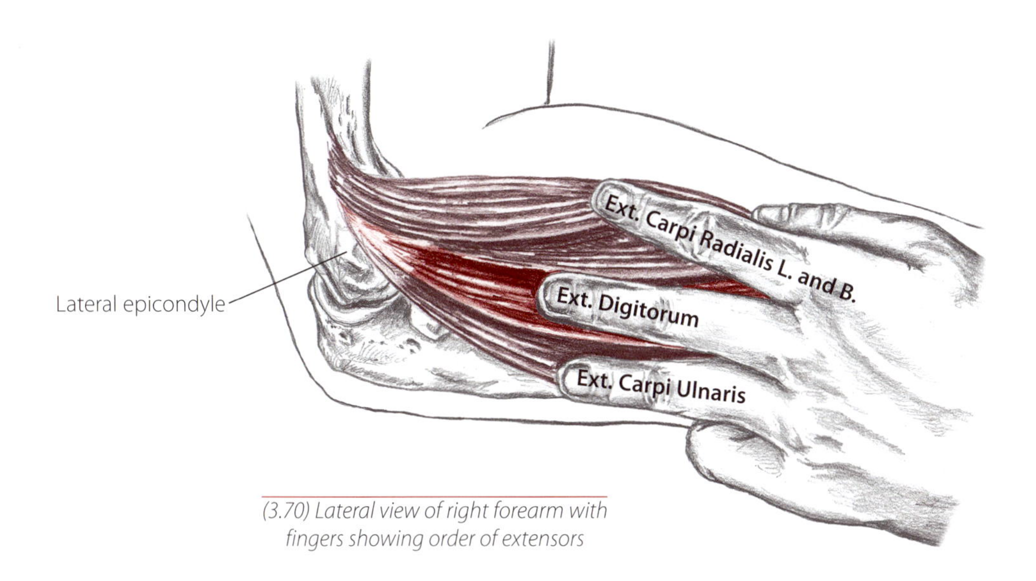

(3.70) Lateral view of right forearm with fingers showing order of extensors

Extensor group

1) Shake hands and flex the elbow to 90˚. Locate the brachioradialis and shaft of the ulna.
2) Lay the flat of your hand between these landmarks and ask your partner to alternately extend and relax her wrist against your resistance (see p. 142).
3) Explore the slender, sinewy fibers of these muscles and note how they contract upon extension. Access their origin at the lateral epicondyle (3.71).

Are you between the brachioradialis and ulnar shaft? Do the muscles contract on extension of the wrist?

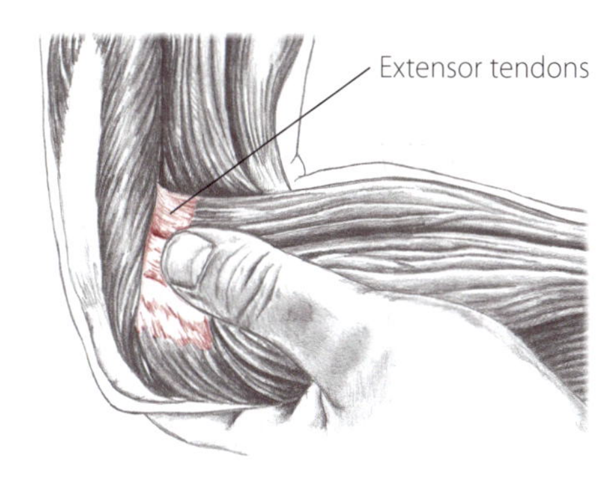

(3.71) Lateral view of right elbow, palpating common tendon of the extensors at lateral epicondyle

Extensor carpi radialis longus and brevis

1) Shake hands and flex the elbow to 90˚. Locate the brachioradialis. Slide laterally off its belly onto the extensor carpi radialis fibers.
2) Ask your partner to alternately abduct and relax her wrist against your resistance. Sense how the fibers tighten with this movement (3.72).
3) Follow their muscle fibers distally as far as possible to where they become tendinous.

Differentiate between the extensor carpi radialis muscles and brachioradialis by asking your partner to alternately abduct and relax her wrist against your resistance. The brachioradialis, which does not cross the wrist joint, will remain slack throughout this action while the extensor carpi radialis muscles will contract.

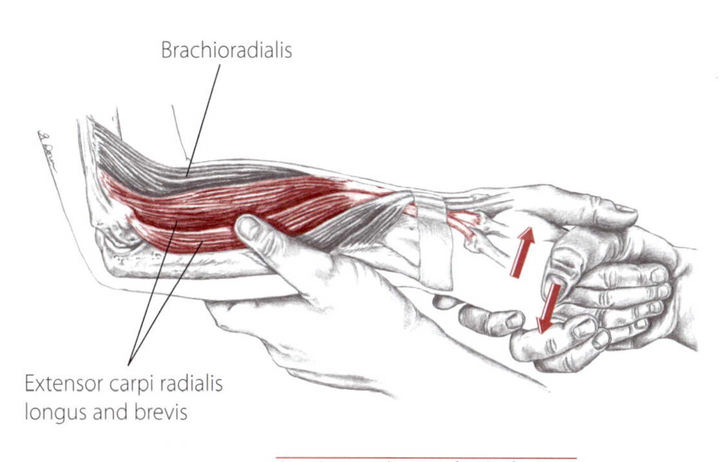

(3.72) Lateral view of right forearm

Extensor digitorum

1) Shake hands with your partner and flex his elbow to 90°. Slide laterally off the extensor carpi radialis fibers.
2) As you move around the forearm, palpate the digitorum's flat surface and roll across its fibers.
3) Isolate its belly by asking your partner to extend her wrist and fingers. Follow the belly distally as it transforms into tendons. The tendons will be palpable as they pass beneath the extensor retinaculum and continue along the dorsal surface of the hand.

Ask your partner to wiggle her fingers as if she were typing (3.73). Do you feel an undulating contraction of the extensor digitorum?

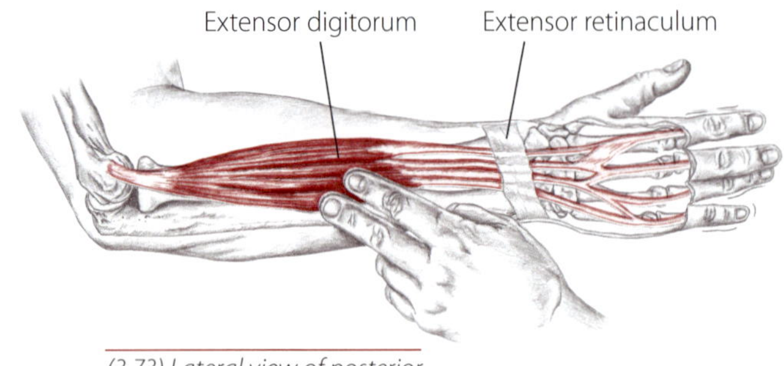

(3.73) Lateral view of posterior surface of right forearm

Extensor carpi ulnaris

1) Shake hands with your partner and flex the elbow to 90°. Locate the shaft of the ulna.
2) Slide laterally off the shaft onto the slender belly of the extensor carpi ulnaris.
3) Ask your partner to adduct her wrist against your resistance (3.75). Note how the tissue directly lateral to the ulna tightens with this movement.
4) Follow the tendon distally past the head of the ulna.

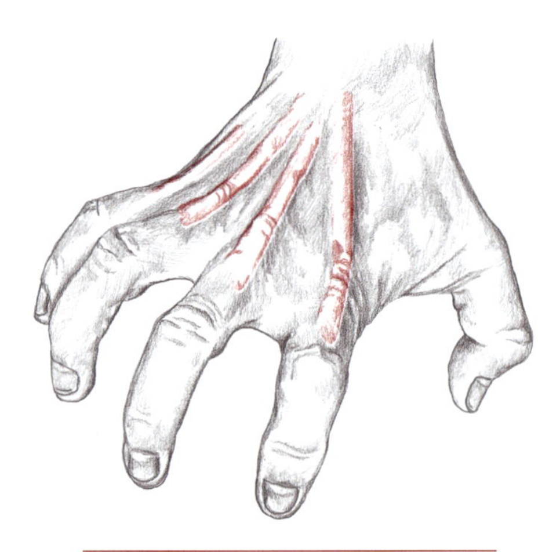

(3.74) Making the infamous "death claw" gesture will bring your extensor digitorum tendons to the surface.

Brachioradialis and extensor carpi radialis longus and brevis are sometimes known collectively as the "wad of three." Together they form a long mass of muscle which extends distally from the lateral supracondylar ridge of the humerus.

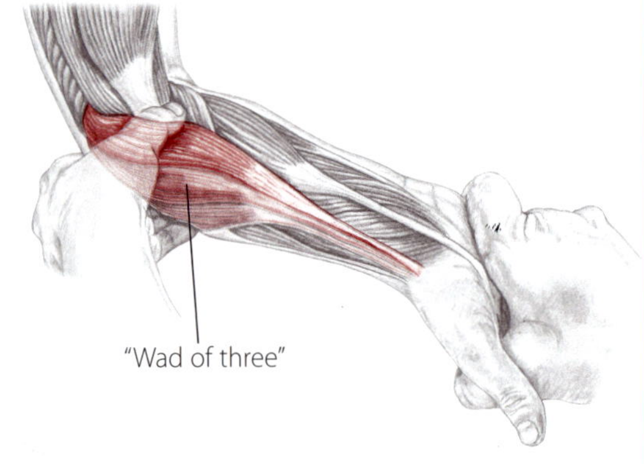

To locate the wad, shake hands with your partner and palpate just lateral to the inner elbow. The wad will be the thick, mobile tissue which can be easily grasped between your fingers and thumb. Follow it distally as far as possible.

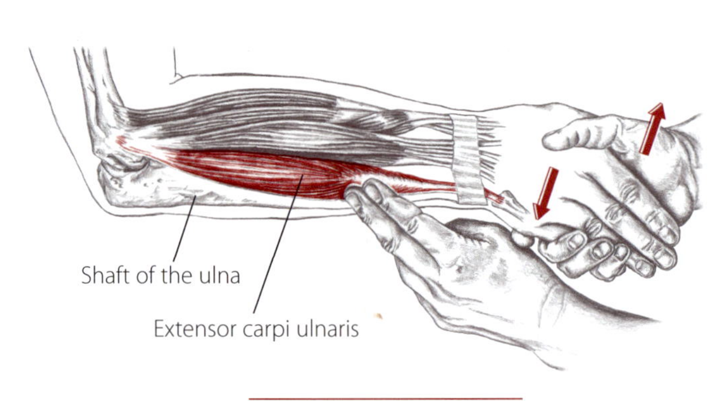

(3.75) Lateral view of posterior surface of right forearm

Anconeus

The anconeus is a weak elbow extensor located lateral to the olecranon process. Triangular-shaped, it originates at the lateral epicondyle of the humerus and fans out to attach on the shaft of the ulna (3.76). The anconeus is superficial, yet can be difficult to differentiate from the surrounding extensors.

A Extend the elbow (humeroulnar joint)

O Lateral epicondyle of the humerus

I Olecranon process and lateral edge of ulnar shaft

N Radial

1) Locate the olecranon process, the proximal shaft of the ulna, and the lateral epicondyle of the humerus.
2) Then lay your index finger along the proximal ulna and the tip of your middle finger upon the lateral epicondyle. The "V" formed by your fingers is the outline of the anconeus (3.77).

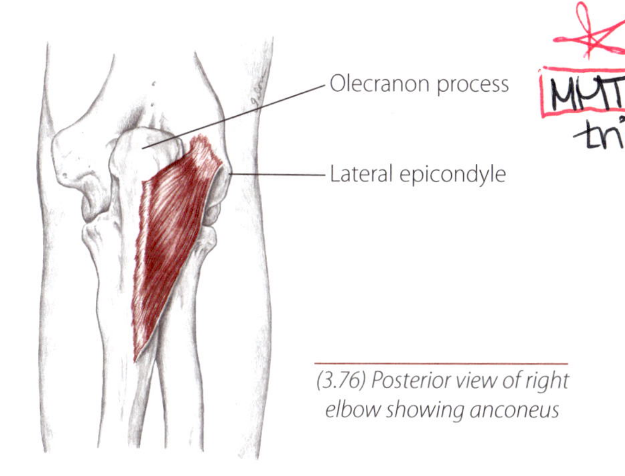

(3.76) Posterior view of right elbow showing anconeus

MMT: Same as triceps.

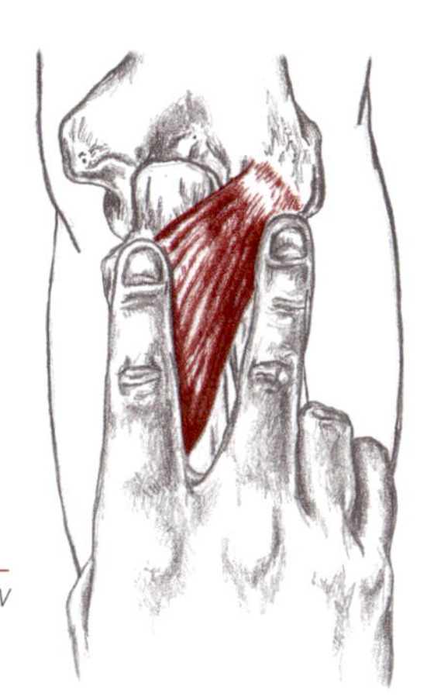

(3.77) Posterior view of right elbow

Extensor Indicis

Located deep to the extensor digitorum and extensor carpi ulnaris, this small but crucial muscle assists the digitorum to specifically extend the index finger. Its tendon runs diagonally across the wrist and hand, sheathed with the extensor digitorum tendon (3.78). If you extend your index finger at the metacarpophalangeal joint you may see two side-by-side tendons passing across the top of the knuckle (3.79). The medial tendon (closer to the middle finger) is the extensor indicis tendon.

A Extend the second finger (metacarpophalangeal joint)

Assist to adduct the second finger

O Posterior surface of distal shaft of ulna and interosseous membrane

I Tendon of the extensor digitorum at the level of the second metacarpal

N Radial

(3.78) Dorsal view of right hand showing extensor indicis

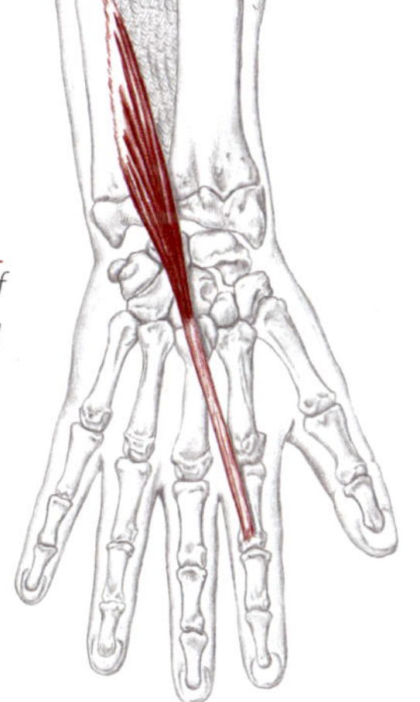

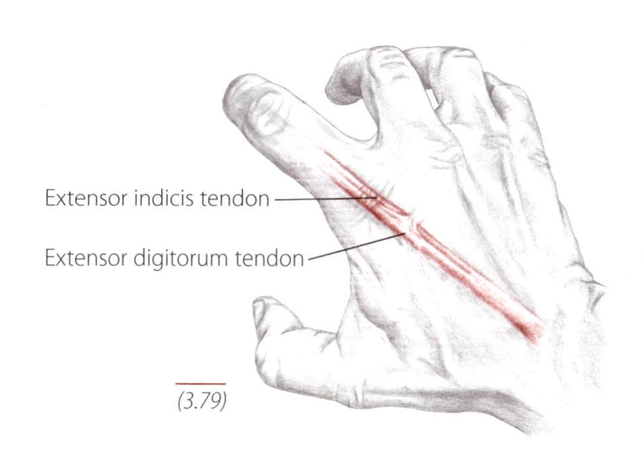

(3.79)

anconeus	an-**ko**-nee-us	Grk. elbow
indicis	**in**-di-kis	

Flexors of the Wrist and Hand

Flexor Carpi Radialis
Palmaris Longus
Flexor Carpi Ulnaris
Flexor Digitorum Superficialis
Flexor Digitorum Profundus

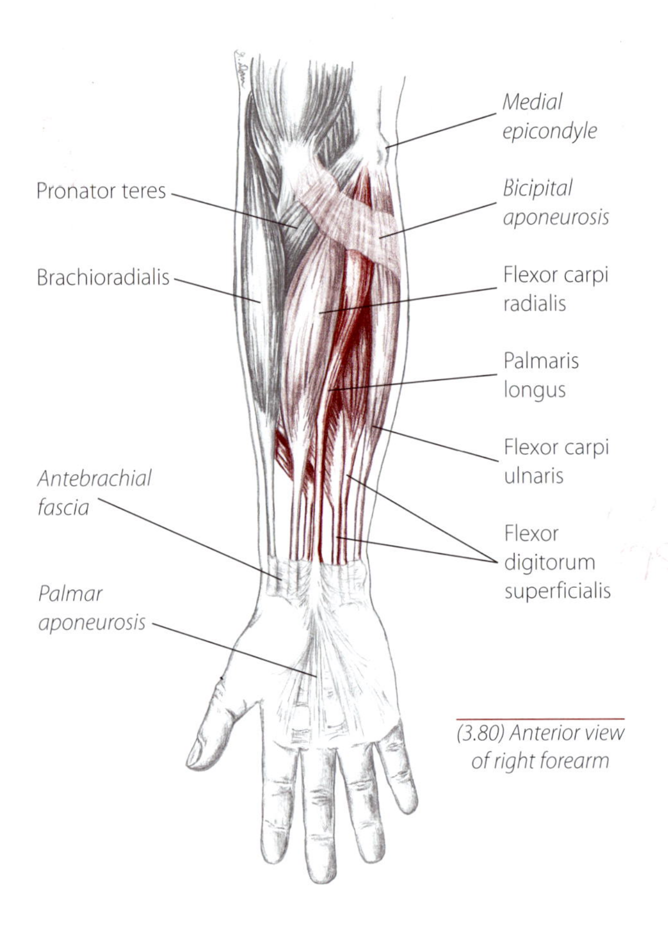

(3.80) Anterior view of right forearm

The five flexors included in this section create flexion primarily at the wrist or fingers (3.80). They are located on the forearm's anterior/medial surface between the brachioradialis and the ulnar shaft. Most of the flexors originate as one mass from the common flexor tendon at the medial epicondyle of the humerus (3.82). The bellies of the flexors extend down the forearm, becoming thin tendons roughly two inches proximal to the wrist.

As a group, the flexors are thicker and more pliable than the extensors. Although the flexors are easily accessed as a group, isolating specific muscle bellies can be challenging.

The flexors are arranged in three layers. The superficial layer is formed by the long bellies of flexor carpi radialis, palmaris longus and flexor carpi ulnaris (3.81). **Flexor carpi radialis** is medial to the pronator teres and brachioradialis. **Flexor carpi ulnaris** lies close to the ulnar shaft and has a distinct tendon attaching to the pisiform. The **palmaris longus**, which is sometimes absent, runs between flexor carpi radialis and flexor carpi ulnaris and attaches to the palmar aponeurosis (p. 169). Portions of all three muscles can be isolated for palpation purposes.

The middle and deep layers contain the wide bellies of **flexor digitorum superficialis** and **flexor digitorum profundus**, respectively (3.87, 3.88). Each digitorum muscle has four thin tendons which pass through the carpal tunnel (p. 169) and attach at the phalanges. The bellies of the digitorums are difficult to access directly, but their density can be felt beneath the superficial flexors.

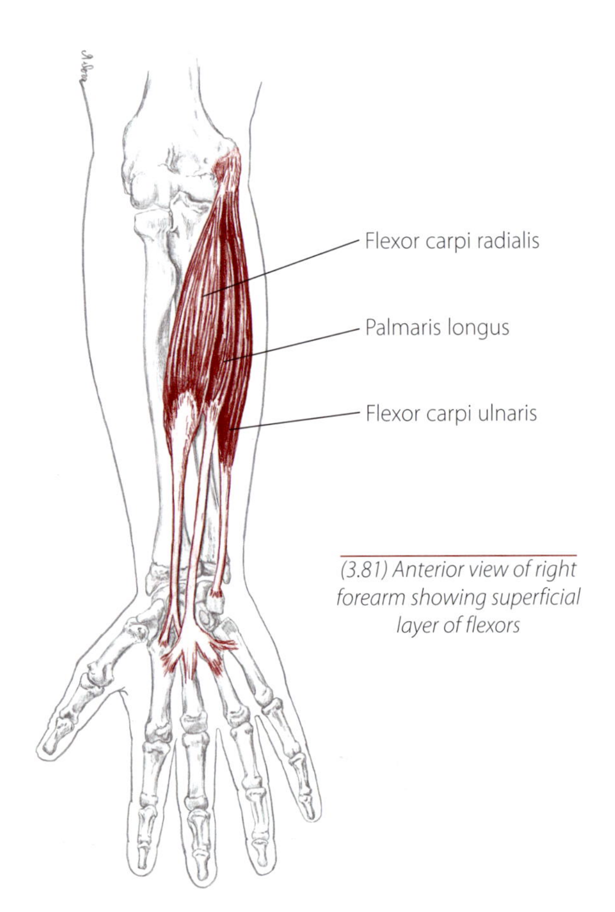

(3.81) Anterior view of right forearm showing superficial layer of flexors

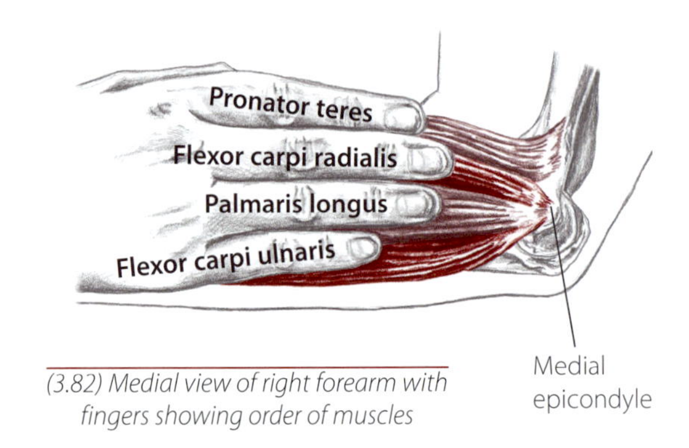

(3.82) Medial view of right forearm with fingers showing order of muscles

superficialis — **soo**-per-**fish**-ee-**a**-lis — L. on the surface
profundus — pro-**fun**-dus — L. deep

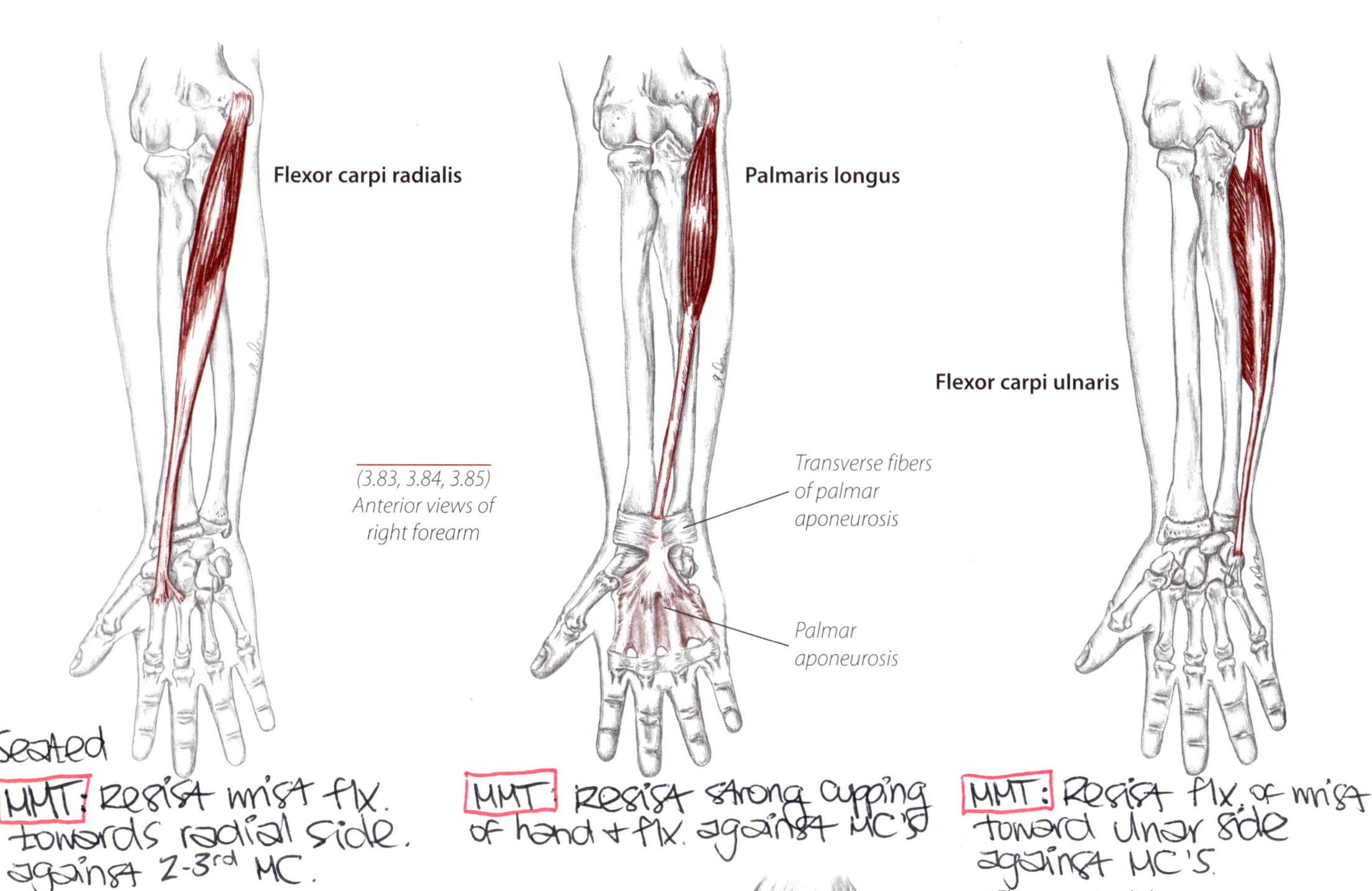

(3.83, 3.84, 3.85) Anterior views of right forearm

Seated
MMT: Resist wrist flx. towards radial side. against 2-3rd MC.

MMT: Resist strong cupping of hand + flx. against MC's

MMT: Resist flx. of wrist toward ulnar side against MC'S.

Flexor Carpi Radialis

A Flex the wrist (radiocarpal joint)
Abduct the wrist (radiocarpal joint)
Flex the elbow (humeroulnar joint)

O Common flexor tendon from medial epicondyle

I Bases of second and third metacarpals

N Median

Palmaris Longus

A Tense the palmar fascia
Flex the wrist (radiocarpal joint)
Flex the elbow (humeroulnar joint)

O Common flexor tendon from medial epicondyle

I Flexor retinaculum and palmar aponeurosis

N Median

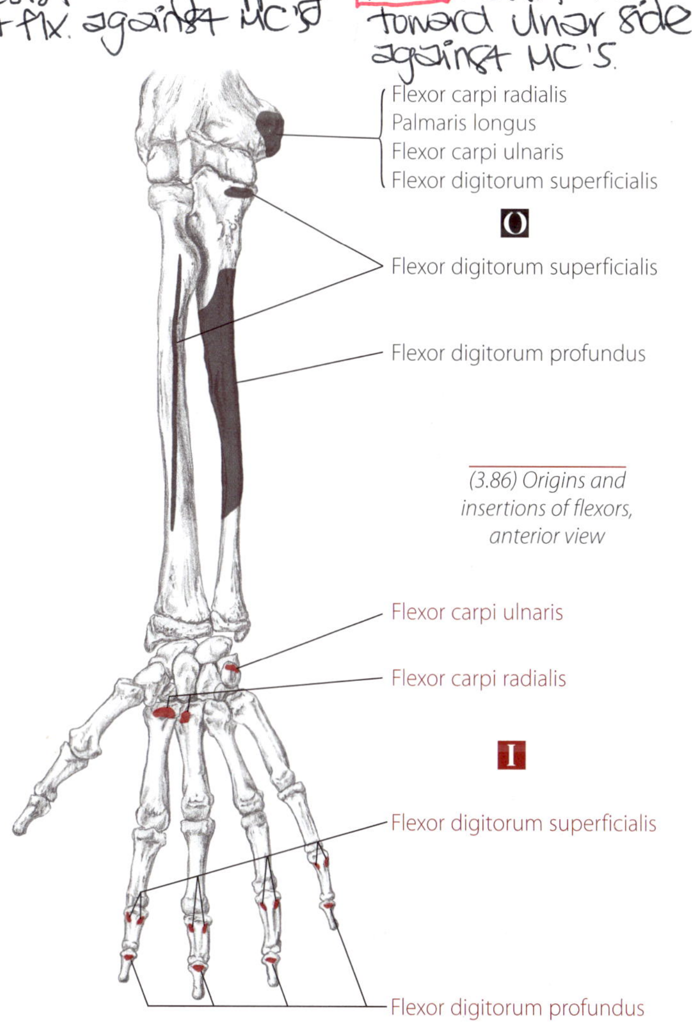

(3.86) Origins and insertions of flexors, anterior view

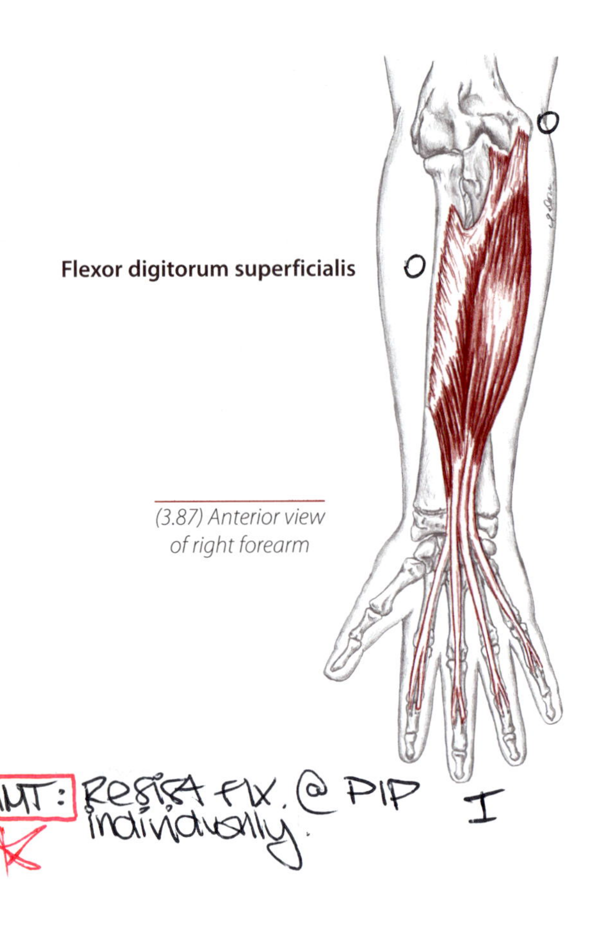

(3.87) Anterior view of right forearm

Flexor Carpi Ulnaris

A Flex the wrist (radiocarpal joint)

Adduct the wrist (radiocarpal joint)

Assist to flex the elbow (humeroulnar joint)

O *Humeral head:*
Common flexor tendon from medial epicondyle of humerus
Ulnar head:
Posterior surface of proximal half of ulnar shaft

I Pisiform

N Ulnar

Flexor Digitorum Superficialis

A Flex the second through fifth fingers (metacarpophalangeal and proximal interphalangeal joints)

Flex the wrist (radiocarpal joint)

O Common flexor tendon from medial epicondyle of humerus, ulnar collateral ligament, coronoid process of ulna and shaft of radius

I By four tendons into sides of middle phalanges of second through fifth fingers

N Ulnar

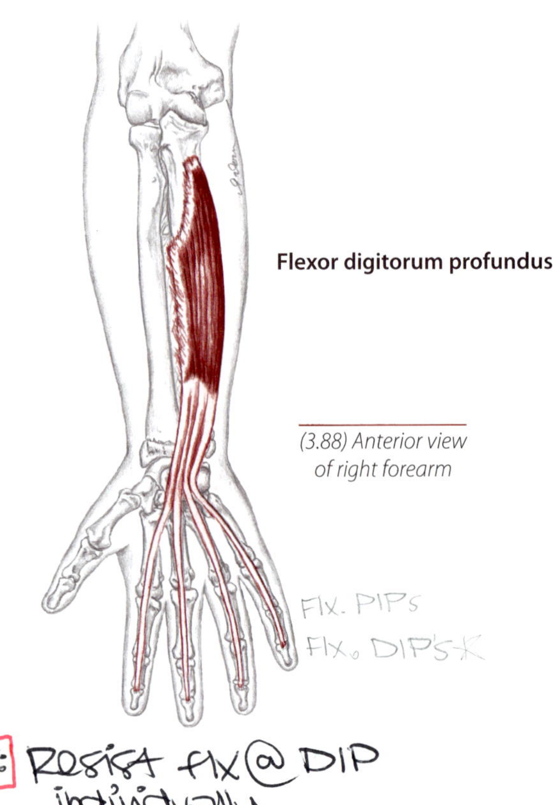

(3.88) Anterior view of right forearm

Flexor Digitorum Profundus

A Flex the second through fifth fingers (metacarpophalangeal and distal interphalangeal joints)

Assist to flex the wrist (radiocarpal joint)

O Anterior and medial surfaces of proximal three-quarters of ulna

2/3

I By four tendons into bases of distal phalanges, palmar surface of second through fifth fingers

N Ulnar

Flexor group

1) Shake hands with your partner and flex the elbow to 90˚. Locate the brachioradialis and shaft of the ulna (see p. 124).
2) Lay the flat of your hand between these landmarks on the forearm's anterior surface and ask your partner to alternately flex and relax her wrist against your resistance (see p. 142).
3) Explore the chubby bellies from their origin at the medial epicondyle to their distal tendons at the wrist (3.89).

Are you between the brachioradialis and ulnar shaft? Do the muscles contract on flexion of the wrist?

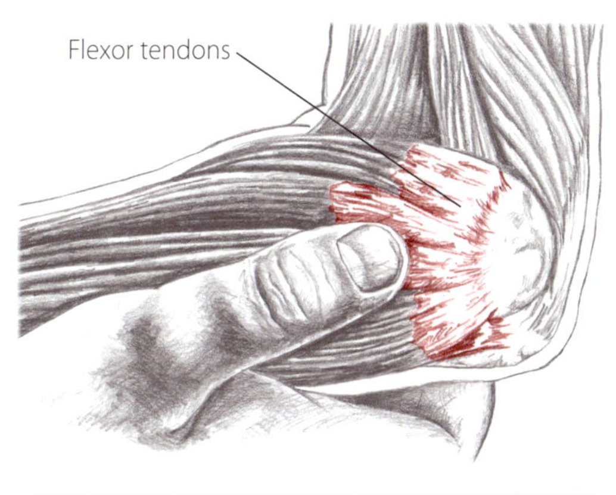

(3.89) Medial view of right elbow, palpating common tendon of the flexors at the medial epicondyle

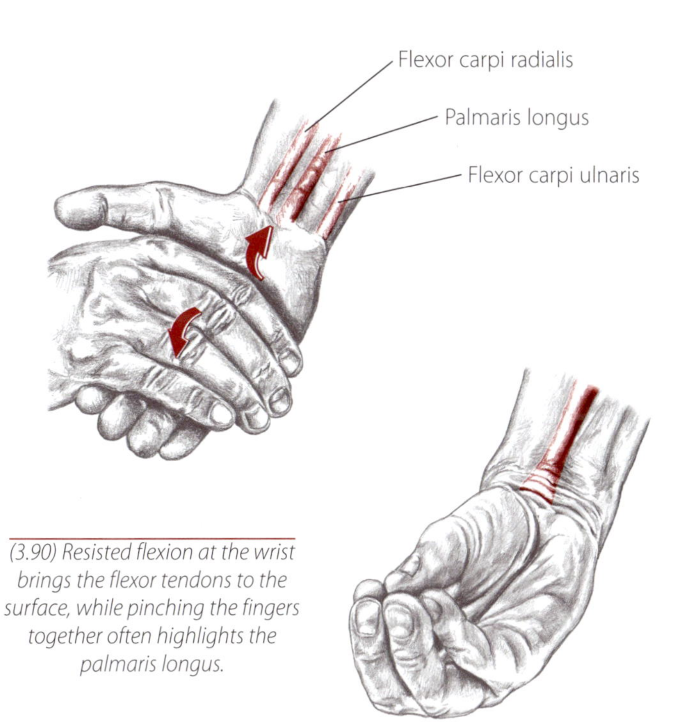

(3.90) Resisted flexion at the wrist brings the flexor tendons to the surface, while pinching the fingers together often highlights the palmaris longus.

Flexor carpi radialis and palmaris longus

1) Flex your partner's elbow to 90˚ and supinate the forearm. Begin at the distal tendons. Ask your partner to flex her wrist against your resistance.
2) At the center of the wrist will be two superficial tendons, flexor carpi radialis and palmaris longus (3.90). The palmaris longus may be absent, but if both tendons are present, the palmaris will be the most medial.
3) As your partner contracts, roll across the tendons and follow them proximally as they expand into muscle bellies (3.91). Ask your partner to alternately abduct and relax her wrist to create a distinct contraction of flexor carpi radialis (3.92).

Are the tendons/muscle bellies superficial? If you palpate the belly of flexor carpi radialis, is it superficial and medial to the pronator teres (p. 154)? Is the palmaris longus medial to the flexor carpi radialis? Follow the bellies toward the elbow. Do they merge at the medial epicondyle of the humerus?

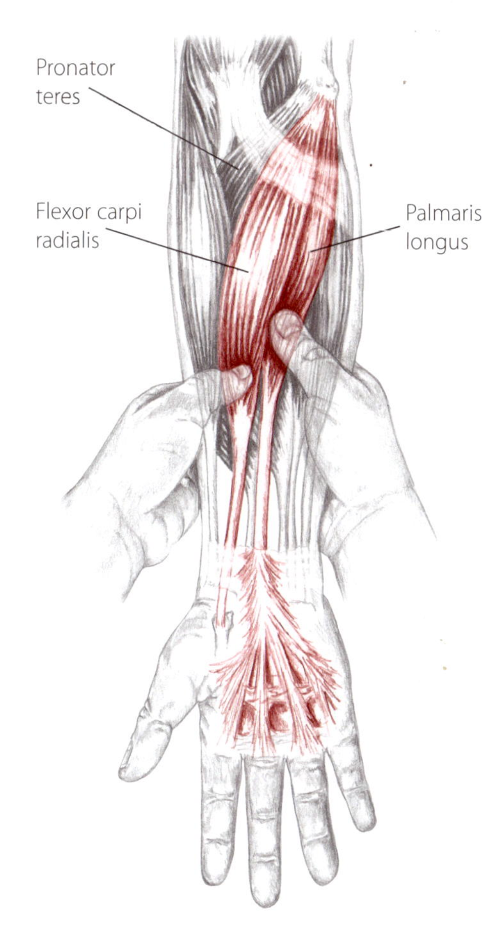

(3.91) Anterior view of right forearm

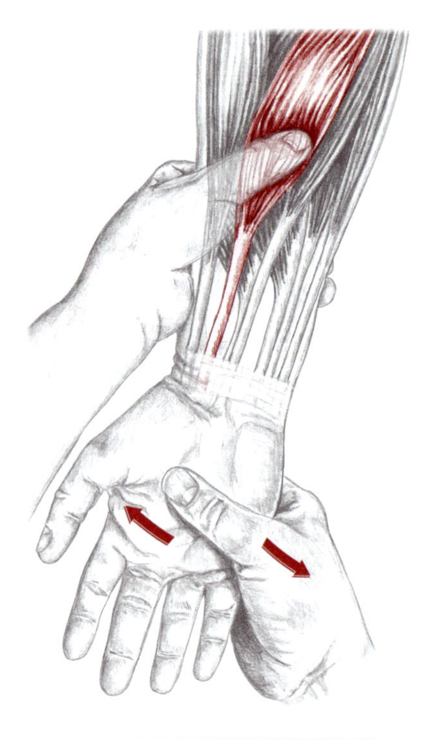

(3.92) Feeling the flexor carpi radialis by resisting abduction of the right wrist

The palmaris longus is absent in about 11% of the population. The palmar aponeurosis, however, is always present. The palmaris longus may vary from a mere tendinous band to a distal muscle belly with a long, proximal tendon. On occasion, there may be two palmaris longus muscles. The insertion site is also variable. It may attach to the fascia of the forearm, the tendon of the flexor carpi ulnaris, the flexor retinaculum, the pisiform or the scaphoid.

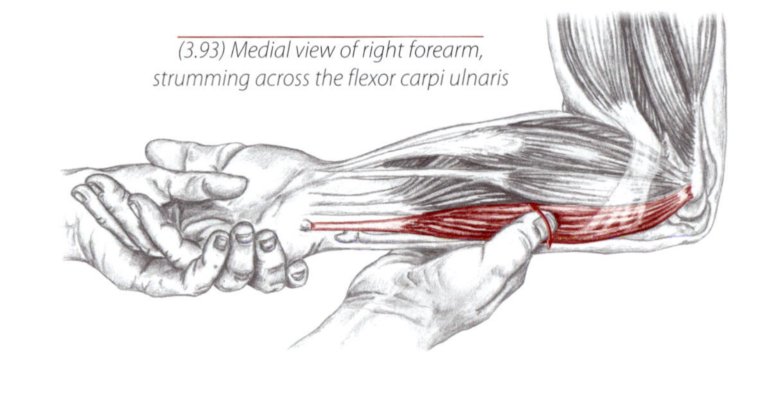

(3.93) Medial view of right forearm, strumming across the flexor carpi ulnaris

Flexor carpi ulnaris

1) Shaking hands with your partner, flex the elbow to 90° and supinate the forearm. Begin at the distal tendon by locating the pisiform (p. 129).
2) Slide proximally off the pisiform to the slender, superficial tendon of flexor carpi ulnaris (3.94).
3) As your partner alternately adducts and relaxes her wrist against your resistance, follow the tendon proximally, strumming across its surface (3.93). Feel how it widens into a muscle belly and heads toward the medial epicondyle. (Note: Unlike the extensor carpi ulnaris, the flexor carpi ulnaris lies roughly a finger's width away from the ulnar shaft.)

Do you feel the muscle contract upon adduction? Is the tendon/muscle belly superficial and along the forearm's anterior/medial surface? Is it medial to the palmaris longus?

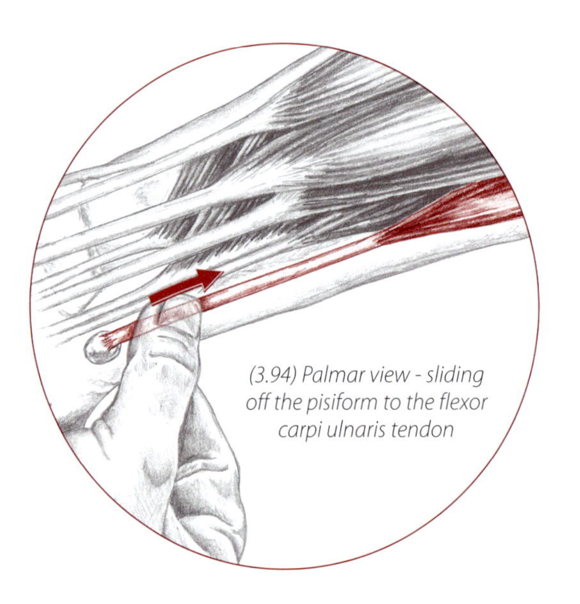

(3.94) Palmar view - sliding off the pisiform to the flexor carpi ulnaris tendon

Flexor digitorum superficialis and profundus

1) Beginning at the wrist, locate the tendons of the superficial flexors (carpi ulnaris, carpi radialis and palmaris longus). Passive flexion of the wrist will soften the tendons and allow for easier access.
2) Slowly work your thumb between the superficial tendons for the deeper digitorum tendons and bellies (3.95).

Is the tissue you are accessing deep to the first layer of flexors? If your partner wiggles the tips of her fingers can you detect any small, undulating contractions in the forearm?

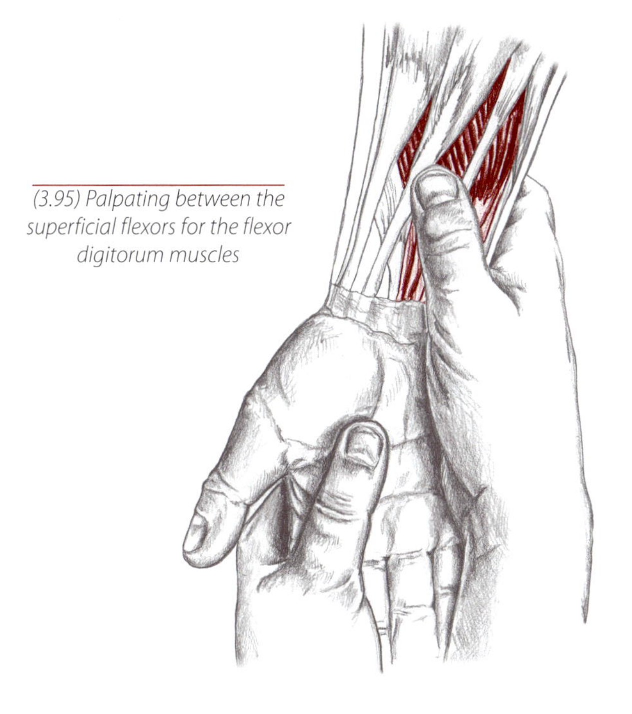

(3.95) Palpating between the superficial flexors for the flexor digitorum muscles

1) Although the digitorum muscles are deep, their contractions are palpable along the medial side of the ulnar shaft. Ask your partner to simultaneously flex her elbow and wrist to 90˚.
2) Locate the ulnar shaft, sliding just off its edge into the flexors. Ask your partner to squeeze the tips of her fifth finger and thumb together and then relax. You should feel the small, but distinct contraction of the digitorum muscles as they bulge into your fingers (3.96).
3) Try squeezing the ring, middle and index fingers together with the thumb and note how this changes the contraction.

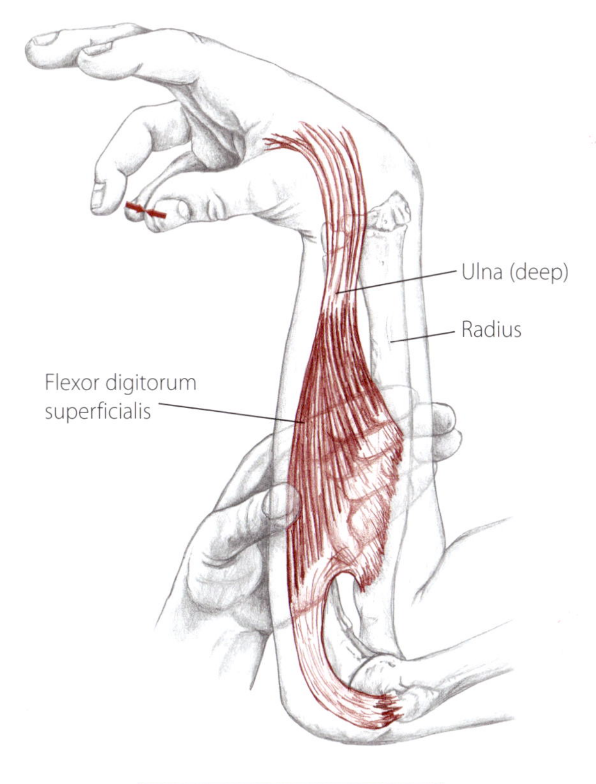

(3.96) Medial view of right forearm

There are two primary supinators (biceps brachii and supinator) and two primary pronators (pronator teres and pronator quadratus). You might assume that this structural symmetry would mean a balance of strength between the pronators and supinators, but in reality, the size and power of the biceps brachii cause the scales to tip in favor of the supinators.

The expression "righty-tighty, lefty-loosey" is not only a reminder of the direction in which to turn a screw, but also applies to the hand that holds the screwdriver. We have more power to supinate than to pronate and the world is dominated by right-handed individuals, so screws have been designed to be tightened by right forearm supination. This, of course, leaves "south paws" to tighten with either weak pronators or the undeveloped supinators of the right forearm.

Pronator Teres

Located on the anterior surface of the forearm, the round pronator teres is tucked between the brachioradialis and forearm flexors (p. 142). It is partially superficial and the only muscle in this vicinity with oblique fibers (3.97). The pronator teres is an antagonist to the biceps brachii and supinator ("carrying a bowl of soup") muscles and creates pronation of the forearm ("prone to spill it"). The distal tendon of biceps brachii, situated just lateral to the pronator teres, is a good landmark for locating its fibers.

(3.97) Anterior view of right forearm showing pronator teres

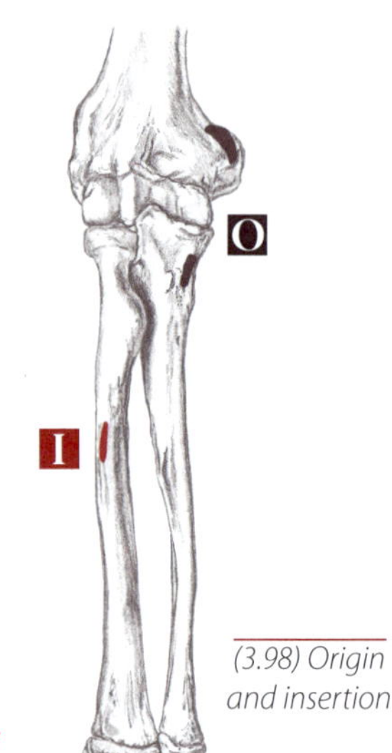

(3.98) Origin and insertion

A Pronate the forearm (radioulnar joints)

Assist to flex the elbow (humeroulnar joint)

O Medial epicondyle of the humerus, common flexor tendon and coronoid process of the ulna

I Middle of lateral surface of the radius

N Median

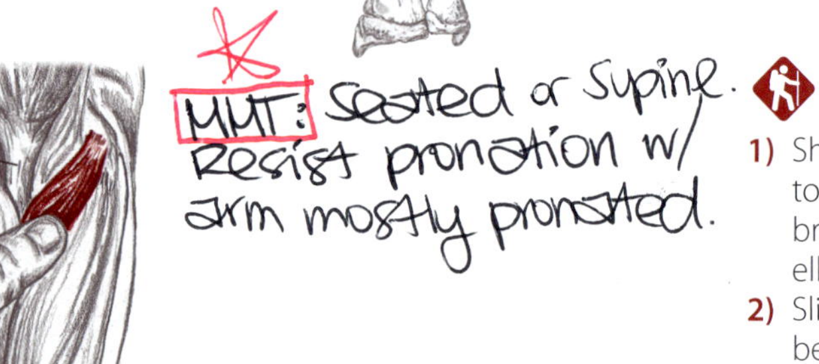

1) Shake hands with your partner and flex the elbow to 90˚. Locate the distal tendon of the biceps brachii. For assistance, ask your partner to flex her elbow against your resistance.
2) Slide distally off the tendon into the valley between the brachioradialis and forearm flexors. Sink your thumb into this space.
3) Explore for the finger-wide pronator belly running obliquely from the medial elbow across to the radius. Strum across its oblique fibers (3.100).
4) Follow it toward the medial epicondyle (noting how it blends into the other flexors) and the middle radius (feeling how it tucks under the brachioradialis).

Biceps brachii tendon

(3.99) Anterior view of right forearm

Shake hands and ask your partner to pronate against your resistance (3.99). Does the belly of the muscle you are palpating form a solid contraction? Do the fibers you are palpating run diagonally toward the middle of the radius?

(3.100) Anterior/medial view of right elbow, strumming across the pronator belly

pronate **pro**-nate L. bent forward

Pronator Quadratus

Although it lacks the pronator teres' power and speed, the small quadratus is still a capable pronator. It has transverse fibers that lie deep to the flexor tendons and the major nerve and blood vessels of the anterior forearm (3.101). The majority of the muscle is inaccessible, except for its most lateral portion. This small palpatory window, however, is also the location of the radial artery - so explore gently.

A Pronate the forearm (radioulnar joints)

O Medial, anterior surface of distal ulna

I Lateral, anterior surface of distal radius

N Median

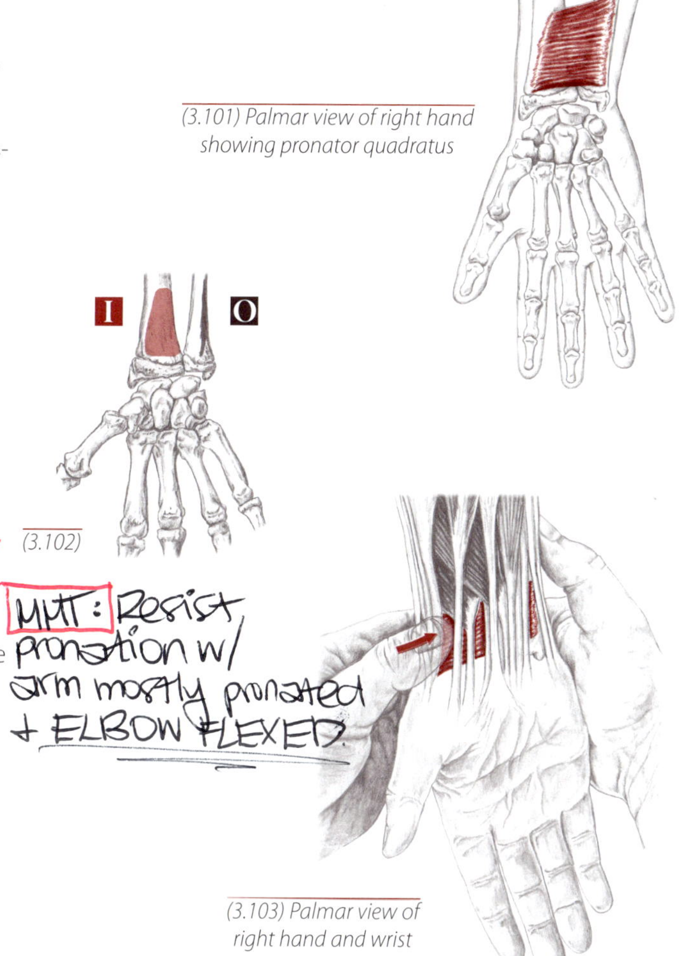

(3.101) Palmar view of right hand showing pronator quadratus

(3.102)

(3.103) Palmar view of right hand and wrist

1) Shake hands with your partner. First, isolate the pulse of the radial artery (p. 170). Then locate the radius' styloid process, sliding around to its anterior surface.
2) Before accessing the quadratus, flex and pronate the wrist slightly, softening the overlaying flexor tendons. Then use your thumb to explore the thin band of tissue between the radius and the tendons (3.103).
3) You might not feel the fibers specifically, but asking your partner to pronate ever so gently can elicit a small contraction.

Supinator

Located on the lateral side of the elbow, the short supinator is deep to the forearm extensors and superficial to the head of the radius (3.104, 3.105). As its name suggests, it supinates the forearm and is an antagonist to the pronator teres. It has a slender muscle belly which can be difficult to truly isolate.

MMT: Seated w/ Elbow + GH extended. Resist supination.

Joint capsule

(3.104) Anterior view of right forearm showing supinator

The deep branch of the radial nerve penetrates the supinator's belly and can illicit a sharp, shooting sensation down the forearm when compressed.

supinate **su**-pi-nate L. bent backward

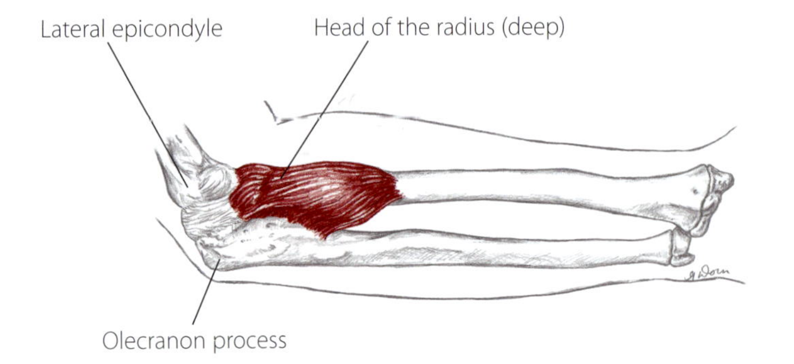

(3.105) Lateral view of right forearm showing supinator

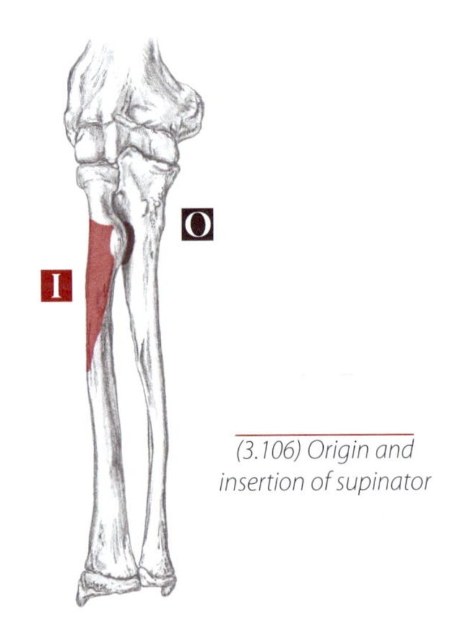

(3.106) Origin and insertion of supinator

A Supinate the forearm (radioulnar joints)

O Radial collateral ligament, annular ligament and supinator crest of the ulna

I Lateral surface of proximal shaft of the radius

N Radial

Supinator

1) Shake hands with your partner and flex the elbow to 90°. Locate the lateral epicondyle of the humerus and the proximal shaft of the radius.
2) Place your fingerpads between these landmarks and palpate through the extensor fibers for the deep supinator belly (3.107).
3) Ask your partner to alternately supinate and relax her forearm against your resistance. The brachioradialis may contract with this movement, but it will be felt superficially, while the supinator is deep to the extensors.

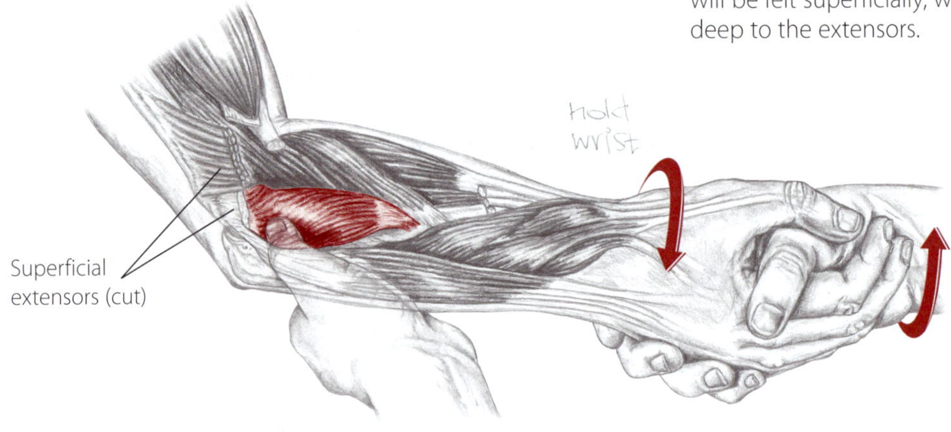

(3.107) Lateral view of right forearm, accessing supinator as your partner supinates against your resistance

Muscles of the Thumb

The eight muscles that move the thumb can be divided into two groups: short and long muscles. The four **short** muscles are located at the thenar eminence (the fleshy mass at the thumb's base).

Opponens Pollicis
Adductor Pollicis
Abductor Pollicis Brevis
Flexor Pollicis Brevis

The four **long** muscles of the thumb are tendinous and originate along the shafts of the radius or ulna.

Abductor Pollicis Longus
Extensor Pollicis Longus
Extensor Pollicis Brevis
Flexor Pollicis Longus

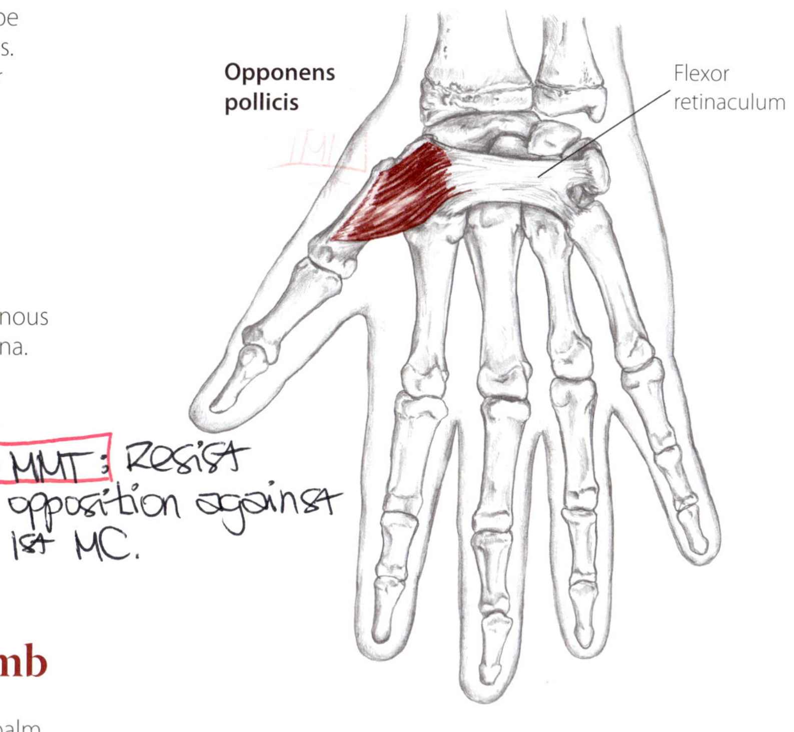

(3.108) Palmar surface of right hand

* MMT: Resist opposition against 1st MC.

Short Muscles of the Thumb

The **opponens pollicis** (3.108) performs the important task of pulling the thumb across the palm (opposition). It is located deep in the thenar eminence and is difficult to isolate. The strong **adductor pollicis** (3.109) lies deep along the palmar surface and draws the thumb toward the index and middle fingers.

Opponens Pollicis

A Opposition of the thumb at the carpometacarpal joint (bringing the pads of the thumb and fifth finger together)

O Flexor retinaculum and tubercle of the trapezium

I Entire length of first metacarpal bone, radial side

N Median

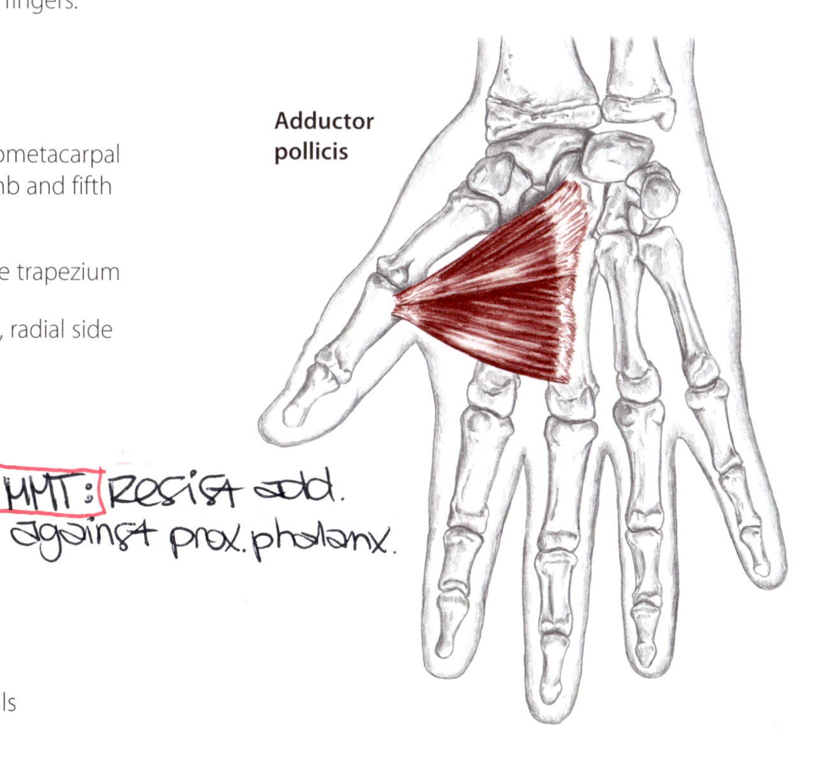

(3.109) Palmar surface of right hand

Adductor Pollicis

* MMT: Resist add. against prox. phalanx.

A Adduct the thumb (carpometacarpal and metacarpophalangeal joints)

Assist in flexion of the thumb (metacarpophalangeal joint)

O Capitate, second and third metacarpals

I Base of proximal phalange of thumb

N Ulnar

opponens	o-**po**-nens	L. opposing
pollicis	**pol**-li-sis	L. thumb

Short Muscles of the Thumb

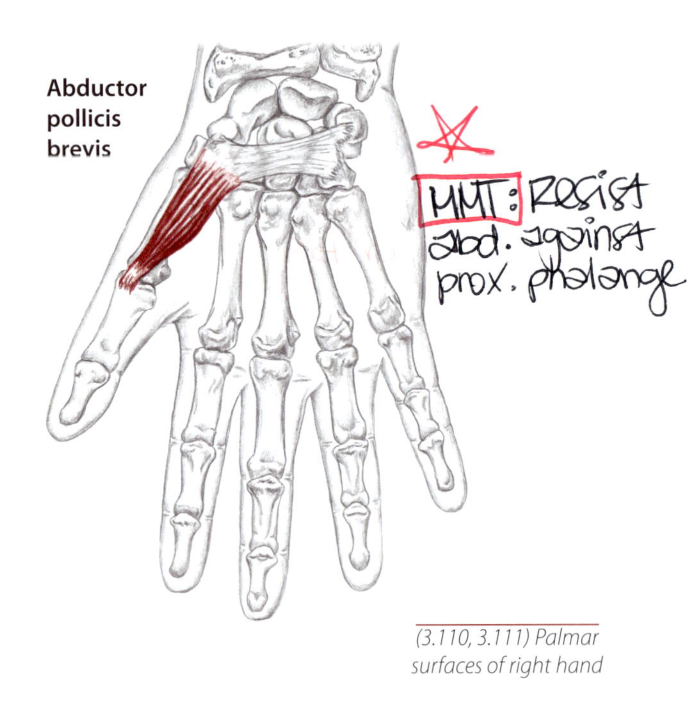

(3.110, 3.111) Palmar surfaces of right hand

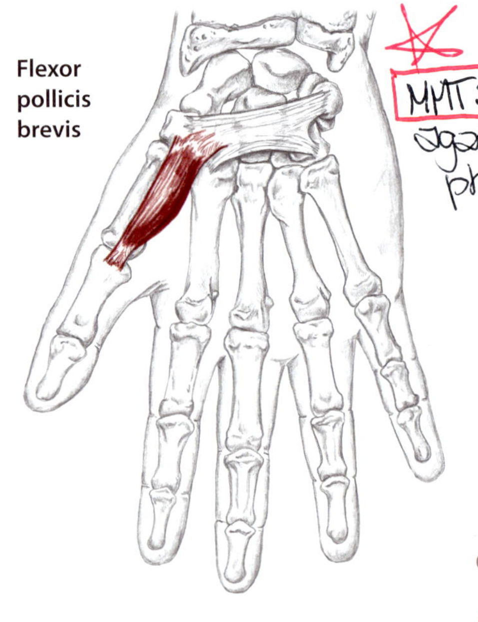

Abductor Pollicis Brevis

A Abduct the thumb
(carpometacarpal and metacarpophalangeal joints)

Assist in opposition of the thumb

O Flexor retinaculum, trapezium and scaphoid tubercles

I Base of proximal phalange of thumb

N Median

Flexor Pollicis Brevis

A Flex the thumb
(carpometacarpal and metacarpophalangeal joints)

Assist in opposition of the thumb

O *Superficial head:*
Flexor retinaculum
Deep head:
Trapezium, trapezoid and capitate

I Base of proximal phalange of thumb

N Median and ulnar

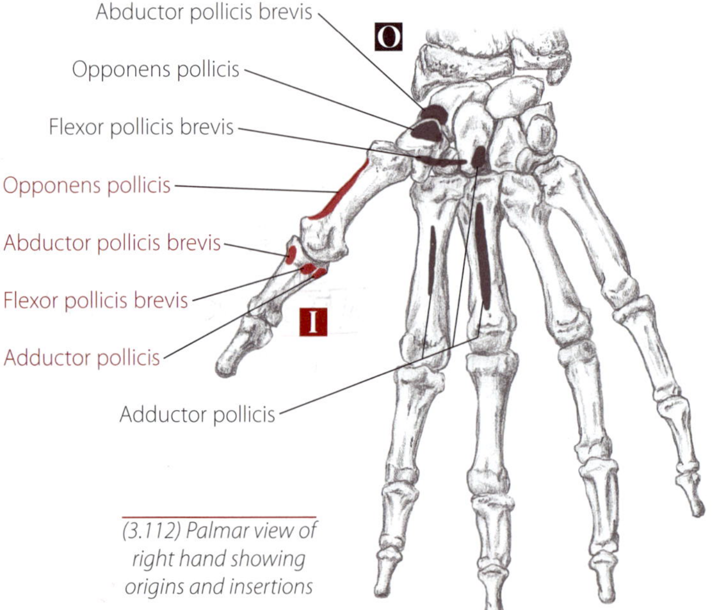

(3.112) Palmar view of right hand showing origins and insertions

thenar **thee**-nar Grk. palm, flat of the hand

Short muscles of the thumb

1) Locate the base of your partner's thumb and explore all sides of the thenar eminence's thick, movable tissue. Palpate from the shaft of the first metacarpal to the webbing between the thumb and finger (3.113).
2) Ask your partner to gently squeeze her thumb and fifth fingerpads together. Note how the thenar eminence becomes dense and compact (3.114).

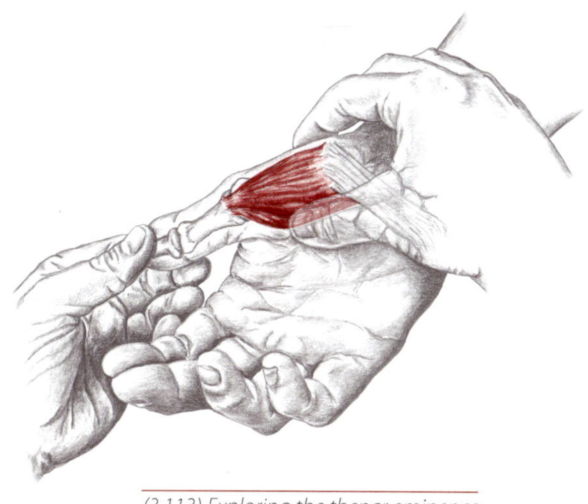

(3.113) Exploring the thenar eminence

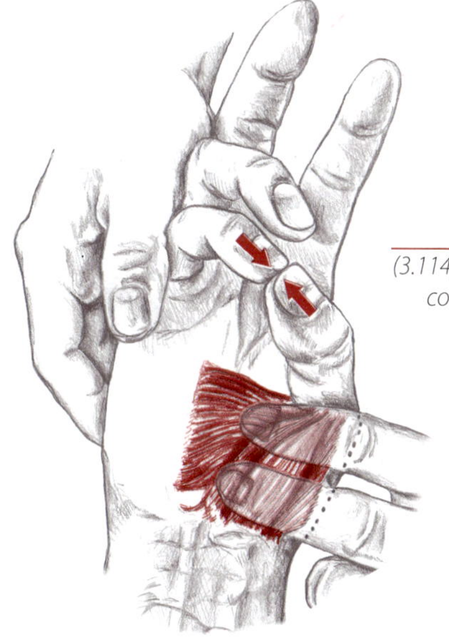

(3.114) Feeling the thenar muscles contract during opposition

When your partner performs an action with her thumb, be sure the contractions are small and repetitive. More forceful contractions will tighten all the surrounding muscles.

The human thumb has several unique qualities that distinguish it from the thumbs of other primates. One characteristic that is *not* distinctly human is the saddle joint of the first carpometacarpal joint. The joint's shape allows for opposition of the thumb and fingers, a skill shared by many higher primates including chimpanzees, orangutans and gorillas.

One reason for the dexterity of the human thumb is the separation between the flexor pollicis longus and flexor digitorum profundus muscles. In other primates these muscles are united, restricting the ability of the fingers and thumb to move independently.

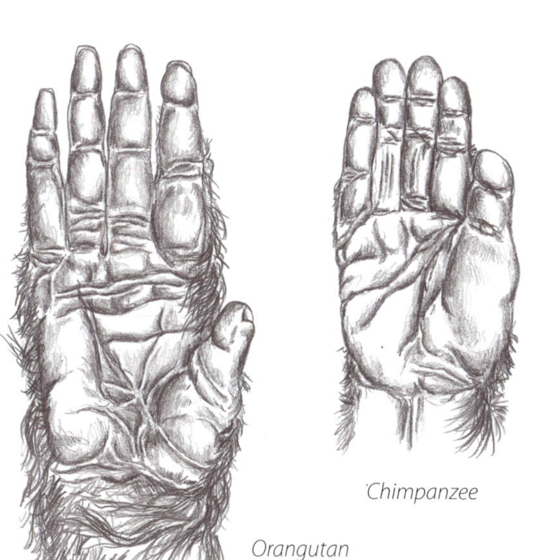

A human is also capable of applying a strong, precise grip with the thumb and fingertips as when one tightens the lid on a jar. A human's thumb is quite long in proportion to the fingers whereas the thumbs of many primates are much shorter than the fingers. Additionally, the muscles on a human's thenar eminence are larger than those on a primate's thenar pad which is typically flat and lacking in thick musculature.

Chimpanzees, gorillas and other primates can grasp with terrific force by curling their digits around an object, but opposing the thumb and finger for a specific, detailed task is something only humans can do.

Long Muscles of the Thumb

The bellies of **abductor pollicis longus** and **extensor pollicis longus and brevis** lie along the posterior aspect of the forearm, deep to the wrist extensors (3.115-3.117). Their distal tendons, however, are superficial and form the "anatomical snuffbox." Used historically as a platform for inhaling a variety of substances, this small cavity is located along the dorsal surface of the hand, just distal to the styloid process of the radius.

The belly of the **flexor pollicis longus** lies on the forearm's anterior surface, deep to the wrist flexors, and is inaccessible. Its long, distal tendon travels through the carpal tunnel between the thenar eminence muscles to the distal phalange of the thumb (3.119).

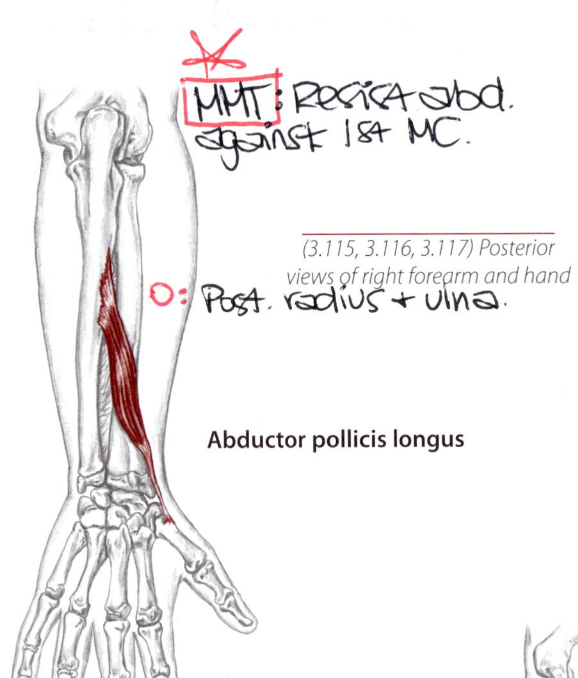

(3.115, 3.116, 3.117) Posterior views of right forearm and hand

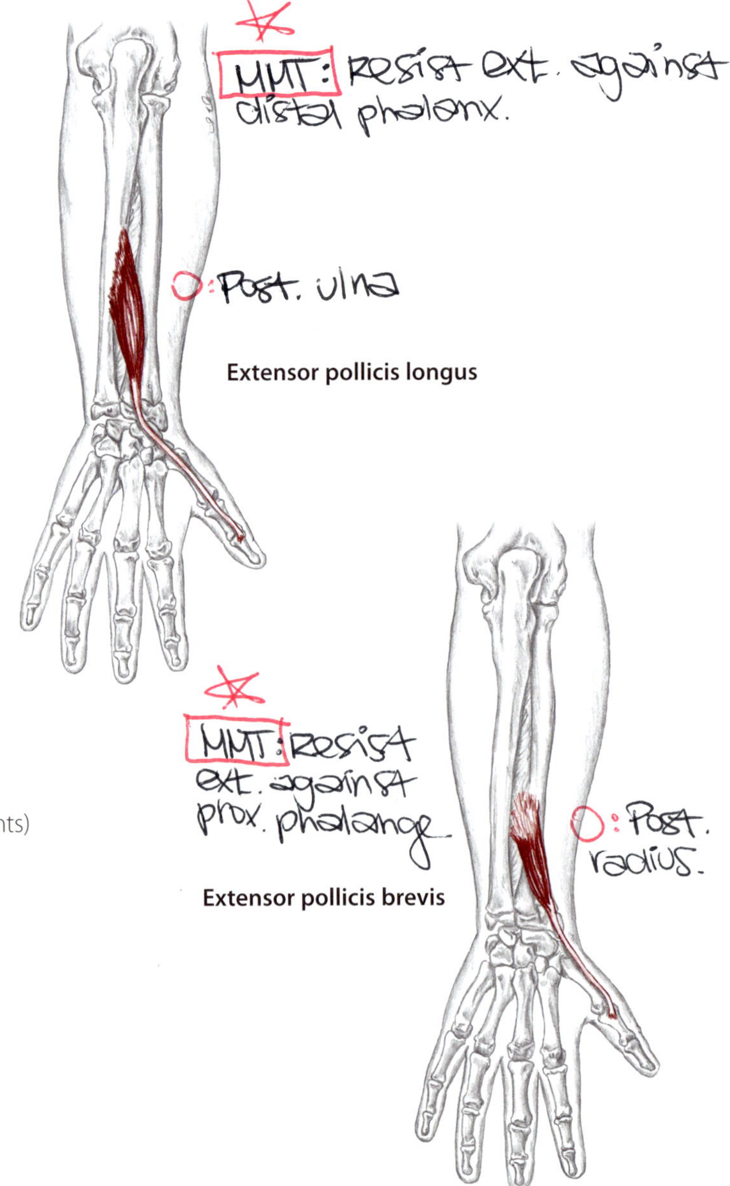

Abductor Pollicis Longus

A Abduct the thumb (carpometacarpal joint)
Extend the thumb (carpometacarpal joint)

O Posterior surface of radius and ulna, and interosseous membrane

I Base of first metacarpal

N Radial

Extensor Pollicis Longus and Brevis

A Extend the thumb (interphalangeal joint)
Assist to extend the thumb (metacarpophalangeal and carpometacarpal joints)

O *Longus:* Posterior surface of ulna and interosseous membrane
Brevis: Posterior surface of radius and interosseous membrane

I *Longus:* Distal phalange of thumb
Brevis: Proximal phalange of thumb

N Radial

Long Muscles of the Thumb

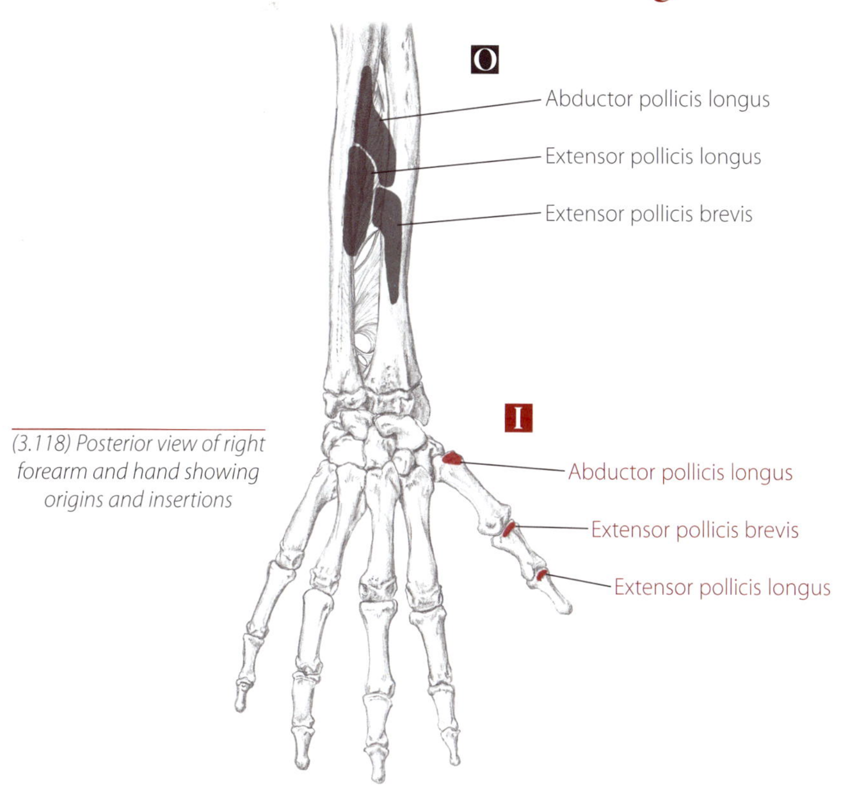

(3.118) Posterior view of right forearm and hand showing origins and insertions

Flexor Pollicis Longus

A Flex the thumb (interphalangeal joint)

Assist to flex the thumb (metacarpophalangeal and carpometacarpal joints)

O Anterior surface of radius and interosseous membrane

I Distal phalange of thumb

MMT: Resist flx. @ dist. phalanx.

N Median

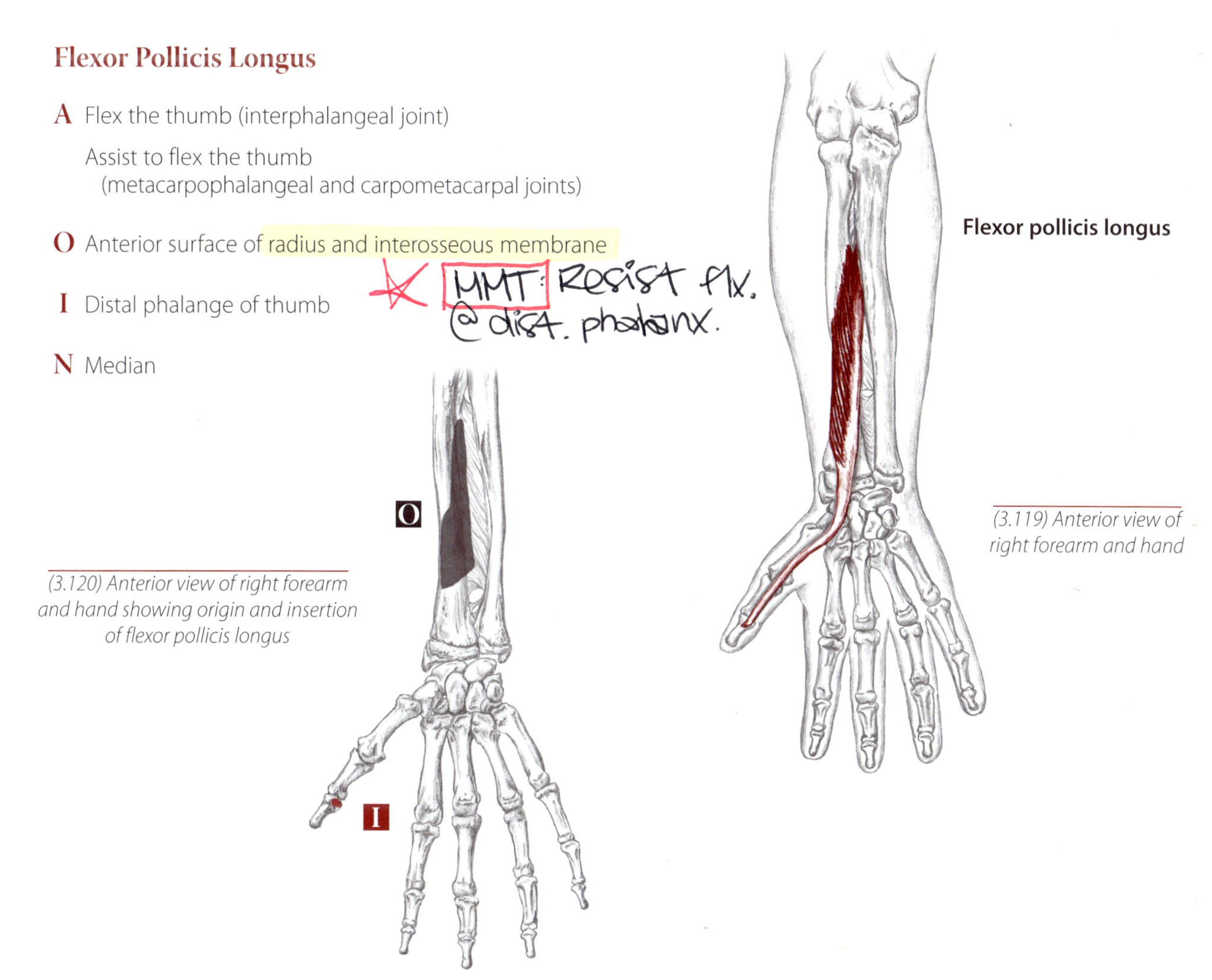

(3.119) Anterior view of right forearm and hand

(3.120) Anterior view of right forearm and hand showing origin and insertion of flexor pollicis longus

Long Muscles of the Thumb

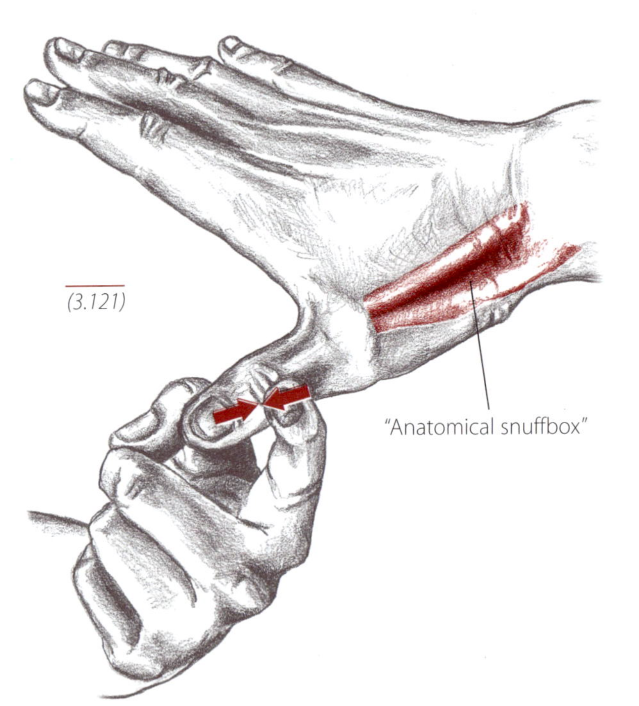

(3.121)

Anatomical snuffbox and long muscles of the thumb

1) With your partner's wrist in a neutral position, ask her to extend her thumb: "Bring your thumbnail toward your inner elbow." (3.121)
2) Just distal to the styloid process of the radius will be a small trough formed by the surrounding tendons. This is the anatomical snuffbox. If not seen immediately, change the angle of the thumb.
3) Follow the tendons that form the snuffbox (extensor pollicis longus, brevis and abductor pollicis) proximally as they slide over the posterior surface of the radius. Lay your fingers along the posterior surface of the radius as your partner circumducts her thumb in order to feel a portion of these muscles contract (3.122).

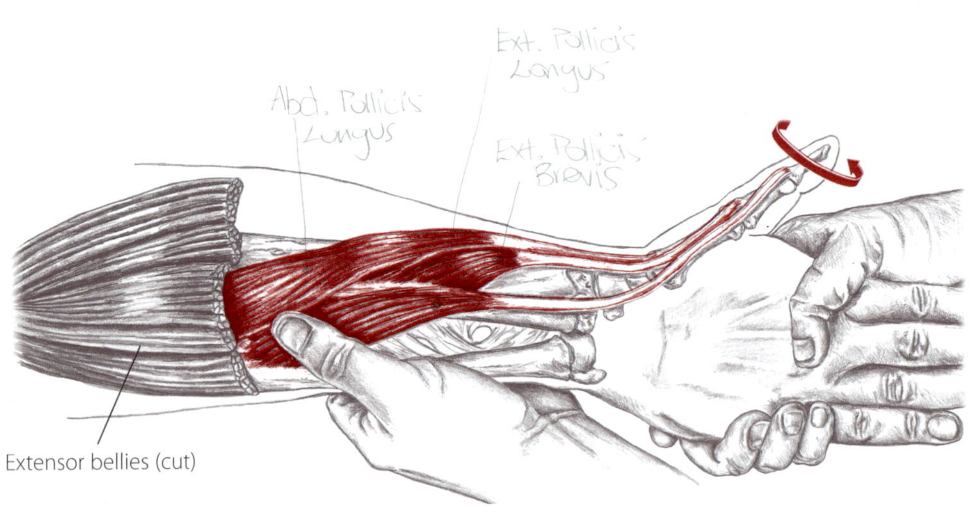

(3.122) Lateral view of right forearm. Exploring the bellies of the thumb muscles, deep to the extensor group, while your partner circumducts her thumb.

Muscles of the Hand

Lumbricals and Interossei

The **lumbricals** sprout from the sides of the flexor digitorum profundus tendons on the palmar side of the hand (3.123). Deep to the lumbricals, the **palmar interossei** (3.124) are located between the metacarpals and hence are difficult to access. The **dorsal interossei** (3.125), however, are accessible between the metacarpals from the hand's dorsal surface (3.126).

MMT: ① Resist Ext. @ mid + dist. phal. ② Resist flx. against prox. phalanges.

Flexor digitorum profundus tendons

Lumbricals

(3.123) Palmar view of right hand

Lumbricals of the Hand

A Extend the second through fifth fingers at the interphalangeal joints

Flex the second through fifth fingers at the metacarpophalangeal joints

O Surfaces of the flexor digitorum profundus tendons

I Extensor aponeurosis on dorsal surface of phalanges

N *Second and third:* Median
Fourth and fifth: Ulnar

MMT: Resist add. of 1, 2, 4 + 5 individually.

Palmar Interossei

A Adduct the thumb, second, fourth and fifth fingers toward the third finger

Assist in flexion of the thumb, second, fourth and fifth fingers at the metacarpophalangeal joints

Assist in extension of the thumb, second, fourth and fifth fingers at the interphalangeal joints (via extensor aponeurosis)

O Base of first, second, fourth and fifth metacarpals

I Base of the proximal phalange of the related finger and the extensor aponeurosis

N Ulnar

Palmar interossei

(3.124) Palmar view of right hand

MMT: Resist abd. of 2, 3 + 4 individually.

Dorsal Interossei

A Abduct the second, third and fourth fingers at the metacarpophalangeal joints

Assist in flexion of the second, third and fourth fingers at the metacarpophalangeal joints

Assist in extension of the second, third and fourth fingers at the interphalangeal joints (via extensor aponeurosis)

O Shafts of all five metacarpals

I Base of the proximal phalange of the second, third and fourth fingers and the extensor aponeurosis

N Ulnar

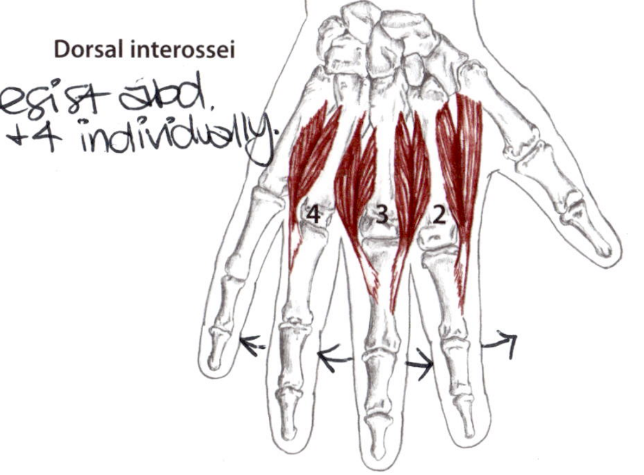

(3.125) Dorsal view of right hand

lumbrical	**lum**-bri-kal	L. earthworm
interroseus	**in**-ter-**ah**-see-us	L. between bones

Muscles of the Hand

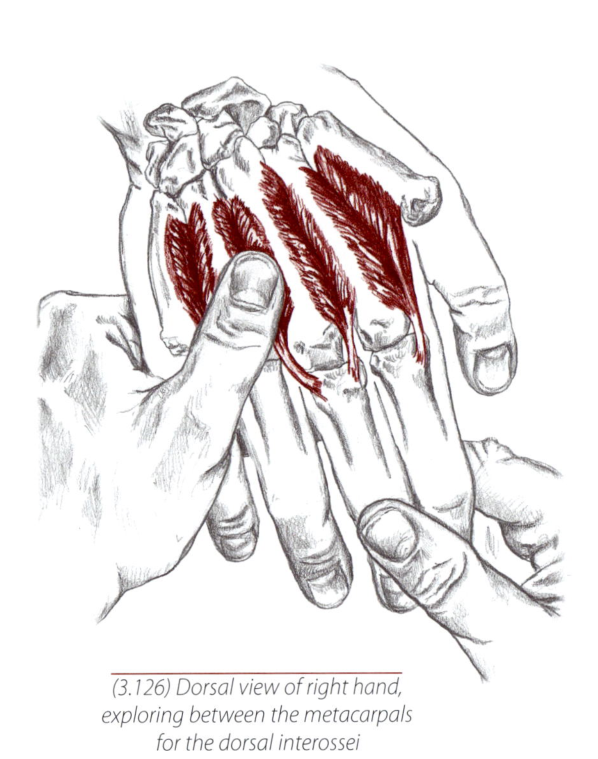

(3.126) Dorsal view of right hand, exploring between the metacarpals for the dorsal interossei

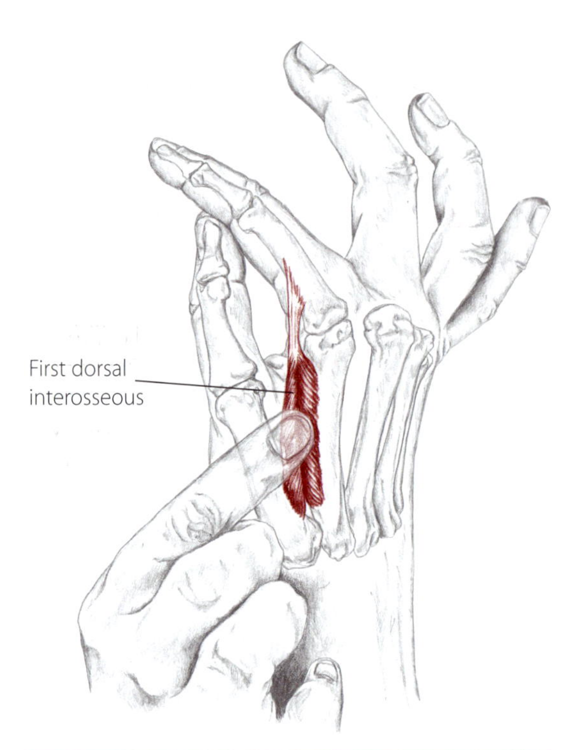

(3.127) Dorsal/radial view of right hand. Ask your partner to adduct her thumb: "Press the side of your thumb into the side of your index finger." Note how the muscles of the thenar eminence may soften, but the muscles in the hand's webbing (adductor pollicis and first dorsal interosseous) become taut.

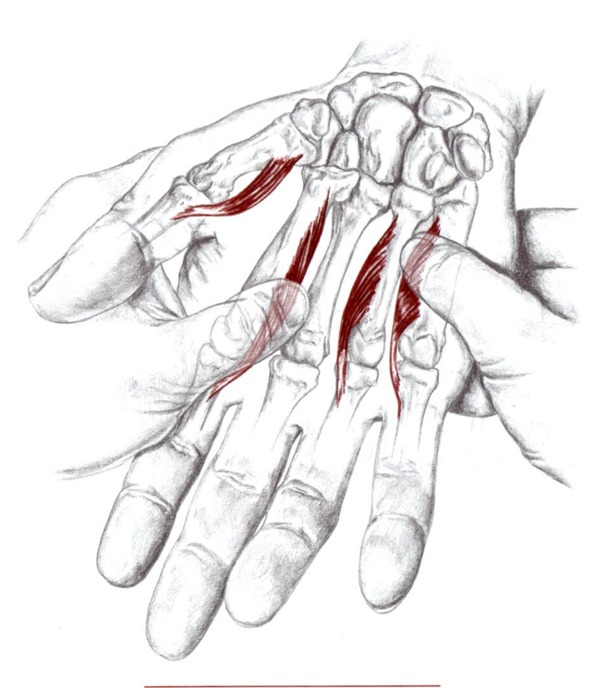

(3.128) Palmar view of right hand, accessing the palmar interossei muscles

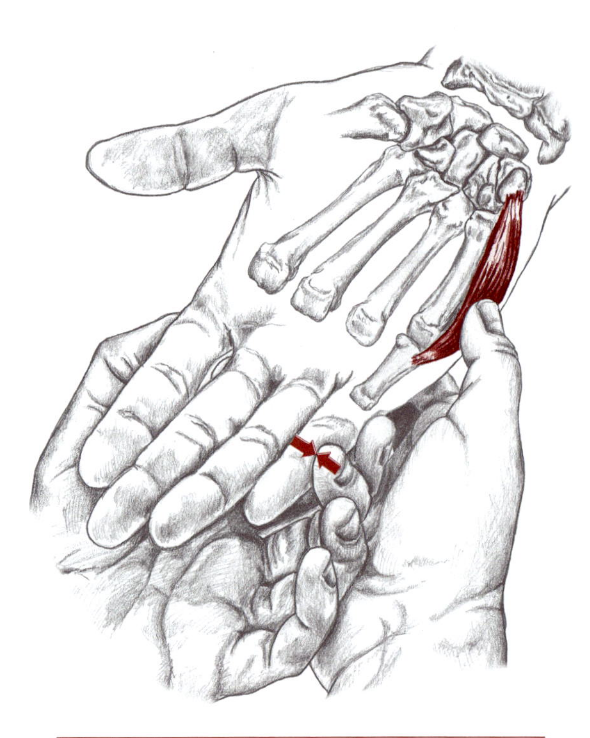

(3.129) Feel the hypothenar muscles contract when your partner abducts her fifth finger against your resistance.

Hypothenar Eminence

Opposite the thenar eminence, along the ulnar side of the palm, is the hypothenar eminence. This oblong mound of flesh is composed of three short muscles: **abductor digiti minimi**, **flexor digiti minimi brevis** and **opponens digiti minimi** (3.130 - 3.132).

The abductor digiti minimi is superficial and extends from the pisiform to the base of the fifth finger. To isolate it, ask your partner to abduct her fifth finger as you apply a little resistance. The solid belly of the digiti minimi will become immediately apparent next to the shaft of the fifth metacarpal (3.129).

Abductor Digiti Minimi

A Abduct the fifth finger (metacarpophalangeal joint)

Assist in opposition of the fifth finger toward the thumb (metacarpophalangeal joint)

O Pisiform and flexor retinaculum

I Base of proximal phalange of fifth finger, ulnar surface

N Ulnar

Flexor Digiti Minimi Brevis

A Flex the fifth finger (metacarpophalangeal joint)

Assist in opposition of the fifth finger toward the thumb

O Hook of hamate and flexor retinaculum

I Base of proximal phalange of fifth finger, palmar surface

N Ulnar

Opponens Digiti Minimi

A Opposition of the fifth finger at the carpometacarpal joint (as when cupping your palm)

O Hook of hamate and flexor retinaculum

I Shaft of fifth metacarpal

N Ulnar

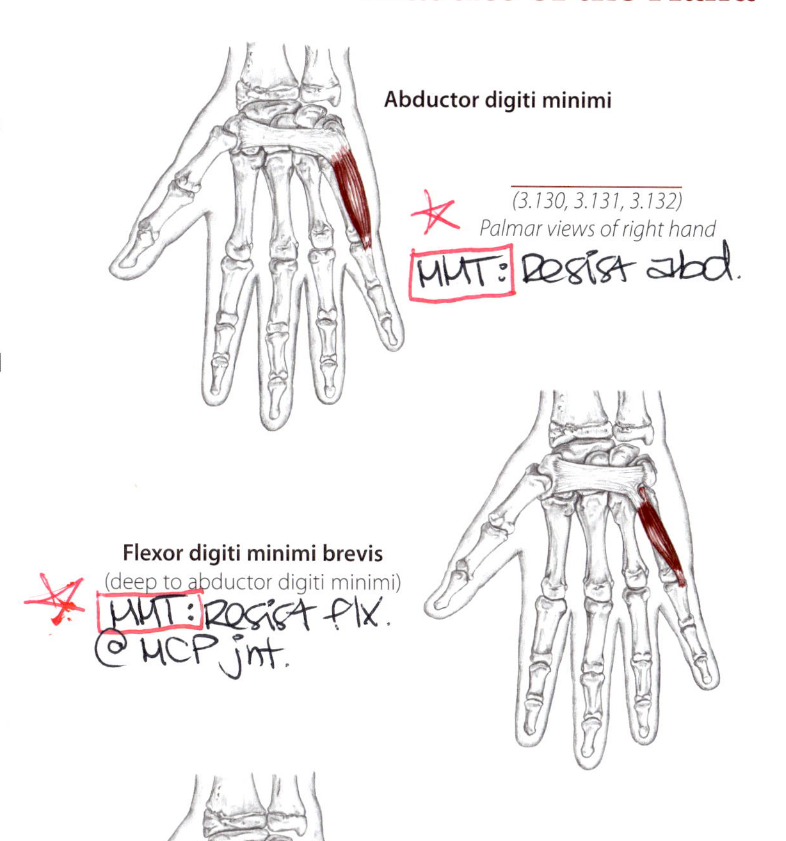

(3.130, 3.131, 3.132) Palmar views of right hand

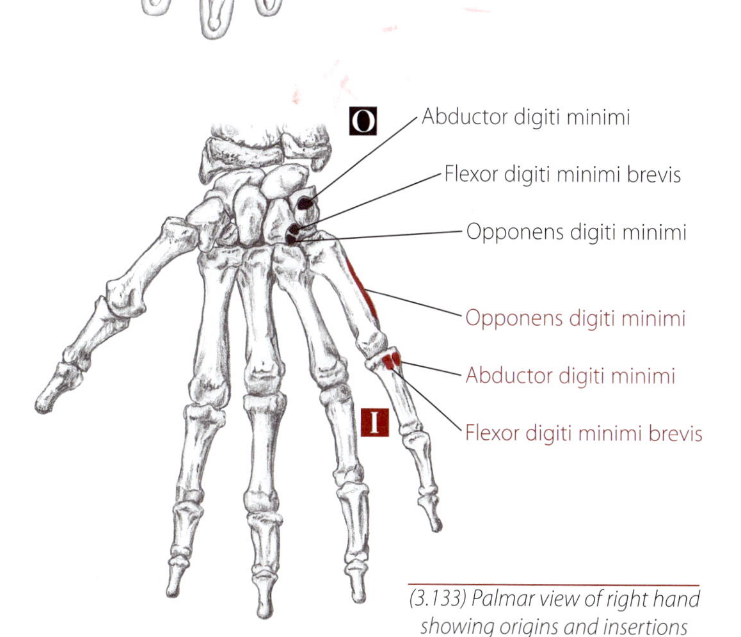

(3.133) Palmar view of right hand showing origins and insertions

hypothenar — **hi**-po-**thee**-nar — Grk. *hypo*, under or below; Grk. *thenar*, palm, flat of the hand

Other Structures of the Forearm and Hand

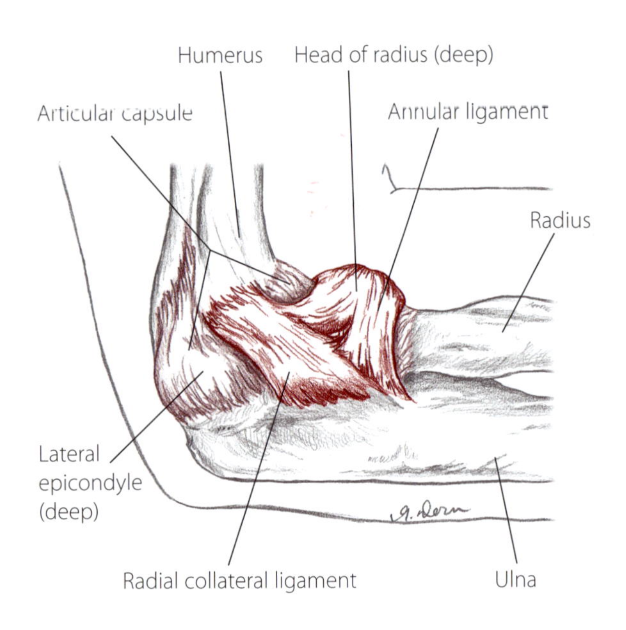

(3.134) Lateral view of right elbow showing humeroulnar and proximal radioulnar joints

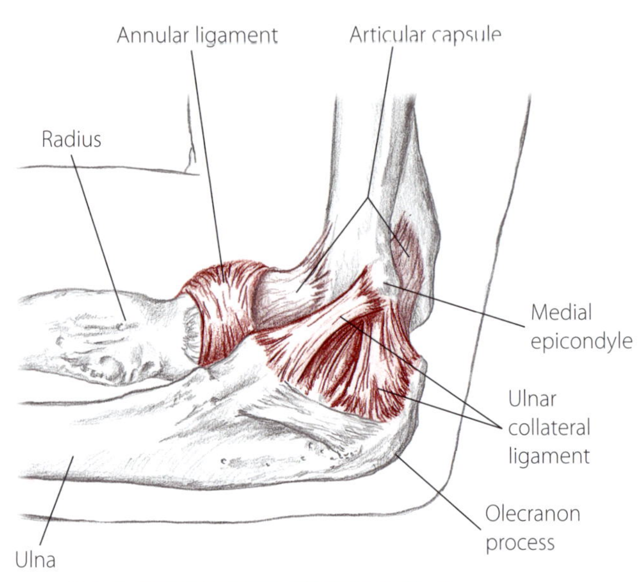

(3.135) Medial view of right elbow showing humeroulnar and proximal radioulnar joints

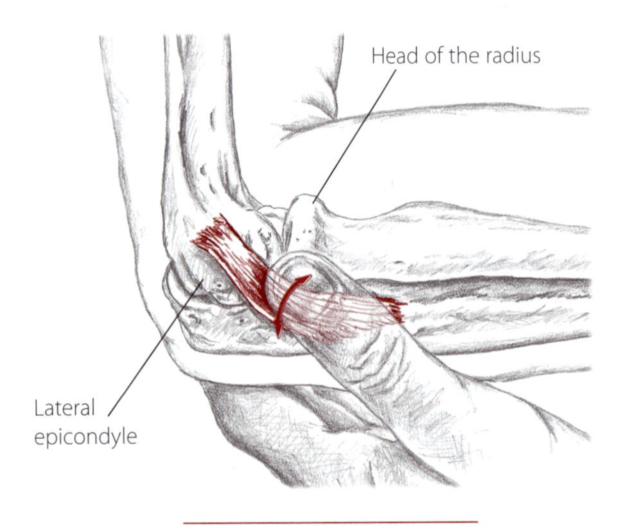

(3.136) Lateral view of right elbow

Radial Collateral Ligament

The radial collateral ligament is a cordlike band that stretches from the lateral epicondyle of the humerus to the annular ligament and olecranon process (3.134). The ligament is deep to the supinator and extensors of the forearm.

1) Shaking hands with your partner, locate the lateral epicondyle of the humerus and head of the radius.
2) Between these landmarks will be a slight ditch. Place your fingertip in this space. Visualize the ligament spanning across the ditch and gently roll your finger across the ligament's slender surface. It may feel like a thin strip of duct tape (3.136).

Are you between the head of the radius and the lateral epicondyle? With the elbow flexed, do the fibers of the ligament run parallel with the forearm?

annular	**an**-u-ler	L. ringlike
collateral	ko-**lat**-er-al	L. of both sides

Annular Ligament

The annular ligament wraps around the head and neck of the radius, stabilizing the proximal radius against the ulna during pronation and supination (3.135). It lies deep to the supinator and the extensor muscles of the forearm. Although the annular ligament cannot be palpated separately, its location can be isolated.

1) With your partner's elbow flexed, place your thumbpad on the head of the radius.
2) While passively pronating and supinating the forearm, allow the head and neck of the radius to pivot under your thumb (3.137, 3.138). You may not feel the annular ligament separately, but visualize it stabilizing the head of the radius to the ulna.

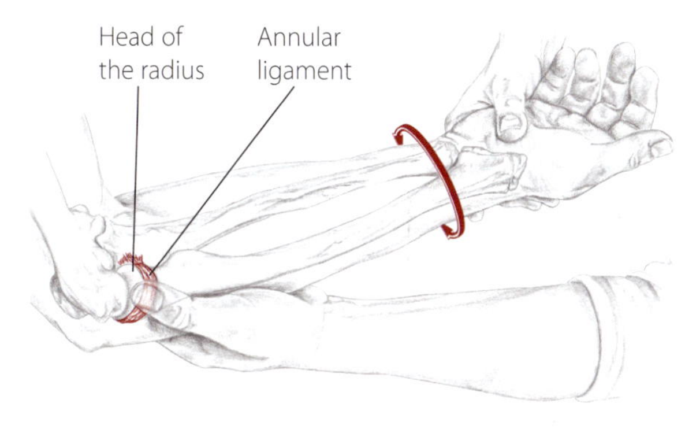

(3.137) Posterior/lateral view of right forearm

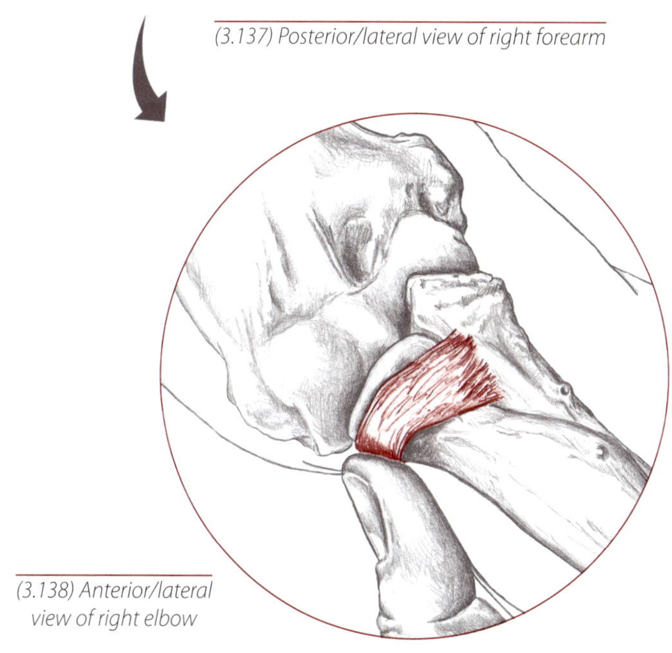

(3.138) Anterior/lateral view of right elbow

Ulnar Collateral Ligament

The ulnar collateral ligament is a strong, triangular-shaped ligament that originates on the humerus' medial epicondyle (3.135). Its fibers spread out and attach to the coronoid process of the ulna and to the olecranon process. The collateral ligament is deep to the common flexor tendon, but superficial to the ulnar nerve.

1) With the elbow flexed, locate the medial epicondyle of the humerus and the medial aspect of the olecranon process.
2) Place your thumb between these landmarks.
3) Palpating through the overlying muscle tissue, explore the ligament's thin fibers which run transversely to the fibers of the muscles (3.139). You may not feel something distinct, but if you are between the stated landmarks you are in the right location.

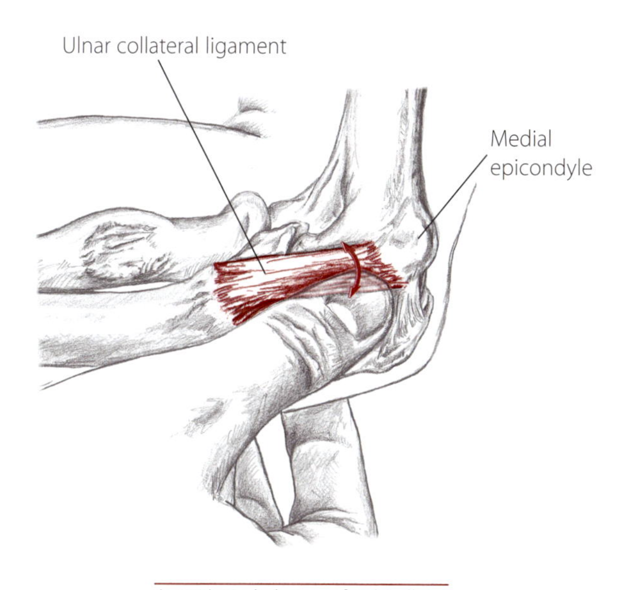

(3.139) Medial view of right elbow

Ulnar Nerve

The ulnar nerve passes between the medial epicondyle and olecranon process as it extends down the forearm. Between these two landmarks the nerve is superficial and easily accessible. Hence, bumping your elbow can irritate the ulnar nerve and create the annoying "funny bone" sensation down the forearm.

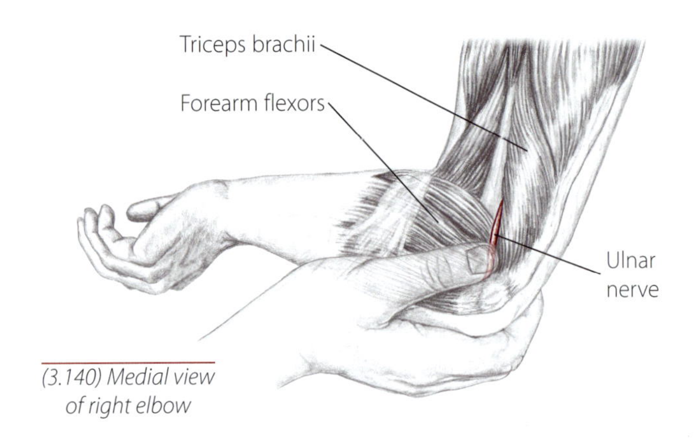

(3.140) Medial view of right elbow

Ulnar nerve

1) With the elbow flexed, locate the medial epicondyle and olecranon process.
2) Using gentle pressure, slide your finger into the space between these landmarks and palpate for the tube-shaped nerve (3.140).
3) Explore its location in relation to the triceps brachii tendon and common flexor tendon.

Is the structure you feel soft and movable? Are you palpating the ulnar nerve or the triceps brachii's fibrous tendon? Ask your partner to extend her elbow. The tendon will tighten and the nerve will "disappear" into the tissue.

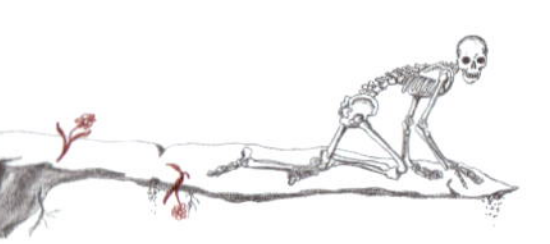

When palpating be sure not to press too hard and impinge on the ulnar nerve which can create tingling or numbness in the forearm or hand.

Olecranon Bursa

Just distal to the triceps brachii tendon, this small bursa pads the space between the olecranon process and skin of the elbow (3.141). Due to its location, the bursa can become inflamed when the elbow is irritated or struck by an external object. This condition, olecranon bursitis, (or "student's elbow") is readily observable by the distinct, localized ballooning of the elbow.

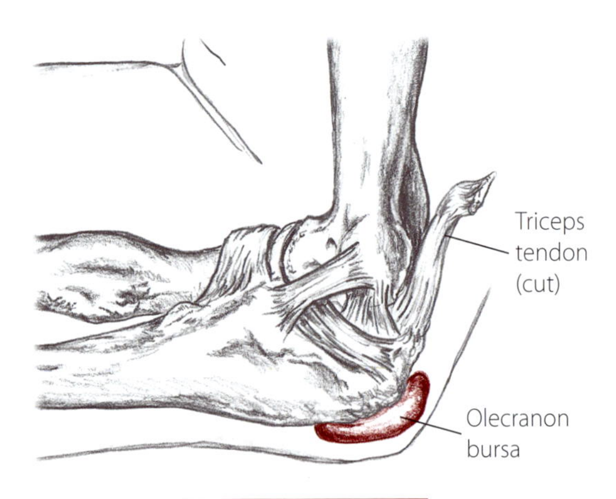

(3.141) Medial view of right elbow

1) With the elbow flexed at 90°, locate the olecranon process.
2) Palpating just distally to the process, gently explore the elbow's thin, malleable tissue. Then let the elbow extend and note how the skin and fascia become even more lax.
3) If the bursa is inflamed, the elbow will present a "goose egg" swelling with localized tenderness. In a healthy state the bursa is not palpable.

Interosseous Membrane

This thin but strong fibrous sheet binds together the forearm bones and serves as an attachment site for several muscles (3.142). Its oblique cord strengthens the membrane's proximal end. During stress, it is not uncommon for the bones to fracture before the interosseous membrane tears.

Because of its deep location, the membrane is not directly accessible. Exploration between the bones at the distal half of the forearm, however, can give you a sense of its presence and tensile firmness.

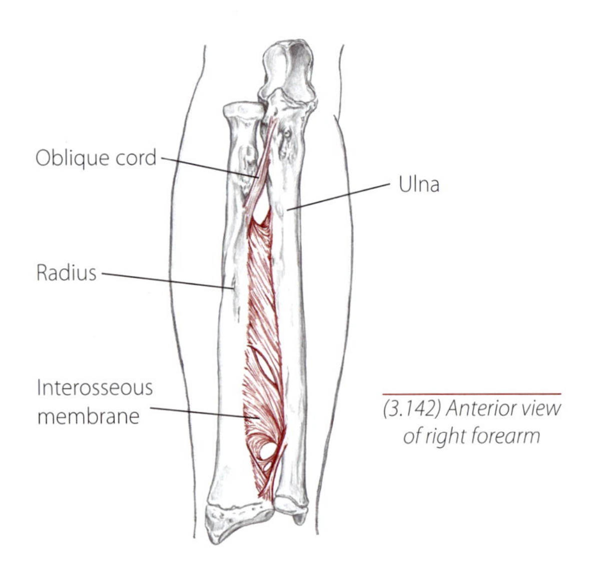

(3.142) Anterior view of right forearm

Retinacula of the Wrist and Palmar Aponeurosis

The **flexor retinaculum** is located on the palmar surface of the wrist just distal to the flexor crease. Its transverse fibers lie deep to the palmaris longus tendon and superficial to the other flexor tendons and median nerve. The flexor retinaculum and carpal bones form the carpal tunnel, in which the flexor tendons and median nerve pass (3.143).

Isolating the thin flexor retinaculum can be difficult, but its transverse fibers (which are perpendicular to the deeper tendons) help to distinguish it. Also, if the retinaculum is "tight" the tissue of the anterior wrist may have an inflexible feel.

The thick **palmar aponeurosis** is a continuation of the antebrachial fascia that stretches superficially across the palm of the hand and is an attachment site for the palmaris longus tendon. It is shaped in a similar way to the plantar aponeurosis (p. 393) on the sole of the foot. Although it may not be easy to palpate, nevertheless, like the flexor retinaculum, its tensile quality can be felt.

The **extensor retinaculum** is superficial and located on the posterior wrist. Like the flexor retinaculum, it is a thickening of fascia that has transverse fibers stretching across the wrist to attach to underlying bones. It stabilizes the wrist and thumb extensors. It is roughly three-quarters of an inch wide and located distal to the head of the ulna and the styloid process of the radius.

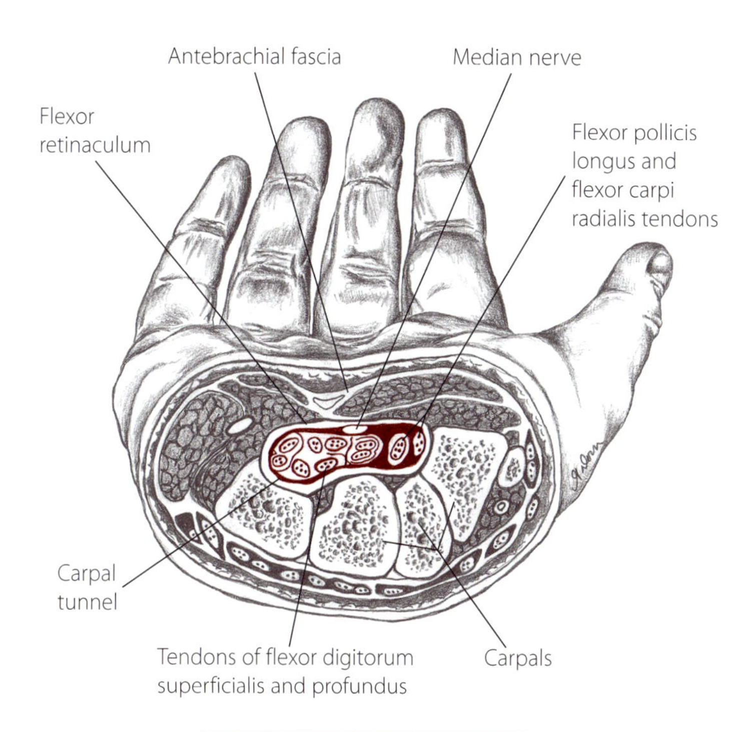

(3.143) Cross section of the right wrist

Flexor retinaculum and palmar aponeurosis

1) Cradle your partner's hand so your thumbpad is on the flexor crease of the wrist. Slide half an inch distally to the crease and sink into the thick tissues of the "heel" of the hand (3.144).
2) As you explore the carpal space, visualize the retinaculum spanning across the carpals. Passively flex and extend the wrist and feel the tension in the retinaculum change.
3) Slide distally onto the palm of the hand and palpate for the thick, superficial palmar aponeurosis.

When palpating the flexor retinaculum, are you distal to the level of the pisiform (p. 129)? To highlight the palmar aponeurosis, ask your partner to tighten his hand as if he were "palming a basketball." (3.145) Note how this action also brings the palmaris longus tendon into view.

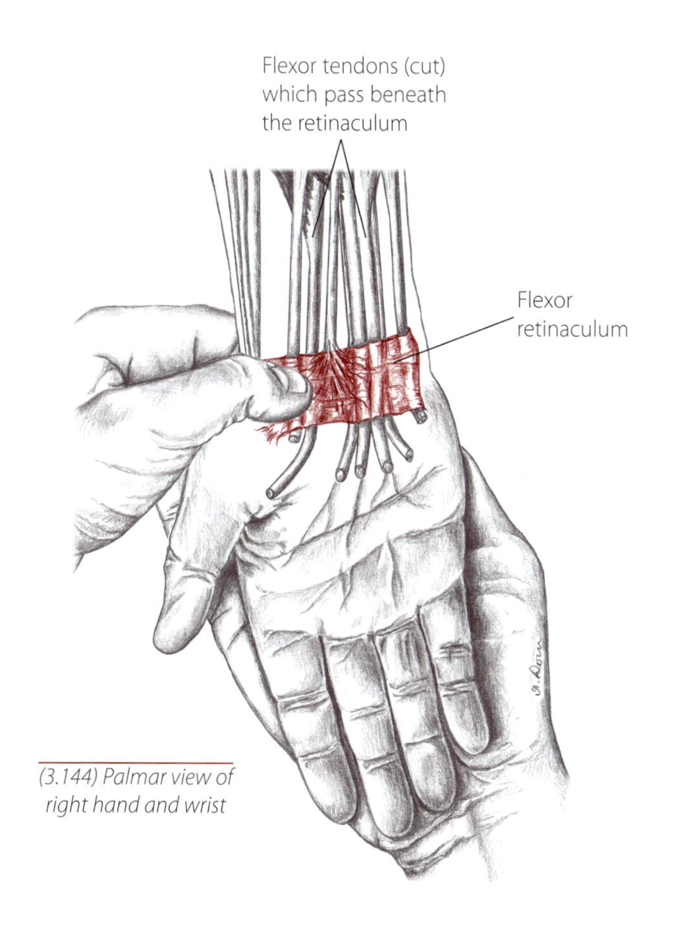

(3.144) Palmar view of right hand and wrist

retinaculum	**ret**-i-**nak**-u-lum	L. halter, band, rope
aponeurosis	**ap**-o-nu-**ro**-sis	Grk. *apo*, from + *neuron*, nerve or tendon

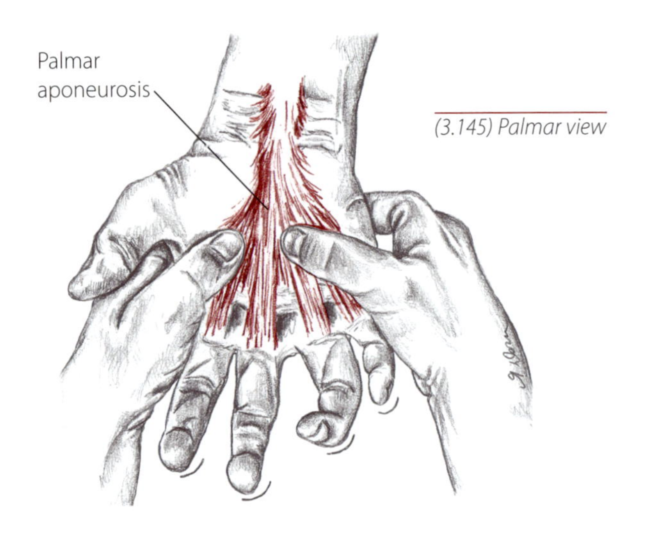

(3.145) Palmar view

Extensor retinaculum

1) Ask your partner to extend her fingers and wrist. The pressure from the bulging extensor tendons will make the retinaculum more distinct.
2) Locate the head of the ulna and the styloid process of the radius.
3) Palpate just distal to these landmarks by sliding across the transverse fibers of the thin retinaculum (3.146).

Are you distal to the head of the ulna and the styloid process of the radius? Can you distinguish superficial, transverse fibers?

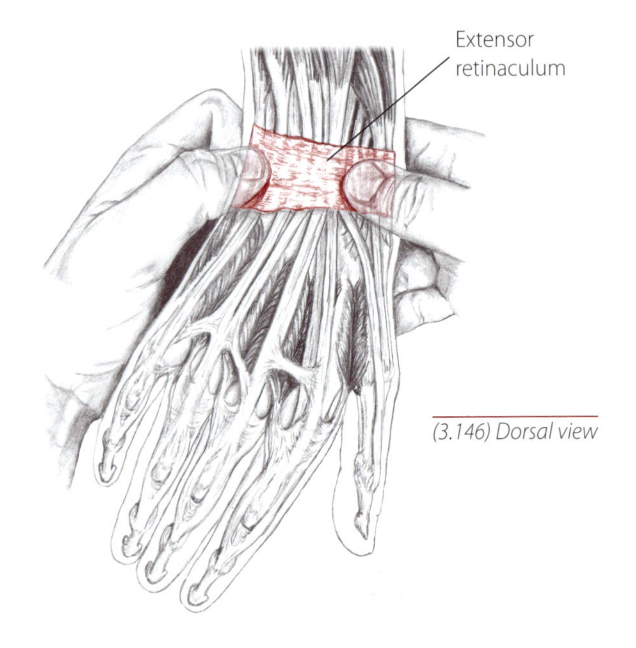

(3.146) Dorsal view

Radial and Ulnar Arteries

The radial and ulnar arteries branch off the brachial artery and travel down the forearm to the hand. The **radial artery** is often used for taking a pulse. It is detectable on the anterior wrist between the tendon of the flexor carpi radialis and the shaft of the radius.

The **ulnar artery** is found proximal to the pisiform and medial to the palmaris longus tendon. Its pulse may not be as easily accessible as the radial pulse.

1) Locate the radial pulse by placing two fingerpads on the flexor side of the wrist. Move laterally and gently press to feel the pulse (3.147).
2) Locate the ulnar pulse by moving your fingerpads to the medial side of the flexor surface (3.148).

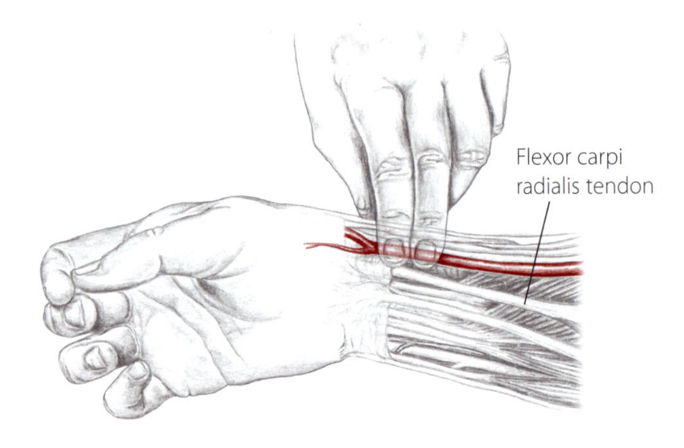

(3.147) Feeling the pulse of the radial artery

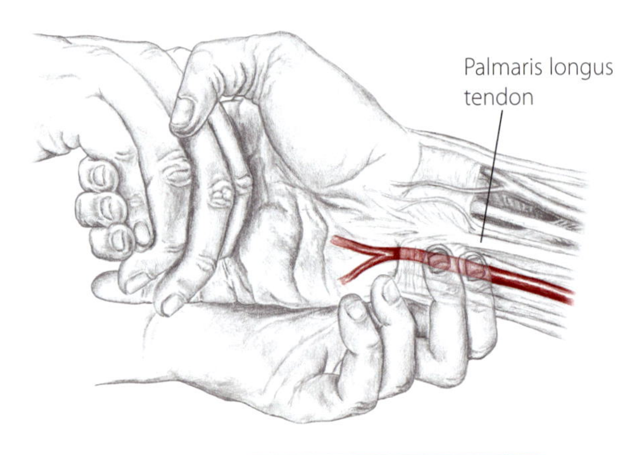

(3.148) Feeling the pulse of the ulnar artery

Ligaments of the Wrist and Hand

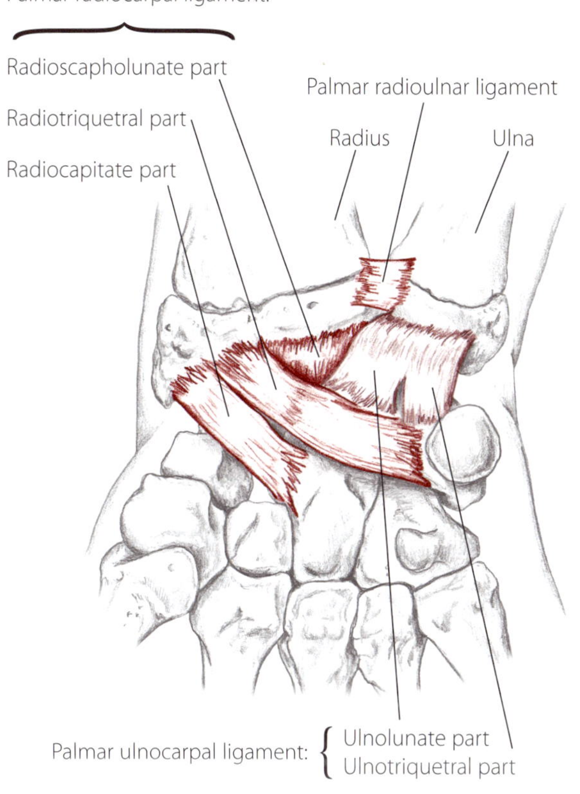

(3.149) Palmar view of right wrist showing ligaments of radiocarpal joints

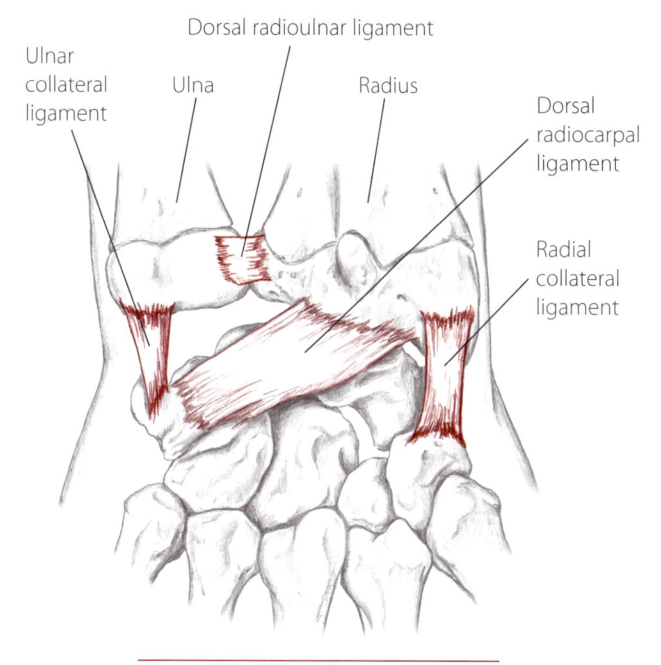

(3.150) Dorsal view of right wrist showing ligaments of radiocarpal joints

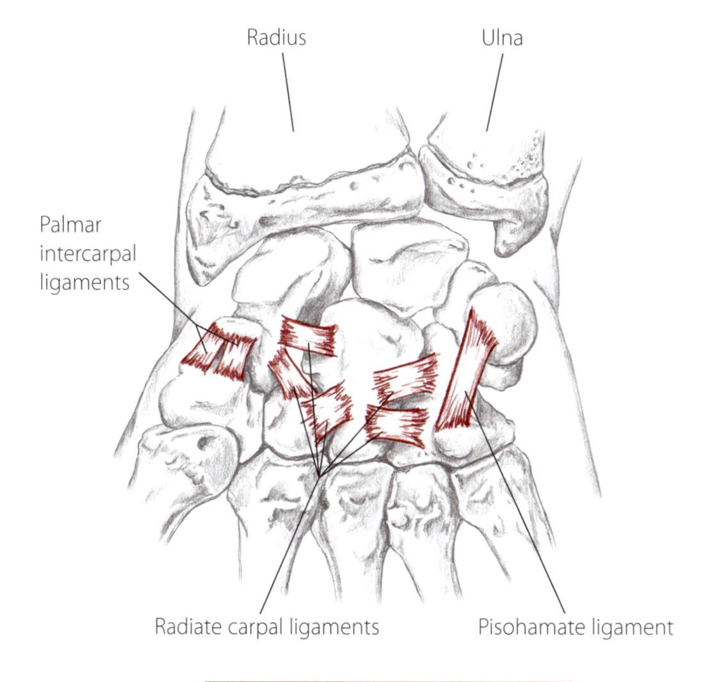

(3.151) Palmar view of right wrist showing ligaments of intercarpal joints

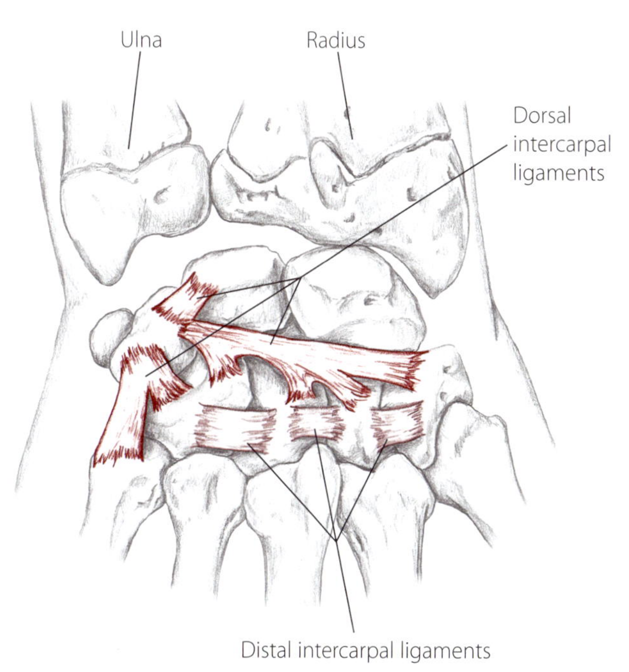

(3.152) Dorsal view of right wrist showing ligaments of intercarpal joints

radioscapholunate	**ray**-dee-o-**skaf**-o-**loo**-nate
radiotriquetrum	**ray**-dee-o-tri-**kwe**-trum
radiocapitate	**ray**-dee-o-**kap**-i-tate

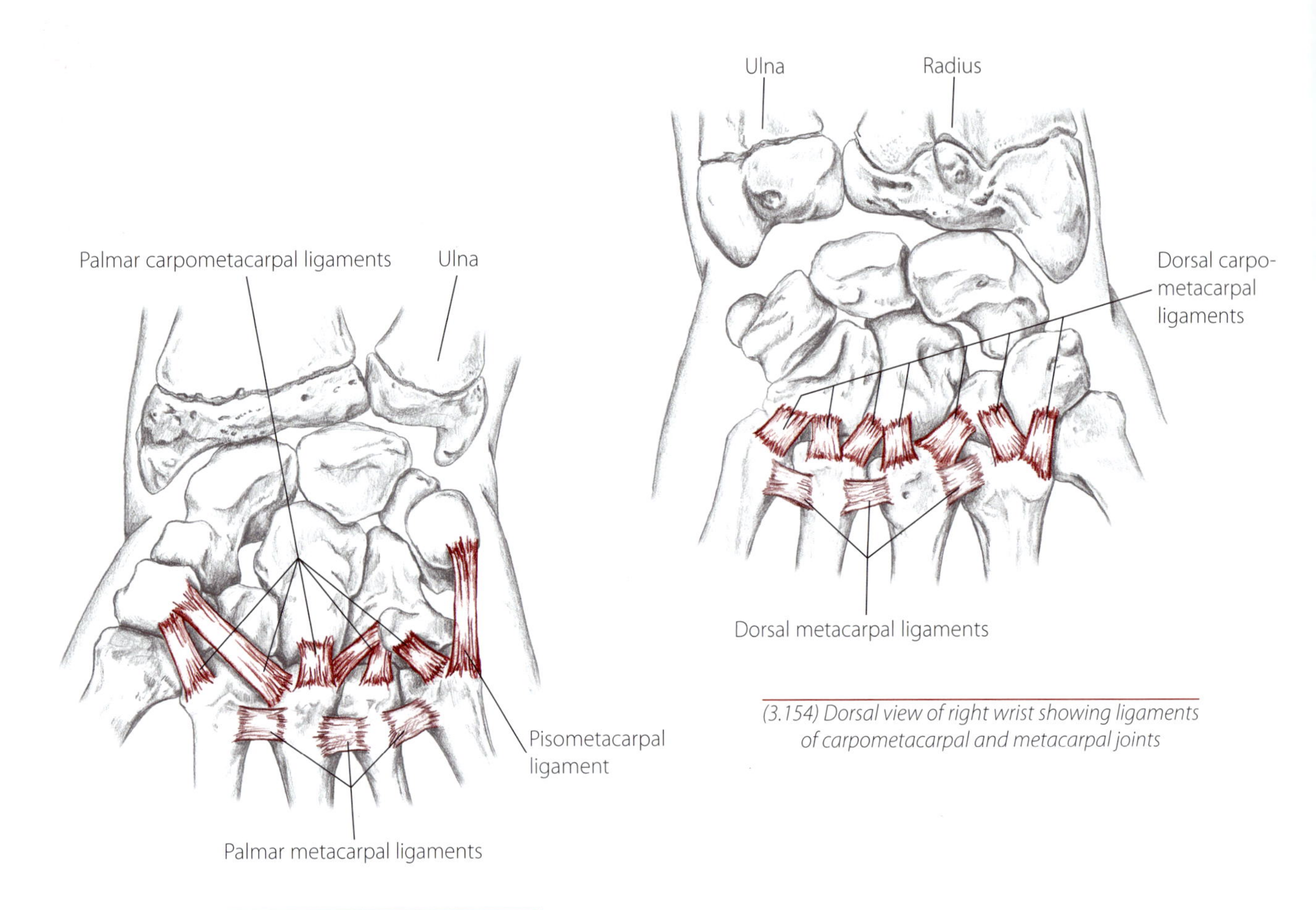

(3.153) Palmar view of right wrist showing ligaments of carpometacarpal and metacarpal joints

(3.154) Dorsal view of right wrist showing ligaments of carpometacarpal and metacarpal joints

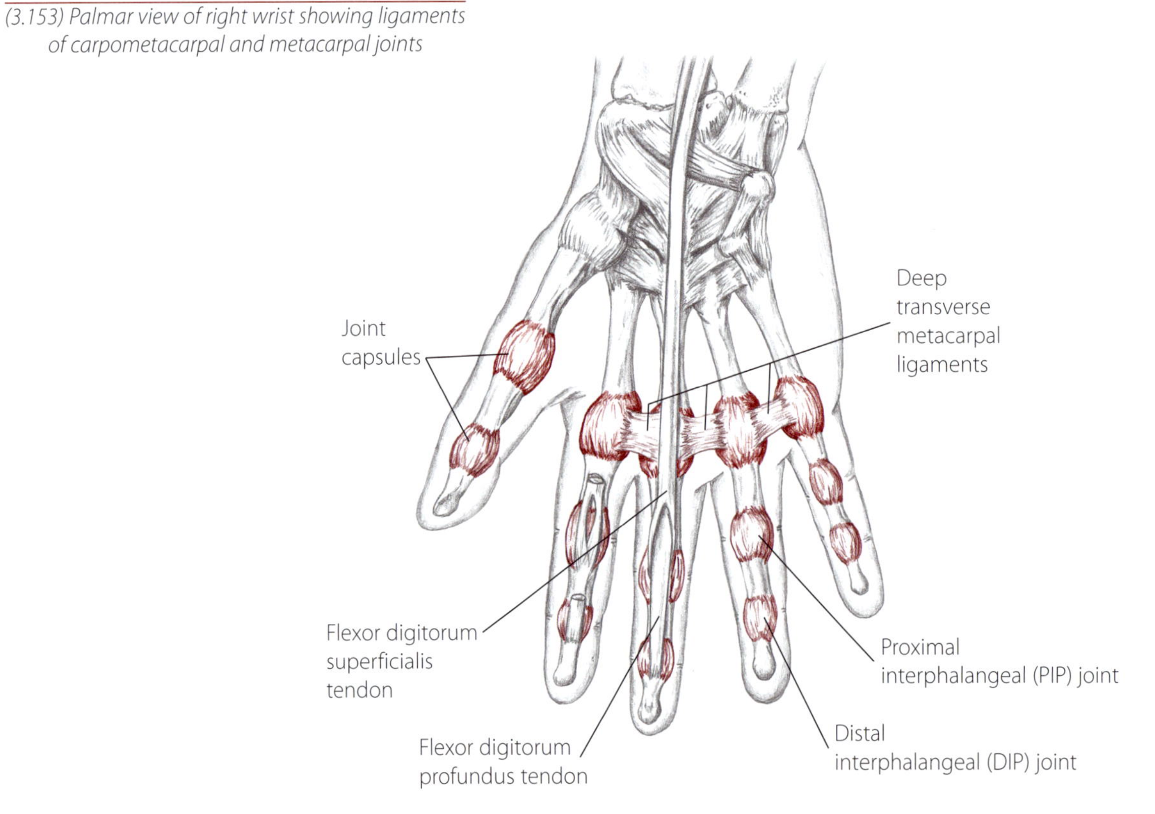

(3.155) Palmar view of right hand showing interphalangeal joints

ulnolunate **ul**-no-**lu**-nate
ulnotriquetrum **ul**-no-tri-**kwe**-trum

4

Spine & Thorax

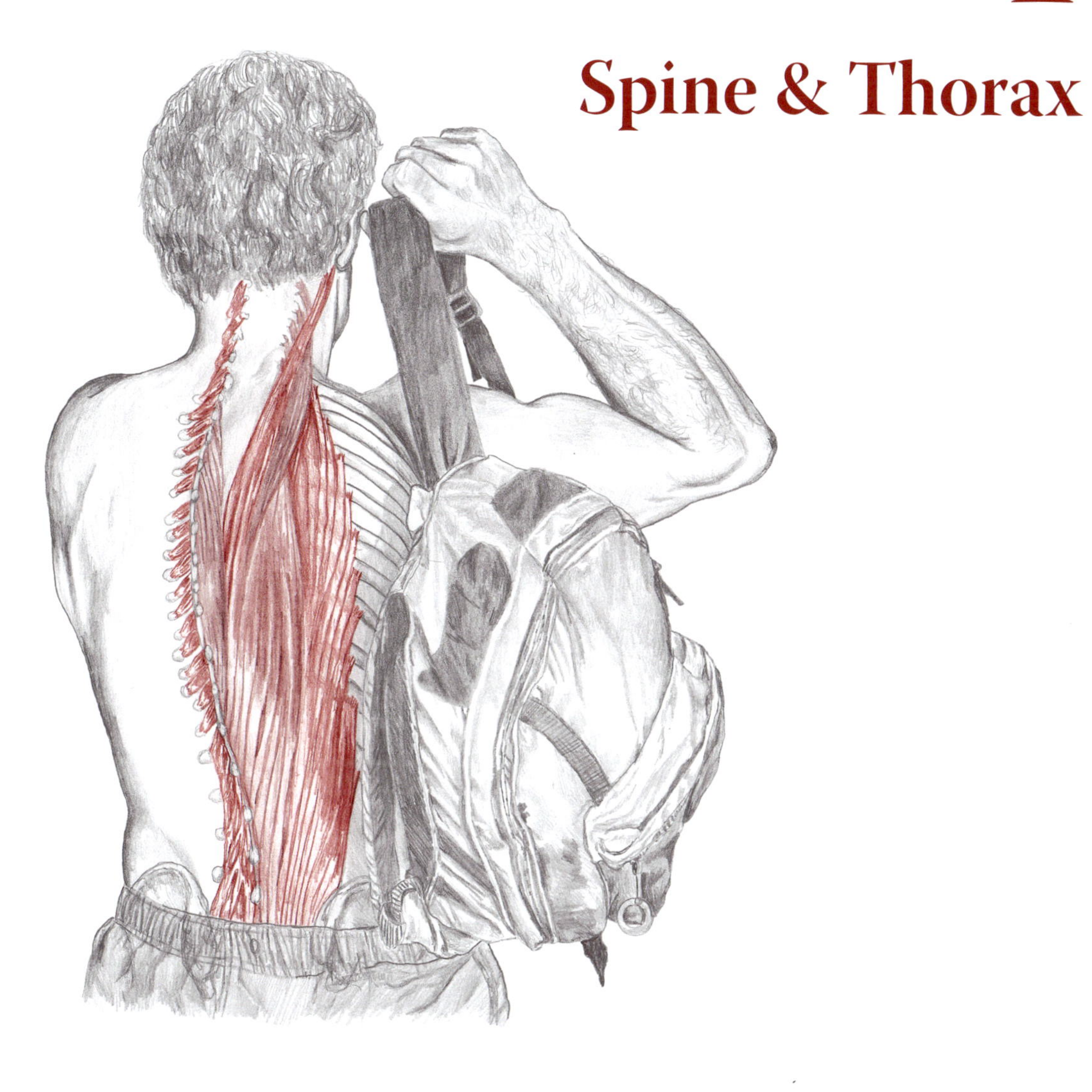

When we are standing, the entire weight of our trunk, head and arms is transferred through the bodies of the vertebrae. The lumbar vertebrae at the base of the spine bear the brunt of this weight.

Fortunately, between the bodies of each vertebrae are intervertebral discs that cushion some of this shock. The discs are composed of a tough outer layer, the annulus fibrosus, and a liquid center called the nucleus pulposus. When weight is placed on a disc, the annulus fibrosus supports the nucleus pulposus in compressing and distributing the pressure. The nucleus is mostly water, some of which is squeezed out in the course of the day.

When you are asleep, the pressure is off the spine and the discs are able to fully restore themselves, so that you wake up in the morning half an inch taller than you were the night before.

Topographical Views

A vertebrate is an animal that has a spinal column. Vertebrates include fishes, amphibians, reptiles, birds and humans. An insect or a mollusk has no spinal column and is therefore called an invertebrate. Animals that walk on four legs are called quadrupeds whereas humans are bipeds.

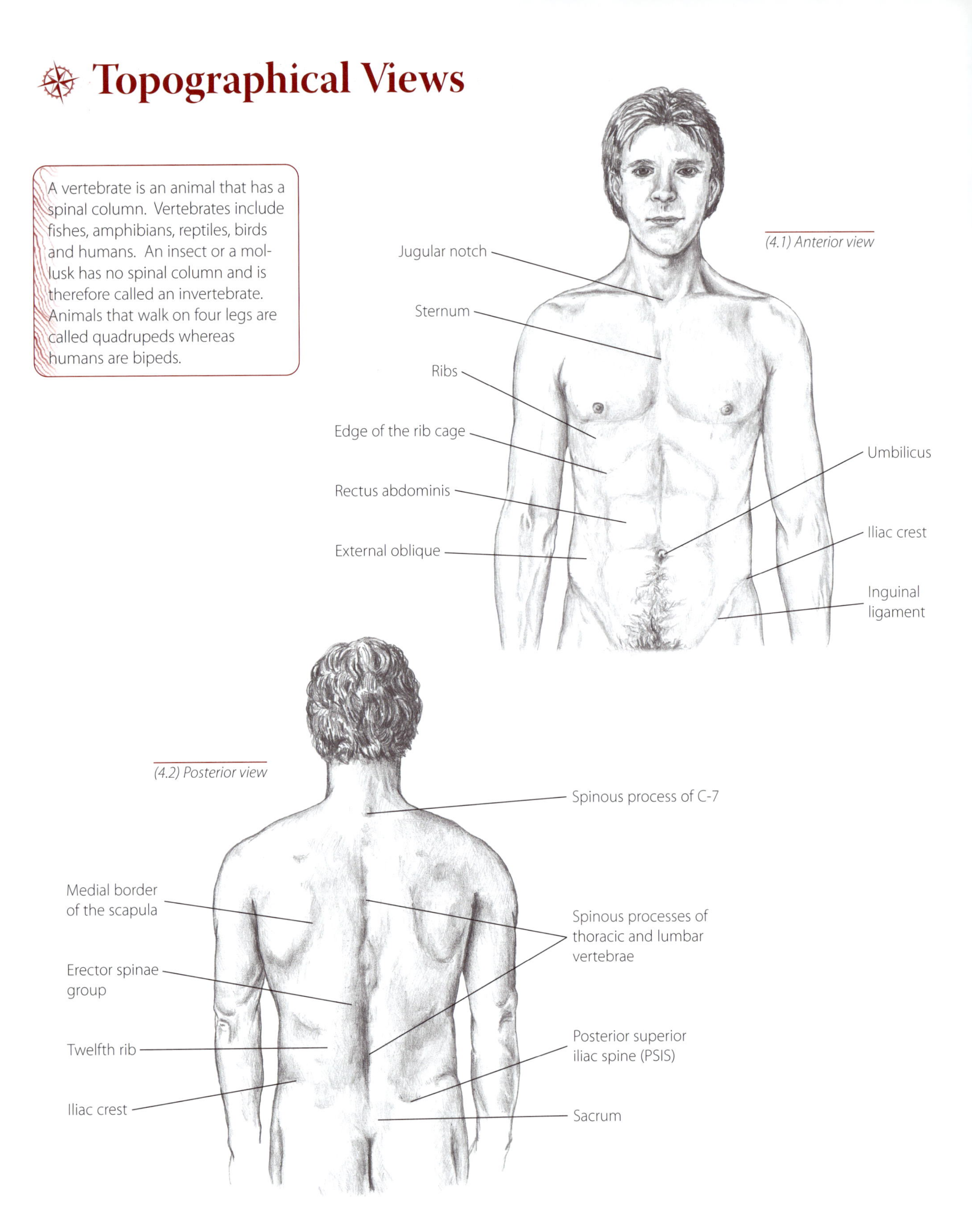

(4.1) Anterior view

(4.2) Posterior view

The vertebrae, sternum and pelvis of a bird are usually air filled or "pneumatized." A bone becomes "pneumatized," it is believed, when its surface comes in contact with an air sac. The bone tissue that rests against the air sac becomes thin before it disappears entirely, leaving behind a cavity that is then penetrated by the air sac. In such a way, the tiny air-filled outgrowths or sacs that extend off the lungs of a bird fill its bones and body cavities, thereby reducing its total body weight.

Exploring the Skin and Fascia

1) Partner prone. Begin by laying your hands on your partner's middle and lower back and sensing the temperature of the tissue. Explore the sides of the torso as well.
2) Begin to gently lift the skin and fascia superficial to the spinal column in the low back region (4.3). Oftentimes this tissue can be quite dense and impliable. Move a few inches laterally and compare it to the tissue superficial to the large erector muscles.
3) Continue further laterally to the sides of the trunk (between the axilla and pelvis). As you move more laterally can you detect any differences in the tissue's elasticity or thickness?

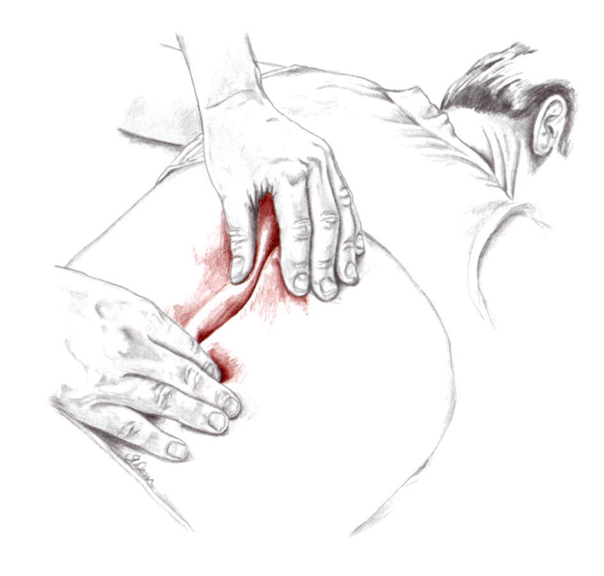

(4.3) Partner prone

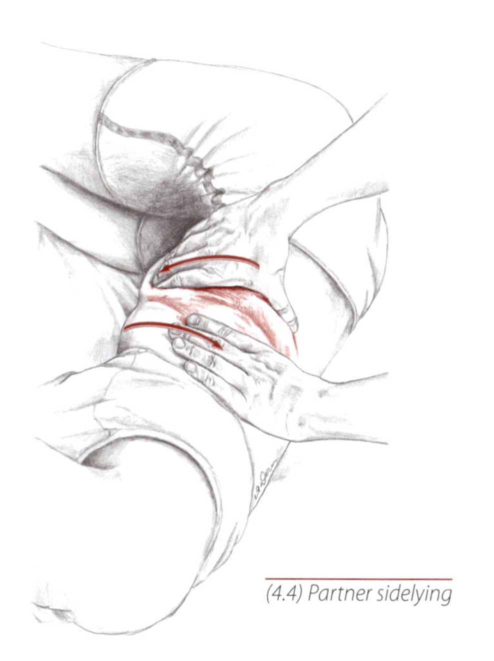

(4.4) Partner sidelying

When thinking about the thorax (the trunk of the body), most of us consider the "belly and back" surfaces and neglect the sides. Sidelying position will allow you to see that the sides of the thorax connect the "belly and back" together and that the thorax is actually one three-dimensional unit.

1) Partner sidelying. Lay both hands on the side of the thorax. See and feel how the anterior, lateral and posterior sides of the thorax form one continuous surface.
2) Gently wring your hands in opposite directions (4.4), sensing the tissue's pliability or resistance. Try to move it in all directions.

The abdomen can be ticklish or sensitive when palpated. Be sure to move slowly and gently, checking in with your partner.

1) Partner supine. Begin with your hands on the sides of the abdomen to sense the temperature of the tissue. Then explore toward the center of the abdomen, up to the edge of the ribs and just below the umbilicus.
2) Begin by gently lifting the skin and fascia of the lateral abdomen and proceed further toward the body's midline (4.5). If some areas may be particularly challenging to grasp, it may be an indication that such action is unwelcome by the body at this time.
3) While grasping a portion of tissue, ask your partner to perform a small "crunch." As the deeper abdominal muscles contract, notice how the tissue may come right out from between your fingers.

(4.5) Partner supine

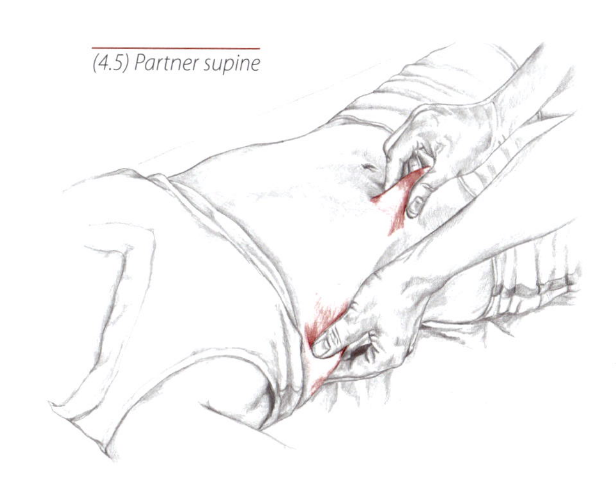

Bones of the Spine and Thorax

The **vertebral column** (or spine) consists of twenty-four vertebrae: seven **cervical** in the neck, twelve **thoracic** of the thorax and five **lumbar** in the lower back (4.6). The sacrum and coccyx are composed of fused vertebrae and are also considered part of the vertebral column. For the purposes of palpation and clarity, the sacrum and coccyx are included in Chapter Six, *Pelvis and Thigh*.

The **cervical vertebrae** are the most mobile and accessible of the twenty-four spinal bones. The twelve **thoracic vertebrae** articulate with the twelve pairs of ribs. Designed for minimal movement, they help to stabilize the thoracic area and protect the internal organs. In contrast, the larger, stockier **lumbar vertebrae**, located between the twelfth rib and posterior iliac crest, are designed to support the weight of the upper body.

As you palpate along the back, all twenty-four vertebrae will be deep to the layers of muscle tissue. However, the spinous and transverse processes protrude from each vertebra and can be helpful location points.

The **thorax** includes the sternum and rib cage. The superficial **sternum** ("breastbone") is located along the midline of the chest. The **rib cage** consists of costal cartilage and twelve pairs of ribs. The costal cartilage is identical in shape and size to the ribs and serves as a bridge between them and the sternum.

Ribs 1 through 7 are known as "true ribs" because they attach directly to the sternum. Ribs 8 through 12 are referred to as "false ribs" because they attach indirectly to the sternum by means of the costal cartilage. Aside from being "false ribs," the eleventh and twelfth ribs are also considered "floating ribs" as they do not attach to the sternum or costal cartilage at all.

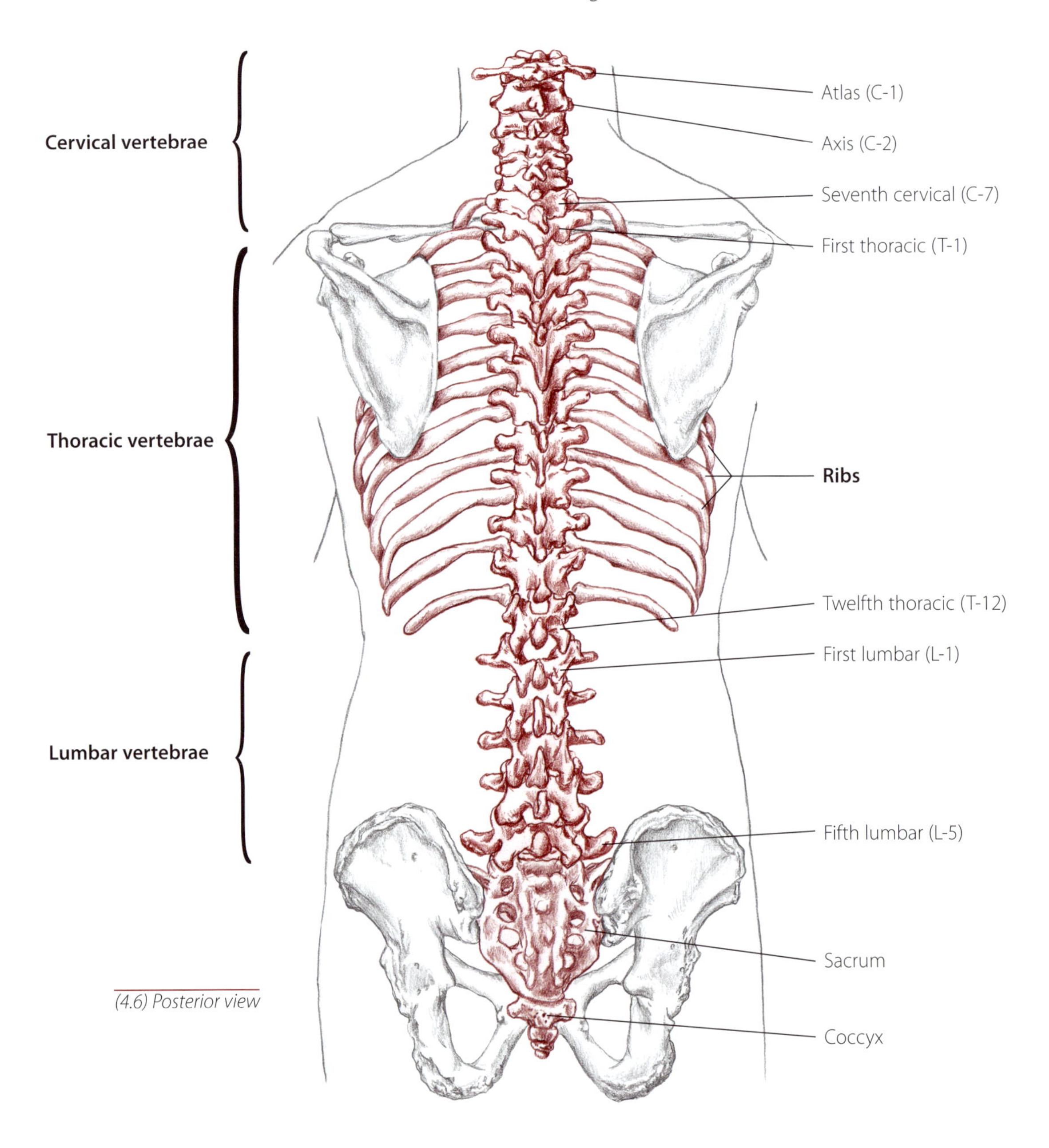

(4.6) Posterior view

cervical	**ser**-vi-kal	L. referring to the neck
chest		AS. box
lumbar	**lum**-bar	L. loin

Bony Landmarks

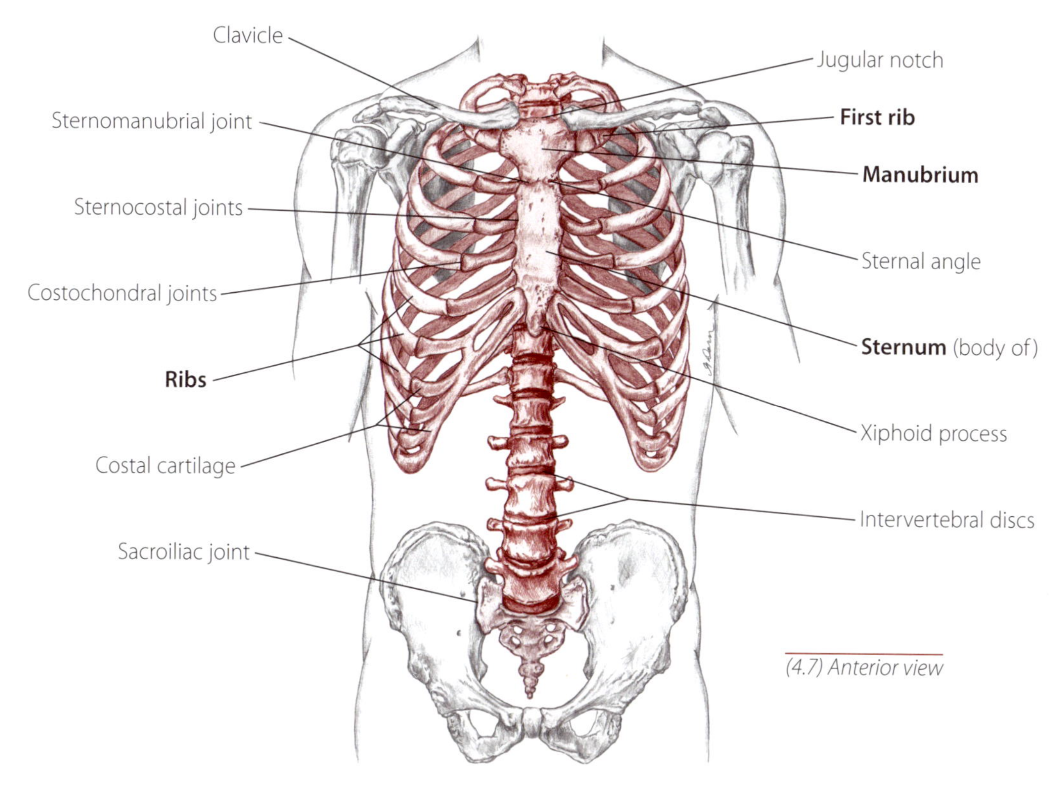

(4.7) Anterior view

Atlas (C-1)

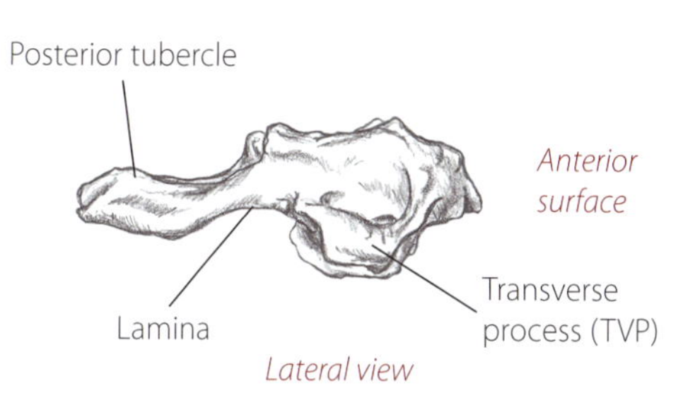

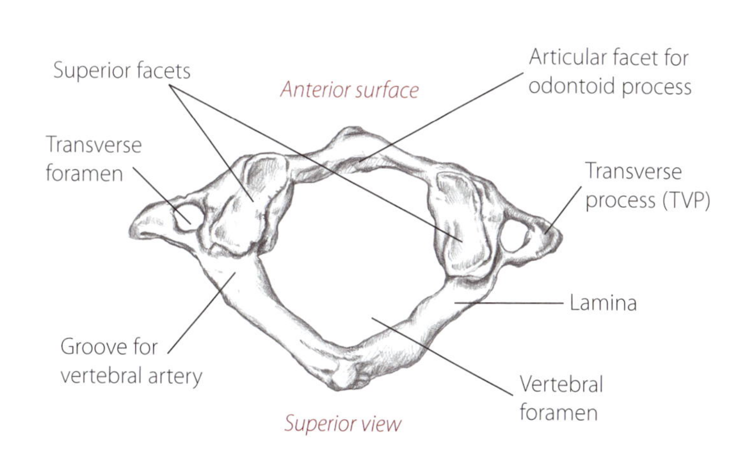

Axis (C-2)

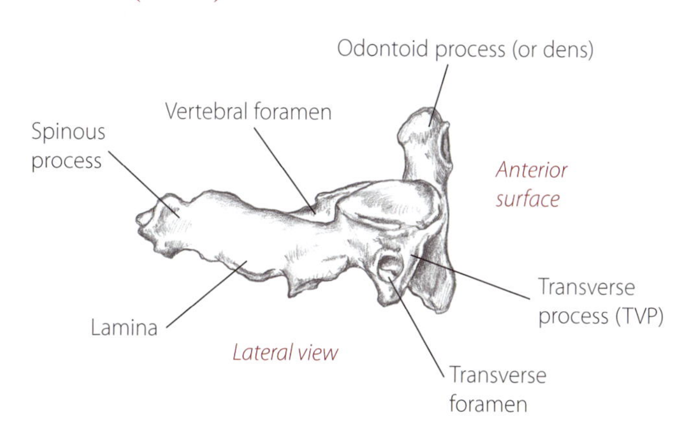

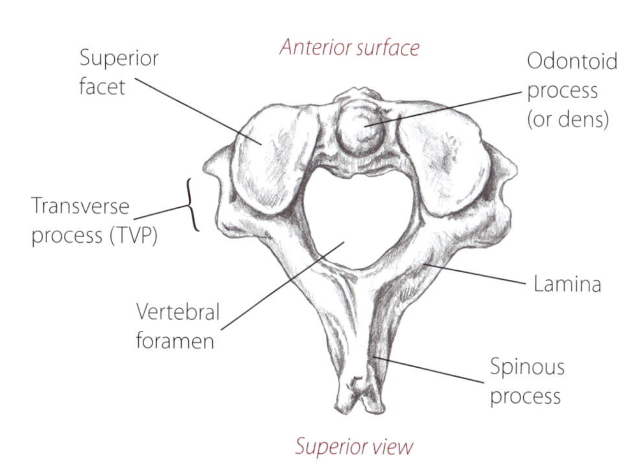

spine		L. thorn
thoracic	tho-**ras**-ik	Grk. chest
vertebra	**ver**-ta-bra	L. joint

Cervical

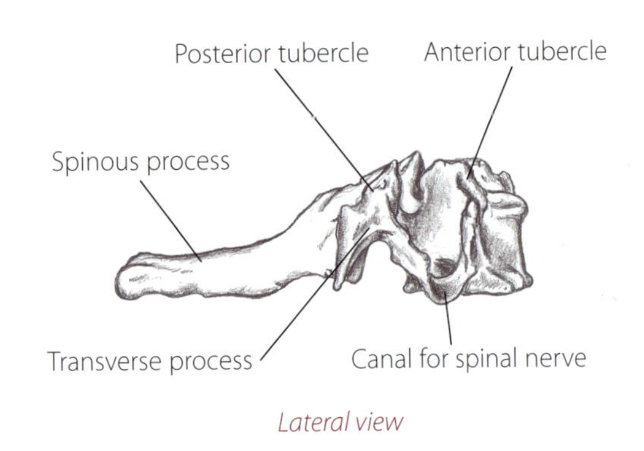

Lateral view

Superior view

Thoracic

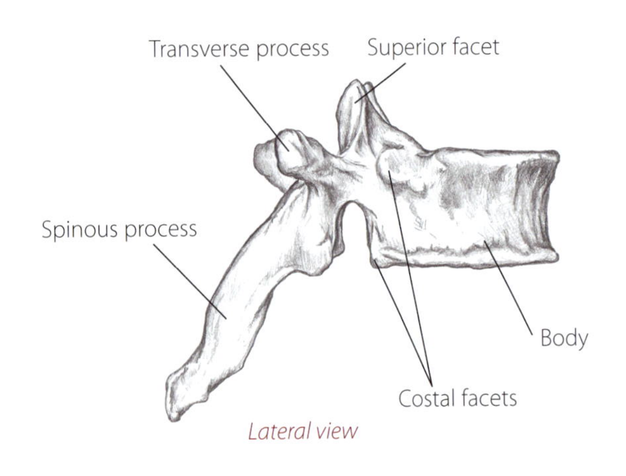

Lateral view

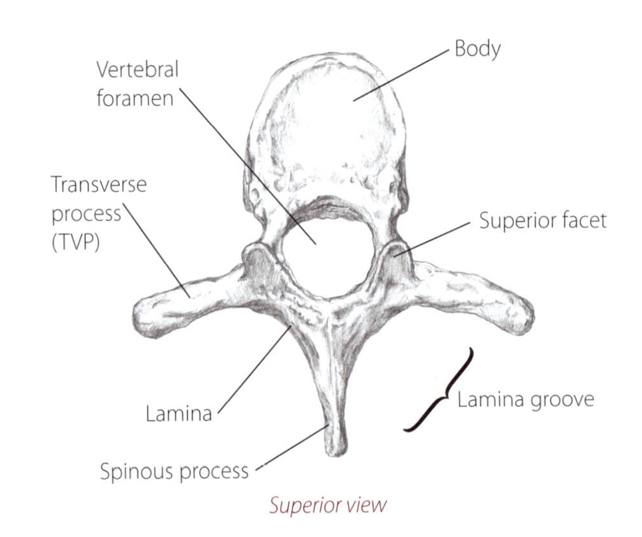

Superior view

Lumbar

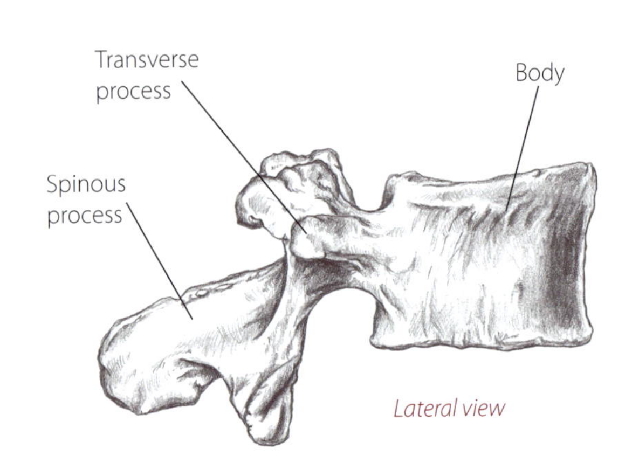

Lateral view

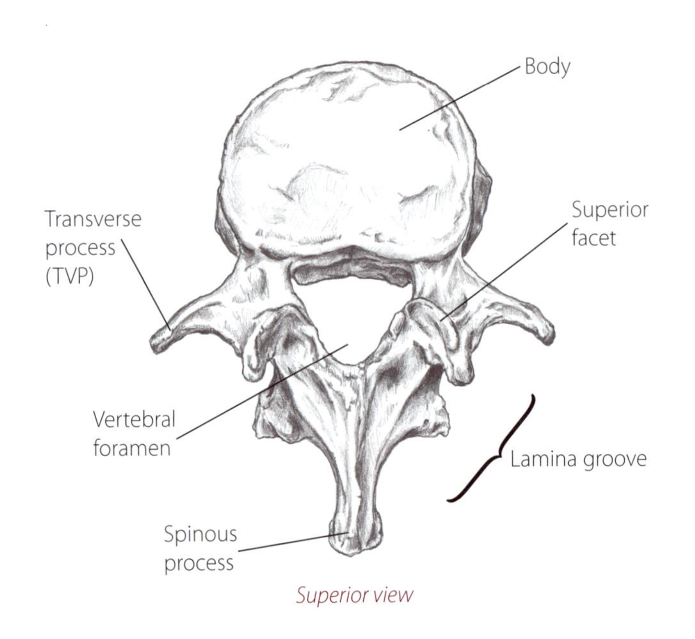

Superior view

facet	**fas**-et	Fr. small face
foramen	for-**aye**-men	L. a passage or opening
odontoid	o-**don**-toyd	Grk. toothlike

Bony Landmarks

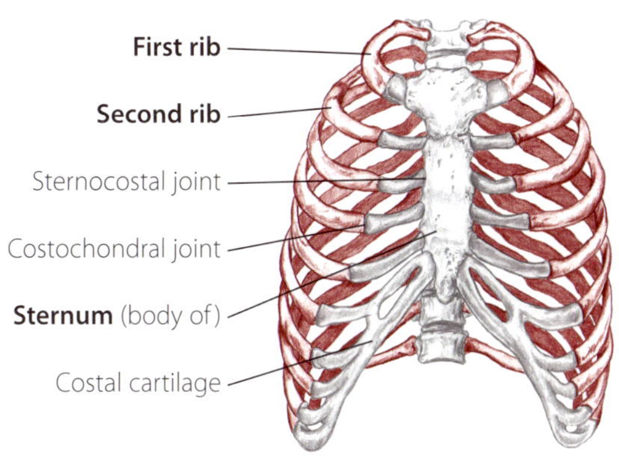

(4.8) Anterior view of thorax

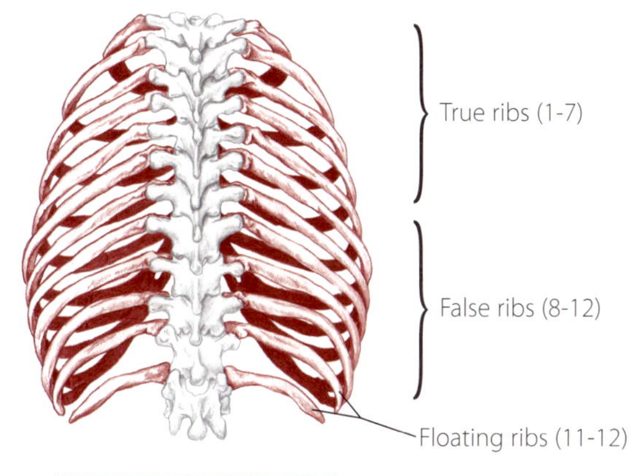

(4.9) Posterior view of thorax

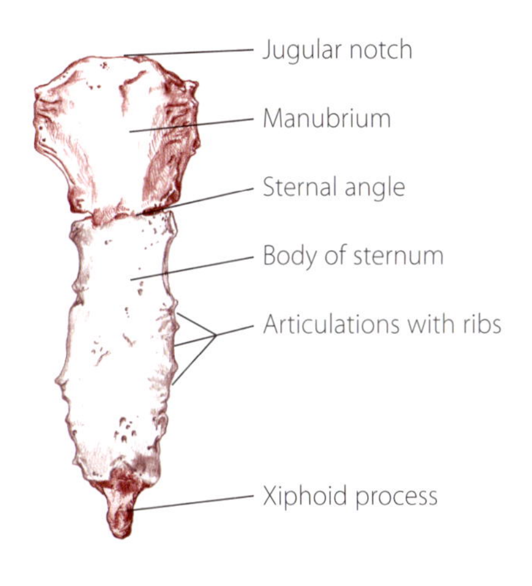

(4.10) Anterior view of sternum

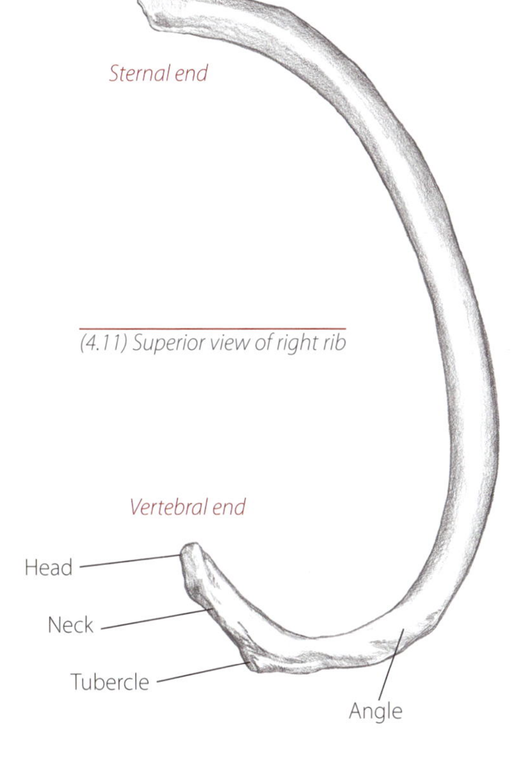

(4.11) Superior view of right rib

Groove for subclavian artery and vein

Sternal end

Vertebral end

Head

Neck

Tubercle

(4.12) Superior view of right first rib

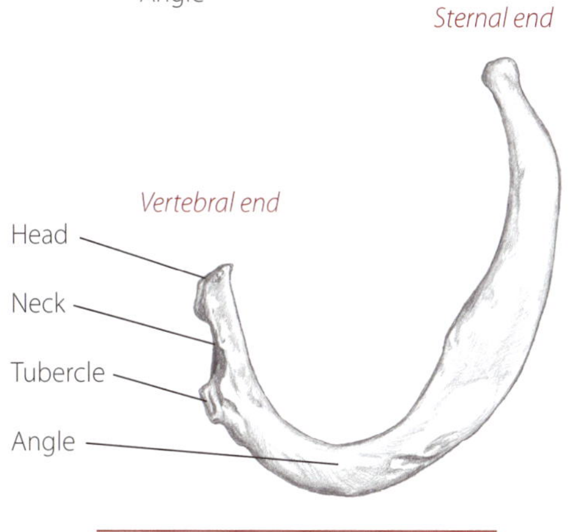

(4.13) Superior view of right second rib

Bony Landmark Trails

Trail 1 "Midline Ridge" explores the spinous processes of the vertebrae and the spaces between them as they run down the middle of the back.

Trail 2 "Crossing Paths" describes surrounding bony landmarks that intersect with specific spinous processes.

C-7 and base of the neck

T-2 and superior angle of the scapula

T-7 and inferior angle of the scapula

T-12 and the twelfth rib

L-4 and top of the iliac crest

Posterior view, Trail #2

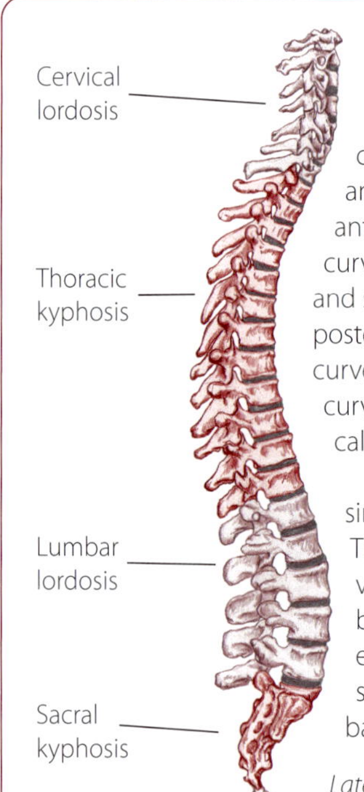

When viewed from the side, the vertebral column has four natural curvatures. The cervical and lumbar regions bend anteriorly to form lordotic curves, while the thoracic and sacral sections bow posteriorly creating kyphotic curves. An abnormal lateral curvature of the spine is called scoliosis.

At birth, the spine has a single kyphotic curvature. The cervical lordosis develops when an infant begins to hold his head erect. As he begins to stand and walk, the lumbar lordosis evolves.

Lateral view

kyphosis	ki-**fo**-sis	Grk. bent, curved, or stooped
lordosis	lor-**doh**-sis	Grk. bent backward

Bony Landmark Trails

Trail 3 "Nape Lane" locates the landmarks of the cervical vertebrae.

a Spinous processes of the cervicals
b Transverse processes of the cervicals
c Lamina groove of the cervicals

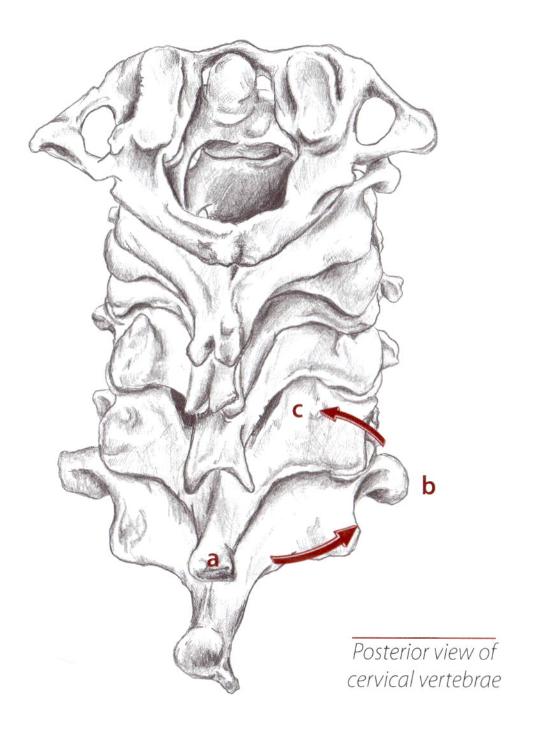

Posterior view of cervical vertebrae

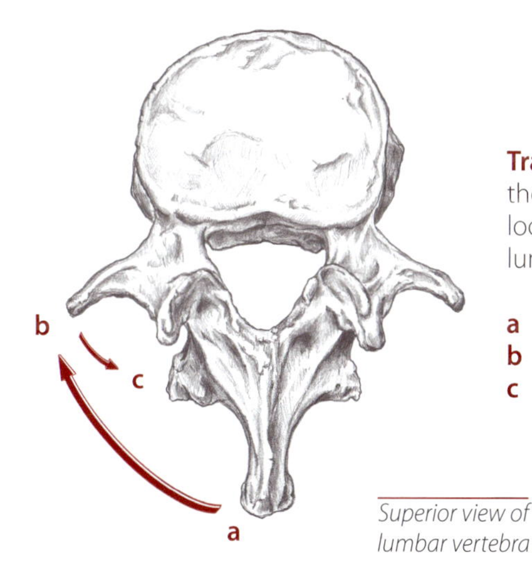

Superior view of lumbar vertebra

Trail 4 "Buried Boulevard" delves into the middle and low back regions to locate landmarks of the thoracic and lumbar vertebrae.

a Spinous processes
b Transverse processes
c Lamina grooves

Trail 5 "Breastbone Ridge" explores the sternum and its landmarks.

a Jugular notch
b Manubrium
c Body of the sternum
d Xiphoid process

Trail 6 "One Bumpy Road" explores the ribs, the rib cage and the costal cartilage.

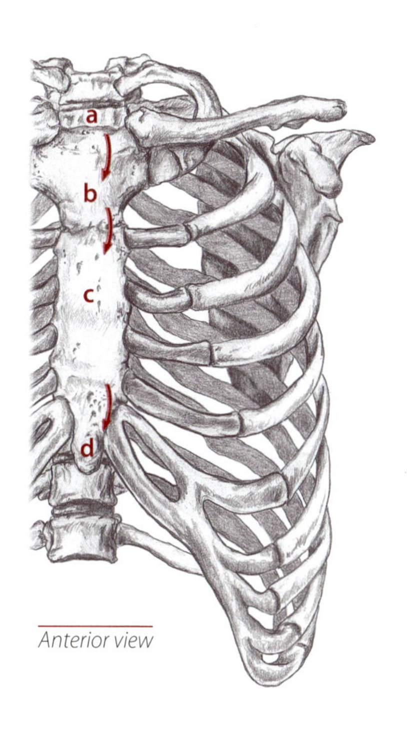

Anterior view

Trail 1 "Midline Ridge"

Spinous Processes of the Vertebrae

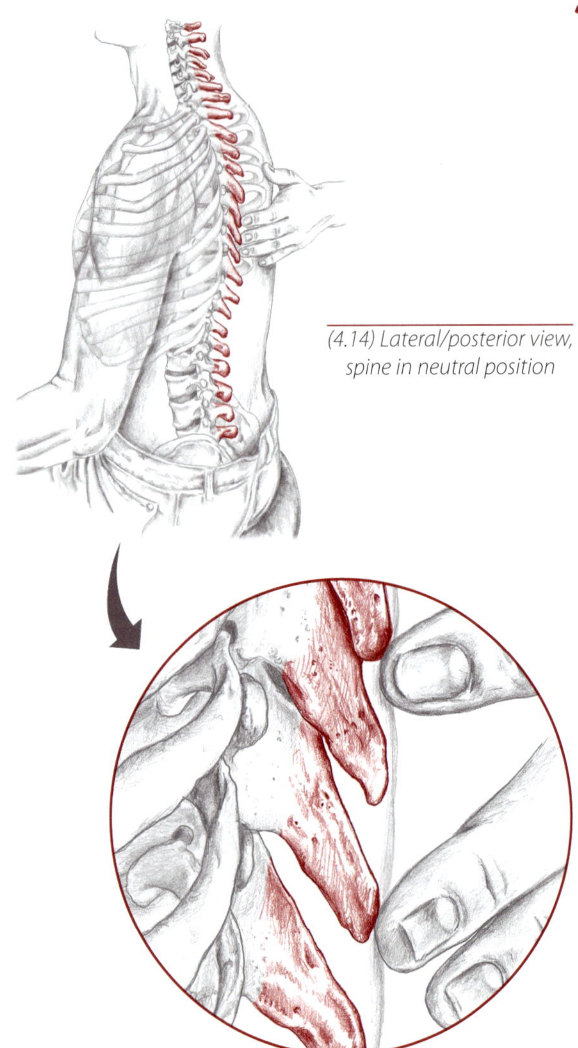

(4.14) Lateral/posterior view, spine in neutral position

A spinous process is a vertebra's posterior projection. As a group, the spinous processes form the visible row of bumps that run down the center of the back. They are designed as attachment sites for layers of muscles, ligaments and fasciae.

The lumbar, thoracic and cervical spinous processes differ from one another in several respects. The **spinous processes of the lumbar vertebrae** are much larger than the thoracic or cervical processes. Tall and stocky, the tips of the lumbar processes may feel more like short strips than points. The bodies of these vertebrae are quite massive and tall; they may have a finger's width of space between their processes. The **thoracic spinous processes** are smaller and closer together than their lumbar counterparts and angle downward.

The **cervical spinous processes** are shorter and smaller than the thoracic processes. Because of the lordotic curve in the cervical spine and the overlying ligamentum nuchae, the cervical spinous processes are actually deeper than the thoracic and lumbar vertebrae. The first cervical vertebra (C-1), or atlas as it is called, is the only vertebra that does not have a spinous process.

1) Partner seated with trunk and neck slightly flexed (this will stretch the overlying tissues and allow the processes to move posteriorly for easier access). Place your fingers along the midline of the back and locate the long line of processes (4.14, 4.15).
2) Slide your fingers slowly up and down the spine, palpating the different sizes, prominences and spaces between processes. Some processes may present themselves immediately while others may be more difficult to find. Ask your partner to slowly flex and extend his spine, noting the movement of the processes.
3) Try this same method with your partner prone.

Can you sculpt out the sides of the processes as you palpate them? Is there a dip superior and inferior to the point you feel? Can you line up three fingers on a series of processes or the spaces between the processes?

Partner prone or seated. Palpate the entire spinal column and count the spinous processes. How many can you feel? All the vertebrae (except C-1) have spinous processes, making a total of twenty-three. Use the intersecting spinous processes, such as C-7, T-12 and L-4, to check your accuracy.

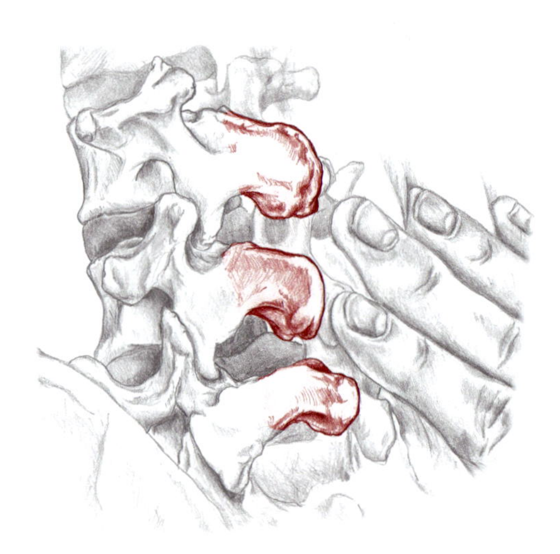

(4.15) Posterior/lateral view, palpating the lumbar spinous processes

Trail 2 "Crossing Paths"

Several spinous processes can be located with the help of intersecting bony landmarks. For example, a line drawn between the tops of the two iliac crests will cross the spinous process of L-4, which, in turn, leads you to its neighboring processes. Because everyone's body is unique, these intersecting landmarks are not definitive; they are best used as guides.

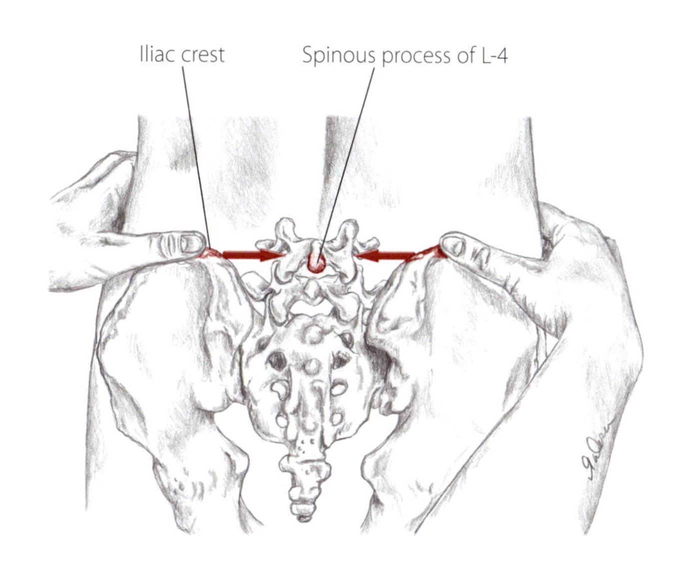

(4.16) Posterior view, partner standing

 L-4 and Top of the Iliac Crest

1) With your partner either prone or standing, locate the lateral aspects of both iliac crests (p. 283).
2) With your index fingers along the top of the crests, slide your thumbs medially, meeting at the spine (4.16).
3) Isolate the large knob of L-4. Explore superiorly and inferiorly for the adjacent lumbar processes.

Are you at the level of the iliac crests? Can you feel a firm protuberance at the midline of the body?

T-12 and the Twelfth Rib

The eleventh and twelfth ribs do not attach to the costal cartilage and are therefore considered "floating ribs." The twelfth rib has a slender, spear-like shape and angles inferiorly. It may vary in length between three to six inches and can be used as a locator for the spinous process of T-12. (See p. 193 for more information about the twelfth rib.)

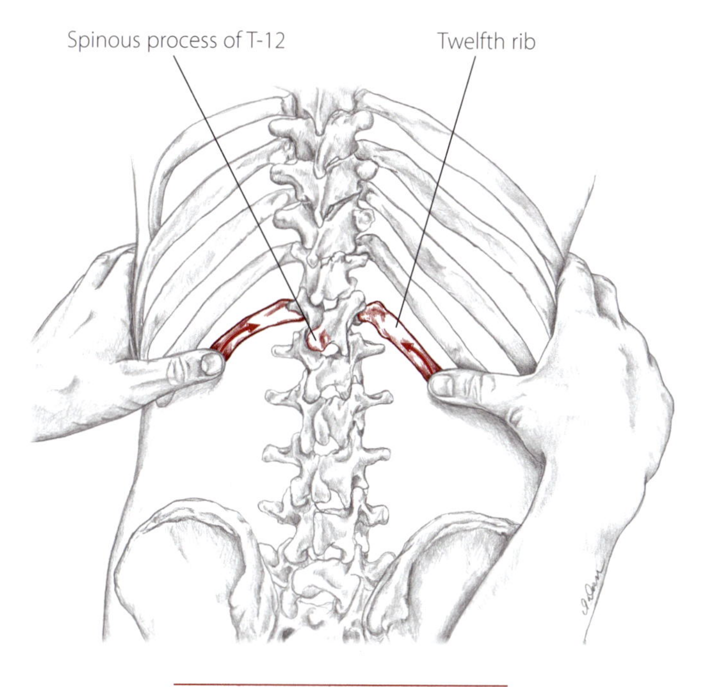

(4.17) Posterior view, partner standing

1) Partner prone or standing. The strategy is to locate the tip of the twelfth rib and follow its shaft to the spinous process. Reaching across to the opposite side of the body, place your hand along the lateral edge of the ribs.
2) Slide inferiorly to the bottom of the rib cage and explore for the tip of the twelfth rib (4.17).
3) With the tip isolated, gently follow the shaft of the rib medially, noting how it lies at an angle. As the rib lies deep to the erector spinae muscles, you may lose contact with its most medial portion. Continue to slide your fingers in the direction of the shaft, palpating for the spinous process.

 *If you have located L-4, can you now count the processes up to T-12?*

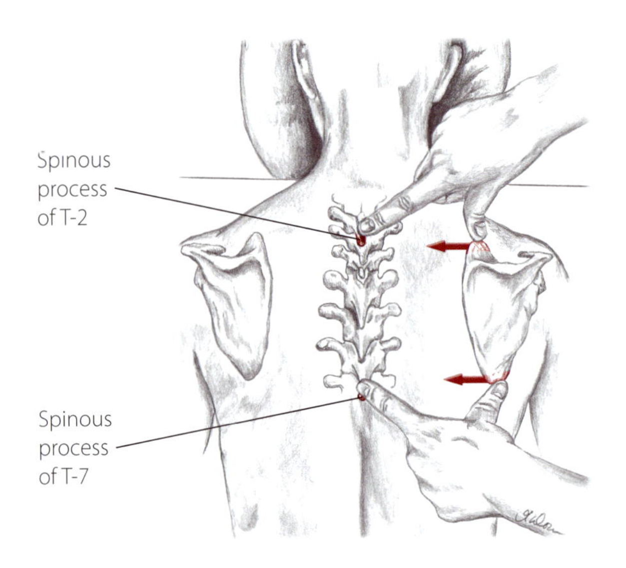

(4.18) Posterior view, partner prone

T-7 and Inferior Angle
T-2 and Superior Angle

Body type, muscular contraction and other factors affect the positioning of the scapulae. The inferior angle of the scapula will generally lie at the level of the spinous process of T-7, while the superior angle is at the level of T-2.

1) Partner prone or standing. Locate the inferior angle (p. 61). Keep one hand at the angle while sliding the other medially to the vertebral column.
2) Locate the superior angle. Keep one hand at the angle while sliding the other medially toward the vertebrae (4.18).

From T-7 can you count the processes down to T-12? Can you count them up to T-2? From T-2 can you count down to T-7?

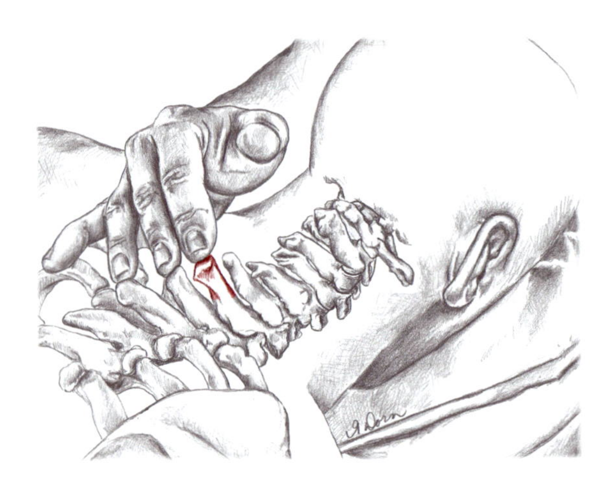

(4.19) Partner prone, locating the spinous process of C-7

C-7 and Base of the Neck

The spinous process of C-7 is located at the base of the neck. It protrudes further than C-6, C-5 and C-4 - a helpful distinction when locating structures in the upper back and neck.

1) Prone. Place your fingerpad superior to the base of the neck along the midline of the body.
2) Slide inferiorly. At the base of the neck, your thumb will bump into the process of C-7 (4.19).
3) Explore its edges and neighboring processes and then try to locate it from a supine position (4.20).

Are you at the base of the neck? Is the process superior to your finger smaller than the process you are palpating? Is there an equally protruding process (T-1) immediately inferior?

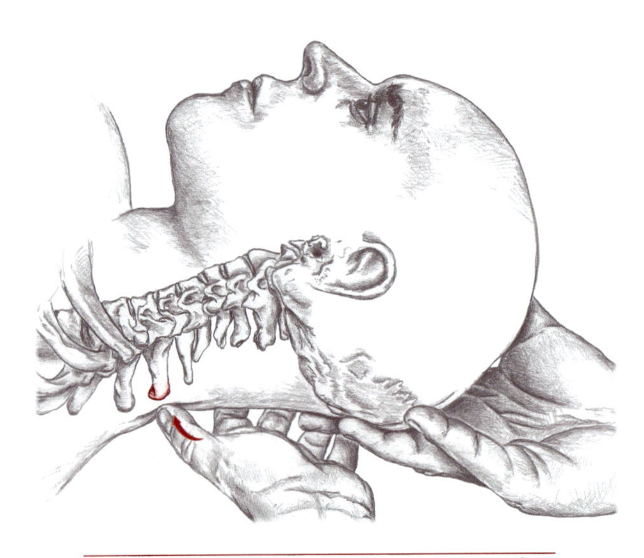

(4.20) Partner supine, palpating spinous process of C-7

When the neck is flexed, the spinous process of C-7 shifts superiorly. T-1, however, is buckled in by the first ribs and does not move. With your partner seated, place a finger on the spinous processes of C-7 and T-1. Have your partner slowly flex his neck and observe both processes. Does C-7 tilt superiorly while T-1 is stationary?

Trail 3 "Nape Lane"

Spinous Processes of Cervicals

The spinous processes of C-6, C-5, C-4 and C-3 protrude posteriorly to approximately the same extent. The process of C-2, however, is larger and more distinct. The tips of the cervical spinous processes are all deep to the ligamentum nuchae (p. 224), a flat ligament attaching to the processes and running superiorly to the occiput (p. 237).

1) Partner supine. Locate the spinous process of C-7.
2) Using gentle pressure, explore the tips and sides of the other cervical processes (4.21). Strum transversely across the dense ligamentum nuchae that spans the tips of the spinous processes.
3) Continue superiorly until you reach the prominence of C-2. As you explore the spinous processes, passively flex, extend and rotate the neck.

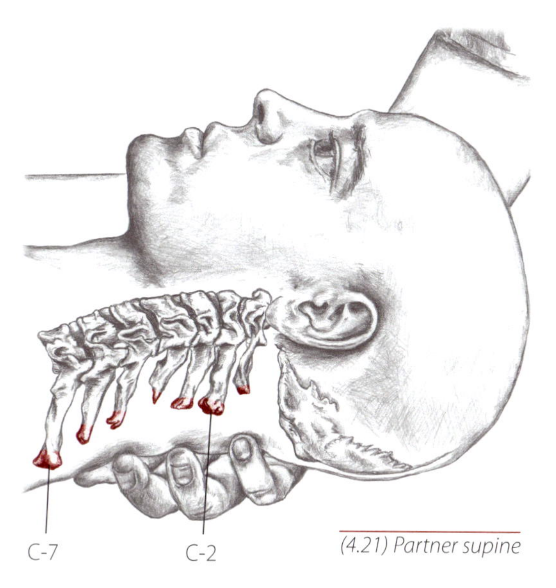

(4.21) Partner supine

Can you feel the subtle ridge formed by the processes along the back of the neck? When exploring the spinous process of C-2, are you inferior to the level of the ear lobes? Is the process of C-2 larger and more pronounced than that of the other cervicals?

1) Being able to differentiate between the spinous process of C-2 and the external occipital protuberance (p. 238) can be helpful when navigating the posterior neck. Begin by laying your fingers horizontally along the base of your partner's head.
2) Place your ring finger at the external occipital protuberance while locating the spinous process of C-2 with your index finger (4.22). Your middle finger will lie between these two structures at the level of C-1. Explore the distance between these prominent landmarks.

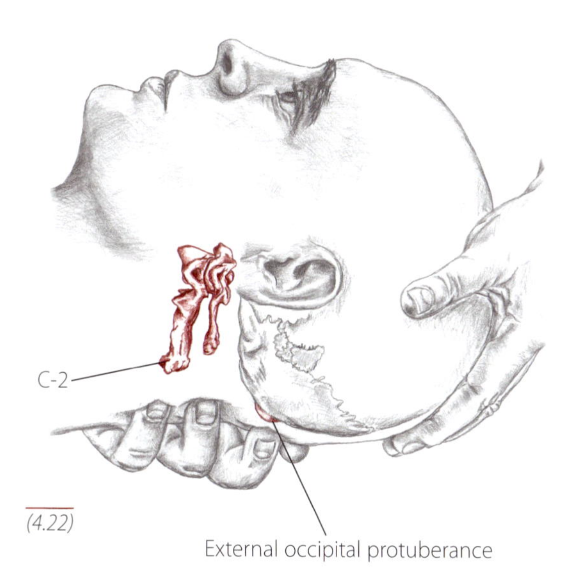

(4.22)

Some bony landmarks serve as attachment sites for multiple tendons and connective tissues. Whether palpating with your fingers or dissecting with a scalpel, these tissues are often difficult to distinguish from one another. The TVPs of the cervical vertebrae are a case in point: Tendons come from several different directions to attach to their surfaces while nerves pass between the tendons, complicating matters further.

To coordinate the tendons and the spinal nerves of the cervical and brachial plexuses, the TVPs of C-2 through C-7 have anterior and posterior tubercles (left). The tubercles are small tips situated on either side of the canal (or sulcus) that channels the cervical nerves. The anterior tubercle is an attachment site for the anterior scalene and other muscles. The middle and posterior scalenes, levator scapula and other posterior muscles attach to the posterior tubercles. It can be difficult initially to palpate individual tubercles, but with experience you will be able to detect them more easily.

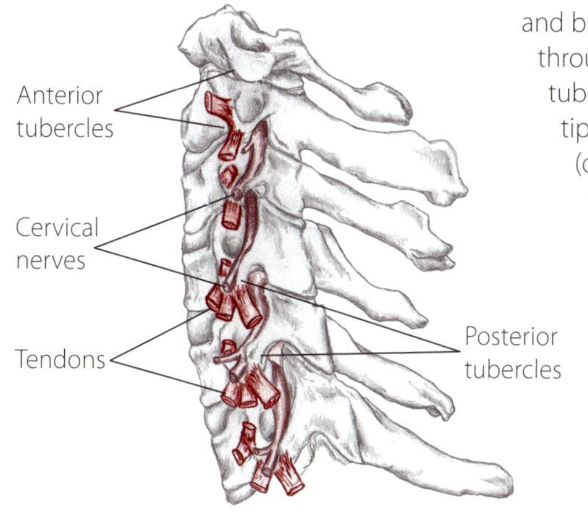

nape ME. the back of the neck

Compression or impingement of the brachial plexus (p. 271) or one of its nerves can create a sharp, shooting sensation down the arm. If this should occur, immediately release and adjust your position posteriorly. As always, ask your partner for feedback.

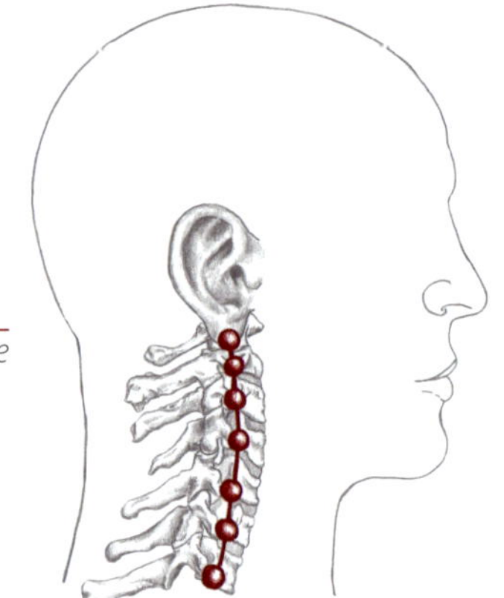

(4.23) The arrangement of the cervical TVPs simulate that of a long dangling earring

Transverse Processes of the Cervicals

The transverse processes (TVPs) of the cervical vertebrae are located on the side of the neck. Old Hollywood films put Frankenstein's neck bolts into his TVPs!

The TVPs extend inferiorly from the mastoid process and many are deep to the sternocleidomastoid muscle (p. 250). All of the TVPs are the same width except for the TVPs of C-1, which are much wider.

The TVPs of C-1 are located just distal and anterior to the tip of the mastoid process (p. 239) and are relatively accessible.

All of the TVPs serve as attachment sites for various muscles, including the scalenes and levator scapulae. The brachial plexus, a large group of nerves that innervates the arm, exits between the TVPs. When first accessing the TVPs, use the flat of your thumb or fingerpads. As your palpation skills improve, explore the TVPs' surfaces more specifically.

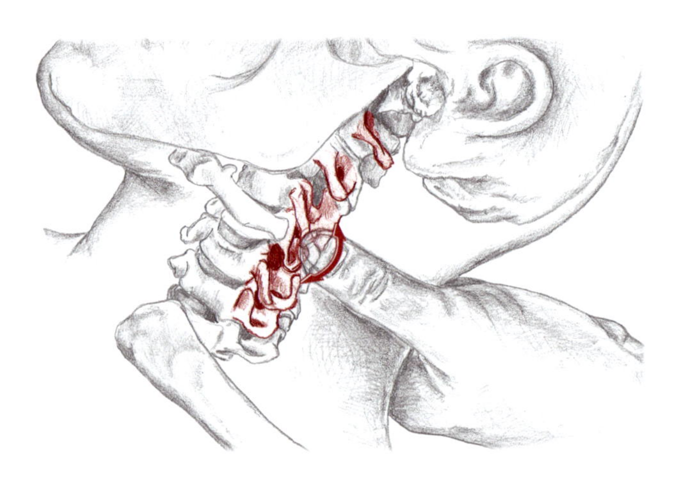

(4.24) Anterior/lateral view, partner supine

TVPs of cervicals

1) Partner supine. Place your fingers on the side of the neck below the earlobes.
2) Using your flat thumbpads, slide anteriorly and posteriorly to feel the ridge of TVPs. Explore the length of the neck (4.24).
3) You may not feel the tips of individual processes, but the ridge formed by the TVPs beneath the overlying tissue instead.

Are you palpating inferior to the earlobe? Do you feel a subtle ridge running down the side of the neck? If you passively flex, laterally flex or rotate the neck, can you feel the TVPs move individually?

1) Partner supine. Rotate the head 45° to the right. With the head in this position, the TVPs form a line from the left mastoid process to the center of the shaft of the clavicle (4.25).
2) Draw an imaginary line from these two landmarks and visualize and palpate the TVPs along the path.

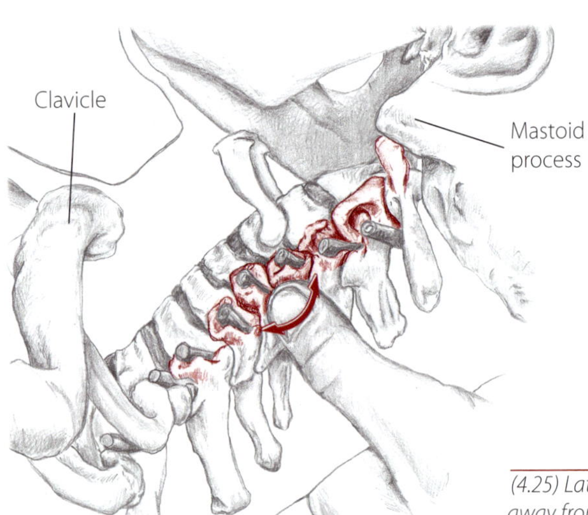

(4.25) Lateral view with head rotated away from the side you are palpating

Transverse process of C-1

1) Partner supine or seated. Locate the left mastoid process of the temporal bone (p. 239) and rotate the head 45° to the right.
2) Using your broad fingerpad, slide slightly inferior and anterior from the mastoid process. Explore deep to the sternocleidomastoid muscle for the solid bump of the transverse process of C-1 (4.26). Even pressing gently on these points may be uncomfortable for your partner, so use a soft touch.
3) For reference, locate the other transverse process.

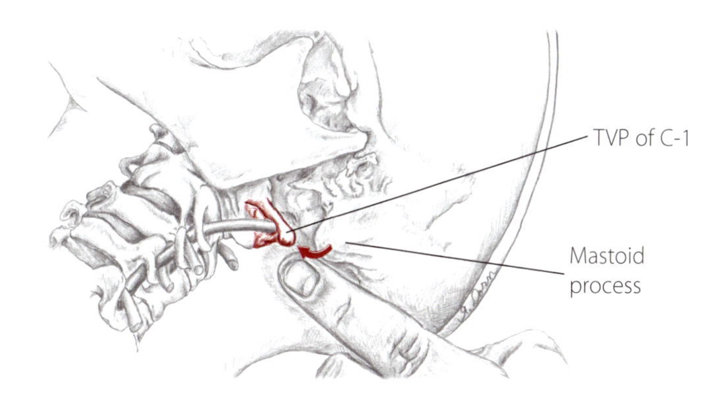

(4.26) Partner supine, anterior/lateral view with head rotated away from the side you are palpating

Cervical Lamina Groove

The lamina groove is the troughlike space between the spinous and transverse processes of the vertebrae. Although sizable on a skeleton, the lamina groove of your partner is filled with layers of muscles which render it almost inaccessible. The lamina groove is best thought of as a helpful region for locating muscle bellies.

1) Partner supine. Scoop the head with one hand, and with your other hand, locate the cervical transverse processes.
2) Slide posteriorly off the TVPs. Explore the space between the transverse and spinous processes which constitutes the lamina groove of the cervical vertebrae (4.28). Again, since the groove is filled with muscles, the bone that forms the groove is impalpable.

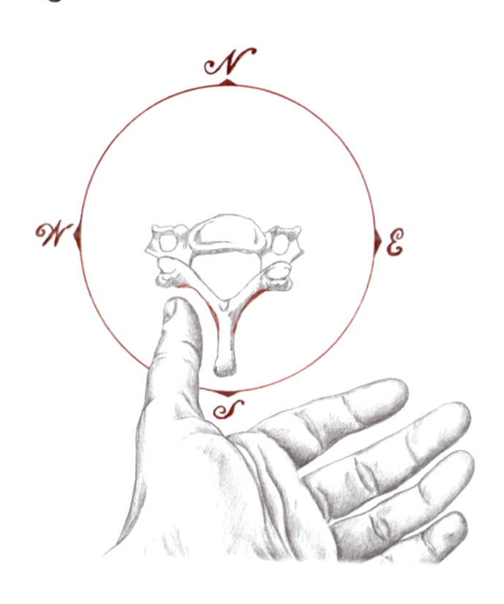

(4.27) A schematic cross section of the neck as if your partner were supine. The lamina grooves can be thought of as being in the neck's southeast and southwest quadrants.

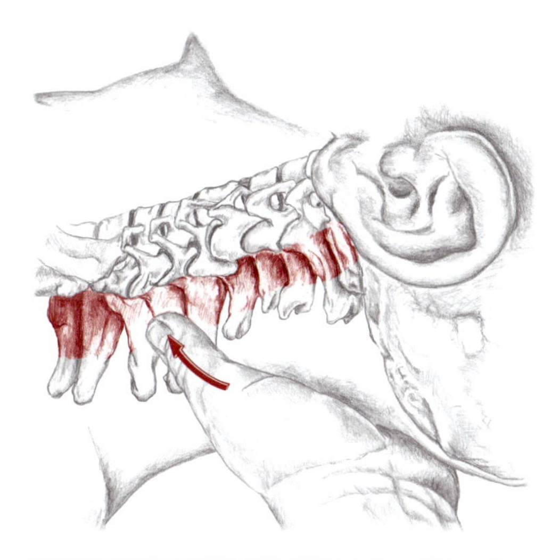

(4.28) Lateral view, palpating in the lamina groove

C-6 has a large anterior tubercle called the carotid tubercle. Its name corresponds to the carotid artery (p. 268) that passes immediately lateral to it. Although you would not want to, you can occlude the carotid artery by placing your finger lateral to the cricoid cartilage and pressing in a posterior direction against the carotid tubercle. Long ago this dramatic maneuver was used in emergency rooms as a last-ditch effort to stem hemorrhaging inside the skull.

lamina **lam**-i-na L. thin plate, leaf

Trail 4 "Buried Boulevard"

Transverse Processes of the Thoracic and Lumbar Vertebrae

The TVPs of the thoracic vertebrae are shorter and do not extend as far laterally as the TVPs of the lumbar vertebrae. They are palpable deep to the erector spinae muscles (p. 202) and superficial to the connecting aspect of the ribs.

The TVPs of the lumbar vertebrae are also deep to the erector spinae. Extending an inch or two laterally, their solid presence can be felt beneath the overlying muscle tissue.

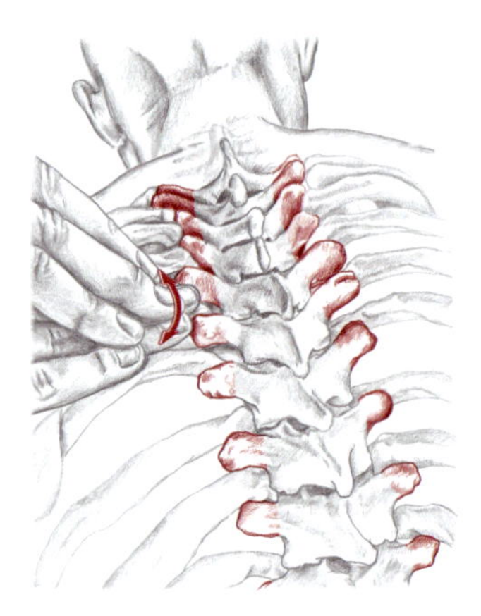

(4.29) Partner prone, isolating the TVPs of the thoracic vertebrae

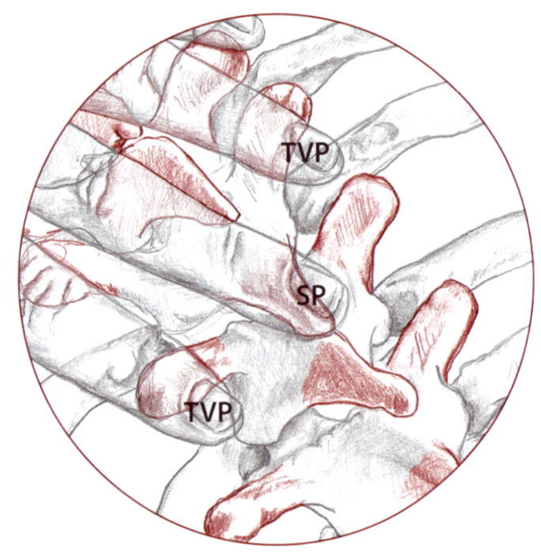

(4.30) Try using three fingerpads to span between the TVPs with the middle finger falling on the spinous process

Thoracic transverse processes

1) Partner prone. Locate a portion of the thoracic spinous processes. Move roughly one inch laterally and sink your fingers through the thick erector spinae muscles.
2) Roll your fingers superiorly and inferiorly, palpating for the TVPs subtle, knobby shape (4.29).

Slide further laterally from the thoracic transverse processes and onto the posterior ribs. Can you determine where the ribs and transverse processes meet? Can you feel the short processes beneath the erector spinae fibers?

Lumbar transverse processes

1) Partner prone. Locate the lumbar spinous processes. Slide roughly two inches laterally to avoid the thick mound of the erector spinae (p. 202).
2) Slowly sink your fingers through the muscle tissue. Directing your pressure at a medial/anterior angle (as if toward the navel), explore for the tips of the TVPs (4.31). Because of the thick overlying tissue, the individual processes may not be directly palpable, but try to sense the solid ridge they form.

Ask your partner to raise her feet slightly to determine whether you are lateral to the erector spinae muscles. Can you feel the hard surface of the processes running horizontally?

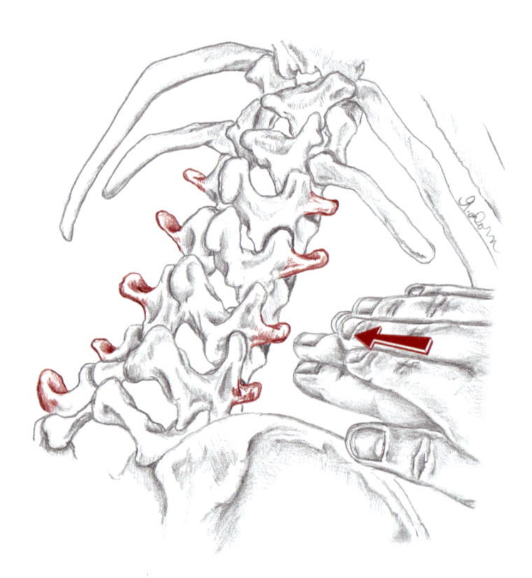

(4.31) Partner prone, palpating the TVPs of the lumbar vertebrae

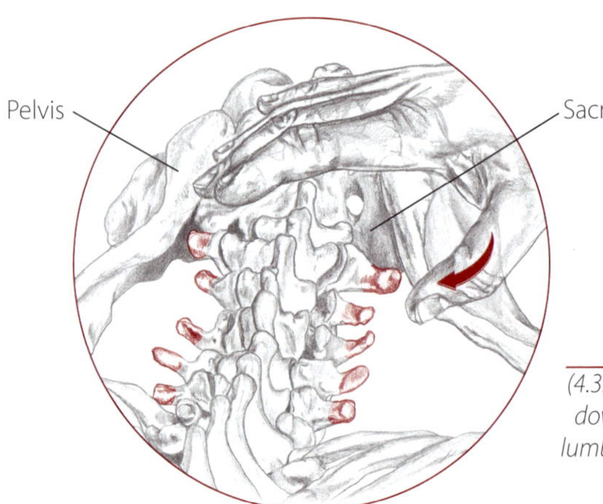

(4.32) Superior view looking down the spine, accessing lumbar TVPs with the thumb

Lamina Groove of the Thoracic and Lumbar Vertebrae

The lamina groove of the thoracic and lumbar vertebrae is located between their spinous and transverse processes. Shaped like a long, vertical trough, the lamina groove expands in depth and width as it progresses down the spine. In the thoracic and lumbar vertebrae the lamina groove is filled with the layers of the erector spinae and transversospinalis muscles. Because of this overlying tissue, the lamina groove is difficult to access directly, but its borders (the spinous and transverse processes) are palpable.

1) Partner prone. Locate the spinous processes of the thoracic vertebrae. With the other hand, locate the TVPs of the thoracic vertebrae.

2) Using firm pressure, explore between these landmarks in the lamina groove (4.33). Note the thick muscle tissue that lies in this groove.

3) Try this same method in the lumbar region (4.34). Observe how the lamina groove widens and deepens and how the muscle tissue is thicker in this region in comparison to the muscle tissue in the thoracic region.

Are you between the transverse and spinous processes of the vertebrae? Can you slide your fingers between the muscle fibers and sink into the lamina groove?

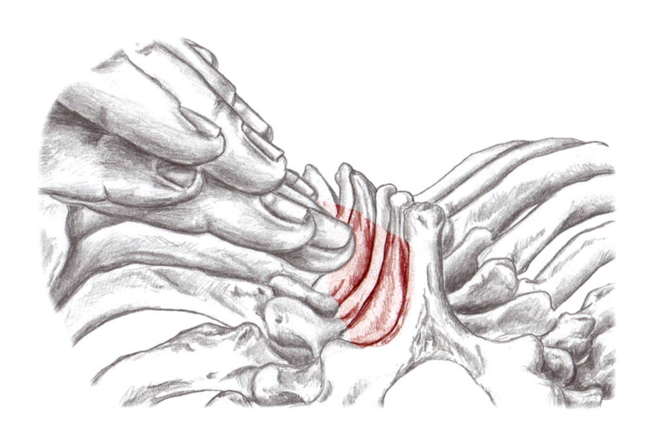

(4.33) Superior view, accessing the lamina groove of the thoracic vertebrae. Muscles overlying the groove (below).

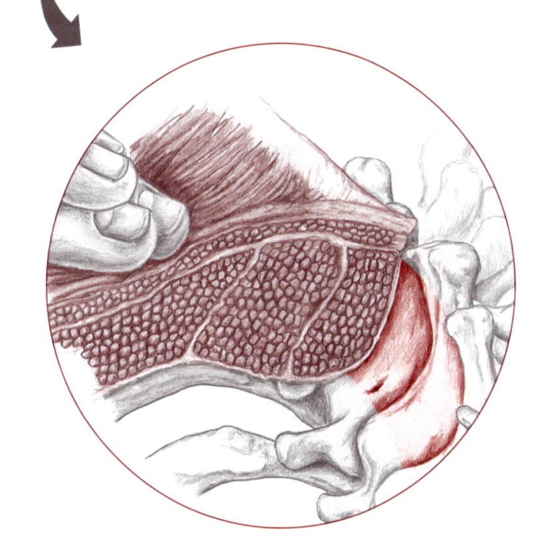

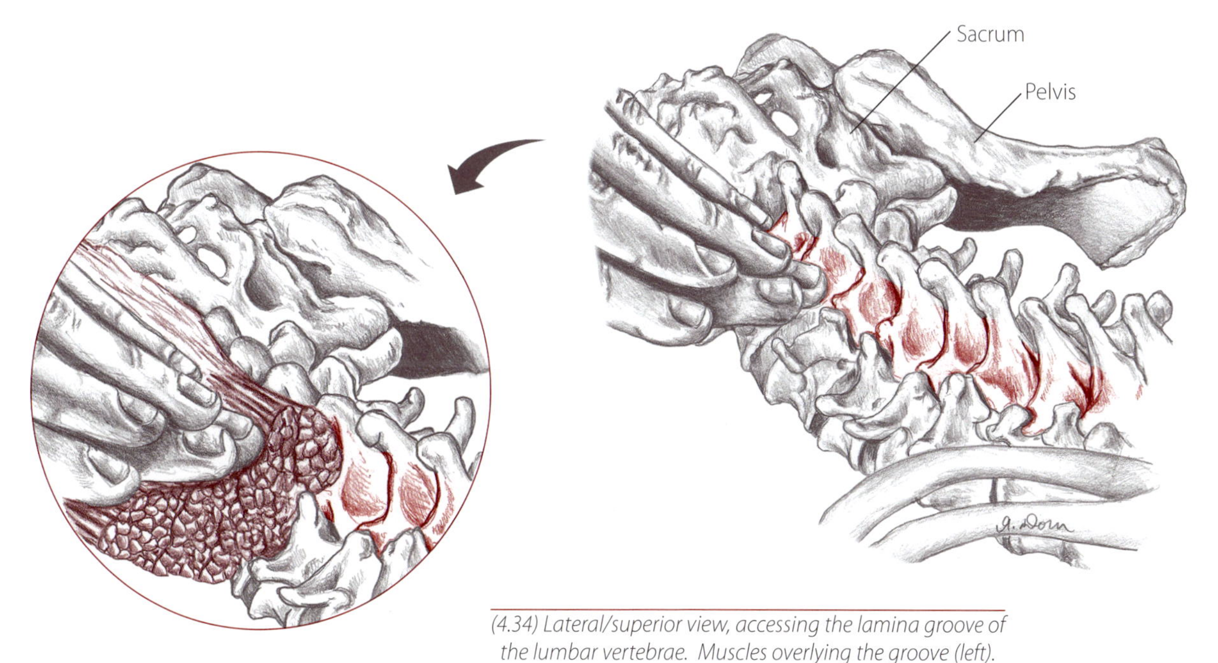

(4.34) Lateral/superior view, accessing the lamina groove of the lumbar vertebrae. Muscles overlying the groove (left).

Trail 5 “Breastbone Ridge”

Sternum

The sternum features several landmarks (p. 179). At the top of the sternum, the **jugular notch** is between the sternal heads of the clavicles. It may be flat or bowl-shaped, and although no muscles attach directly to it, the sternocleidomastoids pass superficially to it while the infrahyoids attach deep to it.

The **manubrium**, the superior portion of the sternum, articulates with the clavicles, the first rib and the second rib. The **body of the sternum** is located inferior to the manubrium and forms the major portion of the sternum. The junction between the manubrium and body of the sternum is called the **sternal angle**.

Extending off the bottom of the sternum, the **xiphoid process** can be an inch in length or completely absent. It is an attachment site for the abdominal aponeurosis. The manubrium, body and xiphoid process of the sternum are superficial, covered only by fasciae and the pectoralis major tendon.

Jugular notch, manubrium and sternum

1) Partner supine. Place your finger upon the sternum at the center of your partner's chest.
2) Slide superiorly until you reach the jugular notch at the top of the sternum. Explore the notch and its location next to the sternoclavicular joints.
3) Move your fingers inferiorly off the jugular notch onto the manubrium and body of the sternum. Explore any crevices or hills upon this “flat” bone. Also, palpate laterally toward its attachments with the costal cartilage.

Xiphoid process

1) Slide your fingers inferiorly until they drop off the sternum and fall into the muscles of the abdomen. Now backtrack to the most inferior tip of the sternum which will be the xiphoid process (4.35). Gently sculpt this tip.

Are you at the most inferior point of the sternum?

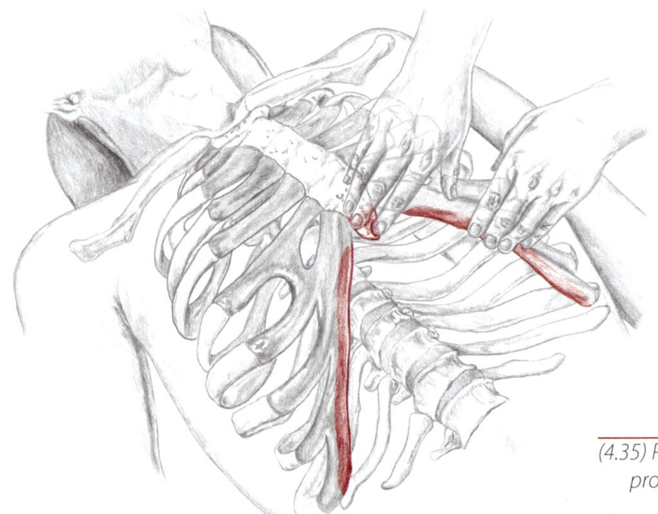

(4.35) Partner supine, palpating the xiphoid process and the edge of the rib cage

Partner supine. The sternal angle is the junction point between the manubrium and sternum. Stretching horizontally, it may feel like a small speed bump or a dip. Locate the jugular notch and glide inferiorly along the surface of the manubrium. Palpate within an inch or two for a ridge or ditch that stretches horizontally across the sternum.

The second rib attaches to the sternum at the level of the sternal angle. Slide your fingers laterally off the angle. Can you feel the round surface of this rib?

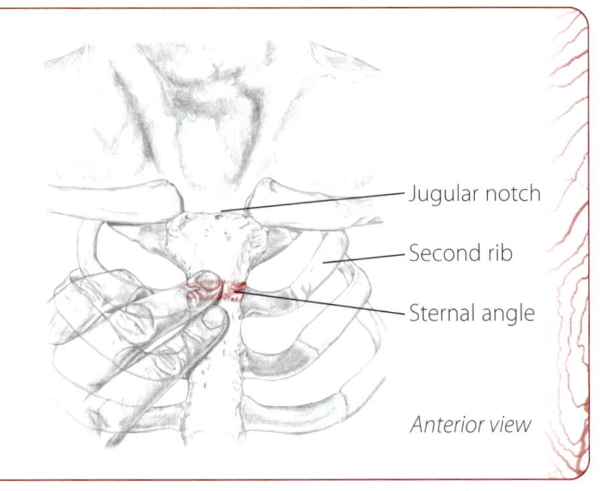

abdomen	**ab**-do-men	L. belly
abdominis	ab-**dah**-min-is	
costal	**kos**-tal	L. rib

Trail 6 "One Bumpy Road"

Ribs and Costal Cartilage

The **ribs** articulate posteriorly with the thoracic vertebrae and then curve around the thorax to the anterior chest (p. 177). Extending off the ribs is the **costal cartilage** that attaches them to the sternum. There are six or seven costal branches, all of which are identical in shape and similar in feel to the ribs. The ribs, with their costal cartilage, run at varying angles around the trunk of the body.

The entire rib cage is deep to muscle tissue; however, the ribs along the sides of the trunk are easily accessed. The spaces between the ribs are filled with thin intercostal muscles that can be easily palpated.

When exploring the thorax, consider its three-dimensional quality. Often the thorax is viewed as having only a front and back, leaving its lateral portions neglected. As you explore the body's trunk, try to connect all its sides together in your mind and with your hands. Note how several muscles, such as the deeper abdominals and intercostals, literally enwrap the thorax.

As you explore the thorax, avoid accessing mammary (breast) tissue. Ask your partner, male or female, whether you may palpate the surrounding areas.

1) Supine. Slide laterally from the sternum onto the costal cartilage. Use your fingertips to locate one costal branch and palpate its rounded surface.
2) Roll off the cartilage into the space between the branches. Explore this groove as it extends laterally. Continue along the length of the sternum, locating and exploring each branch of rib/cartilage and the spaces between them (4.36).

Can you determine how the angle of the ribs changes as you move around the body? Can you differentiate the round shafts of the ribs from the ditchlike spaces between them? Ask your partner to breathe deeply and note any change in the amount of space you can feel between the ribs.

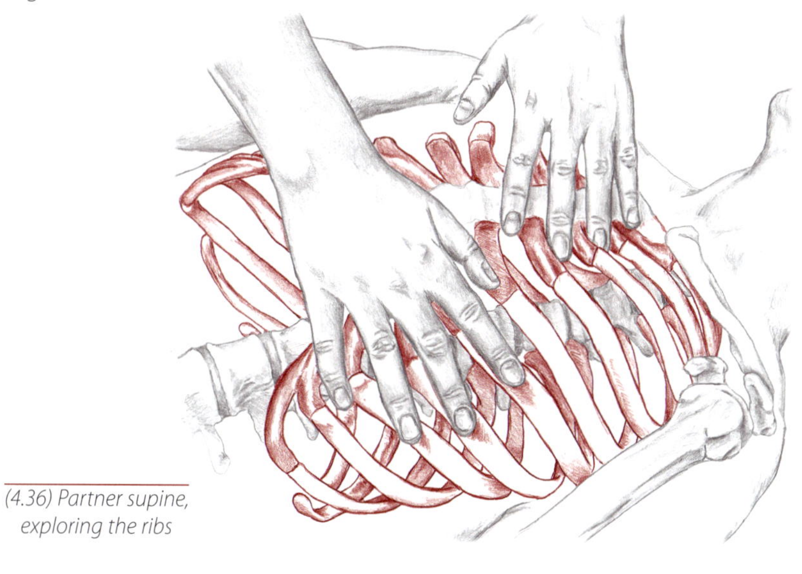

(4.36) Partner supine, exploring the ribs

The jugular notch, sternal angle and xiphoid process can be guidemarks along the vertebral column. The jugular notch lies on the same transverse plane as the spinous process of T-2 **(a)**. The sternal angle lines up with the spinous process of T-4 **(b)**, while the xiphoid process is directly across from the body of T-10 **(c)**. Of course, many factors, such as posture and body type, will affect the placement of the ribs, so use these correlations only as guides.

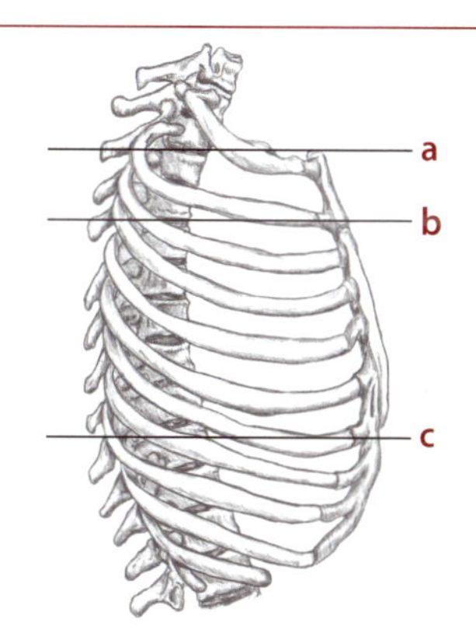

Stand beside your partner and palpate the jugular notch with one hand while your other hand locates the spinous process of T-2. Note whether you can see or feel a difference in the levels of these landmarks. Follow the same procedure for T-4 and T-10.

cricoid	**kri**-koyd	Grk. ring-shaped
jugular	**jug**-u-lar	L. throat
manubrium	ma-**nu**-bree-um	L. handle

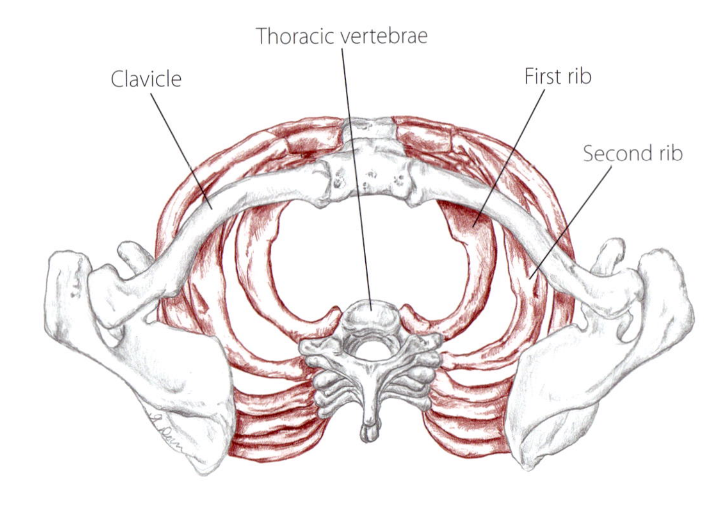

(4.37) Superior view of thorax

Compression or impingement of the brachial plexus or one of its nerves can create a sharp, shooting sensation down the arm. If this should occur, adjust your position to one side. Also, ask your partner for feedback.

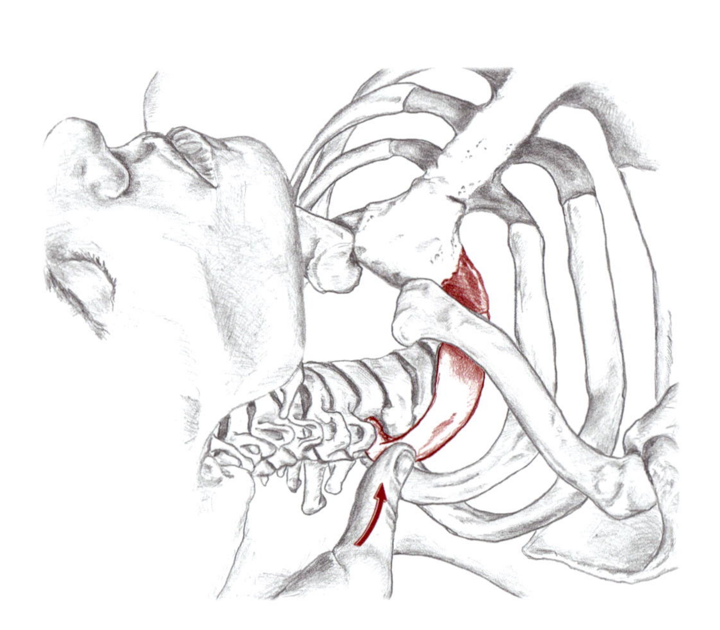

(4.38) Partner supine, palpating the first rib

First Rib

Unlike its cohorts, the first rib is difficult to isolate along the anterior thorax. It lies directly beneath the clavicle and then quickly curves toward the back (4.37). It can, however, be accessed in the posterior triangle (p. 232) of the neck which is formed by the clavicle and the sternocleidomastoid and trapezius muscles.

The scalene muscles (p. 252) fan across the posterior triangle and attach to the first and second ribs. To access the first rib, you must palpate through the scalenes. The brachial plexus (p. 271) and subclavian artery pass between the first rib and the clavicle.

1) Partner supine. Soften the overlying tissue by passively elevating your partner's shoulder.

2) Locate the clavicle and upper flap of the trapezius to identify the posterior triangle. Place your thumbpad between these structures.

3) Slowly sink into the tissue of the scalene muscles, directing your fingers straight in an inferior direction toward your partner's feet (4.38). As your fingers sink into the tissue, you will meet the solid resistance of the shaft of the first rib.

✔ *Ask your partner to take a slow, deep breath into the upper chest. Can you feel the rib rise?*

Partner supine. When palpating in the posterior triangle of the neck, the posterior aspect of the first rib can sometimes be confused with the superior angle of the scapula. Distinguish between these structures by palpating in the posterior triangle and locating what you believe to be the first rib. "Check it" by passively elevating and depressing the scapula. The first rib should remain stationary during this movement.

It is not uncommon for the number of ribs to vary from person to person. There are ordinarily twelve pairs of ribs, but on some people eleven or thirteen pairs can be found. If there is an extra rib, it may be bilateral or merely unilateral and will be found either in the cervical or lumbar areas. A cervical rib often articulates with C-7 and can be felt in the posterior triangle region of the neck at the level of the clavicle. An extra rib in the lumbar will extend off L-1.

sternum	**ster**-num	Grk. chest
thorax	**tho**-raks	Grk. chest
xiphoid	**zif**-oyd	Grk. sword-shaped

Ideally, the ribs are designed to expand in three directions during inhalation: anterior/posterior, lateral and superior. Yet, for reasons ranging from posture to emotional trauma, few people truly breathe in this manner. Often the breath becomes restricted to a portion of the thorax, and the ribs will move in only one or two directions.

1) With this in mind, ask your partner to stand and breathe normally. Observe any changes in the shape or movement of the thorax, shoulders and abdomen.

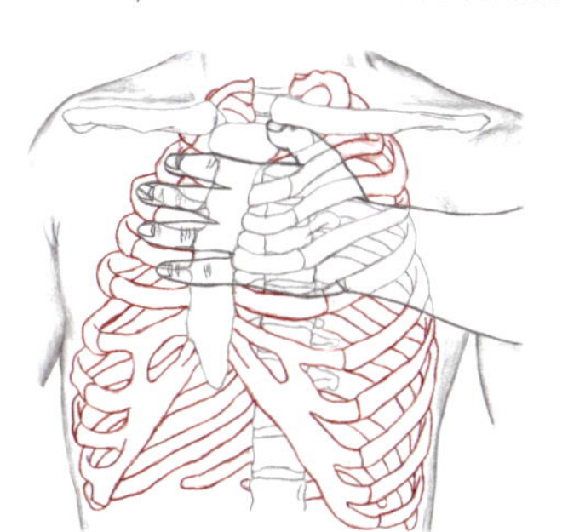

Position of the ribs during exhalation

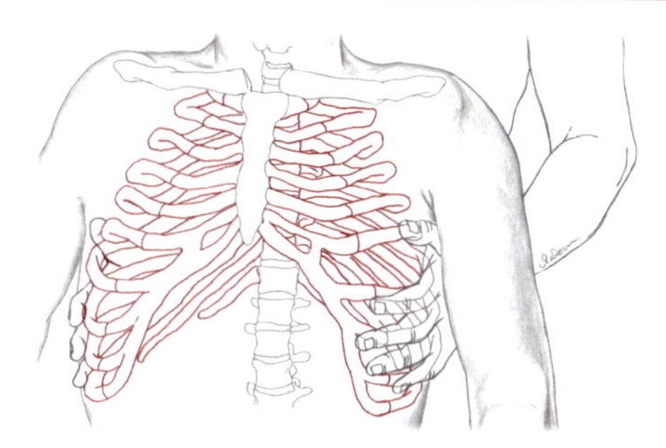

Position of the ribs during inhalation

2) Then lay your hands on all sides of the rib cage and feel for activity in the thorax. Do the ribs move in all three directions? Do some move individually?

3) Ask your partner to inhale deeply and exhale fully. Explore the ribs and anterior neck muscles (scalenes and sternocleidomastoid) during inhalation. These muscles will tighten to elevate the upper ribs. Try these exercises with your partner seated, prone and supine.

Eleventh and Twelfth Ribs

The eleventh and twelfth ribs are called "floating ribs" because they do not attach to the costal cartilage. Both ribs have a slender, spearlike shape and lie roughly at a 45° angle. Their medial portions lie deep to the thick erector spinae muscles; however, their lateral aspects and tips are palpable.

The eleventh rib is six to eight inches in length and extends halfway around the body. The twelfth rib may vary in length from three to six inches. Since anomalies are common in either the length or number of ribs, your partner's ribs may not match this description. (See p. 183 for more information about finding the twelfth rib.)

1) Prone. Reaching across to the body's opposite side, place your hand along the lateral portion of the ribs.

2) Slide inferiorly to the bottom of the rib cage, allowing your hand to sink into the soft abdominal tissue. Compressing your fingerpads into the side of the thorax, explore this region for the tips of the eleventh and twelfth ribs (4.39).

3) With the tips isolated, gently follow the shafts of the ribs medially, noting how they run at an angle.

Can you feel two tips, one of which is more lateral than the other? Ask your partner to take a slow, deep breath and note whether the tips or the bodies of the ribs press into your hand.

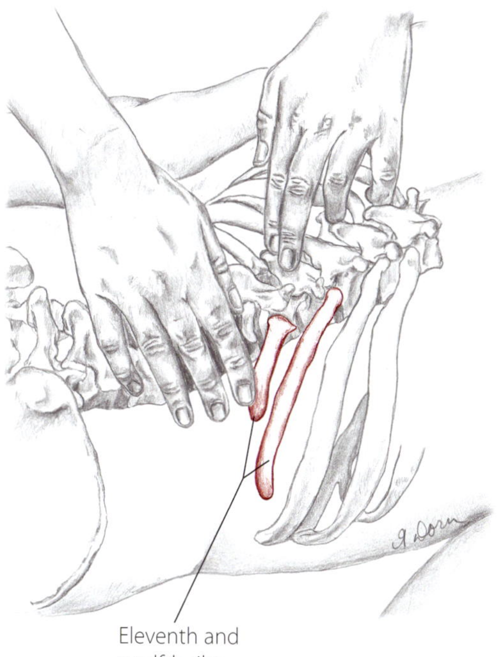

(4.39) Partner prone

Muscles of the Spine and Thorax

The muscles of the spine and thorax are situated along the posterior and abdominal regions and create movement of the vertebral column and rib cage (4.40).

The muscles of the spine are uniquely arranged. Unlike the limb muscles that can often be distinguished individually, the spinal muscles are composed of numerous bands of densely interwoven fibers that make it difficult to isolate a particular portion of muscle.

The muscles of the spine may be divided into small, individual sections or separated into a few major groups. For our purposes, the muscles of the spine will be divided into four groups:

1) The large **erector spinae group** is the most superficial of the spinal muscles and has three major branches.
2) The smaller **transversospinalis group** also has three branches but lies deep to the erectors. Its name refers to its muscle fibers, which extend at varying lengths from the transverse and spinous processes of the vertebrae.
3) The two **splenii** muscles are located along the posterior neck, deep to the trapezius.
4) The eight short **suboccipitals** are the deepest muscles. They are located at the base of the head.

Other muscles affecting the thorax, most notably the sternocleidomastoid and scalenes, are presented in Chapter Five, *Head, Neck and Face*.

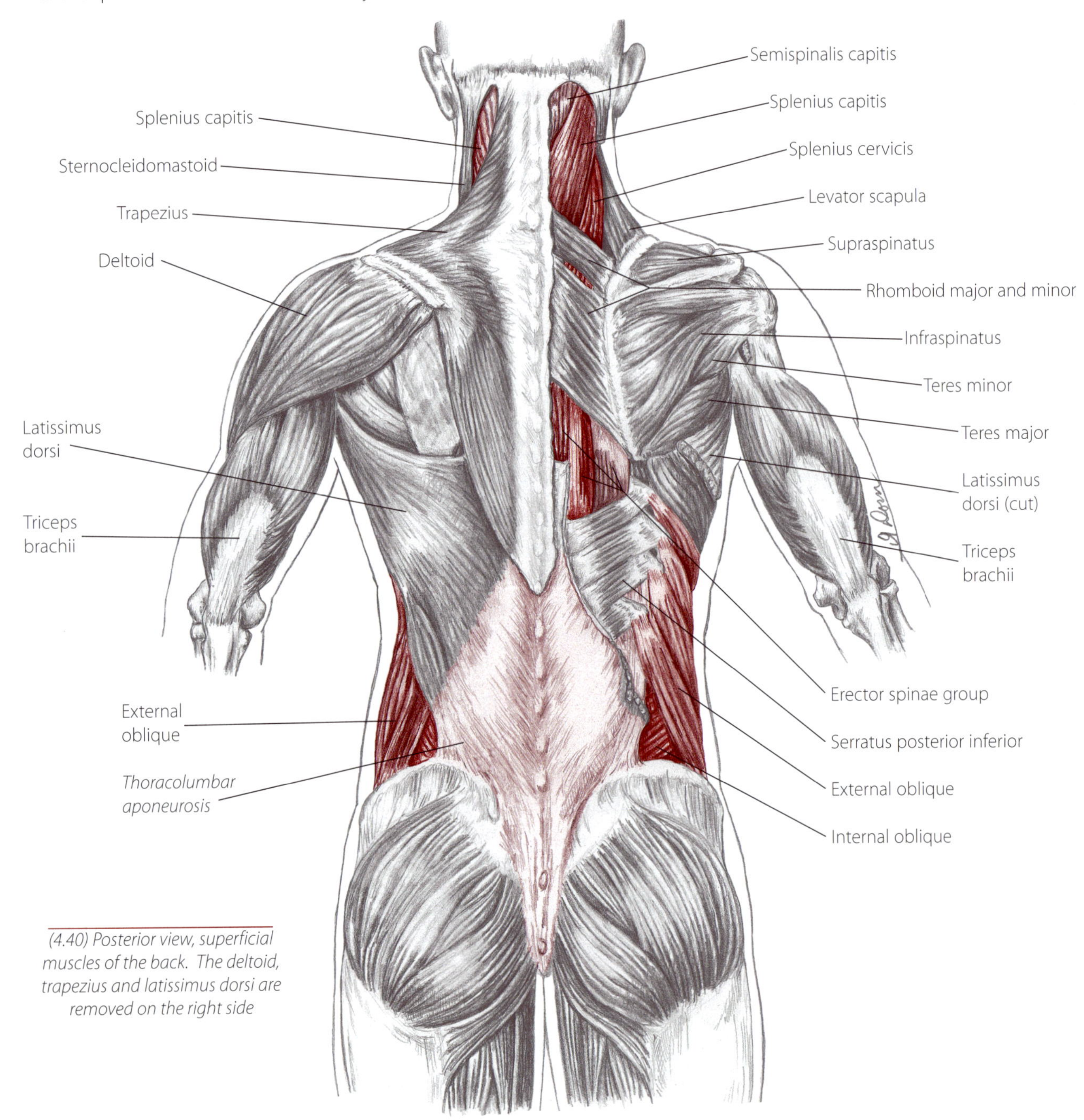

(4.40) Posterior view, superficial muscles of the back. The deltoid, trapezius and latissimus dorsi are removed on the right side

Muscles of the Spine and Thorax

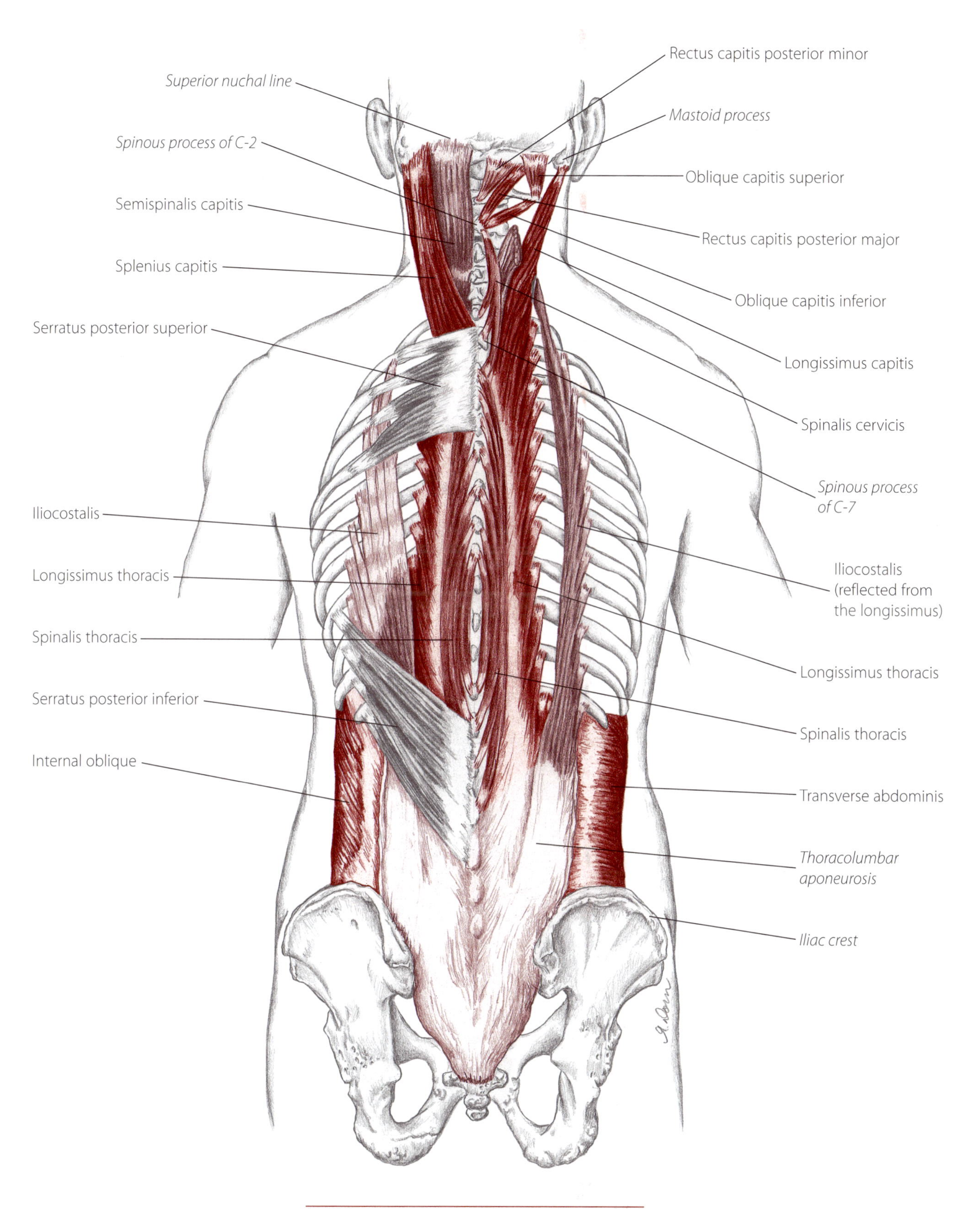

(4.41) Posterior view, intermediate muscles of the back

Muscles of the Spine and Thorax

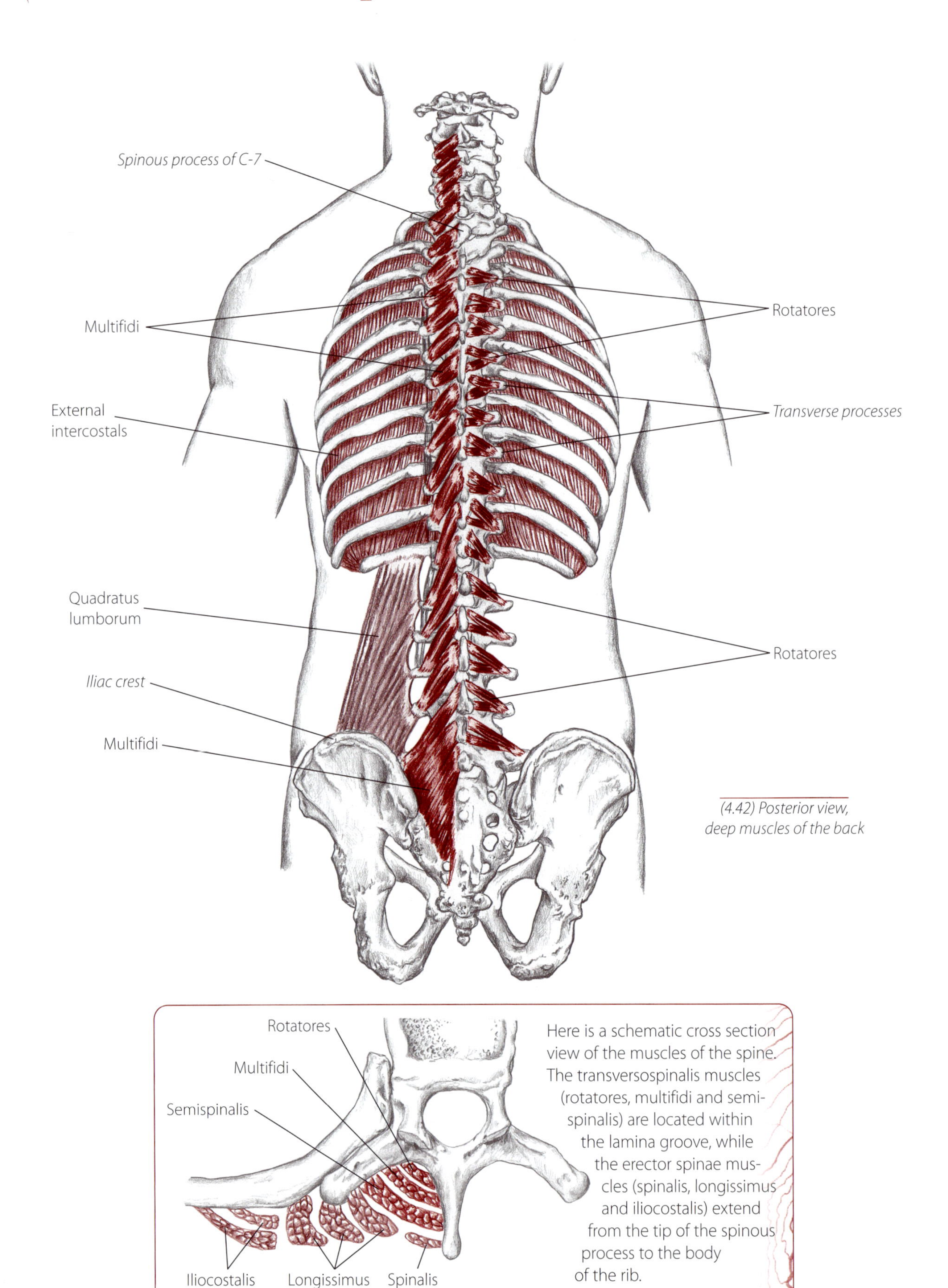

(4.42) Posterior view, deep muscles of the back

Here is a schematic cross section view of the muscles of the spine. The transversospinalis muscles (rotatores, multifidi and semispinalis) are located within the lamina groove, while the erector spinae muscles (spinalis, longissimus and iliocostalis) extend from the tip of the spinous process to the body of the rib.

Muscle Layers of the Posterior Neck

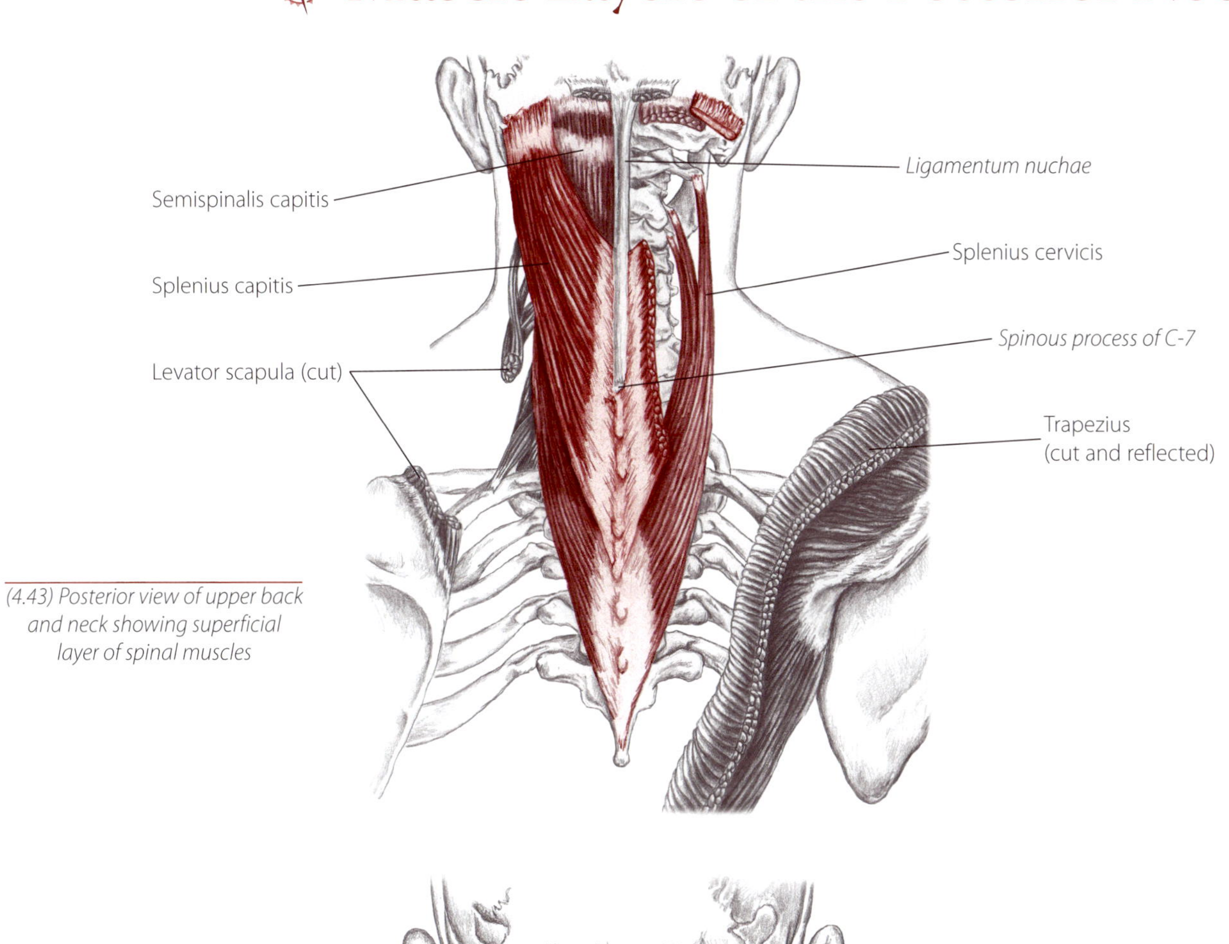

(4.43) Posterior view of upper back and neck showing superficial layer of spinal muscles

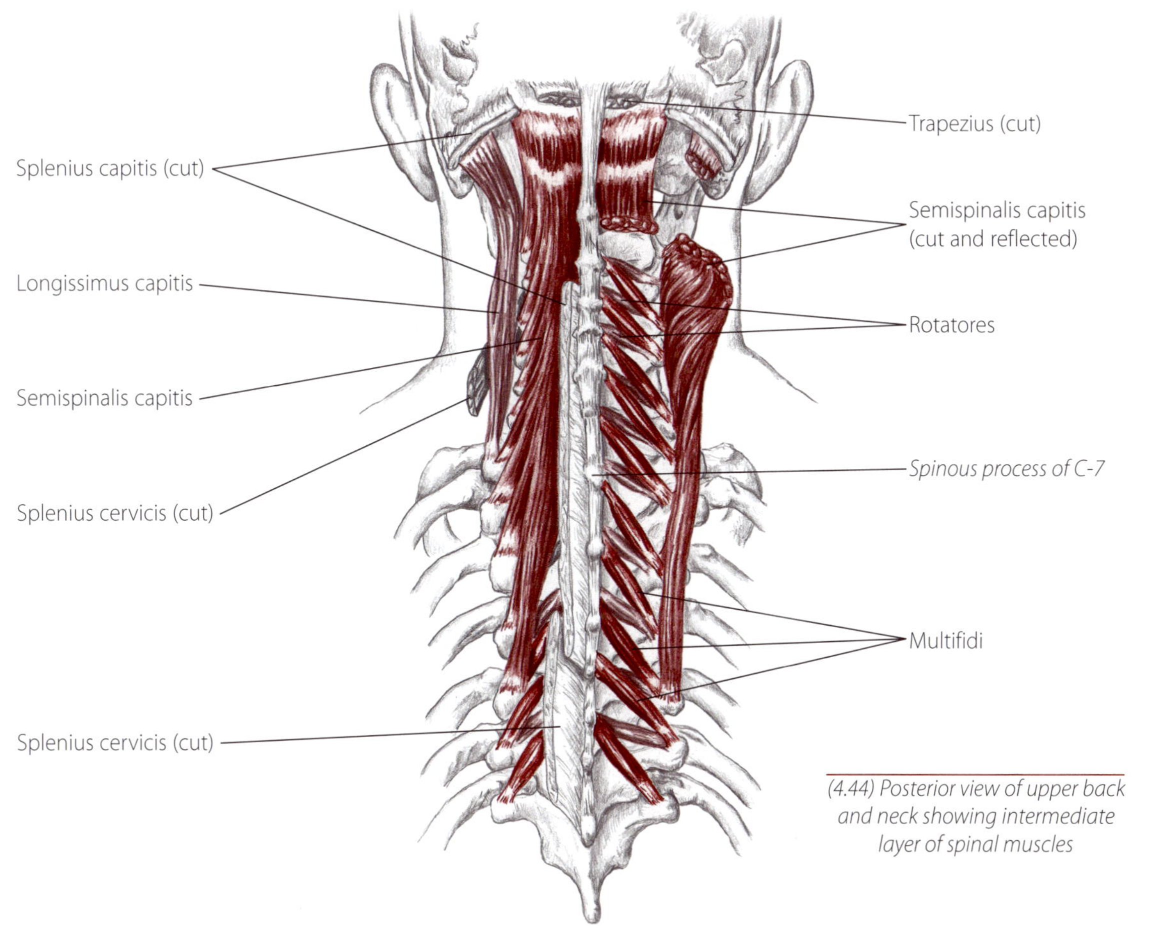

(4.44) Posterior view of upper back and neck showing intermediate layer of spinal muscles

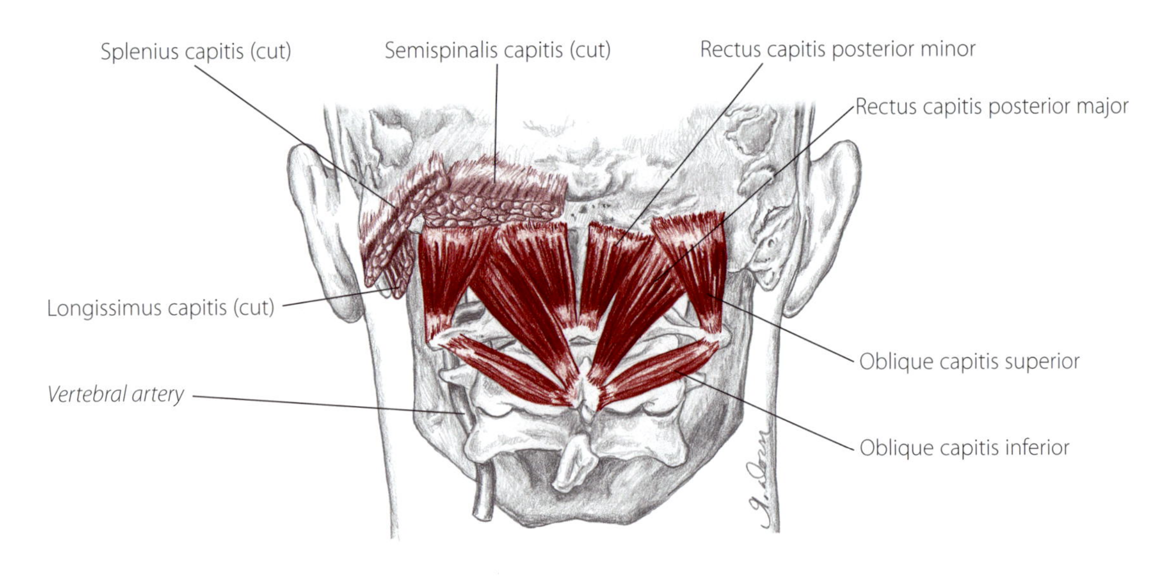

(4.45) Posterior view of upper neck showing deepest layer of spinal muscles

Cross Sections of the Neck and Thorax

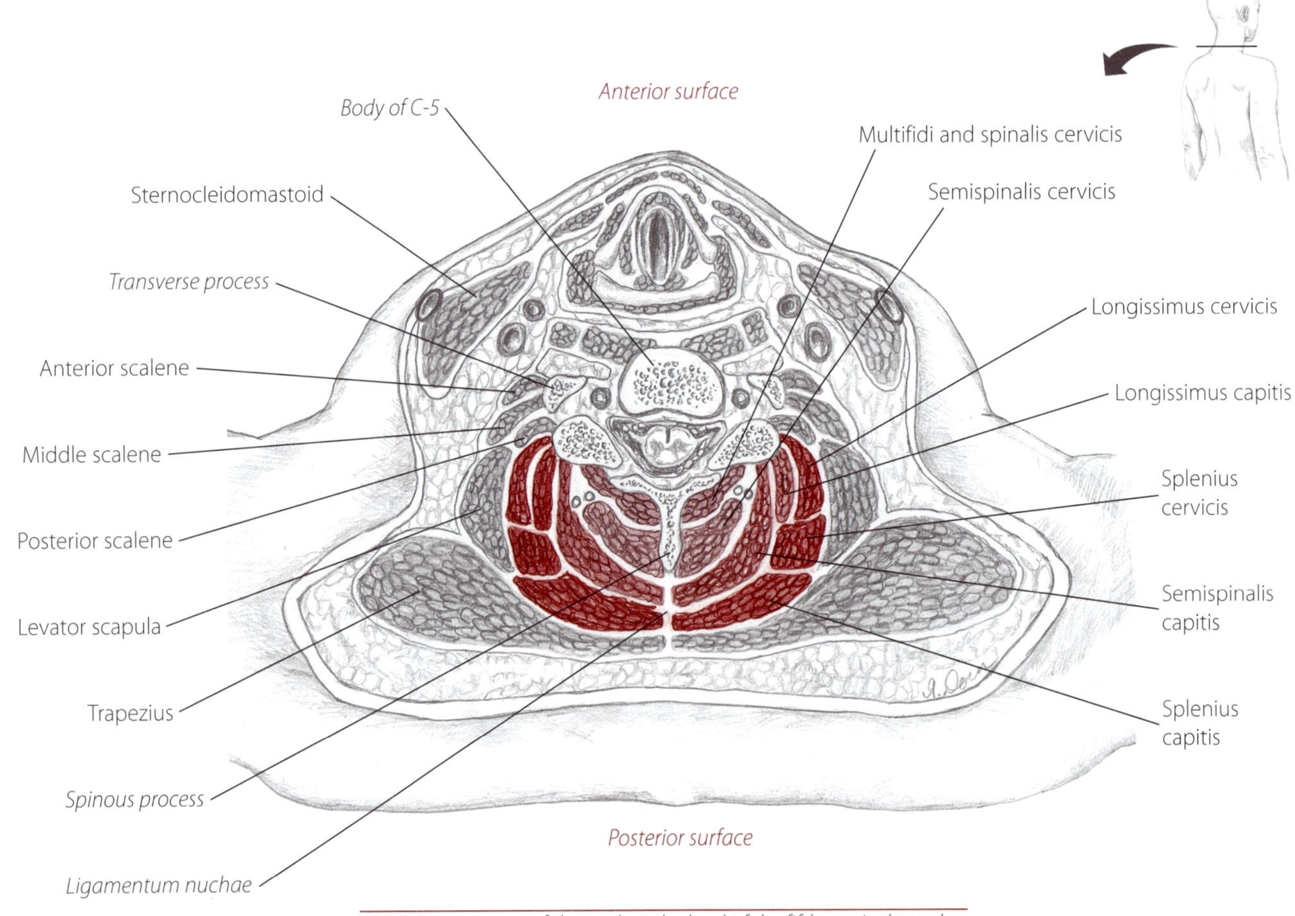

(4.46) Cross section of the neck at the level of the fifth cervical vertebra

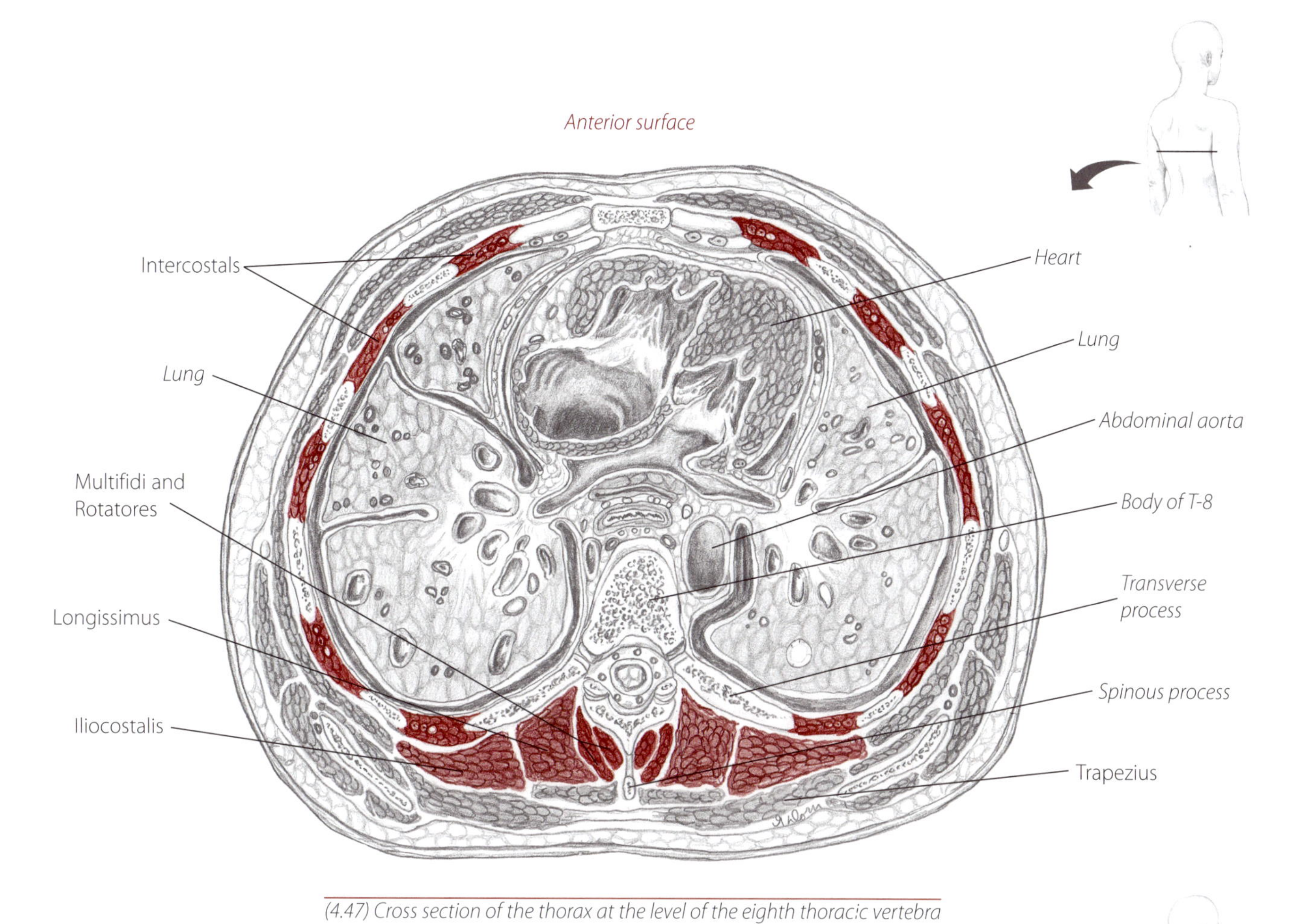

(4.47) Cross section of the thorax at the level of the eighth thoracic vertebra

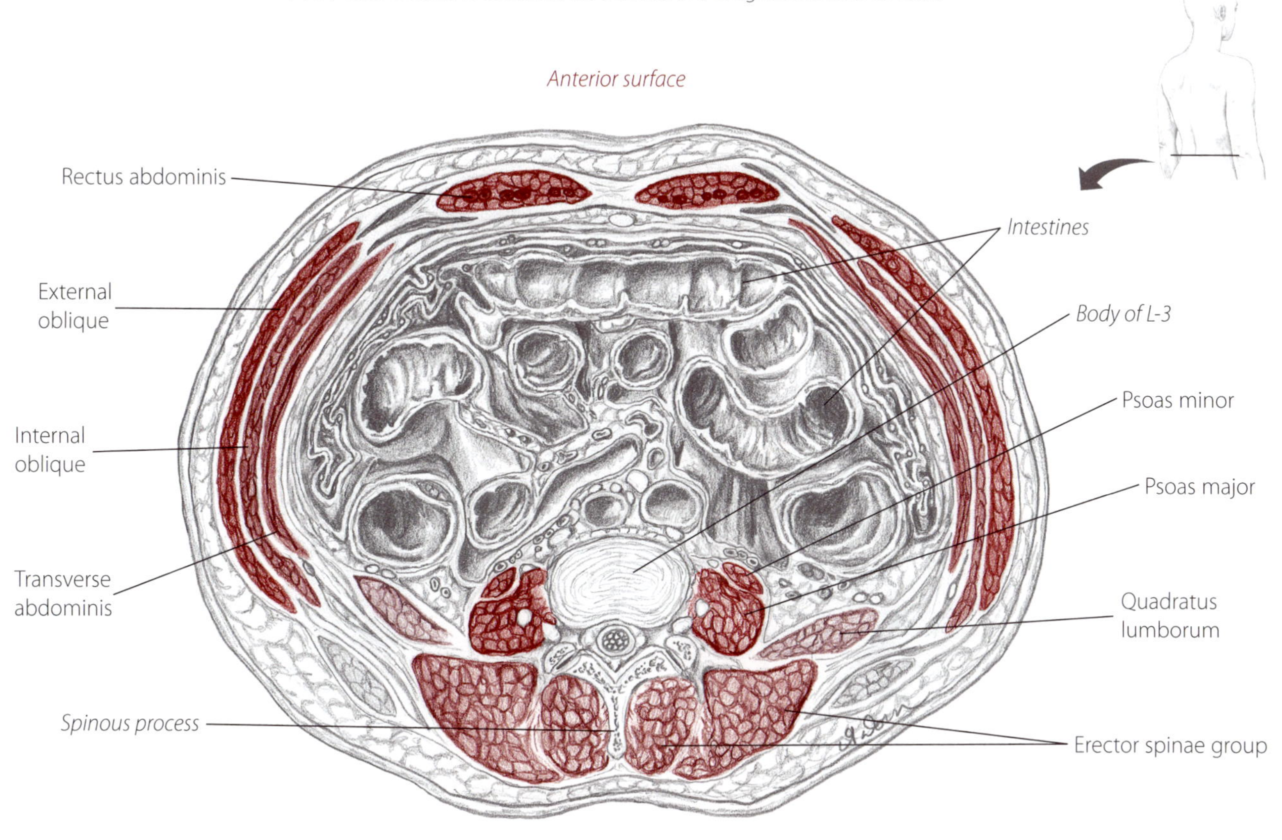

(4.48) Cross section of the abdomen at the level of the third lumbar vertebra

Synergists - Muscles Working Together

*muscles not shown

Vertebral Column

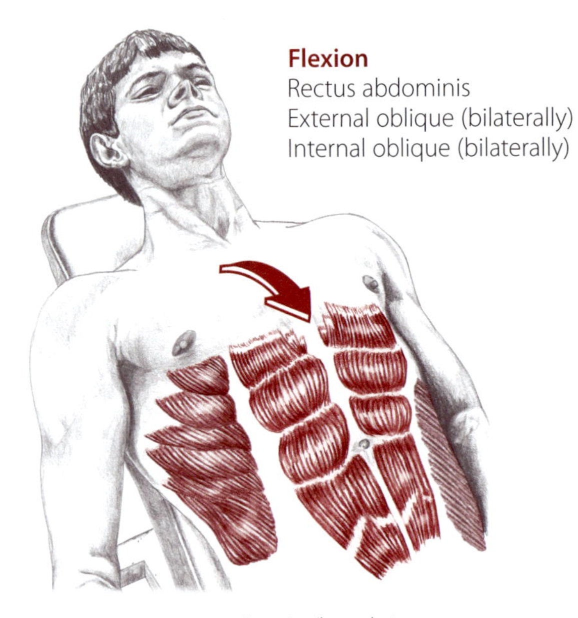

Flexion
Rectus abdominis
External oblique (bilaterally)
Internal oblique (bilaterally)

Anterior/lateral view

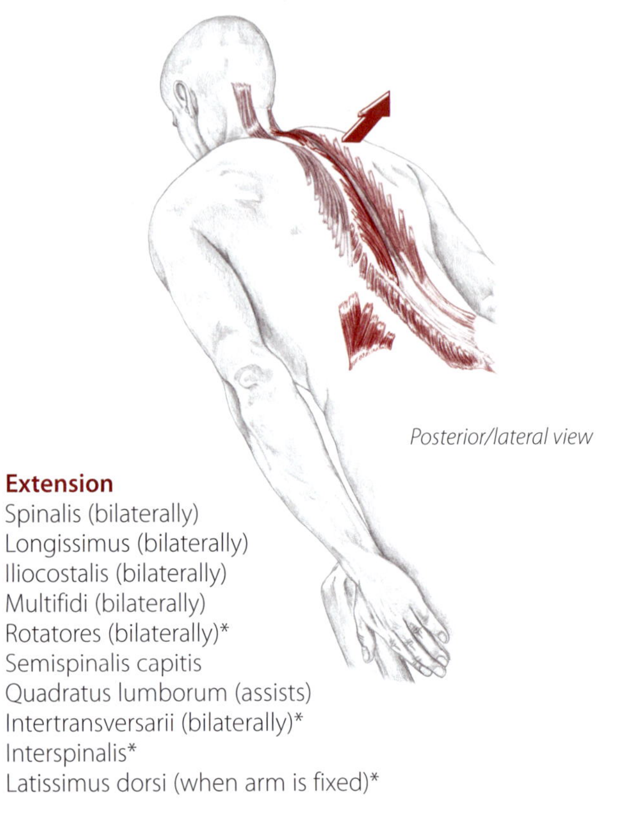

Posterior/lateral view

Extension
Spinalis (bilaterally)
Longissimus (bilaterally)
Iliocostalis (bilaterally)
Multifidi (bilaterally)
Rotatores (bilaterally)*
Semispinalis capitis
Quadratus lumborum (assists)
Intertransversarii (bilaterally)*
Interspinalis*
Latissimus dorsi (when arm is fixed)*

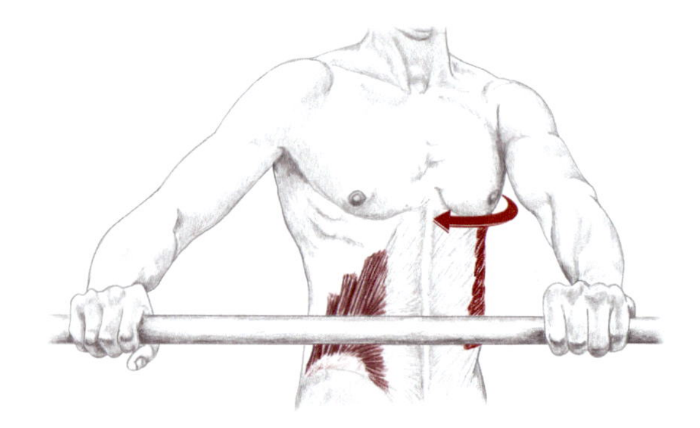

Anterior/lateral view

Rotation *(all unilaterally)*
Multifidi (to the opposite side)
Rotatores (to the opposite side)
External oblique (to the opposite side)
Internal oblique (to the same side)

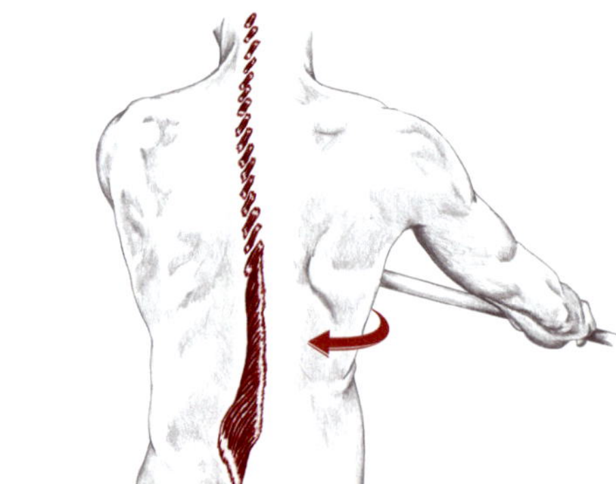

Posterior view of multifidi

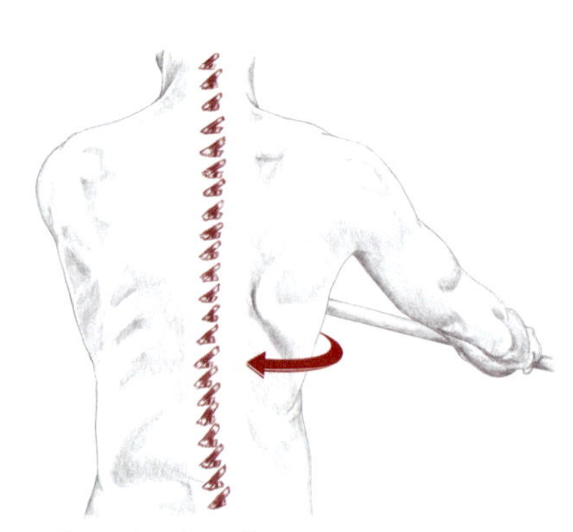

Posterior view of rotatores

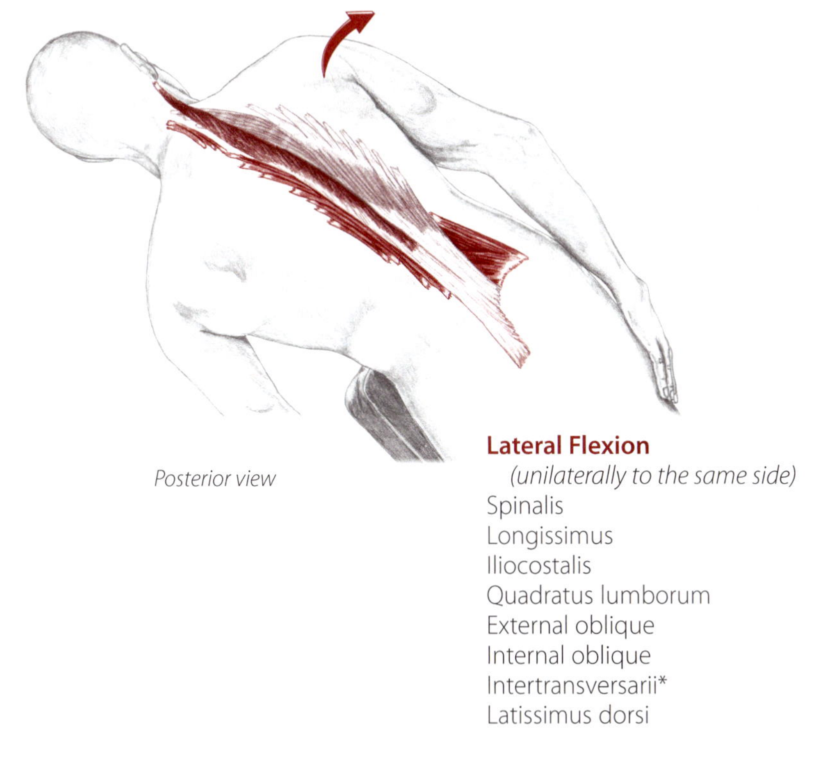

Posterior view

Lateral Flexion
(unilaterally to the same side)
Spinalis
Longissimus
Iliocostalis
Quadratus lumborum
External oblique
Internal oblique
Intertransversarii*
Latissimus dorsi

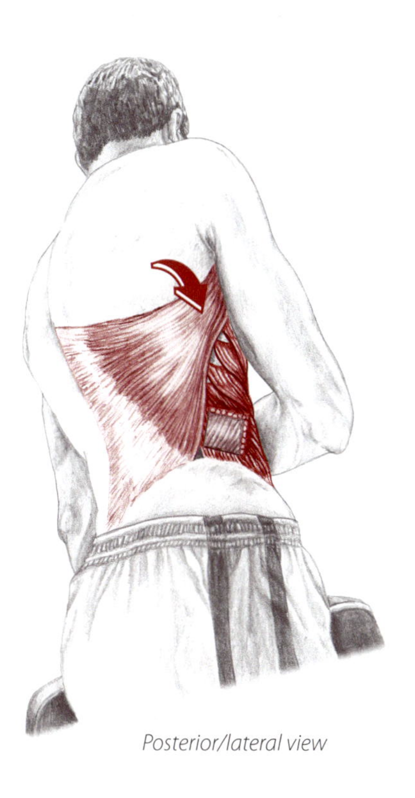

Posterior/lateral view

Ribs/Thorax

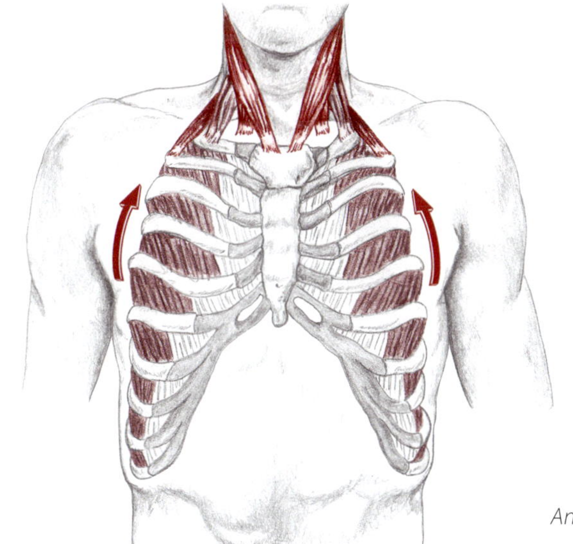

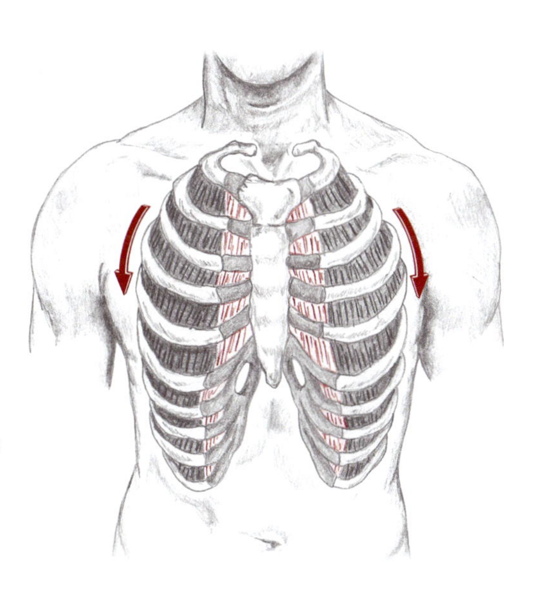

Anterior views

Elevation/Expansion
Anterior scalene (bilaterally)
Middle scalene (bilaterally)
Posterior scalene (bilaterally)
Sternocleidomastoid (assists)
External intercostals (assists)
Serratus posterior superior*
Pectoralis major (may assist if arm is fixed)*
Pectoralis minor (if scapula is fixed)*
Serratus anterior (if scapula is fixed)*
Subclavius (first rib)*

Depression/Collapse
Internal intercostals (assists)
Serratus posterior inferior*

See p. 398 for a list of the muscles of respiration

Erector Spinae Group

Spinalis
Longissimus
Iliocostalis

The erector spinae group runs from the sacrum to the occiput along the posterior aspect of the vertebral column. Its musculature has a dense, layered arrangement that can be difficult to visualize. It might simplify matters if you imagine the erector spinae muscles as a tall poplar tree (4.50) with three main branches - the spinalis, longissimus and iliocostalis (4.49). These branches can then be subdivided into numerous, smaller branches such as spinalis thoracis, longissimus capitis, iliocostalis lumborum and more.

The **spinalis** is the smallest of the three muscles and lies closest to the spine in the lamina groove (4.51). The thick **longissimus** and lateral **iliocostalis** form a visible mound alongside the lumbar and thoracic spine (4.53, 4.54). The long tendons of iliocostalis extend laterally beneath the scapula.

In the lumbar region, the erectors lie deep to the thin but dense thoracolumbar aponeurosis (p. 226). In the thoracic and cervical areas, they are deep to the trapezius, the rhomboids and the serratus posterior superior and inferior. As a group, the erectors are easily palpated along the entire length of the back and neck; locating a specific branch of the erectors, however, can be challenging.

C-7
T-12
L-5
Spinalis
Longissimus
Iliocostalis (reflected)
Thoracolumbar aponeurosis

(4.49) Posterior view of right side showing erector spinae group

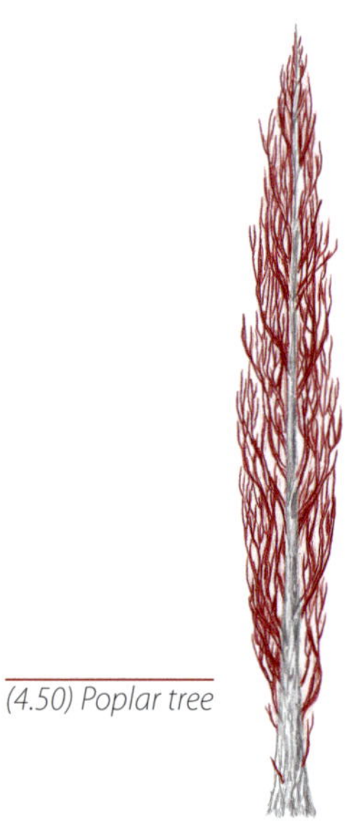

(4.50) Poplar tree

The upper fibers of longissimus and iliocostalis muscles (longissimus cervicis and capitis, iliocostalis cervicis) assist in extension, lateral flexion and rotation of the head and neck to the same side.

iliocostalis	**il**-ee-o-kos-**ta**-lis	L. from hip to rib
longissimus	lon-**jis**-i-mus	L. longest

Erector Spinae Group

A *Unilaterally:*
Laterally flex vertebral column to the same side

Bilaterally:
Extend the vertebral column

O Common tendon (thoracolumbar aponeurosis) that attaches to the posterior surface of sacrum, iliac crest, spinous processes of the lumbar and last two thoracic vertebrae

I Various attachments at the posterior ribs, spinous and transverse processes of thoracic and cervical vertebrae, and mastoid process of temporal bone

N Dorsal primary divisions of spinal nerves

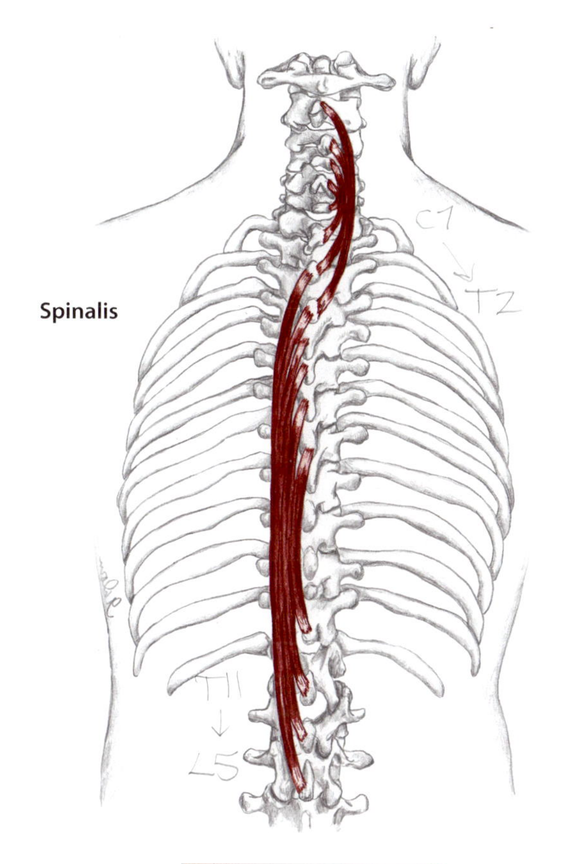

(4.51) Posterior view of thorax

Branches of the Erector Spinae Group

Spinalis

O Spinous processes of the upper lumbar and lower thoracic vertebrae **(thoracis)**
Ligamentum nuchae, spinous process of C-7 **(cervicis)**

I Spinous processes of upper thoracic **(thoracis)**
Spinous processes of cervicals, except C-1 **(cervicis)**

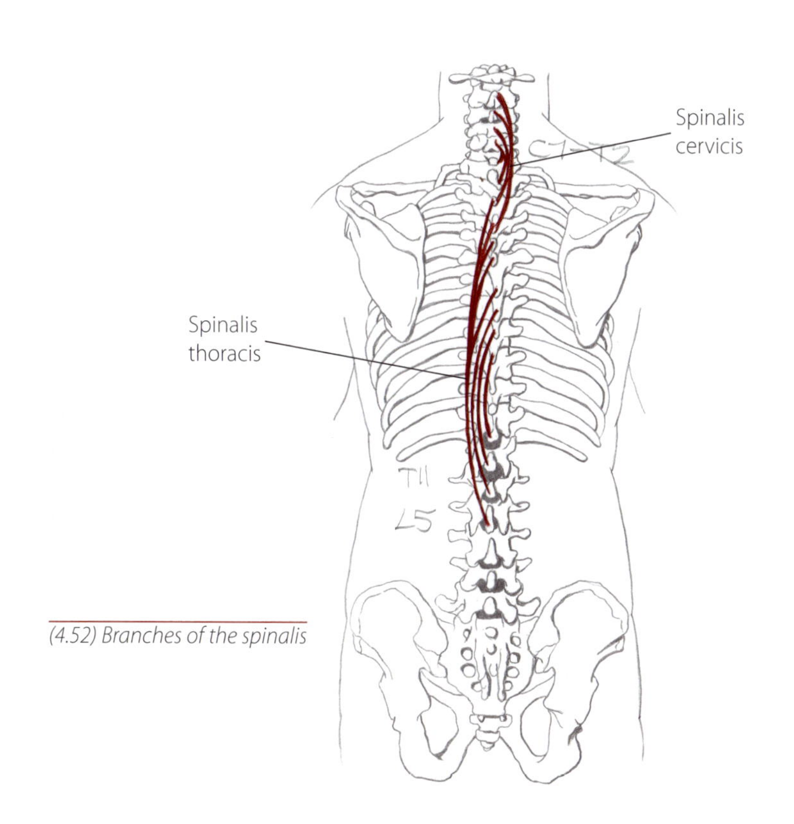

(4.52) Branches of the spinalis

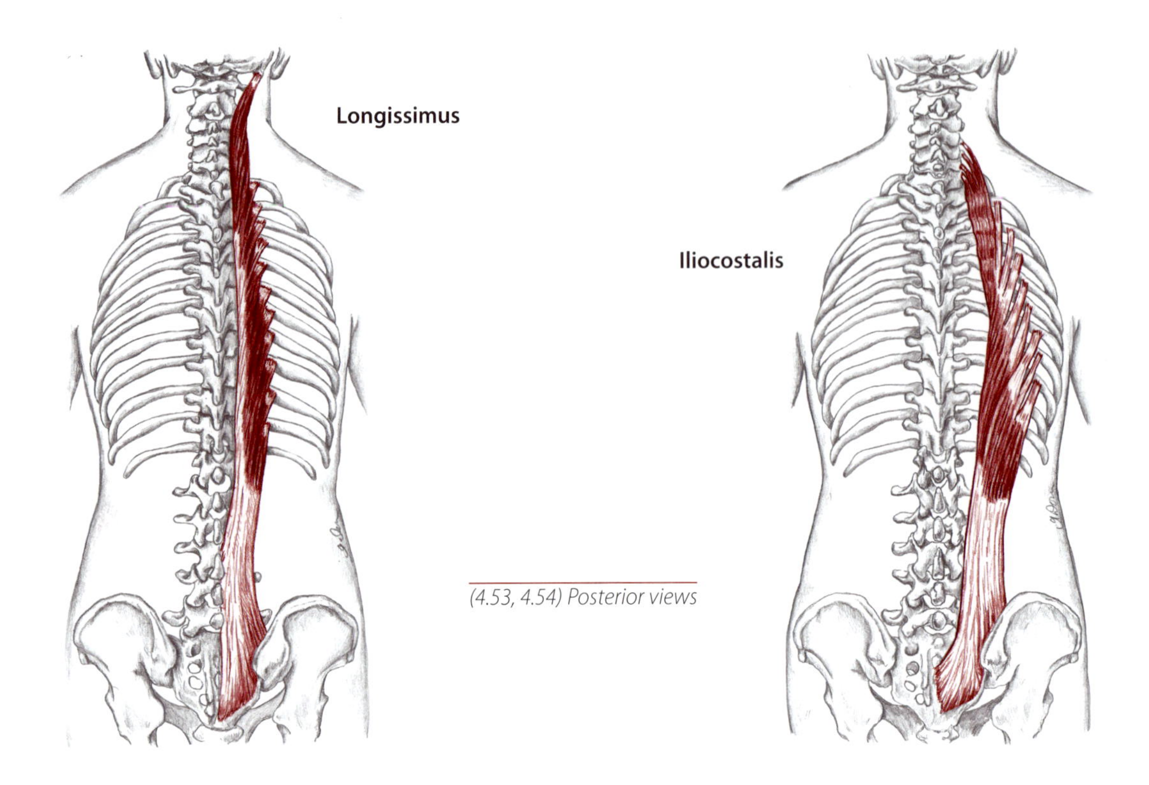

(4.53, 4.54) Posterior views

Longissimus

O Common tendon **(thoracis)**
Transverse processes of upper five thoracic vertebrae **(cervicis and capitis)**

I Lower nine ribs and transverse processes of thoracic vertebrae **(thoracis)**
Transverse processes of cervical vertebrae **(cervicis)**
Mastoid process of temporal bone **(capitis)**

Iliocostalis

O Common tendon **(lumborum)**
Posterior surface of ribs 1-12 **(thoracis and cervicis)**

I Transverse processes of lumbar vertebrae 1-3 and posterior surface of ribs 6-12 **(lumborum)**
Posterior surface of ribs 1-6 **(thoracis)**
Transverse processes of lower cervicals **(cervicis)**

Longissimus capitis
Longissimus cervicis
Longissimus thoracis

(4.55) Branches of the longissimus

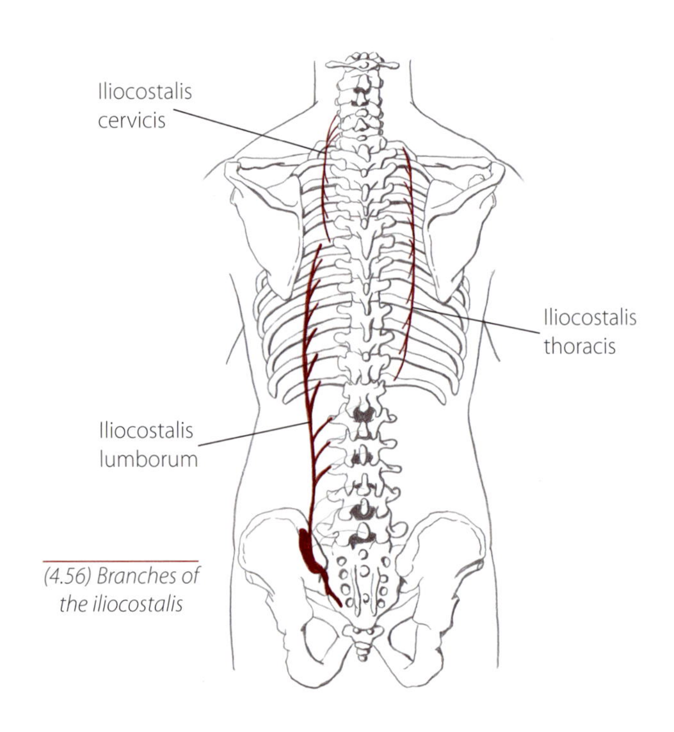

(4.56) Branches of the iliocostalis

Erector spinae group

1) Partner prone. Lay both hands along either side of the lumbar vertebrae. Locate the region of the lower erectors by asking your partner to alternately raise and lower his feet slightly. The erectors do not, of course, raise the feet, but they will contract in order to stabilize the pelvis. Notice how the strong, rounded erector fibers tighten and relax with this action (4.57).
2) As your partner maintains this contraction, palpate inferiorly onto the sacrum and then superiorly along the thoracic vertebrae. Ask your partner to extend his spine and neck slightly in order to contract the erectors in the thoracic region (4.58).
3) Follow the ropy fibers of the erectors between the scapulae and along the back of the neck. These fibers are smallest in the cervical region and are primarily situated lateral to the lamina groove.
4) With your partner relaxed, sink your fingers into the erector fibers, feeling their ropy texture and vertical direction.

Is the tissue you are palpating directly beside the spinous processes of the vertebrae? Do the fibers run parallel to the spine? When the muscles are contracted, can you locate the lateral edge of the erector group? Can you distinguish the fiber direction of the middle trapezius, rhomboids and erectors between the scapulae?

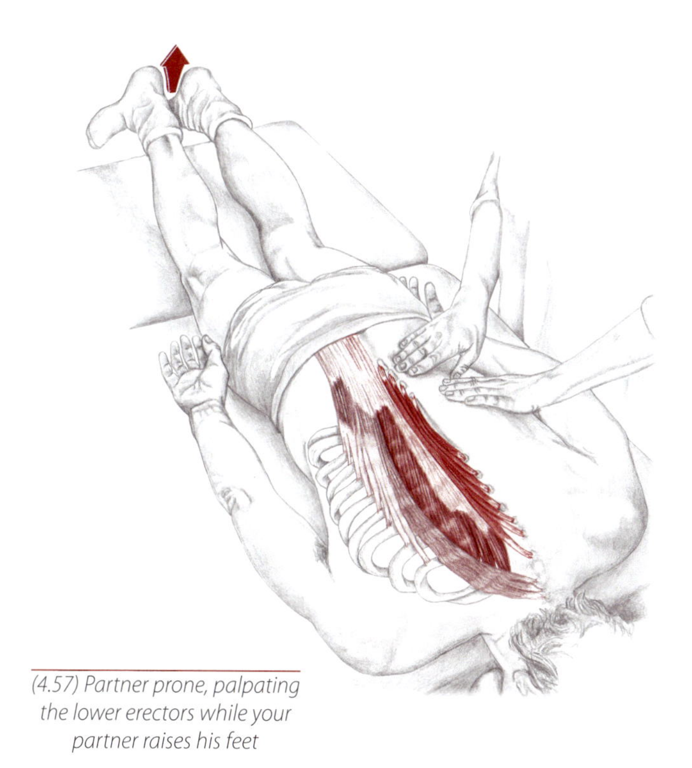

(4.57) Partner prone, palpating the lower erectors while your partner raises his feet

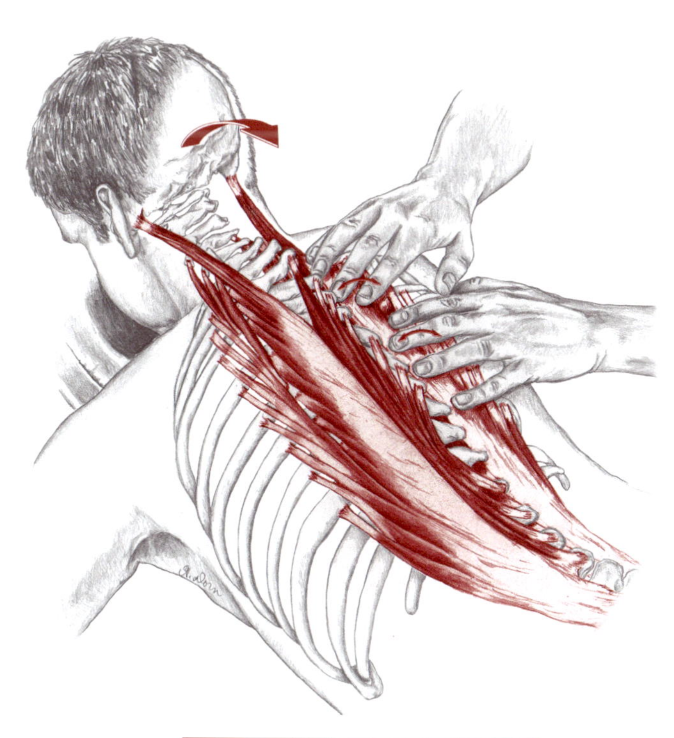

(4.58) Partner prone, palpating the upper erectors while your partner extends his spine

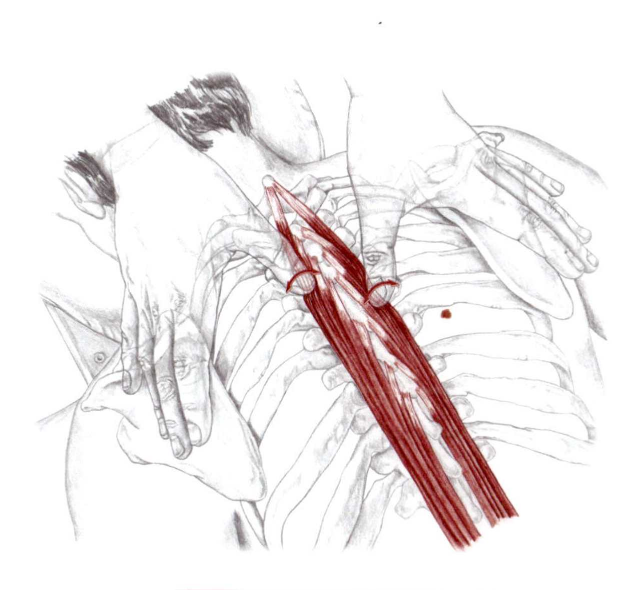

(4.59) Partner prone, strumming your thumbs across the fibers of the spinalis

multifidi	mul-**tif**-i-di	L. *fidi*, to split
rotatores	**ro**-ta-**tor**-ays	L. plural for rotators

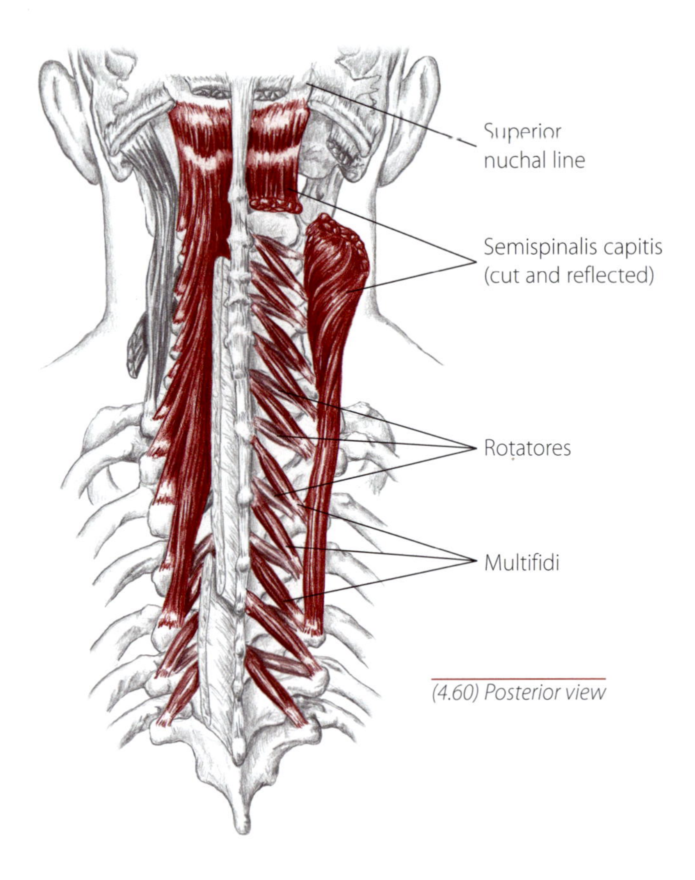

(4.60) Posterior view

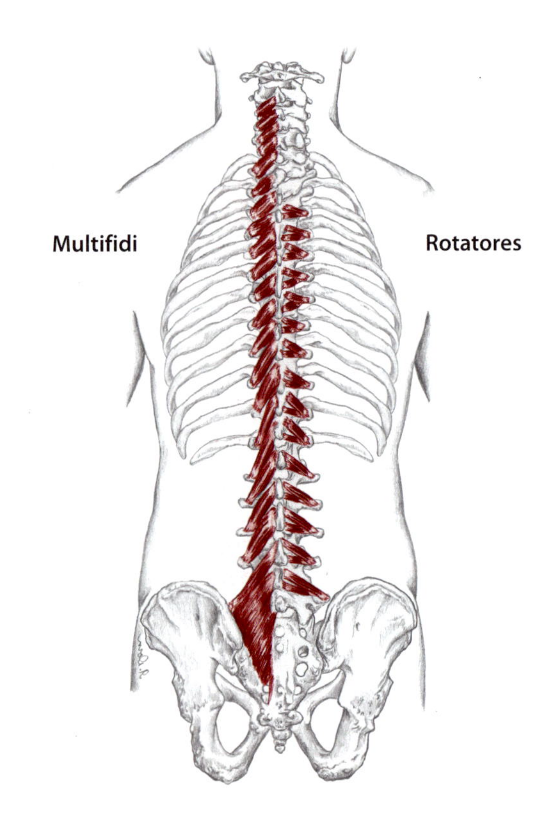

(4.61) Posterior view with the multifidi on the left and the thoracic and lumbar portions of the rotatores on the right

Transversospinalis Group

Multifidi
Rotatores
Semispinalis Capitis

Deep to the erector spinae muscle group is the transversospinalis muscle group. The transversospinalis is composed of three branches - multifidi, rotatores and semispinalis - and extends the length of the vertebral column. Unlike the long, vertical erector fibers, the branches of the transversospinalis consist of many short, diagonal fibers. These fibers form an intricate stitchlike design that links the vertebrae together. The name "transversospinalis" refers to the fact that the fibers of this muscle extend at varying lengths from the transverse and spinous processes of the vertebrae.

The surprisingly thick **multifidi** are directly accessible in the lumbar spine. They are the only muscles with fibers that lie across the posterior surface of the sacrum. The shorter, smaller **rotatores** lie deep to the multifidi (4.61). The **semispinalis capitis** is located along the thoracic and cervical vertebrae and ultimately reaches the cranium (4.60). Both semispinalis muscles form the twin "speed bumps" easily seen on the posterior neck when the neck is extended against resistance.

It can be difficult to isolate the individual bellies of the transversospinalis muscles as they are closely interwoven; however, as a group, their mass or density can be easily felt along the lamina groove of the thoracic and lumbar vertebrae.

Multifidi and Rotatores

A *Unilaterally:*
Rotate the vertebral column to the opposite side

Bilaterally:
Extend the vertebral column

O *Multifidi:*
Sacrum and transverse processes of lumbar through cervical vertebrae

Rotatores:
Transverse processes of lumbar through cervical vertebrae

I Spinous processes of lumbar vertebrae through second cervical vertebra
(Multifidi span two to four vertebrae)
(Rotatores span one to two vertebrae)

N Dorsal primary divisions of spinal nerves

semispinalis — **sem**-eye-spi-**na**-lis — L. half spinal
transversospinalis — trans-**ver**-so-spi-**nal**-is

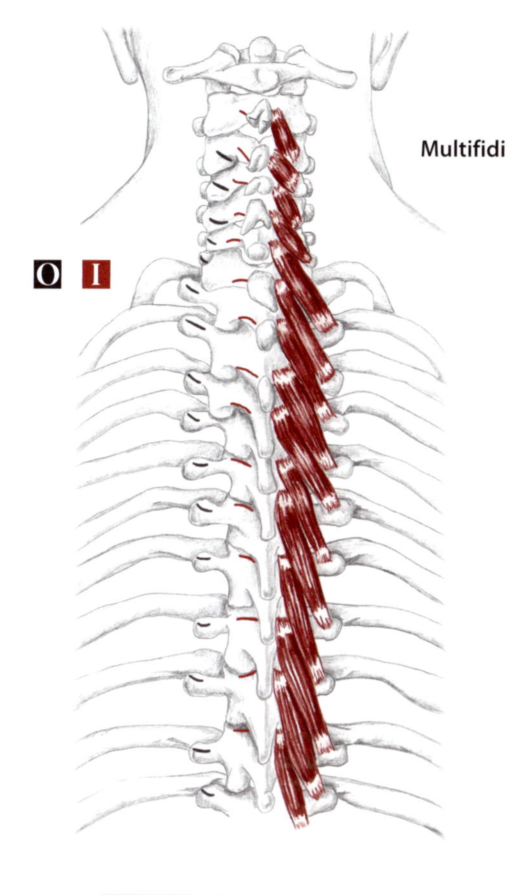

(4.62) Posterior view, showing origins and insertions of the upper multifidi

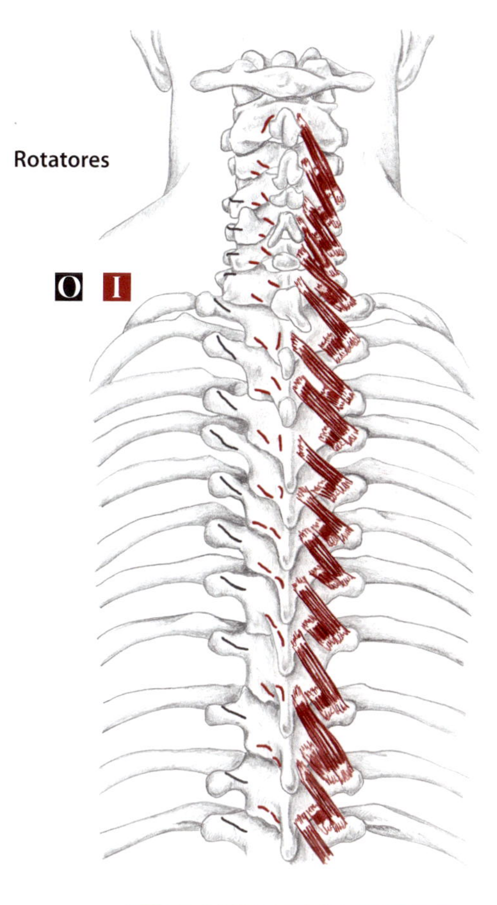

(4.63) Posterior view, showing origins and insertions of the upper rotatores

Semispinalis Capitis

A Extend the vertebral column and head

O Transverse processes of thoracic vertebrae, articular processes of lower cervicals

I Spinous processes of upper thoracic and cervicals (except C-1), and superior nuchal line of occiput

N Dorsal primary divisions of spinal nerves

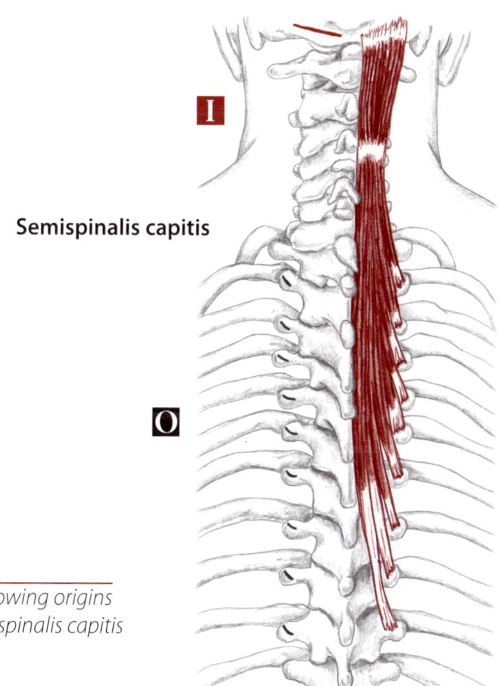

(4.64) Posterior view, showing origins and insertion of the semispinalis capitis

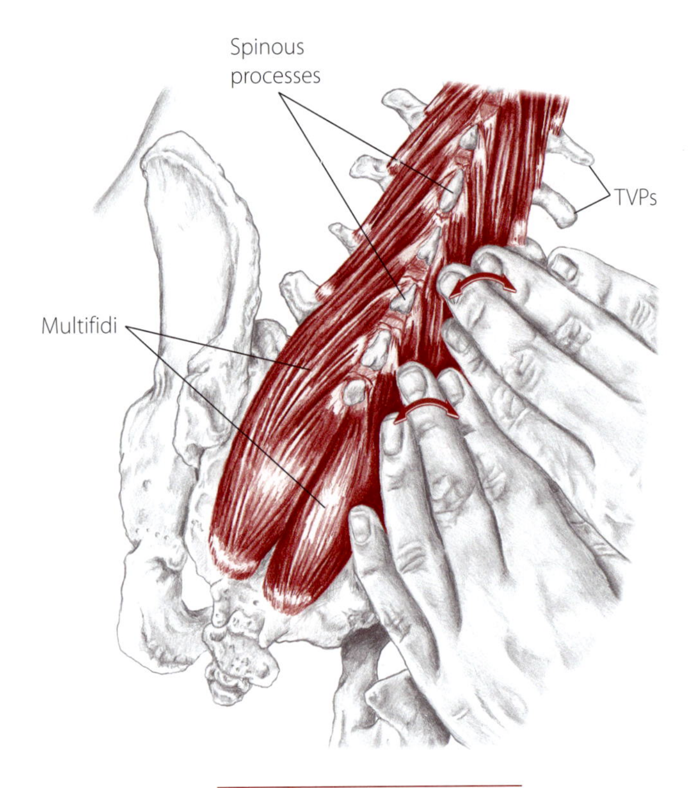

(4.65) Posterior view, partner prone

Transversospinalis group

1) Partner prone. Locate the spinous processes of the lumbar vertebrae. Slide your fingers laterally off the spinous processes, sinking between them and the erector spinae fibers.
2) Pushing the erectors laterally out of the way, explore deeply for the dense, diagonal fibers of the multifidi. Progress inferiorly to the sacrum, rolling your fingers in a perpendicular direction to the multifidi fibers (4.65).
3) Move superiorly, exploring the lamina groove of the thoracic and cervical areas. Then turn your partner supine and palpate the cervical region.

Are you between the spinous and transverse processes? Can you get a sense of these smaller, deeper fibers that stretch at an oblique angle?

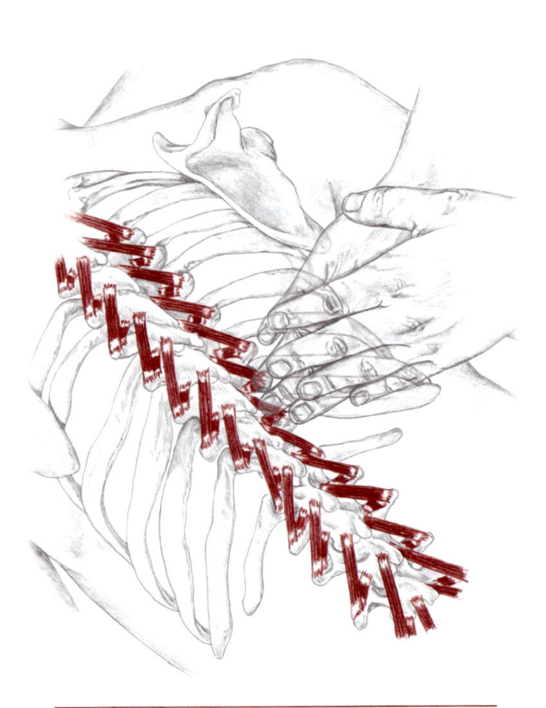
(4.66) Partner prone, posterior view with superficial muscles removed. Accessing the rotatores by directing your fingers **toward** the vertebrae.

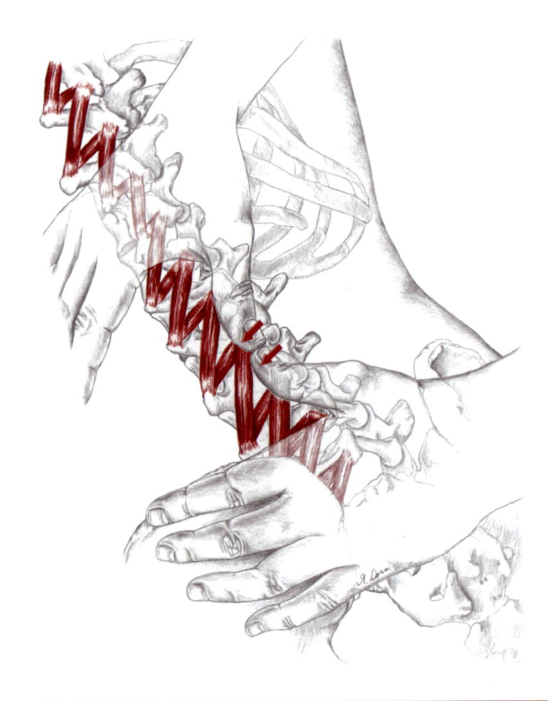
(4.67) Partner prone, posterior/lateral view with superficial muscles removed. Palpating into the lamina groove on opposite side of the body with the thumbs sliding **away** from the vertebrae.

Splenius Capitis and Cervicis

The long splenius capitis and splenius cervicis muscles are located along the upper back and posterior neck (4.68). In contrast to the other back muscles that run parallel to the spine, the splenii fibers run obliquely. The **splenius capitis** is deep to the trapezius and rhomboids. Its fibers angle toward the mastoid process and are superficial between the trapezius and sternocleidomastoid (4.70).

The **splenius cervicis** is deep to the splenius capitis and not as easily isolated; however, its general location can be outlined in the lamina groove of the upper thoracic and cervical spine.

A *Unilaterally:*
Rotate the head and neck to the same side
Laterally flex the head and neck

Bilaterally:
Extend the head and neck

O *Capitis:*
Ligamentum nuchae, spinous processes of C-7 to T-3
Cervicis:
Spinous processes of T-3 to T-6

I *Capitis:*
Mastoid process and lateral portion of superior nuchal line
Cervicis:
Transverse processes of the upper cervical vertebrae

N Branches of dorsal division of cervical

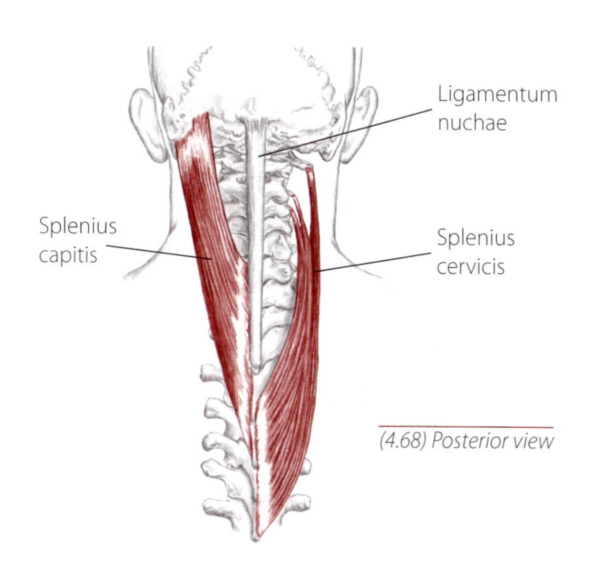

(4.68) Posterior view

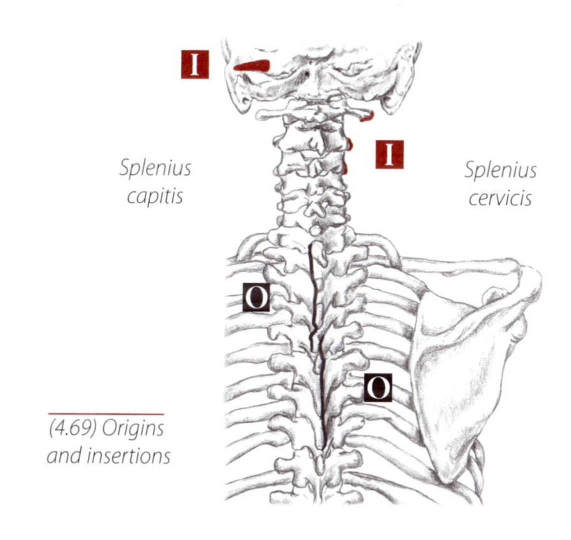

(4.69) Origins and insertions

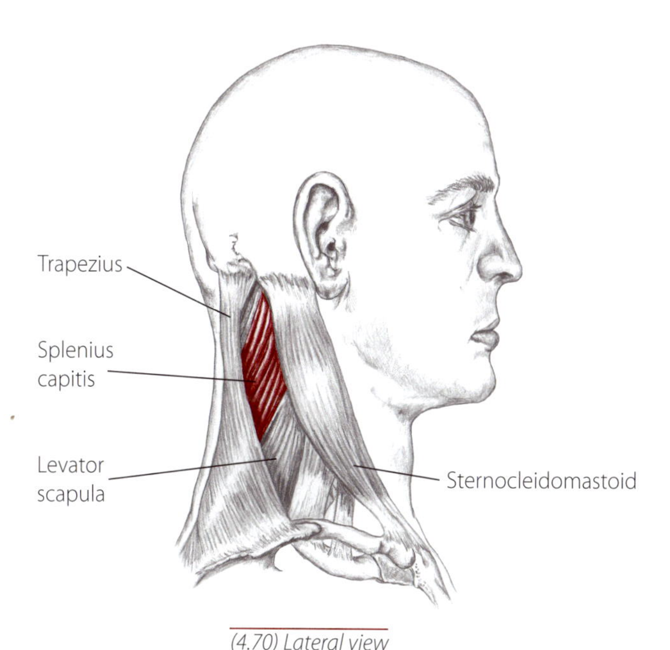

(4.70) Lateral view

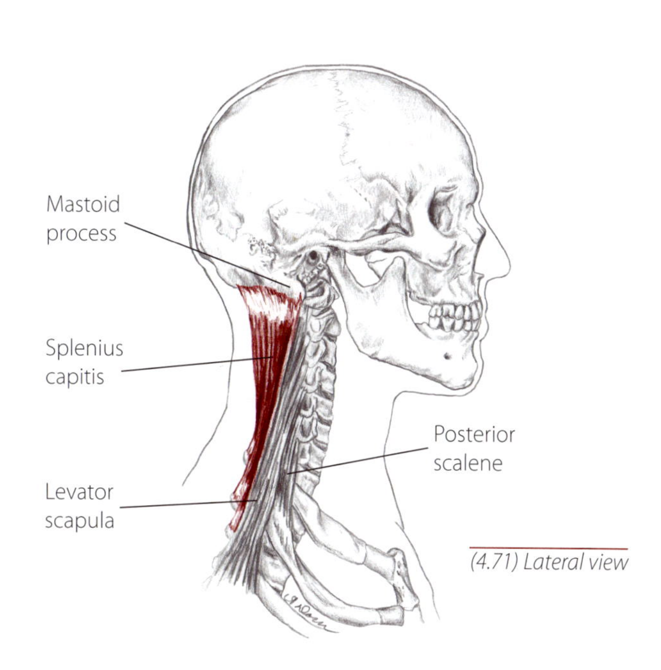

(4.71) Lateral view

splenius capitis **sple**-nee-us **kap**-i-tis
splenius cervicis **sple**-nee-us **ser**-vi-sis

L. bandage-like (muscle) of the head

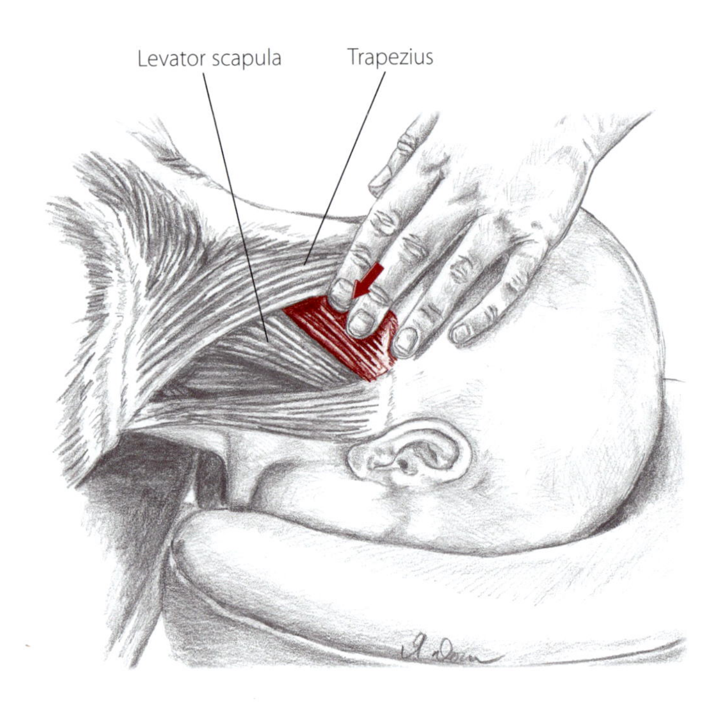

(4.72) Partner prone, locating splenius capitis

Splenius capitis

1) Prone. Locate the upper fibers of the trapezius.
2) Isolate the lateral edge of the trapezius by having your partner extend his head slightly.
3) Ask your partner to relax. Palpate just lateral to the trapezius for the splenius capitis' oblique fibers, following them up to the mastoid process and inferiorly through the trapezius (4.72).

Do the fibers you feel lead toward the mastoid process? Distinguish the trapezius fibers from the splenius capitis fibers by asking your partner to rotate his head slightly toward the side you are palpating. Do you feel these oblique fibers contract while the trapezius remains passive?

Locate the mastoid process and slide medially and inferiorly onto the superficial capitis fibers.

Both splenii muscles

1) Partner supine with the head rotated 45° away from the side you are palpating. Cradle the head with one hand while the other hand locates the lamina groove of the upper cervical and thoracic vertebrae (4.73).
2) Passively extend the neck slightly to shorten the tissue and palpate through the overlying trapezius fibers. These bellies will not be particularly distinct; however, the density of both splenii can be felt in the lamina groove.

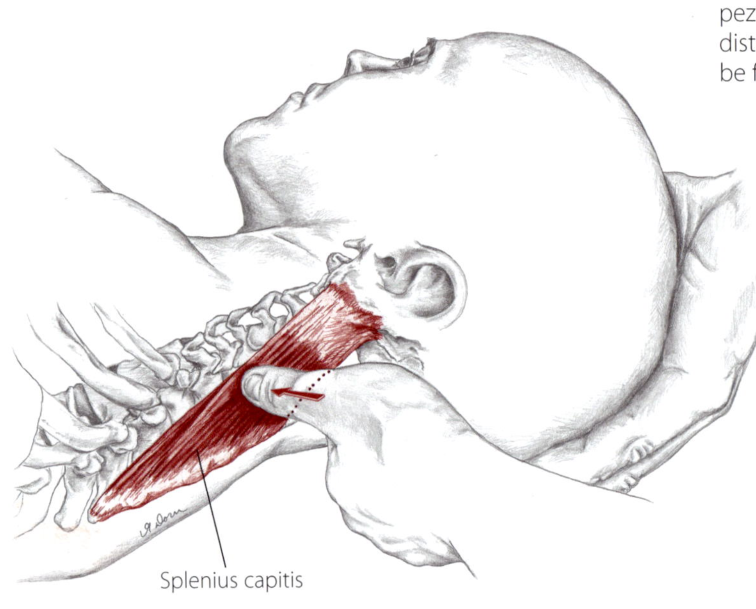

(4.73) Partner supine, with head rotated 45° to the right, away from the side you are palpating

capitis	**kap**-i-tis	L. of the head
cervicis	**ser**-vi-sis	L. neck
splenius	**sple**-nee-us	Grk. bandage

Suboccipitals

Rectus Capitis Posterior Major
Rectus Capitis Posterior Minor
Oblique Capitis Superior
Oblique Capitis Inferior

The eight small suboccipitals are the deepest muscles of the upper posterior neck (p. 198). They are involved in stabilizing the axis and atlas and in creating intrinsic movements such as rocking and tilting of the head. To outline the suboccipitals' location, find the spinous process of C-2, the transverse processes of C-1 and the space between the superior nuchal line of the occiput and C-2 (4.74 - 4.77).

The upper fibers of the trapezius can also be used as a marker. The lateral edge of the muscle is the same width as the suboccipitals. The density of the suboccipital bellies can be felt, but accessing specific muscle bellies may be challenging.

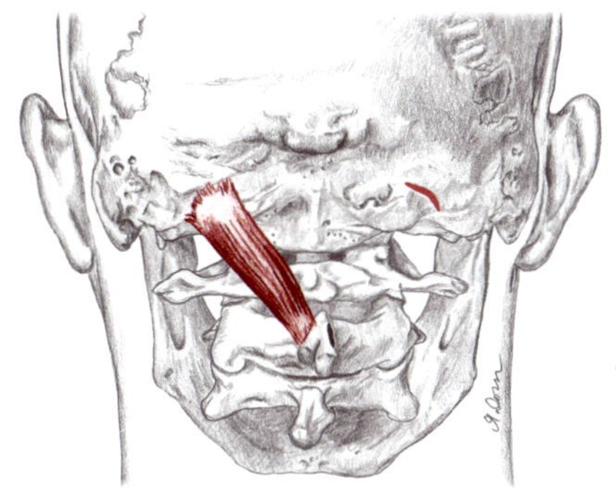

Posterior views with origins (black) and insertions (red)

(4.74) Rectus capitis posterior major

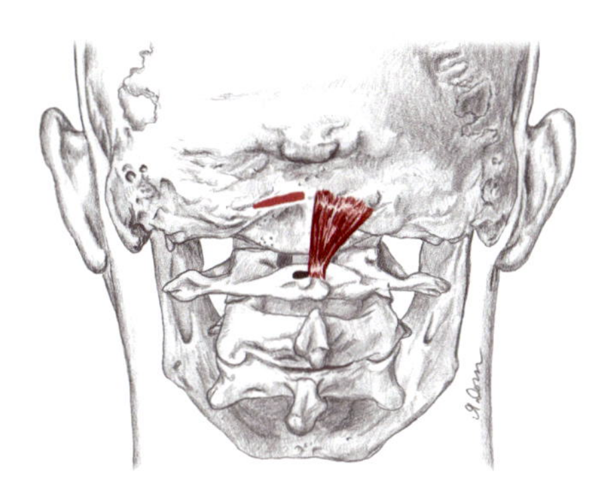

(4.75) Rectus capitis posterior minor

All Suboccipitals:

A *Rectus Capitis Posterior Major*
Rectus Capitis Posterior Minor
Oblique Capitis Superior

Rock and tilt the head back into extension

Rectus Capitis Posterior Major
Oblique Capitis Inferior

Rotate the head to the same side

Rectus Capitis Posterior Major

O Spinous process of the axis (C-2)

I Inferior nuchal line of the occiput

N Suboccipital

Rectus Capitis Posterior Minor

O Tubercle of the posterior arch of the atlas (C-1)

I Inferior nuchal line of the occiput

N Suboccipital

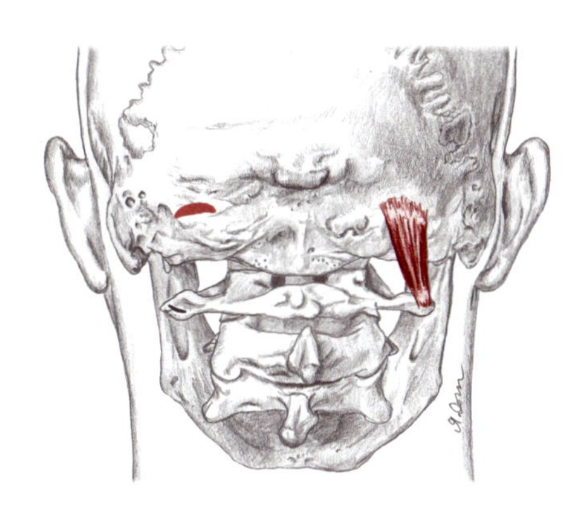

(4.76) Oblique capitis superior

Oblique Capitis Superior

O Transverse process of the atlas (C-1)

I Between the nuchal lines of the occiput

N Suboccipital

Oblique Capitis Inferior

O Spinous process of the axis (C-2)

I Transverse process of the atlas (C-1)

N Suboccipital

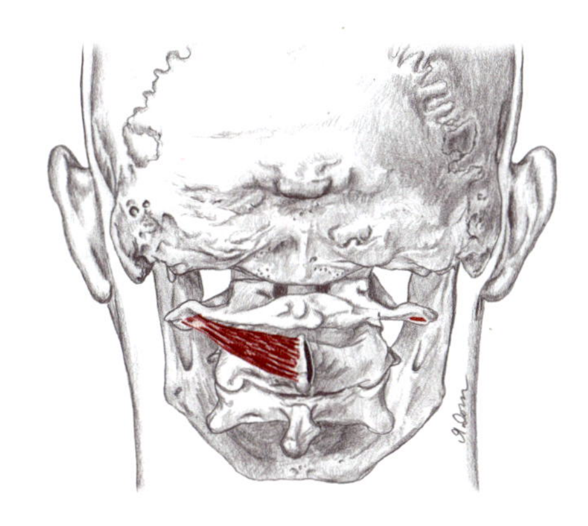

(4.77) Oblique capitis inferior

occiput **ok**-si-put L. the back of the skull

Suboccipitals

1) Partner supine. Cradle the head in both hands. Passively extending the neck a bit will soften the overlying tissue. Locate the superior nuchal line of the occiput and the spinous process of C-2. The suboccipitals span the area between these two landmarks.
2) Cradle the head with one hand while two fingertips of the other hand palpate slowly through the trapezius, splenius capitis and semispinalis capitis fibers (4.78).
3) Roll your fingers across the suboccipitals' small, short bellies. Again, you may initially feel only the density of these muscles rather than the individual bellies.

Are you between the spinous process of C-2 and the superior nuchal line of the occiput? If you ask your partner to tilt his head back ever so slightly, do you feel some contraction in the deepest layer of tissue?

Partner prone. Locate the lateral edge of the trapezius' upper fibers (4.79). Palpating beside the level of C-1, place one finger at the lateral edge of the trapezius. Slowly sink medially into the suboccipitals.

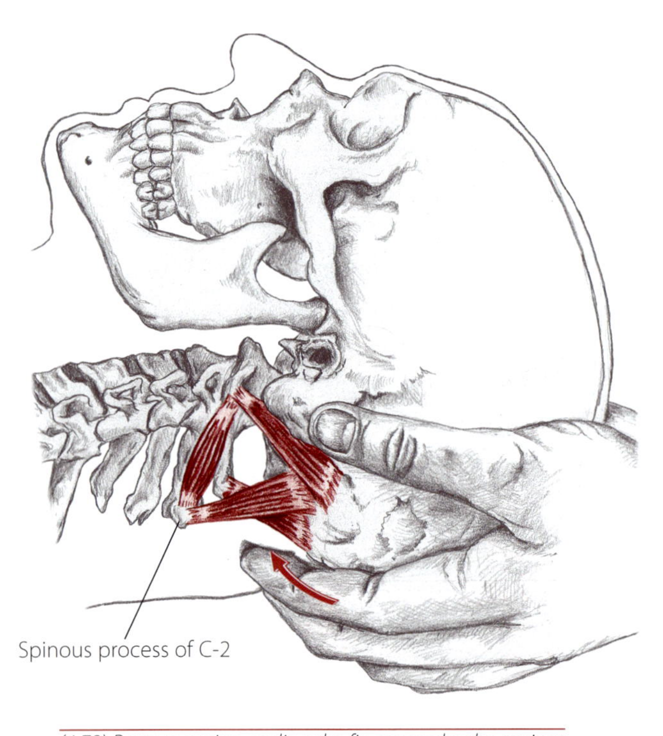

(4.78) Partner supine, curling the fingers under the occiput

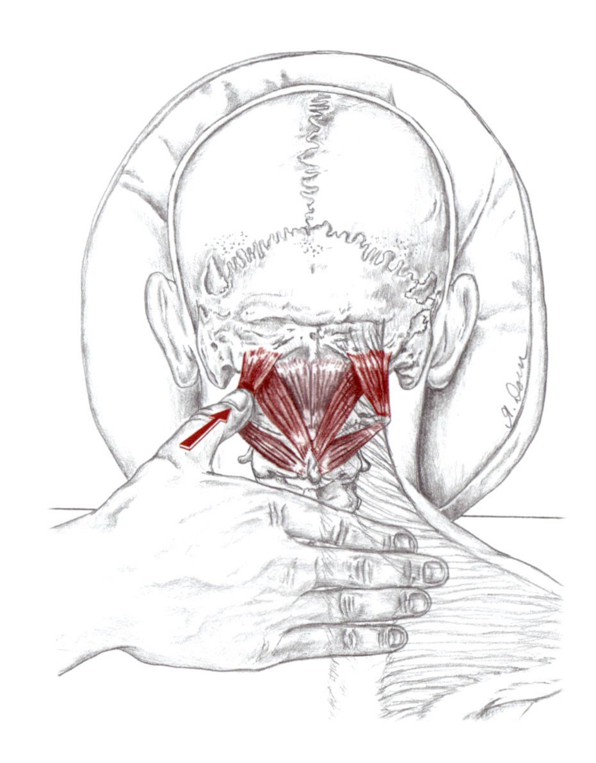

(4.79) Partner prone, sinking your thumb medially, just lateral to the edge of the trapezius

Researchers have discovered that the rectus capitis posterior minor not only attaches to the occiput, but also to the dura mater, the connective tissue that surrounds the spinal cord and brain. Because of this connection between the rectus capitis posterior minor and the dura mater, this muscle may cause headaches by disrupting normal cerebrospinal fluid fluctuations and hence the functioning of the vertebral artery and suboccipital nerve.

dura mater **dyoo**-ra **ma**-ter L. tough mother

Quadratus Lumborum

Although it would seem to be the deepest muscle of the low back, the quadratus lumborum is, strangely enough, the deepest muscle of the abdomen (4.81). Stretching from the posterior ilium to the transverse processes of the lumbar vertebrae and twelfth rib, this squat muscle is simply an abdominal muscle located on the posterior surface of the thorax.

While the medial portion of the quadratus lumborum is buried beneath the thoracolumbar aponeurosis and the thick erector spinae (4.80), its lateral edge is accessible from the side of the torso.

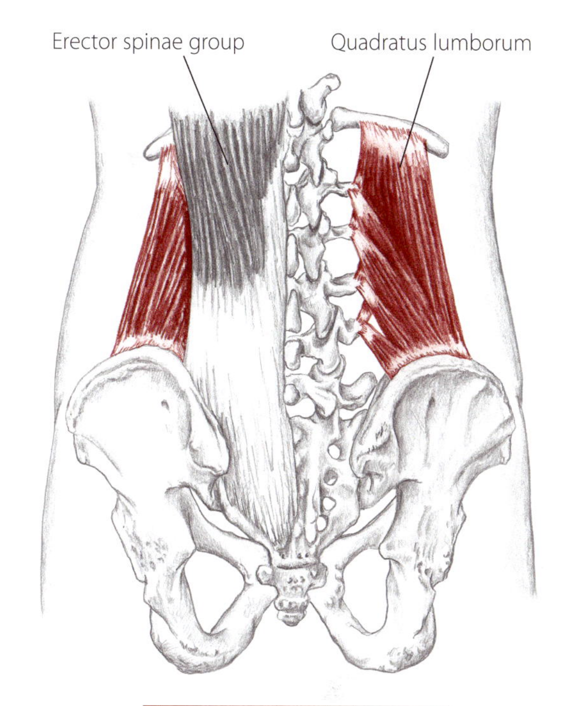

(4.80) Posterior view, erector spinae group removed on right side

A *Unilaterally:*
Laterally tilt the pelvis
Laterally flex the vertebral column to the same side
Assist to extend the vertebral column

Bilaterally:
Fix the last rib during inhalation and forced exhalation

O Posterior iliac crest

I Last rib and transverse processes of first through fourth lumbar vertebrae

N Branches of first lumbar and twelfth thoracic

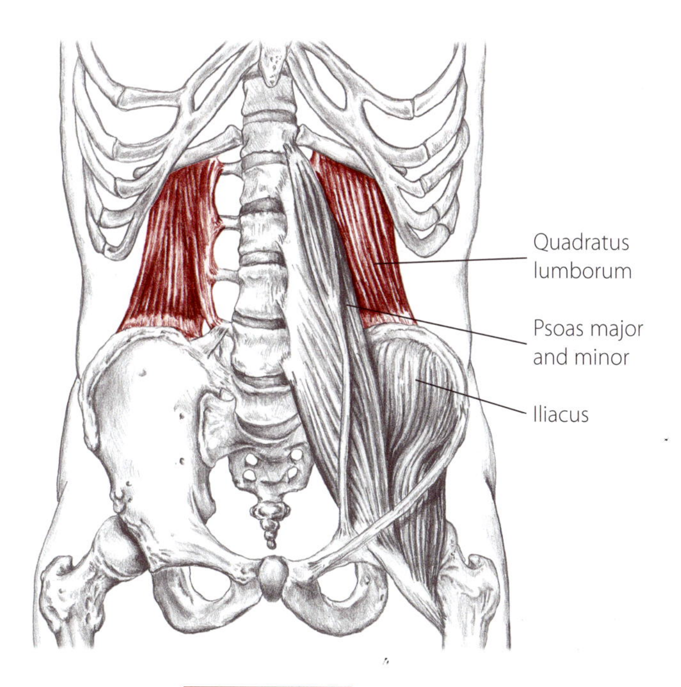

(4.81) Anterior view

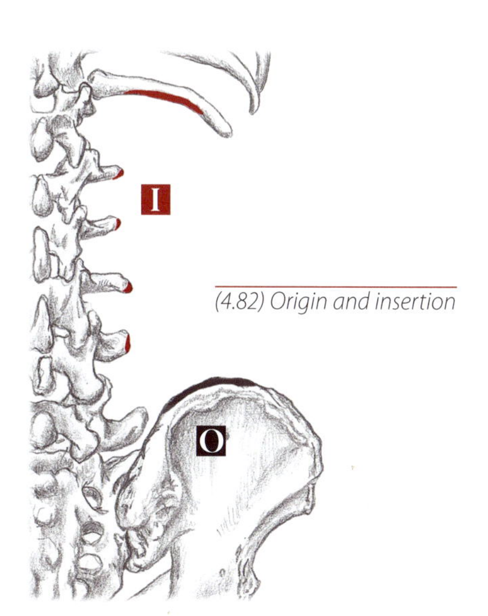

(4.82) Origin and insertion

The quadratus lumborum is sometimes known as the "hip hiker" because of its capacity to laterally tilt (elevate) the hip.

quadratus lumborum **kwod**-rait-us lum-**bor**-um L. four-sided muscle of the lumbar region

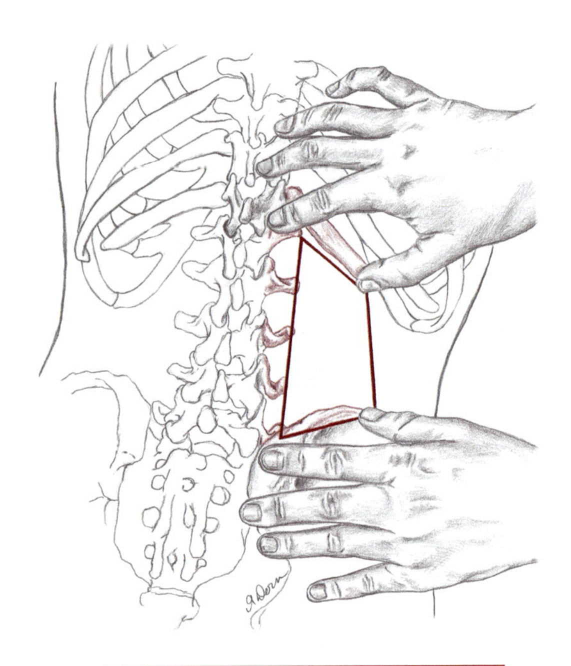

(4.83) Partner prone, laying your fingers along the corners of the quadratus lumborum

Quadratus lumborum

1) Prone. Isolate the borders of the quadratus by locating the twelfth rib, posterior iliac crest and transverse processes of the lumbar vertebrae.
2) Lay your fingers along these landmarks to outline the edges of the quadratus (4.83).
3) Lay your thumbpad along the lateral edge of this square. Using slow, firm pressure, sink your thumb medially toward the lumbar vertebrae and into the edge of the quadratus (4.84).
4) Ask your partner to laterally tilt (elevate) his hip toward his shoulder in order to feel its solid contraction. The hip should remain on the table.

As you palpate, be sure you are accessing the deeper tissue in the low back and not just the superficial external oblique fibers. When your partner hikes his hip, can you feel the lateral edge of the quadratus contract? Can you distinguish between the edge of the erector spinae and the quadratus?

Follow the above instructions, only this time with your partner sidelying (4.85). Placing a bolster between his knees will balance the pelvis and soften the tissue around the quadratus. This position will also allow the abdominal contents to shift away from where you are accessing.

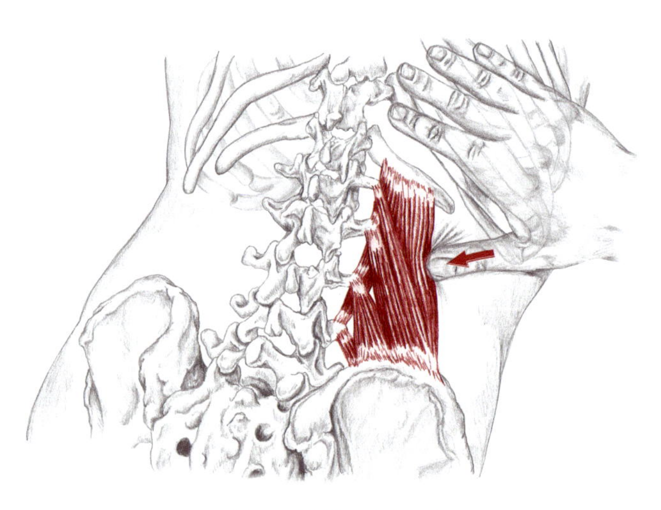

(4.84) Partner prone, accessing the quadratus lumborum

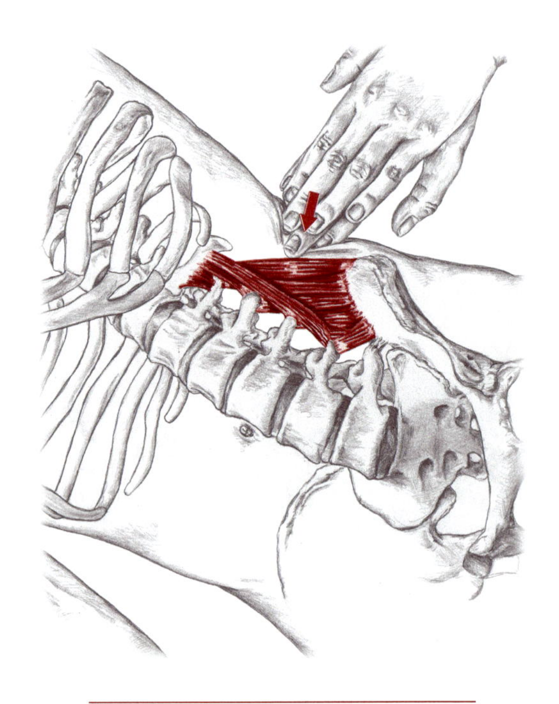

(4.85) Anterior/lateral view, partner sidelying

Abdominals

Rectus Abdominis
External Oblique
Internal Oblique
Transverse Abdominis

The four abdominal muscles expand far beyond the "stomach" region. In fact, they form a muscular girdle that reaches around the sides of the thorax to the thoracolumbar aponeurosis, superiorly to the middle ribs and inferiorly to the inguinal ligament. The immense span of these muscles, with its unique overlapping arrangement and varying fiber direction, helps to stabilize the entire abdominal region.

The revered "washboard belly" is formed by the multiple, superficial bellies of the **rectus abdominis** (4.86, 4.87). Lateral to the rectus abdominis is the **external oblique** (4.89). Unlike the round bellies of the rectus abdominis, the external oblique is a broad, superficial muscle best palpated at its attachments to the lower ribs.

The thin **internal oblique** fibers are deep and perpendicular to the external oblique fibers and can be difficult to distinguish (4.91). The **transverse abdominis**, the deepest muscle of the group, plays a major role in forced exhalation. It cannot be specifically palpated (4.93).

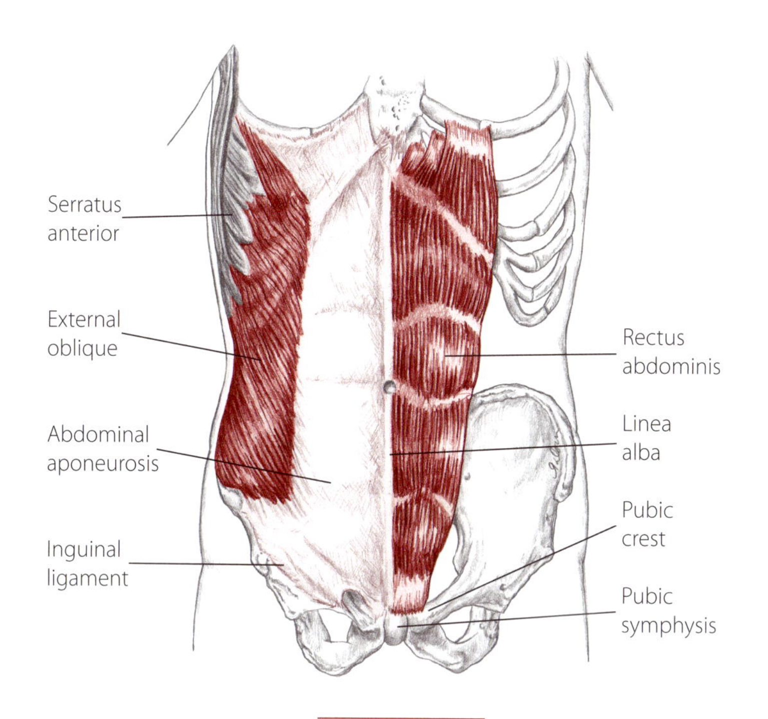

(4.86) Anterior view

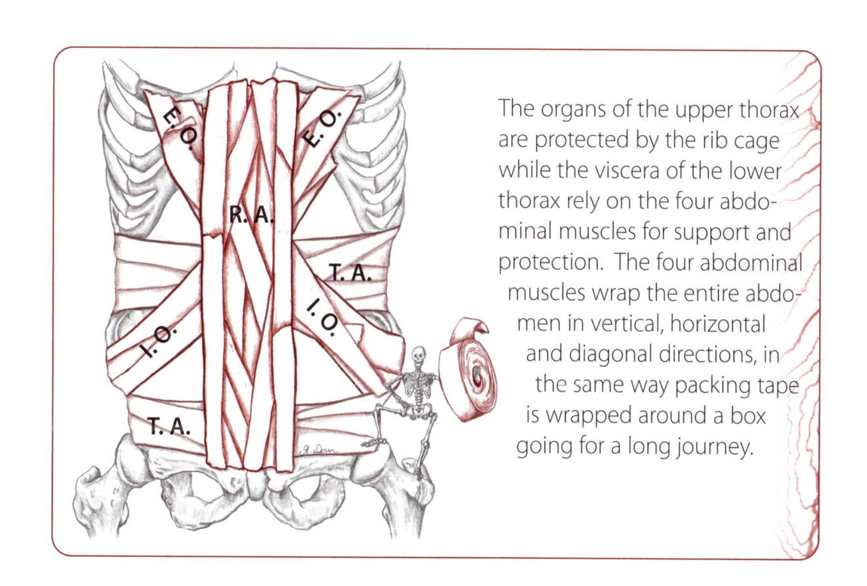

The organs of the upper thorax are protected by the rib cage while the viscera of the lower thorax rely on the four abdominal muscles for support and protection. The four abdominal muscles wrap the entire abdomen in vertical, horizontal and diagonal directions, in the same way packing tape is wrapped around a box going for a long journey.

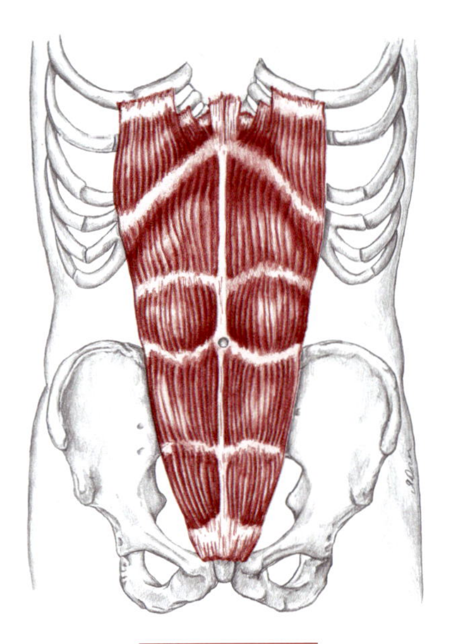

(4.87) Anterior view of rectus abdominis

Rectus Abdominis

A Flex the vertebral column

O Pubic crest, pubic symphysis

I Cartilage of fifth, sixth and seventh ribs and xiphoid process

N Branches of intercostals

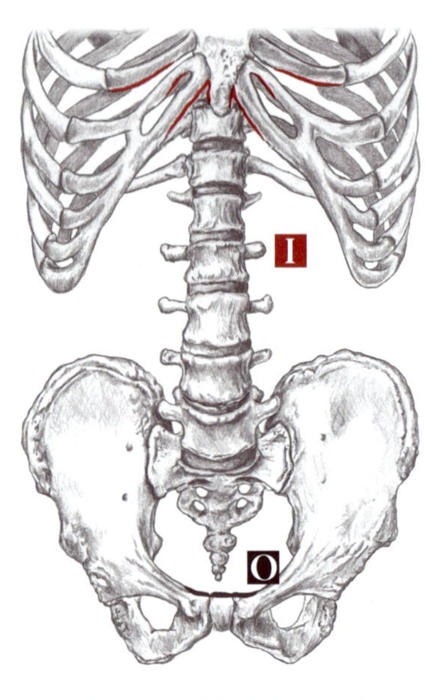

(4.88) Anterior view showing origin and insertion

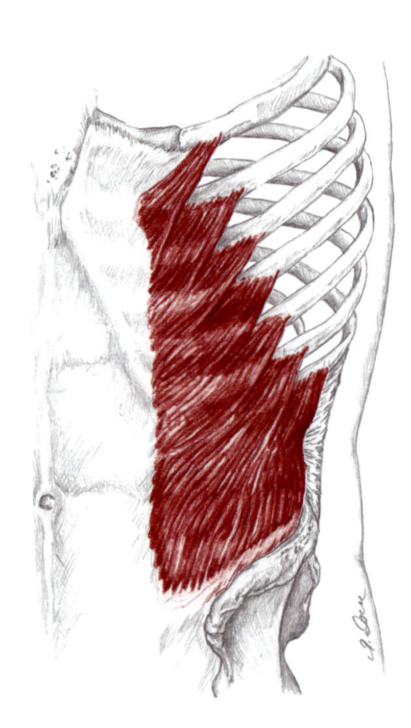

(4.89) Anterior/lateral view of external oblique

External Oblique

A *Unilaterally:*

Laterally flex vertebral column to the same side

Rotate vertebral column to the opposite side

Bilaterally:

Flex the vertebral column

Compress abdominal contents

O Lower eight ribs

I Anterior part of the iliac crest, abdominal aponeurosis to linea alba

N Branches of intercostals

O: Ribs 5-12
I: Via broad aponeurosis

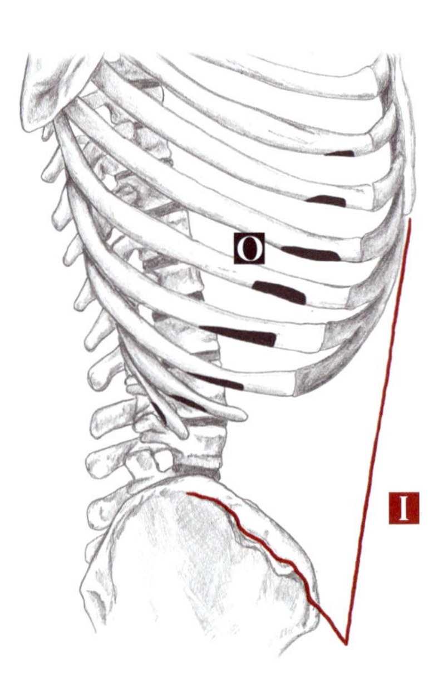

(4.90) Lateral view showing origin and insertion

oblique	o-**bleek**	L. diagonal, slanting
rectus	**rek**-tus	L. straight

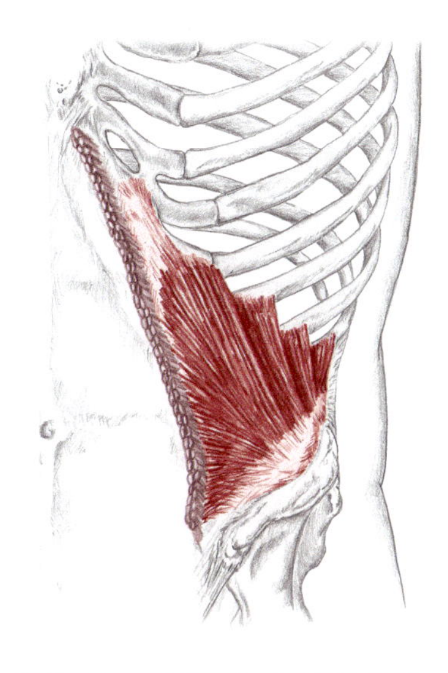

(4.91) Anterior/lateral view of internal oblique (external oblique cut and reflected)

Internal Oblique

A *Unilaterally:*
Laterally flex vertebral column to the same side
Rotate vertebral column to the same side

Bilaterally:
Flex the vertebral column
Compress abdominal contents

O Lateral inguinal ligament, iliac crest and thoracolumbar fascia

I Internal surface of lower three ribs, abdominal aponeurosis to linea alba

N Branches of intercostals

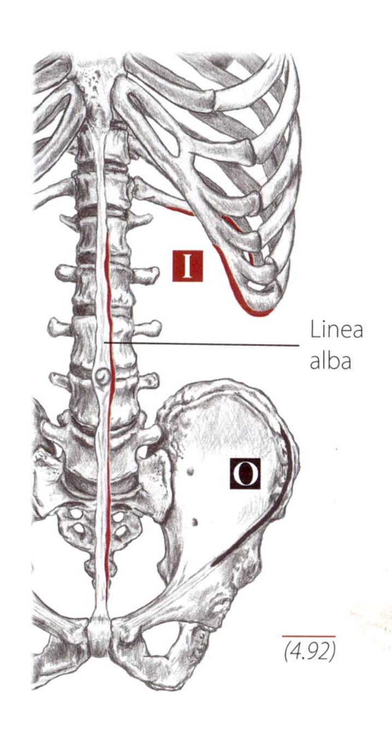

(4.92)

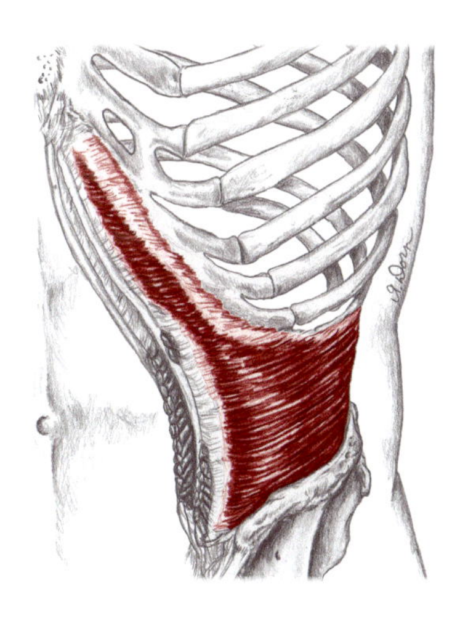

(4.93) Anterior/lateral view of transverse abdominis (both obliques cut and reflected)

Transverse Abdominis

A Compress abdominal contents

O Lateral inguinal ligament, iliac crest, thoracolumbar fascia and internal surface of lower six ribs

I Abdominal aponeurosis to linea alba

N Branches of intercostals

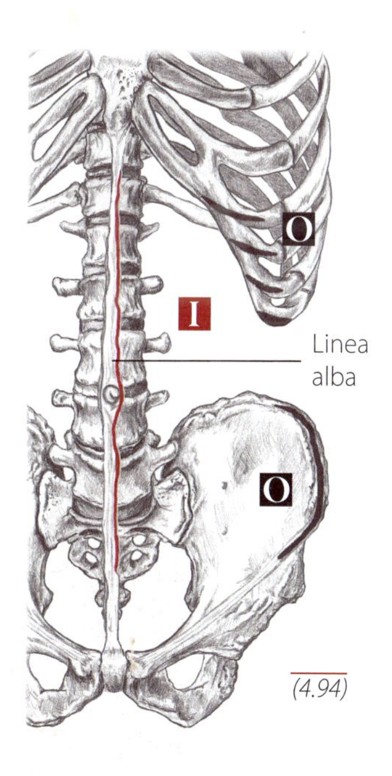

(4.94)

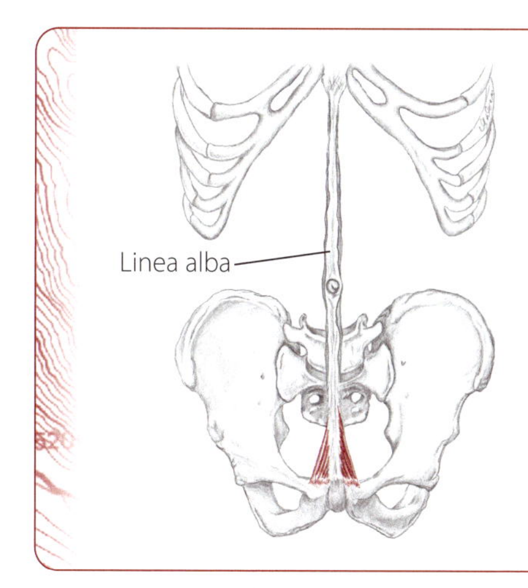

Pyramidalis

Located superficial to the rectus abdominis, this small muscle is absent in roughly 20% of the population.

A Tenses the linea alba

O Pubic symphysis

I Linea alba

N Twelfth thoracic nerve

O: Pubic crest + symphysis.
I: Ext'l surface of xiphoid process + 5th-7th costal cartilages.

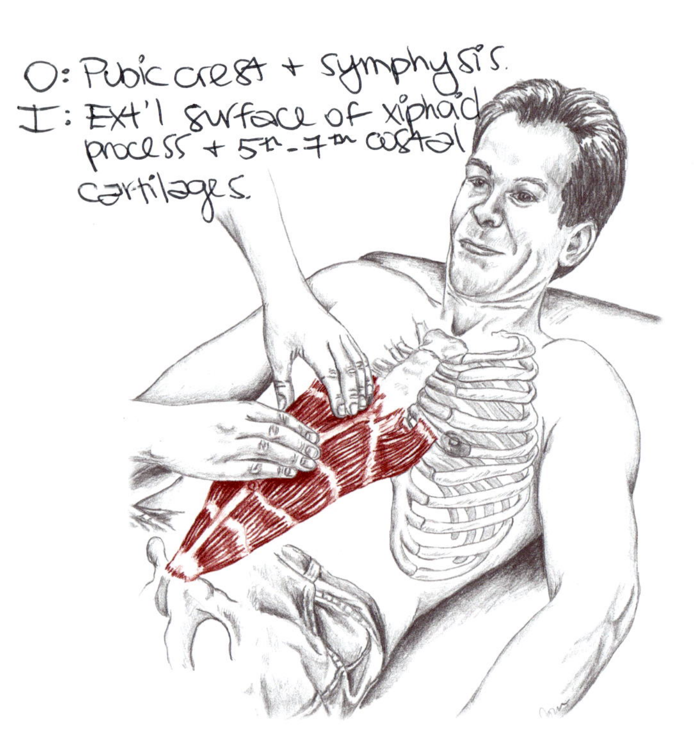

(4.95) Palpating rectus abdominis while your partner flexes his trunk

Rectus abdominis

1) Partner supine with knees flexed. Locate the xiphoid process and the ribs just lateral to the xiphoid. Also locate the pubic crest (p. 284).
2) Place your hand between these landmarks and ask your partner to alternately flex and relax his trunk slightly. "Do a small sit-up."
3) Explore the entire length of the rectus and sculpt between its rectangular muscle bellies (4.95).

As your partner flexes his trunk, can you palpate the lateral edges of the rectus abdominis?

External oblique (left side)

1) Partner supine with his knees flexed. Lay your hand on the left side of the abdomen and lower ribs. Ask your partner to raise his left shoulder toward his right hip (rotating his trunk).
2) Palpate across the superficial fibers of the external oblique, noting their diagonal direction (4.96).
3) With the trunk still rotated, follow the fibers superiorly to where they interdigitate with the serratus anterior, then inferiorly to the abdominal aponeurosis and, finally, laterally to the iliac crest.

Are you palpating lateral to the edge of rectus abdominis? Are the fibers superficial and running at an angle? Palpate lateral to the rectus abdominis with the abdomen relaxed. Can you distinguish between the fibers of the external oblique and the deeper internal oblique? Their fibers should be virtually perpendicular to each other.

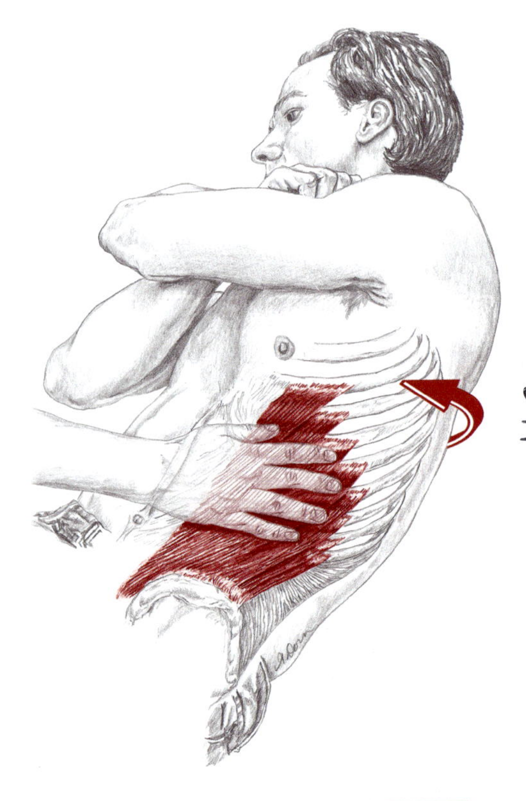

(4.96) Accessing the external oblique while your partner rotates his trunk toward the opposite side

O: Ribs 5-12
I: Via broad ap. to xiphoid, linea alba, pubic tubercle + ant. 1/2 of iliac crest.

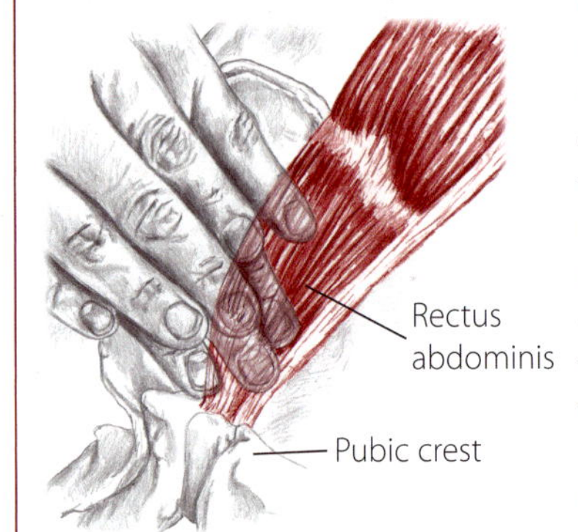

Palpating the inferior attachment of the rectus abdominis requires locating the pubic crest. (See p. 284 for instructions.) Explore the inferior rectus by first locating your partner's umbilicus. As he does a small sit-up, follow the narrowing muscle bellies to the pubic crest. At their insertion site, they are surprisingly slender, perhaps only three inches wide.

interdigitate **in**-ter-**dij**-i-tate L. to interlock, as the fingers of clasped hands

Diaphragm

The diaphragm is the primary muscle of respiration and is unique in both its design and function. Its broad, umbrellalike shape separates the upper and lower thoracic cavities (4.97). The diaphragm's muscle fibers attach to the inner surface of the ribs and the lumbar vertebrae and converge at the central tendon (4.98).

The diaphragm creates inspiration (inhalation) when its muscle fibers contract and pull the central tendon inferiorly. Because the central tendon is attached to the connective tissue that surrounds the lungs, a vacuum is created in the upper thoracic cavity pulling air into the lungs. On exhalation, the muscle fibers of the diaphragm relax, releasing the central tendon and allowing the lungs to deflate.

Although only a small portion of the diaphragm is accessible, the muscle's effect on the thorax and breathing is easily felt.

A Draw down the central tendon of the diaphragm

Increase the volume of the thoracic cavity during inhalation

O *Costal attachment:*
Inner surface of lower six ribs
Lumbar attachment:
Upper two or three lumbar vertebrae
Sternal attachment:
Inner part of xiphoid process

I Central tendon

N Phrenic

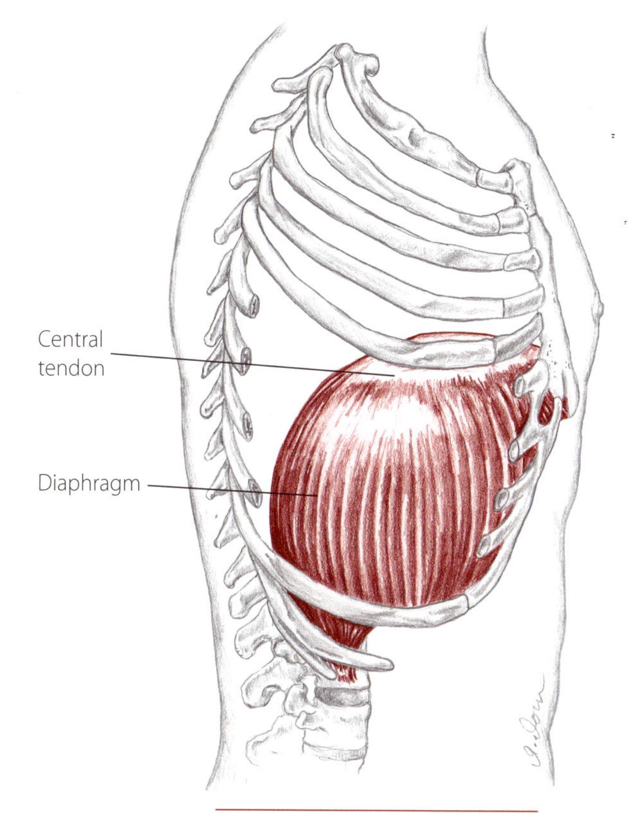

(4.97) Lateral view of thorax showing diaphragm in position of exhalation

An involuntary contraction of the diaphragm will cause air to rush into the lungs and the vocal cords to snap shut. The audible result is a hiccup.

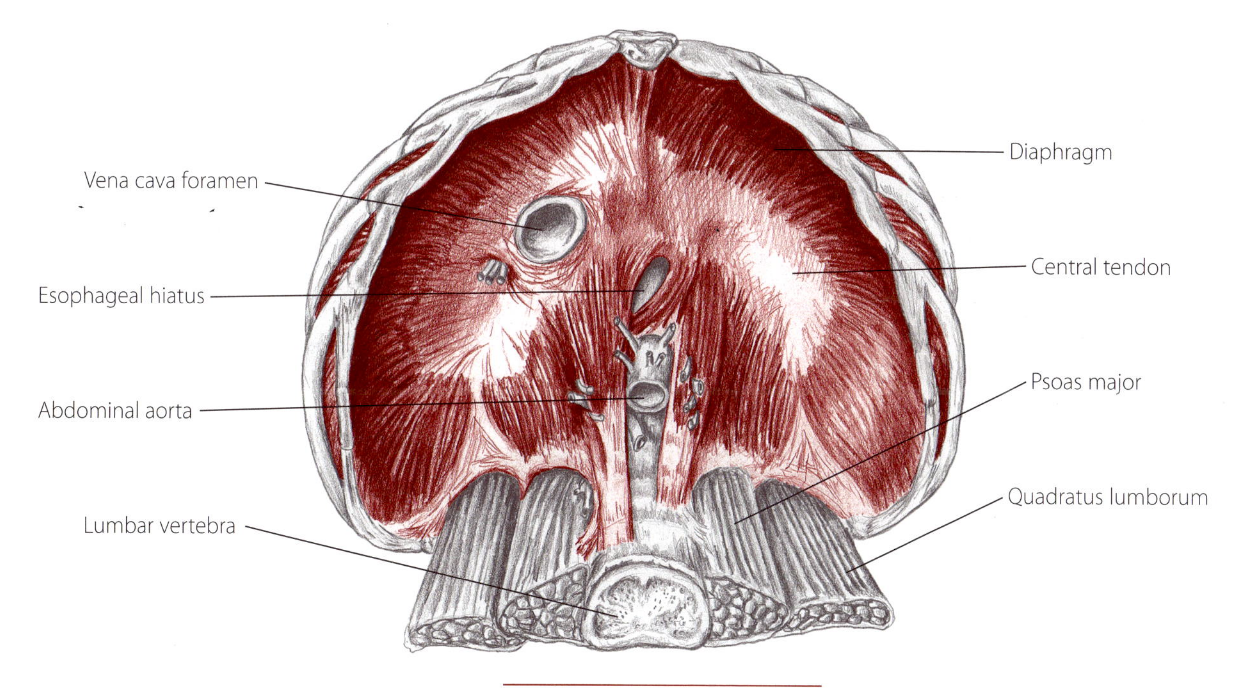

(4.98) Inferior view of the diaphragm

diaphragm **di**-a-**fram** Grk. a partition, wall

Move slowly, communicating with your partner as you palpate. If, at any time, he does not feel safe or comfortable, gently remove your hands.

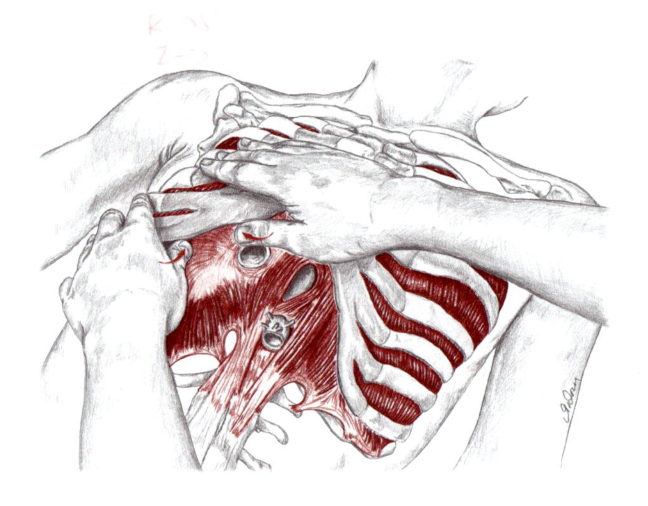

(4.99) Partner supine, palpating the diaphragm

Diaphragm

1) Partner supine, with knees bolstered. Locate the inferior edge of the rib cage, lateral to the xiphoid process.
2) Lay your thumbpads just inferior to the ribs on the abdomen and ask your partner to take slow, deep breaths.
3) Moving only as your partner exhales, slowly press and curl your thumbpads underneath the edge of the ribs (4.99). During inhalation, you may not feel the tissue of the diaphragm, but you will most likely feel its contraction as it pushes other tissues into your thumbpads.

Are your thumbs curling under the ribs rather than sinking into the abdominal organs? Ask your partner to breathe into his belly and notice how the abdominal region expands as the diaphragm contracts.

Try the above procedure with your partner sidelying and his trunk flexed slightly. This position will allow the abdominal contents to shift away from where you are accessing (4.100).

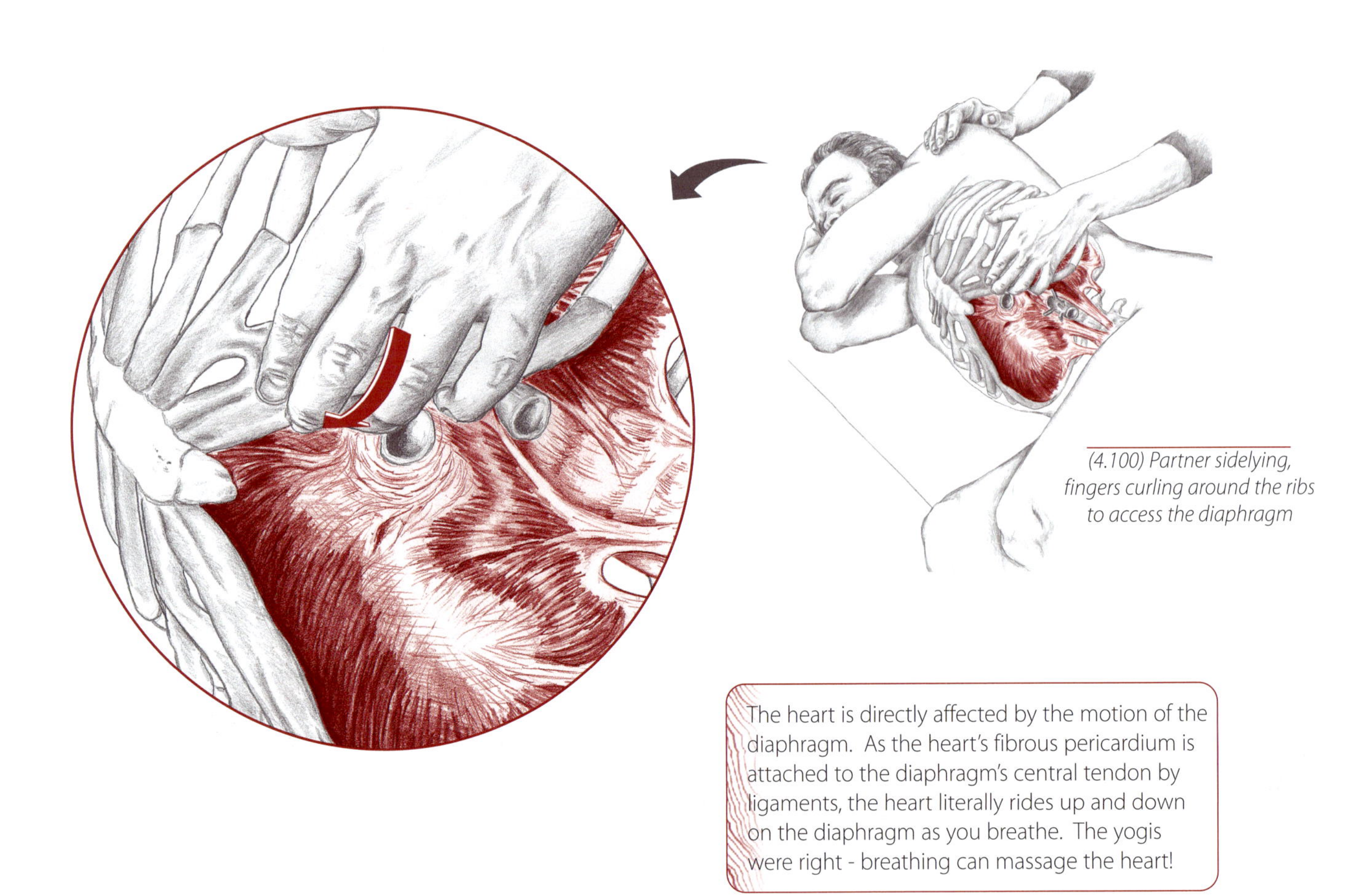

(4.100) Partner sidelying, fingers curling around the ribs to access the diaphragm

The heart is directly affected by the motion of the diaphragm. As the heart's fibrous pericardium is attached to the diaphragm's central tendon by ligaments, the heart literally rides up and down on the diaphragm as you breathe. The yogis were right - breathing can massage the heart!

Intercostals

Better known to carnivores as the meat on spare ribs, the intercostals are the small, slender muscles between the ribs. They are divided into two groups: the external and the internal intercostals (4.101). The fibers of these two groups run perpendicular to each other and can be visualized as extensions of the external and internal oblique muscles (p. 215).

Although the specific role of the intercostals is debatable, their functions include stabilizing the rib cage and assisting in respiration. The entire rib cage lies deep to one or more layers of muscle, but portions of the intercostals are easily accessible. It is not possible, however, to distinguish the external intercostals from the internal intercostals.

Since the ribs and the spaces between them can be sensitive areas to access, use slow, firm hand movements.

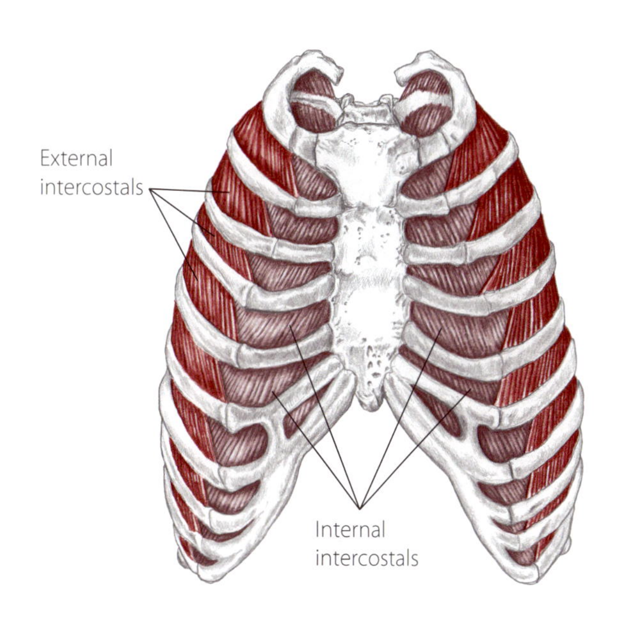

(4.101) Anterior view of rib cage

A *External Intercostals:*
Assist with inhalation by drawing the ribs superiorly, increasing the space of the thoracic cavity

Internal Intercostals:
Assist with exhalation by drawing the ribs inferiorly, decreasing the space of the thoracic cavity

O Inferior border of the rib above

I Superior border of the rib below

N Intercostal

1) Partner supine. Begin just inferior to the pectoralis major on the side of the rib cage. Working across the body, position your fingers in the spaces between the ribs.
2) With one fingerpad, isolate and palpate the tissue between two ribs. Roll your finger along the rib space and palpate the short, dense intercostals that bridge the ribs (4.102).
3) Ask your partner to take several slow, deep breaths. Note any expansion or collapse in the spaces between the ribs. Then turn your partner prone or sidelying and continue to explore the intercostals.

Are you between the ribs or just on the surface? Can you roll your fingers across the small intercostal fibers? Can you sink your fingers through the pectoralis major, latissimus dorsi or external oblique to isolate the intercostals?

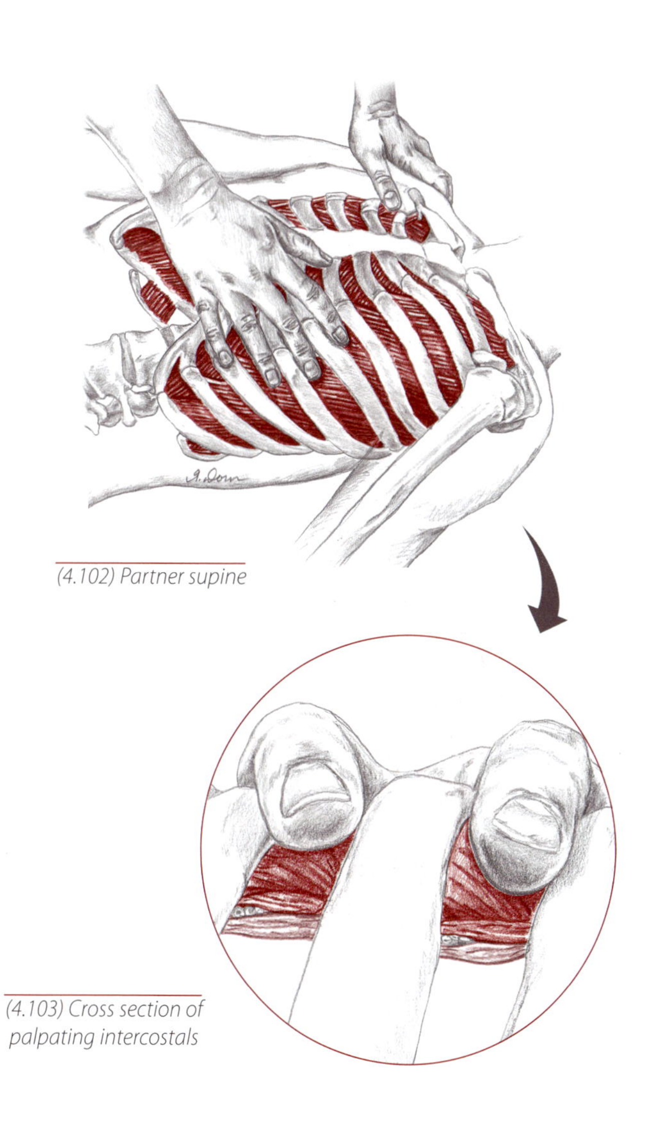

(4.102) Partner supine

(4.103) Cross section of palpating intercostals

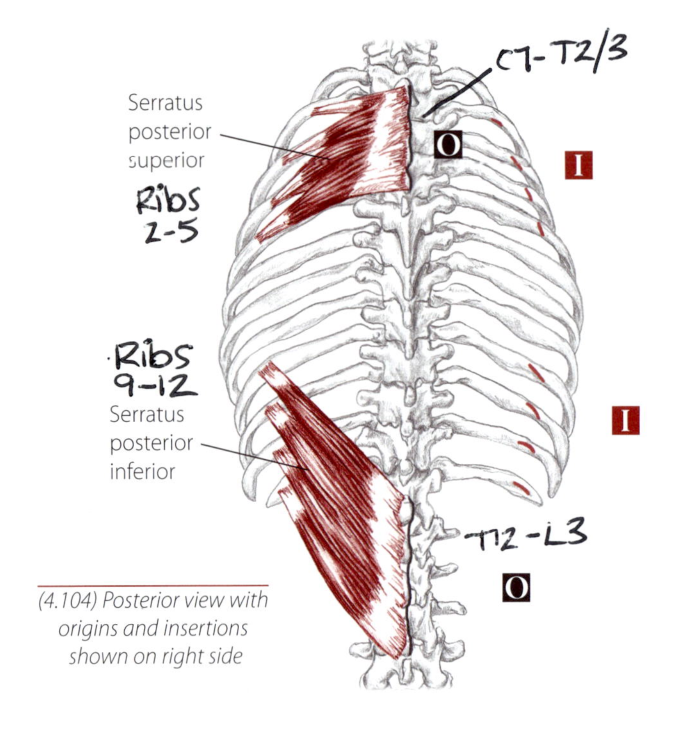

(4.104) Posterior view with origins and insertions shown on right side

Serratus Posterior Superior
Serratus Posterior Inferior

Although they are sandwiched between the shoulder muscles and the erector spinae group, these two broad muscles only affect movement of the ribs. The belly of the **superior** is partially deep to the scapula and has fibers that parallel the superficial rhomboids. The **inferior** is deep to the thoracolumbar aponeurosis (p. 226) and, during exhalation, can stabilize the ribs against the pull of the diaphragm.

Both muscles are superficial enough to be accessed, but due to their thin, tendonous bellies, discerning them is a different story.

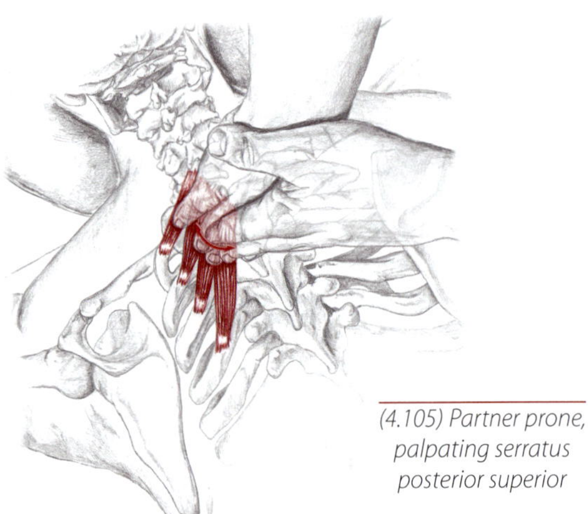

(4.105) Partner prone, palpating serratus posterior superior

Serratus Posterior Superior

A Elevate the ribs during inhalation

O Spinous processes of C-7 to T-3 (T2) + inf. aspect of lig. nuchae

I Posterior surface of second through fifth ribs superior ribs 2-5

N Spinal nerves one through four

Serratus Posterior Inferior

A Depress the ribs during exhalation

O Spinous processes of T-12 to L-3

I Posterior surface of ninth through twelfth ribs

N Spinal nerves nine through twelve

1) **Superior:** Partner prone. With the arm off the side of the table (pulling the scapula laterally), locate the upper portion of the medial border of the scapula.
2) Ask your partner to inhale as you sink your fingers deep to the rhomboid fibers (4.105). Although you may not feel the belly directly, explore this region for its fibers.
3) **Inferior:** Locate the bottom of the rib cage (11th or 12th rib). Ask your partner to slowly exhale as you begin to roll your fingers across the muscle fibers (4.106).

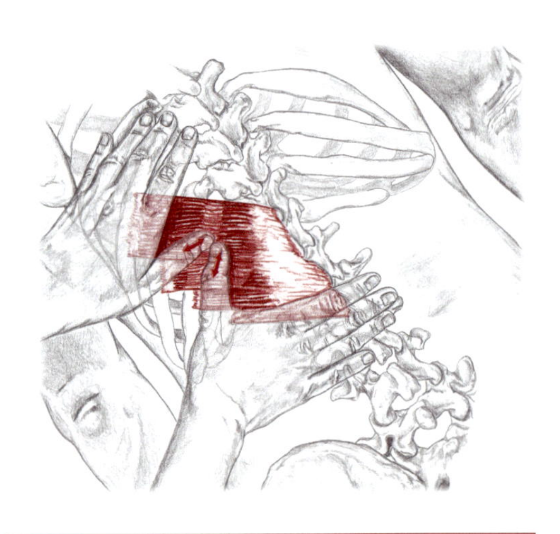

(4.106) Partner prone, palpating serratus posterior inferior

For sandwiched muscles such as these, it can be worthwhile to first identify the muscles that are not the serrati. Then, with a patient and delicate touch, explore the "space between."

Intertransversarii

As their name suggests, these small, short muscles span between the transverse processes. They are the deepest muscles in the cervical and lumbar regions and, for this reason, are nearly impossible to access, let alone detect (4.107).

A *Unilaterally:*
Laterally flex the vertebral column to the same side

Bilaterally:
Extend the vertebral column

O and I
Cervical:
Spanning the transverse processes of vertebrae C-2 to C-7

Lumbar:
Spanning the transverse processes of vertebrae L-1 to L-5

N Dorsal and ventral rami of spinal nerves

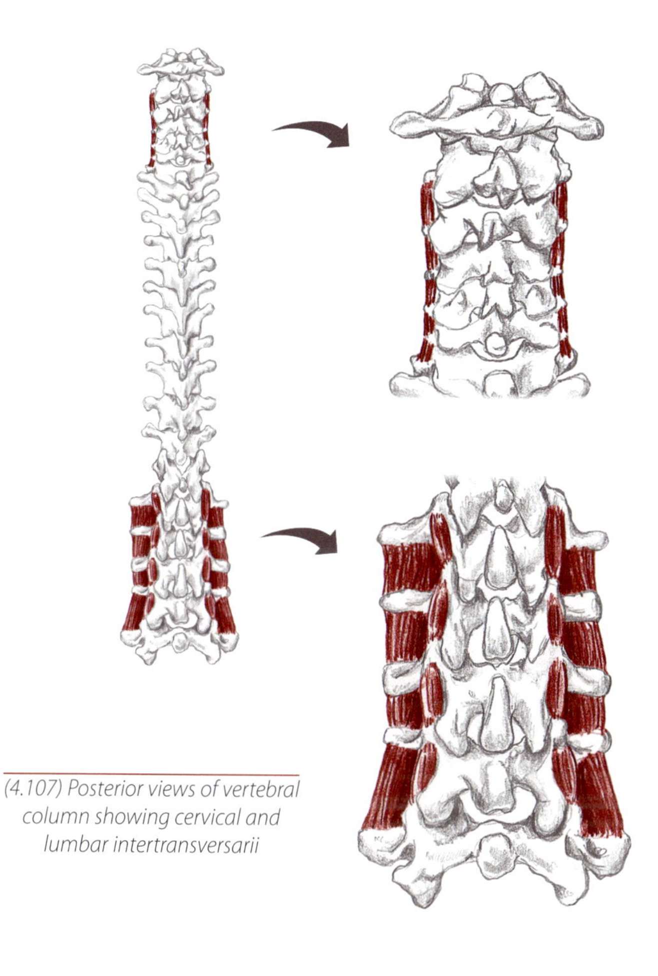

(4.107) Posterior views of vertebral column showing cervical and lumbar intertransversarii

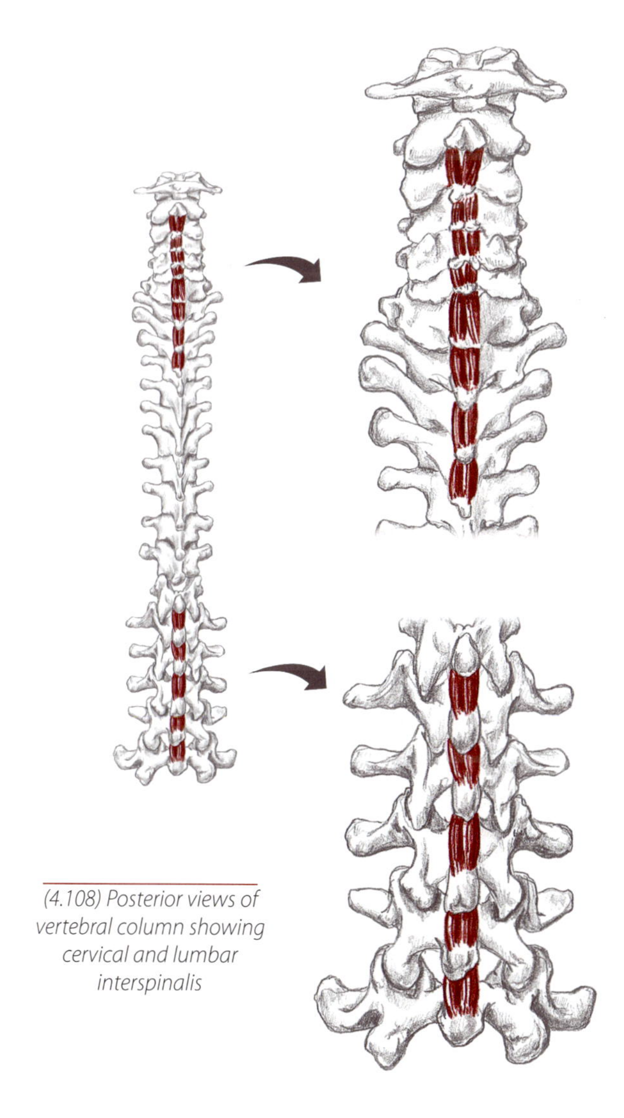

(4.108) Posterior views of vertebral column showing cervical and lumbar interspinalis

Interspinalis

Extending from the spinous processes in the cervical and lumbar regions, these short muscles help extend the spine. The cervical muscles are deep to the ligamentum nuchae while the lumbar muscles are deep to the interspinous ligament (4.108). Like the intertransversarii, these muscles are too deep to isolate.

A Extend the vertebral column

O and I
Cervical:
Spanning the spinous processes of C-2 to T-3

Lumbar:
Spanning the spinous processes of T-12 to L-5

N Dorsal rami of spinal nerves

intertransversarii **in**-ter-trans-**verse**-er-i
interspinalis **in**-ter-spi-**na**-lis

Other Structures of the Spine and Thorax

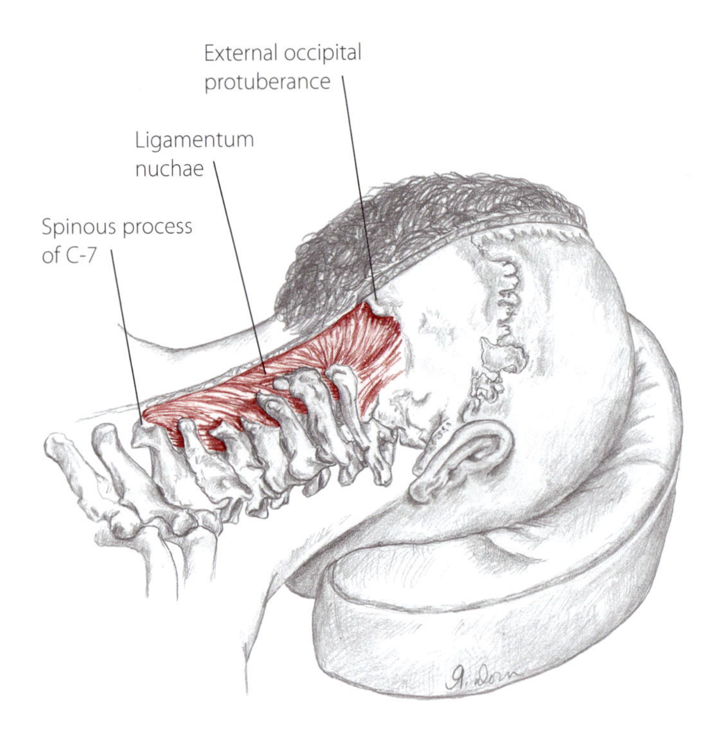

(4.109) Posterior/lateral view with muscles and tissue removed from right side

Ligamentum Nuchae

The ligamentum nuchae is the finlike sheet of connective tissue that runs along the sagittal plane from the external occipital protuberance to the spinous process of C-7 in the neck (4.109).

The chief function of the ligamentum nuchae is to help stabilize the head and neck. It is also an attachment site for the superficial muscles of the posterior neck such as the trapezius and splenius capitis. Since the cervical spinous processes do not extend far enough posteriorly for these superficial muscles to attach, they use the ligamentum nuchae instead.

For palpation purposes, the posterior edge of the ligamentum nuchae is superficial, but can be difficult to discern from the surrounding muscle tissue.

1) Supine. Locate the external occipital protuberance (p. 238) and the spinous process of C-7.
2) Palpate between these landmarks along the midline of the neck. Be sure you are superficial to the spinous processes. It might help to access the ligamentum nuchae if you roll your fingertips across its fiber direction and explore for what may feel like a flap of soft rubber (4.110).
3) Slowly and passively flex and extend the head, rolling your fingers across the fibers of the ligamentum nuchae. Note the changing degree of tension in the ligament as the head is moved.

With your partner seated, ask him to flex his head and neck as far as he can comfortably. In this position, the ligamentum nuchae will stretch and rise to the surface. It should feel like a long, thin "speed bump" on the back of the neck.

Are you superficial to the spinous processes of the vertebrae as you palpate?

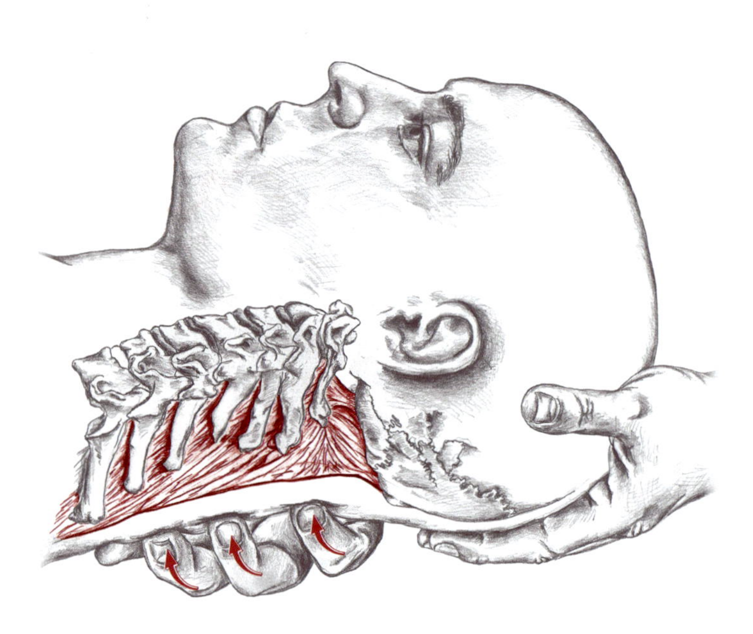

(4.110) Partner supine, palpating the ligamentum nuchae

nuchae **nu**-kay L. nape of neck

Supraspinous Ligament

The long, thin supraspinous ligament extends inferiorly from the ligamentum nuchae. As it continues down the spine, it attaches to the spinous processes of the thoracic and lumbar vertebrae. It is superficial and easily accessed in the spaces between the spinous processes.

1) Partner prone. Locate several thoracic or lumbar spinous processes (4.111).
2) Palpate between the spinous processes. Feel the slender shape and vertical fiber direction of the ligament by rolling your fingertips across its surface.

With your partner seated, ask her to slowly flex and extend her spine. Can you feel any changes in the tension or prominence of the ligament as she moves it?

Abdominal Aorta

Measuring nearly an inch in diameter, the abdominal aorta is the chief artery for carrying blood to the abdominal organs and lower appendages. It lies on the anterior surface of the vertebrae, deep to the small intestines. Lateral to the aorta is the psoas major (p. 326).

1) Partner supine. Locate the umbilicus. Place your fingerpads two inches superior to the umbilicus.
2) Access the pulse of the abdominal aorta by slowly but firmly pressing straight down into the abdomen. Its strong pulse should be easily detectable (4.112).

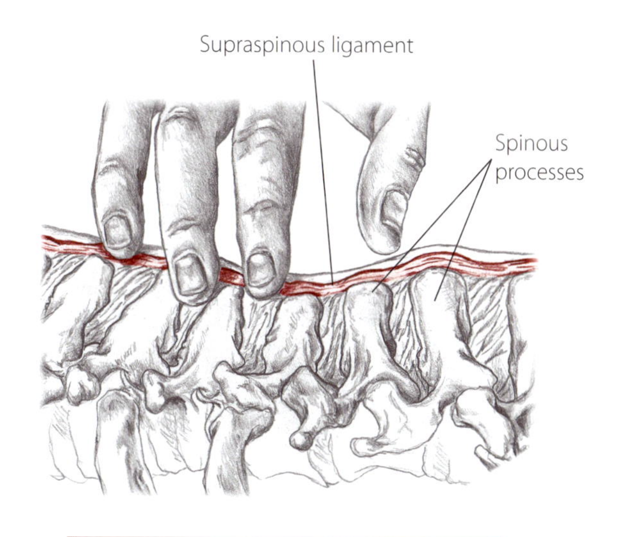

(4.111) Partner prone, lateral view of the vertebrae

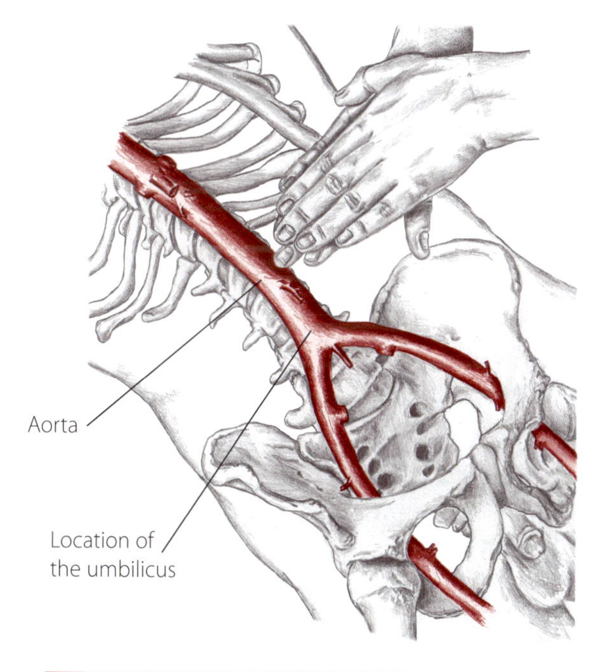

(4.112) Partner supine, feeling the pulse of the abdominal aorta

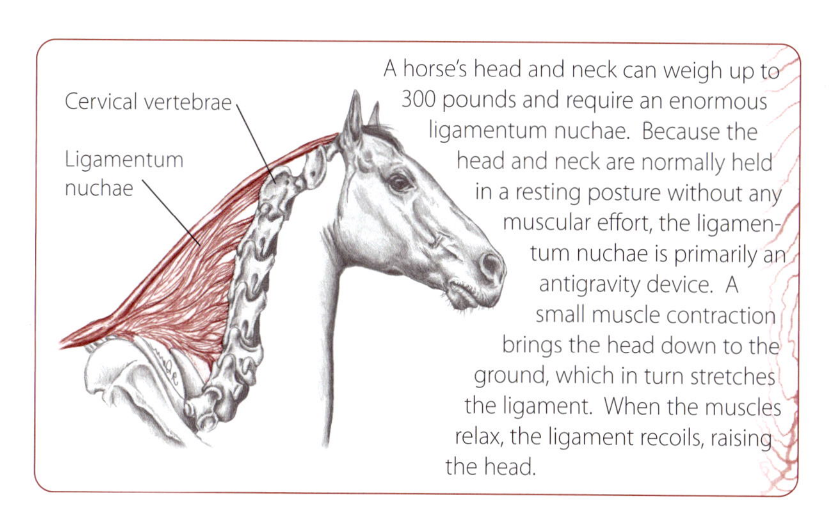

A horse's head and neck can weigh up to 300 pounds and require an enormous ligamentum nuchae. Because the head and neck are normally held in a resting posture without any muscular effort, the ligamentum nuchae is primarily an antigravity device. A small muscle contraction brings the head down to the ground, which in turn stretches the ligament. When the muscles relax, the ligament recoils, raising the head.

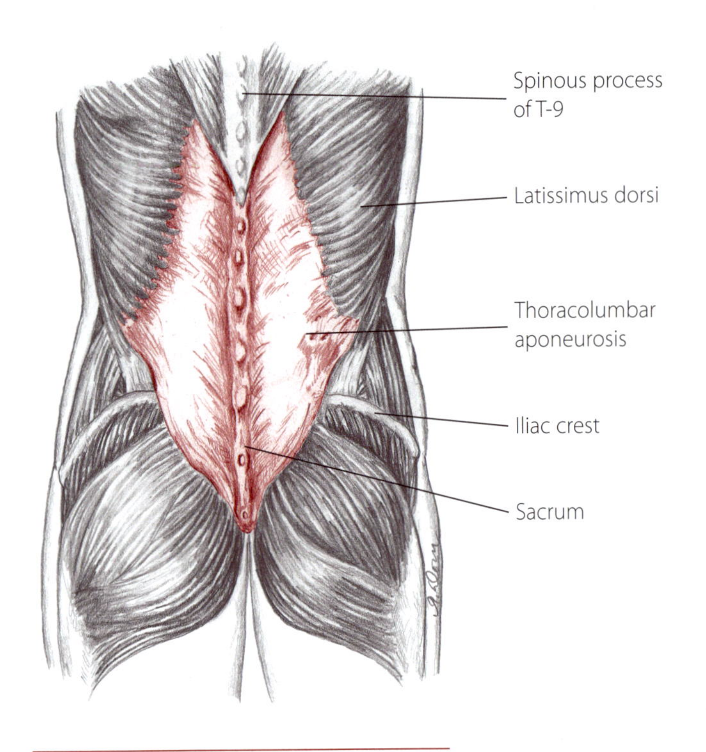

(4.113) Posterior view of lower thorax and pelvis

Thoracolumbar Aponeurosis

Despite its formidable name, the thoracolumbar aponeurosis is just what it says it is: a broad, flat tendon stretching across the thorax and lumbar regions. More accurately, the thoracolumbar aponeurosis is a thick, diamond-shaped tendon that lies superficially across the posterior thorax, stretches across the sacrum to the posterior iliac crest and runs upward to the lower thoracic vertebrae (4.113).

The aponeurosis is an anchor for several muscles in the thorax and hips, including the latissimus dorsi and the erector spinae group. It has a flat, dense texture that is difficult to distinguish from the deeper muscles.

1) Partner prone. Draw out the diamond shape of the aponeurosis by locating the posterior iliac crest, the surface of the sacrum and the lower thoracic vertebrae.

2) Using both hands, firmly grasp and lift the tissue of the low back (4.114). Note the thick layer of connective tissue underneath the skin but superficial to the erector spinae muscles. Do not be surprised if the aponeurosis is so dense that you find it difficult to grasp, let alone lift.

✔ *Ask your partner to alternately raise her elbows slightly and relax (this will contract the latissimus dorsi and tighten the aponeurosis). Do you feel any change in the superficial tissue? Then move your hands laterally off the "diamond" and onto the latissimus dorsi muscle belly. Do you notice any textural differences between these two tissues?*

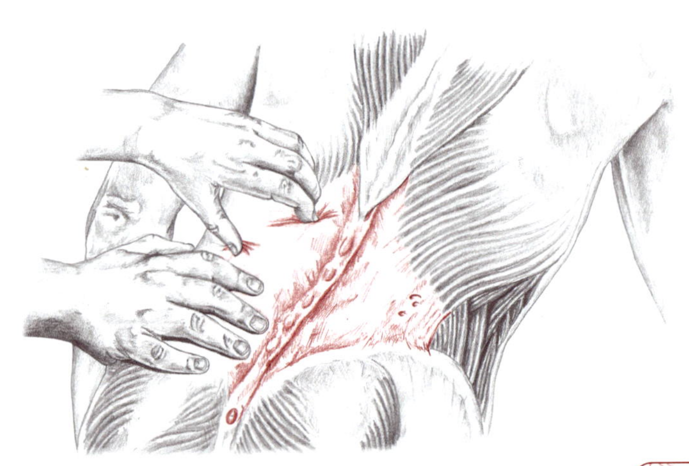

(4.114) Partner prone

A giraffe's neck is more than five feet in length, but nevertheless has just seven cervical vertebrae. The atlas and axis are relatively short, whereas the five cervical vertebrae may measure eleven inches each. The neck and head are stabilized by a massive ligamentum nuchae and an array of short muscles that weave along the posterior surface of the neck. The retractor muscle covers the anterior surface of the cervicals. This muscle extends from the giraffe's sternum all the way up to its hyoid bone and draws back the tongue.

aponeurosis — **ap**-o-nu-**ro**-sis — Grk. *apo*, from + *neuron*, nerve or tendon
thoracolumbar — tho-**rak**-o-**lum**-bar

Craniovertebral Joints - Atlantooccipital and Atlantoaxial Joints

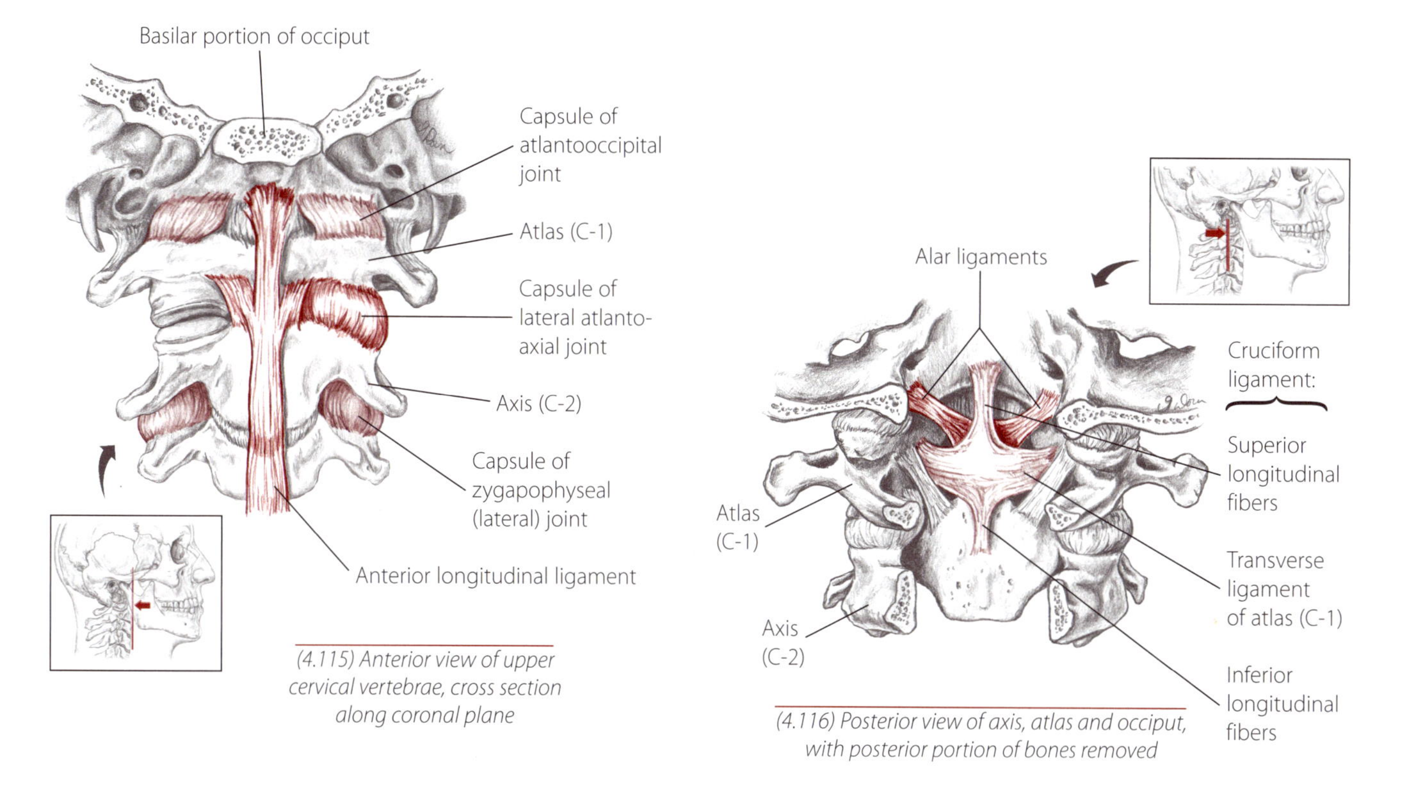

(4.115) Anterior view of upper cervical vertebrae, cross section along coronal plane

(4.116) Posterior view of axis, atlas and occiput, with posterior portion of bones removed

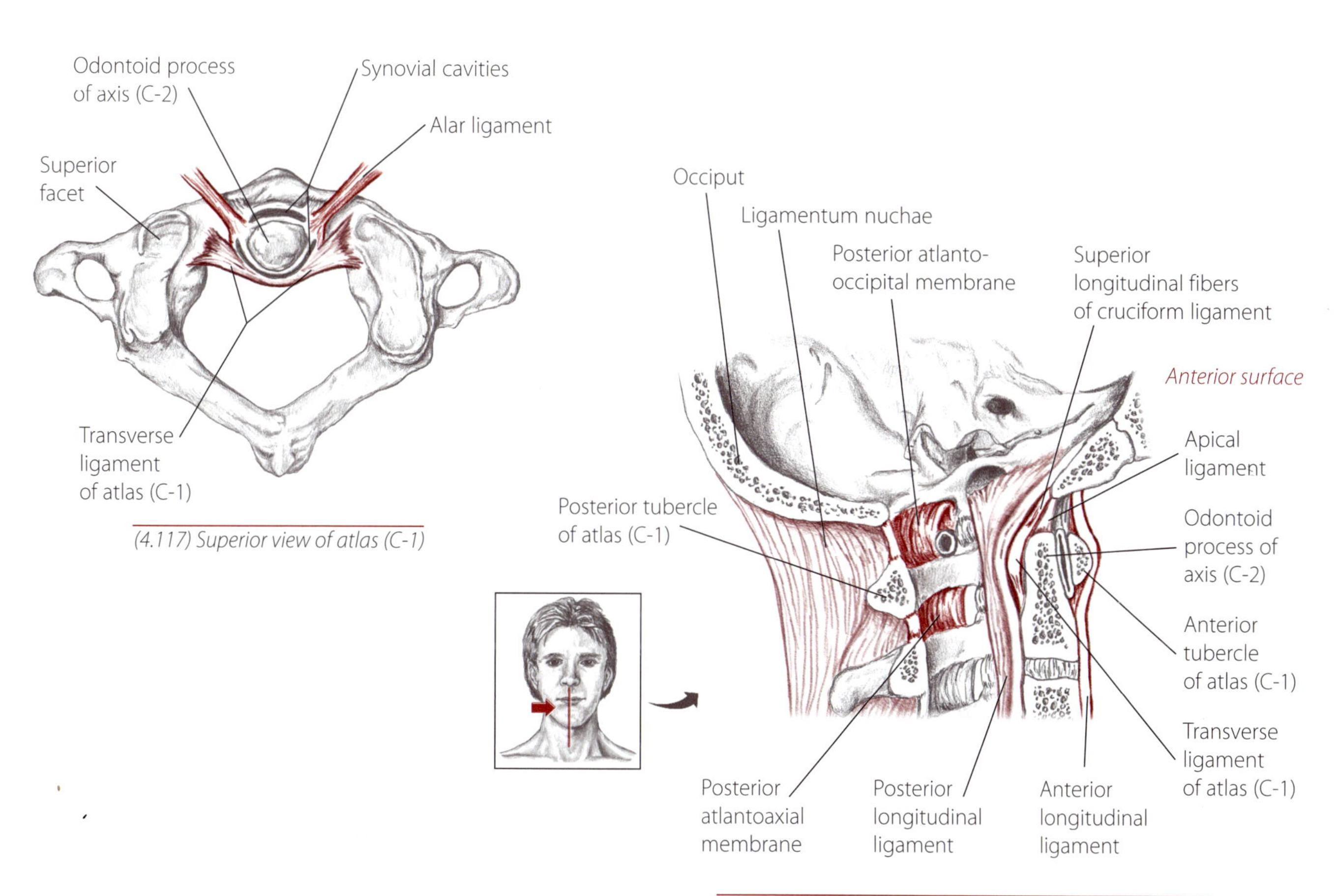

(4.117) Superior view of atlas (C-1)

(4.118) Lateral view, cross section along the sagittal plane

alar	**ay**-lar	atlantoaxial	at-**lan**-to-**ak**-se-al
atlantooccipital	at-**lan**-to-ok-**si**-pi-tal	zygapophyseal	**zy**-gah-**pof**-i-se-al

Intervertebral Joints

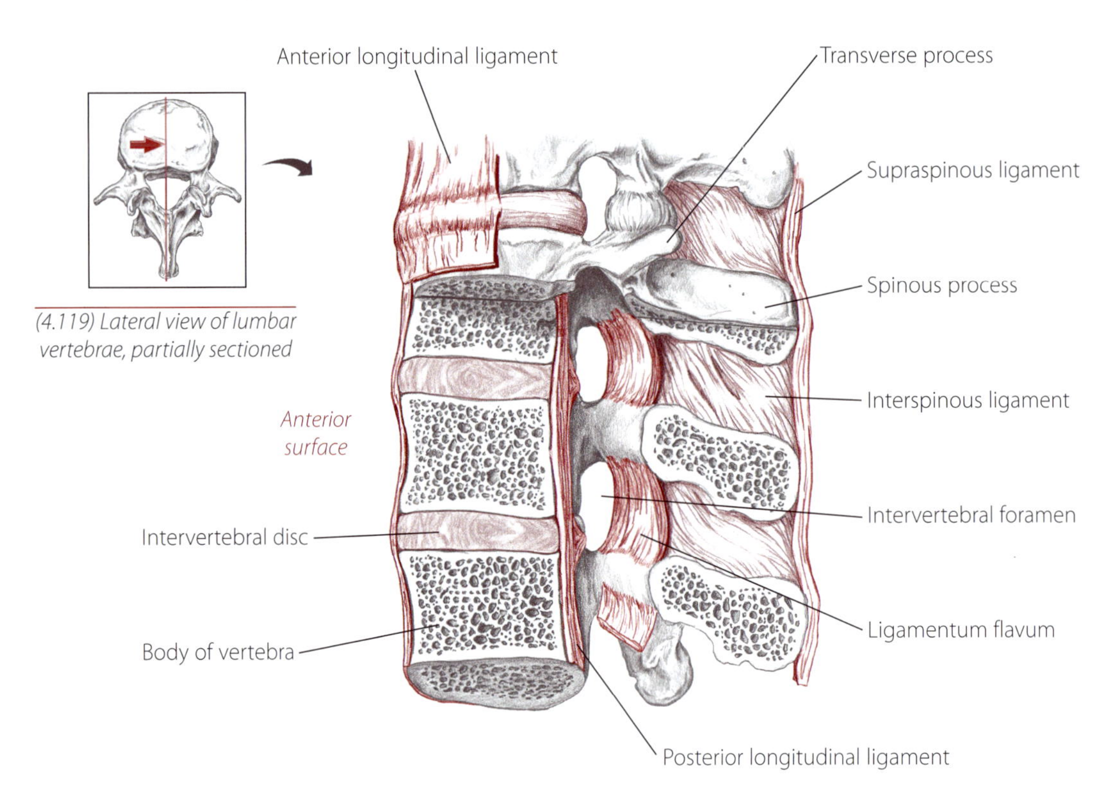

(4.119) Lateral view of lumbar vertebrae, partially sectioned

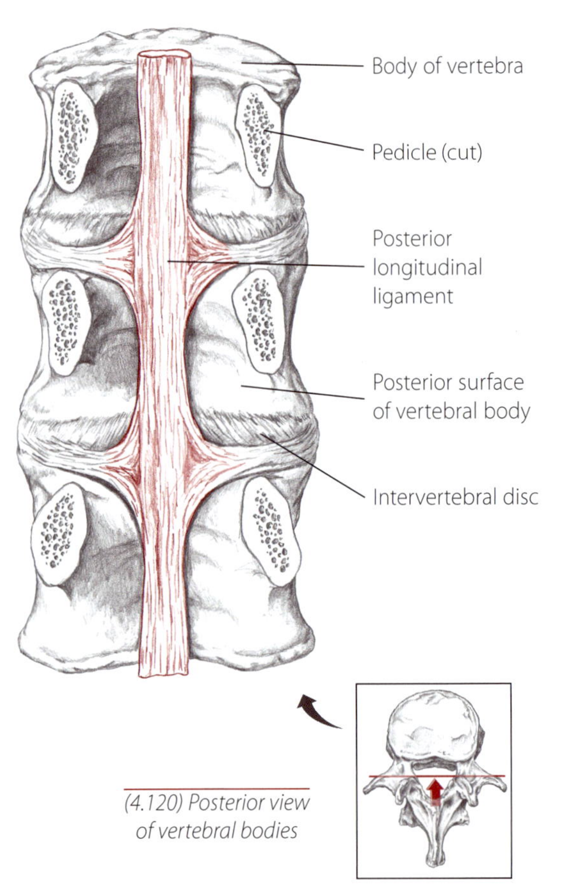

(4.120) Posterior view of vertebral bodies

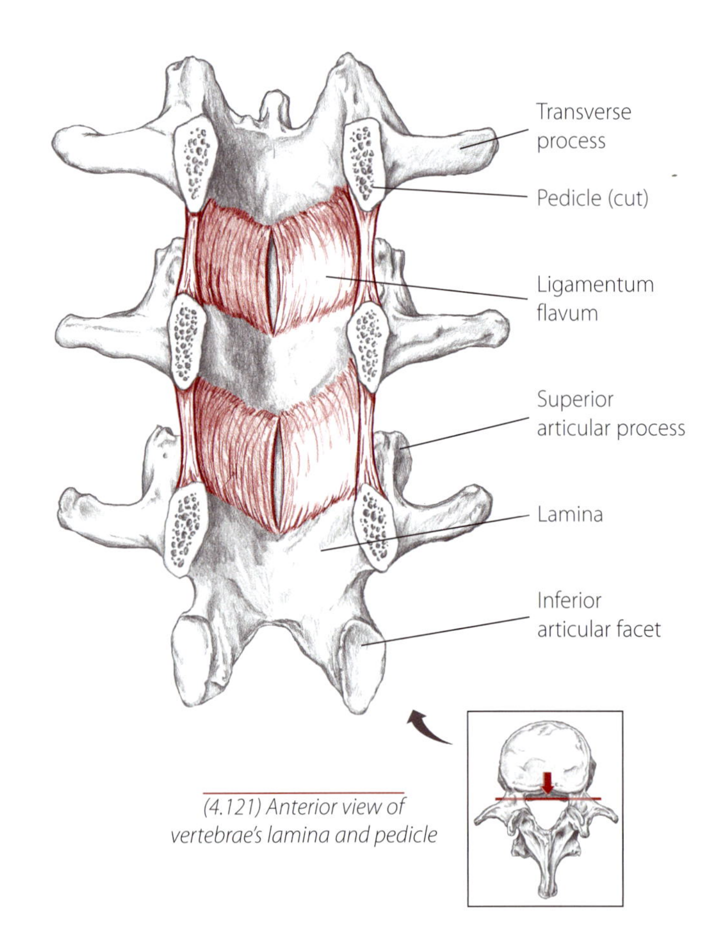

(4.121) Anterior view of vertebrae's lamina and pedicle

flavum **flay**-vum
pedicle ped-i-k'l L. a little foot

Costovertebral Joints

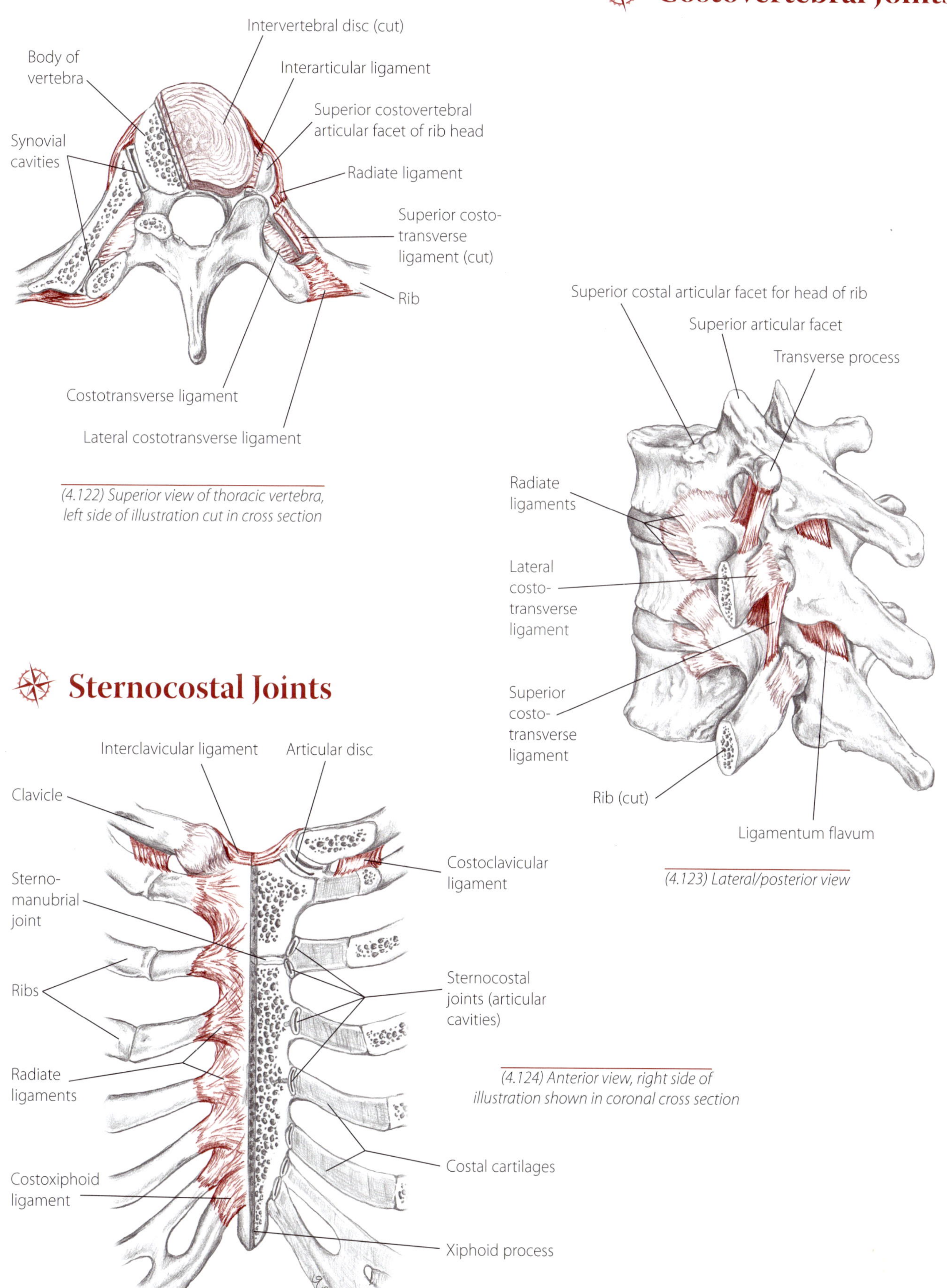

(4.122) Superior view of thoracic vertebra, left side of illustration cut in cross section

(4.123) Lateral/posterior view

Sternocostal Joints

(4.124) Anterior view, right side of illustration shown in coronal cross section

NOTES

Ah, the head, neck and face...

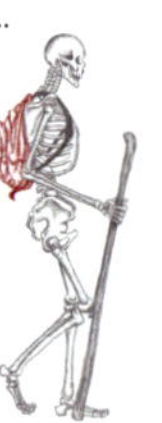

5
Head, Neck & Face

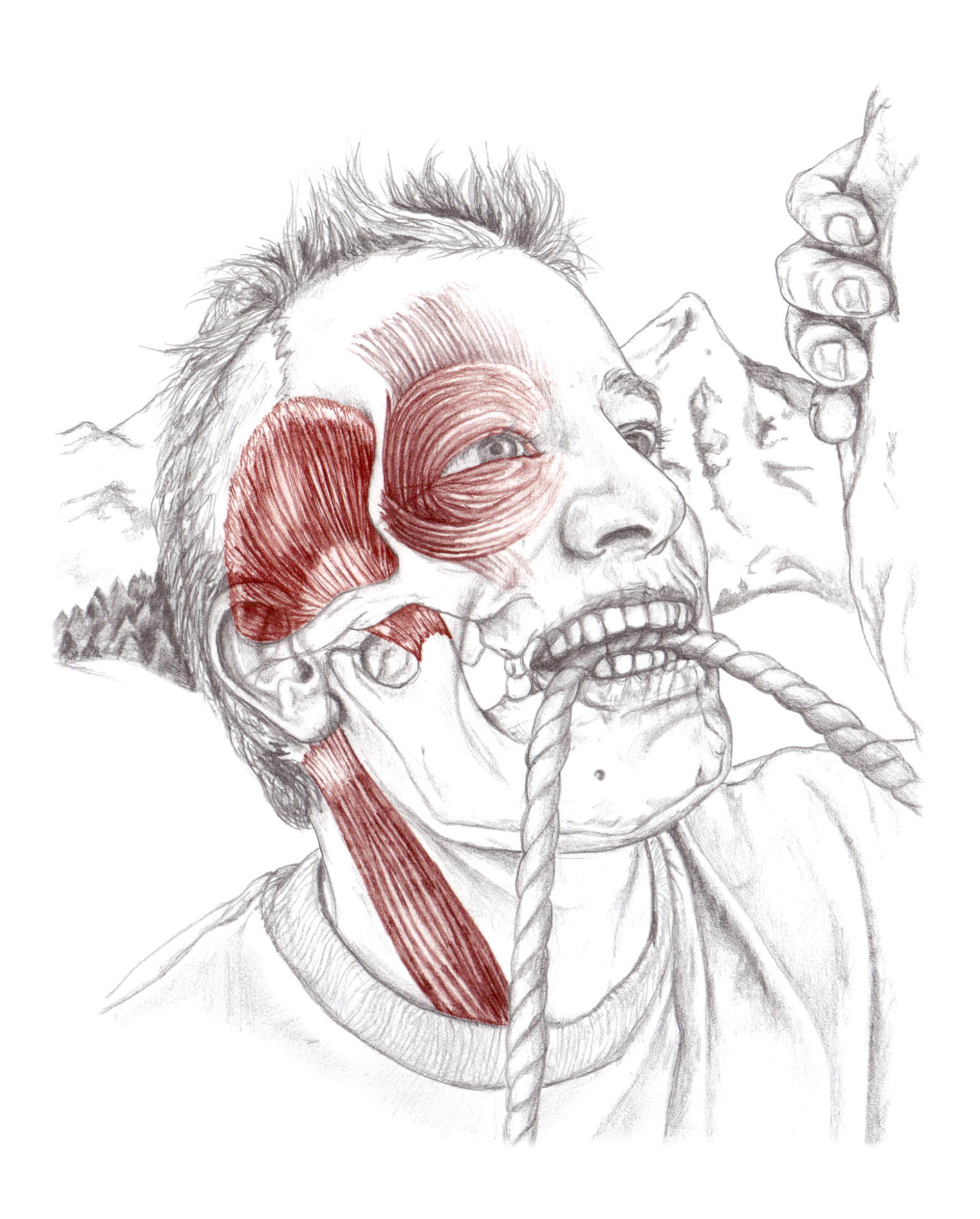

Topographical View

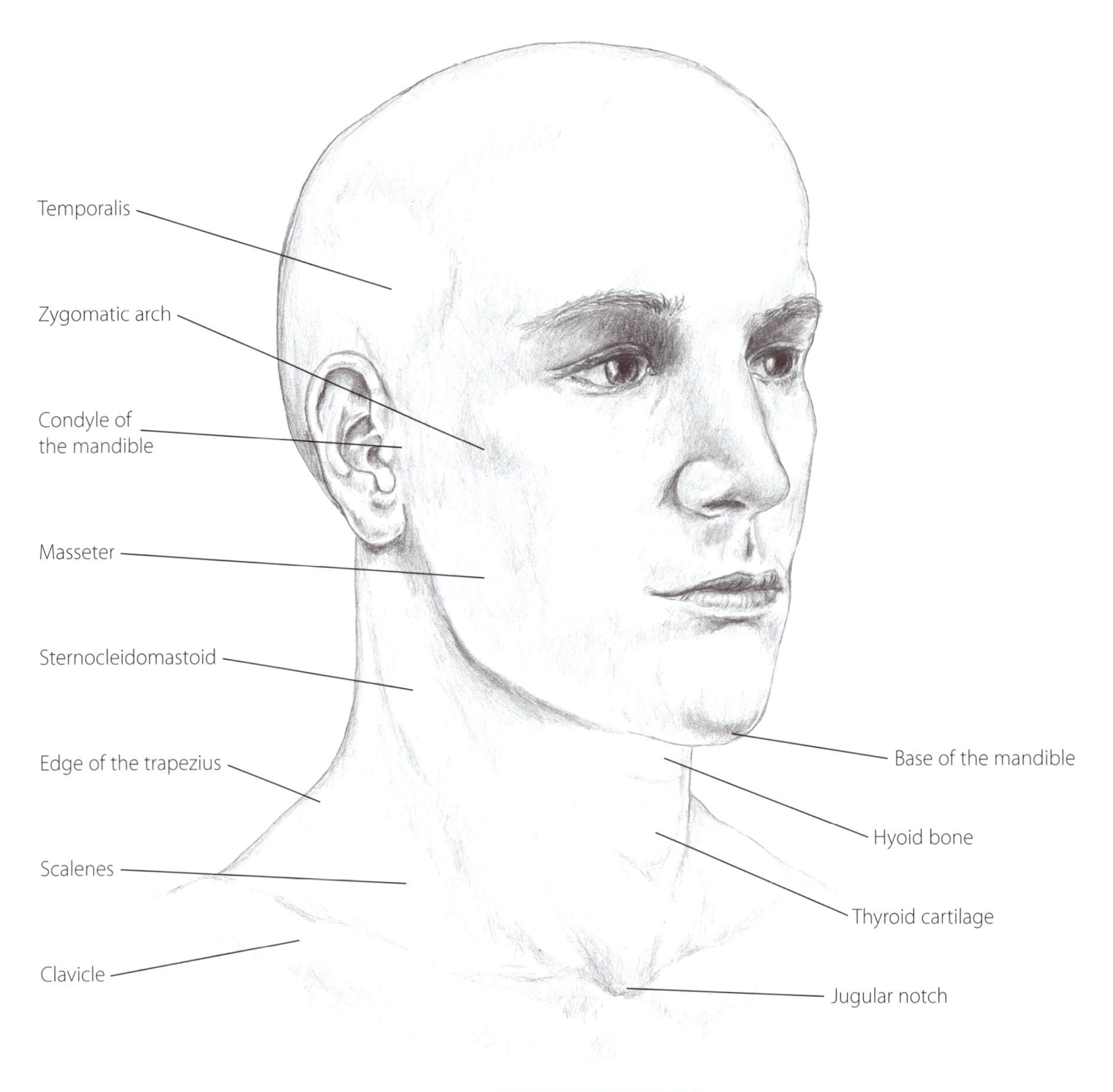

(5.1) Anterior/lateral view

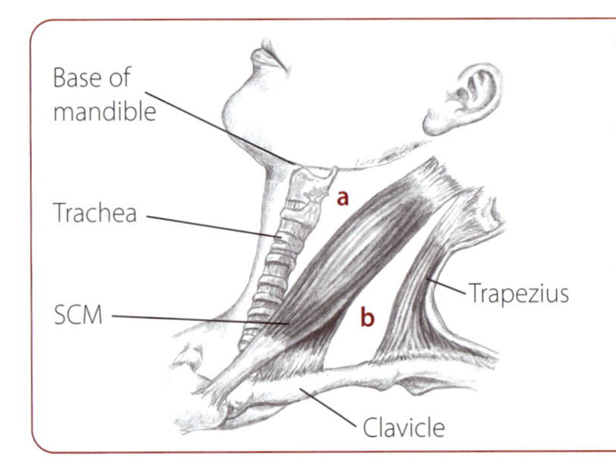

The anterior and lateral sides of the neck can be divided into two triangular regions. The anterior triangle **(a)** is bordered by the sternocleidomastoid (SCM), the base of the mandible and the trachea. The hyoid bone, thyroid gland, carotid artery, submandibular gland and styloid process of the temporal bone are some of the structures within the anterior triangle.

The posterior triangle **(b)** is formed by the sternocleidomastoid (SCM), clavicle and trapezius and contains, among other structures, the brachial plexus and the external jugular vein.

Exploring the Skin and Fascia

1) Partner supine. Sink your fingers into your partner's hair and onto her scalp. Note the temperature and moisture or oiliness.
2) Using your fingerpads for contact, gently tug the scalp in all directions (5.2). Rotate the head to the side to reach behind the ears and at the base of the skull. As you sense the tissue's thickness and mobility, do you notice any regions of the scalp that are more pliable than others?

(5.2) Partner supine

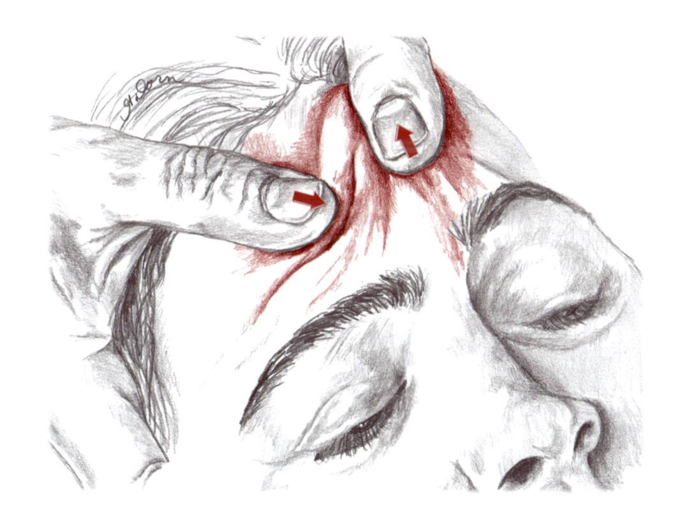

(5.3)

1) Using your thumbpads, gently torque the skin and fascia of the forehead and face. Sense the tissue's elasticity and thickness (5.3).

1) Moving to the neck, gently grasp and lift the skin and fascia of the lateral neck (5.4). Oftentimes the tissue here will be quite thin, almost delicate. Turn your hand 90˚ and try lifting the tissue in a horizontal direction. Is it more challenging to lift? Next explore the anterior neck, including the tissue under the mandible. Do you notice any restrictions in the skin?
2) Resting the head on the heel of your hand, explore the skin and fascia of the posterior neck. Oftentimes this tissue is thicker and denser than the anterior tissue. Is this true?

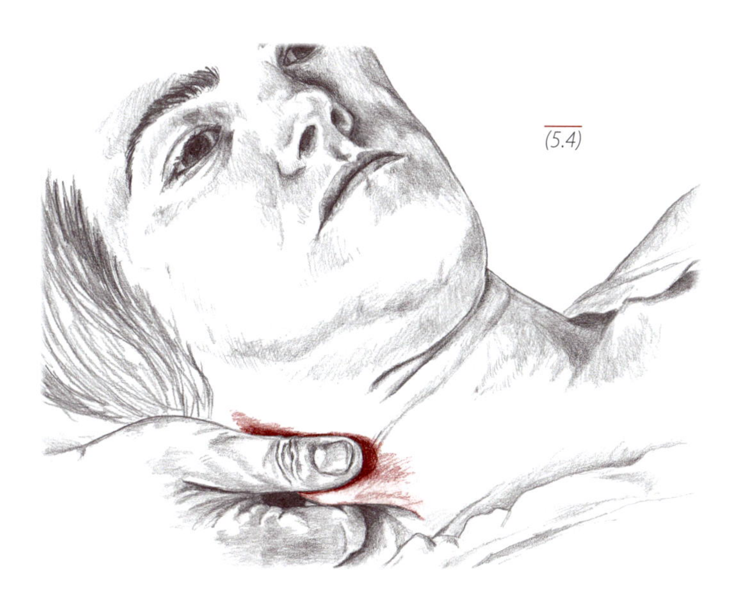

(5.4)

Bones & Bony Landmarks of the Head, Neck and Face

The **skull** is composed of twenty-two bones: eight in the cranium and fourteen in the facial region. Seven of the eight **cranial** bones are directly accessible. The eighth, the ethmoid, is accessible only by way of the nasal cavity. Most of the cranial bones are superficial. Seven of the fourteen **facial** bones are palpable, as are the numerous bony landmarks of the **mandible** (jaw) (5.5, 5.6).

The articulations of the cranial bones are different from the articulations of the appendages. The joints of the arms and legs have a synovial (mobile) joint structure. The cranial bones, in contrast, have fibrous joints that are woven together to form tight-fitting sutures.

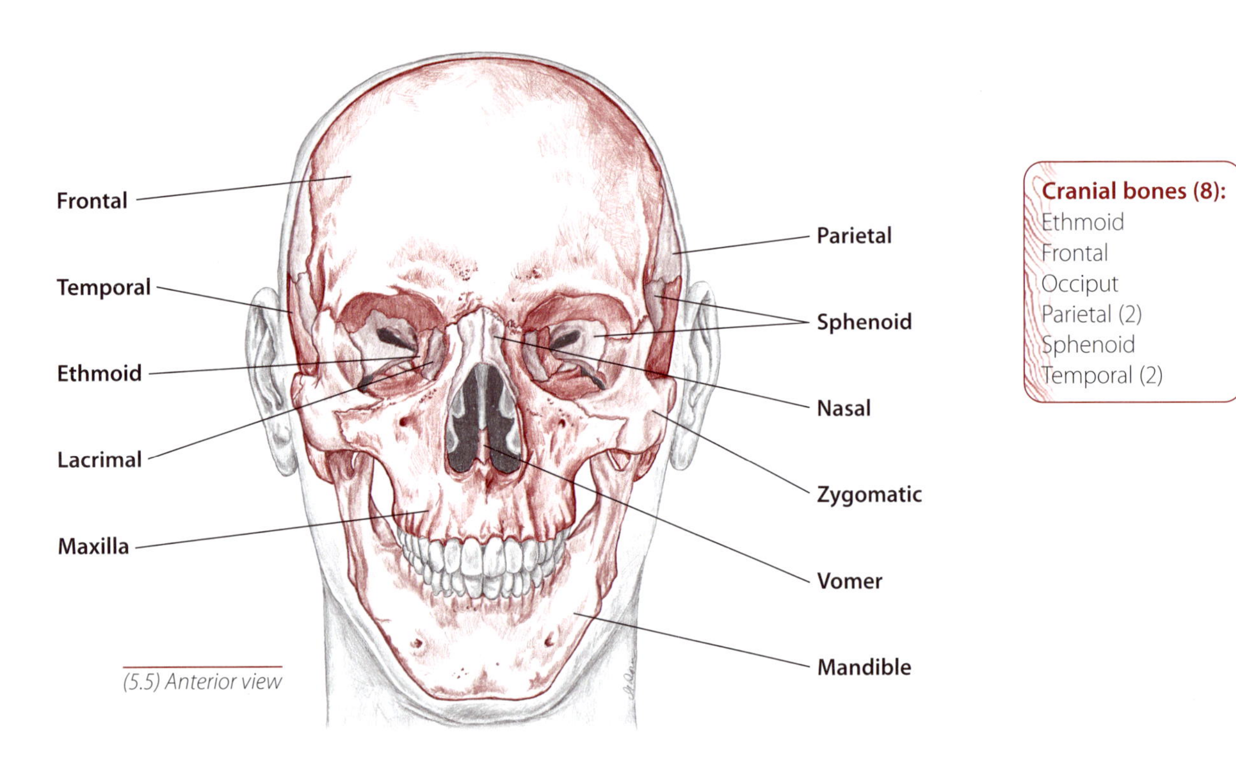

(5.5) Anterior view

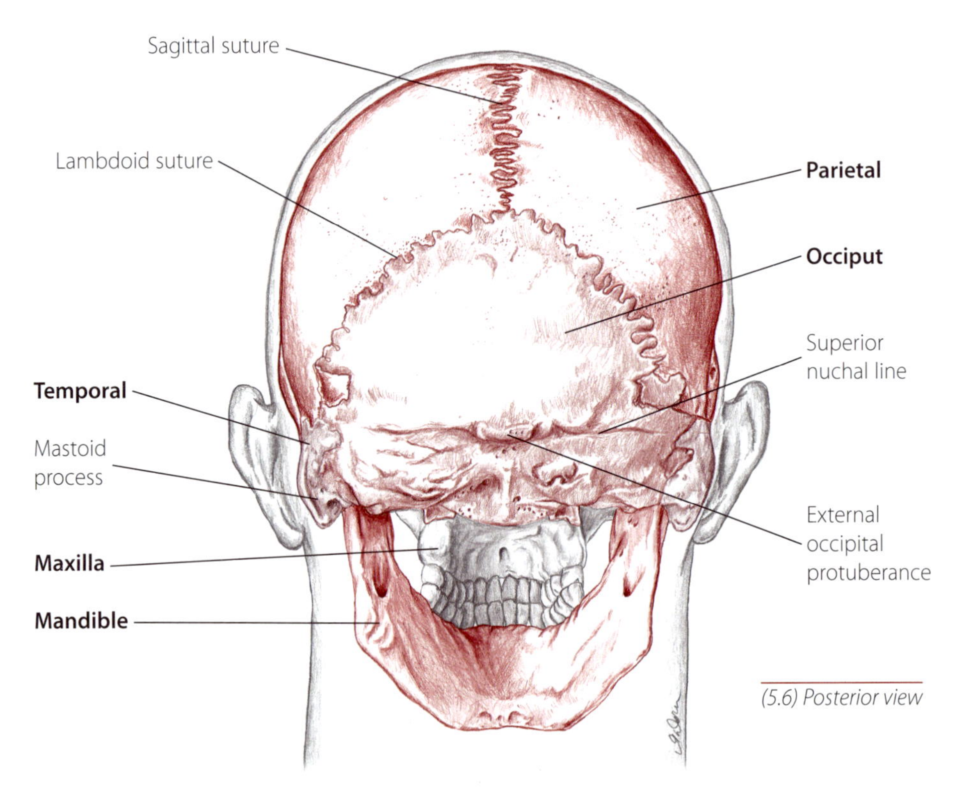

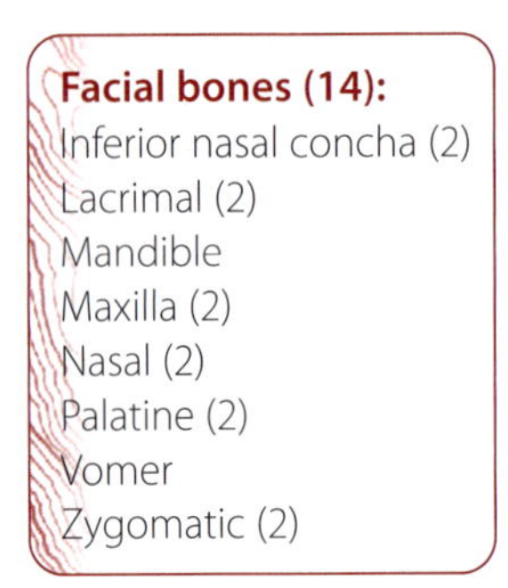

(5.6) Posterior view

cranio- skull	Grk. skull
suture	ME. bow L. a seam

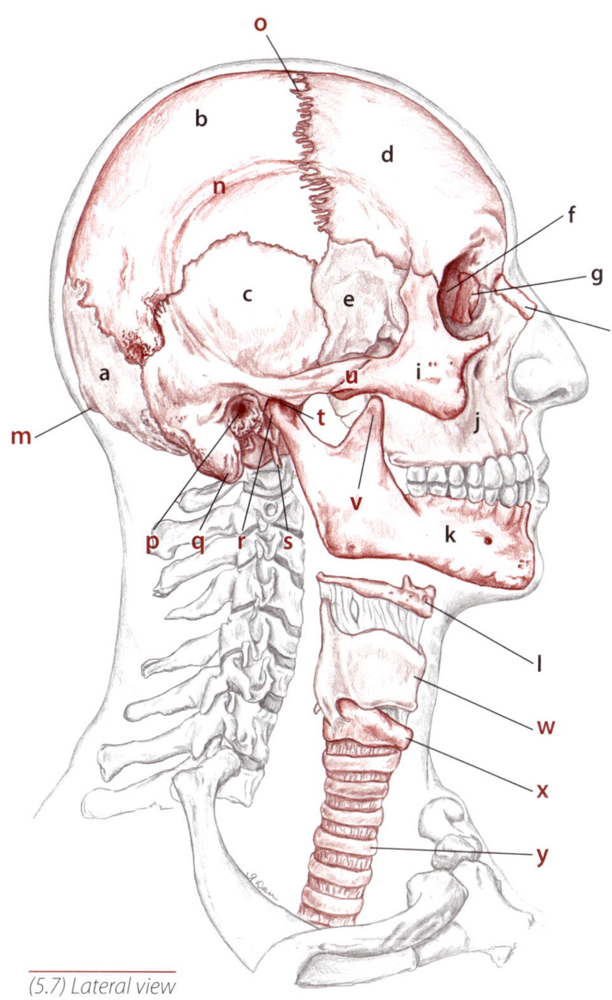

(5.7) Lateral view

a) Occiput
b) Parietal
c) Temporal
d) Frontal
e) Sphenoid
f) Ethmoid
g) Lacrimal
h) Nasal
i) Zygomatic
j) Maxilla
k) Mandible
l) Hyoid
m) External occipital protuberance
n) Temporal lines
o) Coronal suture
p) External auditory meatus
q) Mastoid process
r) Condyle of the mandible
s) Styloid process of the temporal bone
t) Temporomandibular joint
u) Zygomatic arch
v) Coronoid process
w) Thyroid cartilage
x) Cricoid cartilage
y) Trachea

Black letters indicate bones, **red** letters indicate bony landmarks or other structures

Within the medical community it was long believed that the cranial bones did not move. Since the cranial bones, with their tightly woven sutures, are designed to protect the brain, any cursory examination of the skull would seem to support the hypothesis that these bones are immobile.

In the 1920s, however, a young osteopath named William Sutherland was determined to prove that there was an infinitesimal, yet palpable, motion or rhythm of the cranial bones. Using himself as a guinea pig, Sutherland tested his hypothesis by applying a variety of homemade contraptions to his head, including a football helmet with screws drilled through it. While Sutherland monitored his cranial rhythm, his wife quietly detailed his dramatic personality and appearance changes.

Sutherland's research and perseverance lent support to the notion of cranial movement and helped cranial osteopathy to be accepted by the medical establishment.

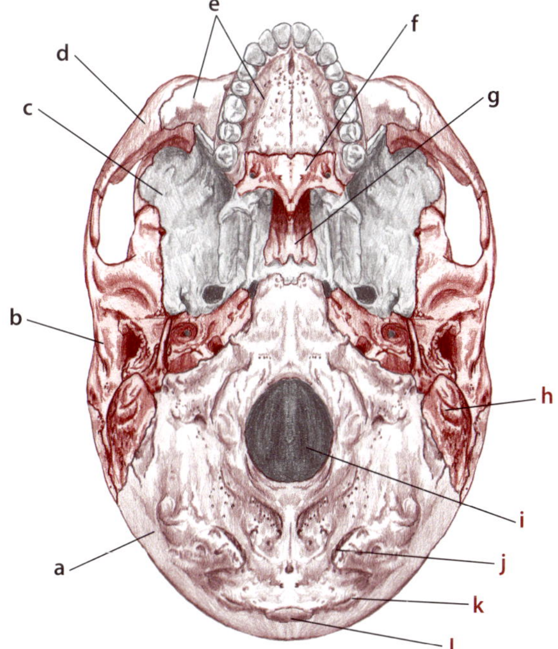

a) Occiput
b) Temporal
c) Sphenoid
d) Zygomatic
e) Maxilla
f) Palatine
g) Vomer
h) Mastoid process
i) Foramen magnum
j) Inferior nuchal line
k) Superior nuchal line
l) External occipital protuberance

(5.8) Inferior view

Bony Landmark Trails

Trail 1 "Around the Globe" palpates the bones and bony landmarks of the cranium and face.

- **a** Occiput
  - External occipital protuberance
  - Superior nuchal lines
- **b** Parietal
- **c** Temporal
  - Mastoid process
  - Zygomatic arch
  - Styloid process
- **d** Frontal
- **e** Sphenoid
- **f** Nasal, zygomatic and maxilla

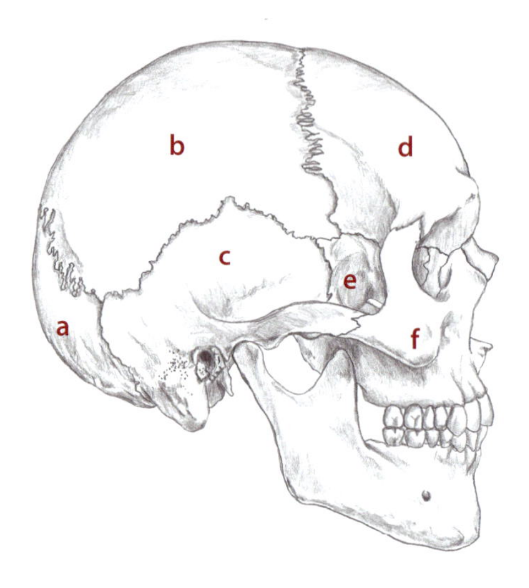

Trail 2 "Jaw Jaunt" explores the mandible.

- **a** Body
- **b** Base
- **c** Submandibular fossa
- **d** Angle
- **e** Ramus
- **f** Coronoid process
- **g** Condyle

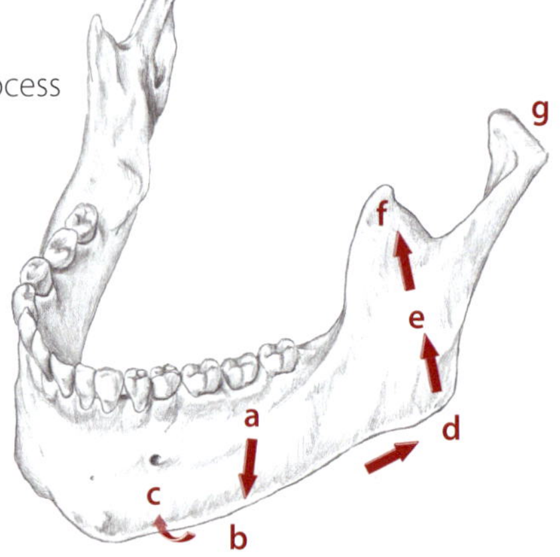

Trail 3 "Horseshoe Trek" locates the cartilaginous structures of the anterior neck and the horseshoe-shaped hyoid bone.

- **a** Trachea
- **b** Cricoid cartilage
- **c** Thyroid cartilage
- **d** Hyoid bone

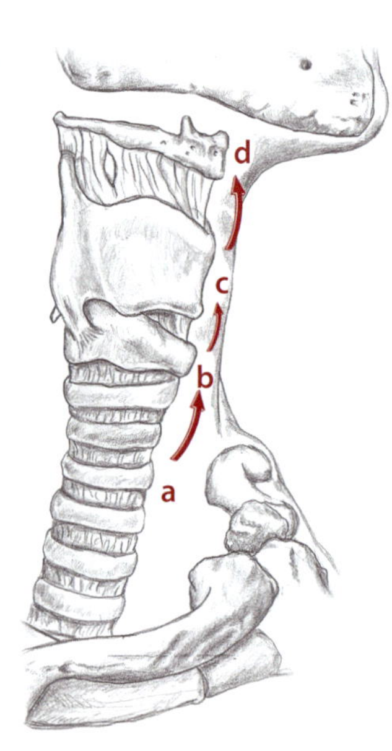

Lateral view of neck

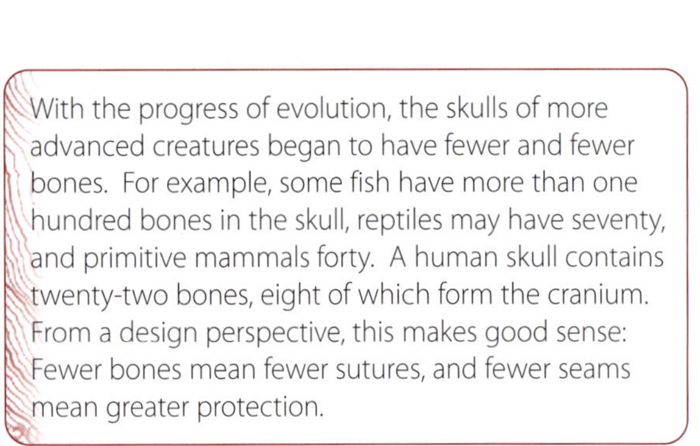

With the progress of evolution, the skulls of more advanced creatures began to have fewer and fewer bones. For example, some fish have more than one hundred bones in the skull, reptiles may have seventy, and primitive mammals forty. A human skull contains twenty-two bones, eight of which form the cranium. From a design perspective, this makes good sense: Fewer bones mean fewer sutures, and fewer seams mean greater protection.

Trail 1 "Around the Globe"

Occiput

External Occipital Protuberance
Superior Nuchal Lines

The occiput is located at the posterior and inferior aspects of the cranium. It extends superiorly from the external occipital protuberance and laterally to the mastoid processes of the temporal bones. The portion of the occiput superior to the occipital protuberance is superficial and easily palpable. The portion inferior to the protuberance curves in and under the head and is covered by layers of tendon and muscle (5.9, 5.10).

Sometimes called the "bump of knowledge," the **external occipital protuberance** is a small, superficial point located along the back of the head at the center of the occiput. It lies between the attachment sites of both trapezius muscles and is the superior attachment for the ligamentum nuchae. Regardless of intelligence, it varies in size.

Located along either side of the occipital protuberance, the **superior nuchal lines** are faint, sometimes bumpy ridges which extend laterally to the mastoid processes. The nuchal lines are attachment sites for the trapezius and splenius capitis muscles.

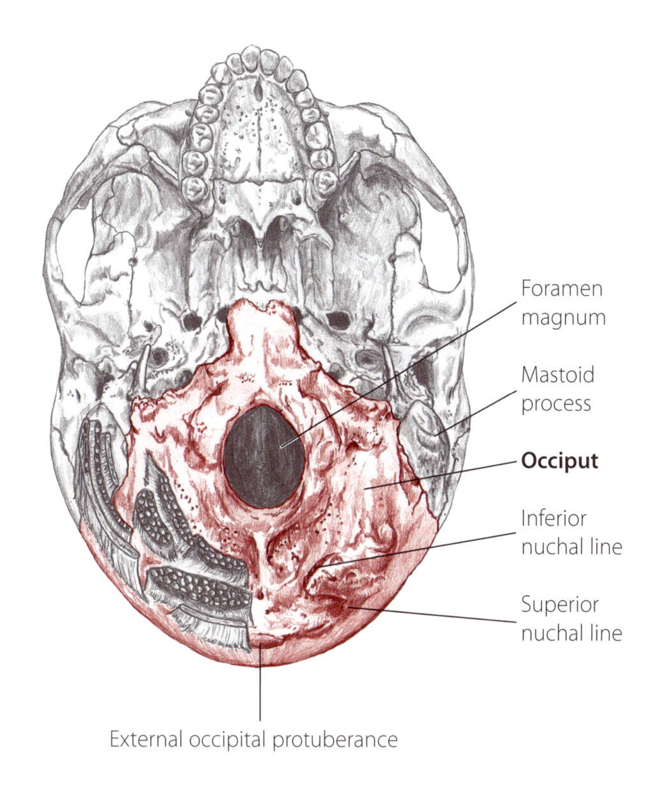

(5.9) Inferior view of cranium; muscle attachment sites on left side identified below

General location of occiput

1) Partner prone. Place your hand on the back of the head between the ears.
2) Explore its surface by sliding your fingers
   - *superiorly* from the occipital protuberance two or three inches;
   - *inferiorly* where the occiput curves and sinks into the muscles of the neck;
   - *laterally* to the mastoid processes behind the ears.

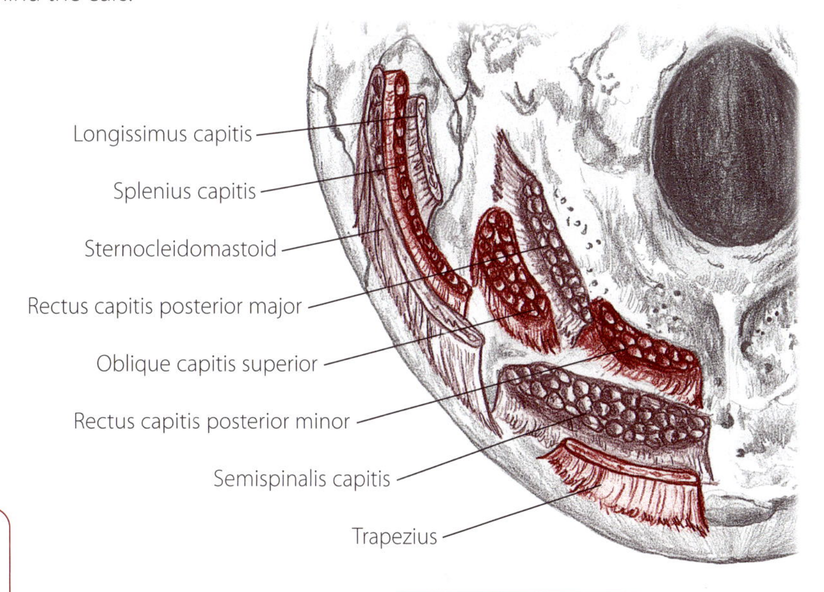

(5.10) Inferior view showing attachment sites

The superior nuchal line is an attachment site for several muscles. Metaphorically, it is the "shoreline" between the dry land of the cranium and the sea of neck muscles.

nuchal	**nu**-kal	L. the back of the neck
occiput	**ok**-si-put	L. the back of the skull

External occipital protuberance

Superior nuchal line

Mastoid process

(5.11) Partner prone, superficial tissue removed on right side of cranium

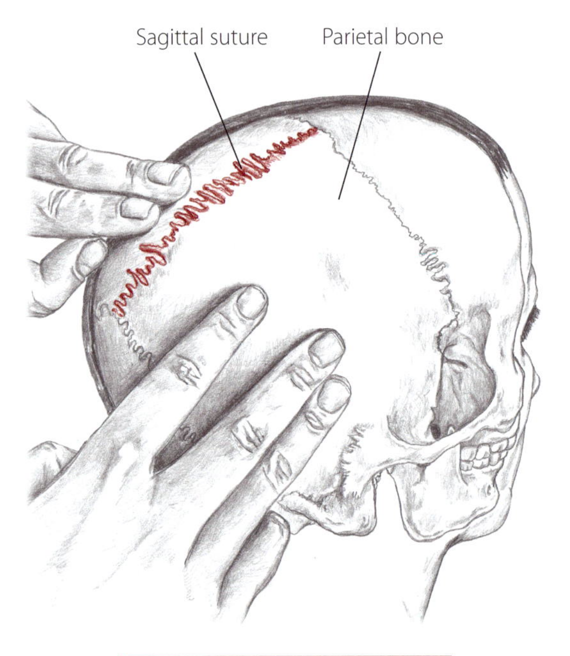

(5.12) Partner seated, superficial tissue removed on right side of cranium

The cranial bones are neither fully developed nor joined at birth. Usually there are six unossified gaps in the skull called fontanels. The name (Old French, *little fountain*) perhaps came from the pulse of the blood vessels felt under the skin that reminded physicians of the spurting of a fountain. The fontanels close over a period of between two and twenty-four months.

External occipital protuberance

1) Partner prone or supine. Place your fingers along the back of the neck at the body's midline (5.11).
2) Slide superiorly onto the bony surface of the cranium. At the "shoreline" between the neck muscles and the cranium will be the protuberance.

Are you level with the top of the ears? If your partner is prone, ask her to extend her head slightly. Is the bump you feel just superior to where the muscles tighten?

Superior nuchal lines

1) Partner prone or supine. Stand at the head of the table and place both index fingers at the external occipital protuberance.
2) Allow the other fingers to fall in place beside them. Glide your fingerpads up and down and palpate the edge of the superior nuchal lines.
3) Follow these ridges laterally as they extend toward the ear and mastoid processes (5.11).

Are you just lateral to the occipital protuberance? Do the ridges lead toward the back of the ears? Can you find them from a prone position? Are you on the cranium as opposed to the muscles of the neck?

Parietal Bones

The large, rectangular parietal bones form the top and sides of the cranium. Positioned between the frontal, occipital and temporal bones, the parietals are saucer-shaped and extend anteriorly to the level of the ear canal. They merge at the body's midline to form the sagittal suture, a slight crest that can often be felt.

1) Partner seated, prone or supine. To access the general area of the parietals, place both hands on top of the cranium.
2) Palpate the sagittal suture between the parietals. If you cannot feel its crest, visualize it along the top of the skull.
3) Follow it anteriorly to the level of the ear canal and posteriorly to the occiput (5.12).

parietal puh-**ri**'e-tul L. wall

Temporal Bone

Mastoid Process
Zygomatic Arch
Styloid Process

The temporal bone is located on the side of the head, encompassing the area around the ear. It has three important bony landmarks: the mastoid process, the zygomatic arch and the styloid process (p. 235). The temporal bone is superficial, except for its superior aspect which lies deep to the temporalis muscle.

The **mastoid process** forms a large, superficial bump directly behind the earlobe. It is an attachment site for the sternocleidomastoid and other muscles.

The superficial **zygomatic arch** (cheekbone) is formed by the temporal and zygomatic bones. It is an attachment site for the masseter muscle. The space between the zygomatic arch and the cranium is filled by the thick temporalis muscle.

The **styloid process** is located behind the earlobe between the mastoid process and the posterior edge of the mandible. Its fanglike shape serves as an attachment site for several ligaments and muscles. The styloid process is deep to overlying muscles and tissue and is not directly palpable; however, its location can be accessed.

The styloid process of the temporal bone is fragile and is deep to the facial nerve (p. 269), so exploration in this area should be very gentle.

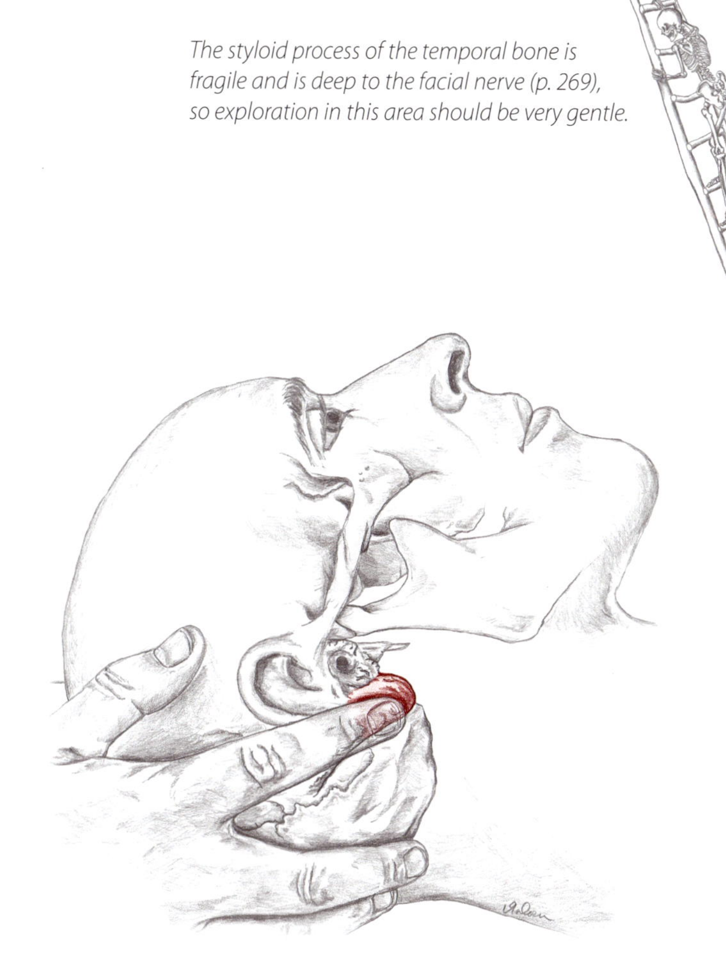

(5.13) Partner supine, accessing the mastoid process, inferior portion of the ear removed

1) Supine. Locate the mastoid process by placing your finger behind the earlobe. Sculpt around its edges, exploring its entire surface (5.13).
2) Explore the zygomatic arch by placing your finger anterior to the ear canal. Move anteriorly along the arch, outlining its sides with your thumb and finger (5.14). Follow it anteriorly as it merges with the orbit of the eye.

When locating the mastoid, are you behind the earlobe? Is the bone you feel round and superficial? Can you palpate posteriorly onto the superior nuchal line of the occiput? Does the ridge of the zygomatic arch run horizontally? Is it level with the ear canal?

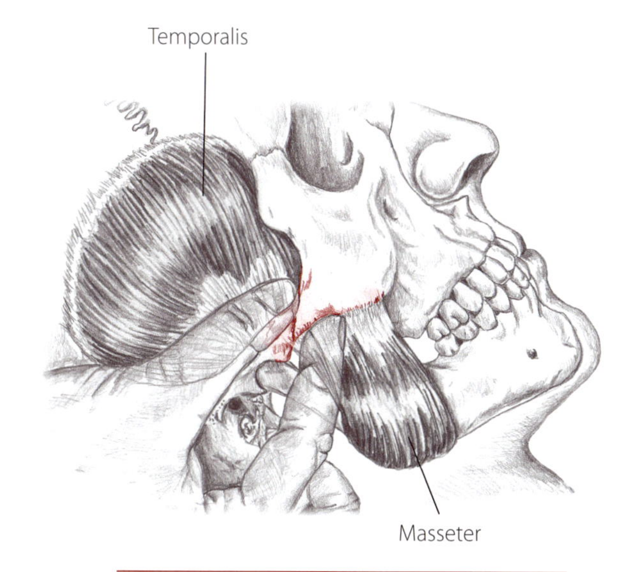

(5.14) Partner supine, palpating the zygomatic arch

mastoid	**mas**-toyd	Grk. breast-shaped
styloid	**sti**-loyd	Grk. a pillar
zygomatic	**zy**-go-**mat**-ik	Grk. cheekbone

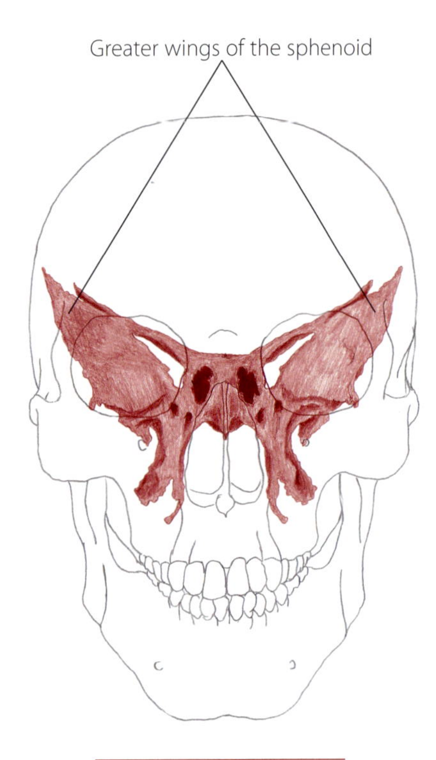

(5.15) Anterior view showing location of the sphenoid bone

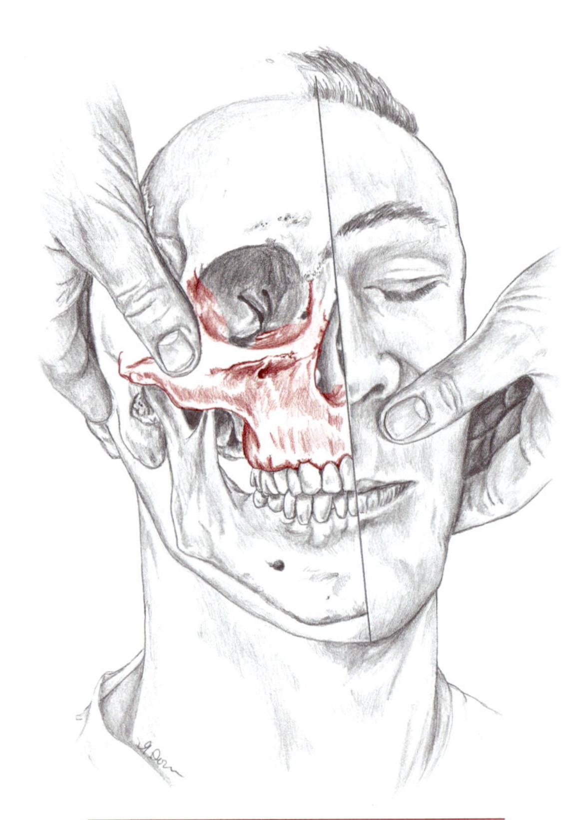

(5.16) Partner supine, exploring the facial bones, superficial tissue removed on left side

Frontal Bone

Located on the anterior aspect of the cranium, the broad frontal bone forms the forehead and upper rim of the eye sockets. It articulates with the parietal bones to form the coronal sutures which are deep to the occipitofrontalis and lateral edge of the temporalis muscles (p. 257).

1) Partner supine. Explore the region of the forehead, moving superiorly to the coronal sutures, inferiorly to the brow and laterally to the anterior edge of the temporalis muscle.

Sphenoid Bone

The sphenoid bone is located inside the cranium and has major articulations with the fourteen bones of the skull. Located behind the eyeballs and superior to the zygomatic arches, the sphenoid is shaped like a butterfly, and its lateral portions are called the greater wings (5.15). The temporalis lies on top of these flat wings, making them inaccessible.

1) Partner supine. Place your fingers at the middle of the zygomatic arch (cheekbone) to locate the greater wings of the sphenoid.

2) Slide your fingers superiorly one inch onto the temporalis muscle belly. Deep to the thick temporalis is the location of the greater wing of the sphenoid.

Facial Bones

Nasal

Located at the bridge of the nose, the small nasal bones are positioned between the frontal and maxillary bones and are virtually indistinguishable from them (p. 234).

Zygomatic

Better known as the cheekbone, the zygomatic bone forms the anterior aspect of the zygomatic arch and the lateral portion of the orbit of the eye (5.16). It serves as an attachment site for the masseter muscle.

Maxilla

The maxillary bones form the center of the face, the inferior portion of the orbit of the eye, the surface around the nose and the upper jaw in which the upper row of teeth articulate.

maxilla	**max**-il-a	L. jawbone
nasal	**na**-zl	L. nose
sphenoid	**sfe**-noyd	Grk. wedge-shaped

Mandible

Body, Base, Submandibular Fossa, Angle, Ramus, Condyle and Coronoid Process

The mandible or "jaw" has numerous landmarks that are superficial and accessible (5.17). The **body** is the flat surface of the mandible inferior to the lower teeth. The **base** or "jaw line" is the edge of the body and an attachment site for the thin platysma muscle. The **submandibular fossa** is located on the underside of the mandible and is an attachment site for the suprahyoid muscles (p. 259).

The superficial **angle** is located at the posterior end of the base. It forms part of the attachment for the masseter. The flat **ramus** is the posterior, vertical portion of the mandible and is deep to the masseter.

The mandible articulates with the cranium at two temporomandibular joints. The superficial **condyle** is located just anterior to the ear canal and inferior to the zygomatic arch. The deeper, inaccessible head of the condyle forms the articulating surface of the mandible at the temporomandibular (TM) joint (5.18).

The **coronoid process** is located an inch anterior to the condyle of the mandible and is the attachment site of the temporalis muscle. When the jaw is closed, the coronoid process lies underneath the zygomatic arch and is inaccessible. Opening the mouth fully, however, will bring the coronoid process out from under the arch and allow the process to be accessed.

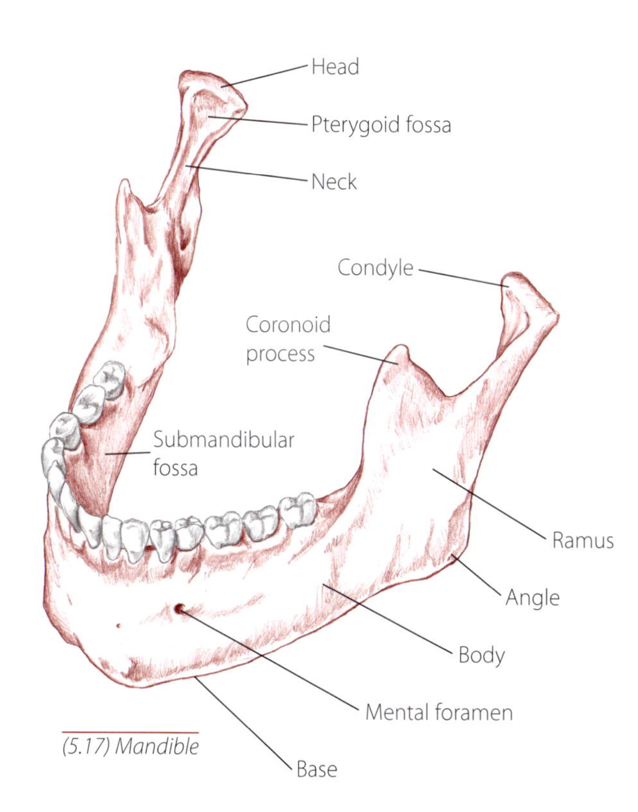

(5.17) Mandible

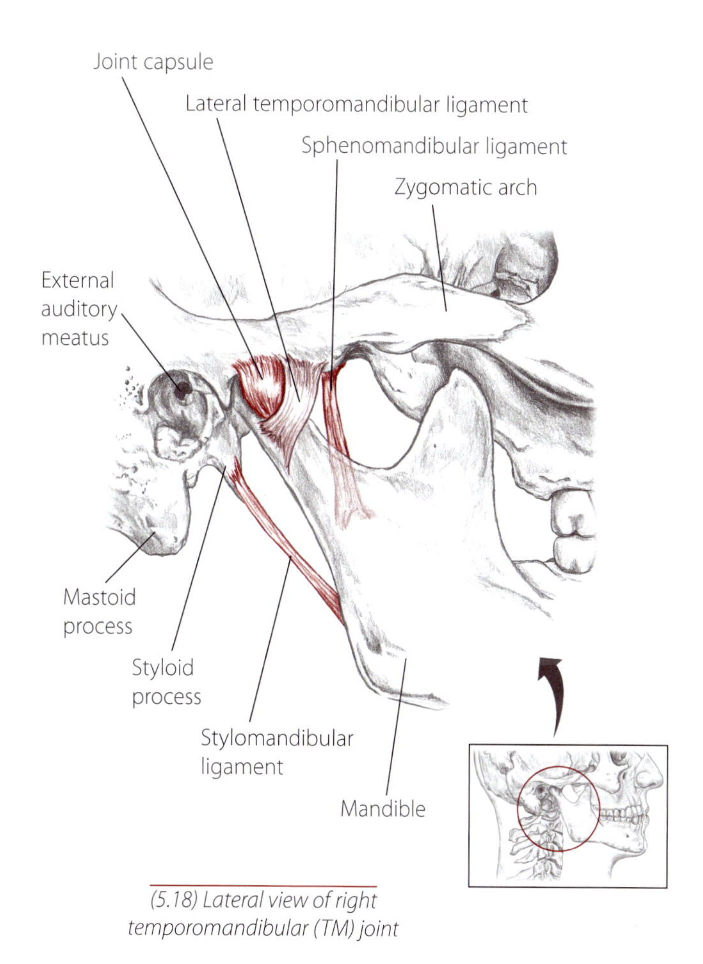

(5.18) Lateral view of right temporomandibular (TM) joint

Exploring in the submandibular fossa can be uncomfortable for your partner because of neighboring glands and nerves. Move slowly, checking in with him or her.

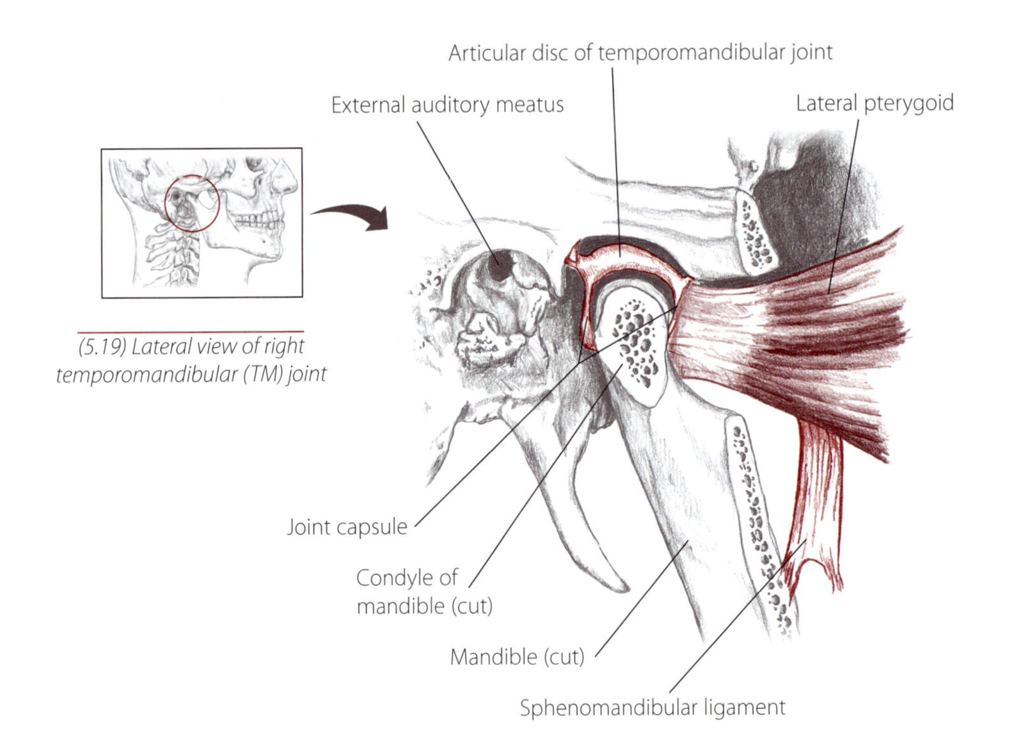

(5.19) Lateral view of right temporomandibular (TM) joint

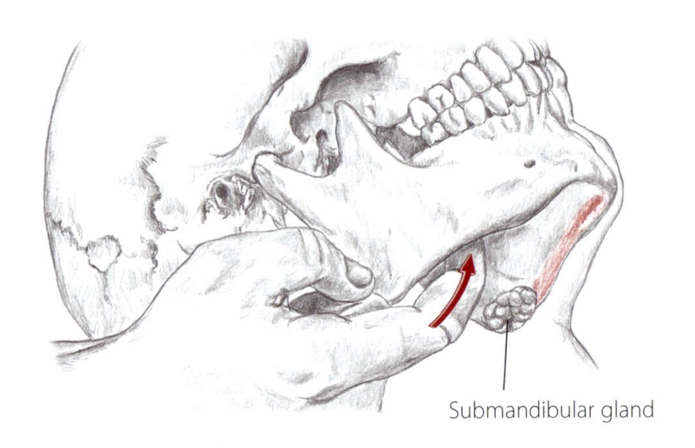

(5.20) Palpating the submandibular fossa

Body, base and submandibular fossa

1) Partner supine. Place your fingers inferior to the bottom teeth and explore the superficial surface of the body.
2) Move inferiorly and palpate the base or edge of the mandible. Explore its entire length from the chin to the angle of the mandible.
3) With one hand stabilizing the mandible, slowly curl a fingertip underneath its edge and into the submandibular fossa (5.20).

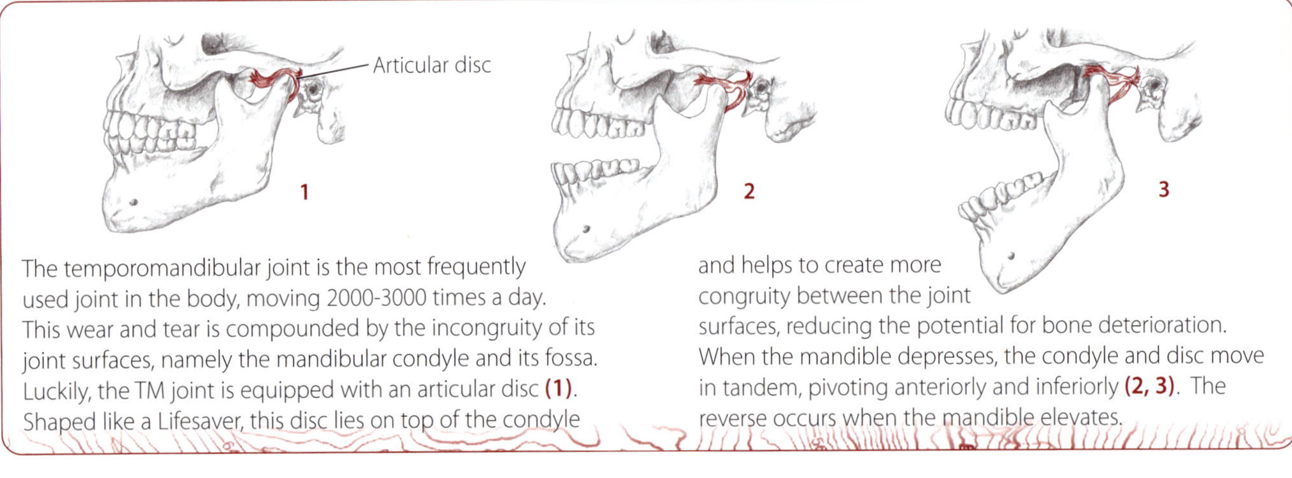

The temporomandibular joint is the most frequently used joint in the body, moving 2000-3000 times a day. This wear and tear is compounded by the incongruity of its joint surfaces, namely the mandibular condyle and its fossa. Luckily, the TM joint is equipped with an articular disc **(1)**. Shaped like a Lifesaver, this disc lies on top of the condyle and helps to create more congruity between the joint surfaces, reducing the potential for bone deterioration. When the mandible depresses, the condyle and disc move in tandem, pivoting anteriorly and inferiorly **(2, 3)**. The reverse occurs when the mandible elevates.

condyle	**kon**-dial	Grk. knuckle
mandible	**man**-di-ble	L. lower jawbone

Angle and ramus of mandible

1) Partner supine. Slide posteriorly along the base of the mandible to the angle. Clarify your location by asking your partner to open her mouth and note the movement of the angle.

2) Slide superiorly from the angle onto the ramus, which is deep to the masseter muscle.

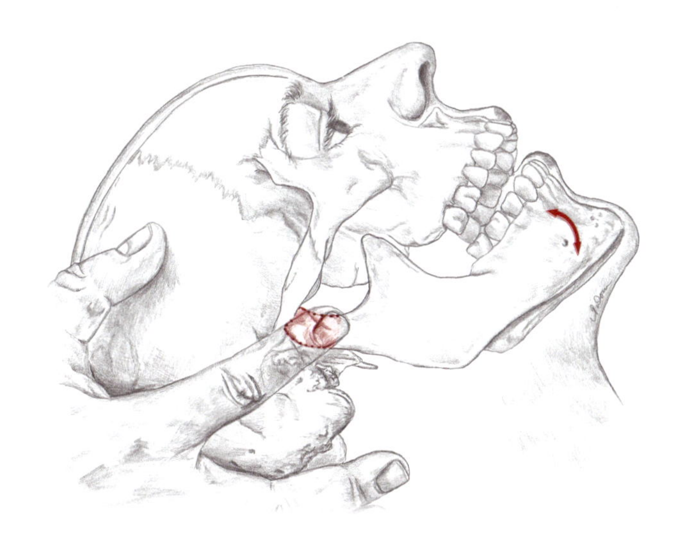

(5.21) Partner supine, feeling the condyle shift as she opens and closes her mouth

Condyle of mandible

1) Place your fingerpad anterior to the ear canal and below the zygomatic arch.

2) Ask your partner to open her mouth fully. With this action, the condyle will become more palpable as it slides anteriorly and inferiorly (5.21).

3) As the jaw closes, follow the condyle to its original position.

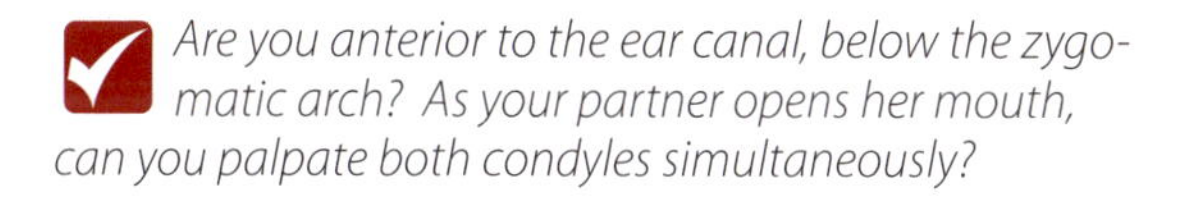

Are you anterior to the ear canal, below the zygomatic arch? As your partner opens her mouth, can you palpate both condyles simultaneously?

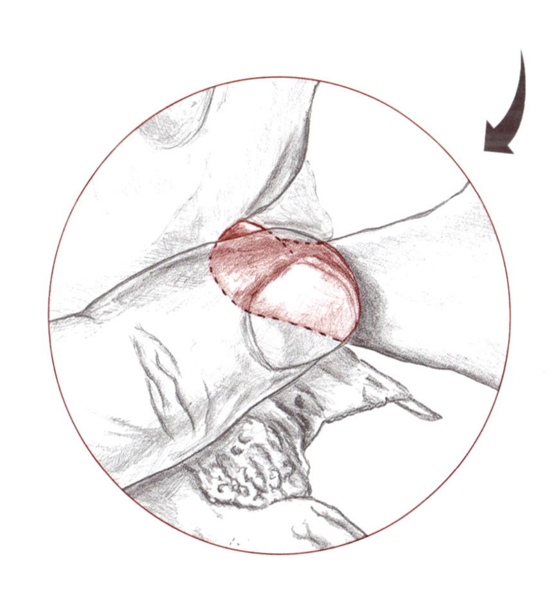

Coronoid process of mandible

1) Place your fingerpad on the middle aspect of the zygomatic arch.

2) Drop half an inch inferiorly and ask your partner to open her mouth fully. As the jaw drops, the large process will press into your finger (5.22).

3) With the mouth still open, explore the surfaces of the process.

Are you inferior to the zygomatic arch? When the mouth is open, can you feel the anterior edge of the process?

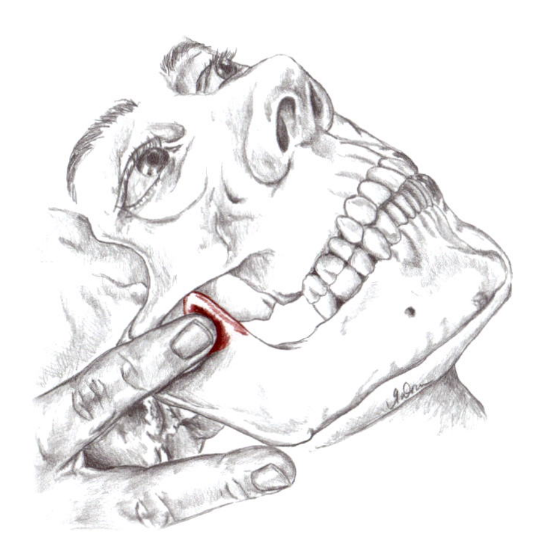

(5.22) Feeling the coronoid process come out from under the zygomatic arch as your partner opens her mouth

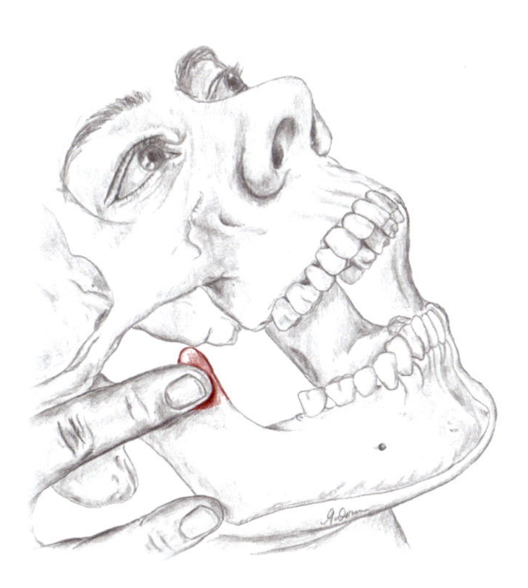

coronoid	**kor**-o-noyd	Grk. crown-shaped
jugular	**jug**-u-lar	L. throat
ramus	**ray**-mus	L. branch

Trail 3 "Horseshoe Trek"

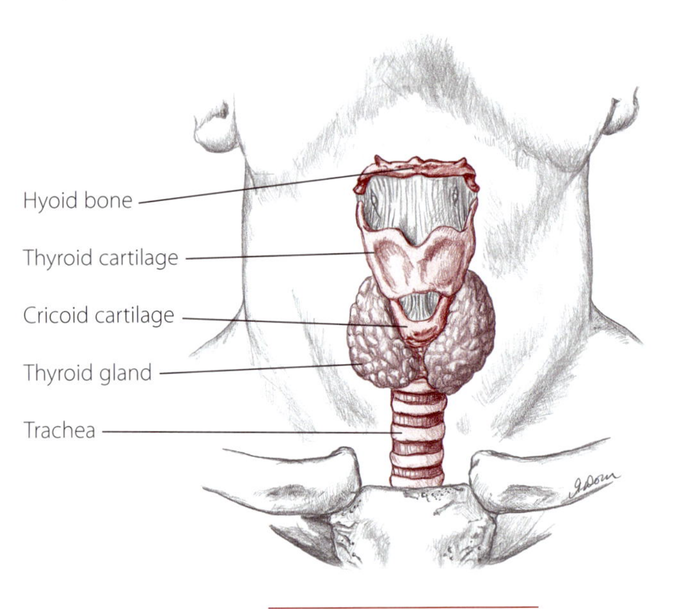

(5.23) Anterior view of the neck

Trachea

Cricoid Cartilage, Thyroid Cartilage and Hyoid Bone

The **trachea** (windpipe) is located at the center of the anterior neck (5.23). It is a ribbed, cartilaginous tube that is roughly an inch in diameter and deep to the thyroid gland. The **cricoid cartilage** is a slightly larger ring of the trachea superior to the thyroid gland. The **thyroid cartilage** (Adam's apple) is superior to the cricoid cartilage below the level of the chin. Present in both sexes, the thyroid cartilage is larger and more visibly protruding on adult males. The three structures are partially deep to the slender infrahyoid muscles (p. 261), yet are easily palpable.

The horseshoe-shaped **hyoid bone** is located superior to the thyroid cartilage (5.26, 5.27). It is roughly an inch in diameter and lies parallel to the base of the mandible (jawline) and the third or fourth cervical vertebrae. The hyoid bone serves as an attachment site for the suprahyoid and infrahyoid muscles. It is accessible and elevates upon swallowing.

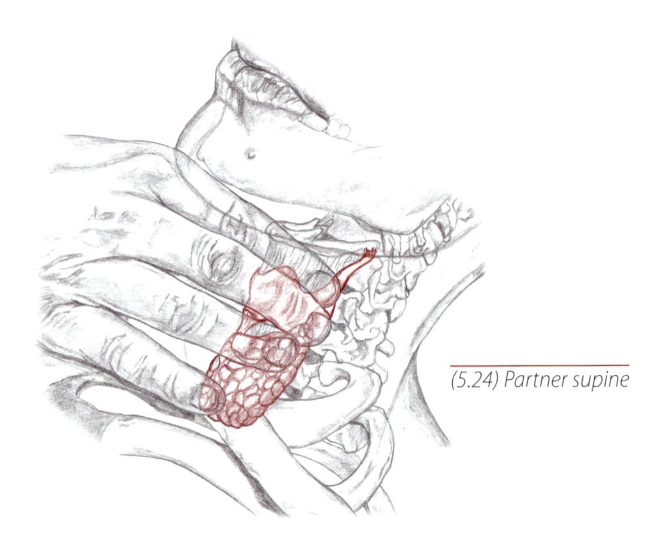

(5.24) Partner supine

Trachea and cartilages

1) Partner supine or seated. Using a fingerpad and thumbpad to palpate, gently explore the anterior surface of the neck for the tubular trachea.
2) Slide your finger up and down to feel the trachea's ribbed surface, and slowly and gently shift it from side to side, noting its pliability (5.24).
3) The cricoid cartilage can be isolated by sliding your finger and thumb superiorly along the trachea to just below the thyroid cartilage. Explore for its large, ringed surface (5.25).
4) Slide superiorly from the cricoid cartilage onto the thyroid cartilage. Palpate its sides and central tip.

Are you at the midline of the neck? Can you distinguish any rings along the trachea's surface? Is the trachea roughly an inch in diameter? With your fingerpad on the thyroid cartilage, ask your partner to swallow. Do you feel it move up and down?

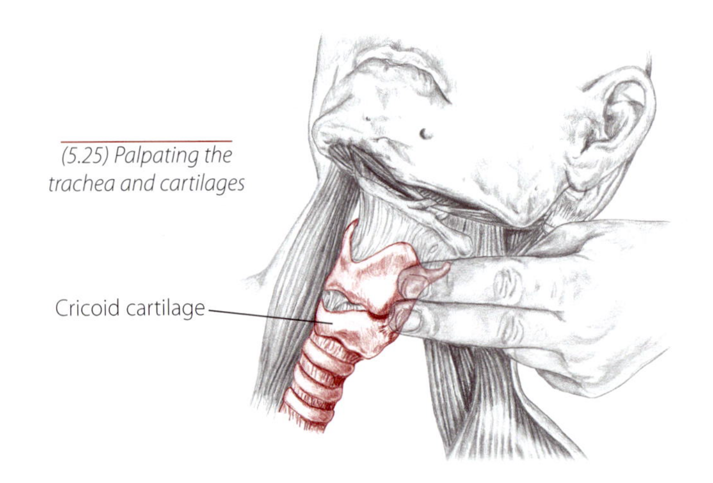

(5.25) Palpating the trachea and cartilages

The thyroid cartilage is sometimes referred to as the "Adam's apple." The name derives from a folk legend that told how the biblical Adam's first bite of apple became stuck halfway down his throat. According to the legend, his male descendants, with their more visibly protruding thyroid cartilage, appear to have carried on his condition.

cricoid	**kri**-koyd	Grk. ring-shaped
hyoid	**hi**-oyd	Grk. U-shaped

Hyoid bone

1) Partner supine or seated. Place your index finger upon the thyroid cartilage.
2) Roll your fingerpad superiorly over the thyroid cartilage, onto the hyoid.
3) Then gently palpate the sides of the hyoid with your first finger and thumb (5.28). The hyoid will be wider than the trachea.
4) Using gentle pressure, explore the surface of the hyoid as well as its small side-to-side movements. If you have difficulty accessing the hyoid, encourage your partner to relax her tongue and jaw.

Are you superior to the thyroid cartilage (Adam's apple)? Can you gently move the hyoid from side to side? With your first finger and thumb on either side of the hyoid, ask your partner to swallow. Do you feel the hyoid rise up and then return (5.29)?

ant. to C3.

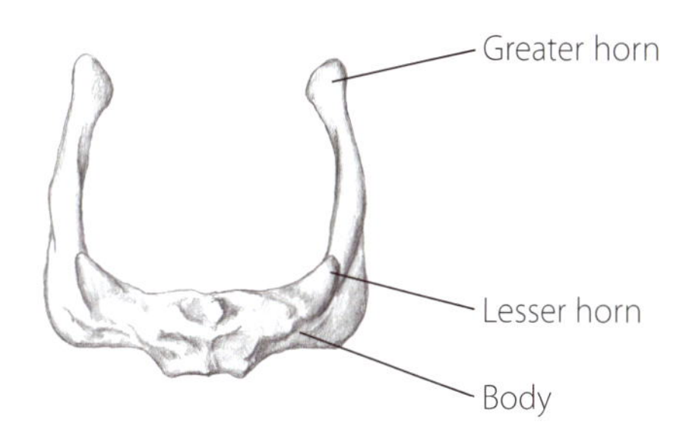

(5.26) Superior view of hyoid

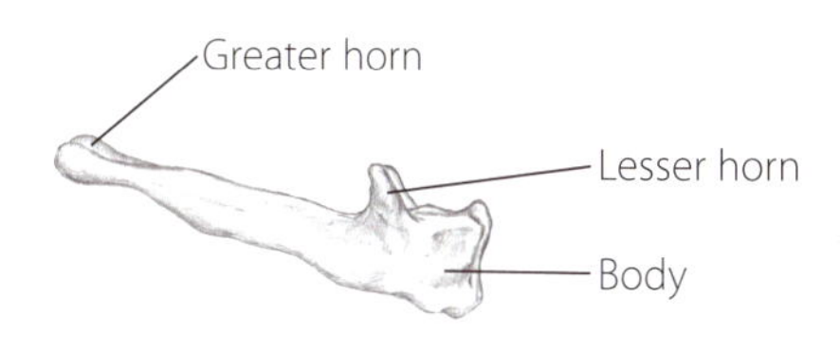

(5.27) Lateral view of hyoid

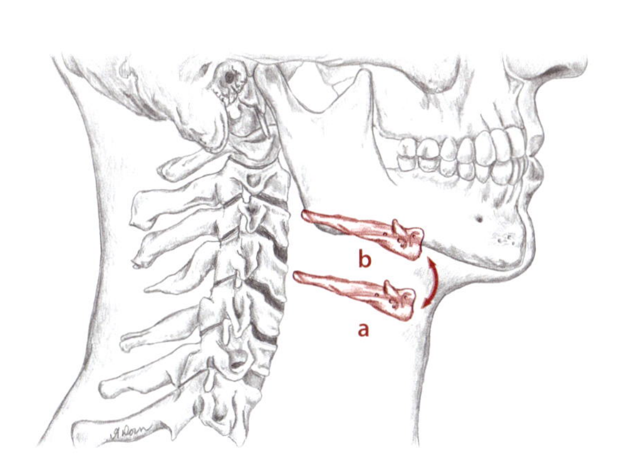

(5.29) The hyoid bone at rest **(a)** *and its placement during swallowing* **(b)**

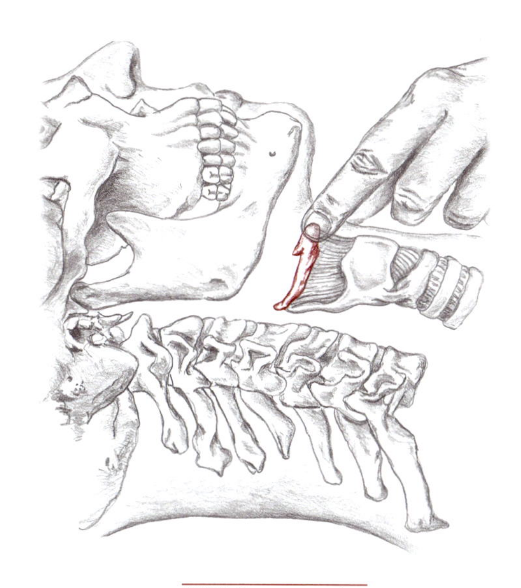

(5.28) Partner supine, isolating the hyoid bone

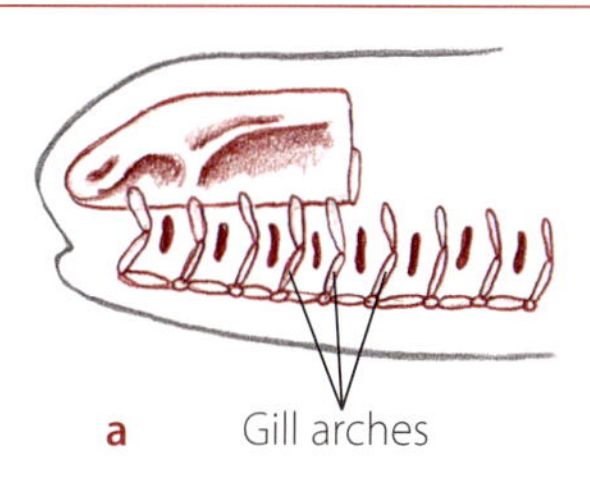

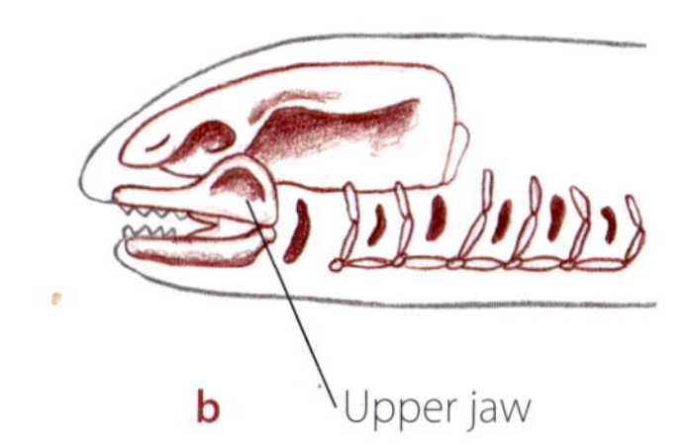

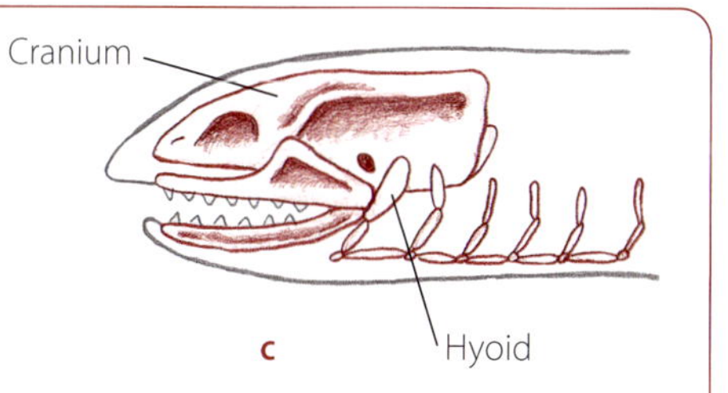

The hyoid bone is an ancestral remnant of the tissue that once formed gills. In the evolution of the jaw, the gill arches (the bones around the gills) **(a)** gravitated toward the head to hold the upper jaw next to the cranium **(b)**. For fish, which do not have the long necks we do, the position of the hyoid provides an important link between the jaw and cranium **(c)**. For humans, the hyoid lost this function and shifted down the neck to become the only non-articulating bone in the body. It is supported instead by the muscles that attach to its surface such as the suprahyoids and infrahyoids.

thyroid	**thi**-royd	Grk. shield
trachea	**tray**-ke-a	Grk. rough

Muscles of the Head, Neck and Face

The head and face contain over thirty pairs of muscles, many of which are small, thin and difficult to isolate. Nevertheless, the several muscles that act upon the mandible are easily accessible on the side of the jaw.

The anterior and lateral neck muscles perform a wide variety of tasks, including moving the head and neck, assistance in swallowing and raising the rib cage during inhalation. The posterior neck muscles, which act primarily upon the cervical spine and head, are detailed in Chapter Four, *Spine and Thorax*.

Before you palpate the following muscles on your partner, you are advised to skip to the back of this chapter in order to familiarize yourself with the arteries, glands and nerves of the head, neck and face (p. 267).

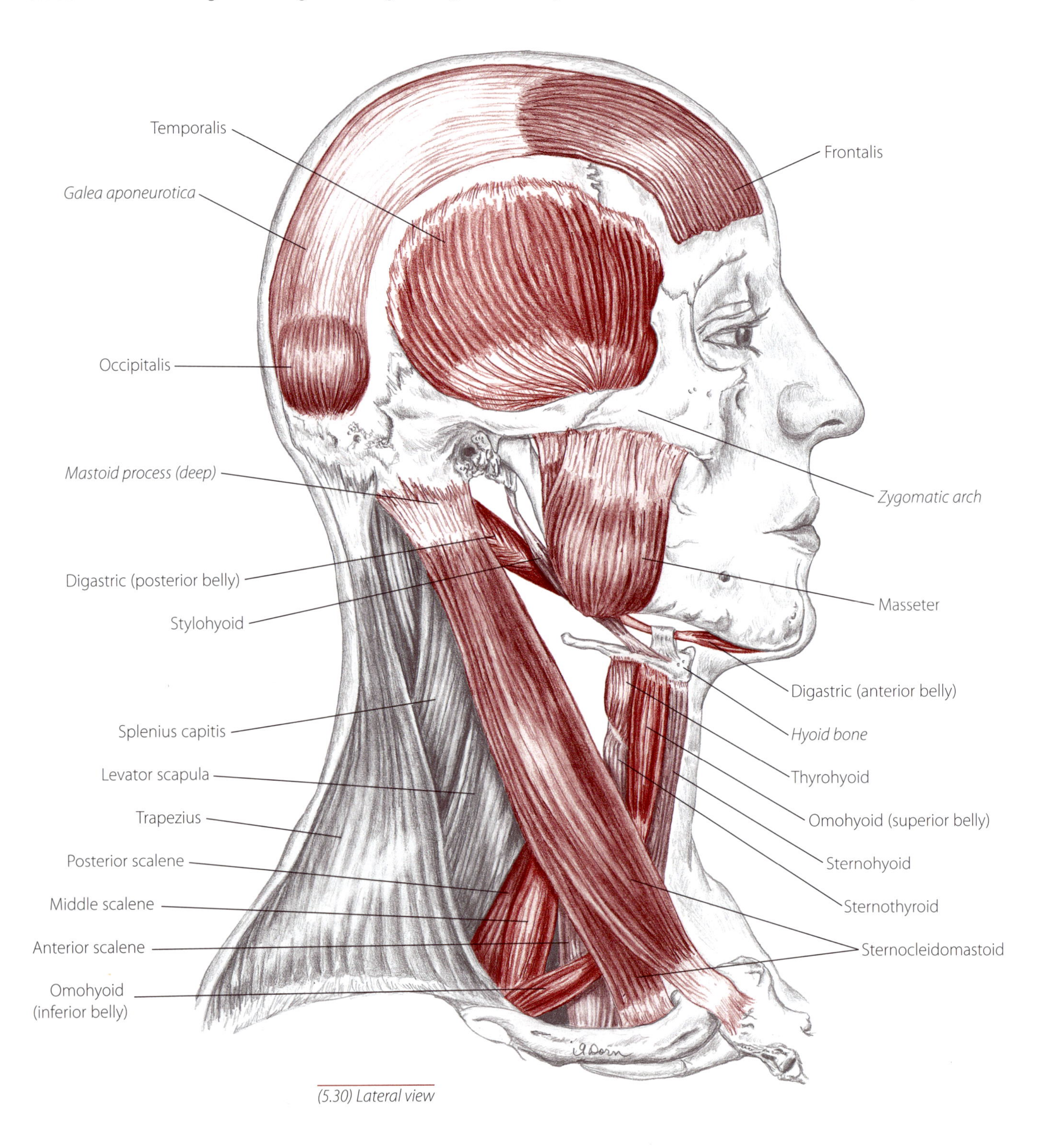

(5.30) Lateral view

Muscles of the Head, Neck and Face

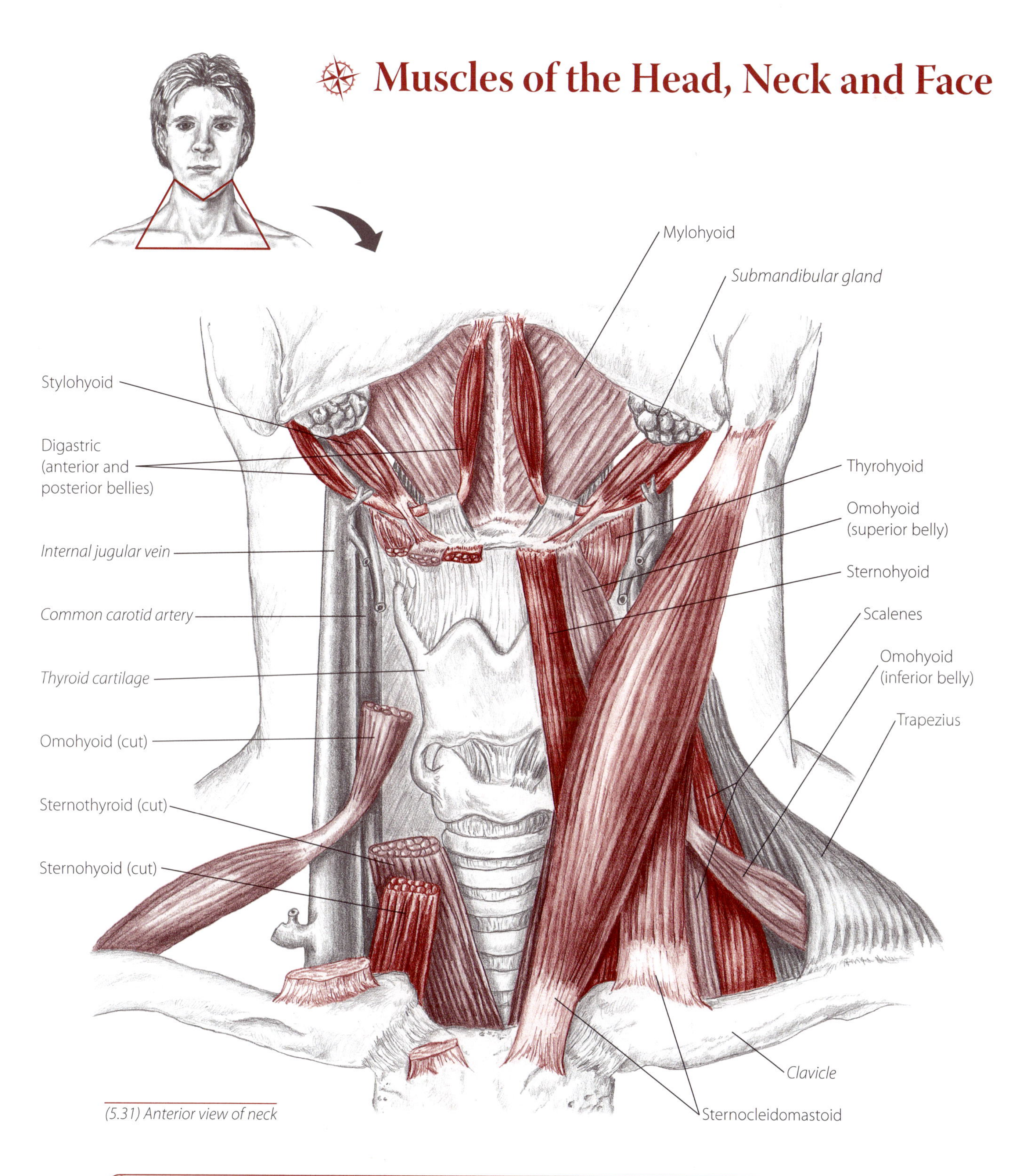

(5.31) Anterior view of neck

The single smallest muscle in the human body is located in the middle ear. The stapedius muscle measures less than 1/20 of an inch, thinner than a U.S. dime. It activates the stirrup, one of the tiny bones of the ear, which sends vibrations from the eardrum into the inner ear.

The stapedius, however, may not be the absolute shortest muscle in the body. A minuscule involuntary muscle called the arrector pili (p. 20) attaches to every hair follicle on the human body. These microscopic muscles nevertheless have a big responsibility: When you are cold or respond to a strong emotion such as fear, the arrector pili muscles raise the hair, producing goose bumps which help to retain body heat. They are also believed to have given our evolutionary ancestors the hair-raising ability of appearing larger to potential enemies.

arrector pili	**a**-rek-tor **pee**-li	L. *arrector*, lifter; *pilus*, hair
stapedius	sta-**pe**-de-us	L. stirrup

Synergists - Muscles Working Together

*muscles not shown

Cervical Spine

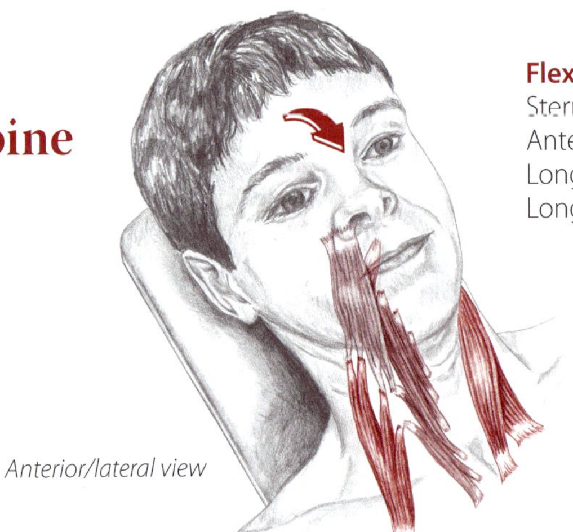

Anterior/lateral view

Flexion

Sternocleidomastoid (bilaterally)
Anterior scalene (bilaterally)
Longus capitis (bilaterally)
Longus colli (bilaterally)

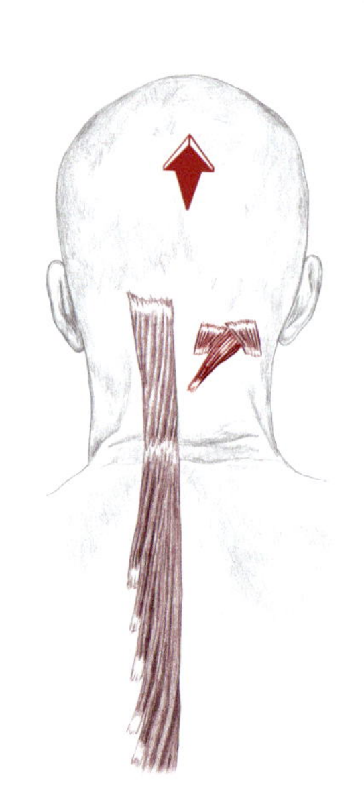

Extension

Trapezius - upper fibers (bilaterally)
Levator scapula (bilaterally)
Splenius capitis (bilaterally)
Splenius cervicis (bilaterally)
Rectus capitis posterior major
Rectus capitis posterior minor
Oblique capitis superior
Semispinalis capitis
Longissimus capitis (assists)*
Longissimus cervicis (assists)*
Iliocostalis cervicis (assists)*

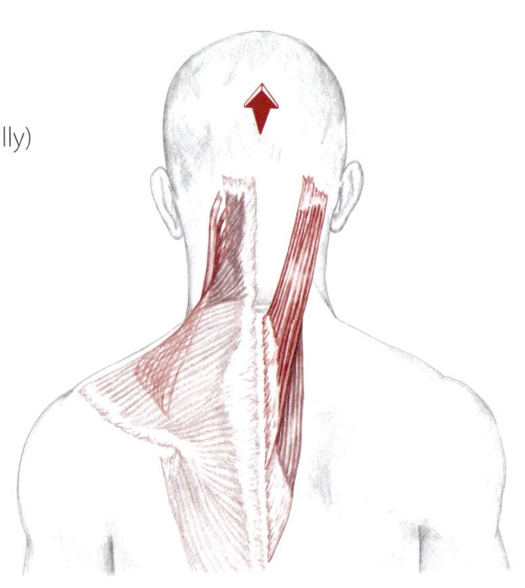

Posterior views

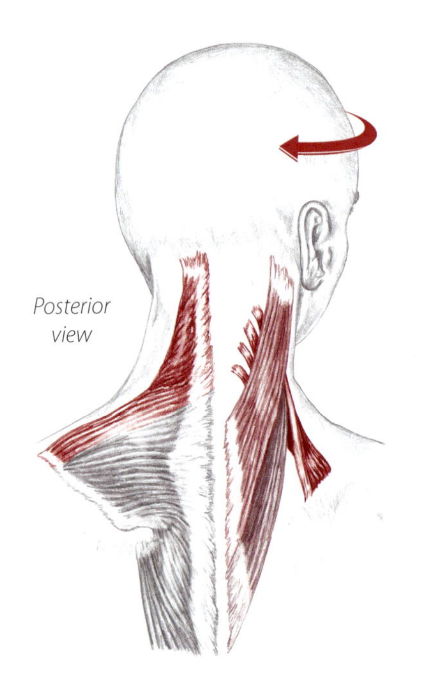

Posterior view

Rotation

(unilaterally to the same side)

Levator scapula
Splenius capitis
Splenius cervicis
Rectus capitis posterior major*
Oblique capitis inferior*
Longus colli*
Longus capitis*
Longissimus capitis (assists)*
Longissimus cervicis (assists)*
Iliocostalis cervicis (assists)*

Rotation

(unilaterally to the opposite side)

Trapezius - upper fibers
Sternocleidomastoid
Anterior scalene
Middle scalene
Posterior scalene

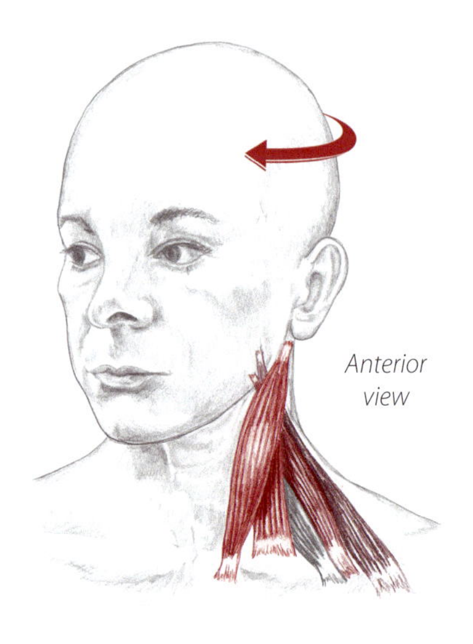

Anterior view

Cervical Spine

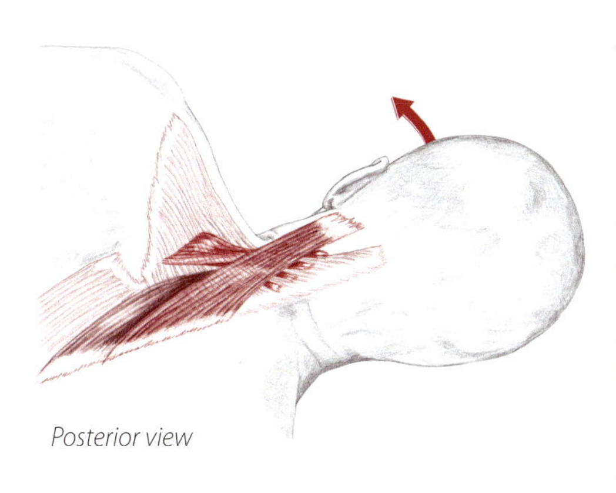
Posterior view

Lateral Flexion
(unilaterally to the same side)
Trapezius - upper fibers
Levator scapula
Splenius capitis
Splenius cervicis
Sternocleidomastoid
Longus capitis
Longus colli
Anterior scalene (with ribs fixed)
Middle scalene (with ribs fixed)
Posterior scalene (with ribs fixed)
Longissimus capitis (assists)*
Longissimus cervicis (assists)*
Iliocostalis cervicis (assists)*

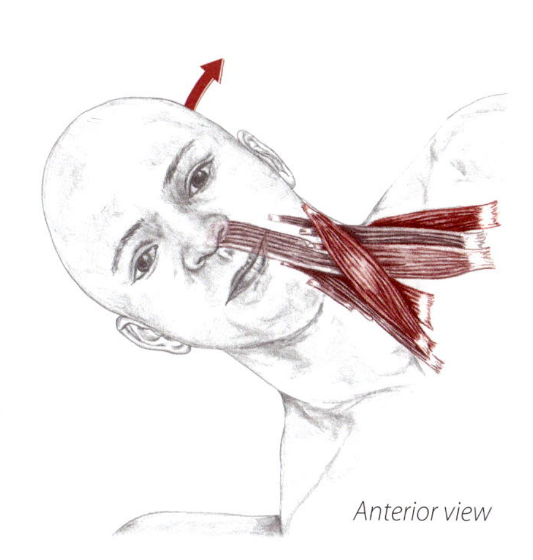
Anterior view

Mandible
(temporomandibular joint)

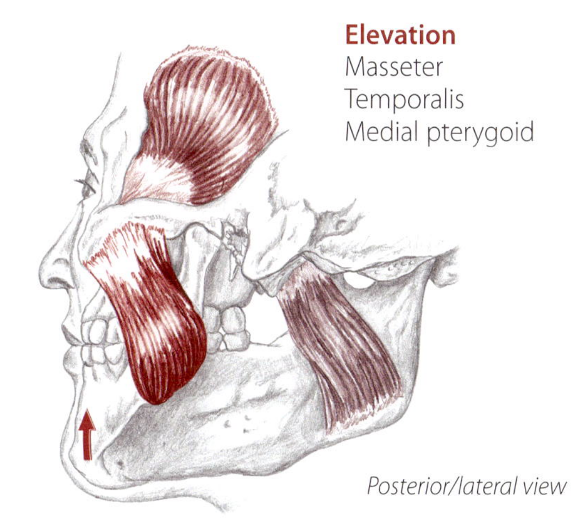
Posterior/lateral view

Elevation
Masseter
Temporalis
Medial pterygoid

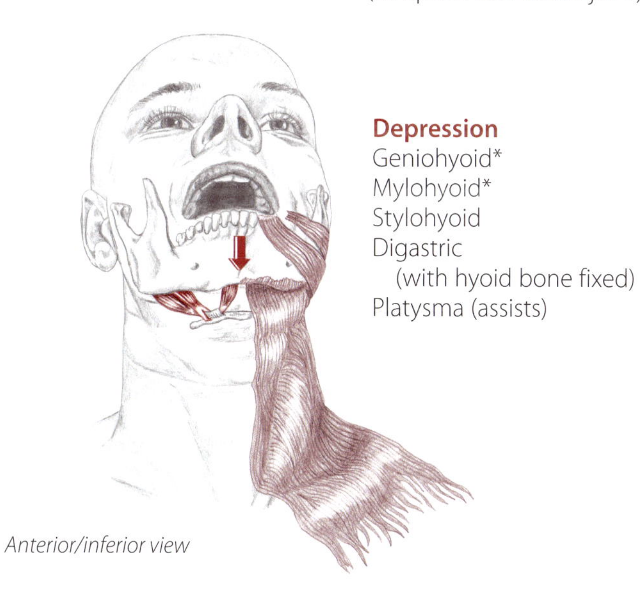
Anterior/inferior view

Depression
Geniohyoid*
Mylohyoid*
Stylohyoid
Digastric
(with hyoid bone fixed)
Platysma (assists)

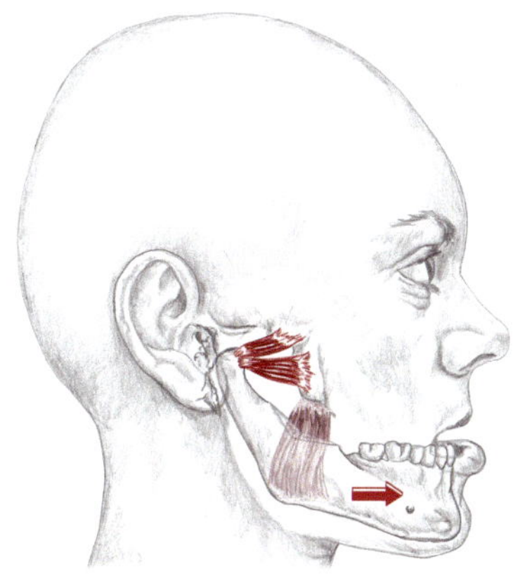

Protraction
Lateral pterygoid (bilaterally)
Medial pterygoid (bilaterally)

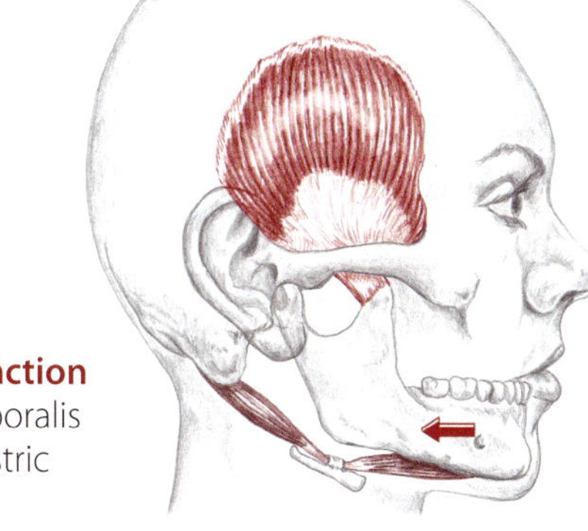

Retraction
Temporalis
Digastric

See p. 398 for synergists of lateral deviation of the mandible

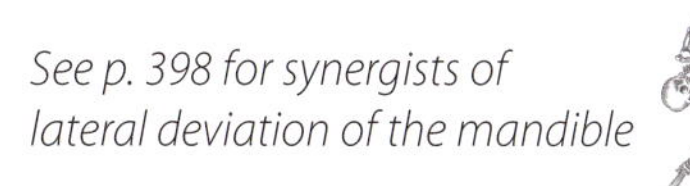

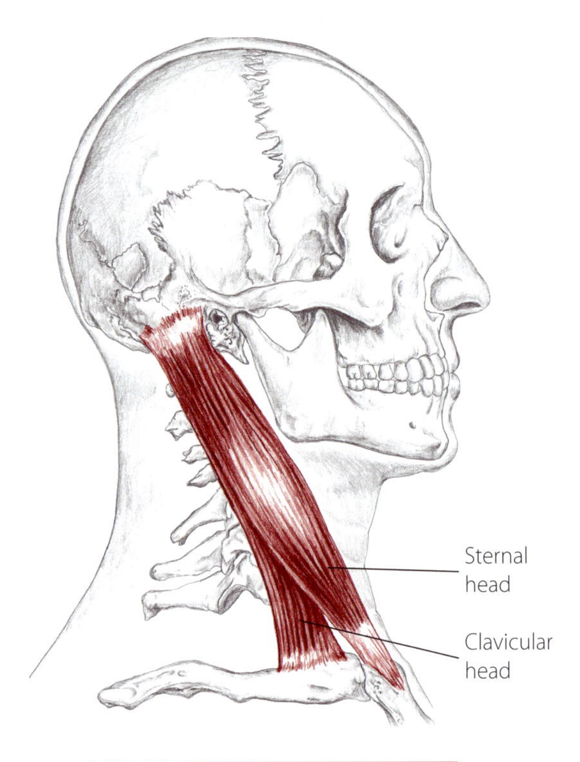

(5.32) Lateral view of sternocleidomastoid

Sternocleidomastoid

The sternocleidomastoid (SCM) is located on the lateral and anterior aspects of the neck. It has a large belly with two heads: a flat, clavicular head and a slender, sternal head (5.32). Both heads merge to attach behind the ear at the mastoid process. The carotid artery (p. 268) passes deep and medial to the SCM; the external jugular vein lies superficial to it. The SCM is also superficial, completely accessible and often visible when the head is turned to the side in Lord Byron-like fashion (5.33).

A *Unilaterally:*
Laterally flex the head and neck to the same side
Rotate the head and neck to the opposite side

Bilaterally:
Flex the neck
Assist in inhalation (elevation of the rib cage)

O *Sternal head:*
Top of manubrium
Clavicular head:
Medial one-third of the clavicle

I Mastoid process of temporal bone and the lateral portion of superior nuchal line of occiput

N Spinal accessory

(5.33) Lord Byron showing off his SCM

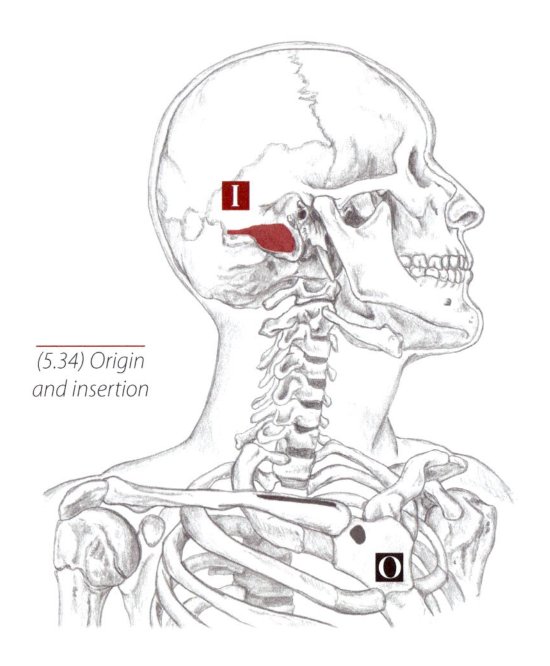

(5.34) Origin and insertion

The sternocleidomastoid and upper fibers of the trapezius (p. 76) begin as one muscle in the embryo and then split later on in development. The location of their attachments hints at their initial relationship: They form an almost continuous tendon along the superior nuchal line and mastoid process. Their other attachments are at either end of the clavicle.

sternocleidomastoid **ster**-no-**kli**-do-**mas**-toyd

Sternocleidomastoid

1) Supine with practitioner at head of the table. Locate the mastoid process of the temporal bone, the medial clavicle and the top of the sternum.
2) Draw a line between these landmarks to delineate the location of the SCM. Note how both SCMs form a "V" on the front of the neck.
3) Ask your partner to raise her head very slightly off the table as you palpate the SCM (5.35). It will usually protrude visibly. (To make the SCM more distinct, rotate the head slightly to the opposite side and then ask her to flex her neck.)
4) Palpate along the borders of the SCM, follow it behind the earlobe and then down to the clavicle and sternum (5.36). Sculpt around the skinny sternal tendon and the wider clavicular tendon.

With your partner relaxed, can you grasp the SCM between your fingers and outline its thickness and shape? How much space is between the clavicular attachments of the SCM and trapezius? It should be roughly two to three inches.

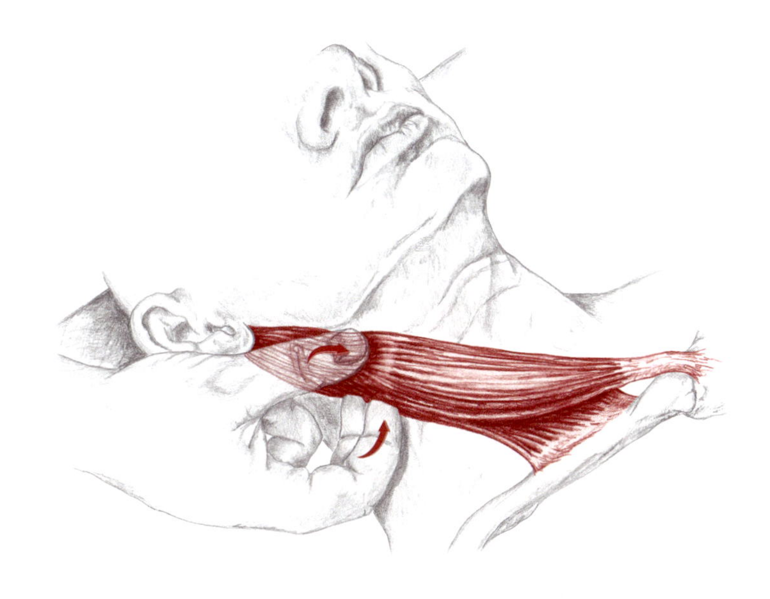

(5.35) Partner supine, flexing her head slightly as you feel the belly of the SCM contract

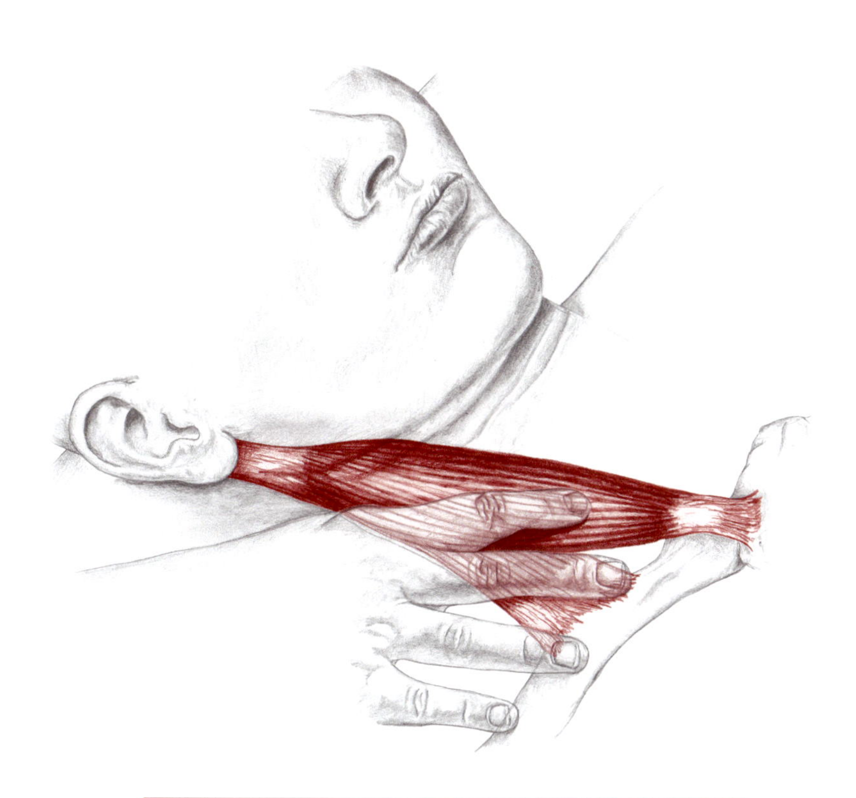

(5.36) Partner supine, following both heads to their separate tendons

Scalenes

Anterior
Middle
Posterior

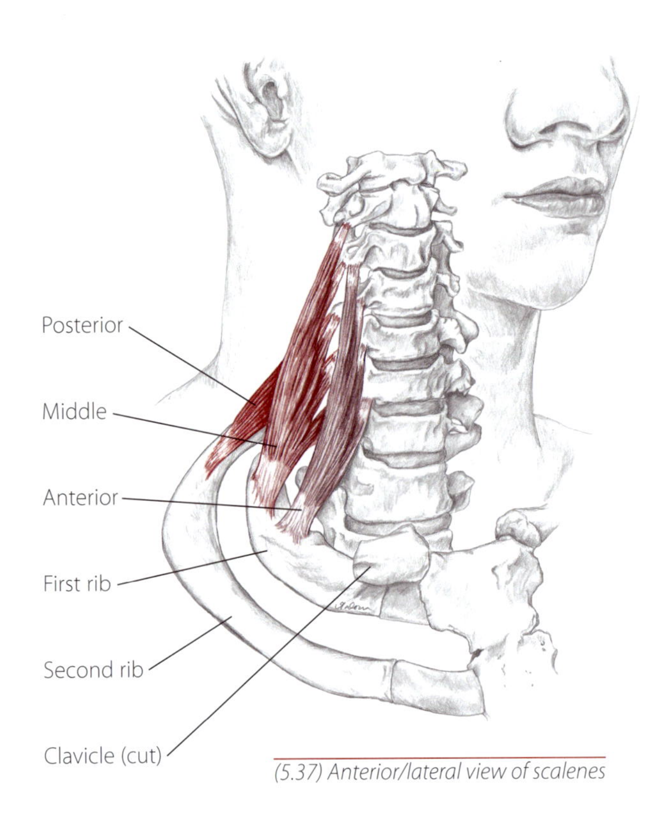

(5.37) Anterior/lateral view of scalenes

The three scalenes are sandwiched between the sternocleidomastoid and the anterior flap of the trapezius on the anterior, lateral neck. Their fibers begin at the side of the cervical vertebrae, dive underneath the clavicle and attach to the first and second ribs (5.37). During normal inhalation, the scalenes perform the vital task of elevating the upper ribs.

The **anterior scalene** (5.40) lies partially tucked beneath the sternocleidomastoid. The **middle scalene** (5.41) is slightly larger and lies lateral to the anterior scalene. Both muscle bellies are directly accessible. The smaller **posterior scalene** (5.42) is located between the middle scalene and levator scapula. The posterior scalene is positioned deeper than the other scalenes. Due to its small size and buried location, the posterior scalene can be difficult to distinguish from the surrounding bellies.

The large branches of the brachial plexus and subclavian artery pass through a small gap between the anterior and middle scalenes. Individual nerves of the brachial plexus may penetrate through or in front of the anterior scalene (5.38).

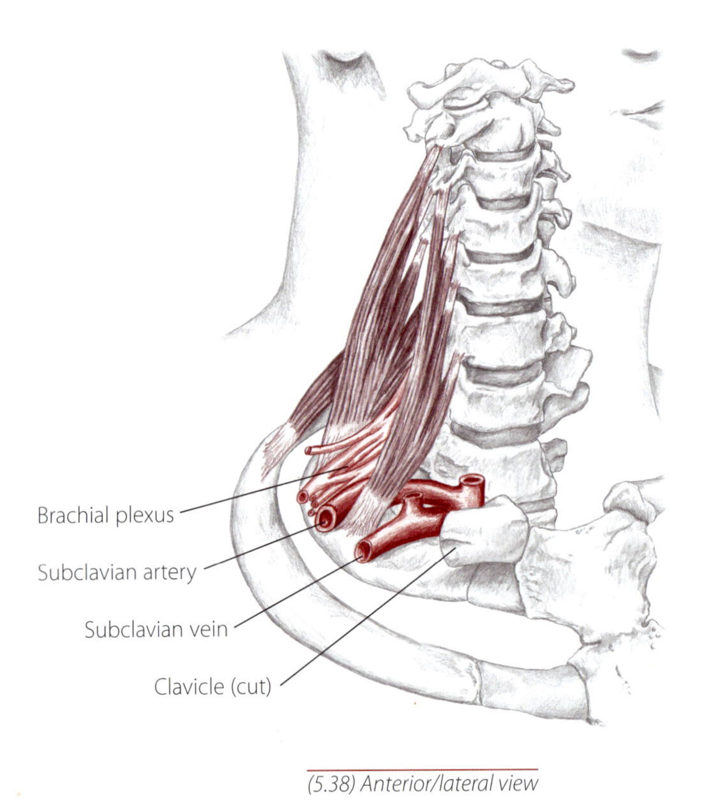

(5.38) Anterior/lateral view

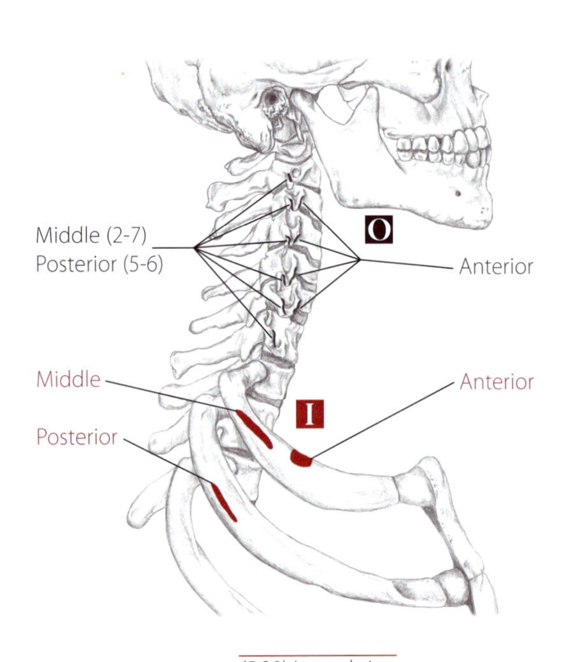

(5.39) Lateral view of origins and insertions

scalene **skay**-leen Grk. uneven

Actions of Scalenes

Unilaterally:

With the ribs fixed, laterally flex the head and neck to the same side (All)

Rotate head and neck to the opposite side (All)

Bilaterally:

Elevate the ribs during inhalation (All)

Flex the head and neck (Anterior)

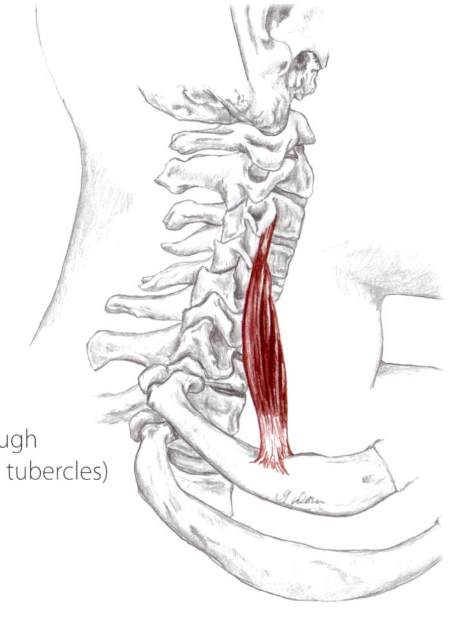

Anterior Scalene

O Transverse processes of third through sixth cervical vertebrae (anterior tubercles)

I First rib

N Cervical and brachial plexuses

(5.40) Lateral view of anterior scalene

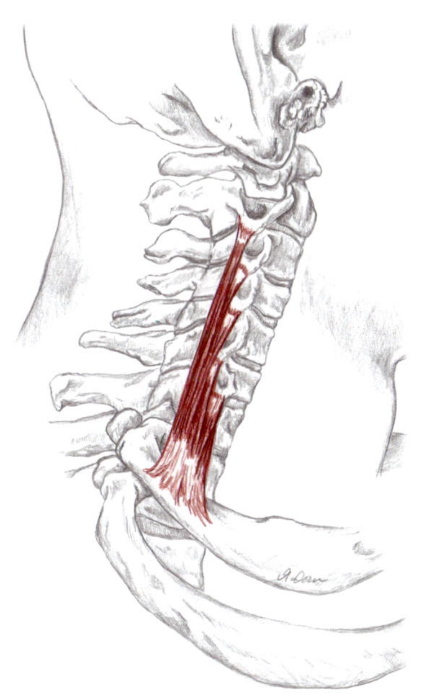

Middle Scalene

O Transverse processes of second through seventh cervical vertebrae (posterior tubercles)

I First rib

N Cervical and brachial plexuses

(5.41) Lateral view of middle scalene

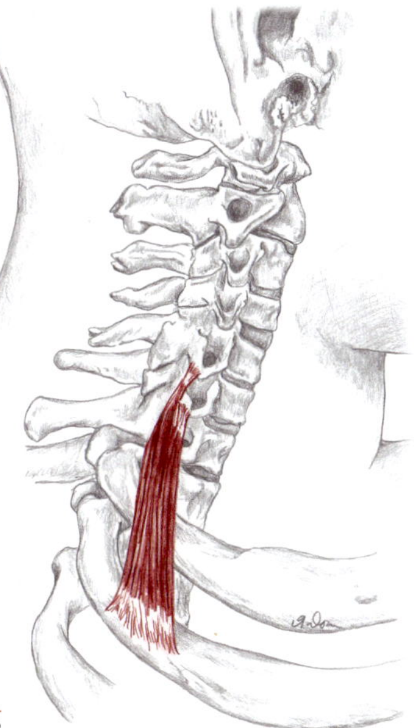

Posterior Scalene

O Transverse processes of fifth and sixth cervical vertebrae (posterior tubercles)

I Second rib

N Brachial plexus

(5.42) Lateral view of posterior scalene

Compression or impingement of the brachial plexus or one of its nerves can send a sharp, shooting sensation or numbness down the arm. If this should occur, immediately release and adjust your position posteriorly. Be sure to ask your partner for feedback while palpating the scalenes.

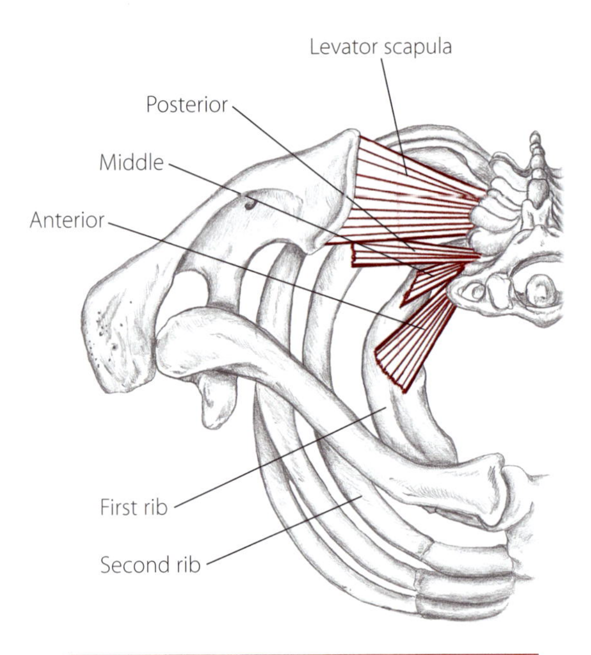

(5.43) Superior view showing the fiber direction of the scalenes and levator scapula. Muscles are not to scale.

Scalenes as a group

1) Partner supine, with practitioner at head of the table. Cradle the head (passively flexing it) to allow for easier palpation. Place your fingerpads along the anterior and lateral sides of the neck between the SCM and trapezius.
2) With the pads of your fingers, use gentle pressure to palpate the stringy, superficial muscle bellies in this triangle.

Are you between the SCM and trapezius? Ask your partner to inhale deeply into her upper chest. As she fully inhales, do you feel the muscles in this triangle contract (5.44)?

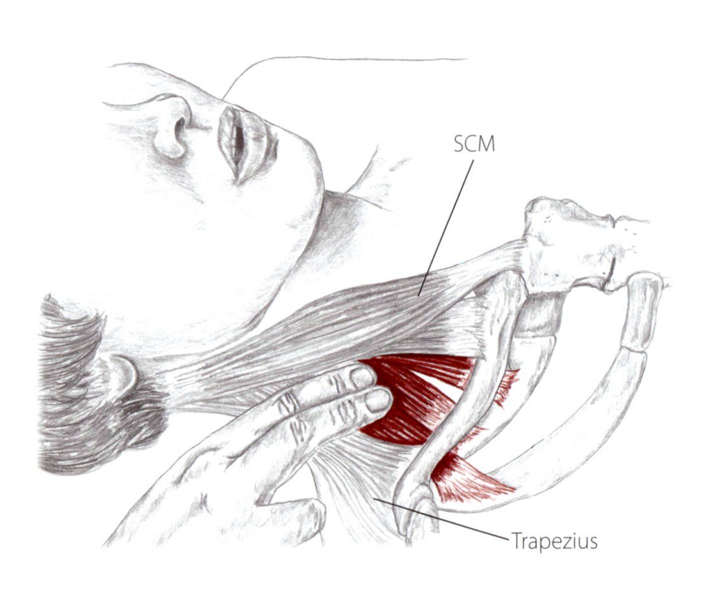

(5.44) Partner supine, feeling the scalenes contract as your partner inhales

Anterior and middle scalenes

1) Partner supine. Since the anterior scalene lies partially deep to the lateral edge of the SCM, rotate the head slightly to the opposite side to better expose it. Gently palpate under the SCM's lateral edge and roll across the belly of the anterior scalene (5.45).
2) Follow it inferiorly as it tucks under the clavicle.
3) Move laterally to explore the middle scalene, noting its similarly shaped belly (5.46).

Do the muscles you feel have a slender, stringy texture? If you follow them inferiorly, do they sink beneath the clavicle (in the direction of the ribs)? Can you follow them superiorly to the transverse processes of the cervical vertebrae? Ask your partner to flex her head slightly. Can you feel the scalenes contract?

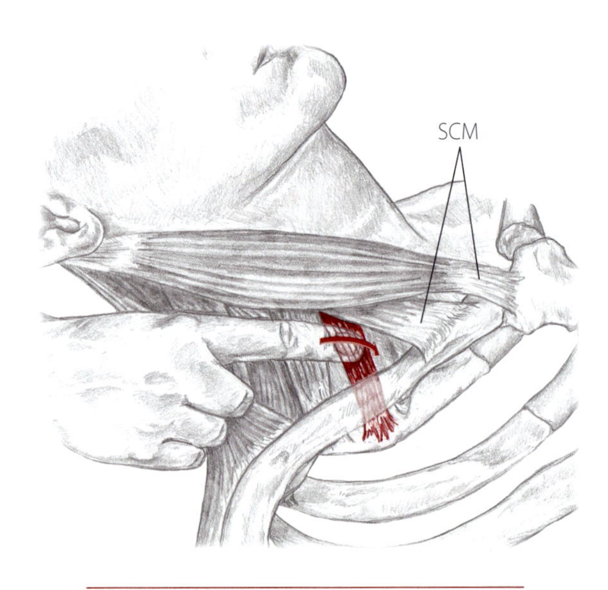

(5.45) Partner supine, locating the anterior scalene

Posterior scalene

1) Partner supine. The posterior scalene extends laterally off the neck and is squeezed between the middle scalene and levator scapula (p. 91).
2) Locate the middle scalene and the levator scapula. Place a finger between these bellies and sink inferiorly (5.47).
3) Slowly strum across the thin band of tissue running laterally from the transverse processes to the second rib.

To distinguish between the posterior scalene and levator scapula, locate the posterior scalene and ask your partner to slowly elevate her scapula. Since the posterior scalene does not create this action, there should be no contraction of its fibers. However, if you ask your partner to slowly inhale into her upper chest, you should feel the posterior scalene contract.

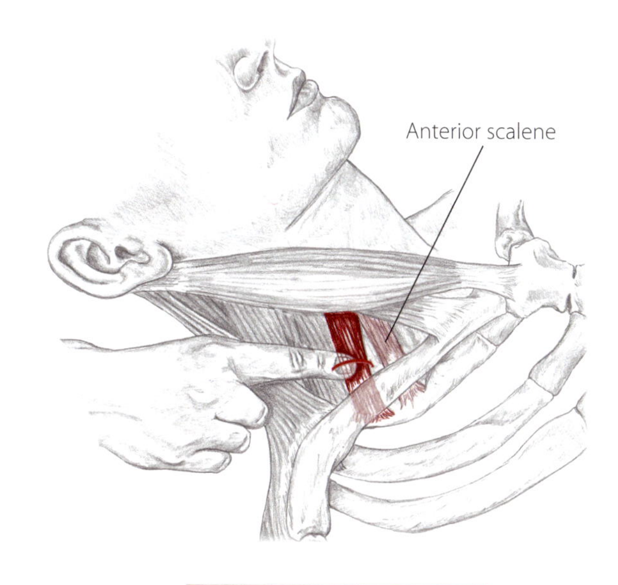

(5.46) Lateral view, partner supine, accessing the middle scalene

All scalenes

Partner prone. Begin by isolating the edge of the upper fibers of the trapezius (p. 77). Then curl your fingers around the anterior edge of the trapezius into the tissue of the lateral neck (5.48). The levator scapula will be just anterior to the trapezius followed by the posterior and middle scalenes.

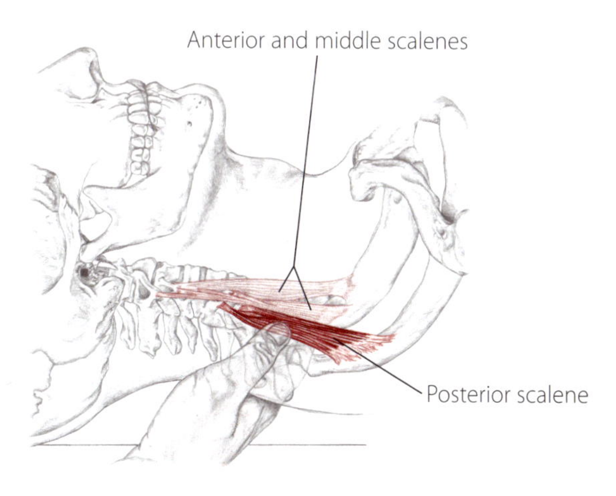

(5.47) Lateral view, partner supine, palpating the posterior scalene

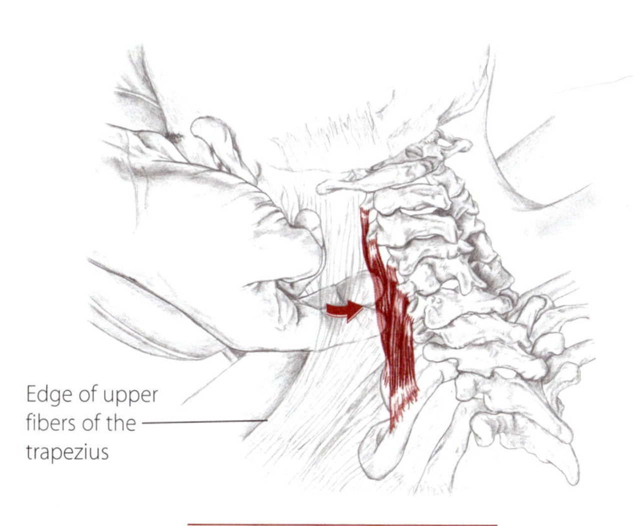

(5.48) Posterior/lateral view, partner prone, palpating the middle scalene

The existence of a fourth muscle, the scalene minimus, is one of several variations on the scalene muscle group. Present in roughly 40% of the population, the minimus often attaches from the sixth and seventh cervical vertebrae to the first rib or pleural dome of the lung. Lying inferior and deep to the anterior scalene, this muscle may nevertheless be quite strong.

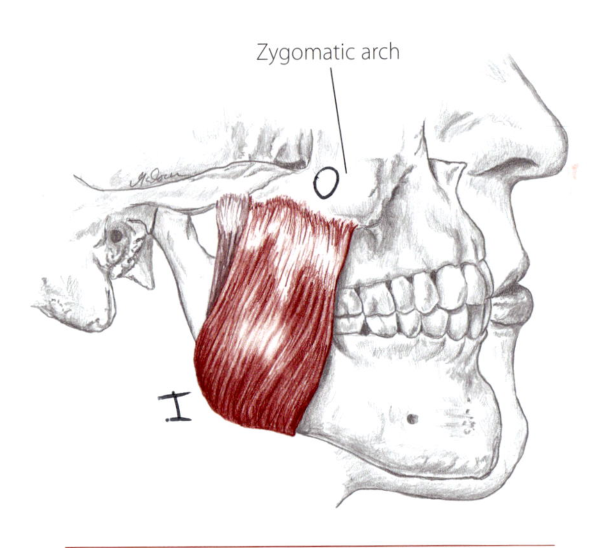

(5.49) Lateral view showing superficial head of masseter

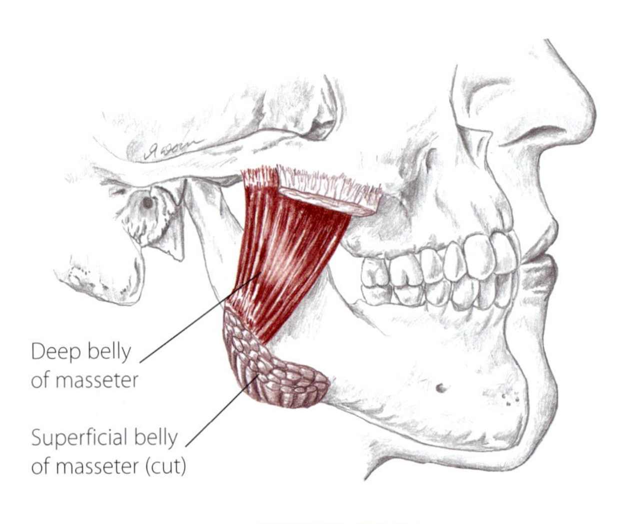

(5.50) Lateral view

Masseter

The masseter is the strongest muscle in the body relative to its size. The two masseters together exert a biting force of nearly one hundred-fifty pounds of pressure - enough to bite off a finger! The masseter is the primary chewing muscle and is used in speaking and swallowing.

Located on the side of the mandible, the square-shaped masseter is composed of two overlapping bellies. The superficial belly can be accessed from the face (5.49); the deep belly is palpable from inside the mouth (5.50). The masseter is situated deep to the parotid gland (p. 269), yet is easily palpable.

A Elevate the mandible (temporomandibular joint)
+ protrudes

O Zygomatic arch

I Angle and ramus of mandible

N Mandibular nerve via masseteric nerve

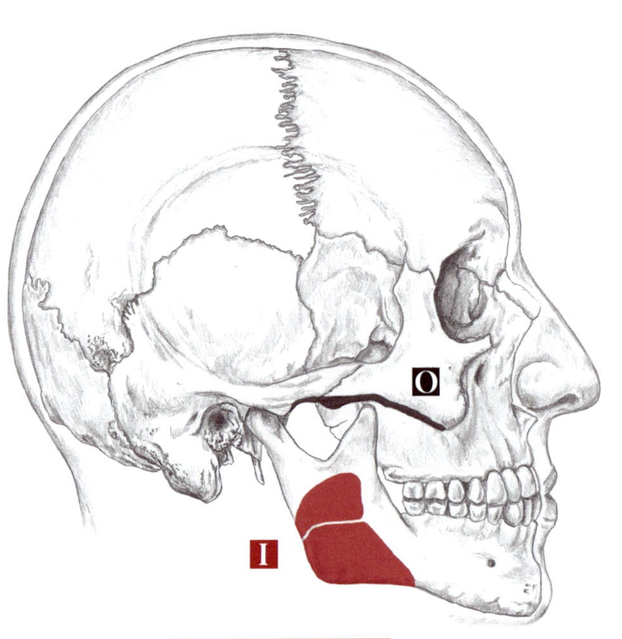

(5.51) Origin and insertion

masseter **mas**-se-ter Grk. chewer

Masseter

1) Partner supine. Locate the zygomatic arch and angle of the mandible.
2) Place your fingers between these bony landmarks and palpate the surface of the masseter.
3) Ask your partner to alternately clench and relax her jaw as you sculpt out the square shape of the belly (5.52). Clarify the masseter's fiber direction by strumming your fingers horizontally across its muscle fibers.
4) Now ask your partner to relax and try grasping the chunky bellies of the masseter (5.53).

As your partner clenches her jaw, can you outline the anterior edge of the masseter? If your partner opens her jaw as wide as possible, can you feel the tissue lengthen?

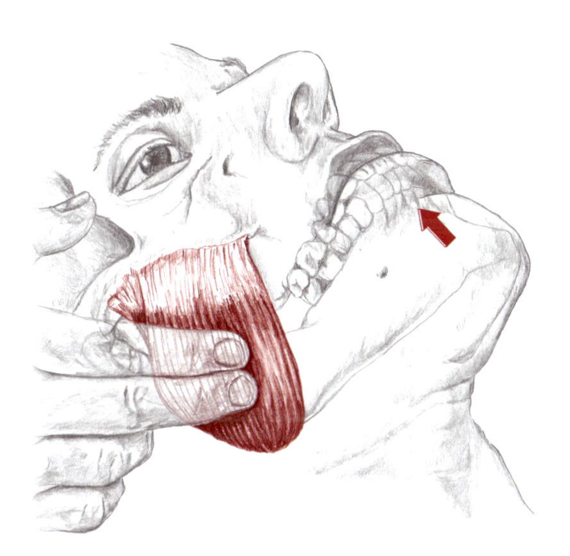

(5.52) Partner supine, clenching her jaw

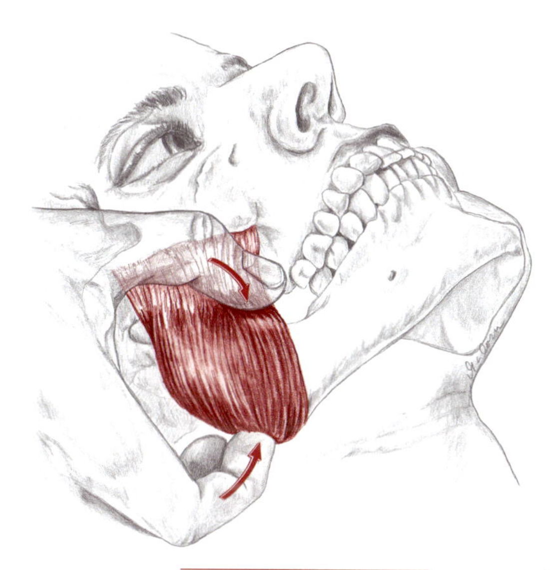

(5.53) Partner relaxes her jaw while you grasp the masseter

Temporalis

The temporalis muscle is located on the temporal aspect of the cranium. Its broad origin attaches to the frontal, temporal and parietal bones (5.54). Its fibers converge in a thick mass, reaching under the zygomatic arch to connect at the coronoid process. Though deep to the temporal fascia and artery, the temporalis is superficial and directly accessible.

A Elevate the mandible (temporomandibular joint)

Retract the mandible (temporomandibular joint)

O Temporal fossa and fascia

I Coronoid process of the mandible

N Deep temporal branch of mandibular nerve

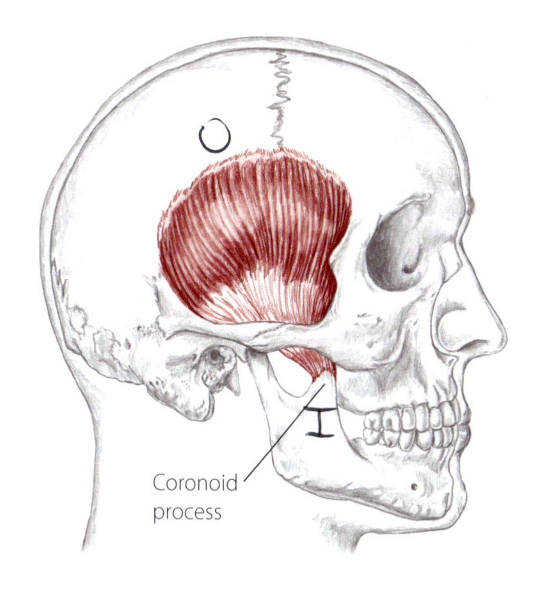

(5.54) Lateral view of temporalis

temporalis — **tem**-po-**ra**-lis — L. time, seen by the graying of hairs in this region

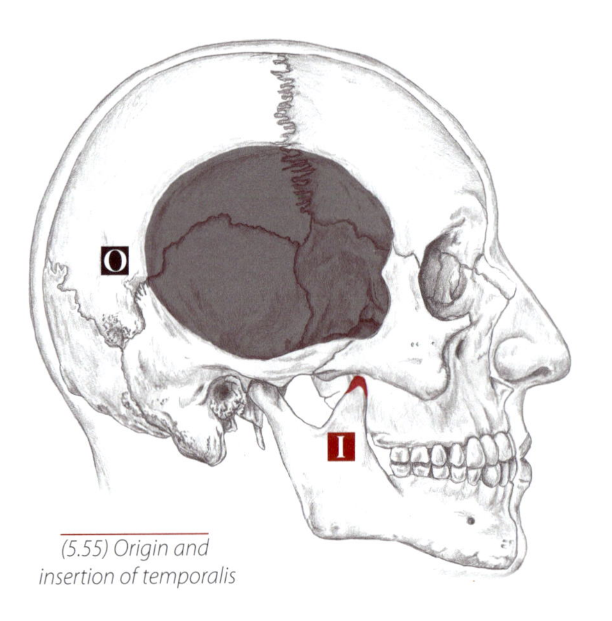

(5.55) Origin and insertion of temporalis

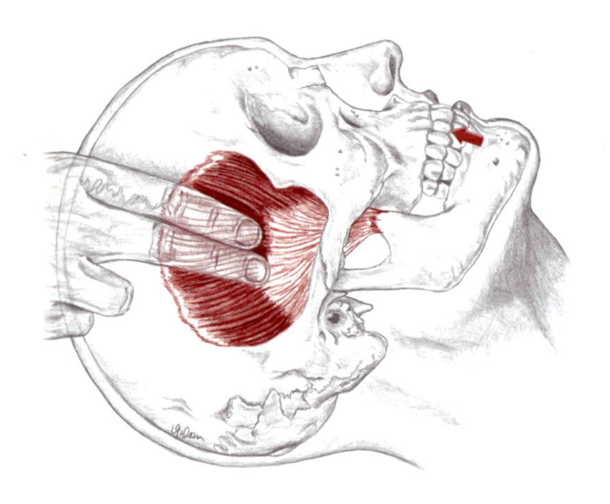

(5.56) Partner supine, feeling the temporalis as your partner clenches her jaw

Temporalis

1) Partner supine with practitioner at head of the table. Locate the zygomatic arch.
2) Place your fingerpads one inch superior to the arch and ask your partner to alternately clench and relax her jaw. Do you feel the strong temporalis contracting beneath your fingers (5.56)?
3) To locate the insertion site of the temporalis tendon, ask your partner to open her mouth wide.
4) Locate and explore the coronoid process (5.57). Although the coronoid process is easily accessed, you may not be able to isolate the tendon of the temporalis.

To outline the wide origin of the temporalis, place your fingers in various positions on the side of the head and ask your partner to alternately clench and relax her jaw. If your fingers are on the muscle, you will feel the temporalis fibers tighten and soften. If you are off the muscle, you will feel nothing.

When exploring the muscle belly, are you superior to the zygomatic arch on the side of the head? Can you discern the direction of the muscle fibers and feel them converge?

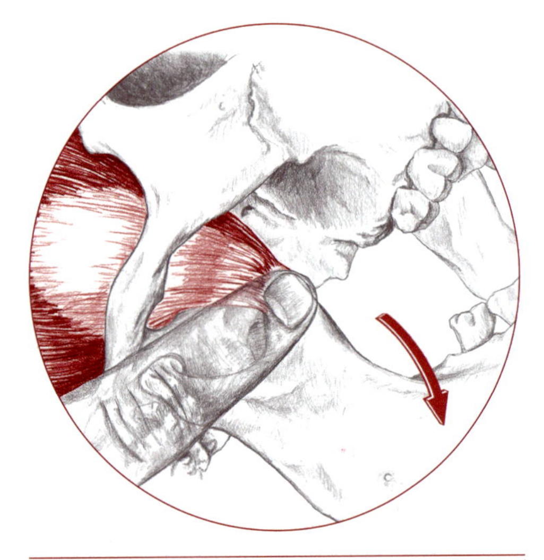

(5.57) Isolating the temporalis tendon at the coronoid process of the mandible, with your partner's jaw open

Suprahyoids and Digastric

The suprahyoids **(geniohyoid, mylohyoid and stylohyoid)** form a wall of muscle along the underside of the jaw (5.58, 5.59). Stretching from the edge of the mandible to the hyoid bone, they lie inferior to the glossus muscles (the muscles of the tongue).

Even though each of the three suprahyoids is quite small, collectively they affect the tongue and hyoid bone and are important in chewing, swallowing and speaking. They are partially deep to the digastric muscle, yet are accessible. The suprahyoid bellies cannot be individually distinguished.

The long, round **digastric** muscle is composed of a posterior and an anterior belly. The posterior belly runs from the mastoid process to the hyoid bone (penetrating through the stylohyoid) and then loops through a tendinous sling on the hyoid's anterior surface. It continues on as the anterior belly to attach at the underside of the mandible (5.60). Both bellies are superficial, yet difficult to distinguish from the deeper suprahyoid muscles.

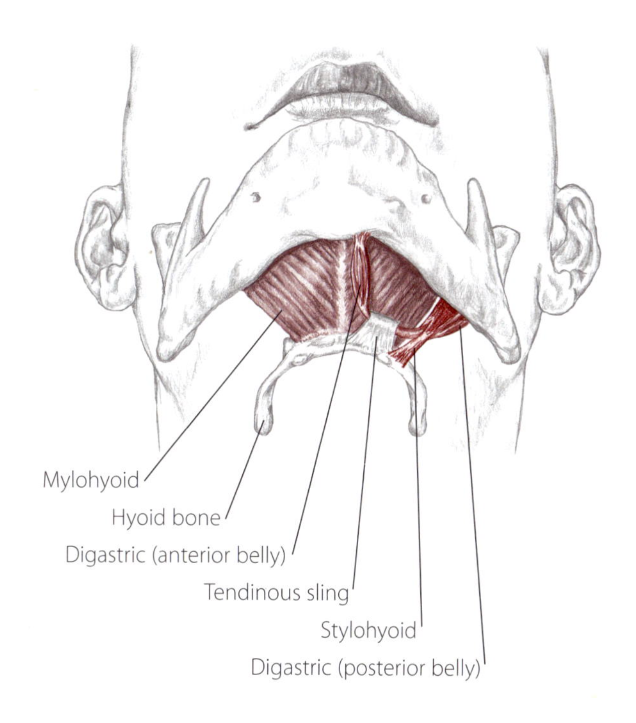

(5.58) Anterior/inferior view; geniohyoid is deep to mylohyoid

Suprahyoids

A Elevate hyoid and tongue

Depress mandible (temporomandibular joint)

O *Geniohyoid, Mylohyoid:*
Underside of mandible

Stylohyoid:
Styloid process

I Hyoid bone

N *Geniohyoid:* Hypoglossal
Mylohyoid: Mylohyoid
Stylohyoid: Facial

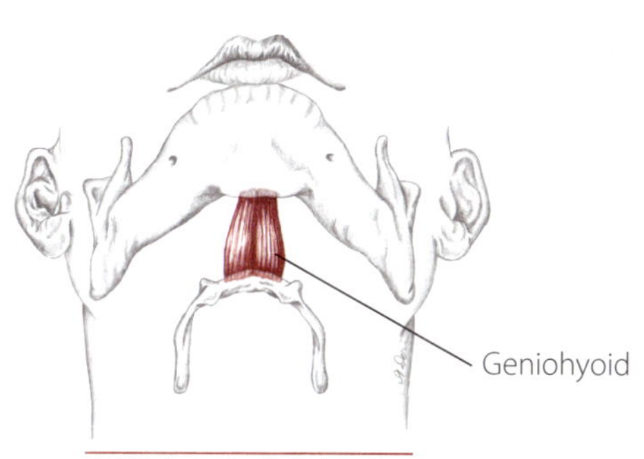

(5.59) Anterior/inferior view

Digastric

A With hyoid bone fixed, it depresses the mandible (temporomandibular joint)

With mandible fixed, it elevates the hyoid bone

Retracts the mandible (TM joint)

O Mastoid process (deep to sternocleidomastoid and splenius capitis)

I Inferior border of the mandible

N Mylohyoid and facial

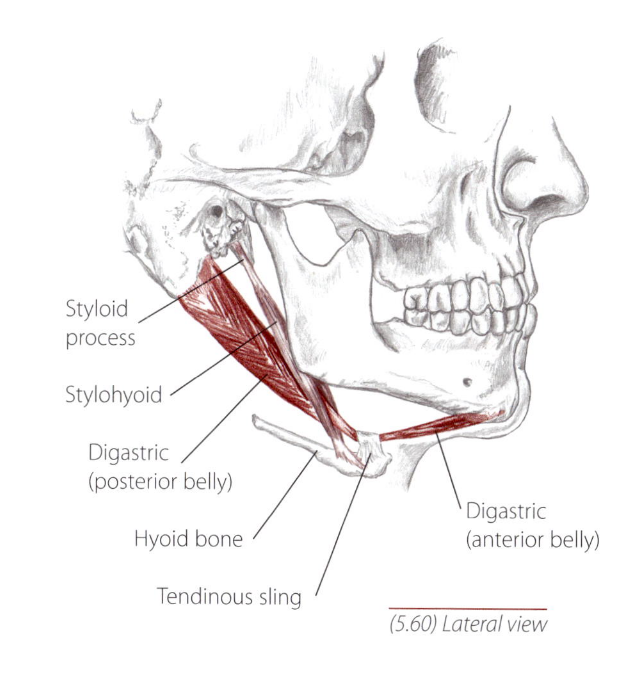

(5.60) Lateral view

geniohyoid	**je**-ne-o-**hi**-oyd	Grk. *genion*, chin
glossus	**glah**-sis	Grk. tongue

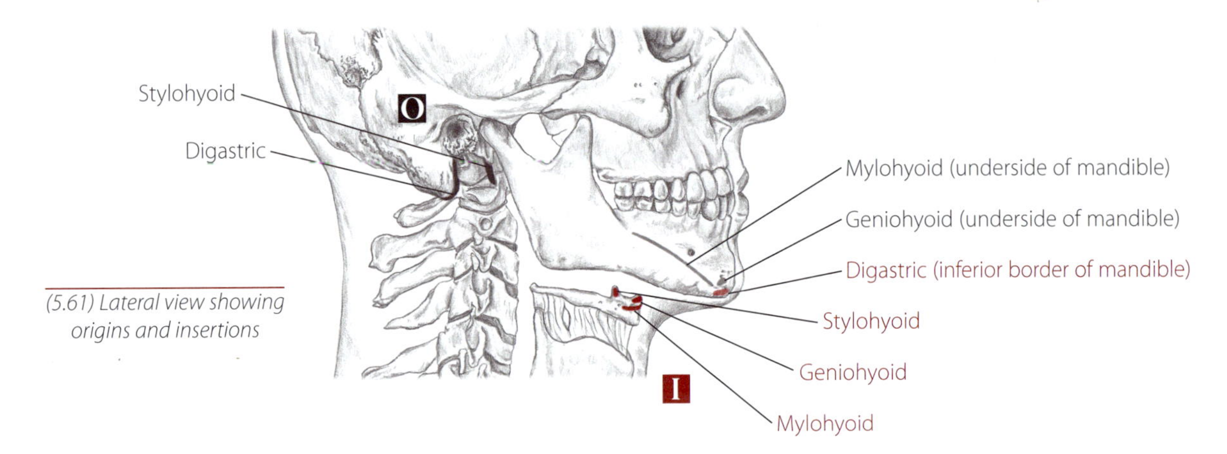

(5.61) Lateral view showing origins and insertions

Suprahyoids

1) Supine. With your partner's jaw closed, place your finger along the underside of the mandible.
2) Contract the suprahyoids by asking your partner to press the tip of her tongue firmly against the roof of her mouth. Note how this action forms a wall of taut muscle along the base of the mandible (jawline). Follow it as it extends down to the hyoid bone (5.62).
3) With the tongue relaxed, palpate the flat surface of the suprahyoid tissues, distinguishing them from the lumpy texture of the submandibular gland (p. 269).

If you place a fingerpad underneath the tip of the chin and ask your partner to gently depress her mandible into your finger, do the suprahyoids contract? Also, ask your partner to swallow as you palpate the suprahyoids. Do these tissues contract?

Digastric

1) Partner supine with practitioner at head of table. Locate the mastoid process of the temporal bone and the hyoid bone.
2) Draw an imaginary line between these points. Using your index finger, palpate along this line for the skinny, posterior digastric (5.63).
3) Draw an imaginary line between the hyoid bone to the underside of the chin and palpate for its anterior belly.
4) To feel the digastric contract, place your finger under the chin and ask your partner to try to open her mouth against your gentle resistance. This contraction will sometimes allow both of the digastric bellies to be located more easily.

Is the muscle you are palpating superficial and pencil-width? Does it extend from the mastoid process to the hyoid bone to the chin?

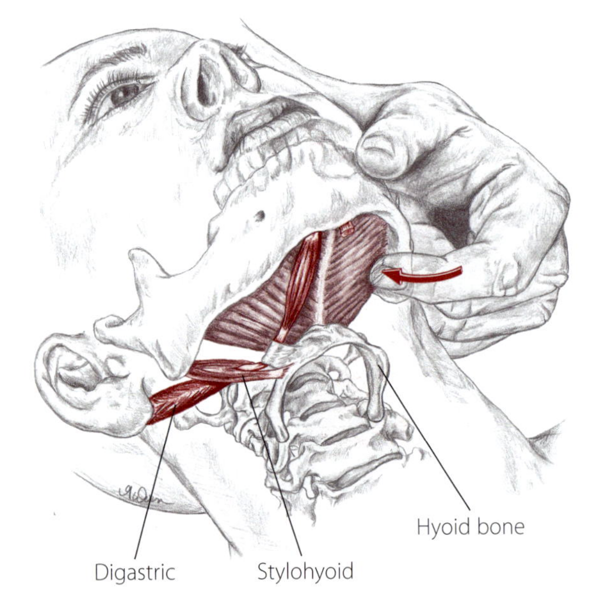

(5.62) Partner supine, curling the finger around the mandible to palpate the mylohyoid

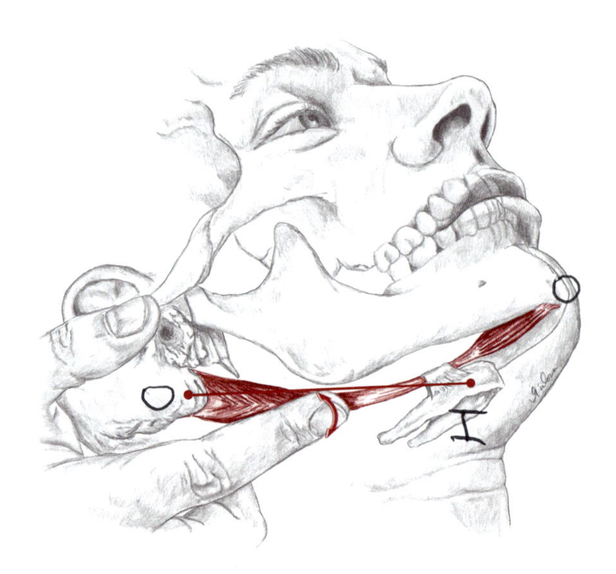

(5.63) Isolating the digastric between the mastoid process and the hyoid bone

digastric	di-**gas**-trik	Grk. double-bellied
mylohyoid	**my**-lo-**hi**-oyd	Grk. *myle*, mill
stylohyoid	**sti**-lo-**hi**-oyd	

Infrahyoids

The infrahyoids are located on the anterior neck, superficial to the trachea (5.64). All four muscles are thin and delicate and function as antagonists to the suprahyoids. The superficial **sternohyoid** and **sternothyroid** are layered just to the side of the trachea and, although difficult to distinguish individually, are directly accessible. Deep to these two muscles is the **thyrohyoid**. As its name suggests, it spans from the thyroid cartilage to the hyoid bone.

The **omohyoid** is perhaps the most bizarre muscle in the body. It has a skinny, ribbonlike belly which begins at the hyoid bone, passes underneath the SCM and scalenes, and attaches to the scapula. Aside from depressing the hyoid, the omohyoid tightens the fascia of the neck and dilates the internal jugular vein. Because of its depth and slender belly, the omohyoid is mostly inaccessible.

A Depress the hyoid bone and thyroid cartilage

O *Sternohyoid and Sternothyroid:* Top of manubrium

Thyrohyoid: Thyroid cartilage

Omohyoid: Superior border of the scapula

I *Sternohyoid, Omohyoid and Thyrohyoid:* Hyoid bone

Sternothyroid: Thyroid cartilage

N Upper cervical

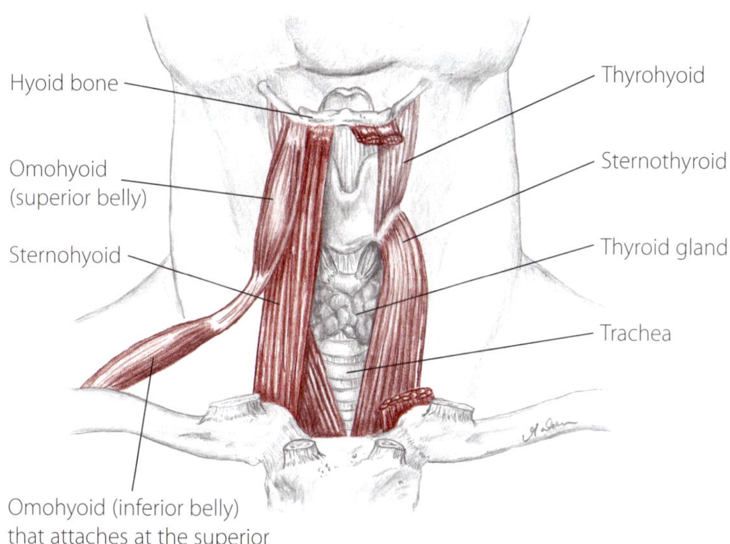

(5.64) Anterior view of neck, SCM removed

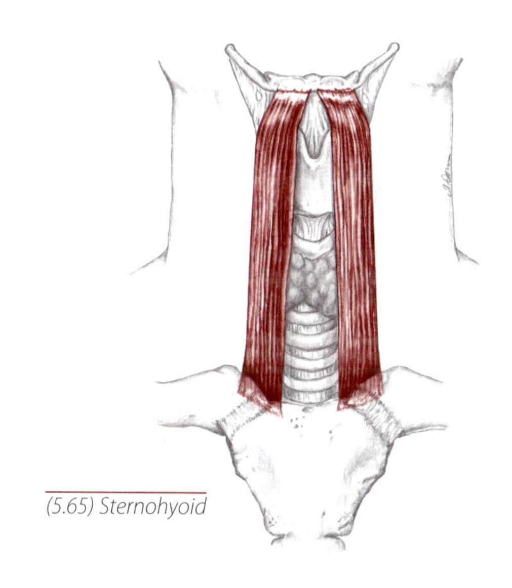

(5.65) Sternohyoid

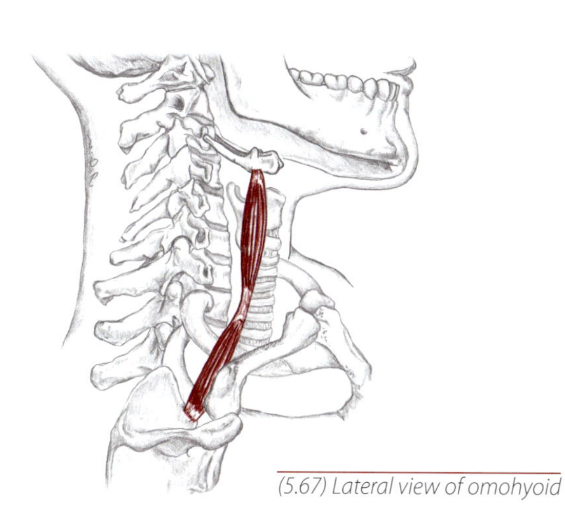

(5.67) Lateral view of omohyoid

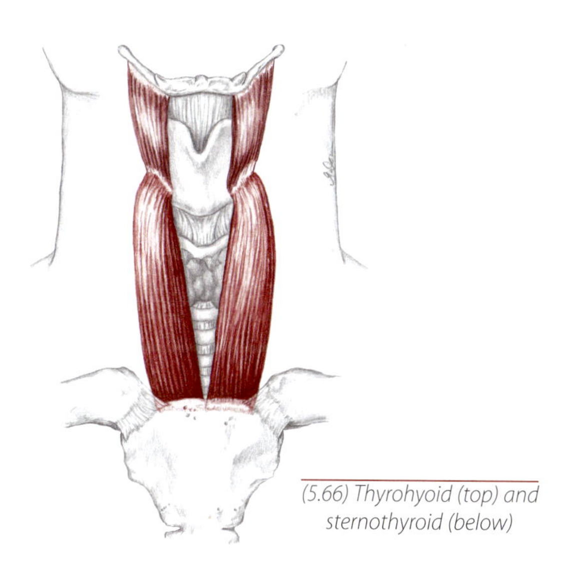

(5.66) Thyrohyoid (top) and sternothyroid (below)

omohyoid	**o**-mo-**hi**-oyd	Grk. *omos*, shoulder
sternohyoid	**ster**-no-**hi**-oyd	
sternothyroid	**ster**-no-**thi**-royd	

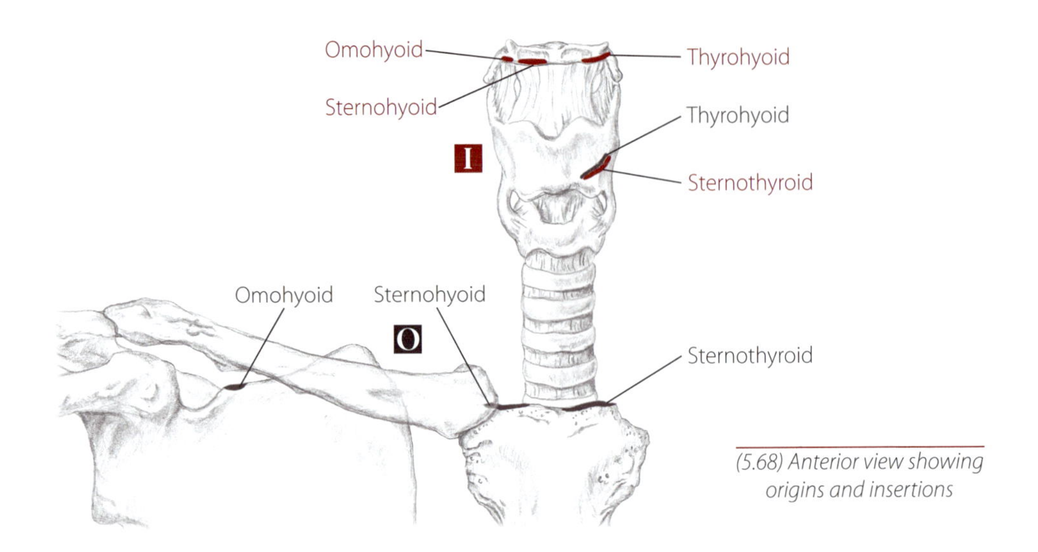

(5.68) Anterior view showing origins and insertions

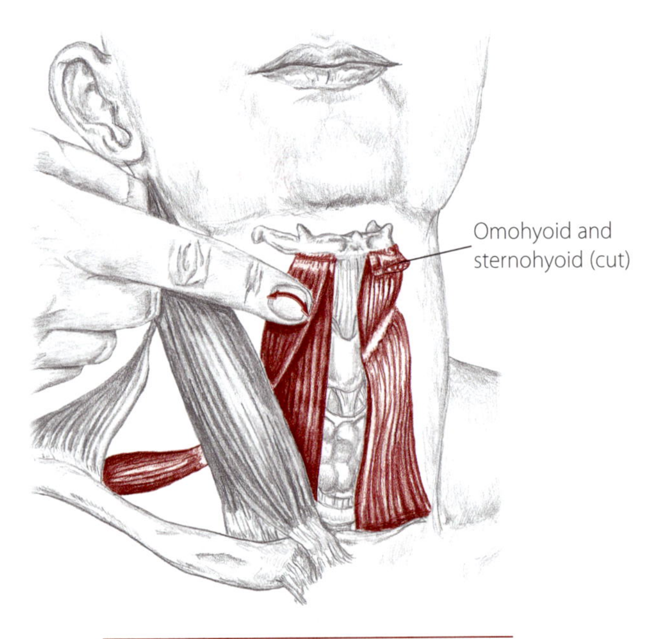

(5.69) Partner supine, palpating the infrahyoids

Sternohyoid and sternothyroid

1) A hand (or two) on the front of the neck can be unnerving, so put your partner at ease by exploring with one hand at a time and then using only one fingerpad. Also, to avoid irritating the thyroid gland, explore only the superior half of these muscles.
2) Locate the surface of the trachea, just below the thyroid cartilage (Adam's apple). With one fingerpad, slide lateral to the trachea and gently explore the thin tissue lying superficial to the windpipe. Try to roll your finger across the thin bellies of the infrahyoids (5.69).
3) Ask your partner to tighten the muscles of the anterior neck. Sometimes this isometric contraction will make the infrahyoids quite solid and easily palpable.

A giraffe's trachea is formed by more than a hundred tracheal rings and is the cause of a unique breathing problem. Because of the windpipe's length, each inhalation brings in nearly two gallons of air that never reach the lungs. (Comparatively, a resting human takes in two gallons of air every minute.) To counteract this anatomical dead space, a giraffe is equipped with a massive lung capacity of nearly twelve gallons. It has also been suggested that a giraffe's trachea may serve as a cooling device. As the trachea is filled with moist air, it cools the nearby blood vessels that travel up to the brain.

Platysma

The platysma is a thin, superficial sheath spanning the anterior neck from the mandible to the chest (5.70). The platysma and other facial muscles are integumentary muscles. Instead of connecting to bones, these muscles are embedded in the superficial fascia and attach to the skin and overlying muscle. The platysma's claim to fame is its ability to create the infamous "Creature from the Black Lagoon" expression.

A Assist in depression of the mandible (temporomandibular joint)

Tighten the fascia of the neck

O Fascia covering superior part of pectoralis major

I Base of mandible, skin of lower part of face

N Facial

1) Partner supine. Ask your partner to jut his head anteriorly and protract his jaw (5.71). Then ask him to tighten the tissue on the front of his neck.
2) Explore this thin sheet of muscle from the mandible down to the upper chest. Note any "flaps" the platysma forms along the lateral side of the neck.

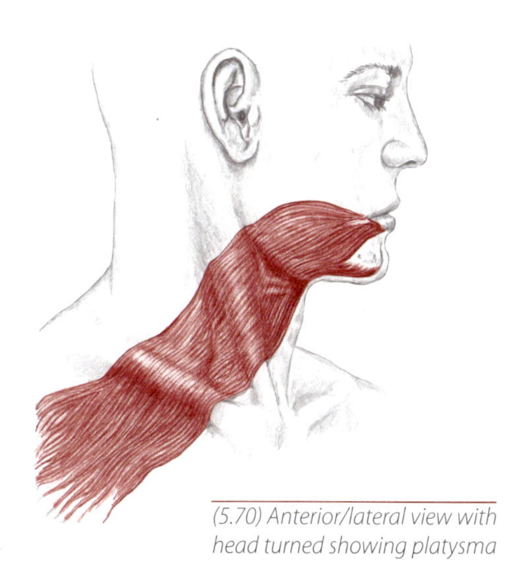

(5.70) Anterior/lateral view with head turned showing platysma

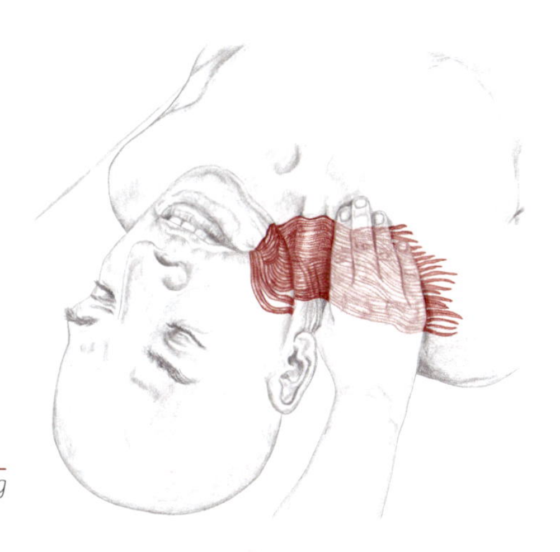

(5.71) Partner supine, contracting the tissue of the anterior neck

Occipitofrontalis

The occipitofrontalis is the muscle primarily responsible for raising the forehead into an expression of surprise. It is a unique muscle with four thin bellies - two occipital bellies located on the back of the head and two frontal bellies on the forehead. The four bellies are joined by the galea aponeurotica, a broad sheath of connective tissue stretching across the top of the cranium (5.72). Although the occipitofrontalis is superficial, its thin fibers cannot be isolated.

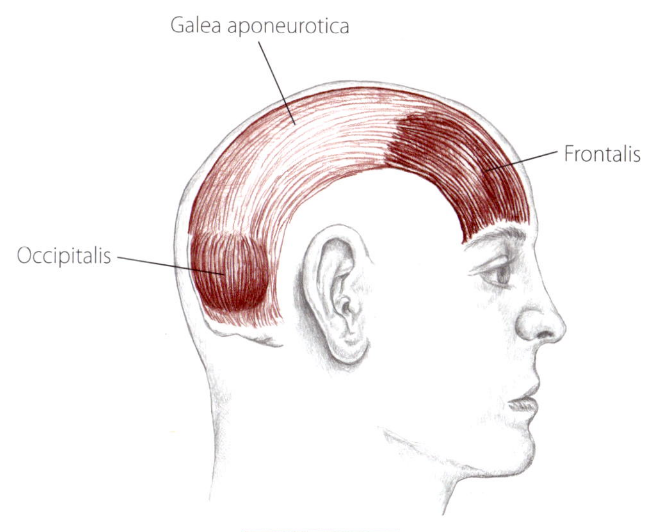

(5.72) Lateral view

panniculus carnosus	pan-**ik**-u-lus car-**no**-sis	L. small, fleshy garment
platysma	pla-**tiz**-ma	Grk. plate

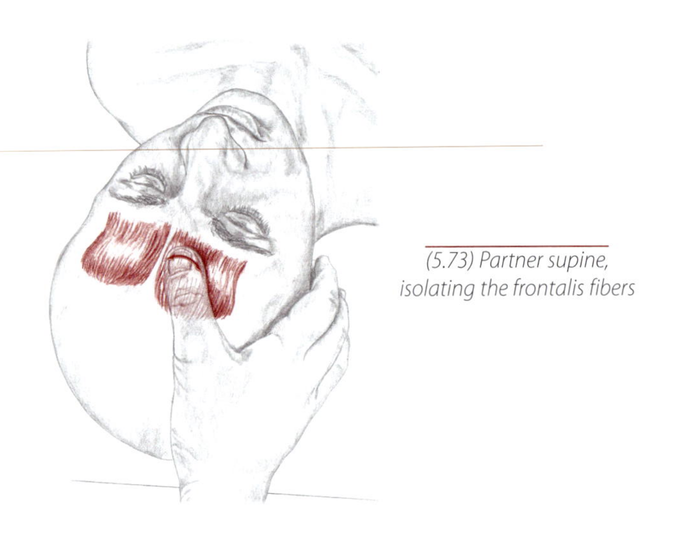

(5.73) Partner supine, isolating the frontalis fibers

A *Frontalis:*
Raise the eyebrows and wrinkle the forehead

Occipitalis:
Anchor and retract the galea posteriorly

O *Both:* Galea aponeurotica

I *Frontalis:* Skin over the eyebrows
Occipitalis: Superior nuchal line of the occiput

N Facial

Frontalis fibers

Partner supine. Place your fingers on the forehead and ask your partner to raise his eyebrows (5.73). Do you feel the tissue of the forehead contract?

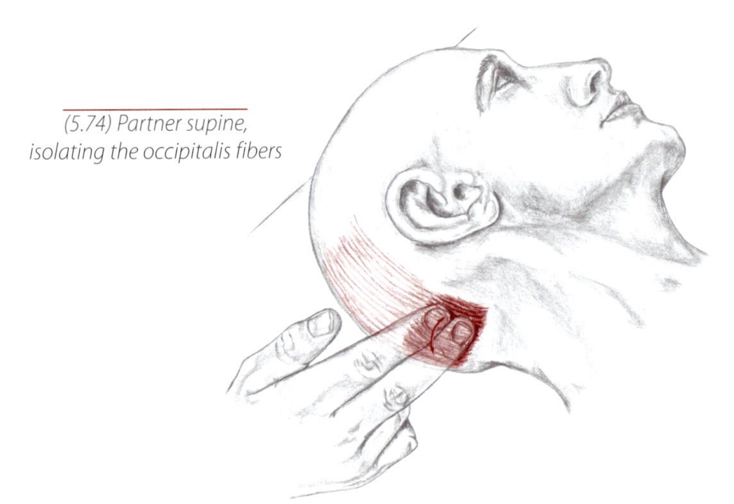

(5.74) Partner supine, isolating the occipitalis fibers

Occipitalis fibers

Supine or prone. Locate the superior nuchal line of the occiput (p. 238) and slide your fingers one inch superiorly to isolate the region of the oval occipital bellies (5.74).

Muscles that express emotion are called mimetic muscles. There are thirty pairs of mimetic muscles on the human face - more than on any other animal. Collectively they can form an incredible range of expressions from crinkling the eyebrows in confusion (corrugator supercili), flaring the nostrils in anger (levator labii superioris), puckering the lips for a kiss (orbicularis oris) or raising the chin to pout (mentalis). Smiling is generated by eight muscles while frowning requires up to twenty.

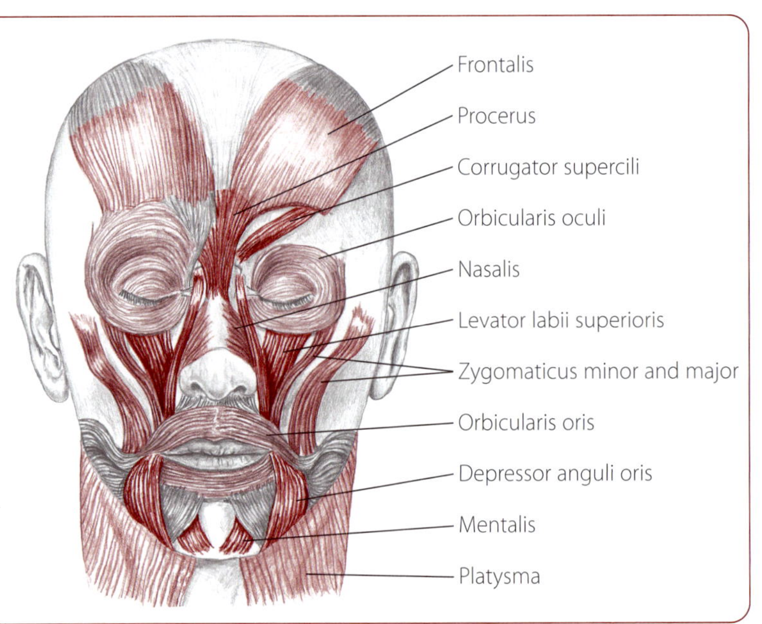

occipitofrontalis	ok-**sip**-i-to-fron-**ta**-lis	
mentalis	men-**tal**-is	L. chin

Medial and Lateral Pterygoids

The medial and lateral pterygoids assist the masseter and temporalis with movement of the mandible. The **medial pterygoid** helps to elevate the mandible, while the lateral pterygoid protracts it. The medial pterygoid is located on the *interior* side of the mandible (5.75) and its shape and position mirror the *exterior* masseter muscle (p. 256).

The **lateral pterygoid** has horizontal fibers that extend from the sphenoid bone to the joint capsule and articular disc of the temporomandibular joint (5.77). Portions of the pterygoids can be accessed from both inside and outside the mouth.

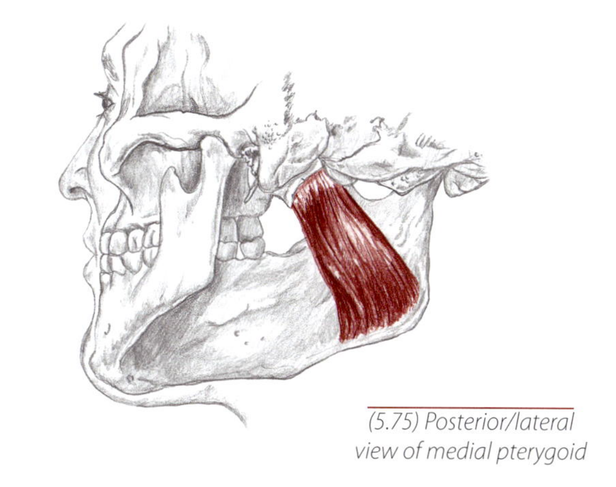

(5.75) Posterior/lateral view of medial pterygoid

Medial Pterygoid

A *Unilaterally:*
Laterally deviate the mandible to the opposite side

Bilaterally:
Elevate the mandible
Protract the mandible

O Medial surface of lateral pterygoid plate of sphenoid bone and tuberosity of maxilla

I Medial surface of ramus of the mandible

N Mandibular

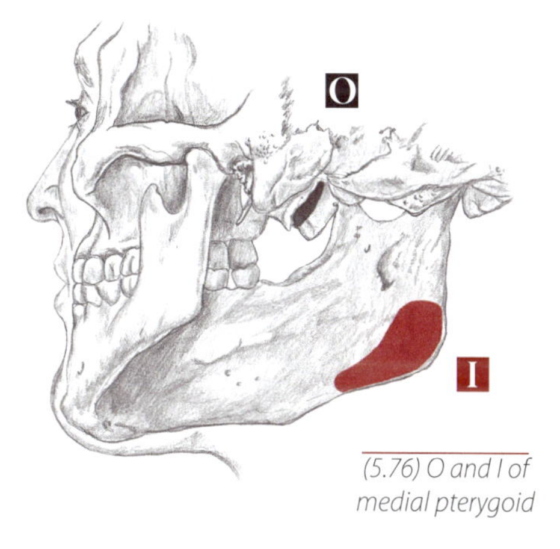

(5.76) O and I of medial pterygoid

Lateral Pterygoid

A *Unilaterally:*
Laterally deviate the mandible to the opposite side

Bilaterally:
Protract the mandible

O *Superior head:*
Infratemporal surface and crest of greater wing of sphenoid bone
Inferior head:
Lateral surface of lateral pterygoid plate of sphenoid bone

I Articular disc and capsule of temporomandibular joint, neck of mandible

N Mandibular

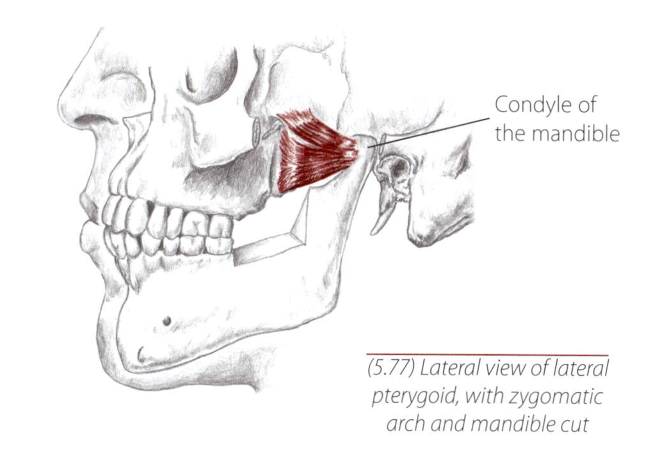

(5.77) Lateral view of lateral pterygoid, with zygomatic arch and mandible cut

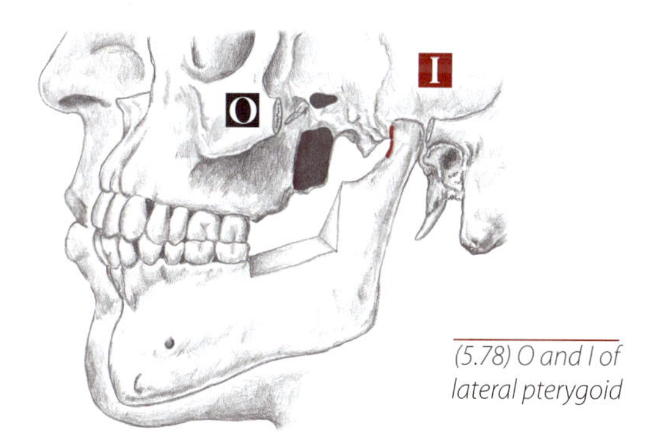

(5.78) O and I of lateral pterygoid

pterygoid ter-i-**goyd** Grk. wing-shaped

Longus Capitis and Longus Colli

Tucked between the trachea and the cervical vertebrae are two small muscles - **longus capitis** and **longus colli** (5.79, 5.80). Attaching from the anterior surface of the cervical vertebrae to the occiput and atlas, they laterally flex, rotate and flex the head and neck. They also help to reduce the lordotic curve of the cervical vertebrae. Each muscle has a multibranched design similar to that of the erector spinae muscles of the back.

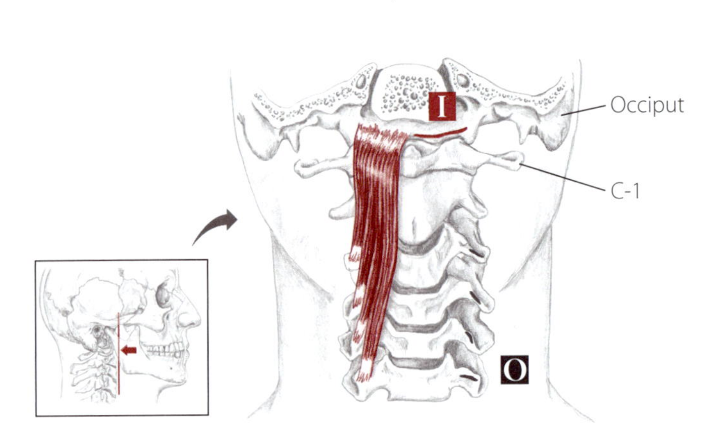

(5.79) Anterior view of cervical vertebrae showing longus capitis

Longus Capitis

A *Unilaterally:*

Laterally flex the head and neck to the same side

Rotate the head and neck to the same side

Bilaterally:

Flex the head and neck

O Transverse processes of C-3 through C-6

I Inferior surface of occiput

N Ventral rami of C-1 to C-4

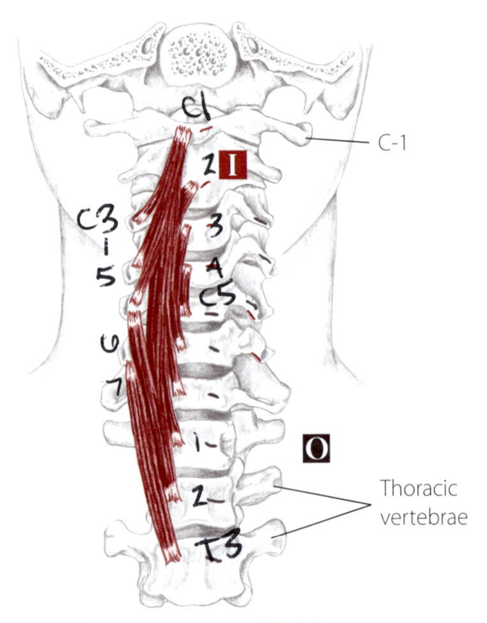

(5.80) Anterior view of cervical vertebrae showing longus colli

Longus Colli

A *Unilaterally:*

Laterally flex the head and neck to the same side

Rotate the head and neck to the same side

Bilaterally:

Flex the head and neck

O Bodies of C-5 through T-3, transverse processes of C-3 through C-5

I Tubercle on anterior arch of the atlas; bodies of the axis, C-3 and C-4; transverse processes of C-5 and C-6

N Ventral rami of C-1 to C-7

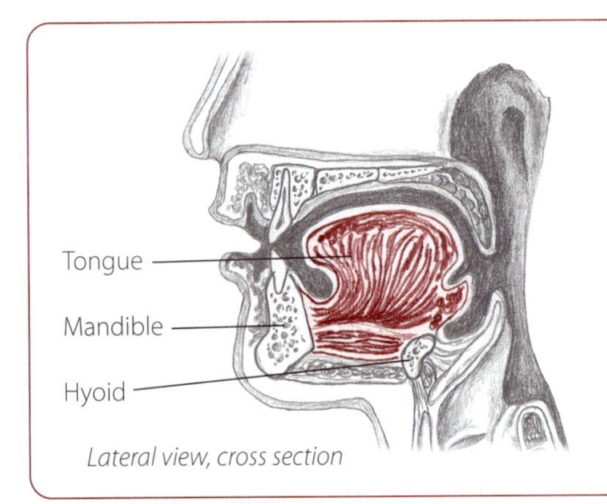

Lateral view, cross section

Muscles of the Tongue

There are two groups of muscles that coordinate the tongue: the glossus muscles and the intrinsic muscles. The three glossus muscles attach to the hyoid and other bones and move the tongue during chewing and swallowing. Three other intrinsic muscles of the tongue interweave with each other and are responsible for changing the tongue's shape during speech. As the tongue is basically a bag of fluid with a constant volume, these intrinsic muscles mold and twist it in the same way you might bend and shape a water balloon.

Other Structures of the Head, Neck and Face

There are several accessible arteries, glands and nerves in the head, neck and face (5.81). Many are superficial and delicate and should therefore be palpated gently. It is advisable to locate and explore these structures on yourself before palpating them on a partner.

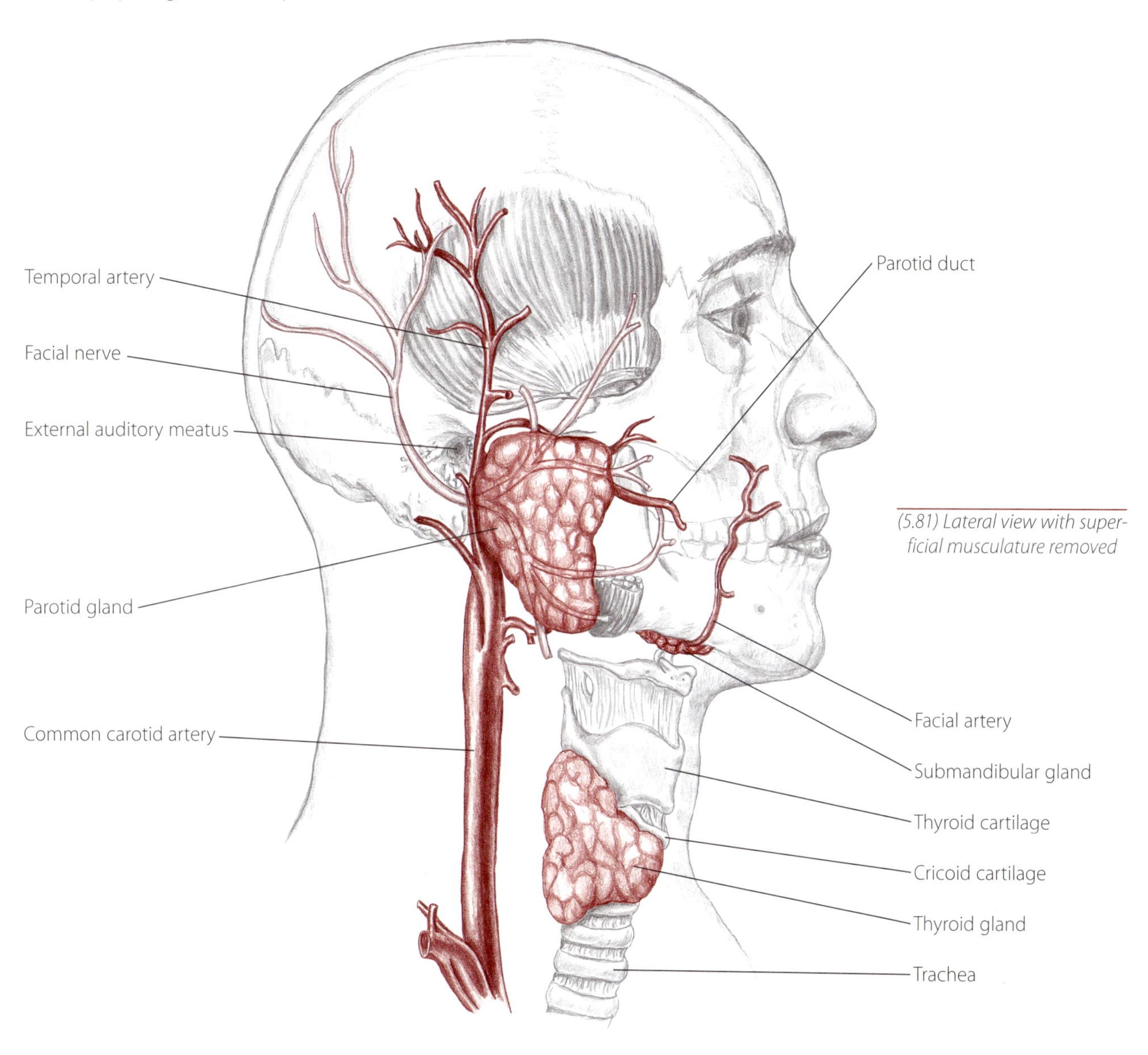

(5.81) Lateral view with superficial musculature removed

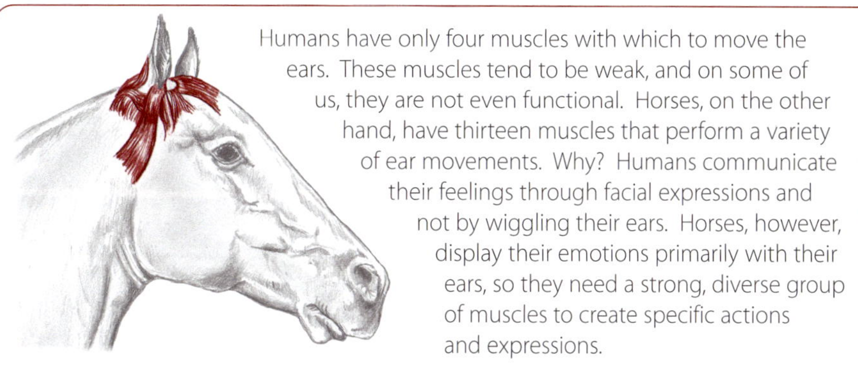

Humans have only four muscles with which to move the ears. These muscles tend to be weak, and on some of us, they are not even functional. Horses, on the other hand, have thirteen muscles that perform a variety of ear movements. Why? Humans communicate their feelings through facial expressions and not by wiggling their ears. Horses, however, display their emotions primarily with their ears, so they need a strong, diverse group of muscles to create specific actions and expressions.

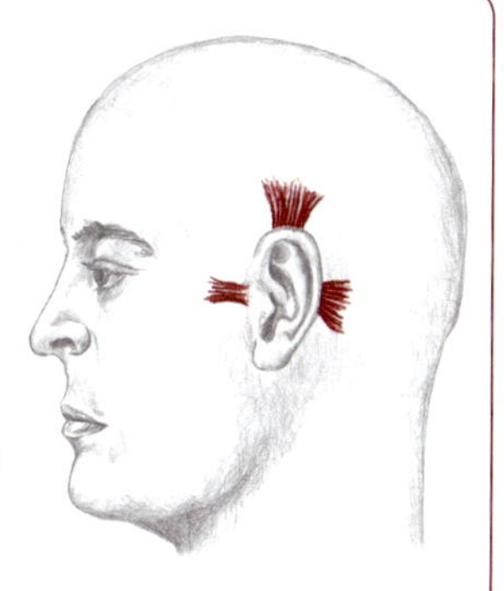

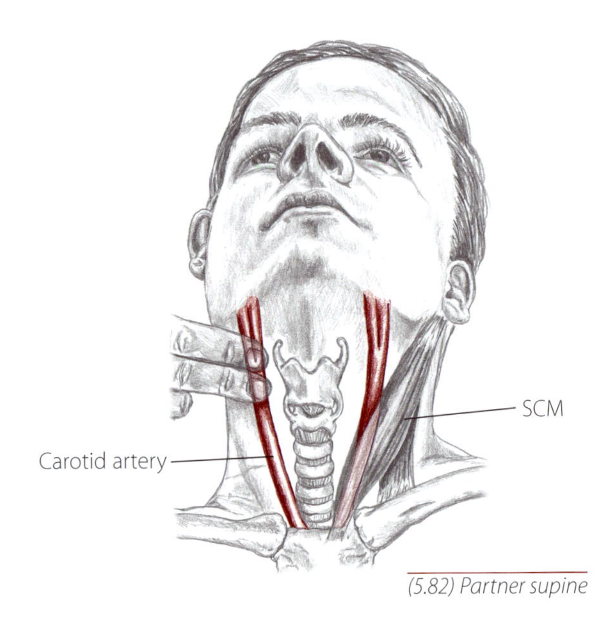

(5.82) Partner supine

Common Carotid Artery

The carotid artery is the primary supplier of blood to the head and neck. It ascends the anterior and lateral sides of the neck and lies deep to the sternocleidomastoid (SCM) and infrahyoid muscles. Its strong pulse can be felt medial to the SCM at the level of the hyoid bone.

1) Partner supine or seated. Place two fingerpads at the angle of the mandible.
2) Slide off the angle in an inferior and medial direction and press gently into the neck (5.82). The strong pulse of the carotid artery should be quite noticeable.

Are you medial to the SCM? Are you under the mandible at the level of the hyoid bone?

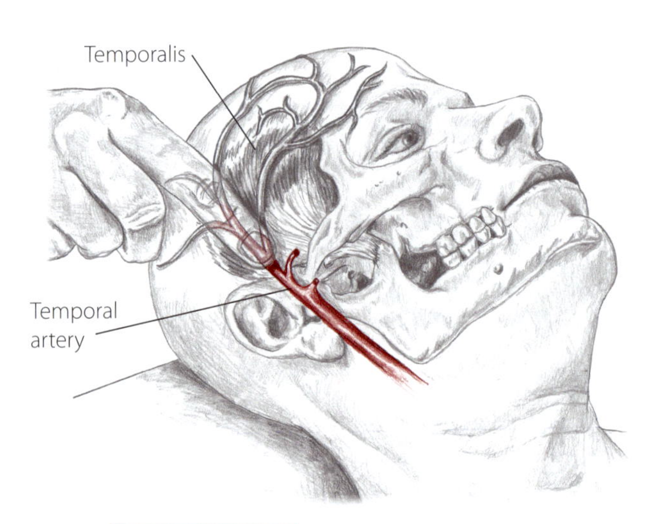

(5.83) Partner supine

Temporal Artery

The temporal artery branches off the external carotid artery and crosses over the top of the zygomatic arch. It continues superiorly along the side of the cranium, lying superficial to the temporalis muscle. The pulse of the temporal artery can be detected in front of the ear along the zygomatic arch.

1) Partner supine or seated. Place your fingerpad in front of the ear at the zygomatic arch (5.83).
2) Gently explore and palpate for the artery's pulse. If you do not feel it, adjust your finger position and make sure your pressure is not too deep.

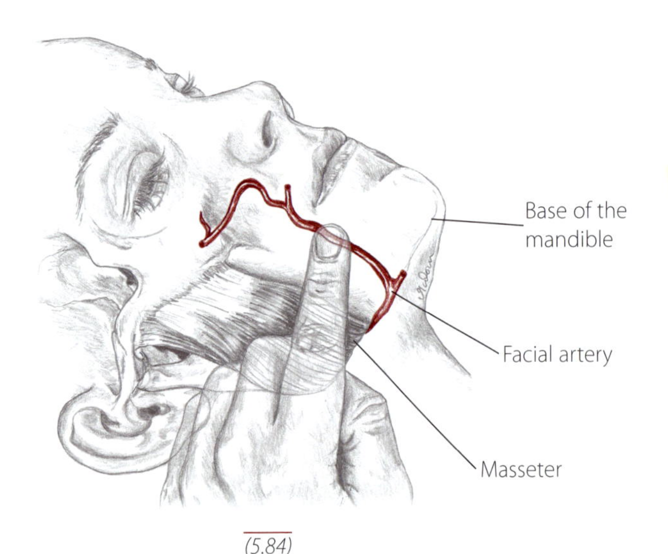

(5.84)

Facial Artery

The small, superficial facial artery branches off from the external carotid artery and curves around the base of the mandible (jawline) toward the mouth and nose. Its pulse may be difficult to detect, but can be felt along the base of the mandible at the anterior edge of the masseter.

1) Partner supine. With your partner clenching her jaw, locate the masseter's anterior edge.
2) Position your finger next to the base of the mandible and gently palpate for the pulse of the artery (5.84).

Are you at the base of the mandible, along the anterior edge of the masseter?

carotid ka-**rot**-id Grk. causing deep sleep

Facial Nerve

The facial nerve is not a structure you will want to palpate specifically, but because of its proximity to other palpable structures on the side of the face, you will need to be aware of its location.

The facial nerve (cranial nerve VII) exits from the cranium and emerges superficially just anterior to the mastoid process. As it passes beneath the parotid gland, the nerve branches off and spreads across the face, scalp and neck (5.85). Often two or more branches of the facial nerve cross superficially over the zygomatic arch.

When exploring the parotid gland, masseter or zygomatic arch, be mindful of the facial nerve's presence. Static pressure on the nerve can cause irritation, inflammation or even induce nausea in your partner.

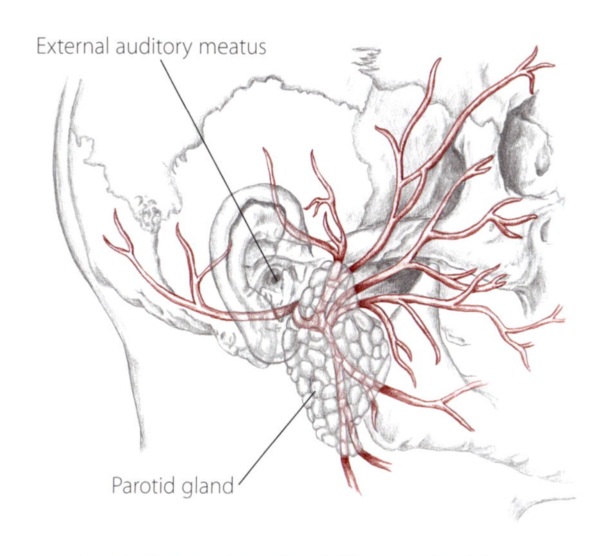

(5.85) Lateral view showing branches of the facial nerve

Vertebral Artery

The vertebral artery branches off the subclavian artery and supplies blood to the brain and spinal cord. It ascends the neck through the transverse foramen of C-6 to C-1 before passing through the foramen magnum of the occiput (5.86). It is inaccessible, of course, but a vital structure to be aware of when palpating and/or passively moving the head and neck.

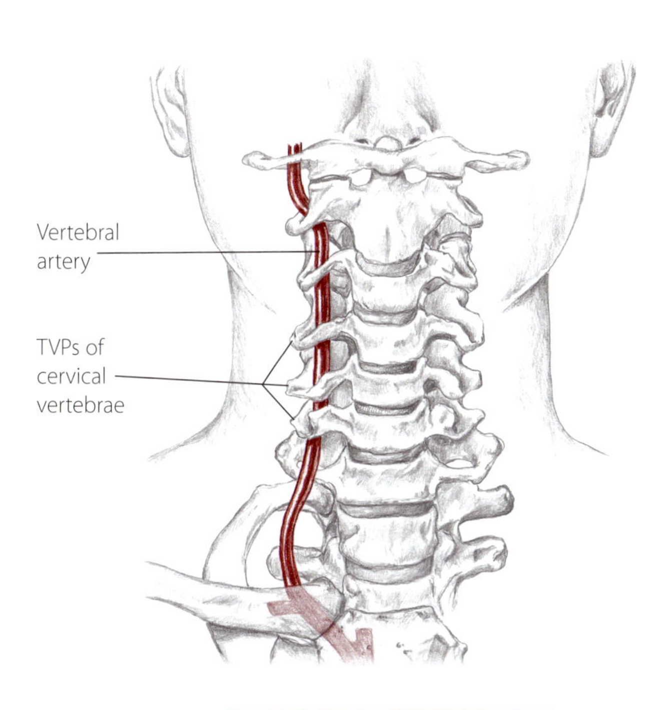

(5.86) Anterior view of neck and head

Parotid Gland and Duct, Submandibular Gland

There are three salivary glands in the neck and face: the parotid, submandibular and sublingual. All are accessible, but be aware that palpation of the salivary glands can stimulate the production of saliva.

The **parotid gland** is located in front of the earlobe, superficial to the masseter muscle (p. 256). It has a soft, lumpy surface and is penetrated by branches of the facial nerve. The **parotid duct** is a spaghetti-sized tube extending anteriorly from the parotid gland. It tucks around the anterior edge of the masseter to funnel saliva to the mouth.

True to its name, the **submandibular gland** is tucked underneath the base of the mandible. Its round shape can be located anterior to the angle of the mandible.

> The flap of tissue that hangs from the roof of the mouth (soft palate) is called the uvula. It is designed to cover the nasal passages during swallowing. When a person sleeps on her back with her mouth open, air passes the uvula and palate, causing these and other tissues to vibrate; this, of course, produces a snore. The loudest snore ever recorded was 69 decibels - frightening when you consider that a pneumatic drill produces 70 to 90 decibels!

parotid	pa-**rot**-id	Grk. beside the ear
thyroid	**thi**-royd	Grk. shield
uvula	**uv**-u-la	L. a little grape

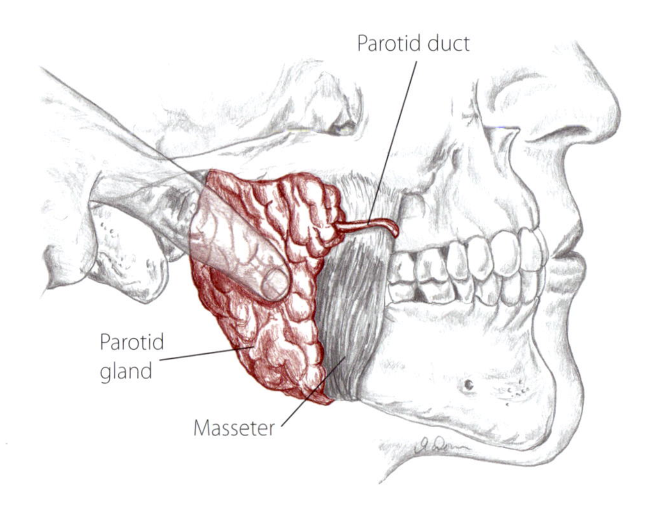

(5.87) Palpating the parotid gland

Parotid gland

1) Partner supine. Place your fingerpads in front of the earlobe on the masseter muscle.
2) Using gentle pressure along the superficial tissue, palpate between the angle of the mandible and the zygomatic arch for the gland's gelatinous texture (5.87).
3) Press deep to the gland in order to feel the striated fibers of the masseter muscle. Compare the different textures of these structures.

Parotid duct

1) Ask your partner to clench her jaw.
2) Place your fingerpads below the zygomatic arch, along the anterior edge of the masseter. Roll your finger back and forth (in a superior/inferior direction) and palpate for the mobile, horizontal tube.

Are you along the anterior edge of the masseter? Is the duct the diameter of a strand of spaghetti, and does it run horizontally?

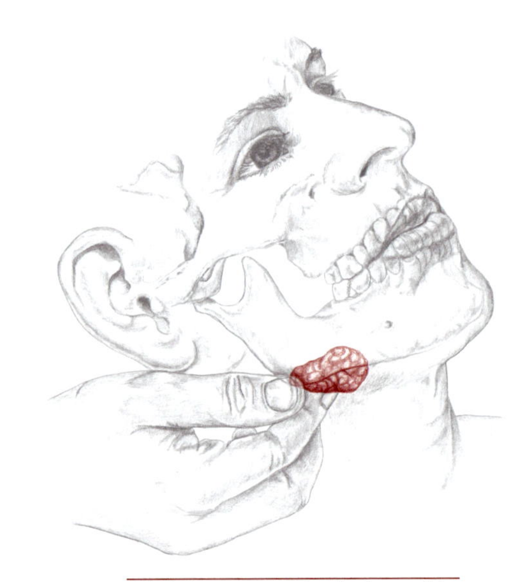

(5.88) Accessing under the base of the mandible for the submandibular gland

Submandibular gland

1) Place a finger along the base of the mandible.
2) Move your fingers medially, underneath the base, to palpate the superficial, marble-sized gland (5.88).

Can you roll your finger along the surface of the gland, outlining its shape?

Thyroid Gland

The left and right lobes of the thyroid gland are located on the anterior surface of the trachea. The gland lies deep to the infrahyoid muscles and has a soft, spongy texture that can be difficult to distinguish from the surrounding tissues.

1) Partner supine or seated. Using one fingerpad, locate the surface of the trachea between the jugular notch and cricoid cartilage.
2) Palpate for the soft texture of the thyroid gland lying on top of the trachea (5.89). Respecting the gland's delicacy, explore gently and briefly.

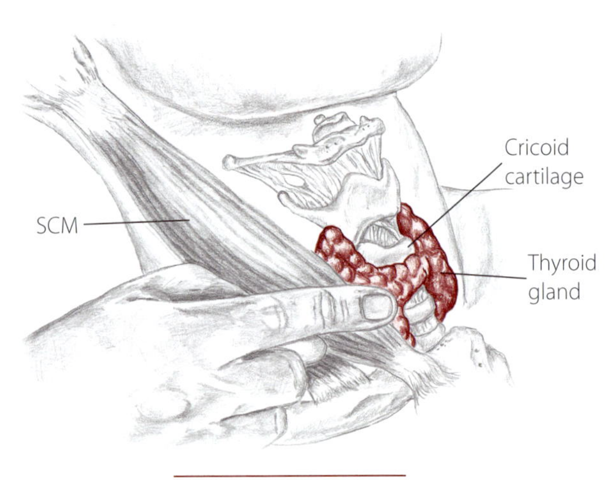

(5.89) Anterior/lateral view

Cervical Lymph Nodes

The numerous bundles of lymph nodes in the cervical region are divided into two groups: superficial and deep. The superficial cervical nodes (5.90) are located primarily along the underside of the mandible, posterior and inferior to the earlobe, and in the posterior triangle (p. 232) between the platysma and the deep fascia. The deep cervical nodes are larger and lie beside several large vessels and glands. Both the superficial and deep lymph nodes are slightly movable and have the size and texture of soft lentils or moist raisins. They are often tender on palpation.

1) Supine or seated. Place your fingers on the lateral side of the neck. Using your broad fingerpads, gently palpate under the skin for the superficial cervical nodes.
2) Explore along the underside of the mandible and in the posterior triangle. Once you have located a node, carefully outline its size and shape.

Are they slightly movable and the size and texture of soft lentils?

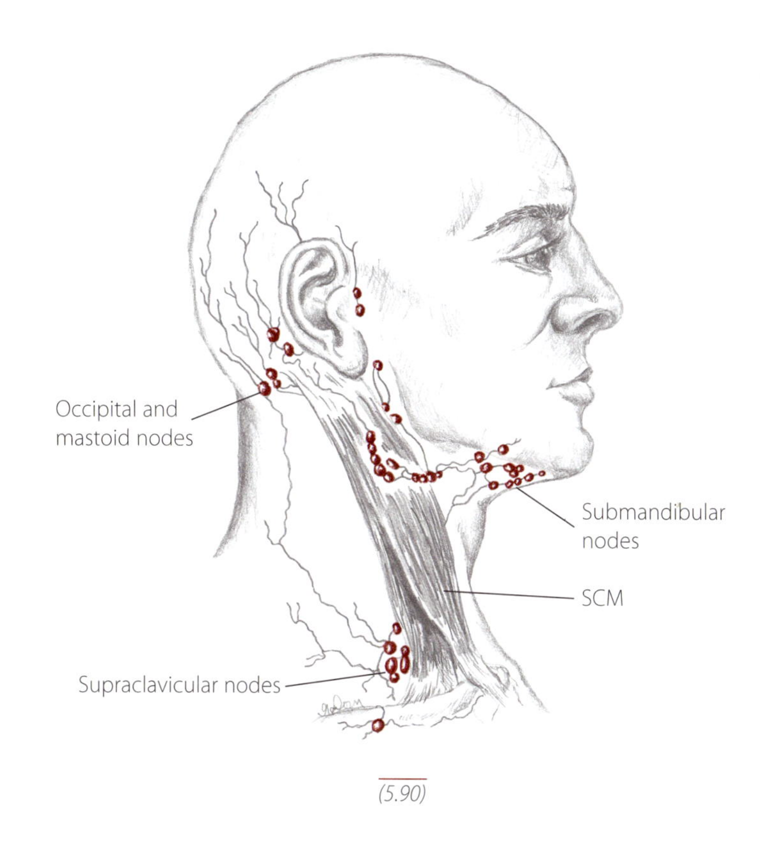

(5.90)

Brachial Plexus

The brachial plexus is a large bundle of nerves that innervates the shoulder and upper extremity. After exiting from the transverse processes of C-5 to T-1, it squeezes between the anterior and middle scalenes, continues inferiorly and laterally, and ducks underneath the clavicle to the axillary region (5.91).

Although the brachial plexus can be accessed, it is best avoided. Compressing or impinging one of its nerves can create a sharp, shooting sensation down the arm.

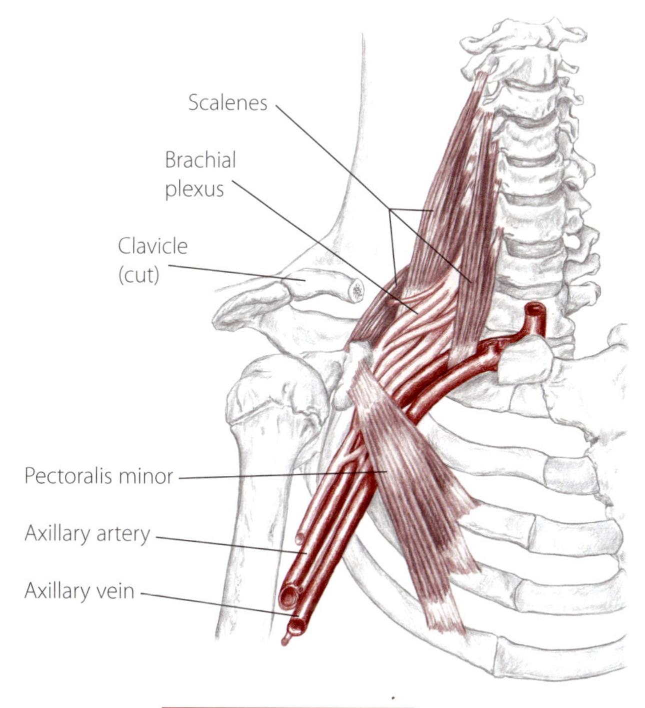

(5.91) Anterior/lateral view of right shoulder and cervical spine

Most mammals have a broad, thin sheet of muscle called the panniculus carnosus. It is an integumentary muscle that attaches to the underside of the skin and, on some species, covers the entire thorax. It enables a horse to shake off flies, an armadillo to roll into a ball and a cat to raise the hair on its back (left). For humans, the platysma is believed to be all that remains of the panniculus carnosus.

plexus **plek**-sus L. interwoven

NOTES

Off to the pelvis and thigh...

6

Pelvis & Thigh

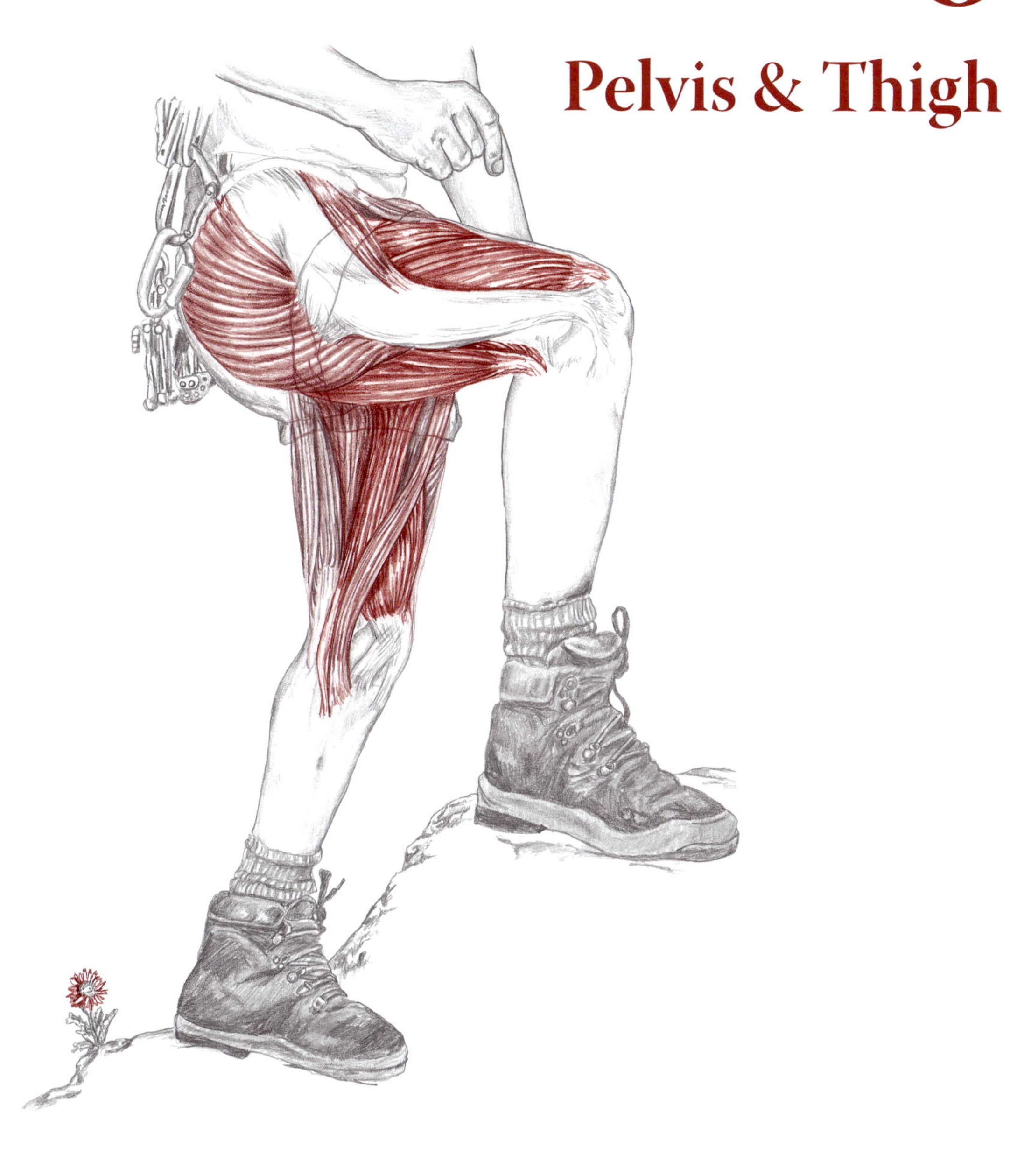

Topographical Views

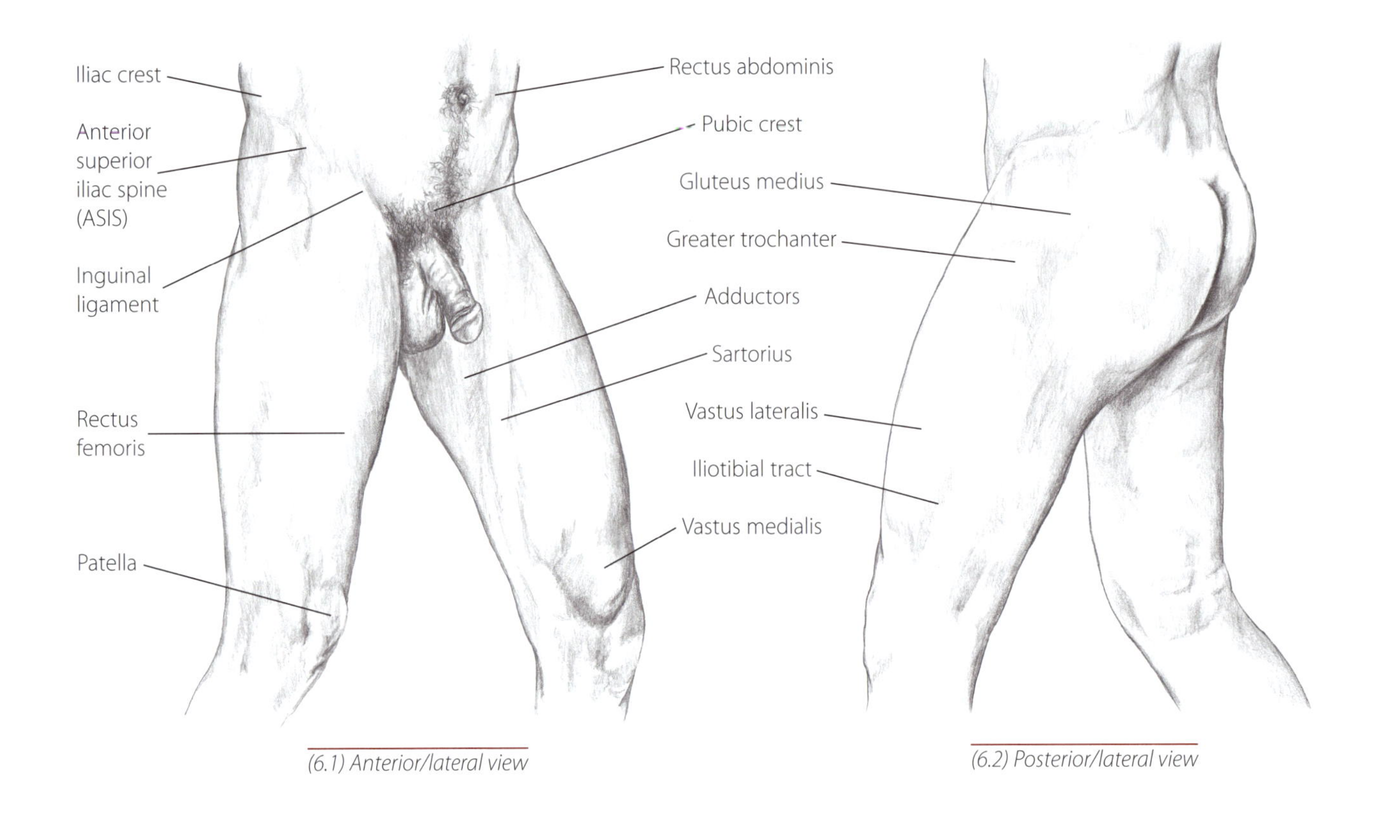

(6.1) Anterior/lateral view

(6.2) Posterior/lateral view

In this chapter, the male genitals have been included in the illustrations that demonstrate palpatory techniques near the base of the pelvis. This is to clarify their location with respect to the structure you are palpating. *See p. 293 for more information.*

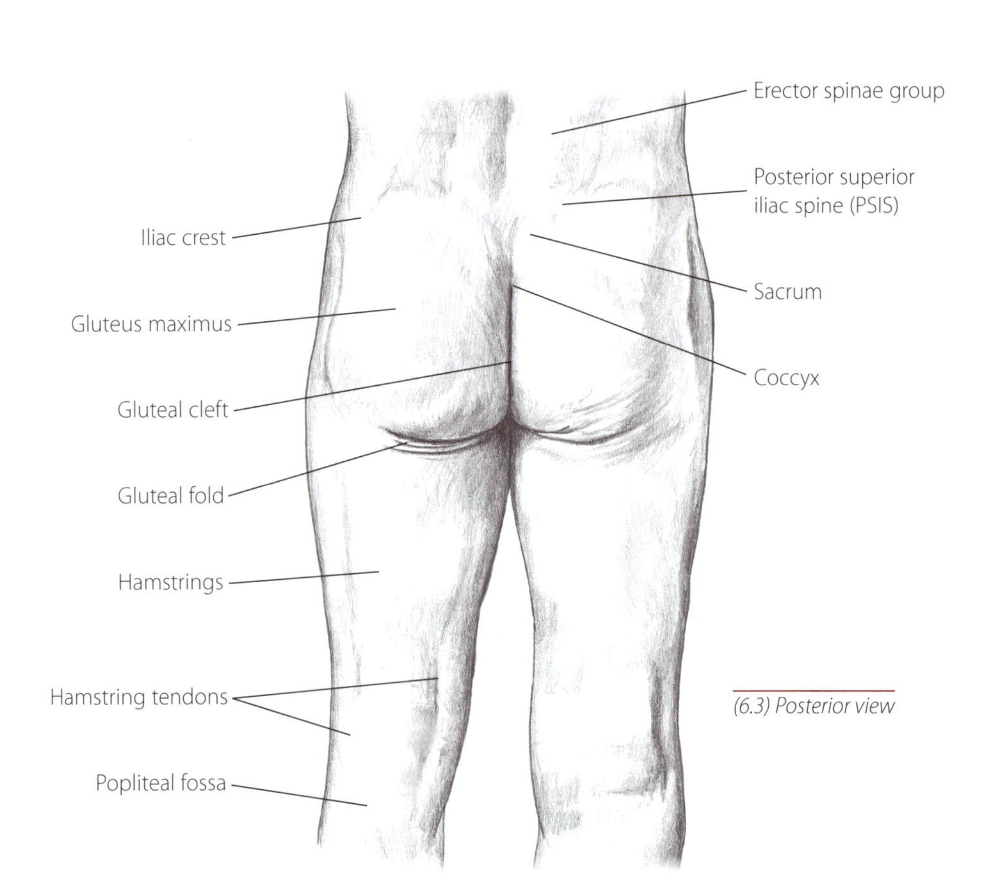

(6.3) Posterior view

Exploring the Skin and Fascia

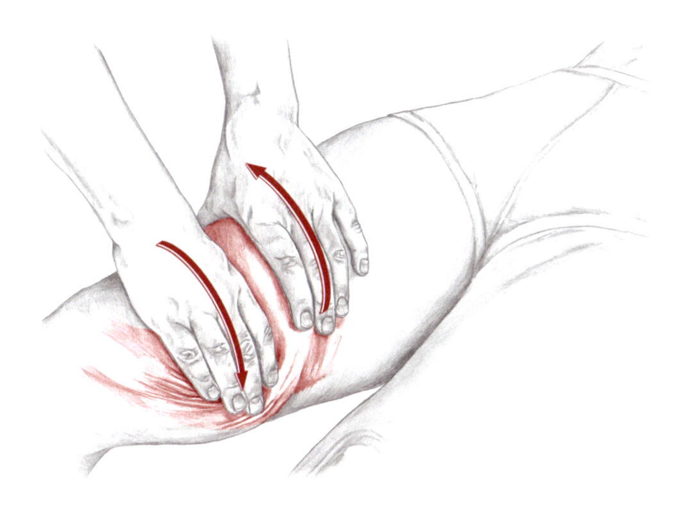

(6.4) Partner supine, medial view of right thigh

1) Partner supine. Begin by placing your hands on your partner's thigh. Explore from the pelvis to the knee, sensing the temperature of the tissue. Be sure to access the thigh's medial and lateral sides as well.
2) Sink your hands into the thigh and gently twist the tissue in opposite directions (6.4). Take particular note of its thickness and elasticity. For example, the skin and fascia just proximal to the knee may be thinner than the tissue near the pelvis.

1) Partner prone. Holding the ankle, passively flex the knee. Use your other hand to gently grasp the skin and fascia proximal to the posterior knee. As you roll it between your fingers, note the tissue's pliability and texture.
2) Compare what you have felt to the thicker tissue of the medial thigh and the denser fascia of the lateral thigh.
3) Still grasping the skin and fascia, passively flex and extend the knee (6.5). Feel the tissue stretch as the knee extends. Continue this same movement while grasping the sides of the thigh. If the tissue is difficult to grasp, you may want to use the flat of your hand to sense the changes in the tissue.

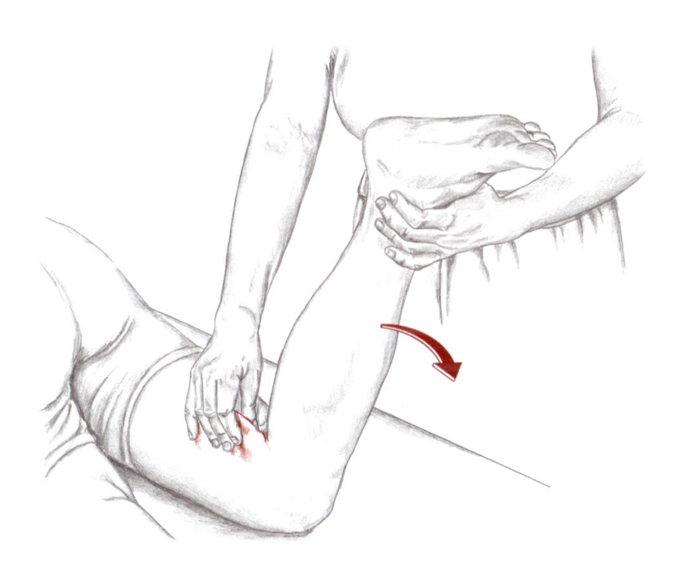

(6.5) Partner prone, posterior/medial view of right thigh

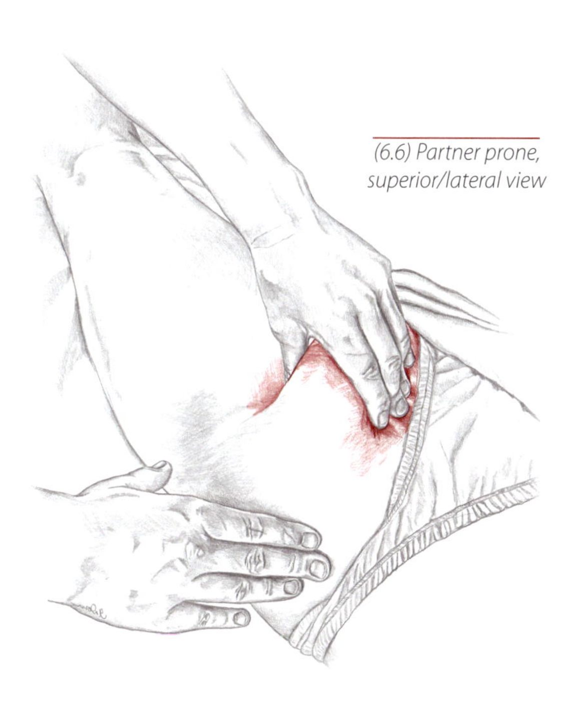

(6.6) Partner prone, superior/lateral view

1) Partner prone. Using the back of your hand, explore the temperature of the posterior and lateral buttocks. It is not uncommon for the tissue to be cooler here than it is on the posterior thigh or low back.
2) Since the buttocks are composed of both large muscles and large quantities of adipose, they are a good area for exploring tissue differences. Setting your thumb on the gluteal fold (see 6.3), gently but firmly grasp the tissue of the buttock.
3) Try grasping just the superficial skin and fascia, sensing its gelatinous quality. Then grasp a bit deeper and feel the thick, striated mass of the gluteal muscles (6.6).

Bones of the Pelvis and Thigh

The **pelvis** (pelvic girdle) consists of the two hip bones, the sacrum and the coccyx (6.7). Each **hip** (coxal) bone is formed by the fusion of three smaller bones: the **ilium**, **ischium** and **pubis** (6.8). Although the pelvis is deep to surrounding muscles, organs and adipose tissue, aspects of it are easily palpable.

The superficial **sacrum** lies posteriorly between the hip bones. The small **coccyx** extends inferiorly from the sacrum. The sacrum and coccyx, both made up of fused vertebrae, are considered part of the vertebral column.

The **femur** is the longest, heaviest and strongest bone in the body. Its proximal end articulates with the hip at the acetabulum to form the ball-and-socket-shaped coxal (hip) joint. Portions of the proximal femur are partially accessible. The femoral shaft is surrounded by the thick muscles of the thigh, while the distal end of the femur is superficial.

The distal femur articulates with the proximal tibia to form the tibiofemoral (knee) joint. The tibiofemoral joint is a modified hinge joint, which means it can flex and extend and, when in a flexed position, can medially and laterally rotate the knee.

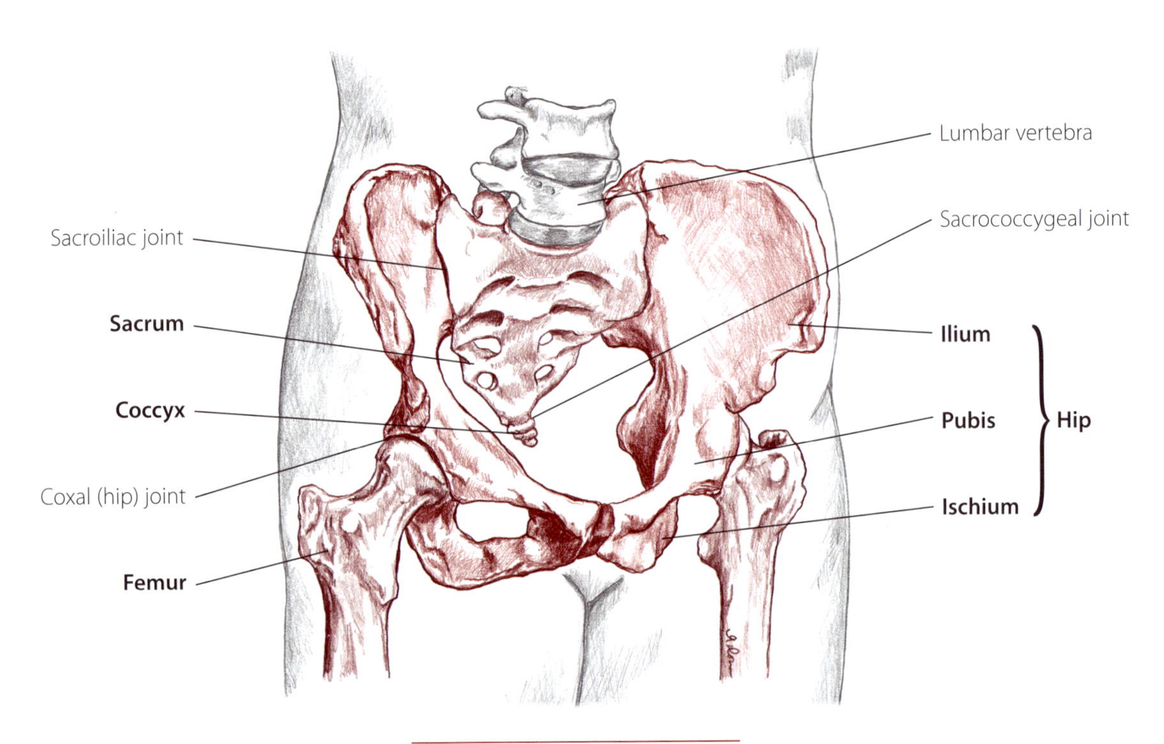

(6.7) Anterior/lateral view of the pelvis

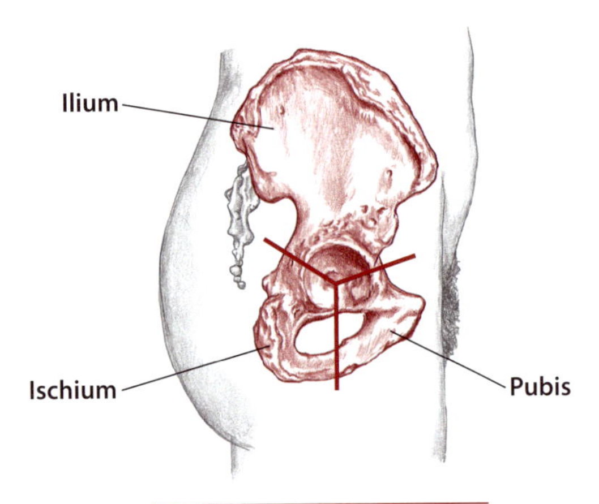

(6.8) Lateral view, bones of the hip

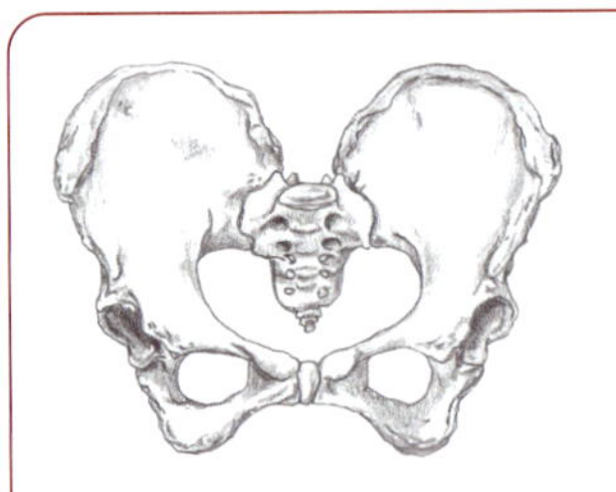

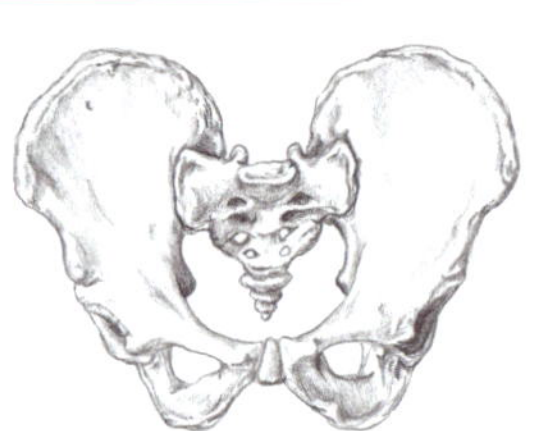

The shape of the pelvis is different in females (left) and males (right). The female pelvis is broader for carrying and delivering a child. It has a wider iliac crest, a larger pelvic "bowl," and a greater distance between ischial tuberosities.

femur	**fee**-mur	L. thigh
pelvis	**pel**-vis	L. basin

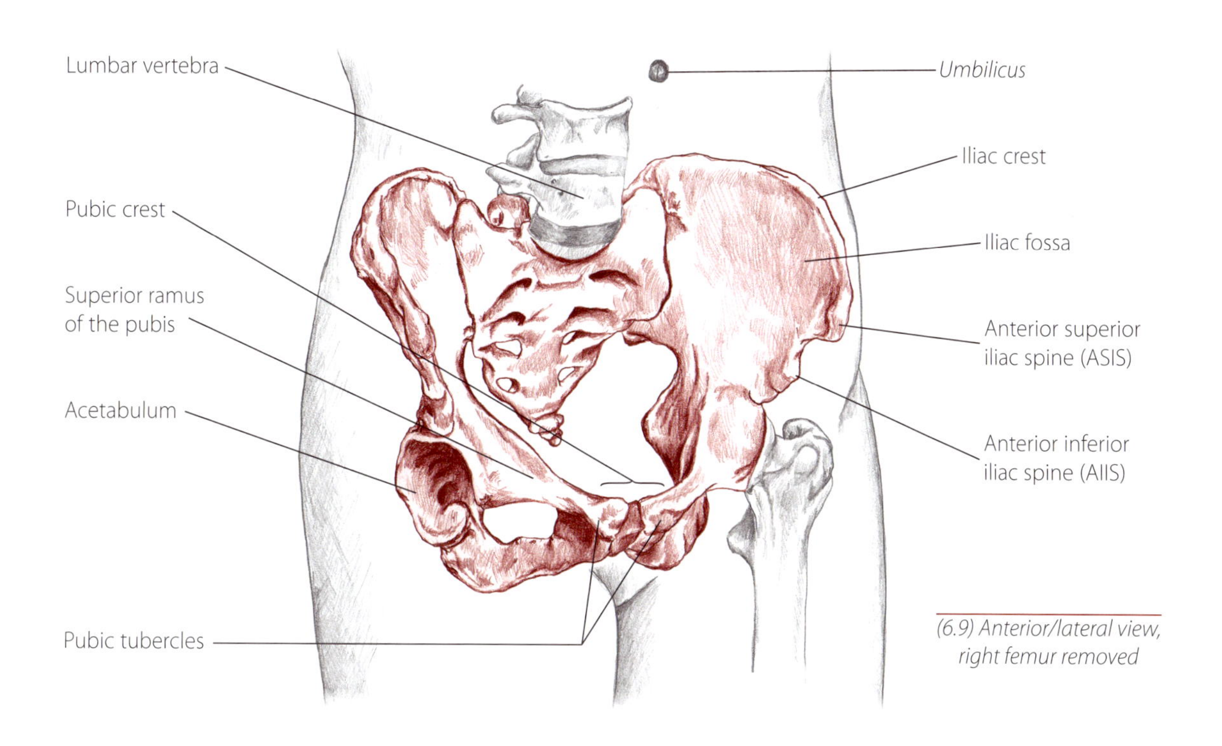

(6.9) Anterior/lateral view, right femur removed

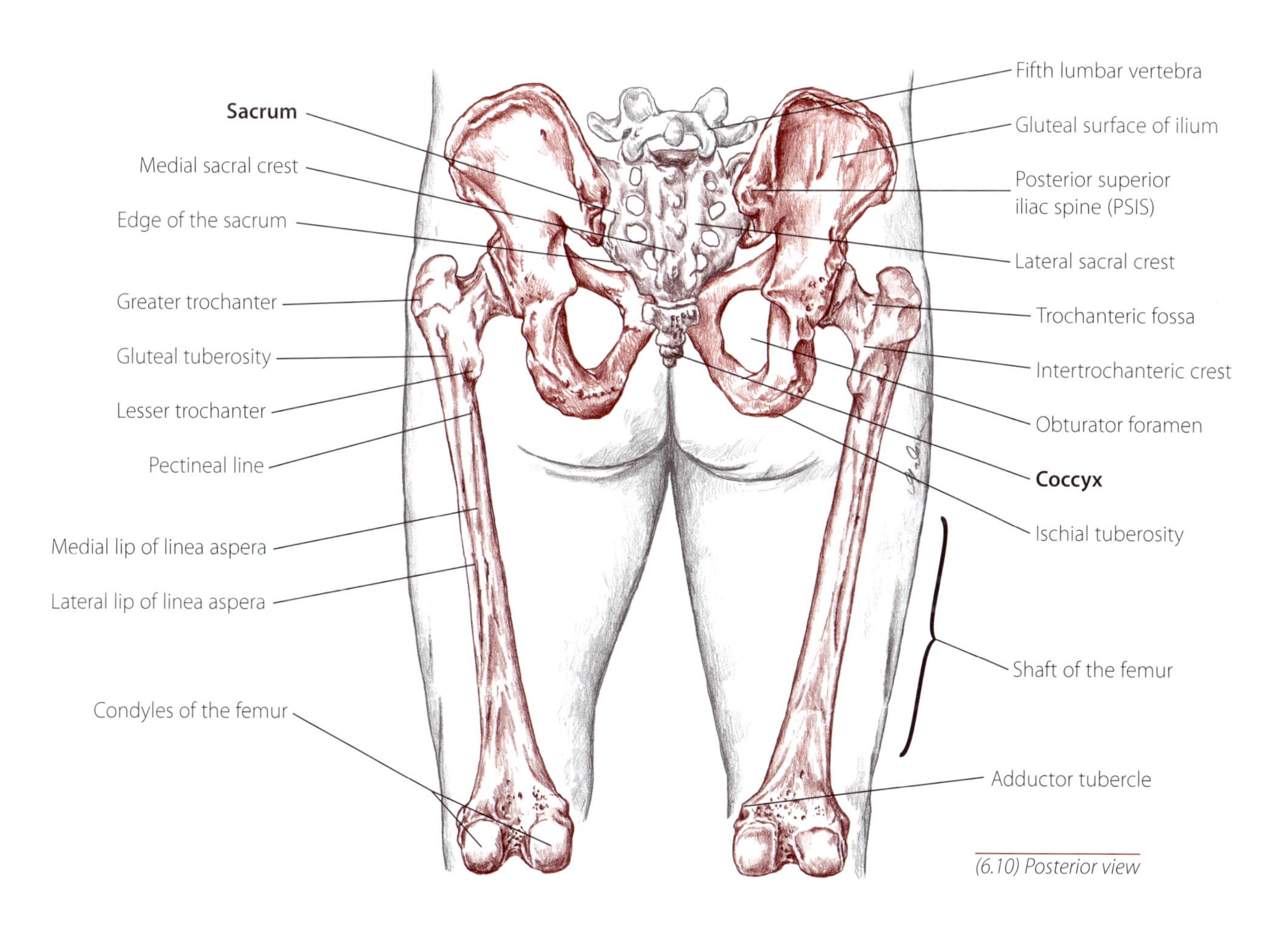

(6.10) Posterior view

ischium	**ish**-ee-um	Grk. hip
ilium	**il**-ee-um	L. groin, flank
pubis	**pu**-bis	NL. bone of the groin

Bony Landmarks

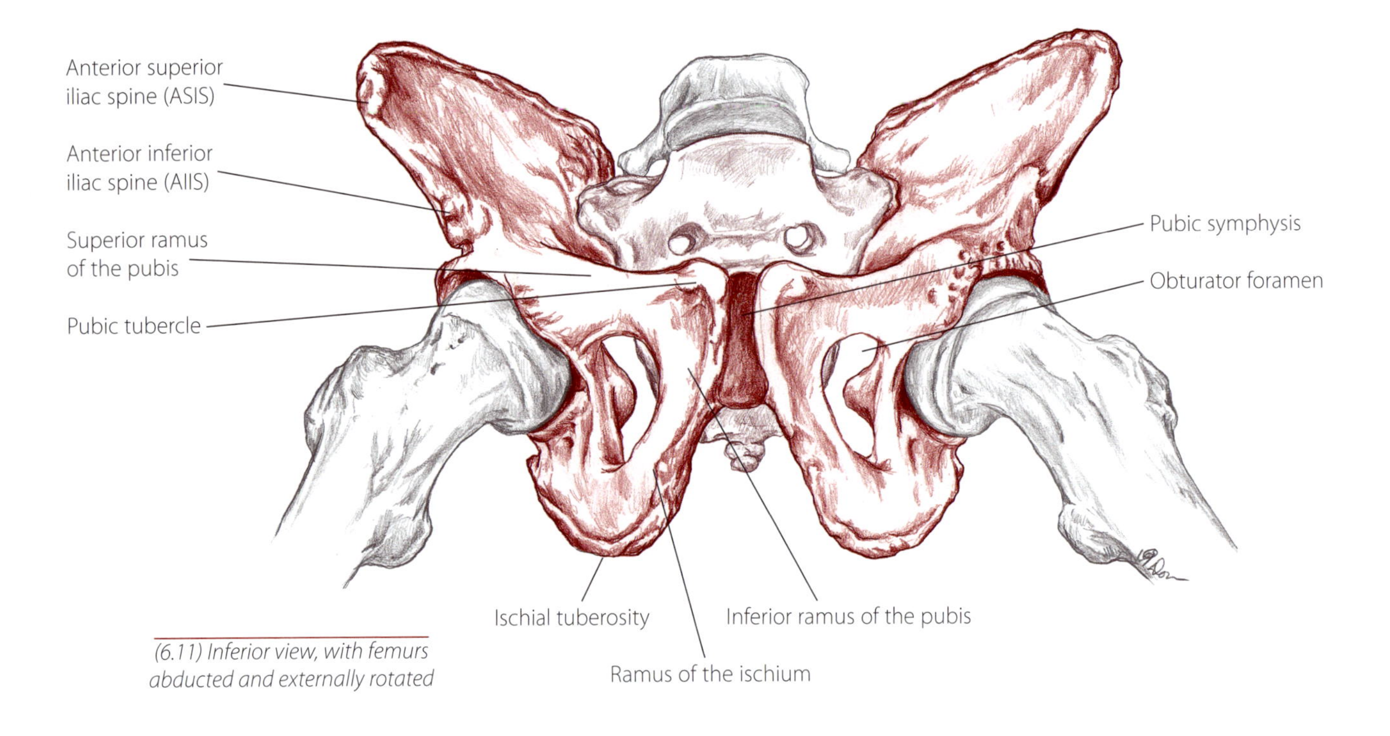

(6.11) Inferior view, with femurs abducted and externally rotated

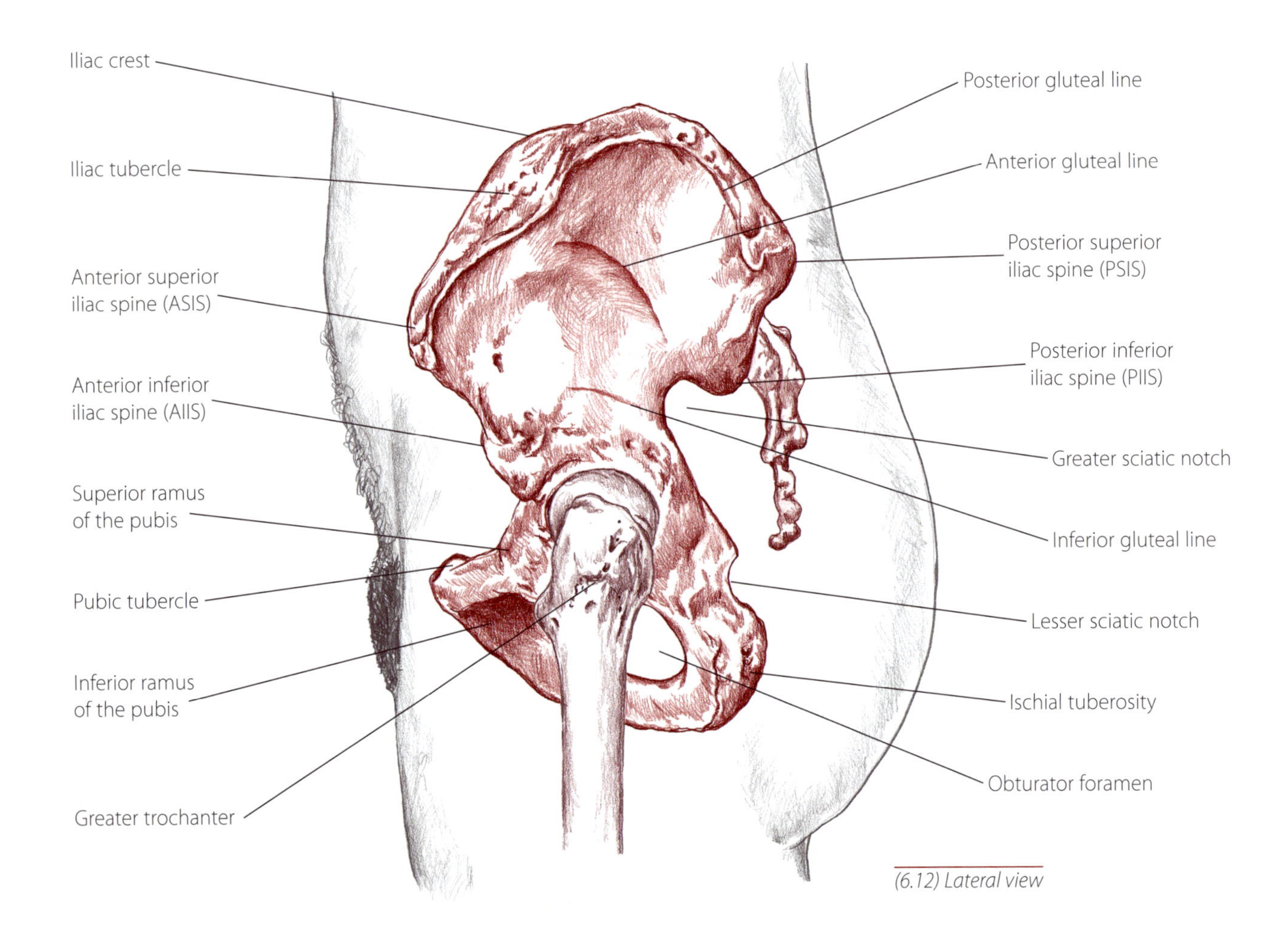

(6.12) Lateral view

acetabulum	**as**-e-**tab**-u-lum	L. a little saucer for vinegar
foramen	for-**aye**-men	L. a passage or opening
linea aspera	**lin**-e-a **as**-per-a	L. rough line

Bony Landmarks

a) Iliac crest
b) Iliac fossa
c) Anterior superior iliac spine (ASIS)
d) Anterior inferior iliac spine (AIIS)
e) Pectineal line
f) Superior ramus of the pubis
g) Pubic tubercle
h) Symphyseal surface
i) Inferior ramus of the pubis
j) Posterior superior iliac spine (PSIS)
k) Articular surface for sacrum
l) Posterior inferior iliac spine (PIIS)
m) Greater sciatic notch
n) Ischial spine
o) Lesser sciatic notch
p) Obturator foramen
q) Ischial tuberosity
r) Ramus of the ischium

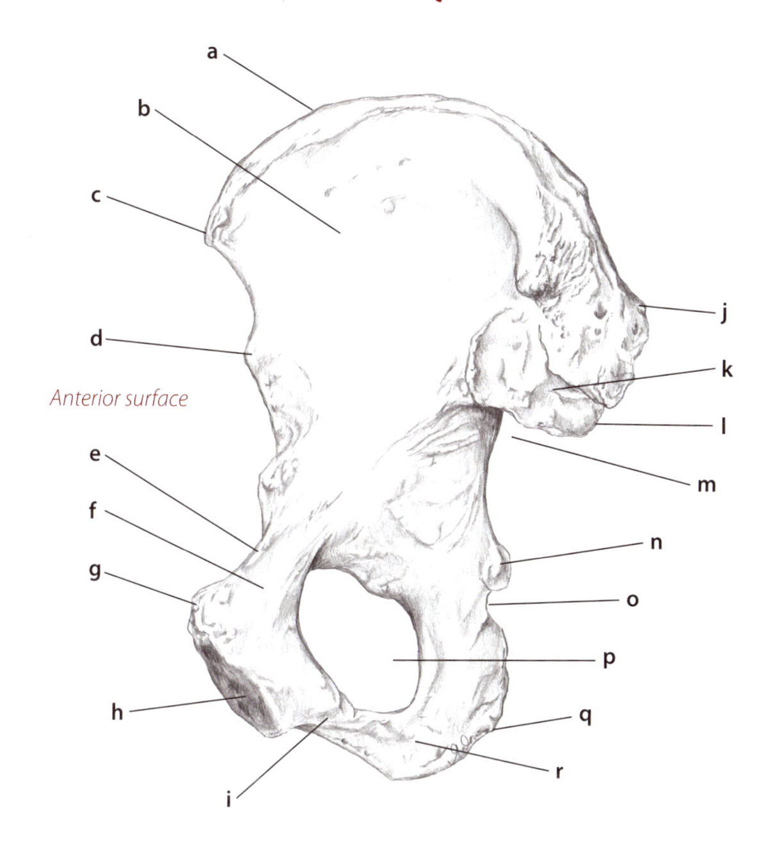

(6.13) Medial view of right hip

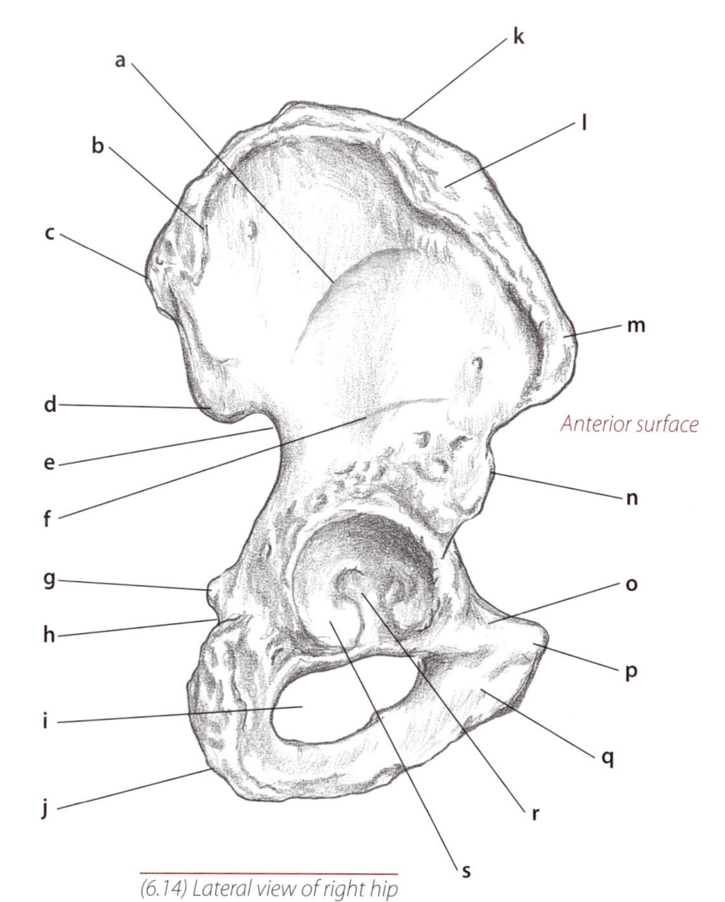

(6.14) Lateral view of right hip

a) Anterior gluteal line
b) Posterior gluteal line
c) Posterior superior iliac spine (PSIS)
d) Posterior inferior iliac spine (PIIS)
e) Greater sciatic notch
f) Inferior gluteal line
g) Ischial spine
h) Lesser sciatic notch
i) Obturator foramen
j) Ischial tuberosity
k) Iliac crest
l) Iliac tubercle
m) Anterior superior iliac spine (ASIS)
n) Anterior inferior iliac spine (AIIS)
o) Superior ramus of the pubis
p) Pubic tubercle
q) Inferior ramus of the pubis
r) Acetabulum
s) Lunate surface of acetabulum

Bony Landmarks

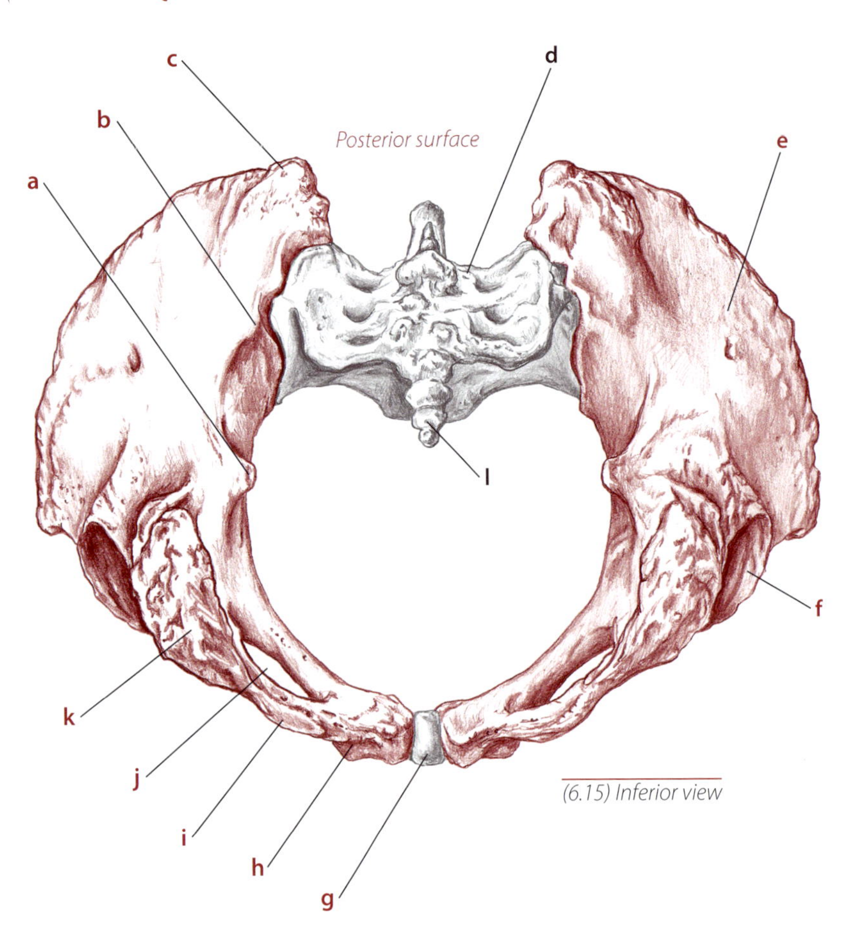

(6.15) Inferior view

a) Ischial spine
b) Posterior inferior iliac spine (PIIS)
c) Posterior superior iliac spine (PSIS)
d) Sacrum
e) Gluteal surface of ilium
f) Acetabulum
g) Pubic symphysis
h) Inferior ramus of the pubis
i) Ramus of ischium
j) Obturator foramen
k) Ischial tuberosity
l) Coccyx

Black letters indicate bones; **red** letters indicate bony landmarks or other structures

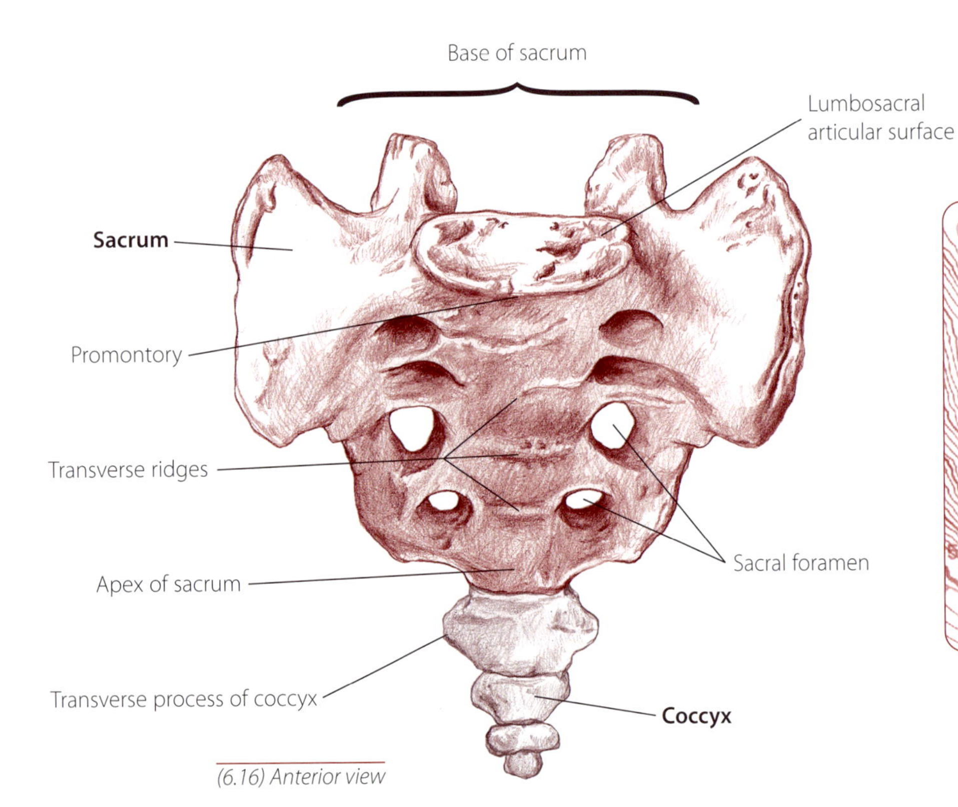

(6.16) Anterior view

Why is there a big hole in the hip? The obturator foramen is located along the inferior part of the pelvis, surrounded by the pubis and ischium. It began quite small - as seen in the skeletal remains of early reptiles. On humans, the foramen is situated between weight-bearing lines, and since bone is heavy and (in a sense) costly to maintain, the foramen evolved into a larger hole covered with a membrane.

Bony Landmarks

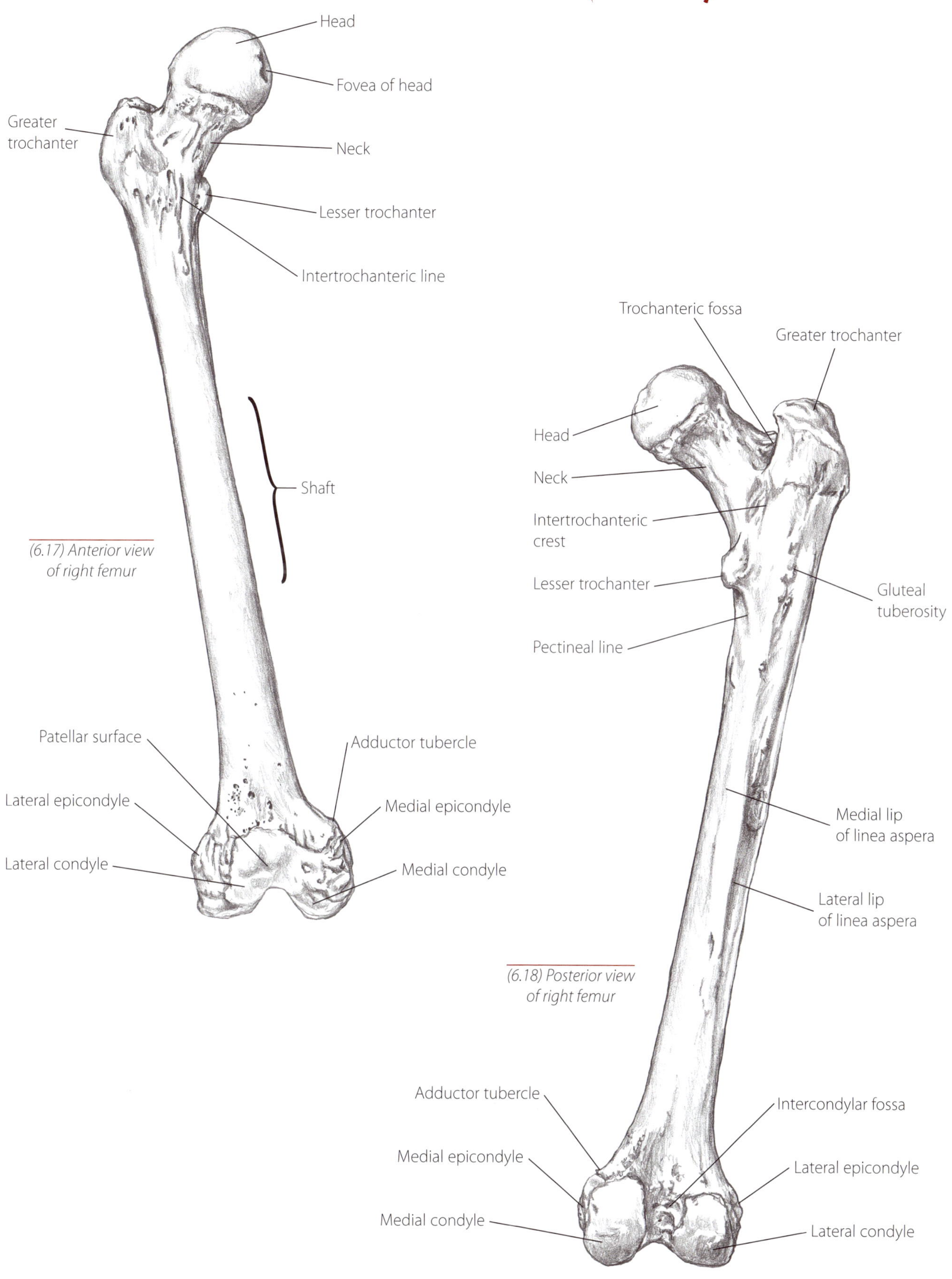

(6.17) Anterior view of right femur

(6.18) Posterior view of right femur

Bony Landmark Trails

Trail 1 "Solo Pass"

Due to the pelvis' multifaceted shape and proximity to sensitive areas, palpating your partner's pelvic region may challenge you initially. **Trail 1** is designed to give you an opportunity to access your own pelvic region first. This will generate the confidence needed to palpate effectively on your partner over the next four trails. These six landmarks can be seen as your "base camps" - they are clearly accessible and will lead you to the other landmarks of the pelvis.

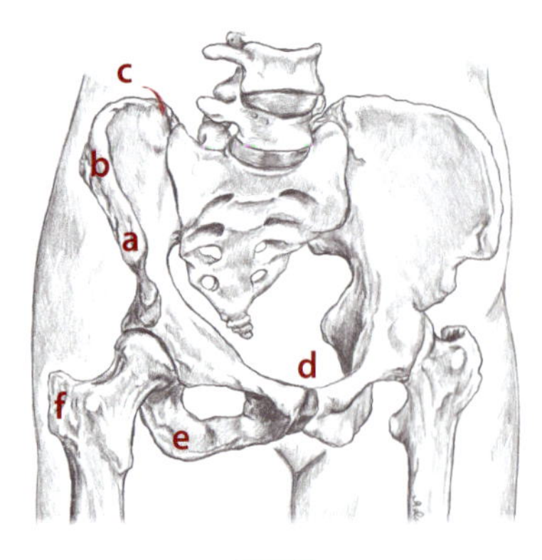

Trail 1

- **a** Anterior superior iliac spine (ASIS)
- **b** Iliac crest
- **c** Posterior superior iliac spine (PSIS)
- **d** Pubic crest
- **e** Ischial tuberosity
- **f** Greater trochanter of the femur

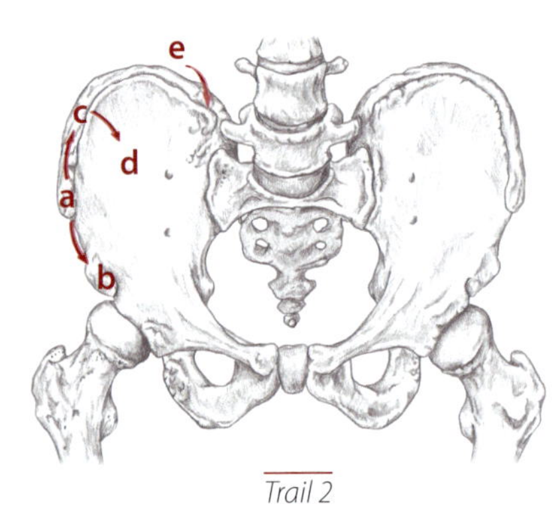

Trail 2

Trail 2 "Iliac Avenue" travels along the superior aspect of the pelvis on the ilium.

- **a** Anterior superior iliac spine (ASIS)
- **b** Anterior inferior iliac spine (AIIS)
- **c** Iliac crest
- **d** Iliac fossa
- **e** Posterior superior iliac spine (PSIS)

Trail 3 "Tailbone Trail" accesses the bones at the base of the spine.

- **a** Posterior superior iliac spine (PSIS)
- **b** Sacrum
- **c** Medial sacral crest
- **d** Edge of the sacrum
- **e** Coccyx
- **f** Sacroiliac joint

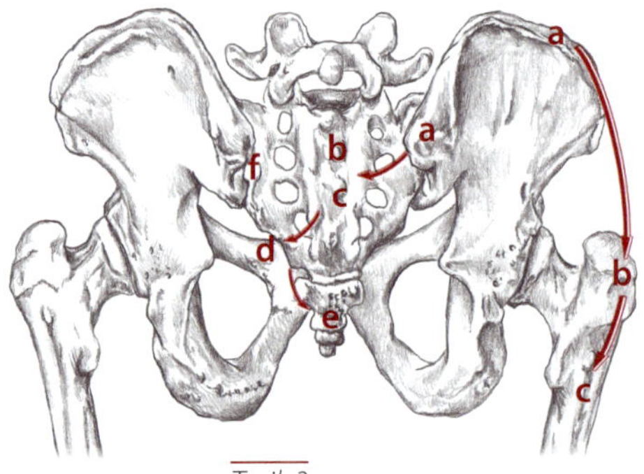

Trail 3

Trail 4

Trail 4 "Hip Hike" explores the lateral hip and landmarks of the proximal femur.

- **a** Iliac crest
- **b** Greater trochanter
- **c** Gluteal tuberosity

Trail 5 "The Underpass" follows around the pubic region to access the landmarks of the medial thigh.

- **a** Umbilicus
- **b** Pubic crest and tubercles
- **c** Superior ramus of the pubis
- **d** Inferior ramus of the pubis and ramus of the ischium
- **e** Ischial tuberosity

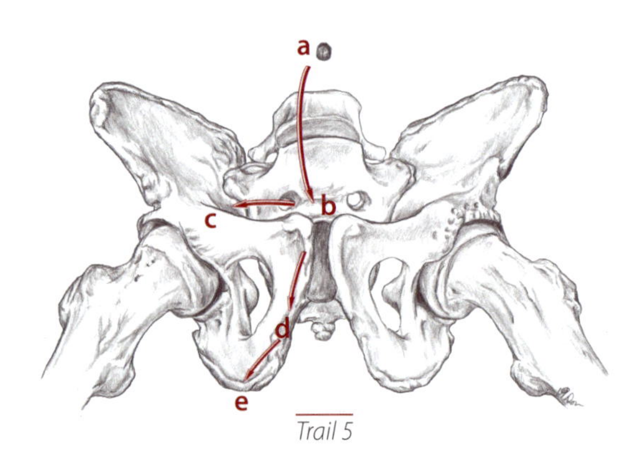

Trail 5

Trail 1 "Solo Pass"

Anterior Superior Iliac Spine (ASIS)

As the name suggests, the ASIS is located on the anterior and superior aspect of the ilium. Both ASISes are the superficial tips located below the waistline underneath the front pants pockets. The ASIS serves as the attachment site for the sartorius muscle and the inguinal ligament.

1) Locate both anterior superior iliac spines by placing your hands upon your hips with fingers in front, thumbs behind. Feel for the tip of the pelvis that sticks out anteriorly (6.19).
2) Explore these points and the surrounding structures of the ilium. Try palpating them on yourself from a seated position so that the overlying tissue will be soft.

Are the bones you feel just beneath the surface of the skin? Are you inferior to the level of the umbilicus?

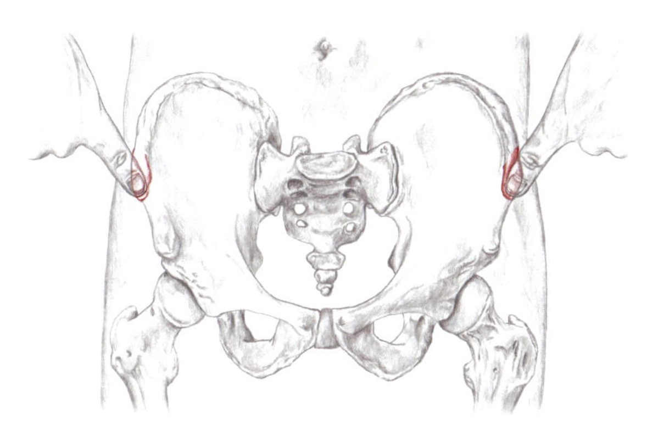

(6.19) Anterior view, palpating your ASISes

Iliac Crest

The iliac crest is the long, superior edge of the ilium. It begins at the ASIS and extends around the side of the torso to end at the PSIS. Besides helping to keep your pants up, the iliac crests serve as attachment sites for the quadratus lumborum (p. 213) and abdominal muscles. Each crest is superficial and easily palpable as the muscles that attach to it do not cross over its edge (6.20).

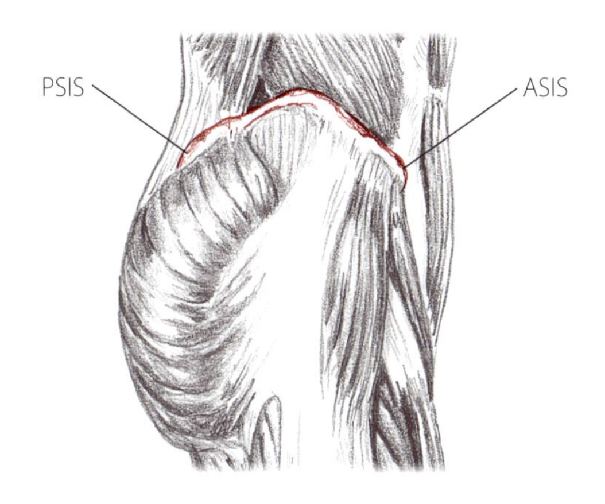

(6.20) Lateral view of pelvis showing the muscles which surround the iliac crest

1) Locate the ASIS. Slowly walk your fingers around the side of your hip, pressing into the wide edge of the crest. Note how the crest rises from the ASIS and soon after may widen laterally (6.21).
2) Follow the crest as it continues around to the posterior side of the body and ends at the PSIS.

Can you sink your fingers into the flesh of the abdomen just above the iliac crest?

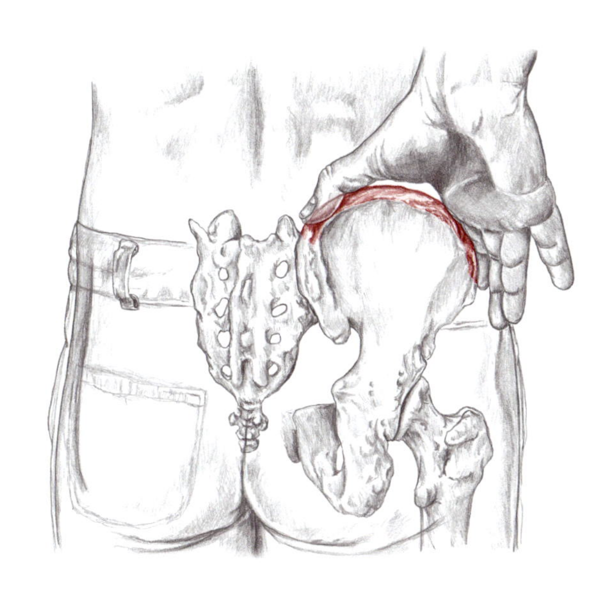

(6.21) Posterior view, sliding your fingers along your iliac crest

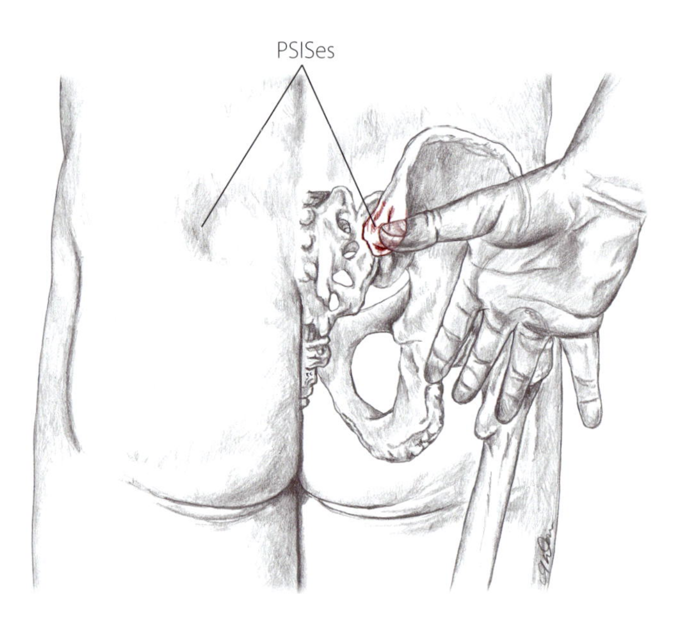

(6.22) Posterior view, locating your PSIS

Posterior Superior Iliac Spine (PSIS)

The superficial PSIS is located at the posterior end of the iliac crest. In most people, both PSISes can be visibly identified by the two small dimples found at the base of the low back. Without the help of a mirror, you may have trouble seeing your own PSISes, but you can still palpate them.

1) Place your thumbs upon your iliac crests. Follow the crests around the posterior hip. Note how they descend as you move medially.
2) The PSISes may feel like small humps surrounded by thicker tissues and are not as pronounced as the ASISes (6.22).

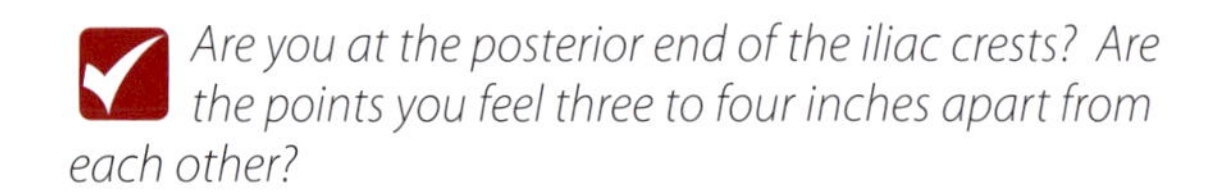

Are you at the posterior end of the iliac crests? Are the points you feel three to four inches apart from each other?

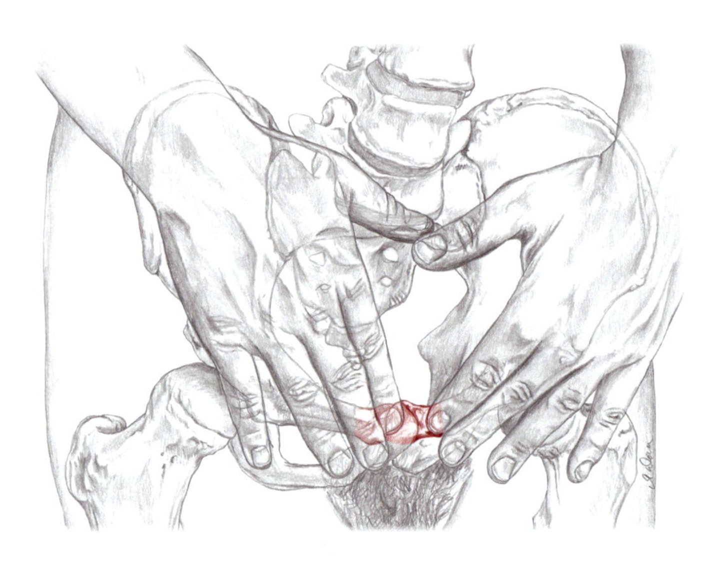

(6.23) Anterior view, accessing your pubic crest

Pubic Crest

The pubic crest is located directly inferior to the navel and superior to the genitals. Formed by the superior, medial edge of both pubic bones, the horizontal crest is roughly two inches wide and clearly palpable. It is an attachment site for the rectus abdominis muscle (p. 215) and the abdominal aponeurosis.

1) Position your fingers at your navel.
2) Slowly slide your fingers down the midline of the body toward the pubic region (6.23). You may travel five to eight inches before you feel the firm ridge of the pubic crest. You will be one to two inches superior to the genitals.

Are you at the midline of the body? Are you inferior to the level of the ASISes? Do you feel a solid, horizontal ridge of bone just above the genital region?

Locate the ASIS. Follow the inguinal ligament (p. 333) inferiorly 45° to the midline of the body until you reach the crest.

Ischial Tuberosity

If you have ever sat through a long musical or sporting event on a metal folding chair, then your ischial tuberosities are no stranger to you. The "sits bones" are located on the most inferior aspect of the pelvis at the level of the gluteal fold (the horizontal crease between the buttocks and thigh). The ischial tuberosity serves as an attachment site for the hamstrings, adductor magnus and the sacrotuberous ligament.

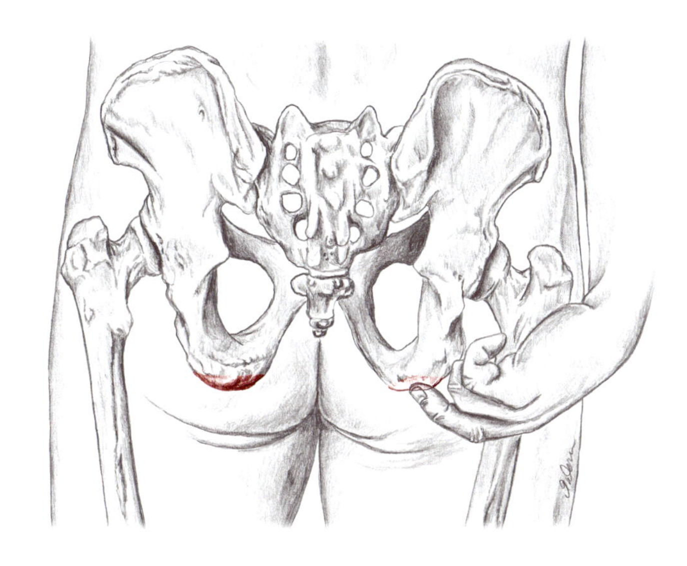

(6.24) Posterior view, palpating your ischial tuberosity

1) Have a seat on a hard chair or surface and rock side-to-side feeling your "sits bones."
2) Stand up and palpate the bone you were sitting on - your ischial tuberosity (6.24). Explore in all directions the large surface of the tuberosity.

Do you feel an identical structure between the other buttock and thigh?

When exploring the area around the sacrum and posterior iliac crest, it is not uncommon to locate small nodules of fibrofatty tissue. Embedded in the superficial fascia, they may vary in size from a pea to a large marble.

Greater Trochanter of the Femur

Located distal to the iliac crest, the greater trochanter is the large, superficial mass located on the side of the hip. It is easily palpable and serves as an attachment site for the gluteus medius, gluteus minimus and lateral rotators of the hip.

1) Locate the middle of the iliac crest.
2) Slide your fingerpads inferiorly four to six inches along the lateral side of the thigh until you reach the superficial mass of the greater trochanter. Explore and sculpt around all sides of its wide hump.

Medially and laterally rotate your hip as you palpate the trochanter. Do you feel its wide, knobby surface swivel back and forth under your fingers (6.25)?

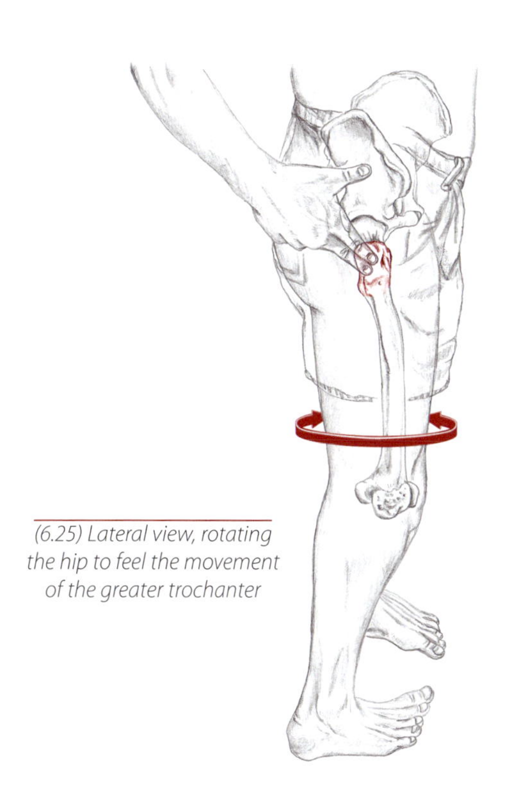

(6.25) Lateral view, rotating the hip to feel the movement of the greater trochanter

trochanter   tro-**kan**-ter   Grk. to run

Trail 2 "Iliac Avenue"

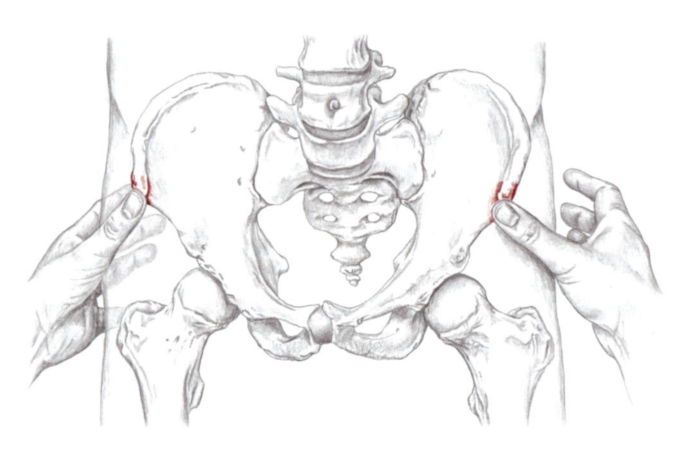

(6.26) Partner standing, locating both ASISes on your partner

Anterior Superior Iliac Spine (ASIS)

(Refer to p. 283 for more information)

1) Partner standing. Place your hand upon the side of the abdomen, below the level of the umbilicus.
2) Gently compress inferiorly until you feel the superficial tip of the ASIS (6.26). Palpate and observe the distance between the two ASISes and their relationship to each other.

Is the bony tip you feel superficial? Are you inferior to the level of the navel? Are you superior to the genital region?

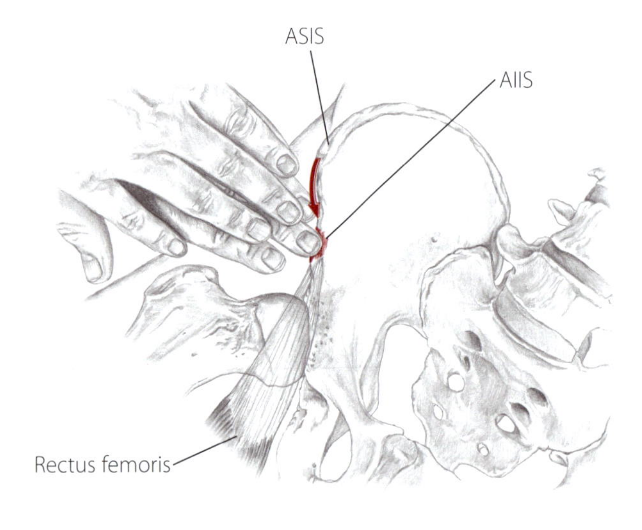

(6.27) Anterior view, partner supine, accessing the AIIS

Anterior Inferior Iliac Spine (AIIS)

The AIIS is located inferior and medial to the ASIS and is the attachment site for the rectus femoris muscle (p. 300). Smaller and flatter than the ASIS, the AIIS is deep to the sartorius muscle and inguinal ligament. Because of its subtle shape and its depth to the sartorius, the AIIS may be difficult to distinguish.

1) Supine. Flex the hip by bolstering under your partner's knee to shorten and soften the overlying tissue.
2) Locate the ASIS. Slide inferiorly and medially approximately one inch.
3) Palpate deep to the overlying tendons and explore for the small mound of the AIIS (6.27).

Are you medial and inferior to the ASIS? If your partner flexes his hip slightly, can you feel the tendon of the rectus femoris tighten under your fingers? (The overlying sartorius tendon will also become taut with this action.)

Along the lateral edge of the iliac crest, there is a subtle widening called the iliac tubercle. It designates the boundary between the origins of the tensor fasciae latae and gluteus medius muscles. **1)** Locate the ASIS. **2)** Slide posteriorly along the iliac crest approximately two inches. Explore the lateral edge of the iliac crest where it swells slightly. This is the iliac tubercle.

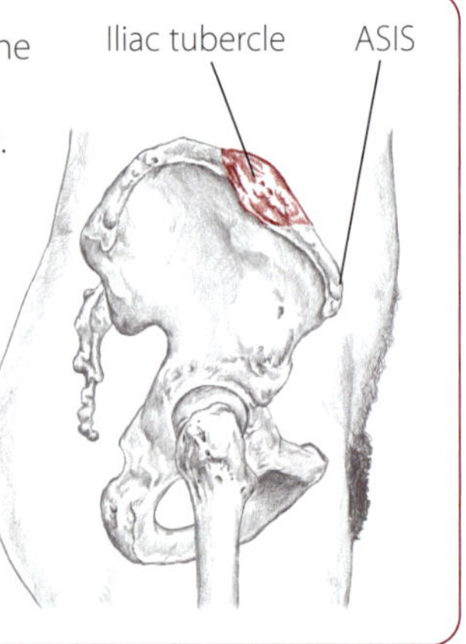

Iliac Crest

(Refer to p. 283 for more information)

1) Partner sidelying. Locate the ASIS.
2) Slide posteriorly along the iliac crest, observing how it widens and rises up along its path.
3) Follow the crest as it continues around the posterior side of the body to the PSIS (6.28).

 *Can you spread the webbing between your finger and thumb along the length of the crest?*

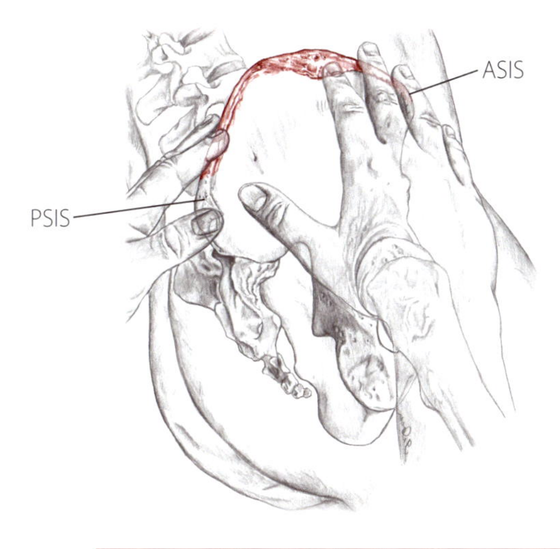

(6.28) Partner sidelying, sculpting out the iliac crest

Iliac Fossa

The bowl-shaped iliac fossa is located on the medial surface of the ilium and is an attachment site for the iliacus muscle. The presence of the iliacus and the abdominal contents makes the majority of the fossa inaccessible; however, you can sink your fingers slowly over the iliac crest and into the fossa to palpate it.

1) Supine. Flex the hip by bolstering under your partner's knee to shorten and soften the overlying tissue.
2) Lay the fingertips of one hand along the iliac crest just posterior and superior to the ASIS.
3) Moving slowly and patiently, curl your fingertips over the lip of the iliac crest into the iliac fossa (6.29). Depending on the firmness of the tissue, you may sink in only a small distance.

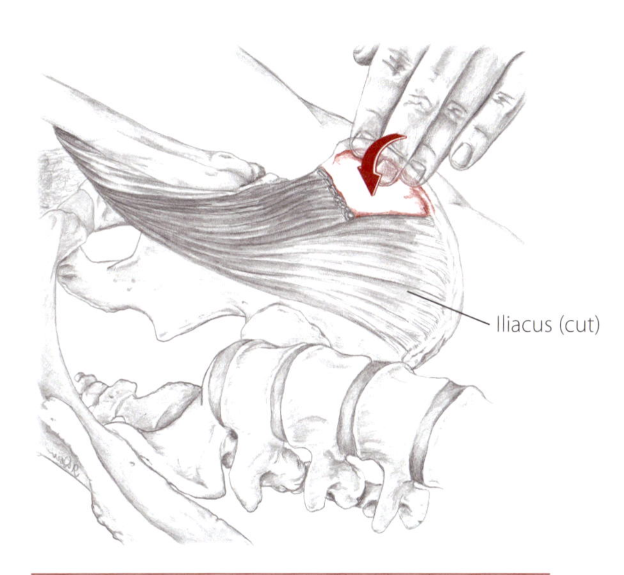

(6.29) Partner supine, curling your fingers into the iliac fossa

Posterior Superior Iliac Spine

(Refer to p. 284 for more information)

1) Partner standing. Follow both iliac crests posteriorly around the waist.
2) Follow the crests as they descend toward the sacrum and end at each PSIS (6.30). The PSIS will feel like a shallow hump surrounded by thicker tissues. It is not as pronounced as the ASIS, but is nevertheless accessible.
3) If possible, visibly locate the dimples of the low back and explore the surrounding region.

Are you at the posterior end of the iliac crest? Are both landmarks roughly horizontal to each other and three to five inches apart?

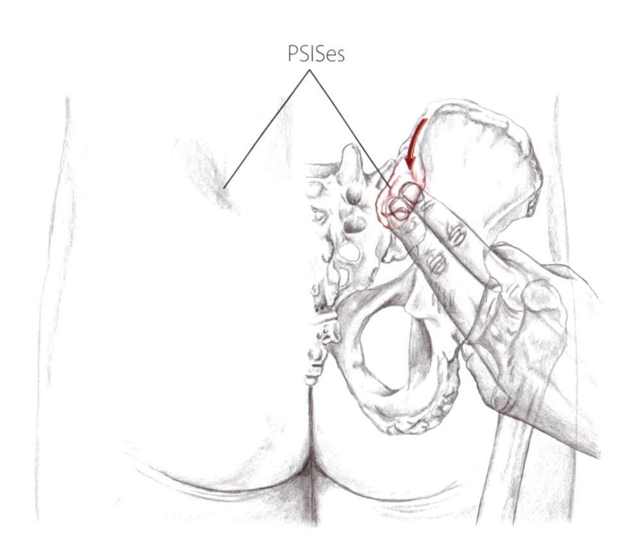

(6.30) Posterior view, isolating the PSIS

fossa **fos**-a L. a shallow depression

Trail 3 "Tailbone Trail"

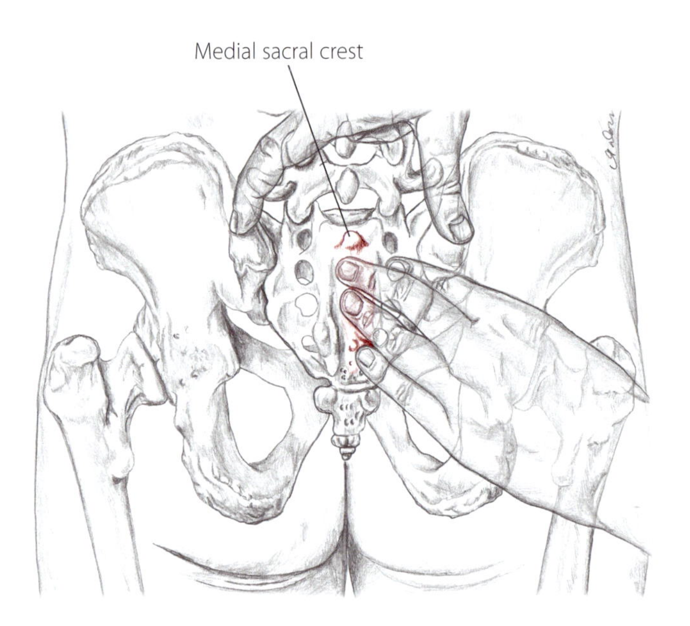

(6.31) Posterior view, using the PSISes as guides to locate the medial sacral crest of the sacrum

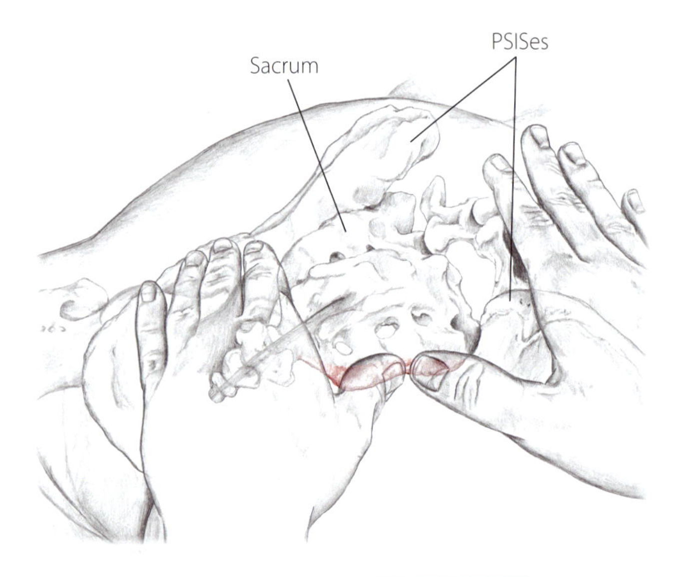

(6.32) Posterior/lateral view, partner prone, exploring the edge of the sacrum

Sacrum

Medial Sacral Crest
Edge of the Sacrum

The **sacrum** is a large, triangular bone at the inferior end of the vertebral column. Situated between the overhanging sides of the pelvis, the sacrum is made up of a series of four or five vertebrae that are fused together.

Running down the center of the sacrum and composed of three to four points is the **medial sacral crest**. On either side of the medial sacral crest is the lateral sacral crest - a smaller series of bony knobs. The **edge of the sacrum** is part of the attachment site for the gluteus maximus and the sacrotuberous ligament. Although the sacrum's bumpy surface lies deep to the thoracolumbar aponeurosis and sacroiliac ligaments, it is easily accessible.

1) Partner prone. Place a thumb and finger upon each PSIS and explore between and below these points for the surface of the sacrum.
2) Locate the midline of the sacrum and explore the points of the sacral crest (6.31). Palpate superiorly to the level of the PSIS and just above the coccyx.
3) Slide your fingers laterally off the side of the sacrum, pressing your fingertips into its solid edge (6.32). Follow the lateral edge up toward the PSIS and down to the coccyx.

How many small tips can you feel along the sacral crest? Can you follow both lateral edges inferiorly to where they converge at the coccyx? If you move laterally from the outer edge of the sacrum, can you feel the mass of the gluteus maximus (p. 309)?

Reptiles and most birds have two sacral vertebrae while mammals have between three and five. Humans in particular have more because, as upright creatures, the entire weight of the upper body is transferred through the sacrum to the pelvis and legs. All that remains of the spinous processes of the sacrum's vertebrae are the time-worn tips of the medial sacral crest.

sacrum **sa**-krum L. sacred or holy thing, from the use of the sacrum in Roman animal sacrifice

Coccyx

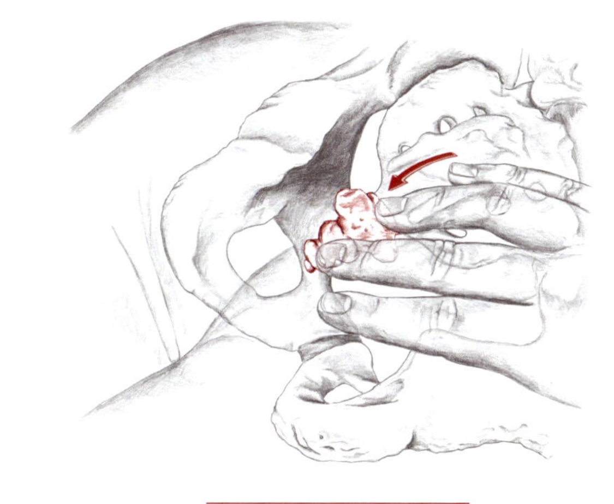

(6.33) Posterior/lateral view, partner prone, palpating the coccyx

The coccyx is located at the top of the gluteal cleft and attaches to the end of the sacrum. Composed of three or four fused bones, it has a segmented, bumpy contour and can be an inch or more in length. Its tip may curve in toward the body or veer slightly to the left or right.

Because of its proximity to the gluteal cleft, palpating the coccyx may be challenging initially for both you and your partner, so palpate your own coccyx before palpating your partner's.

1) Partner prone. Walk your fingers down the medial sacral crest toward the gluteal cleft. At the top of the cleft, the bumpy surface of the coccyx will be felt.
2) Explore the surface and sides of the coccyx, noting how the wide upper part narrows to a tip (6.33). The tip of the coccyx may not be accessible since it curves into the body.

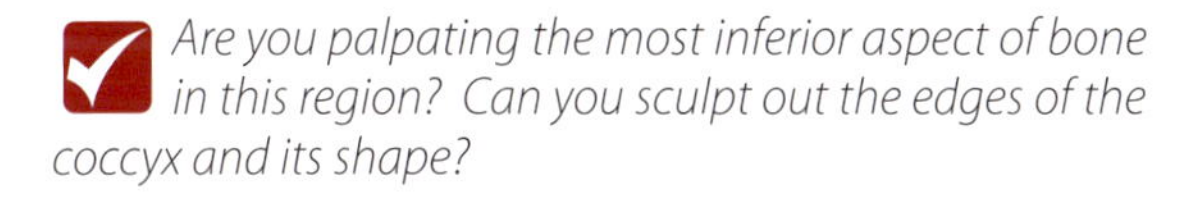

Are you palpating the most inferior aspect of bone in this region? Can you sculpt out the edges of the coccyx and its shape?

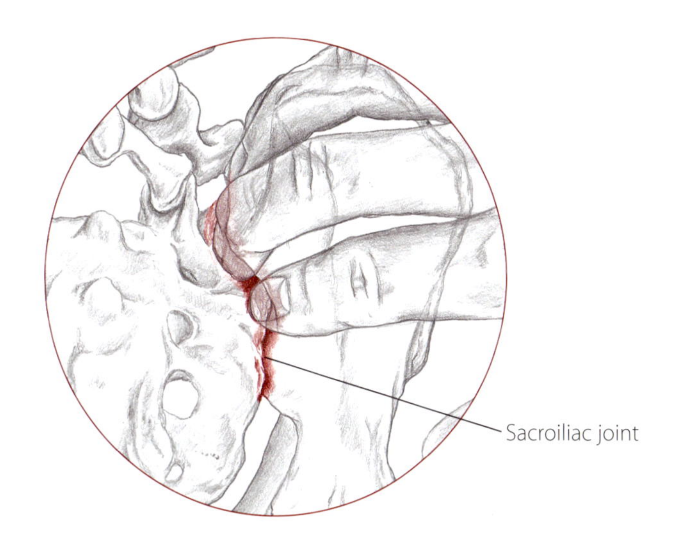

Sacroiliac Joint

The sacroiliac joint is the junction between the sacrum and the ilium. It is located medial to the PSIS and is deep to the thoracolumbar aponeurosis and posterior sacroiliac ligaments (p. 334). The ilium overhangs the sacroiliac joint, leaving only the edge of the joint accessible.

1) Partner prone. Locate the PSIS. Move slightly inferior and medial to locate the sacroiliac joint.
2) Create a small widening at the joint by keeping one hand upon it while the other hand flexes the knee to 90°. Then passively rotate the hip laterally, feeling for a small opening at the joint space (6.34). Also try medially rotating the hip.

Are you just medial and distal to the PSIS? Can you sculpt out the edge of the ilium as it overlaps the sacrum?

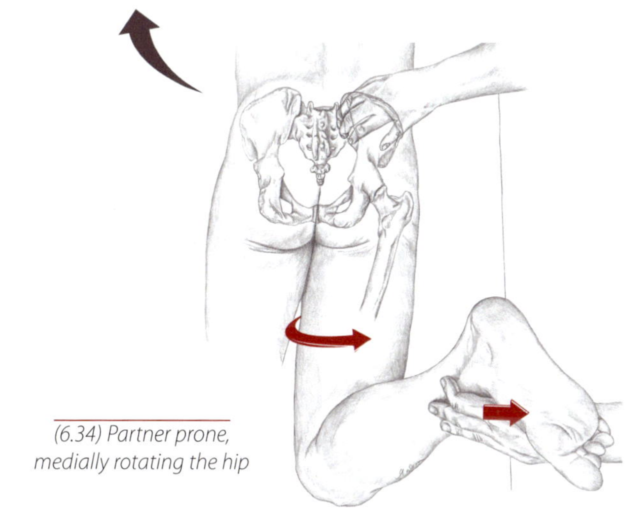

(6.34) Partner prone, medially rotating the hip

The Greek philosopher Herophilus named the last segments of the vertebral column the "kokkyx" since it resembled a cuckoo's beak. However, during the Renaissance, the French anatomist Jean Riolan thought the term referred to the release of gas from the anus that can sound like the cry of a cuckoo. The coccyx is also called the "tailbone" - an appropriate term when referring to the human fetus: During early development a small, distinct tail extends off the sacrum, but by the eighth week it disappears, leaving just what is recognizable as the coccyx.

coccyx **kok**-siks Grk. cuckoo

Trail 4 "Hip Hike"

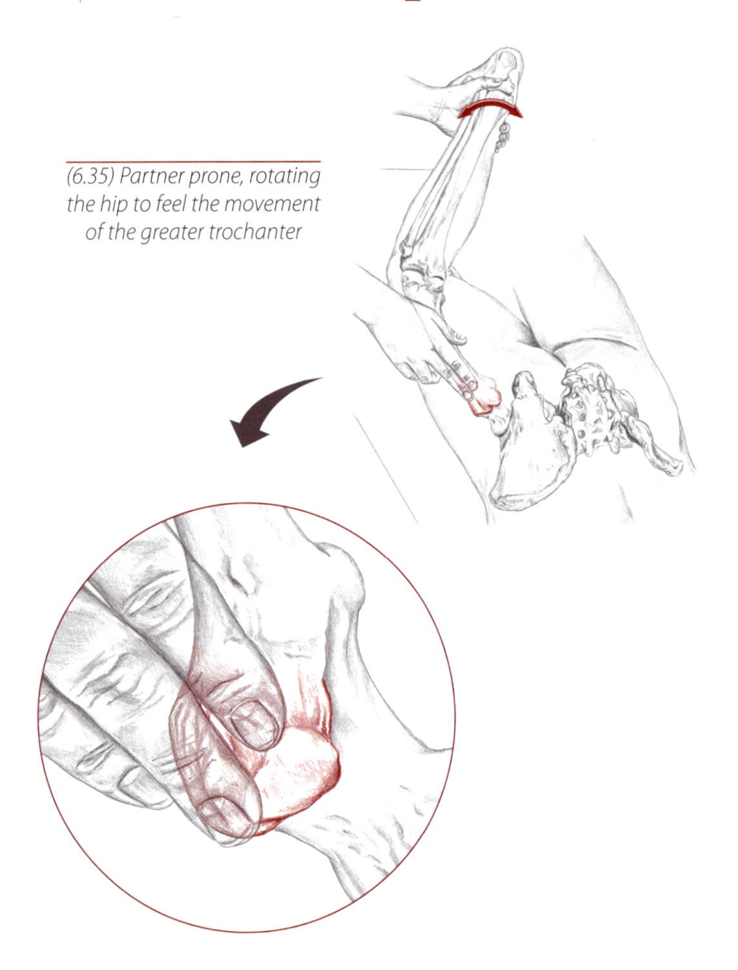

(6.35) Partner prone, rotating the hip to feel the movement of the greater trochanter

Greater Trochanter

(Refer to p. 285 for more information)

1) Partner prone. Locate the middle of the iliac crest.
2) Slide your fingerpads distally four or five inches along the side of the thigh. There you will feel the superficial hump of the greater trochanter.
3) Sculpt around its two-inch-wide surface and explore all of its sides.

Holding the ankle, flex the knee to 90°. As your proximal hand palpates the greater trochanter, use the other hand to medially and laterally rotate the hip (6.35). Do you feel the trochanter swivel back and forth under your fingers?

Gluteal Tuberosity

The gluteal tuberosity is located distal to the posterior surface of the greater trochanter. It is a small ridge serving as an attachment site for the lower fibers of the gluteus maximus muscle. Although it is surrounded by the gluteus maximus tendon and the upper fibers of the vastus lateralis muscles (6.37), the gluteal tuberosity is relatively superficial and accessible.

1) Partner prone. Locate the posterior surface of the greater trochanter.
2) Slide one or two inches distally along the posterior shaft of the femur until you feel the solid surface of the tuberosity (6.36). It may not feel like a ridge, but more like a flat, superficial portion of bone.

Can you press into the area you are palpating and feel the superficial surface of the femur? Are you directly lateral to the ischial tuberosity (p. 285)?

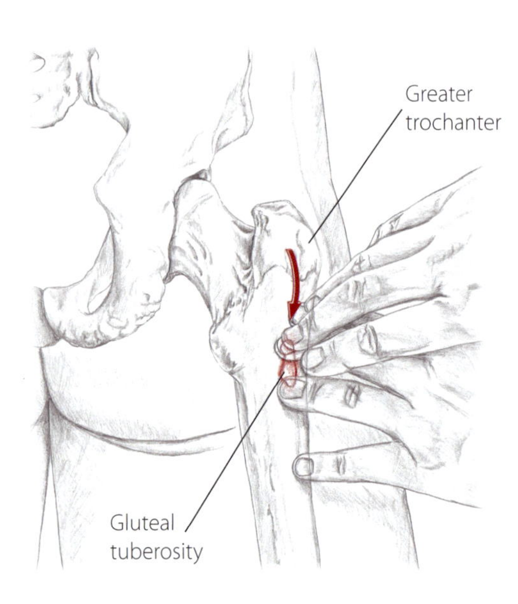

(6.36) Posterior view

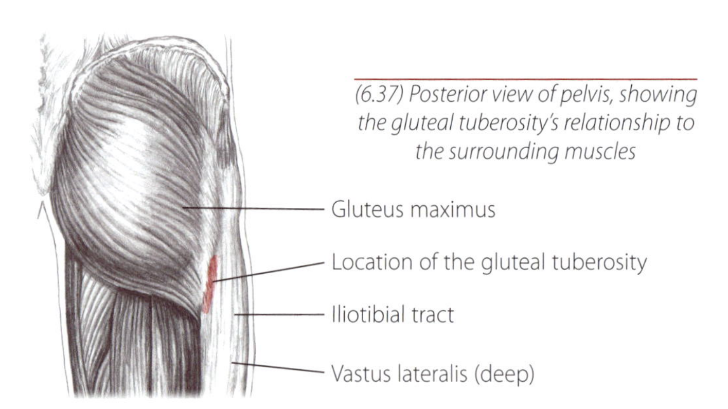

(6.37) Posterior view of pelvis, showing the gluteal tuberosity's relationship to the surrounding muscles

Trail 5 "The Underpass"

Umbilicus

The umbilicus (or navel) will, of course, be visible when the abdomen is undraped. When not exposed, the umbilicus can be felt at the midline of the body, superior to the level of the ASISes.

Pubic Crest and Tubercles

(Refer to p. 284 for more information)

The pubic tubercles are located on the superior aspect of the pubic crest. Each tubercle is shaped like a small horn and serves as an attachment site for the adductor longus muscle and the inguinal ligament. The tubercles may be one to two inches apart and are not always easy to palpable.

1) Face your partner as he lies supine on the table. Set your fingertips on his umbilicus, allowing your palm to rest on the abdomen. The heel of your hand will be on (or just superior to) the pubic crest.
2) Turn your hand and relocate the crest with your fingers (6.38). Explore its horizontal ridge. Remember that the pubic crest is the only horizontal stretch of bone in this vicinity.
3) Move laterally and explore for the tips of the pubic tubercles. Palpate both tubercles, noting the distance between them.

Begin at the ASIS and follow the inguinal ligament (p. 333) inferiorly and medially 45° to the pubic tubercle.

Do you feel a firm, bony prominence inferior and medial to the level of the ASIS? With respect to the pubic tubercles, are the bony prominences you feel on the superior part of the pubic crest? Are the tubercles on the same level as the greater trochanters?

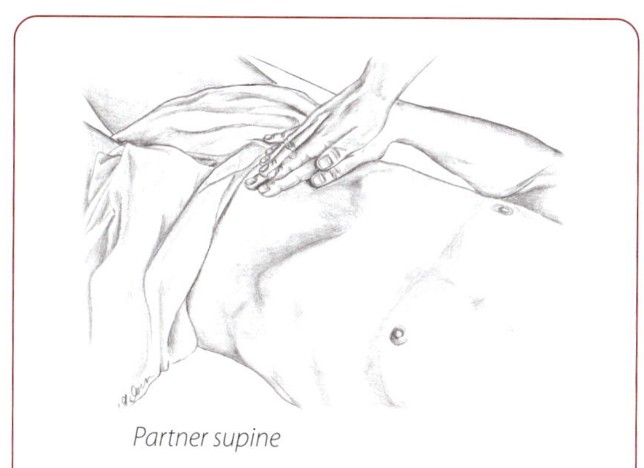

Partner supine

Here are a few suggestions to make sure this route is comfortable for you and your partner: **a)** Explain to your partner what you will be doing and obtain permission to proceed. **b)** If your partner would be more comfortable, use his or her hand to palpate with your hand guiding on top (above).

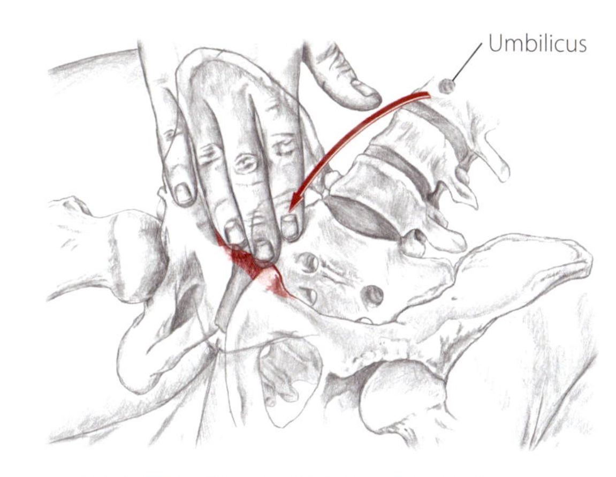

(6.38) Anterior/lateral view, partner supine

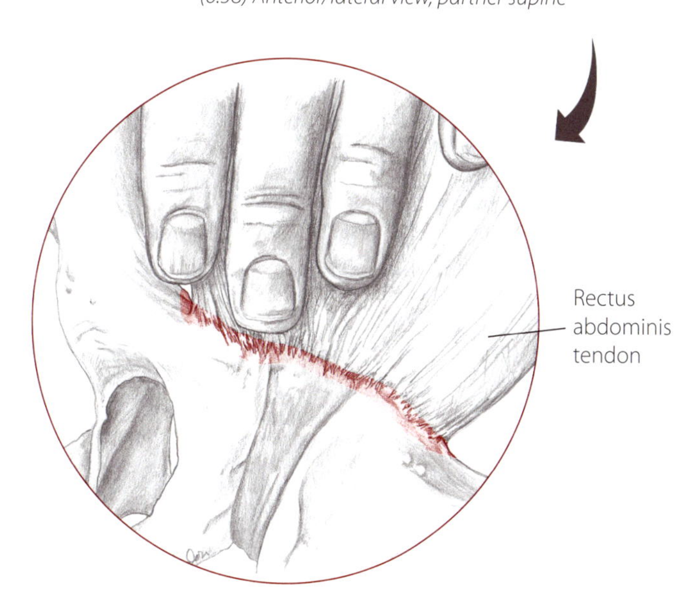

symphysis	**sim**-fi-sis	Grk. growing together
umbilicus	um-**bil**-i-kus	L. navel, center

When palpating the superior ramus, be mindful of the pulse of the femoral artery (p. 333).

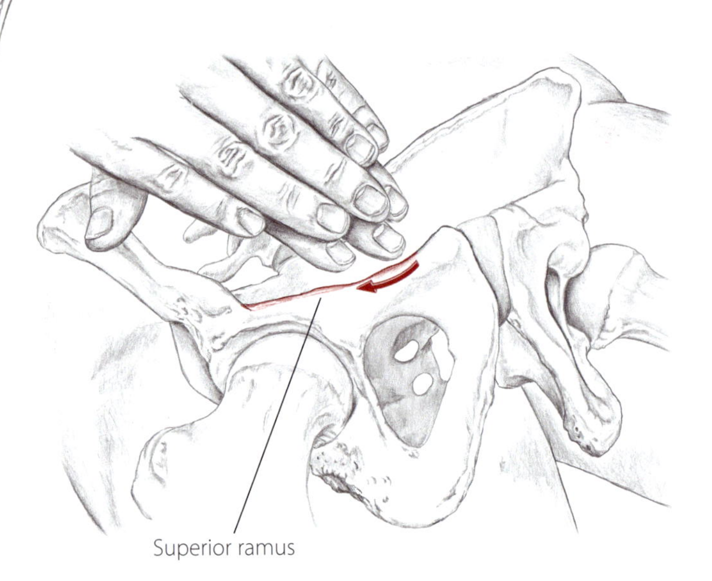

(6.39) Partner supine

Superior Ramus of the Pubis

The superior ramus of the pubis spans 45° from the pubic tubercle toward the AIIS. It forms a ridge that serves as an attachment site for the pectineus muscle (p. 313). Since it is deep to the inguinal ligament and a neurovascular bundle, the superior ramus can be challenging to palpate.

1) Partner supine. Place your flexed knee under your partner's knee. This position will flex and laterally rotate the hip, allowing for easier palpation.
2) Locate the pubic crest. Slide laterally off the crest toward the AIIS. Sink into the tissue, feeling for the buried ridge of the superior ramus (6.39).

Are you lateral and slightly superior to the pubic tubercle? If you cannot feel the edge of the ramus, can you sense its density beneath the superficial tissue?

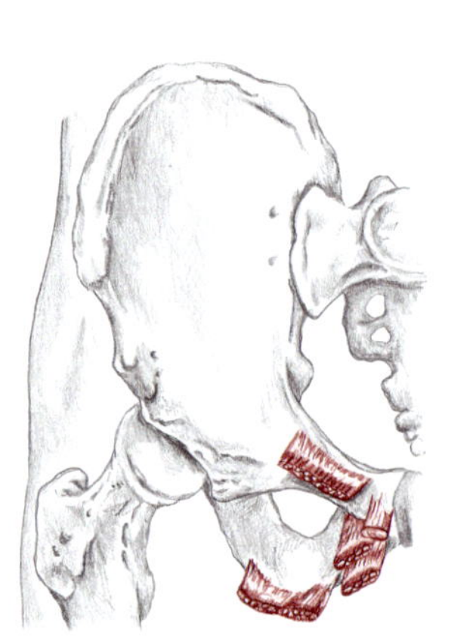

(6.40) Adductor attachment sites along the rami

Inferior Ramus of the Pubis and Ramus of the Ischium

The two rami are located along the inferior aspect of the pelvis and together form a bridge between the pubic crest and the ischial tuberosity. The ramus of the pubis, the anterior half of the bridge, serves as an attachment site for the gracilis and adductor brevis muscles; both rami are attachment sites for the adductor magnus muscle (6.40). When palpating the rami, use your fingertips, keeping them close to the medial thigh. The angle formed by the rami will be wider on females than males.

1) Partner supine. Place your flexed knee under your partner's knee.
2) Locate the pubic crest. Then move to the lateral edge of the pubic crest and slide posteriorly around the medial thigh (6.41). Using slow, but firm pressure, palpate for the hard ridge of the rami. This "bridge of bone" is the only bony mass in the area, so if you are pressing on a solid line of bone, you have found it.
3) Continue around the thigh until you reach the large ischial tuberosity.

As you follow the rami, do they lead you posteriorly around the inside of the thigh? As you move around the thigh, do you feel the rami widen laterally? Can you feel where any of the adductor tendons (p. 313) attach to the rami?

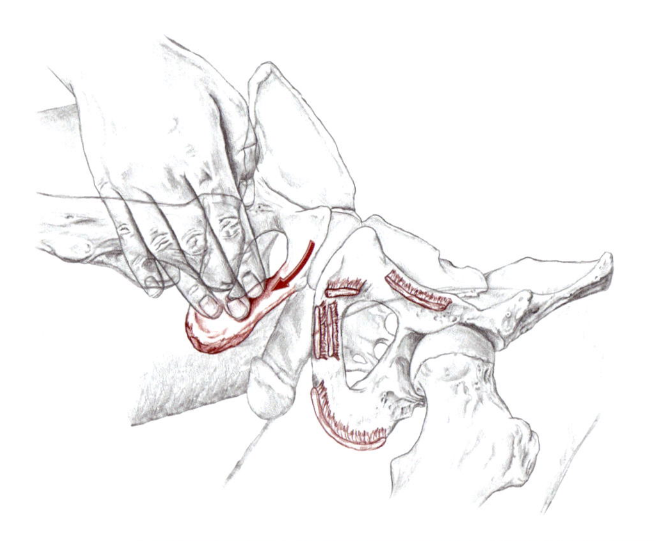

(6.41) Partner supine, palpating from the rami to the ischial tuberosity

ramus **ray**-mus L. branch

Ischial Tuberosity

(Refer to p. 285 for more information)

1) Prone. Locate the gluteal fold, the horizontal line between the buttock and thigh. Place your fingers at the center of the gluteal fold and press superiorly and medially until your fingertips bump into the large surface of the ischial tuberosity (6.42).
2) Explore all sides of its large mass and note its relationship to the greater trochanter.

Are you palpating between the inferior buttock and proximal thigh? Can you feel the large hamstring tendons attach to the ischial tuberosity?

Partner sidelying, with the top hip flexed. Place your hand on the medial thigh. Slide proximally to the gluteal fold and ischial tuberosity (6.43).

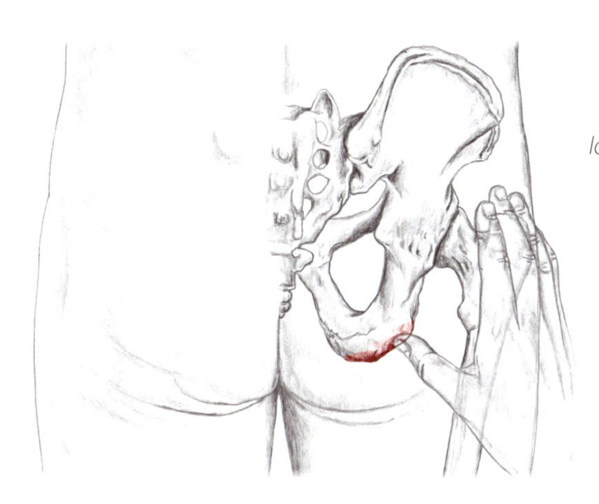

(6.42) Posterior view, locating the ischial tuberosity

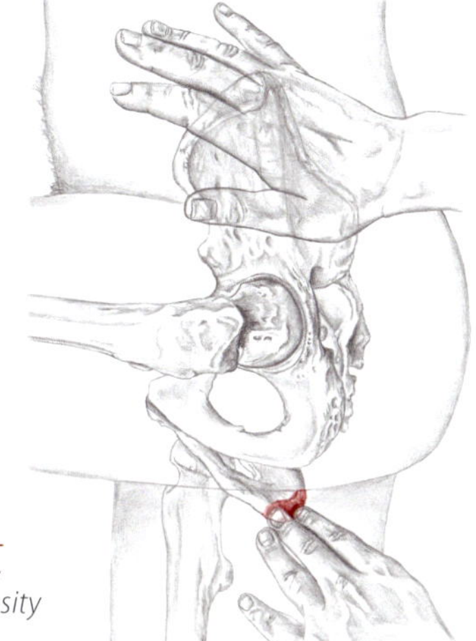

(6.43) Partner sidelying, locating the ischial tuberosity

"How do you access bones and muscles that are close to the genitals?" Actually, all of the bony landmarks, tendons and blood vessels in this region can be easily palpated without contacting the genitals (below). If you follow the instructions given, the comfort of you and your partner will be maintained.

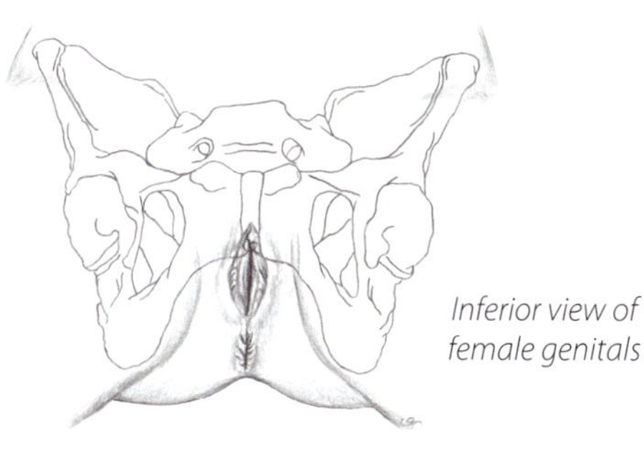

Inferior view of female genitals

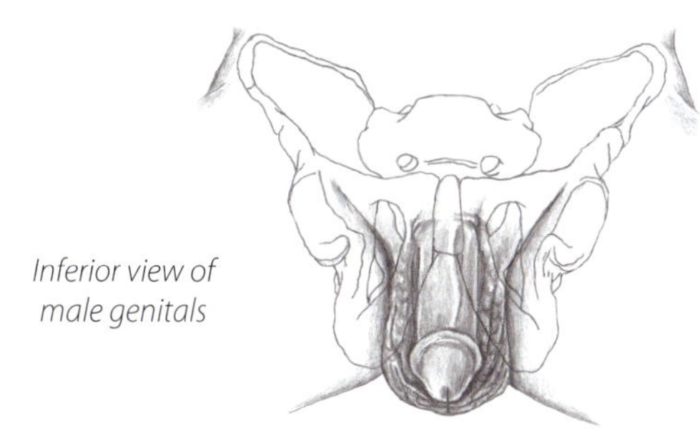

Inferior view of male genitals

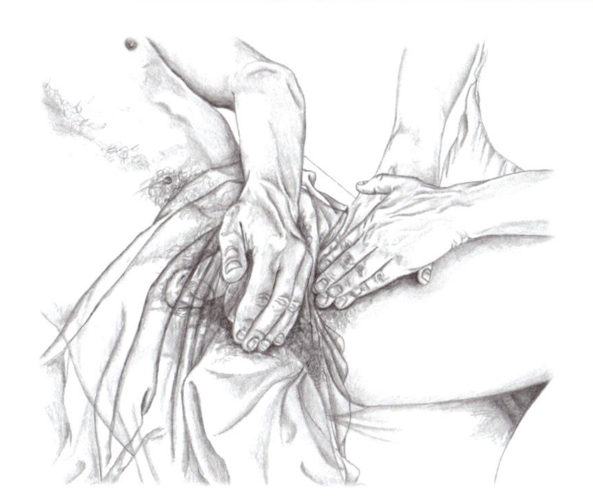

Partner supine

With that said, it should be obvious that palpation on males is complicated by the position of the penis and testicles. In a supine position, flexion and lateral rotation of the thigh will bring it away from the pelvis and allow for easier palpation. (See p. 316, for example.)

The sidelying position allows the genitals to shift away from the base of the pelvis. To make sure that your partner's genitals are away from the area you are trying to access, ask him to shift and hold his genitals away from the side you will be contacting (above).

Muscles of the Pelvis and Thigh

The muscles of the pelvis and thigh primarily create movement at the coxal (hip) and tibiofemoral (knee) joints. Most of the hip and thigh muscles can be divided into five groups. There are two groups in the buttock region while three make up the mass of the thigh:

1) Three **gluteal** muscles give shape to the buttock and lateral hip.
2) Six small **lateral rotators** are deep to the gluteals.
3) Four **quadriceps** are located on the thigh's anterior and lateral surfaces.
4) Three long **hamstrings** lie along the posterior thigh.
5) Five **adductors** are tucked between the quadriceps and hamstrings along the medial thigh.

Additional muscles include the iliopsoas, sartorius and tensor fasciae latae.

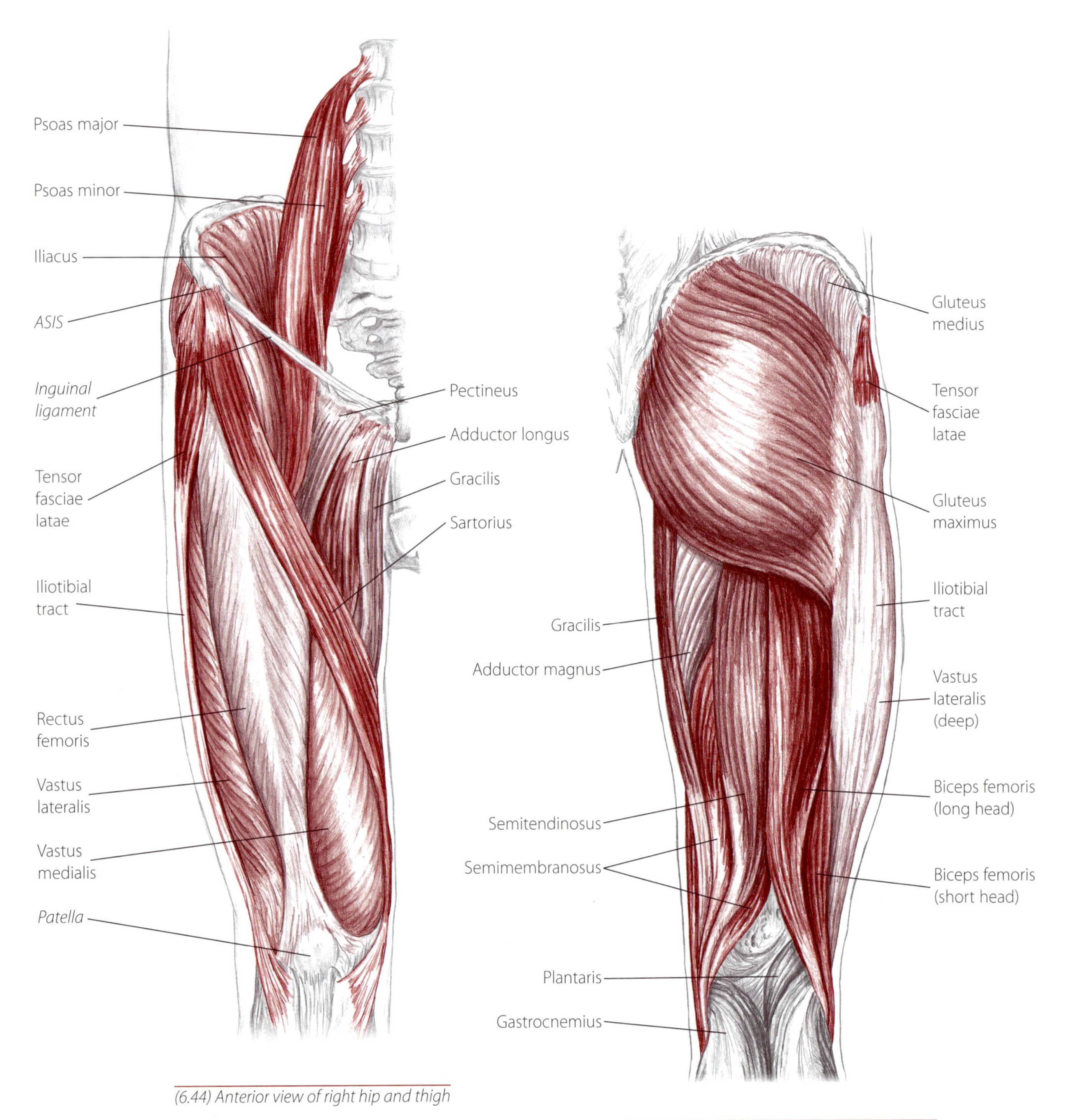

(6.44) Anterior view of right hip and thigh

(6.45) Posterior view of right buttock and thigh

Muscles of the Pelvis and Thigh

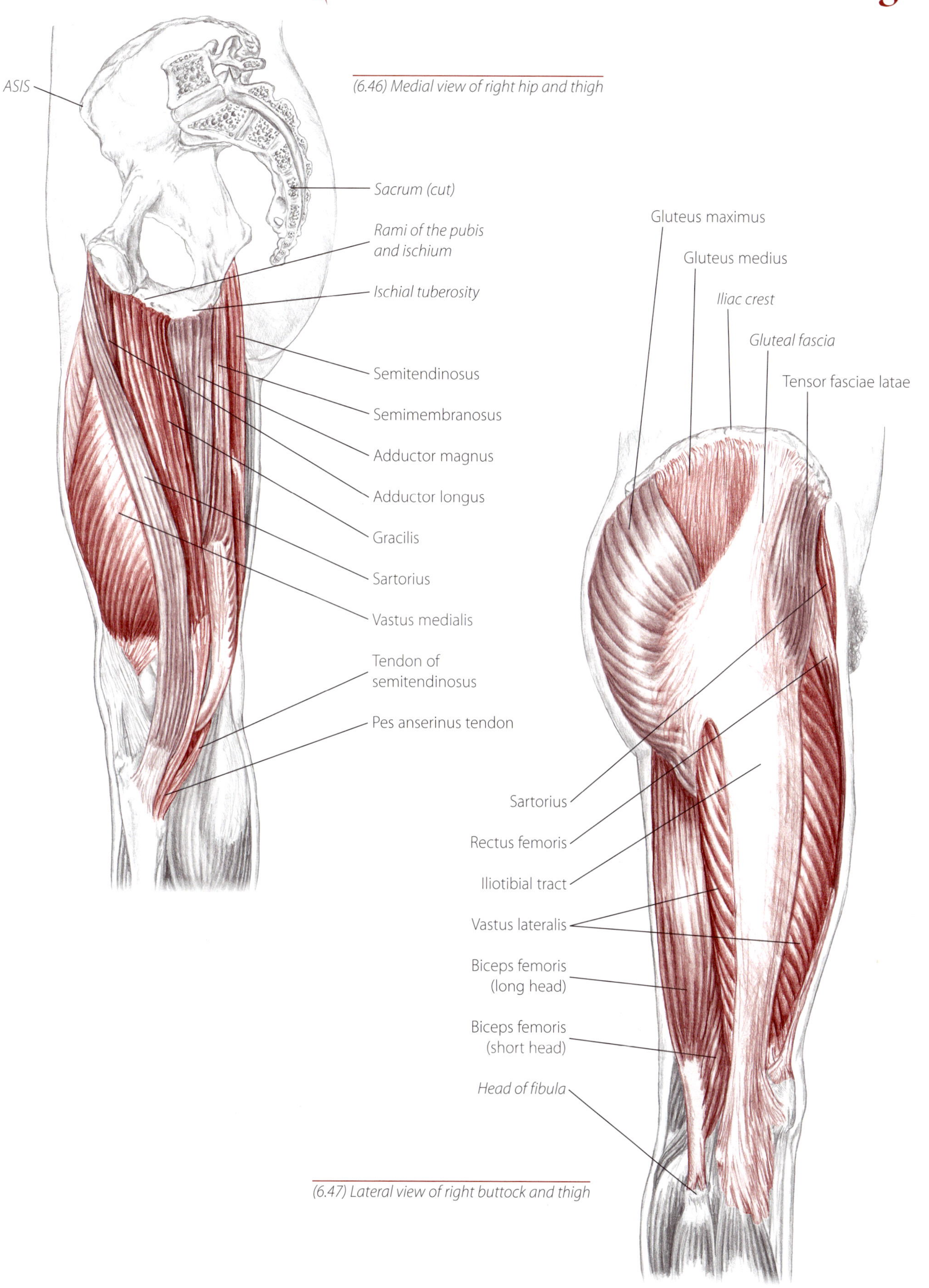

(6.46) Medial view of right hip and thigh

(6.47) Lateral view of right buttock and thigh

Synergists - Muscles Working Together

*muscles not shown

Coxal

(hip joint)

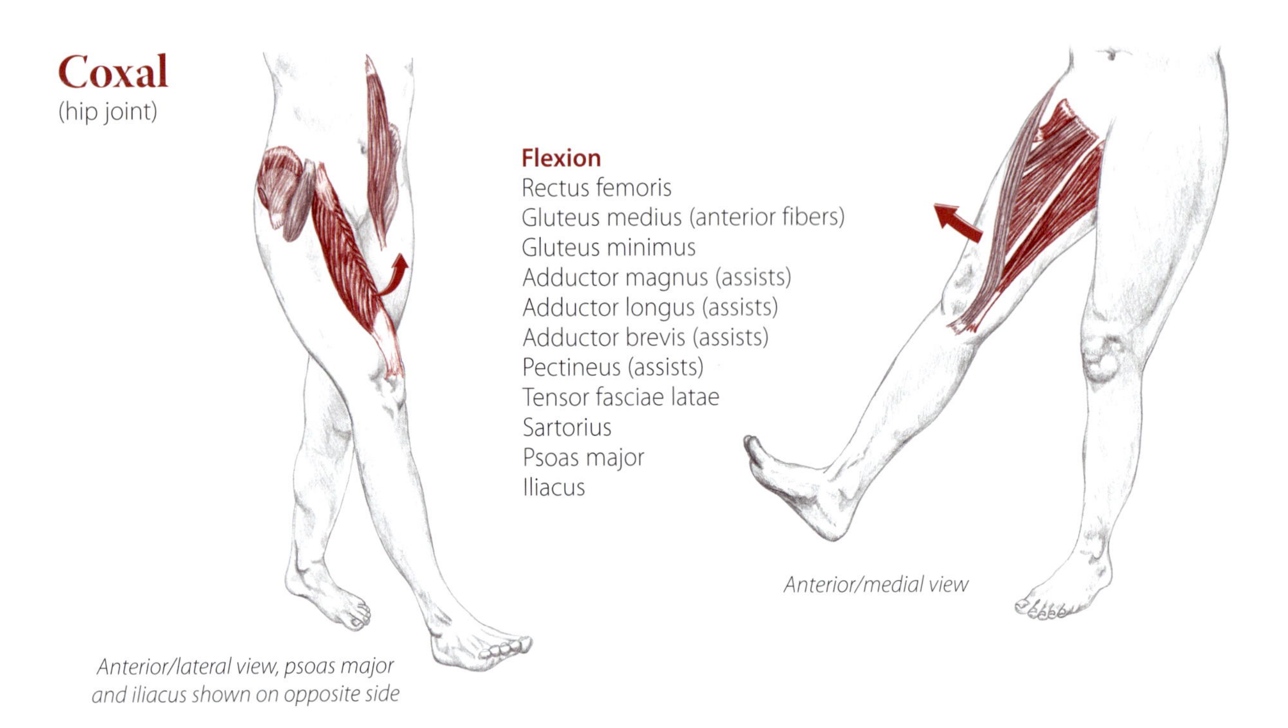

Flexion

Rectus femoris
Gluteus medius (anterior fibers)
Gluteus minimus
Adductor magnus (assists)
Adductor longus (assists)
Adductor brevis (assists)
Pectineus (assists)
Tensor fasciae latae
Sartorius
Psoas major
Iliacus

Anterior/medial view

Anterior/lateral view, psoas major and iliacus shown on opposite side

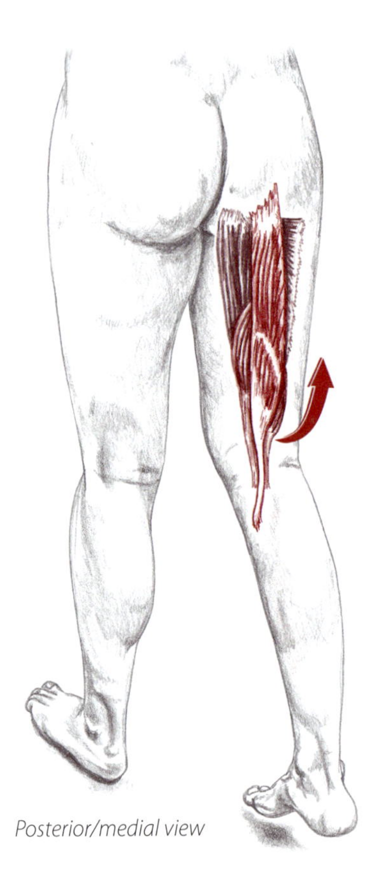

Posterior/medial view

Extension

Biceps femoris
Semitendinosus
Semimembranosus
Gluteus maximus
(all fibers)
Gluteus medius
(posterior fibers)
Adductor magnus
(posterior fibers)

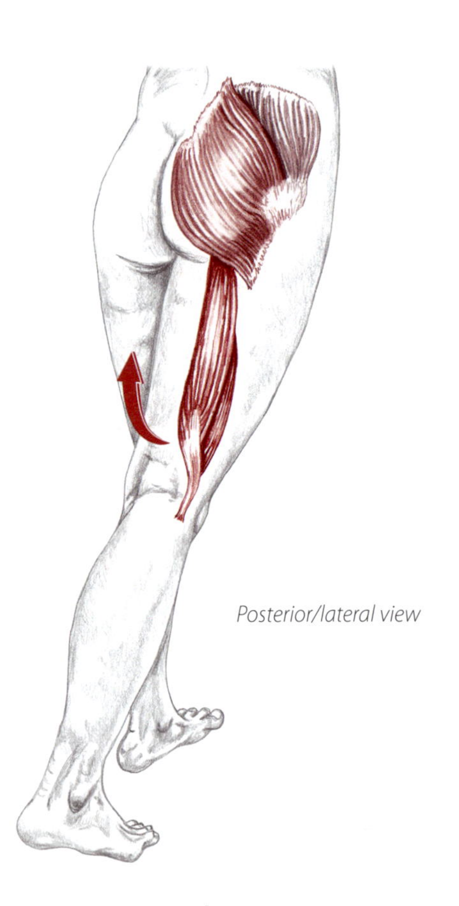

Posterior/lateral view

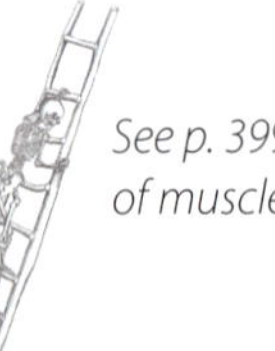

See p. 399 for a complete list of muscles that tilt the pelvis

Medial Rotation (internal rotation)

Semitendinosus
Semimembranosus
Gluteus medius
(anterior fibers)
Gluteus minimus
Adductor magnus
Adductor longus
Adductor brevis
Gracilis
Pectineus
Tensor fasciae latae

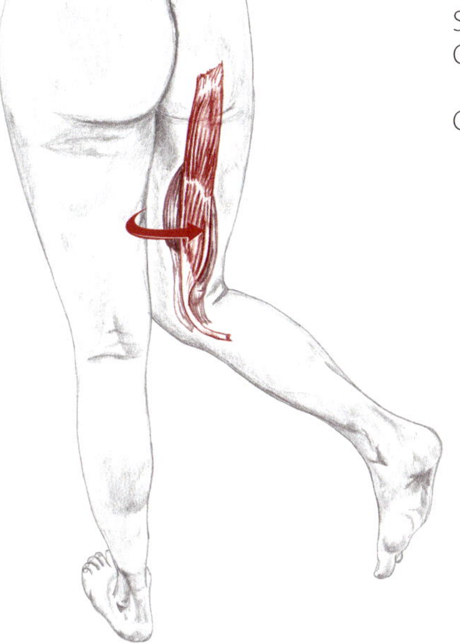

Posterior/medial view

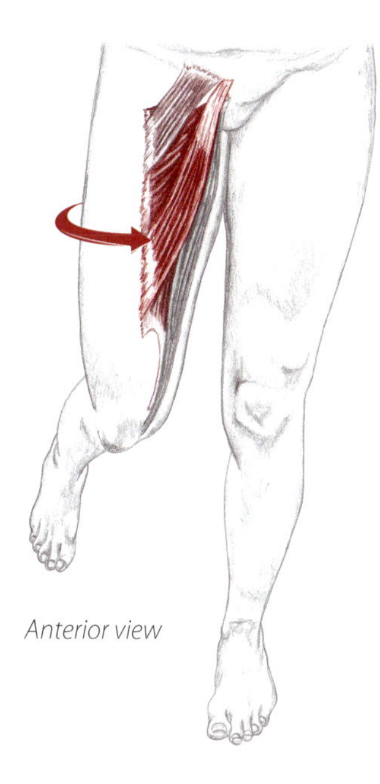

Anterior view

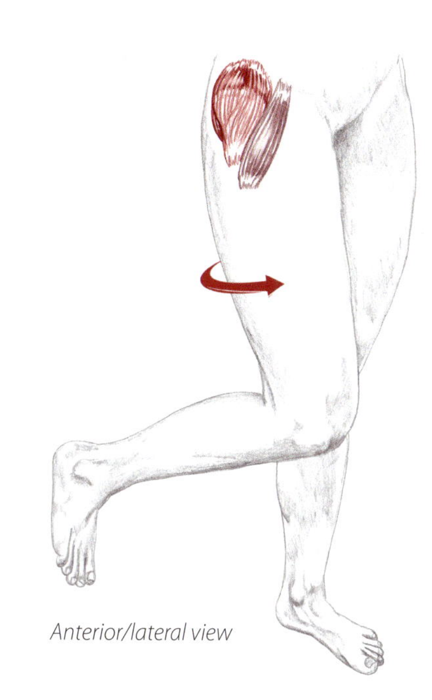

Anterior/lateral view

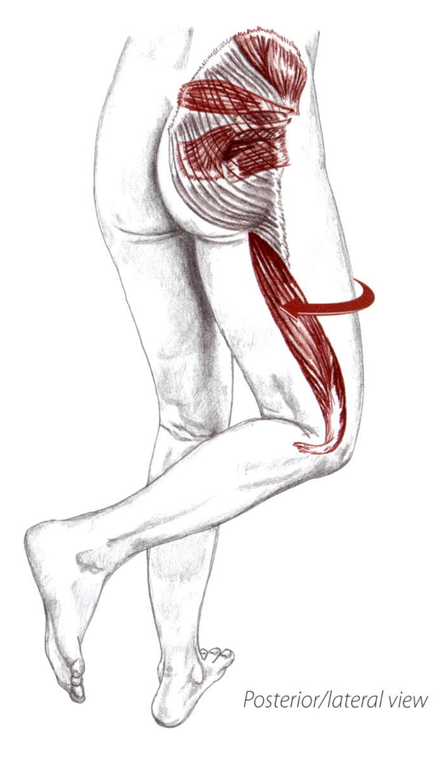

Posterior/lateral view

Lateral Rotation (external rotation)

Biceps femoris
Gluteus maximus
(all fibers)
Gluteus medius
(posterior fibers)
Sartorius
Piriformis
Quadratus femoris
Obturator internus
Obturator externus
Gemellus superior
Gemellus inferior
Psoas major
Iliacus

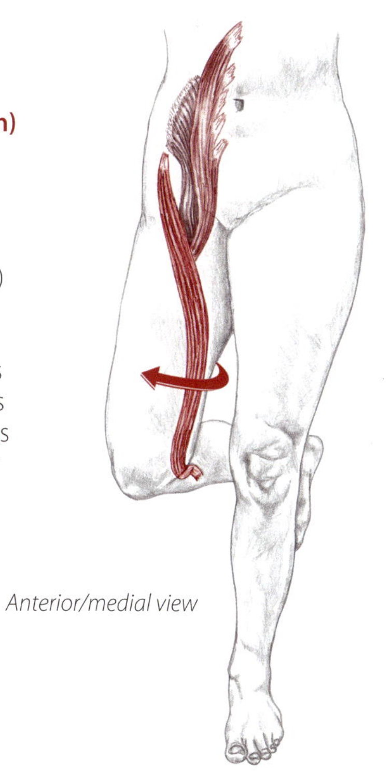

Anterior/medial view

Coxal
(hip joint)

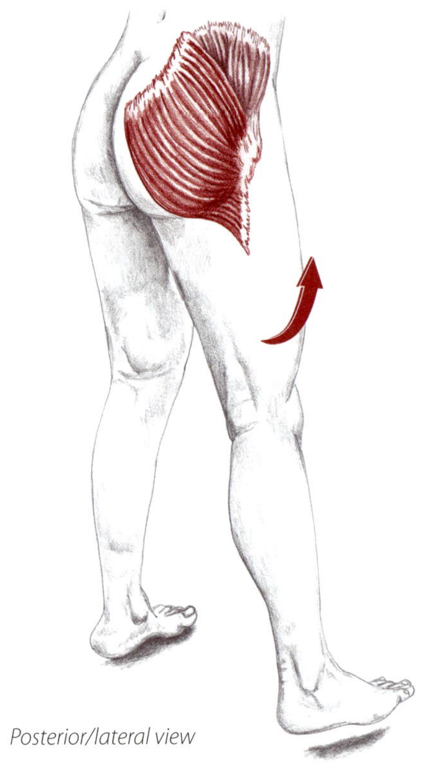
Posterior/lateral view

Abduction
Gluteus maximus (all fibers)
Gluteus medius (all fibers)
Gluteus minimus
Tensor fasciae latae
Sartorius
Piriformis (when the hip is flexed)*

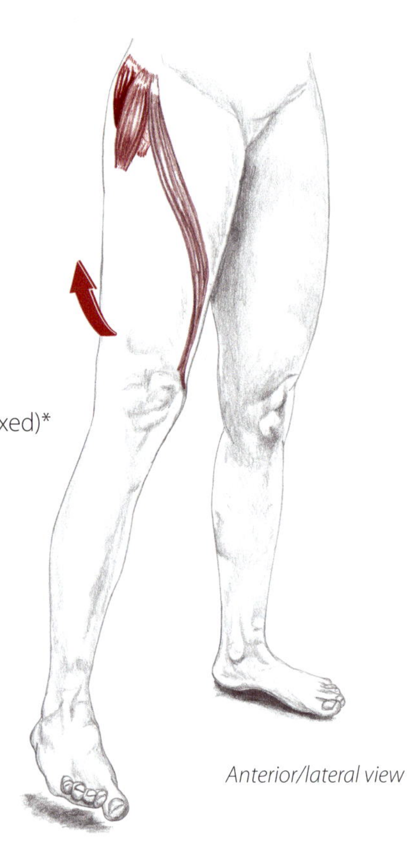
Anterior/lateral view

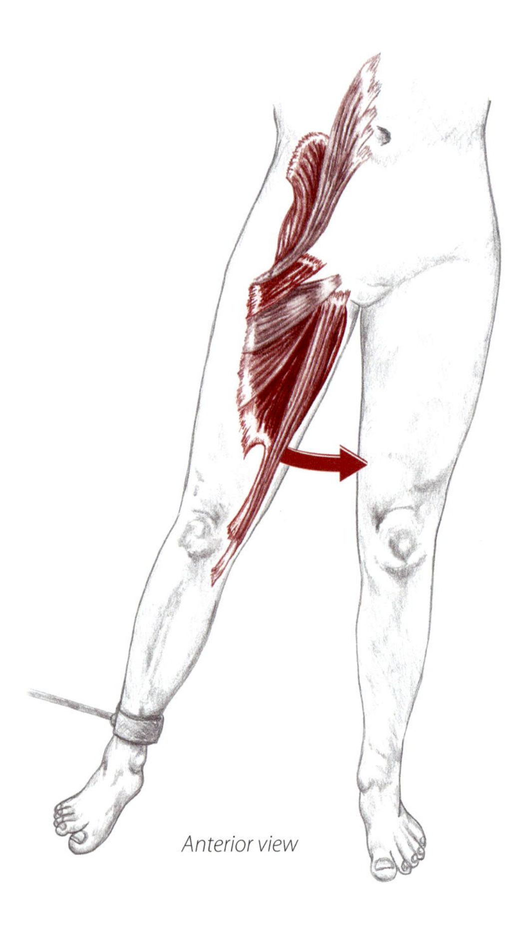
Anterior view

Adduction
Adductor magnus
Adductor longus
Adductor brevis
Pectineus
Gracilis
Psoas major
Iliacus
Gluteus maximus
(lower fibers)

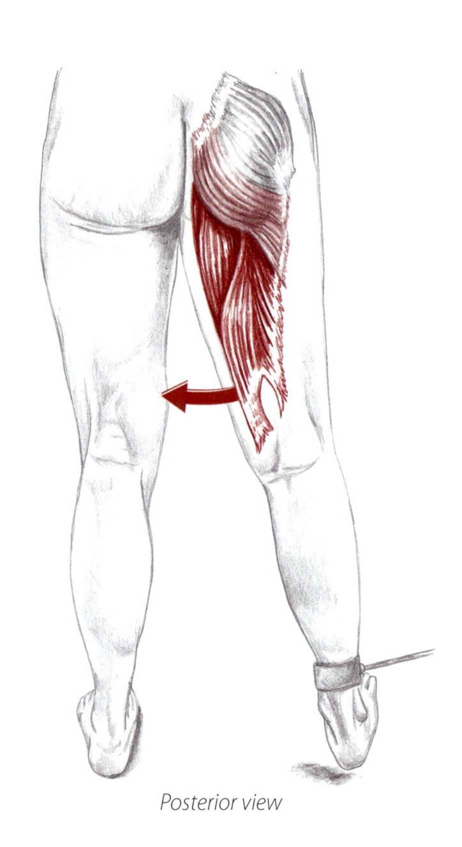
Posterior view

Knee
(tibiofemoral joint)

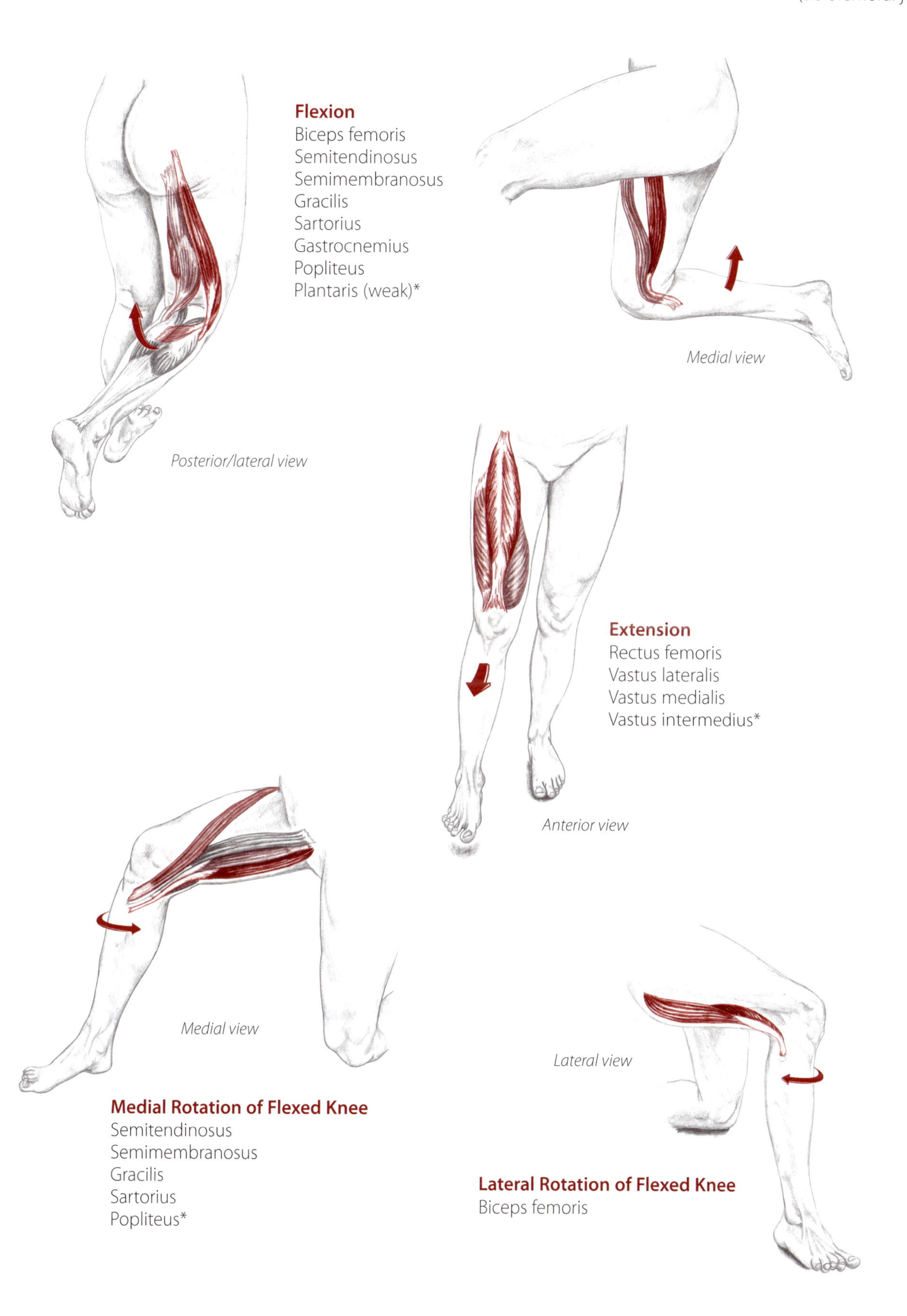

Flexion
Biceps femoris
Semitendinosus
Semimembranosus
Gracilis
Sartorius
Gastrocnemius
Popliteus
Plantaris (weak)*

Extension
Rectus femoris
Vastus lateralis
Vastus medialis
Vastus intermedius*

Medial Rotation of Flexed Knee
Semitendinosus
Semimembranosus
Gracilis
Sartorius
Popliteus*

Lateral Rotation of Flexed Knee
Biceps femoris

Quadriceps Femoris Group

Rectus Femoris
Vastus Medialis
Vastus Lateralis
Vastus Intermedius

The four large quadriceps muscles primarily extend the knee. The cylindrical, superficial **rectus femoris** is located on the anterior thigh and is the only quadriceps that crosses two joints - the hip and knee (6.48). **Vastus intermedius** is deep to the rectus femoris; however, its edges can be accessed if the rectus femoris is shifted to the side (6.49).

The palpable aspect of **vastus medialis** forms a "tear-drop" shape at the distal portion of the medial thigh (6.50) while the **vastus lateralis** is the sole muscle of the lateral thigh. The posterior edge of the lateralis lies next to the biceps femoris, one of the hamstrings. Although vastus lateralis is deep to the iliotibial tract (p. 318), its fibers are easily accessible (6.51).

All four quadriceps muscles converge into a single tendon above the knee. The tendon connects to the top and sides of the patella before attaching to the tibial tuberosity.

A *All:*
Extend the knee (tibiofemoral joint)

Rectus Femoris:
Flex the hip (coxal joint)

O *Rectus Femoris:*
Anterior inferior iliac spine (AIIS)

Vastus Lateralis:
Lateral lip of linea aspera, gluteal tuberosity

Vastus Medialis:
Medial lip of linea aspera

Vastus Intermedius:
Anterior and lateral shaft of the femur

I Tibial tuberosity

N Femoral

AIIS
Rectus femoris
Vastus lateralis
Vastus medialis
Patella
Patellar ligament
Tibial tuberosity

(6.48) Anterior view of right hip and thigh

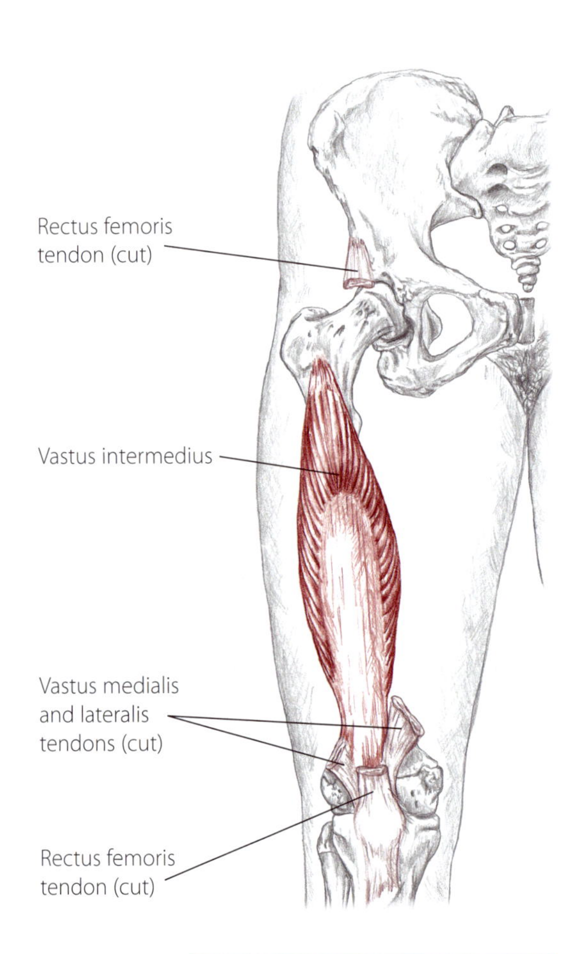

(6.49) Anterior view of right hip and thigh

quadriceps	**kwod**-ri-seps	L. four-headed
rectus	**rek**-tus	L. straight

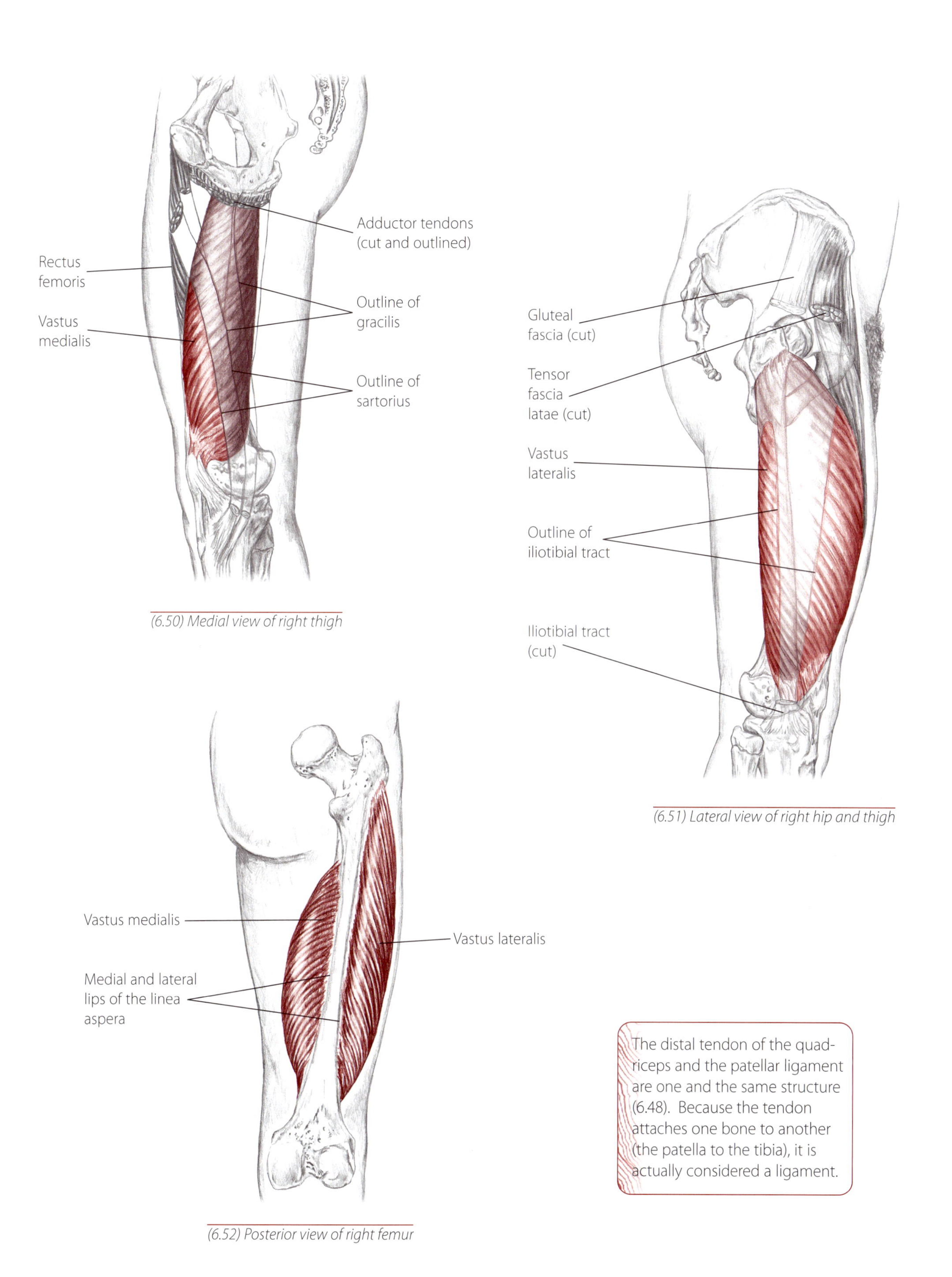

(6.50) Medial view of right thigh

(6.51) Lateral view of right hip and thigh

(6.52) Posterior view of right femur

The distal tendon of the quadriceps and the patellar ligament are one and the same structure (6.48). Because the tendon attaches one bone to another (the patella to the tibia), it is actually considered a ligament.

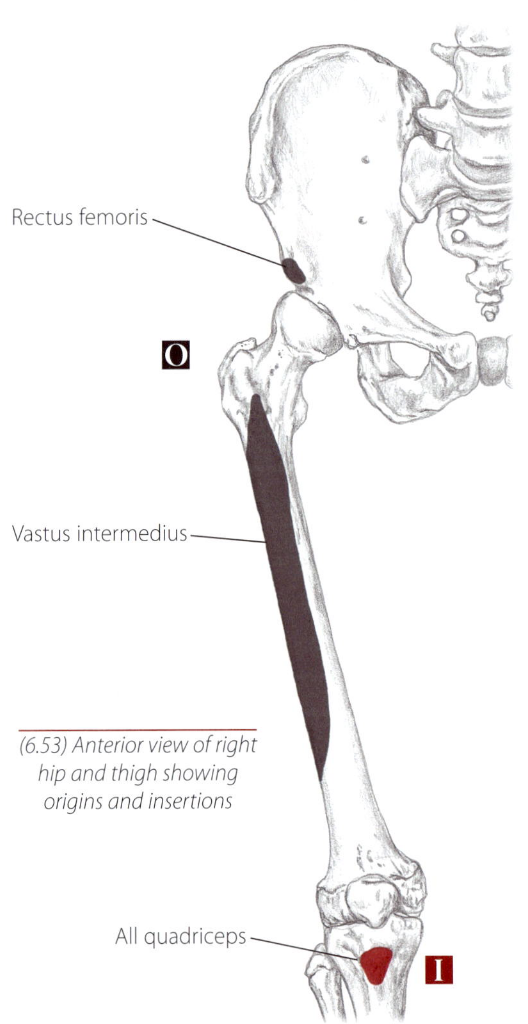

(6.53) Anterior view of right hip and thigh showing origins and insertions

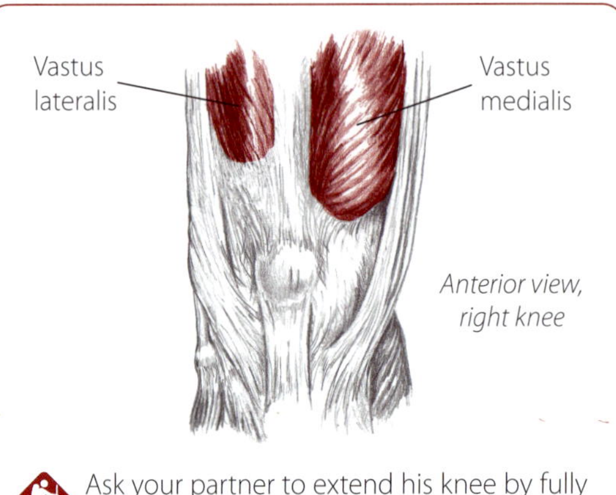

Anterior view, right knee

Ask your partner to extend his knee by fully contracting his quadriceps. Observe and palpate the distal ends of the vastus medialis and vastus lateralis. Do you notice that the vastus medialis extends further distally than the vastus lateralis? The reason for this variance concerns the tracking (or movement) of the patella. The angle of the femur, combined with the pull of the quadriceps, causes the patella to track laterally. This is prevented, however, in two ways: First, the edge of the lateral condyle of the femur (p. 346) is elevated, forming a lateral wall, and secondly, the distal fibers of vastus medialis are set at an angle, pulling the patella medially.

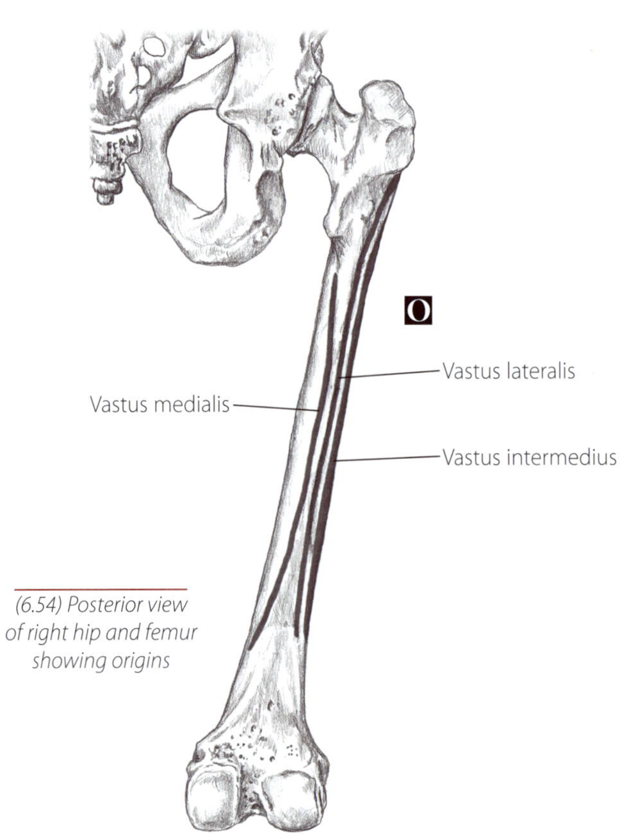

(6.54) Posterior view of right hip and femur showing origins

Quadriceps as a group

1) Partner seated. Lay the flat of your hands on the anterior surface of the thigh.
2) Ask your partner to alternately extend and relax his knee slowly. Explore the lateral and medial sides of the thigh (6.55). Do you feel the quadriceps tighten as the knee extends? For greater contraction, provide a little resistance below the knee as your partner tries to raise it.

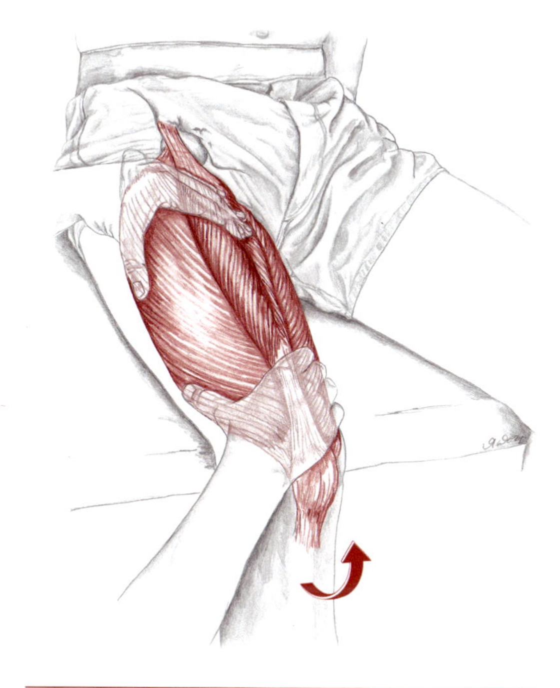

(6.55) Partner seated, palpating the quadriceps as a group

Rectus femoris

1) Supine with knee bolstered. Locate the AIIS (p. 286) and the patella (p. 344).
2) Draw an imaginary line between these two points and follow the path of the rectus femoris (6.56).
3) Palpate along this line and strum across the rectus fibers. (It will be two to three fingers wide.)
4) Ask your partner to flex his hip and hold his foot up off the table (6.57). This position contracts the rectus femoris, making it more pronounced.

Are you on the anterior surface of the thigh? Can you follow the muscle belly to the patella and toward the AIIS? Can you shift it to the side and feel the density of vastus intermedius beneath it?

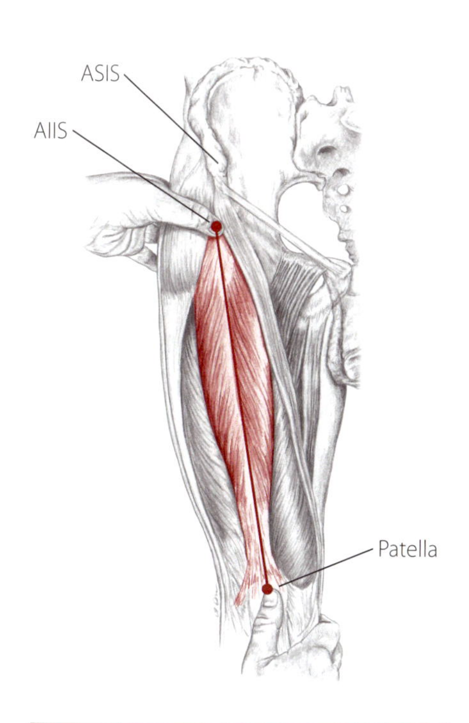

(6.56) Anterior view, drawing a line between the AIIS and patella to isolate the rectus femoris

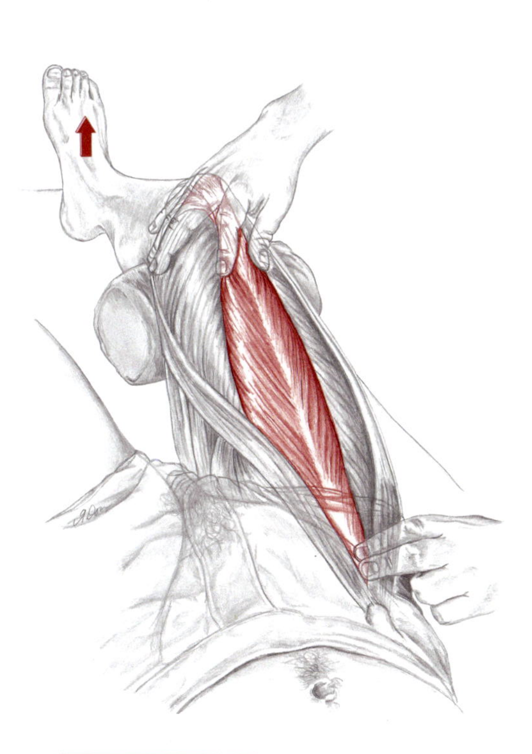

(6.57) Locating the rectus femoris as your partner flexes his hip and holds his foot off the table

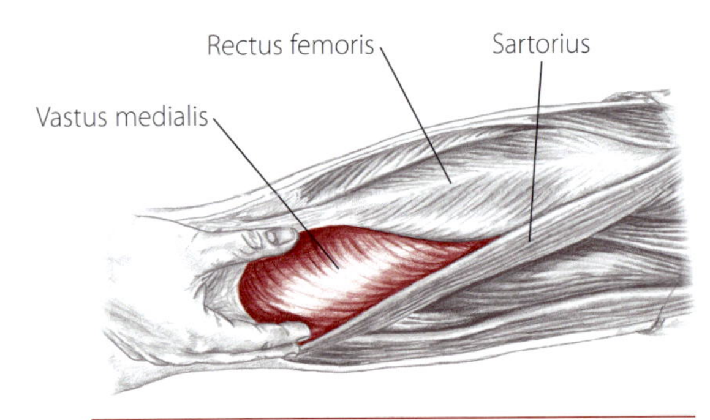

(6.58) Partner supine, anterior/medial view of right thigh

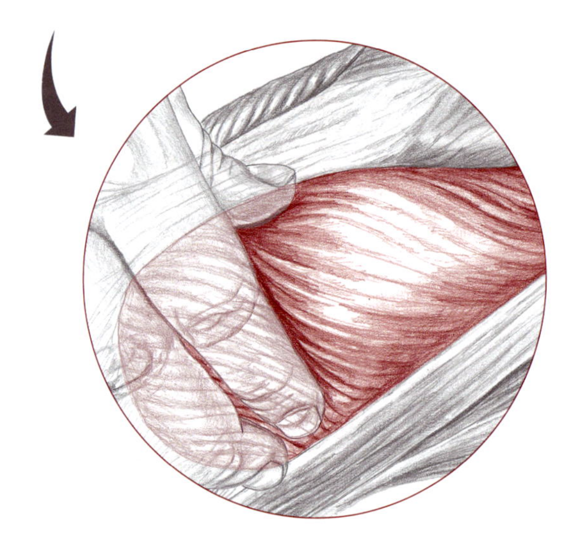

 Vastus medialis

1) Supine with the knee bolstered. Ask your partner to fully contract his quadriceps by extending his knee. Palpate just medial and proximal to the patella for the bulbous shape of the medialis.
2) Locate the rectus femoris and sartorius (p. 320), noting how these muscles surround the medialis to form its long "teardrop" shape (6.58).

Are you medial to the rectus femoris? Can you make out the round shape of the vastus medialis and follow its fibers to the patella?

 Vastus lateralis

1) Partner sidelying. Place the flat of your hand on the lateral side of the thigh while your partner slowly extends and relaxes his knee (6.59). Notice the vastus lateralis contracting and relaxing.
2) Palpate its entire belly - posteriorly to the biceps femoris (p. 305) and proximally to the greater trochanter. With the thigh relaxed, identify the direction and depth of the lateralis fibers and the superficial iliotibial tract (p. 318).

Can you follow its fibers to the patella? Can you differentiate between the vertical fibers of the iliotibial tract and the deeper, oblique fibers of the lateralis (6.60)?

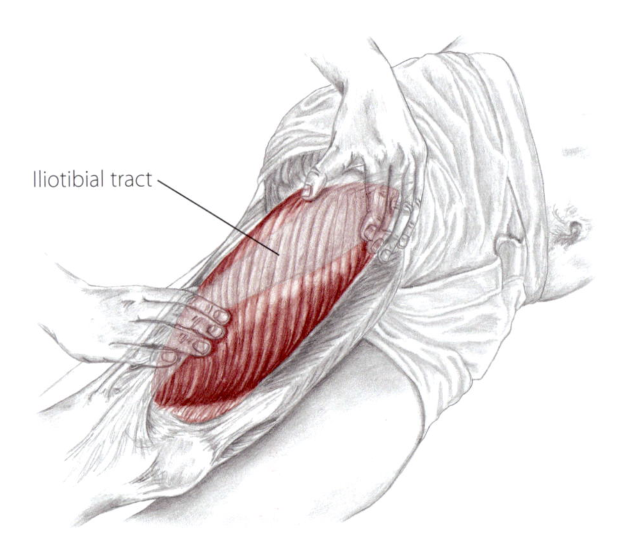

(6.59) Partner sidelying, palpating the vastus lateralis deep to the iliotibial tract

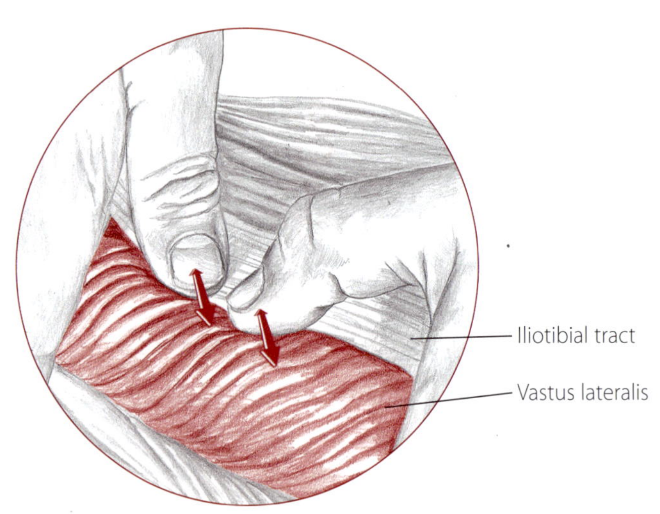

(6.60) Partner sidelying, looking distally down the thigh, rolling your fingers across the fibers of the vastus lateralis

Hamstrings

Biceps Femoris
Semitendinosus
Semimembranosus

The hamstrings are located along the posterior thigh between the vastus lateralis and adductor magnus (6.61). Comparatively, the hamstrings are not as massive as the quadriceps femoris group, but are nonetheless strong hip extensors and knee flexors. All three hamstrings have a common origin at the ischial tuberosity. Their tubular bellies extend superficially down the thigh before becoming long, thin tendons that stretch across the posterior knee. As a group, the hamstrings and their distal tendons are easily palpable.

Biceps femoris is the lateral hamstring. It has two heads - a superficial long head and a deeper, indiscernible short head (6.62, 6.63). The medial hamstrings include the two "semi" muscles. The **semitendinosus** lies superficial to the wider and deeper **semimembranosus** (6.64, 6.65).

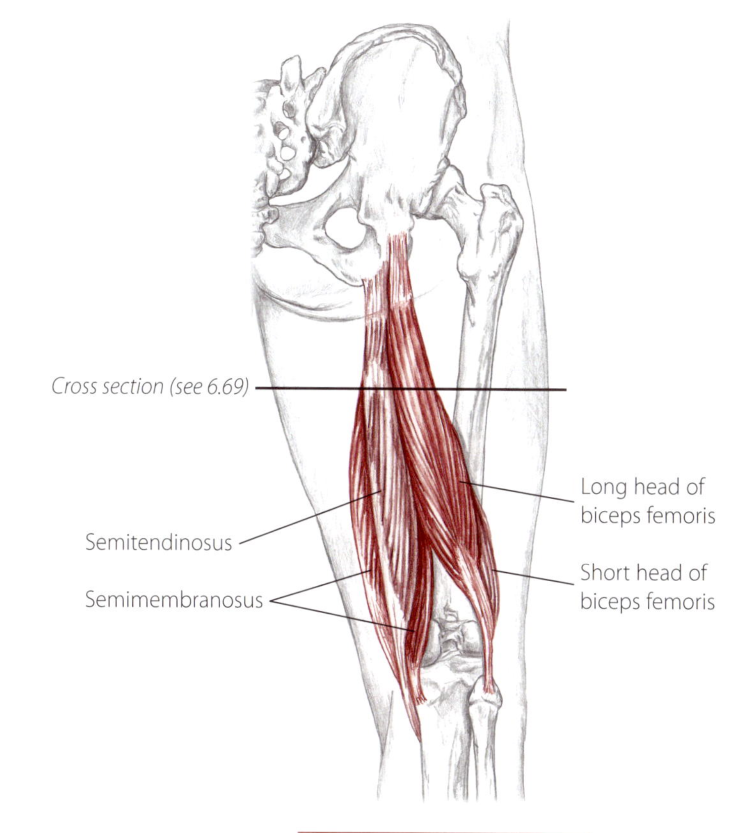

(6.61) Posterior view of right thigh showing superficial hamstrings

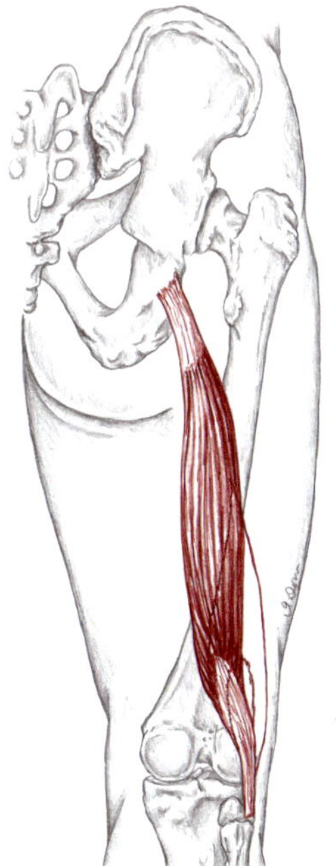

(6.62) Posterior view

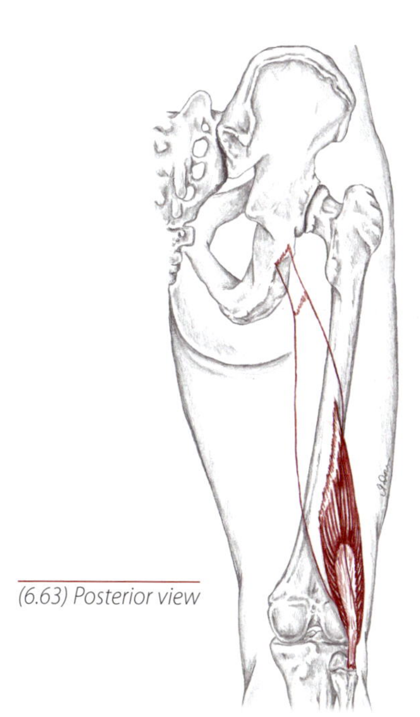

(6.63) Posterior view

The term "hamstring" originated in eighteenth century England. Back then, butchers would display pig carcasses in their shop windows by hanging them from the long tendons at the back of the knee.

semimembranosus	**sem**-eye-**mem**-bra-**no**-sus	L. half membranous
semitendinosus	**sem**-eye-**ten**-di-**no**-sus	L. half tendinous

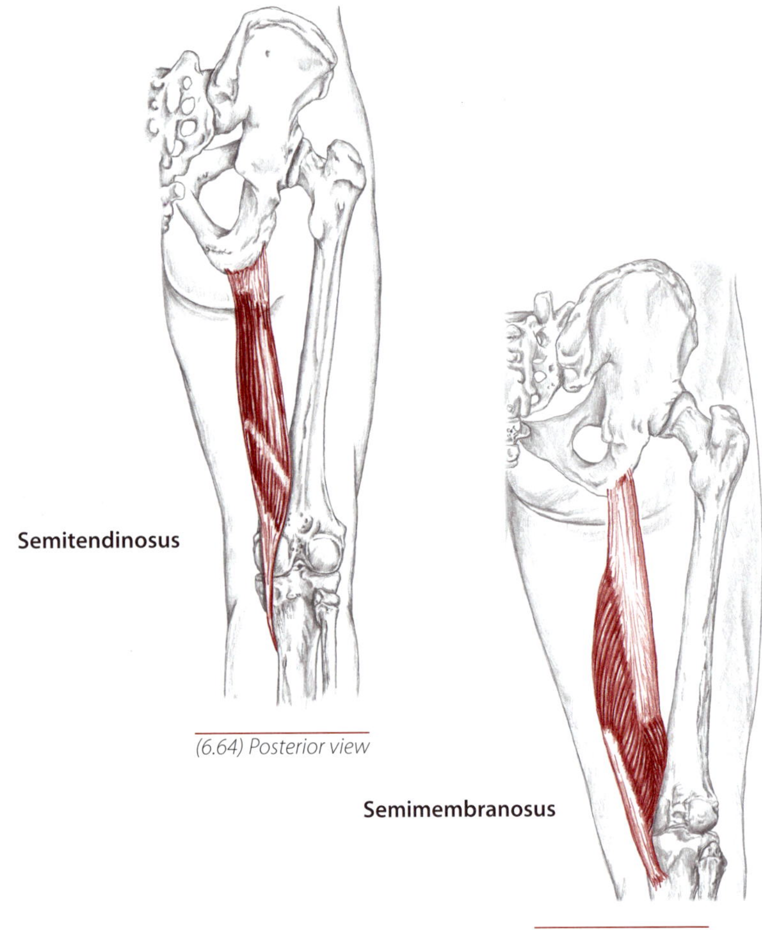

(6.64) Posterior view

(6.65) Posterior view

Biceps Femoris

A Flex the knee (tibiofemoral joint)
Laterally rotate the flexed knee (t/f joint)
Extend the hip (coxal joint)
Laterally rotate the hip (coxal joint)
Tilt the pelvis posteriorly

O *Long head of Biceps Femoris:*
Ischial tuberosity
Short head of Biceps Femoris:
Lateral lip of linea aspera

I Head of the fibula

N Tibial and peroneal

Semitendinosus

A Flex the knee (tibiofemoral joint)
Medially rotate the flexed knee (t/f joint)
Extend the hip (coxal joint)
Medially rotate the hip (coxal joint)
Tilt the pelvis posteriorly

O Ischial tuberosity

I Proximal, medial shaft of the tibia at pes anserinus tendon

N Tibial

Semimembranosus

A Flex the knee (tibiofemoral joint)
Medially rotate the flexed knee (t/f joint)
Extend the hip (coxal joint)
Medially rotate the hip (coxal joint)
Tilt the pelvis posteriorly

O Ischial tuberosity

I Posterior aspect of medial condyle of tibia

N Tibial

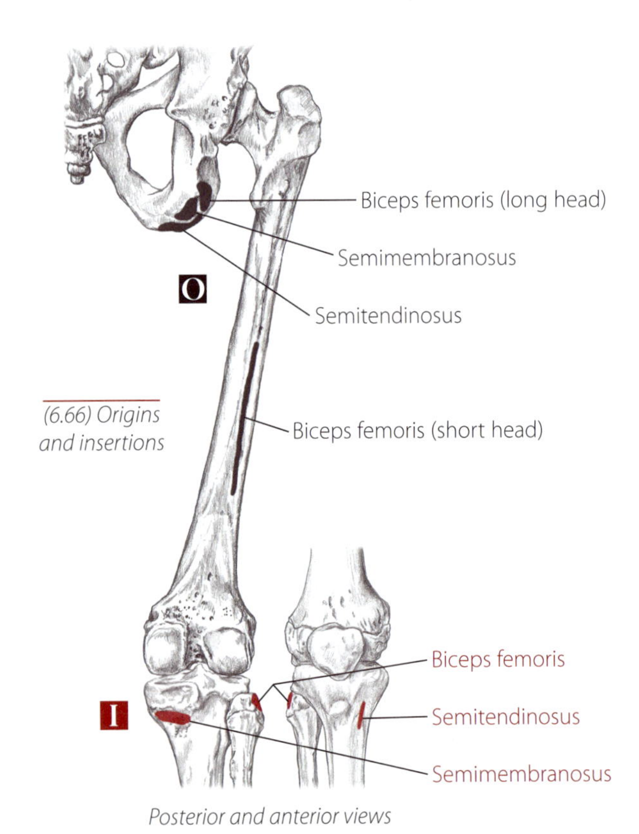

(6.66) Origins and insertions

Posterior and anterior views

Hamstrings as a group

1) Prone. Place a hand on the posterior thigh between the buttocks and knee. Ask your partner to flex his knee, holding his foot off the table. As the hamstrings contract, explore their mass and width (6.67).
2) Locate the ischial tuberosity. Slide your fingertips distally one inch and strum across the large, solid tendon of the hamstrings (6.68).
3) Follow the tendon distally as it spreads out into the separate bellies of the hamstrings.

Follow the bellies proximally. Do they attach to the ischial tuberosity? Follow the bellies distally. Do you feel their skinny tendons along the posterior knee?

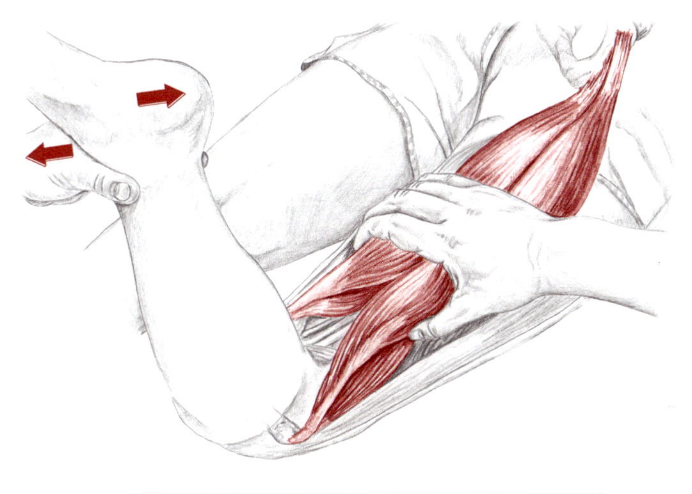

(6.67) Partner prone, grasping the hamstrings group

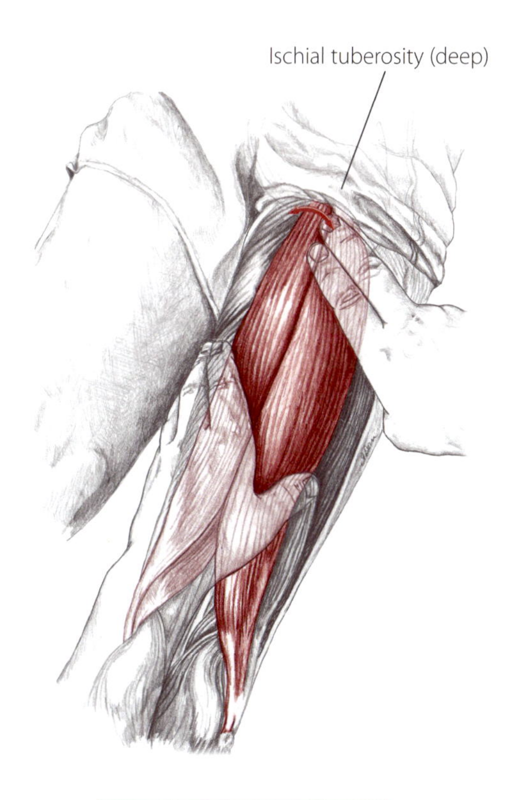

(6.68) Partner prone, isolating the hamstrings tendon at the ischial tuberosity

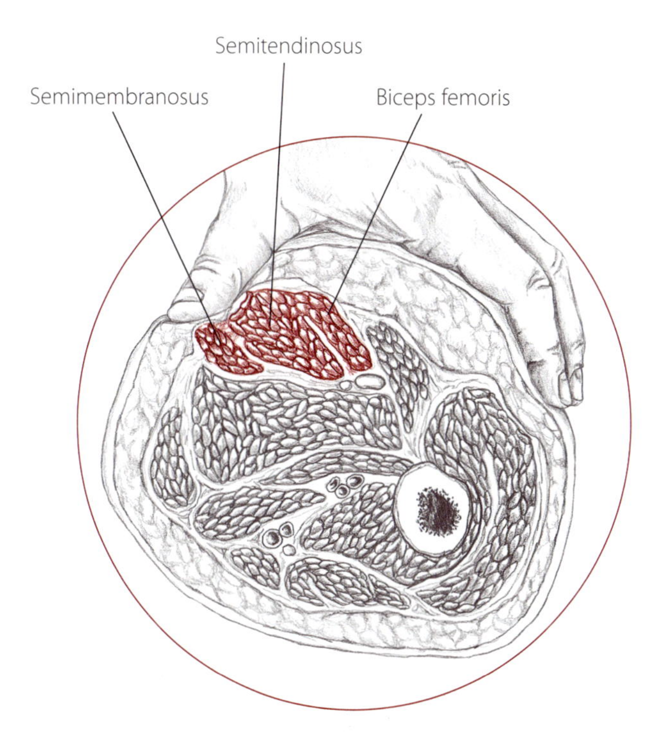

(6.69) Superior view, cross section of right thigh (see 6.61). Sinking your thumb into the medial edge of the hamstrings

biceps femoris	**bi**-seps fe-**mor**-is	Grk. the two-headed (muscle) of the thigh
pes anserinus	pes **an**-ser-**i**-nus	L. *pedis*, foot; *anserinus*, gooselike

Individual bellies and distal tendons

1) Partner prone. Ask him to hold his knee in a flexed position. Again, explore the bellies of the hamstrings.
2) The lateral half of the hamstring belly is the biceps femoris. Its belly will lead toward the head of the fibula. Palpate on the lateral side of the knee for the long, prominent tendon of the biceps femoris and follow it toward the head of the fibula (6.70).
3) The medial half of the hamstrings consists of the layered bellies of the semitendinosus and semimembranosus. Move to the medial side of the knee and palpate for the tendons of these muscles (6.71).
4) The most superficial tendon will be the semitendinosus. Turn your partner supine and follow it distally as it merges with the pes anserinus tendon. The semimembranosus is tucked deep to the semitendinosus and is often difficult to isolate.

Are the tendons along the back of the knee slender and superficial? Does the biceps femoris tendon lead to the head of the fibula? Can you follow the "semis" as they seem to disappear into the medial knee?

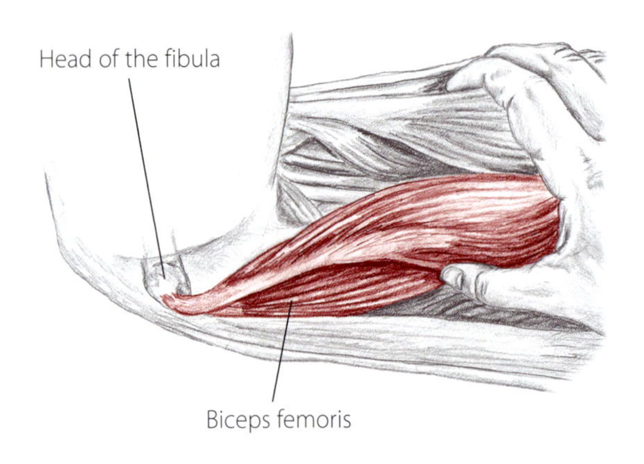

(6.70) Partner prone, posterior/lateral view of flexed right knee

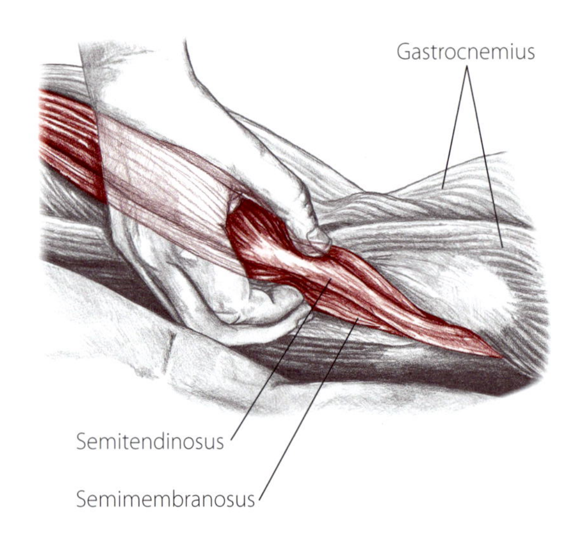

(6.71) Partner prone, posterior/medial view of extended right knee

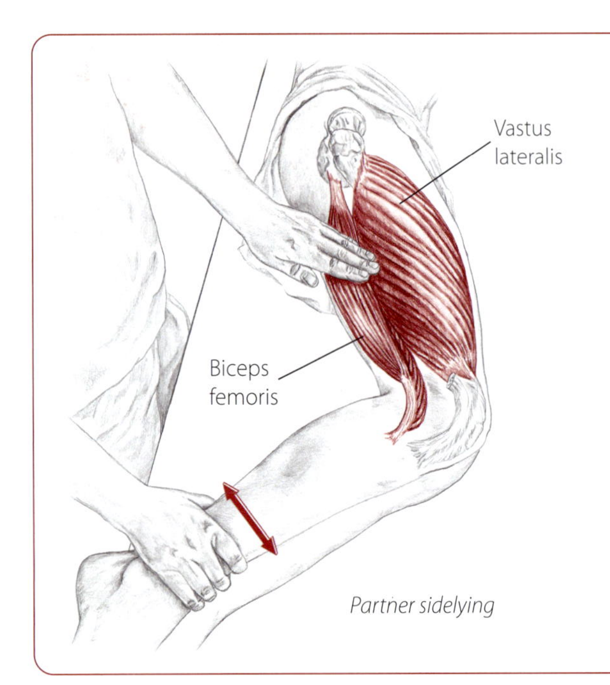

Partner sidelying

How do you differentiate the vastus lateralis from the biceps femoris on the posterior, lateral thigh? Have these muscles do what comes naturally - be antagonists.

1) Partner sidelying. Bend the top knee to 90° and clasp the ankle. Lay your other hand on the lateral thigh.
2) Ask your partner to alternate between flexing and extending his knee ever-so-slightly against your resistance. Sense how the vastus lateralis contracts upon extension while the biceps femoris remains lax. The opposite will happen when flexing the knee.
3) Often there will be a palpable dividing line or depression between the edges of these muscles.

Gluteals

Gluteus Maximus
Gluteus Medius
Gluteus Minimus

The three gluteal muscles are located in the buttock region, deep to the surrounding adipose tissue. The large, superficial **gluteus maximus** is the most posterior of the group and has fibers that run diagonally across the buttock (6.72).

The **gluteus medius** is located on the outside of the hip and is also superficial, except for the posterior portion which is deep to the maximus (6.73). Both the gluteus maximus and medius are strong extensors and abductors of the hip. With its convergent fibers that pull the femur in multiple directions, the gluteus medius could be thought of as the "deltoid muscle of the coxal joint."

The **gluteus minimus** lies deep to the gluteus medius and is inaccessible; however, its dense fibers can be felt beneath the medius (6.74). Because it attaches to the anterior surface of the greater trochanter, the gluteus minimus flexes and medially rotates the hip, thus performing the opposite movement of the other gluteals.

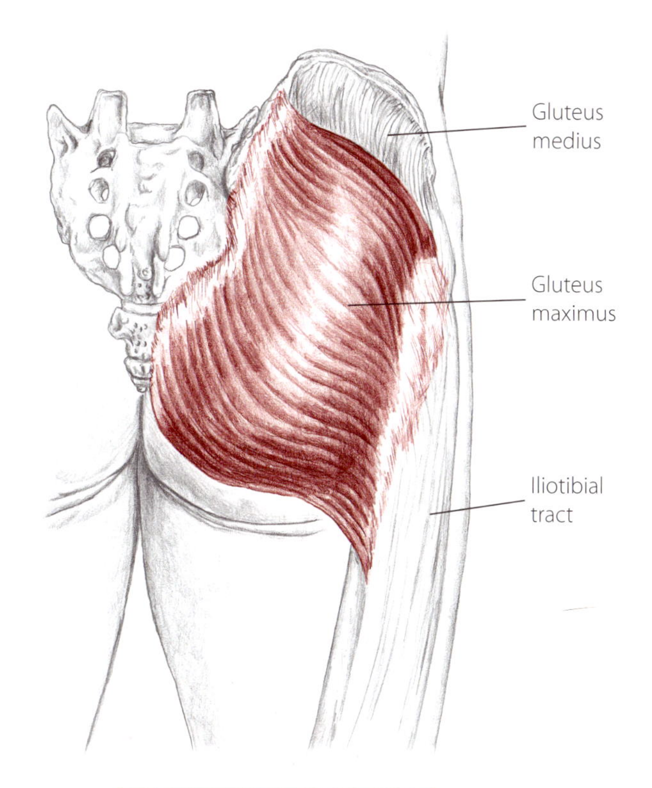

(6.72) Posterior view of right buttock

Gluteus Maximus

A *All fibers:*
Extend the hip (coxal joint)
Laterally rotate the hip (coxal joint)
Abduct the hip (coxal joint)

Lower fibers:
Adduct the hip (coxal joint)

O Coccyx, edge of sacrum, posterior iliac crest, sacrotuberous and sacroiliac ligaments

I Gluteal tuberosity (upper fibers) and iliotibial tract (lower fibers)

N Inferior gluteal

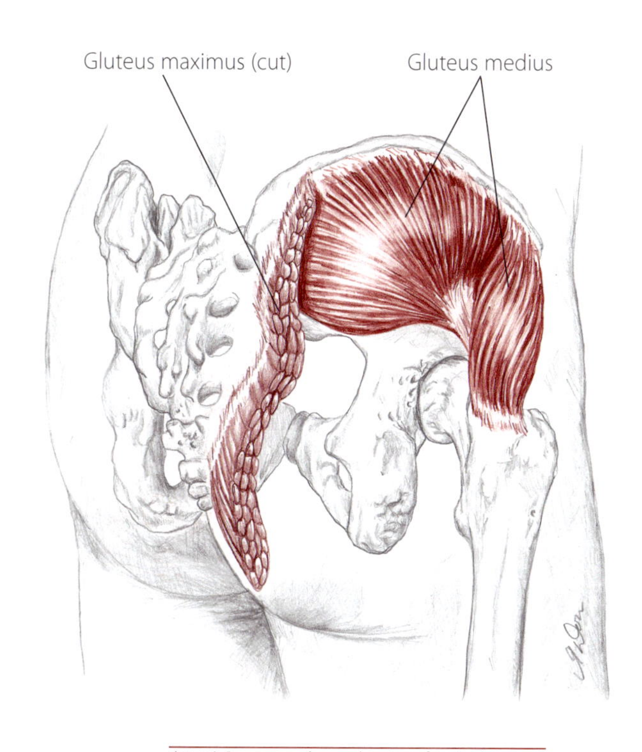

(6.73) Posterior/lateral view of right buttock

Gluteus Medius

A *All fibers:*
Abduct the hip (coxal joint)

Anterior fibers:
Flex the hip (coxal joint)
Medially rotate the hip (coxal joint)

Posterior fibers:
Extend the hip (coxal joint)
Laterally rotate the hip (coxal joint)

O Gluteal surface of the ilium between the iliac crest and the posterior and anterior gluteal lines

I Greater trochanter

N Superior gluteal

gluteus **gloo**-te-us Grk. *gloutos*, buttocks, which in turn is Anglo-Saxon for *buttuc*, meaning end

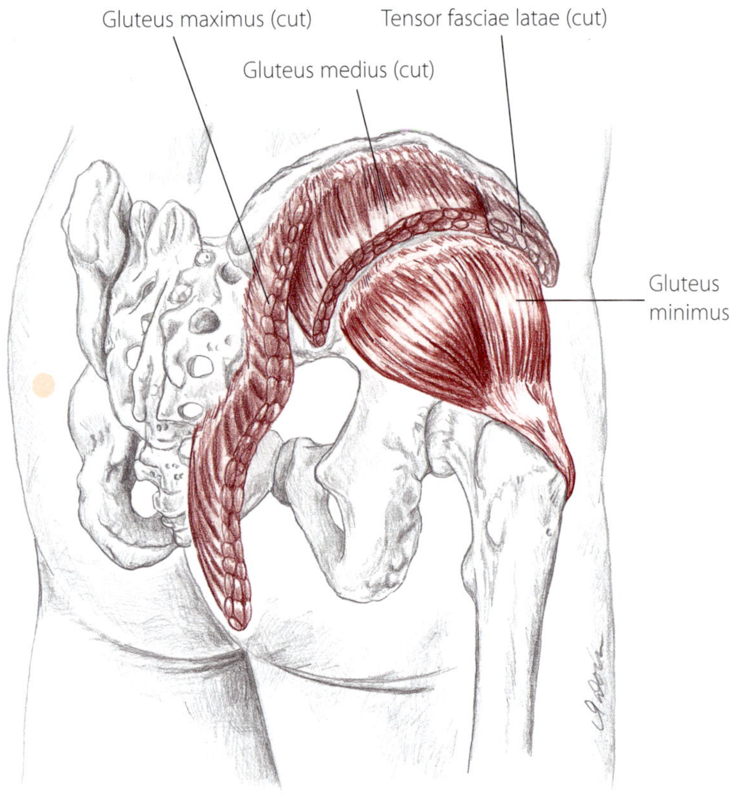

(6.74) Posterior/lateral view of right buttock

Gluteus Minimus

A Abduct the hip (coxal joint)
Medially rotate the hip (coxal joint)
Flex the hip (coxal joint)

O Gluteal surface of the ilium between the anterior and inferior gluteal lines

I Anterior border of greater trochanter

N Superior gluteal

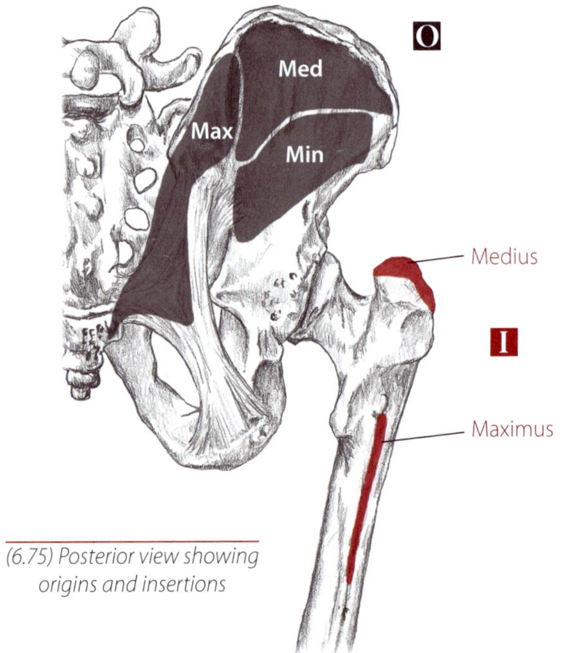

(6.75) Posterior view showing origins and insertions

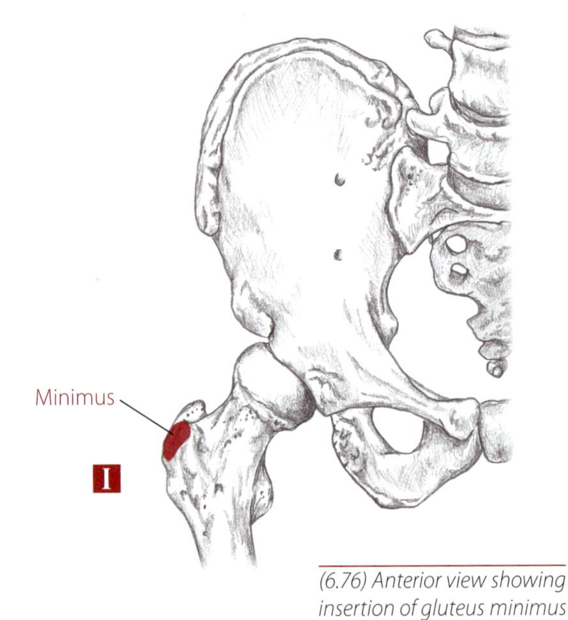

(6.76) Anterior view showing insertion of gluteus minimus

Gluteus maximus

1) Partner prone. Locate the coccyx, the edge of the sacrum, the PSIS and the posterior two inches of the iliac crest in order to isolate the landmarks that form the origin of the maximus (6.77).
2) Locate the insertion of the maximus at the gluteal tuberosity.
3) Connect its origin to its insertion by drawing the fiber direction on your partner. Then palpate its thick, superficial fibers. Also notice differences in texture and depth between the adipose tissue of the buttock and the muscle fibers of the maximus. The adipose is superficial to the maximus and often has a soft, gel-like consistency.

Ask your partner to extend his hip (6.78). Palpate the bulging fibers that lead to the gluteal tuberosity. If this is difficult with the knee extended for your partner while prone, try palpating with the knee flexed or with your partner standing.

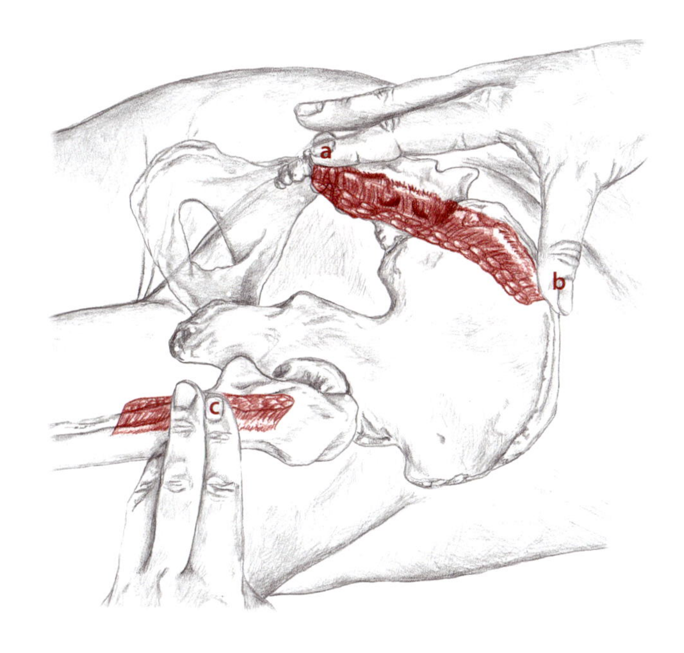

(6.77) Partner prone, isolating the borders of the gluteus maximus: **a)** *coccyx,* **b)** *posterior iliac crest,* **c)** *gluteal tuberosity*

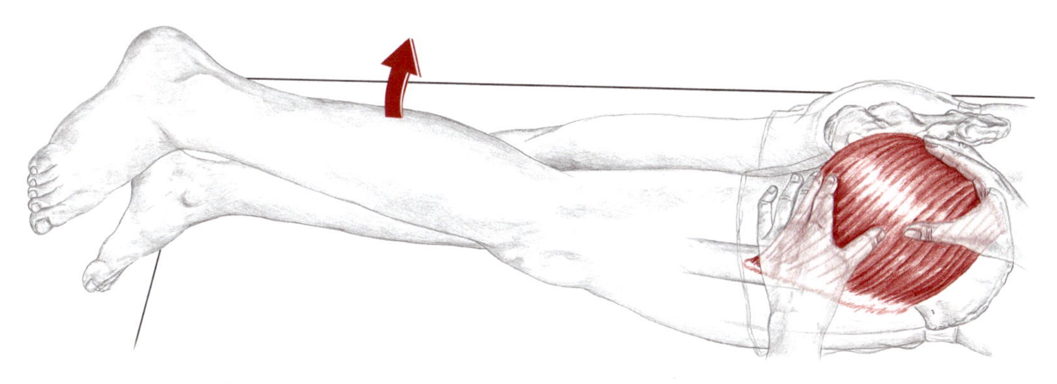

(6.78) Partner prone, extending his hip by contracting his gluteus maximus

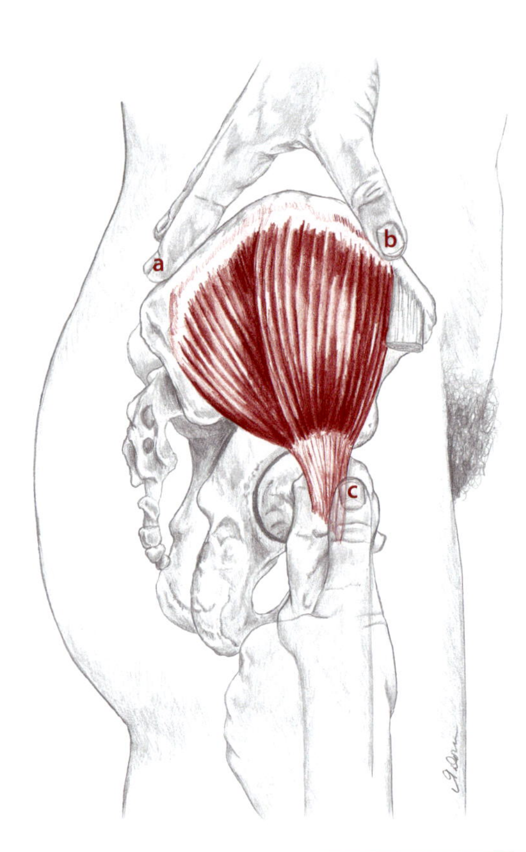

(6.79) Partner sidelying, isolating the borders of the gluteus medius: **a)** *PSIS,* **b)** *iliac crest,* **c)** *greater trochanter*

Gluteus medius and minimus

1) Sidelying. Isolate the shape of the gluteus medius by placing the webbing of one hand along the iliac crest (from PSIS to nearly the ASIS) while the other hand locates the greater trochanter.
2) Your hands will form the pie-shaped outline of the gluteus medius (6.79).
3) Palpate in this area from just below the iliac crest to the greater trochanter for the dense fibers of the gluteus medius.
4) Sink your fingers deep to the gluteus medius in order to explore for the density and mass of the gluteus minimus.

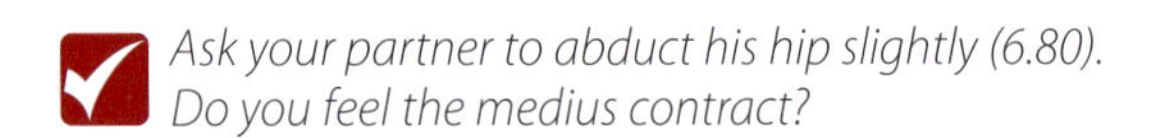
Ask your partner to abduct his hip slightly (6.80). Do you feel the medius contract?

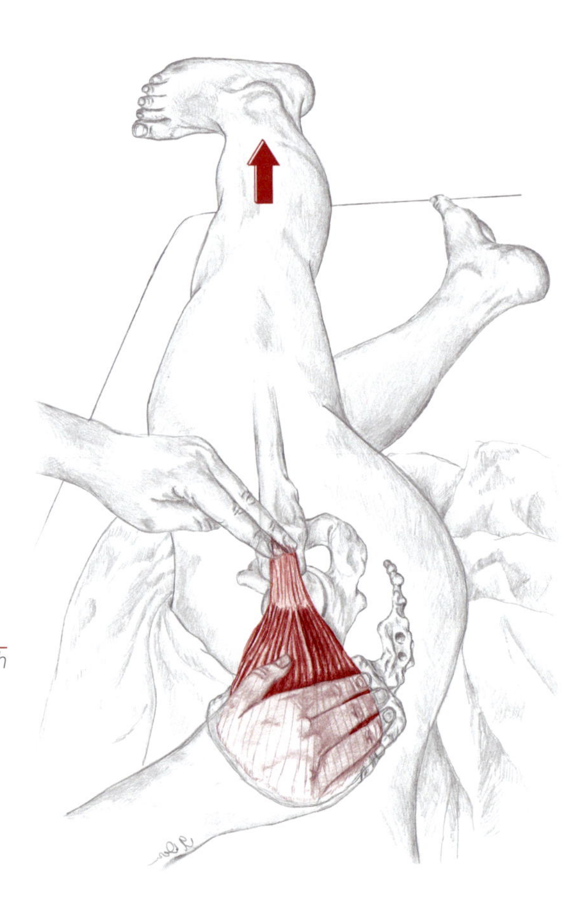

(6.80) Isolating the gluteus medius with partner sidelying, abducting the hip

Humans are unique among mammals not only with respect to their extra-large brains, but also because of their well-padded buttocks. No other mammal has such deposits of adipose tissue in the gluteal region, and no one seems to know *why* we have them. It was thought that the buttocks gave us something to sit upon, but we really sit upon our ischial tuberosities. And for good reason: If we did not, the gluteus maximus and gluteal fascia would be compressed beneath us.

Since women typically have larger buttocks than men, it was conjectured that the buttocks served as fat-storage sites during pregnancy. Not so.

One thing is known: The gluteal fold - the crease between the buttock and thigh - helps localize the subcutaneous adipose at the top of the thigh. Biomechanically, it is easier to swing the thigh back and forth when walking with the tissue situated proximally rather than dispersed down the thigh.

Adductor Group

Adductor Magnus
Adductor Longus
Adductor Brevis
Pectineus
Gracilis

The five adductors are located along the medial thigh between the hamstrings and quadriceps femoris muscles (6.81). Their proximal tendons attach at specific locations along the base of the pelvis. Together these tendons form a connective tissue drape that extends from the superior ramus of the pubis to the ischial tuberosity (6.82, 6.88).

When the thigh is viewed anteriorly, the muscle bellies of the adductors lie in three layers. The **pectineus** and **adductor longus** are most anterior (6.83). Behind them is the **adductor brevis** (6.84), and most posterior is the **adductor magnus** (6.85). The broad span of adductor magnus, known as the "floor of the adductors," lies anterior to the hamstrings (6.86). These four muscles tuck posteriorly to the quadriceps femoris group and insert on the posterior femur. The fifth adductor, **gracilis**, lies superficially on the medial thigh. It is the only adductor that crosses the knee (6.84).

Although their individual bellies may be challenging to isolate, as a group, the adductors are easy to locate. When palpating the adductor tendons near the pubic bone, there will be a prominent tendon that extends off of or nearby the pubic tubercle. The source of this superficial tendon is either the gracilis or adductor longus; in some cases, it is a merging of both tendons.

In either case, the tendon can serve as an important guidepost for locating not only gracilis and adductor longus, but also pectineus and adductor magnus. The pectineus will be located on the anterior side of this tendon while the adductor magnus will be located posterior to it.

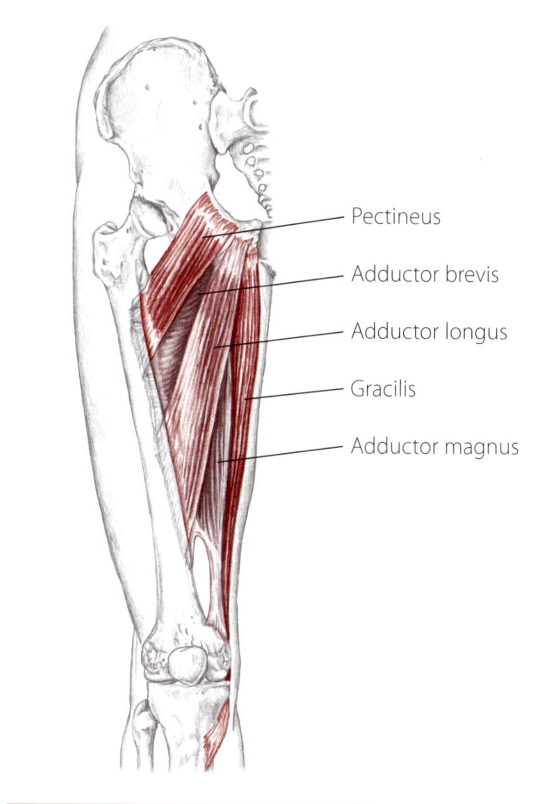

(6.81) Anterior view of right hip and thigh

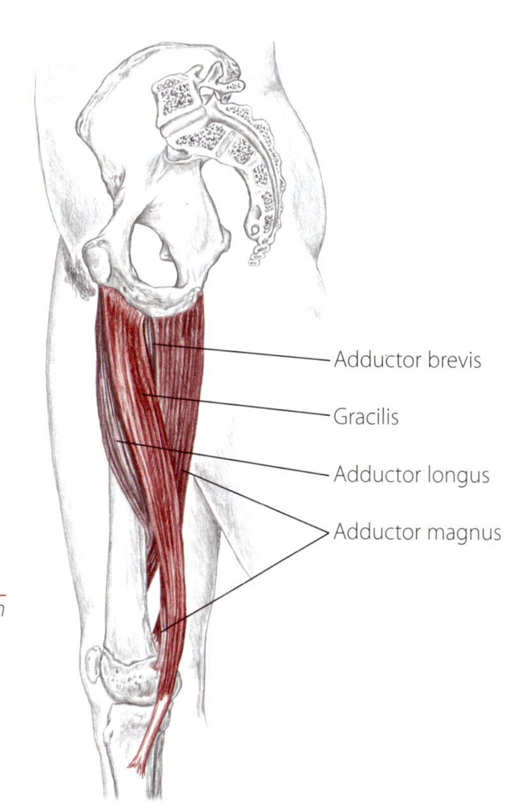

(6.82) Medial view of right hip and thigh

brevis	**breh**-vis	L. short
gracilis	gra-**cil**-is	L. slender, graceful
pectineus	pek-**tin**-e-us	L. comblike

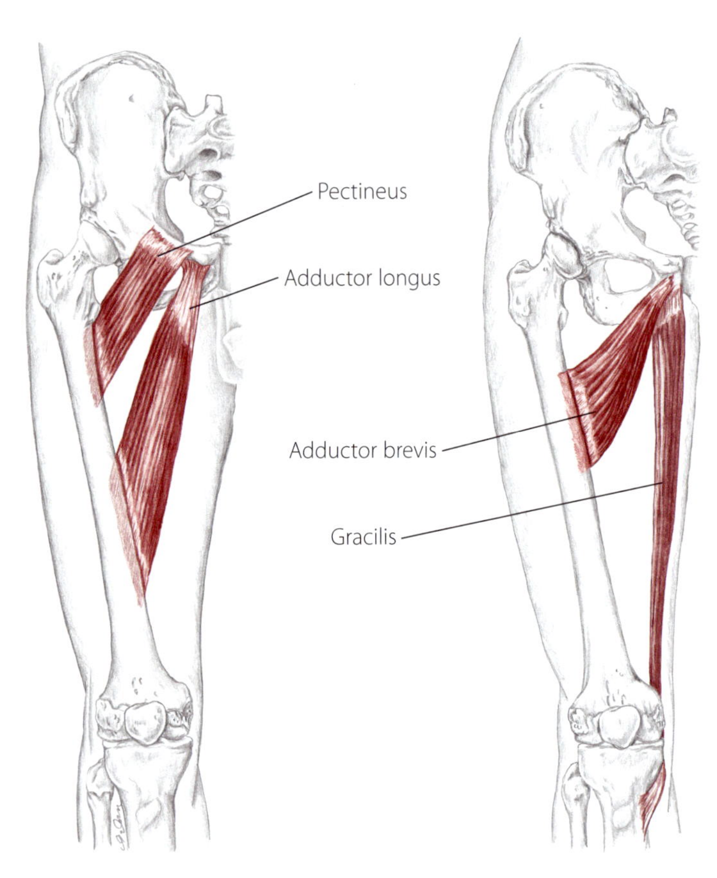

(6.83, 6.84) Anterior views of right hip and thigh

A *All:*
Adduct the hip (coxal joint)
Medially rotate the hip (coxal joint)

All, except Gracilis:
Assist to flex the hip (coxal joint)

Gracilis:
Flex the knee (tibiofemoral joint)
Medially rotate the flexed knee (t/f joint)

Posterior fibers of Adductor Magnus:
Extend the hip (coxal joint)

Adductor Magnus

O Inferior ramus of the pubis, ramus of ischium and ischial tuberosity

I Medial lip of linea aspera and adductor tubercle

N Obturator and tibial

Adductor Longus

O Pubic tubercle

I Medial lip of linea aspera

N Obturator

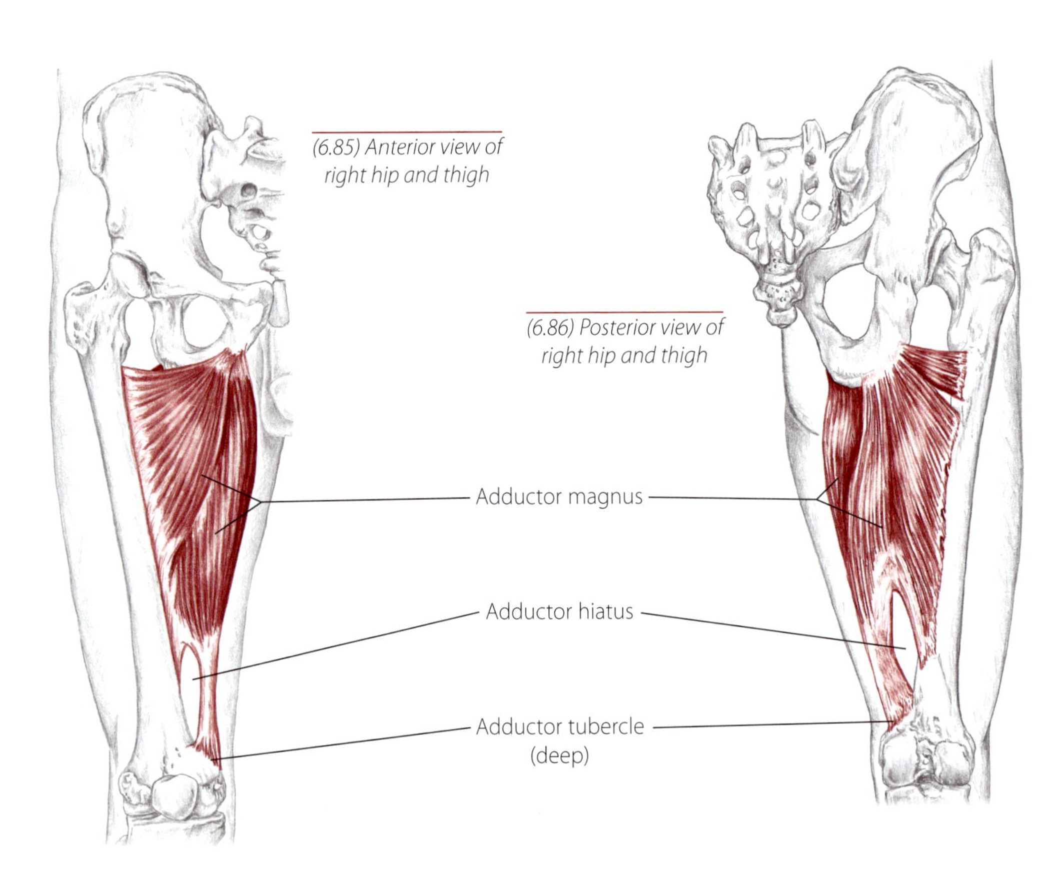

(6.85) Anterior view of right hip and thigh

(6.86) Posterior view of right hip and thigh

Adductor Brevis

O Inferior ramus of pubis

I Pectineal line and medial lip of linea aspera

N Obturator

Pectineus

O Superior ramus of pubis

I Pectineal line of femur

N Femoral and obturator

Gracilis

O Inferior ramus of pubis and ramus of ischium

I Proximal, medial shaft of tibia at pes anserinus tendon

N Obturator

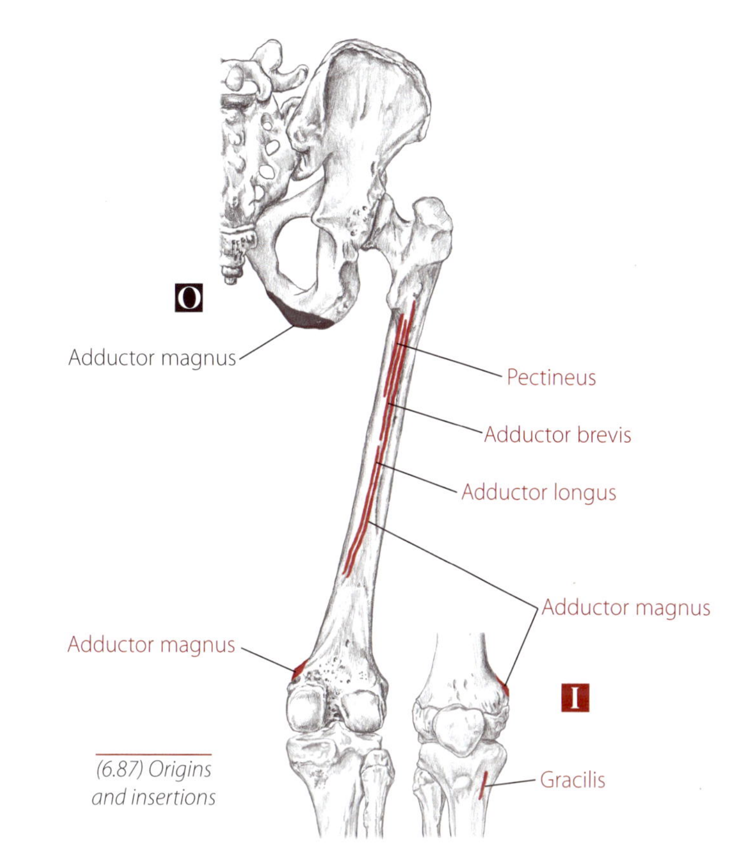

(6.87) Origins and insertions

Posterior and anterior views

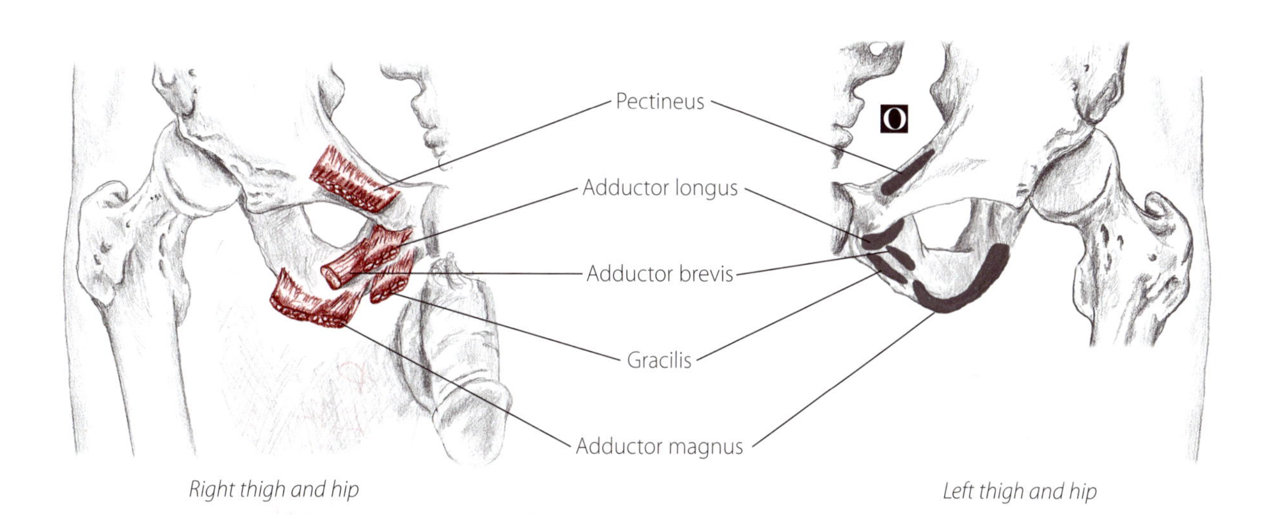

(6.88, 6.89) Anterior views showing the origin sites of the adductors

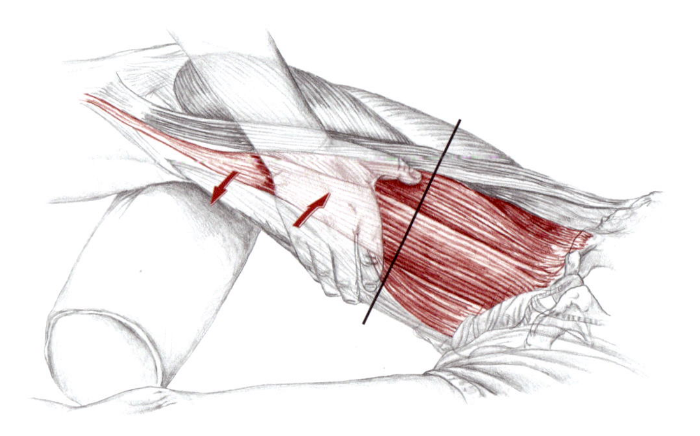

(6.90) Partner supine, palpating the adductors as a group, line indicating cross section

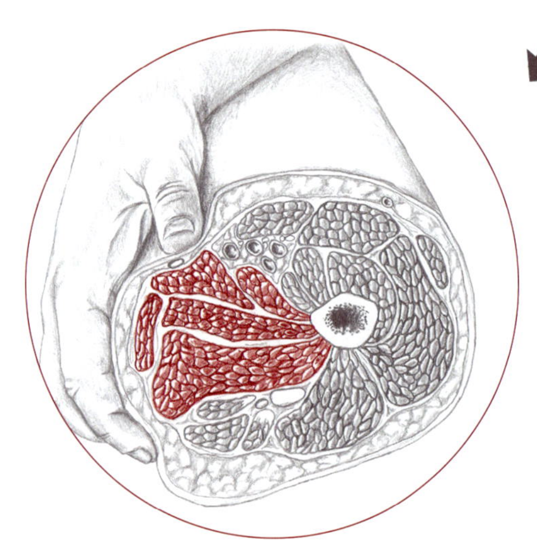

(6.91) Inferior view, cross section of right thigh, grasping the highlighted adductors

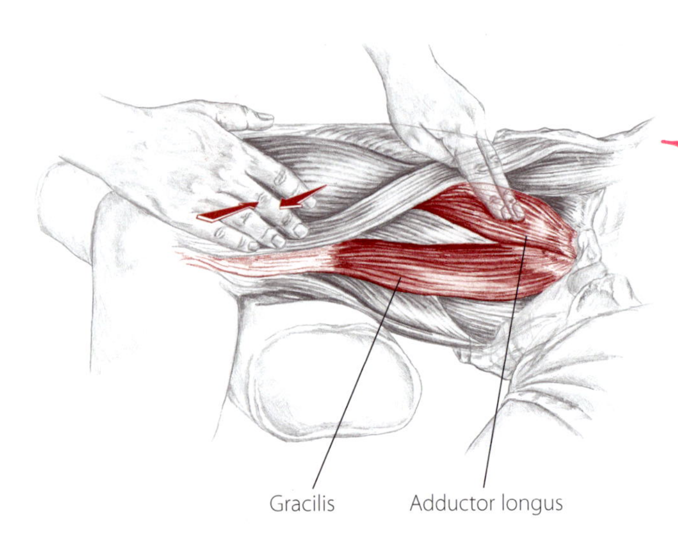

(6.92) Locating the prominent tendon(s) of the adductors

Adductors as a group

1) Partner supine with the hip slightly flexed and laterally rotated. Place your hand along the medial thigh and ask your partner to adduct his hip against your resistance (6.90). Do you feel the adductors tighten?
2) Ask your partner to alternately adduct and relax, as you palpate proximally to the adductor tendons. Then move distally, exploring anterior and posterior to the edges of the adductor bellies.

Are you on the medial side of the thigh? Explore either side of the adductors to determine if you are between the hamstrings and quadriceps femoris group (6.91). You should be.

Gracilis and adductor longus

1) Partner supine with the hip slightly flexed and laterally rotated. Place the flat of your hand at the middle of the medial thigh. Ask your partner to adduct his hip slightly.
2) While your partner contracts, slide your fingers proximally to the pubic bone and locate the taut, prominent tendon(s) of the gracilis and adductor longus extending off of (or nearby) the pubic tubercle.
3) Strum your fingertip across this tendon and follow it distally as it develops into muscle tissue (6.92). If the muscle belly slowly angles into the medial thigh, you are palpating adductor longus. If the belly is slender and continues down the medial thigh toward the knee, you are accessing gracilis.

The shape and location of the sartorius (p. 320) are similar to the shape and location of the gracilis. Distinguish between the two by simply following the muscle you are palpating proximally. If it leads toward the ASIS, it is the sartorius; if toward the pubis, the muscle is the gracilis.

MMT as group:
(sidelying) Lift top leg up, hold it there for clt.
Clt. brings other leg up + you resist.

Pectineus

1) Partner supine with the hip slightly flexed and laterally rotated. Place the flat of your hand on the middle of the medial thigh and ask your partner to adduct his hip slightly.
2) Locate the prominent tendon of the adductor longus or gracilis. Slide off the tendon laterally toward the ASIS. Slowly sink into the belly of pectineus (6.93). You should be inferior to the superior ramus of the pubis (p. 292).
3) Ask your partner to alternately adduct and relax his hip and feel the fibers of pectineus contract.

Are you just anterior to the prominent adductor tendon? Do the fibers you are palpating contract upon adduction?

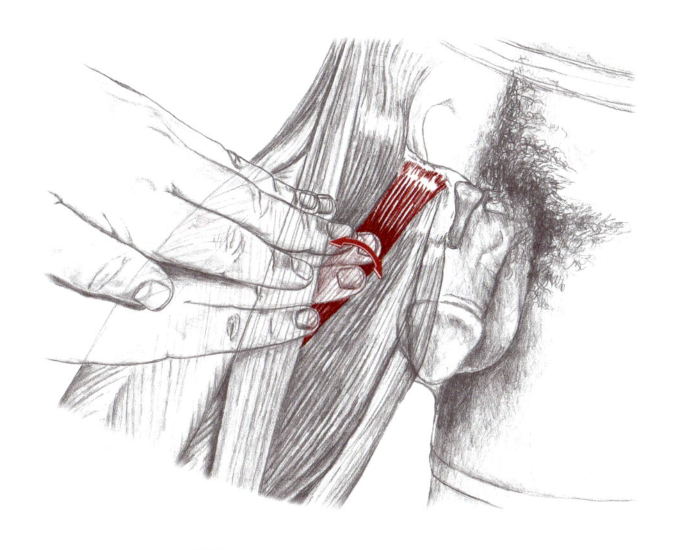

(6.93) Partner supine, rolling your fingers across the pectineus

Adductor magnus

1) Partner sidelying with top hip flexed. Begin by locating the ischial tuberosity.
2) Ask your partner to adduct his hip slightly. Shifting anteriorly, locate the prominent tendon of adductor longus or gracilis. Then slide off the tendon posteriorly. Palpate the wide tendon of adductor magnus as it stretches to the ischial tuberosity (6.94).
3) Follow the fibers of adductor magnus distally by strumming your fingers across its belly. It is difficult to differentiate magnus fibers from semimembranosus fibers. Nevertheless, the thin, distal tendon of the magnus is distinguishable and can be accessed where it attaches onto the adductor tubercle (p. 347).

Be alert to the pulse of the femoral artery (p. 333). If you feel it under your fingers, remove your pressure and shift to one side.

Because the adductors attach on the posterior femur, you might assume that they would rotate the coxal joint *laterally* rather than *medially*. In anatomical position, however, the adductors will medially rotate the femur. With that said, if the femur is in a laterally rotated position, some of the adductors will produce lateral rotation.

Partner supine. Lay your hand on the adductors. Ask your partner alternately to rotate his hip medially and laterally. Grasping the ankle to create a little resistance will verify this movement. Do you feel the adductors contract on medial rotation? What do they do when the hip is laterally rotated?

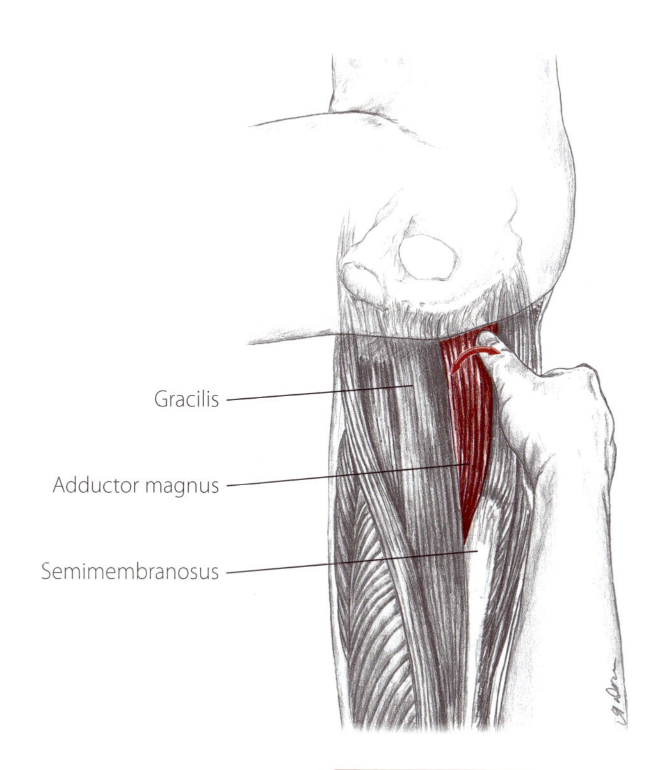

(6.94) Partner sidelying, accessing the adductor magnus

Tensor Fasciae Latae and Iliotibial Tract

The **tensor fasciae latae** (TFL) is a small, superficial muscle located on the lateral side of the upper thigh (6.95). Approximately three fingers wide, the TFL is easily accessible between the upper fibers of the rectus femoris and the gluteus medius. The TFL attaches to the iliotibial tract along with the gluteus maximus.

The **iliotibial tract** is a superficial sheet of fascia with vertical fibers that run along the lateral thigh. It emerges from the gluteal fascia, is wide and dense over the vastus lateralis muscle (p. 300) and funnels into a strong cable along the side of the knee before inserting at the tibial tubercle (p. 345). The fibers of tensor fasciae latae and some fibers of gluteus maximus (p. 309) attach to the proximal aspect of the iliotibial tract. The iliotibial tract has a thick, matted texture (similar to packing tape) that makes it a strong stabilizing component of the hip and knee.

The iliotibial tract is entirely accessible. The distal cable portion, anterior to the biceps femoris tendon, is the easiest part of the iliotibial tract to isolate.

A Flex the hip (coxal joint)
Medially rotate the hip (coxal joint)
Abduct the hip (coxal joint)

O Iliac crest, posterior to the ASIS

I Iliotibial tract

N Superior gluteal

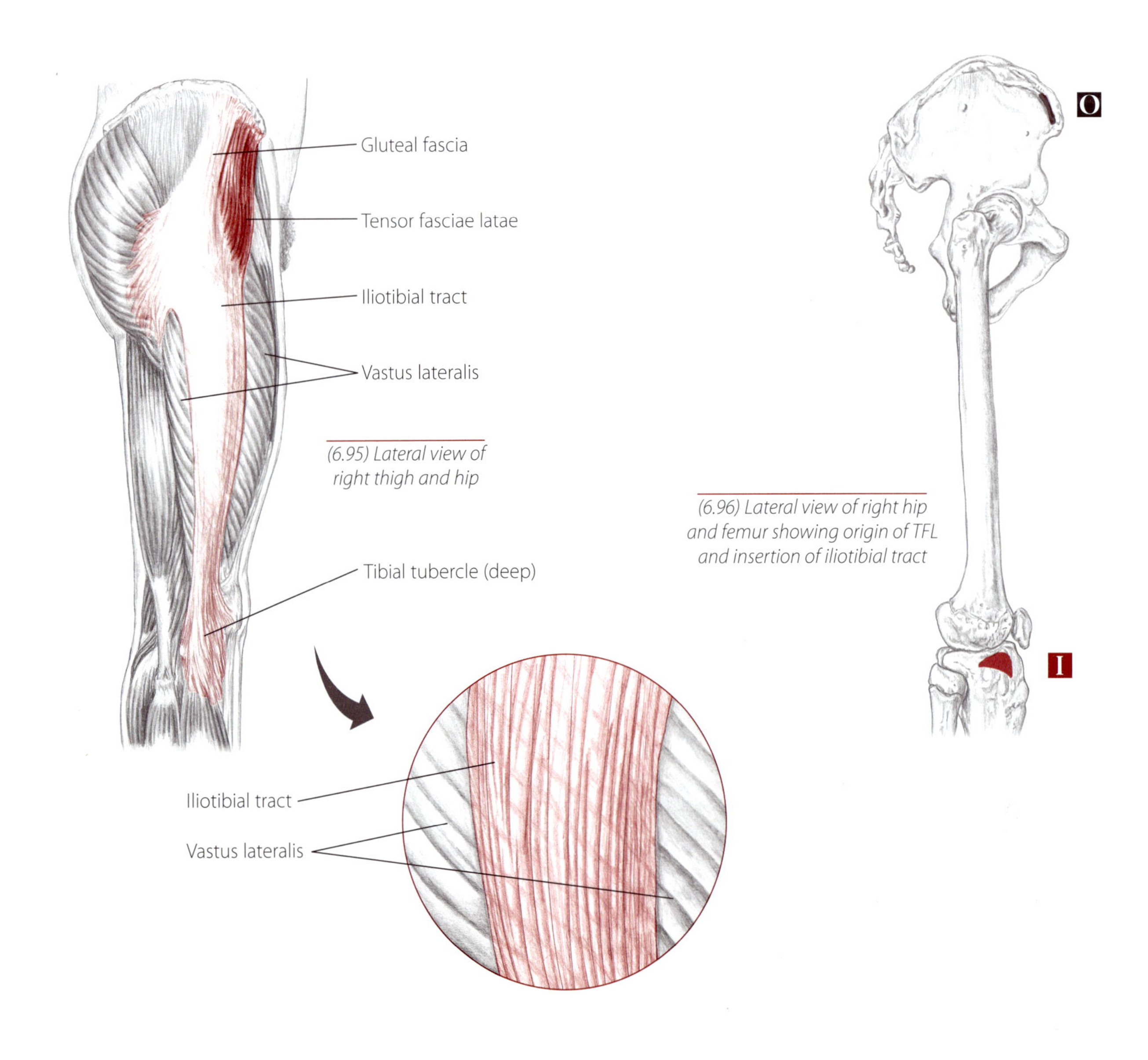

(6.95) Lateral view of right thigh and hip

(6.96) Lateral view of right hip and femur showing origin of TFL and insertion of iliotibial tract

latae	**la**-ta	L. broad
tensor	**ten**-sor	L. a stretcher
tract		L. extent, drawn out

Tensor fasciae latae

1) Supine. Locate the ASIS. Place the flat of your hand posterior and distal to the ASIS and iliac crest.
2) Ask your partner to alternate medial rotation with relaxation of the hip. Upon medial rotation, the TFL will contract into a solid, oval mound beneath your hand (6.97).
3) Palpate its vertical fibers, outline its width and follow it distally until the TFL blends into the iliotibial tract.

Are you posterior and distal to the anterior iliac crest? If you ask your partner to rotate the hip laterally, does the TFL contract? It should not.

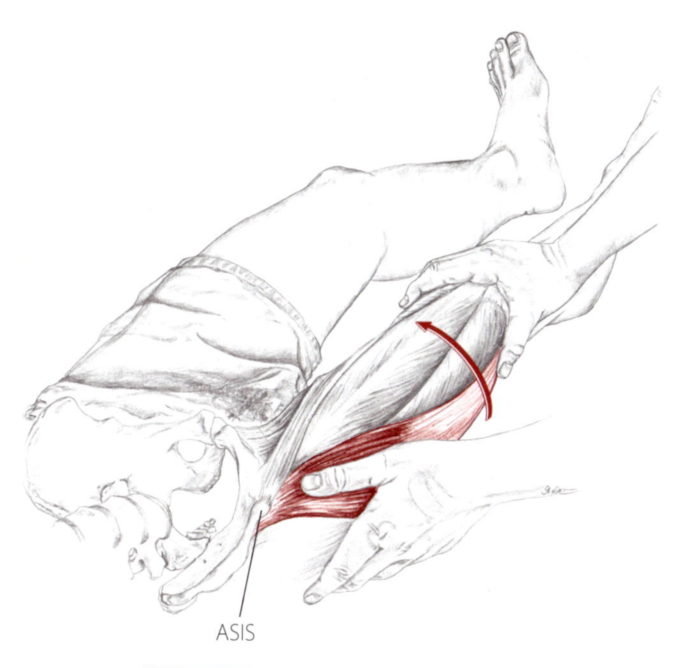

(6.97) Partner supine, feeling the TFL contract as your partner medially rotates his hip

Distal end of the iliotibial tract

1) Sidelying. Locate the biceps femoris tendon (p. 305) just proximal to the back of the knee.
2) Slide anteriorly from the biceps femoris tendon to the lateral thigh. Roll your fingers horizontally across the fibers of the iliotibial tract and explore for its tough, superficial quality. Its most distal aspect may feel similar in size and shape to the biceps femoris tendon.
3) Follow it distally as it disappears toward the tibial tubercle. Explore proximally and note how it becomes broader and thinner as it progresses up the thigh. Feel the tension of the iliotibial tract change by asking your partner to alternately abduct and relax his hip (6.98).

Are the fibers you feel superficial and stringy compared to the deeper, fleshier vastus lateralis fibers? Do the fibers run vertically down the thigh and converge into a thin, cablelike tendon at the tibial tubercle?

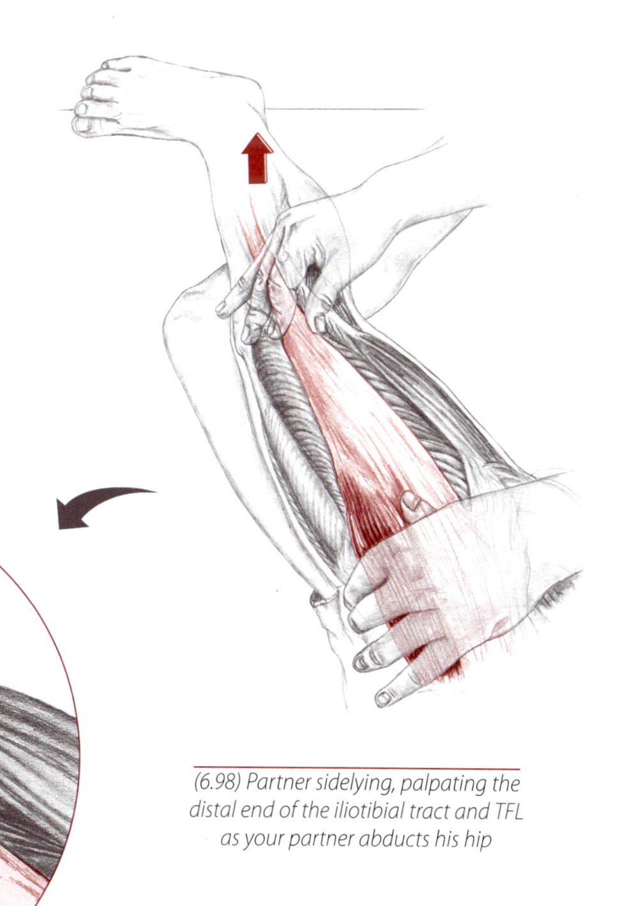

(6.98) Partner sidelying, palpating the distal end of the iliotibial tract and TFL as your partner abducts his hip

Sartorius

The sartorius is the longest muscle in the body, stretching from the anterior superior iliac spine (ASIS), across the thigh, to the medial knee (6.100). Though it is entirely superficial, the slender belly of the sartorius, roughly two fingers wide, can be difficult to isolate. Its proximal fibers are lateral to the femoral artery (p. 333). Its name (L. *sartor*, tailor) refers to its ability to bring the thigh and leg into the position a tailor would use when sewing (6.99).

A Flex the hip (coxal joint)

Laterally rotate the hip (coxal joint)

Abduct the hip (coxal joint)

Flex the knee (tibiofemoral joint)

Medially rotate the flexed knee (t/f joint)

(6.99)

O Anterior superior iliac spine (ASIS)

I Proximal, medial shaft of the tibia at pes anserinus tendon

N Femoral

1) Partner supine. Ask your partner to position his foot so it is resting on his opposite knee. The hip will be flexed and laterally rotated.
2) Place your hand along the middle of the medial thigh. Ask your partner to raise his knee toward the ceiling (contracting the sartorius).
3) Strum your fingers across the slender sartorius, following it proximally to the ASIS and distally to the medial tibia (6.102).
4) Maintain your hand placement and ask your partner to relax his hip. Continue to palpate, noticing how the sartorius curves from the ASIS to the medial side of the thigh.

Is the muscle belly you feel roughly two fingers wide and superficial? When distal to the ASIS, can you strum across its tendon? Are you medial to the vastus medialis belly? The sartorius and gracilis are slender, superficial muscles along the medial thigh. Differentiate between them by following their respective bellies proximally: The sartorius will lead toward the ASIS, the gracilis toward the pubic tubercle.

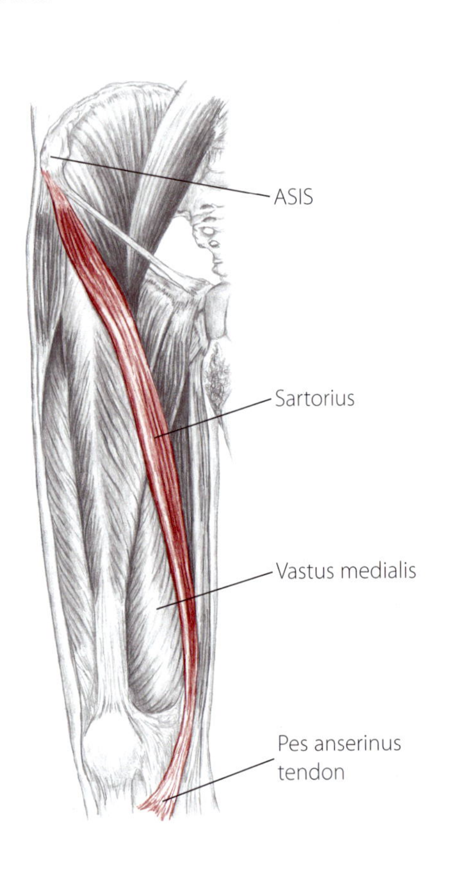

(6.100) Anterior/medial view of right hip and thigh

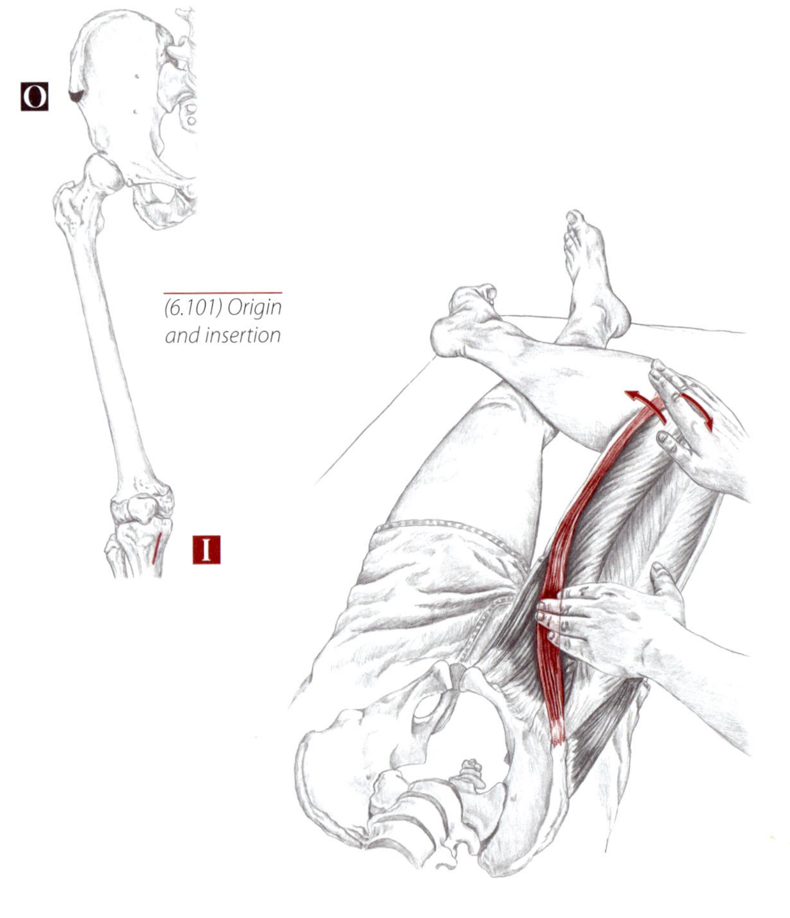

(6.101) Origin and insertion

(6.102) Partner supine

sartorius sar-**tor**-ee-us L. *sartor*, tailor

Tendons of the Posterior Knee

There are five distinct tendons located on the posterior aspect of the knee (6.103). Biceps femoris and the iliotibial tract are located on the lateral/posterior knee; sartorius, gracilis and semitendinosus are bundled together on the medial/posterior knee. These three tendons merge distally at the proximal, medial shaft of the tibia to become the pes anserinus tendon.

Where is the semimembranosus tendon? Its distal tendon is very short and deep to the semitendinosus and gracilis. The distal aspect of semimembranosus can be accessed by palpating between the tendons of semitendinosus and gracilis.

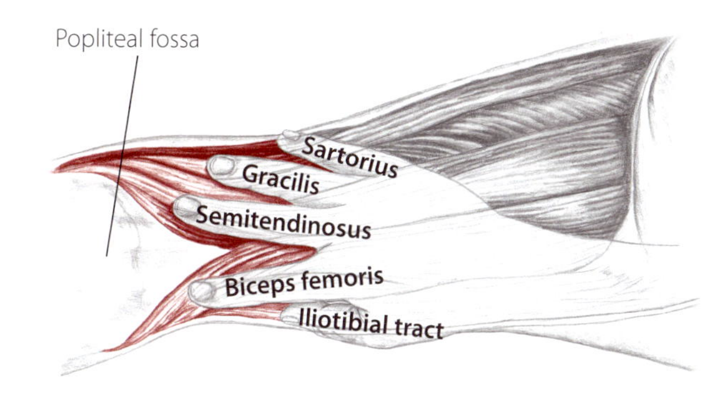

(6.103) Posterior view of right thigh with partner prone; fingers on the posterior knee showing location of distal tendons

Lateral tendons

1) Prone. Ask your partner to flex and hold his knee at 45°. The tendons will become taut in this position. For greater clarity, place your hand on the ankle and give your partner some resistance.
2) The most prominent tendons will be biceps femoris and semitendinosus. Follow the slender biceps femoris tendon as it extends down to the head of the fibula (6.104).
3) Move laterally approximately one inch from the biceps tendon and palpate the iliotibial tract. Unlike the biceps femoris, the iliotibial tract is broader and located on the lateral side of the thigh.

Head of the fibula
Gastrocnemius
Patella
Biceps femoris
Iliotibial tract

(6.104) Partner prone, lateral view of right knee, isolating the biceps femoris and iliotibial tract

Medial tendons

1) Supine. Move to the medial side of the knee and palpate the thin, prominent tendon of the semitendinosus.
2) Slide off semitendinosus anteriorly and palpate the equally slender tendon of gracilis.
3) Situated anterior to gracilis will be sartorius. Compared to the long, skinny tendons of semitendinosus and gracilis, the sartorius tendon is shorter and wider (6.105). For this reason, it may be challenging to isolate.
4) Follow the three tendons distally as they blend together to become the pes anserinus tendon which attaches on the proximal, medial shaft of the tibia.

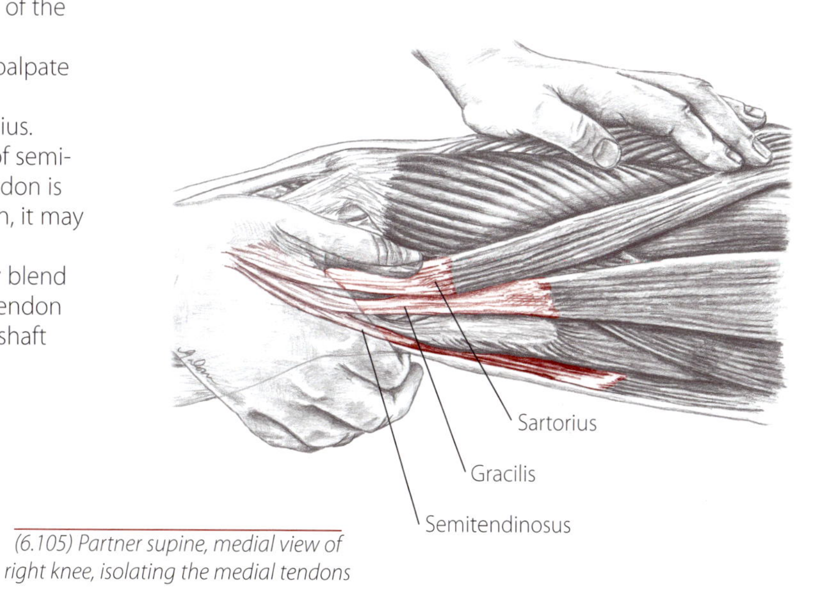

(6.105) Partner supine, medial view of right knee, isolating the medial tendons

Lateral Rotators of the Hip

Piriformis
Quadratus Femoris

Obturator Internus
Obturator Externus

Gemellus Superior
Gemellus Inferior

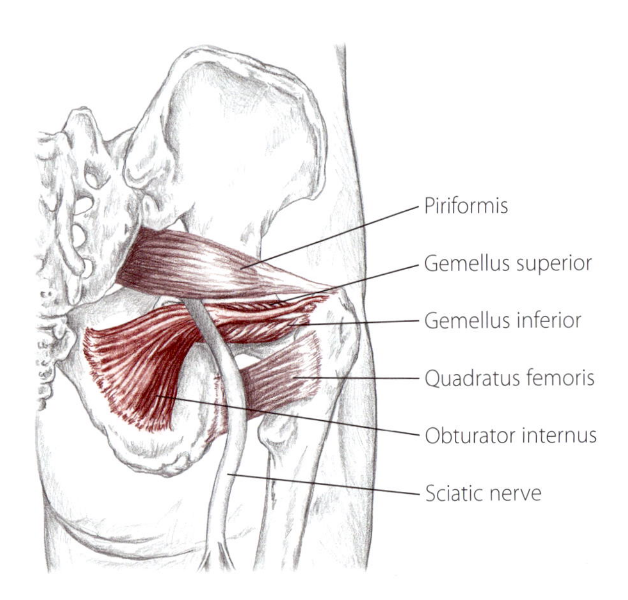

(6.106) Posterior view of right hip with gluteals removed

Sometimes known as the "deep six," these small muscles are located deep to the gluteus maximus and create lateral rotation of the hip. All attach to aspects of the greater trochanter and fan medially to attach to the sacrum and pelvis (6.106 - 6.112).

All of the lateral rotators are deep to the large sciatic nerve (p. 335), except for the piriformis. The piriformis lies superficial to the sciatic nerve and, if overcontracted, can compress it. Nevertheless, the lateral rotators are accessible as a group, with the piriformis and quadratus femoris being the most discernible.

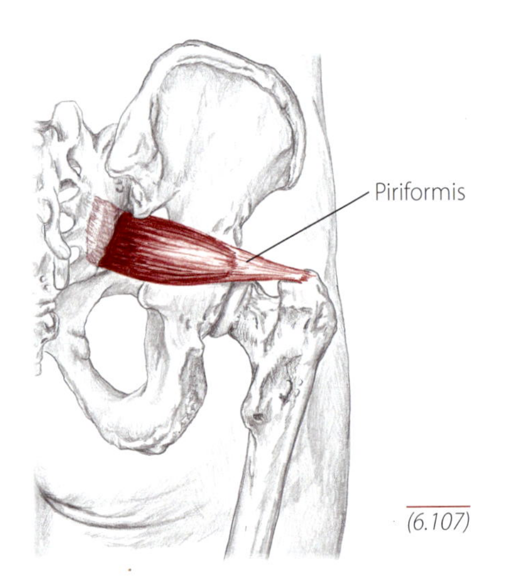

(6.107)

Piriformis

A Laterally rotate the hip (coxal joint)
Abduct the hip when the hip is flexed

O Anterior surface of sacrum

I Greater trochanter

N Branch of sacral plexus

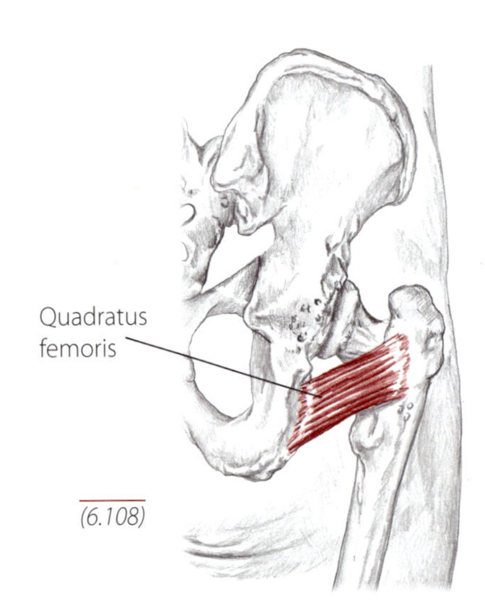

(6.108)

Quadratus Femoris

A Laterally rotate the hip (coxal joint)

O Lateral border of ischial tuberosity

I Intertrochanteric crest, between the greater and lesser trochanters

N Branch of sacral plexus

Obturator Internus

A Laterally rotate the hip (coxal joint)

O Obturator membrane and inferior surface of obturator foramen

I Medial surface of greater trochanter

N Branch of sacral plexus

gemellus	jem-**el**-us	L. twin
obturator	**ob**-tu-**ra**-tor	L. obstructor

Obturator Externus

A Laterally rotate the hip (coxal joint)

O Superior and inferior rami of pubis

I Trochanteric fossa of femur

N Obturator

Gemellus Superior

A Laterally rotate the hip (coxal joint)

O Ischial spine

I Upper border of greater trochanter

N Branch of sacral plexus

Gemellus Inferior

A Laterally rotate the hip (coxal joint)

O Ischial tuberosity

I Upper border of greater trochanter

N Branch of sacral plexus

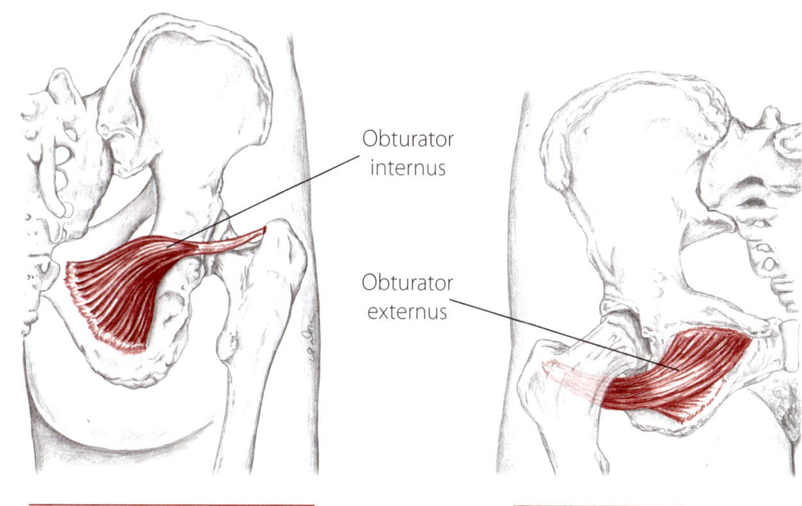

(6.109) Posterior view of right hip

(6.110) Anterior view of right hip

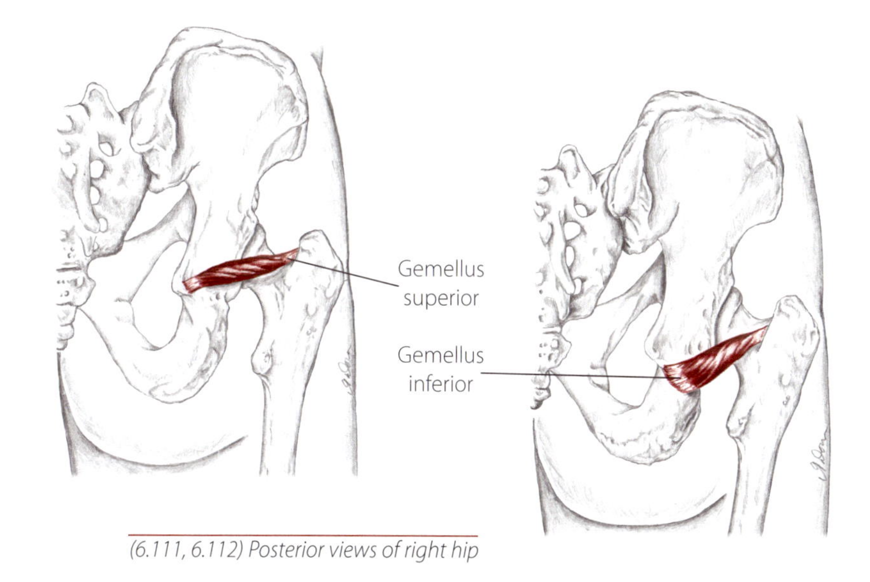

(6.111, 6.112) Posterior views of right hip

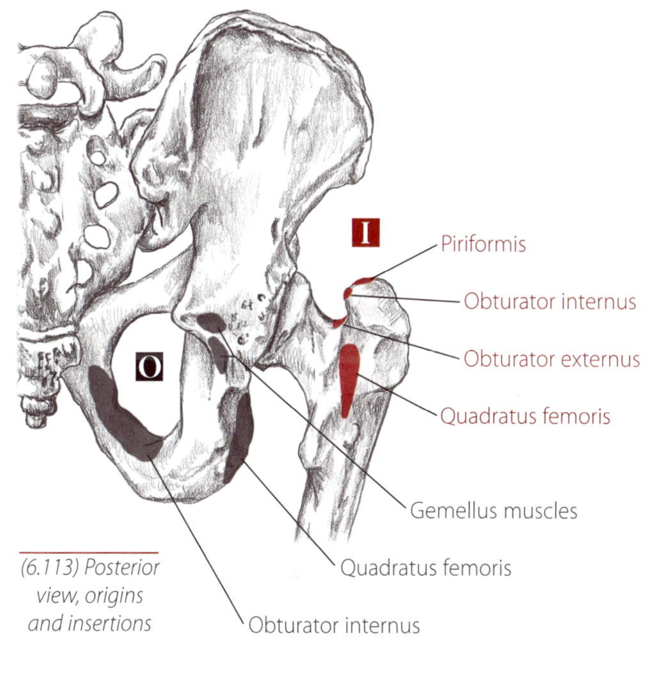

(6.113) Posterior view, origins and insertions

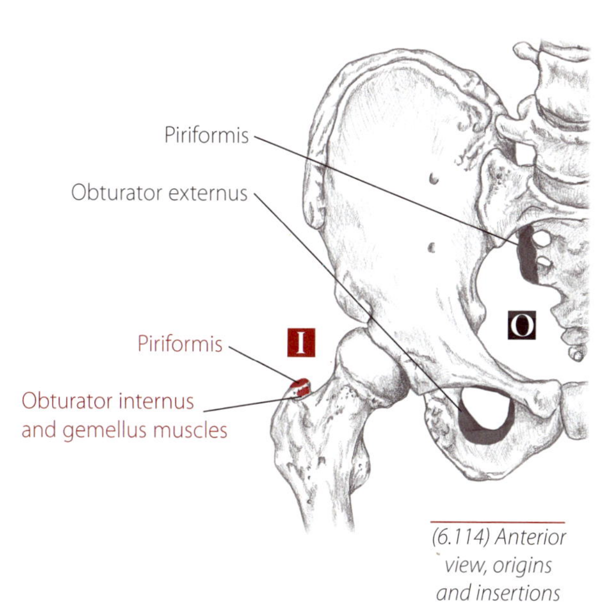

(6.114) Anterior view, origins and insertions

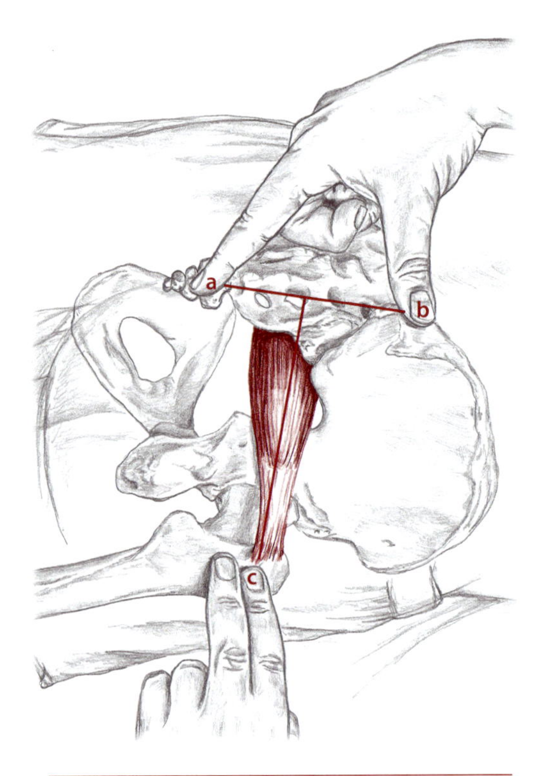

*(6.115) Posterior/lateral view, partner prone, isolating the piriformis by forming a "T," **a)** coccyx, **b)** PSIS, **c)** greater trochanter*

Piriformis

1) Prone. Locate the coccyx, PSIS and greater trochanter. Together, these landmarks form a "T". The piriformis is located along the base of the "T" (6.115).
2) Place your fingers along this line. Working through the thick gluteus maximus, roll your fingers across the belly of the slender piriformis.
3) Strum across the belly to clarify its location, staying mindful of the deeper sciatic nerve (6.116).

Are you compressing through the thick gluteus maximus fibers? With your fingers on the piriformis, bend the knee to 90° and ask your partner to rotate his hip laterally against your gentle resistance (6.117). You may feel gluteus maximus contract, but can you feel piriformis contract beneath it?

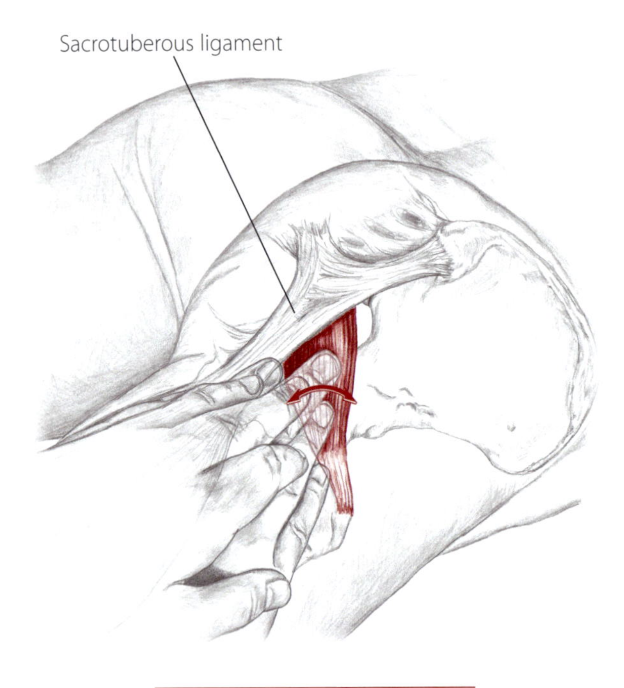

(6.116) Posterior/lateral view, partner prone, rolling over piriformis

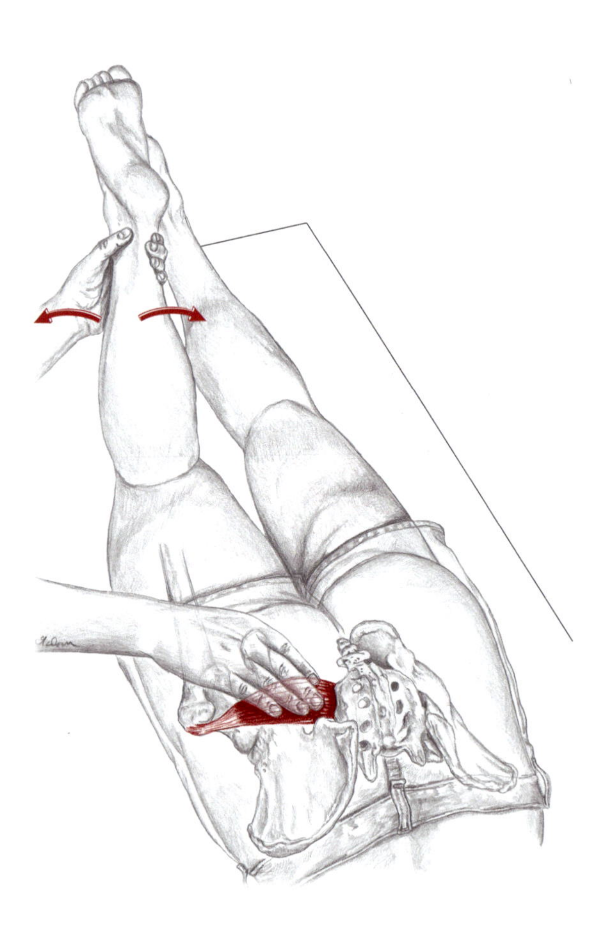

(6.117) Prone, feel the piriformis contract by asking your partner to laterally rotate his hip against your resistance

piriformis	**pir**-i-**form**-is	L. pear-shaped
quadratus	**kwod**-rait-us	L. squared, four-sided

Quadratus femoris

1) Partner prone. Locate the distal, posterior aspect of the greater trochanter and the ischial tuberosity. Place your fingerpads between these two landmarks.
2) Pressing firmly through the gluteus maximus fibers, strum vertically across the fibers of the rectangle-shaped quadratus femoris.

Does the belly stretch between the ischial tuberosity and the distal trochanter? Rolling your fingers over the belly, can you feel its horizontal fibers? Flex the knee zto 90° and passively rotate the hip medially and laterally. Can you sense changes in the muscle's tension as it shortens and lengthens (6.118)?

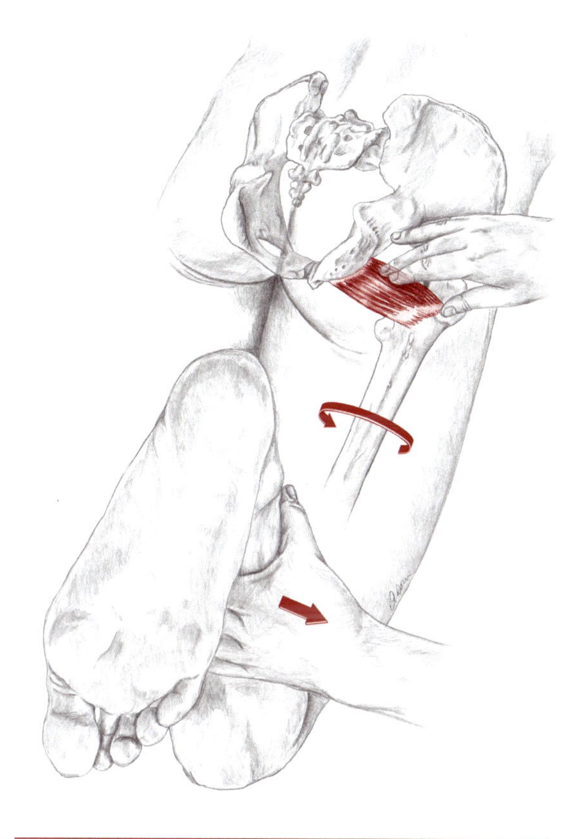

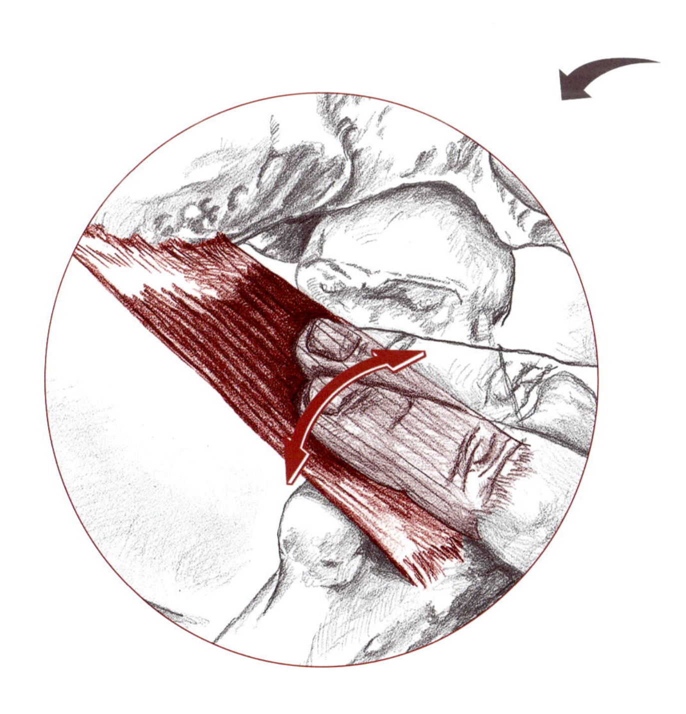

(6.118) Prone, sense the quadratus femoris contract by asking your partner to laterally rotate his hip against your resistance

Compared to its evolutionary precursors, the piriformis is a remnant of its former glory. It is a descendant of the great caudofemoral elevator muscles that can still be seen today extending from a reptile's femur to its tail. These large muscles provide reptiles with the tremendous thrusting force needed to extend the femur while running.

Iliopsoas

Psoas Major
Iliacus

The iliacus and psoas major, together called the iliopsoas, are important hip flexors and low back stabilizers (6.119). Best known to your local butcher as "tenderloin" or "filet mignon," the long, slender **psoas major** is located deep to the abdominal contents (6.120). It stretches from the lumbar vertebrae, underneath the inguinal ligament, to the lesser trochanter.

The stockier **iliacus** is located deep to the abdomen in the iliac fossa (6.121). Because of their respective locations, these muscles are only partially accessible and may be challenging to palpate.

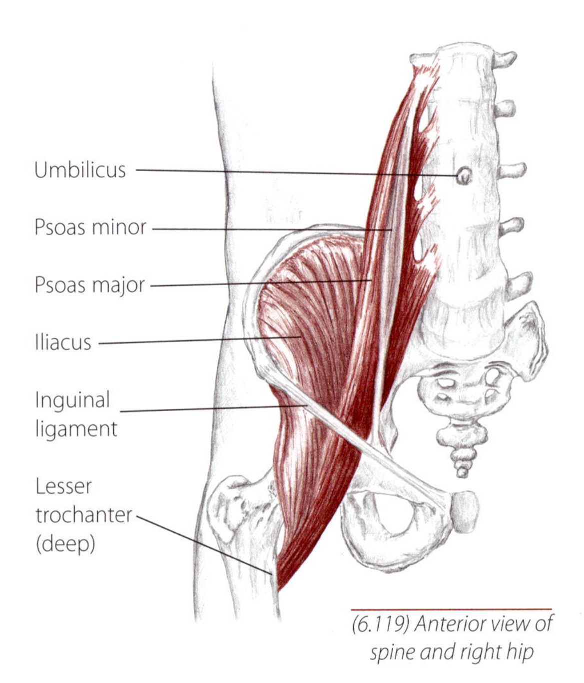

(6.119) Anterior view of spine and right hip

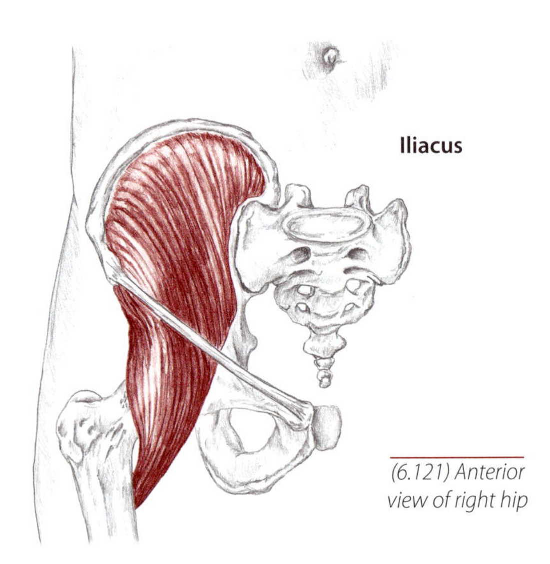

(6.121) Anterior view of right hip

Psoas Major

A Flex the hip (coxal joint)
Laterally rotate the hip (coxal joint)
Adduct the hip (coxal joint)

O Bodies and transverse processes of lumbar vertebrae

I Lesser trochanter

N Lumbar plexus

Iliacus

A Flex the hip (coxal joint)
Laterally rotate the hip (coxal joint)
Adduct the hip (coxal joint)

O Iliac fossa

I Lesser trochanter

N Femoral

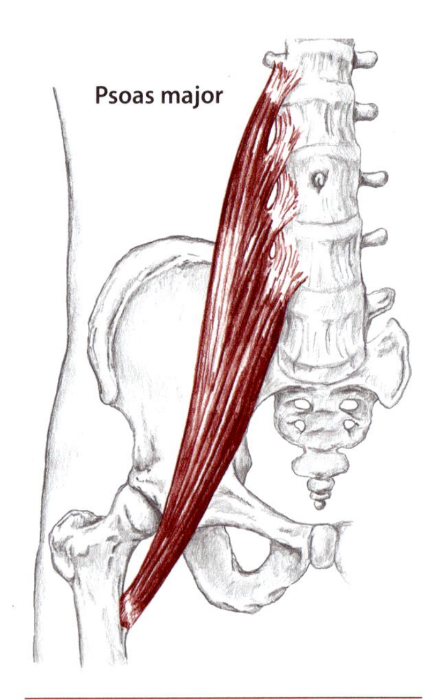

(6.120) Anterior view of spine and right hip

iliacus	i-**lee**-a-cus	L. pertaining to the loin
psoas	**so**-as	Grk. muscle of the loin

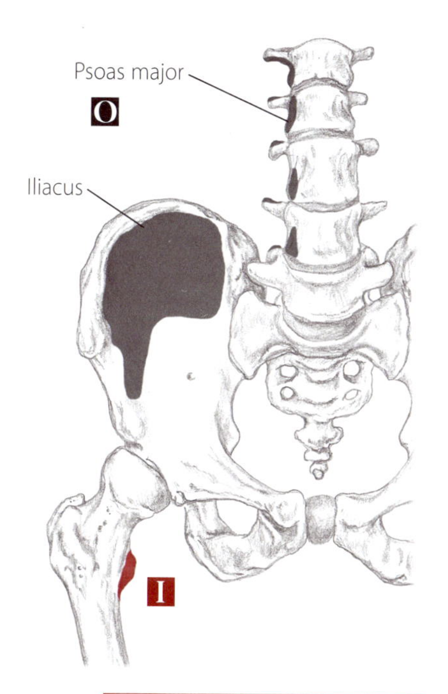

(6.122) Anterior view of right hip and spine showing origins and insertion

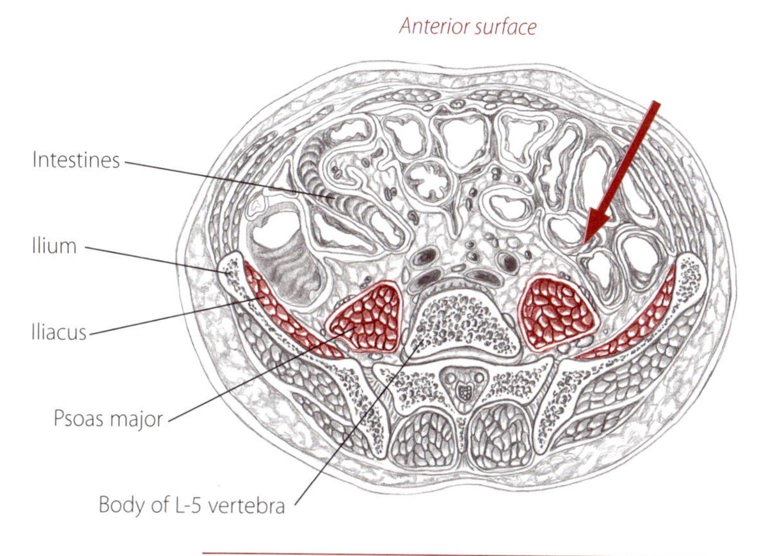

(6.123) Cross section of the trunk at the level of L-5, arrow showing direction of fingers when accessing the psoas major

Psoas Minor

Roughly 40% of the population has a psoas minor. It is a small muscle which extends from the lumbar vertebrae to the superior ramus of the pubis. When present, the psoas minor assists in posterior tilt (upward rotation) of the pelvis - the opposite action of the psoas major (see box, p. 328). Interestingly, the psoas minor is an important muscle of locomotion on a dog or cat because of the relationship of the pelvis to the vertebrae in quadrupeds. In a human biped, however, the psoas minor is a relatively insignificant muscle, except when doing the horizontal rumba!

A Assist to create lordotic curvature in the lumbar spine
Tilt pelvis posteriorly

O Body and transverse process of first lumbar vertebra

I Superior ramus of pubis

N Ventral rami of lumbar

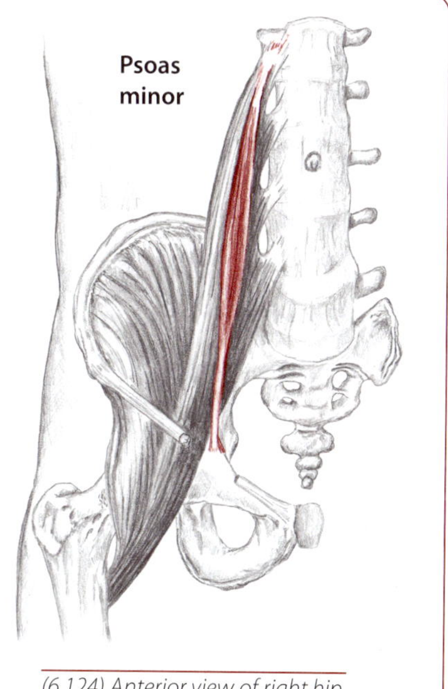

(6.124) Anterior view of right hip, with inguinal ligament cut

Psoas major

When accessing either the psoas or iliacus, palpate slowly and communicate with your partner. If at any point he does not feel comfortable or safe, slowly remove your hands. The psoas major lies just lateral to the abdominal aorta (p. 225). If you feel a strong pulse directly beneath your fingers when accessing the muscle, realign them more laterally.

1) Partner supine, with the hip slightly flexed and laterally rotated. Support your partner's thigh by placing your thigh underneath it. Locate the navel and ASIS, placing your fingerpads hand-on-hand between these points.
2) Slowly compress your fingerpads into the abdomen, moving only when your partner exhales (6.125). (Compressing in small circles upon your partner's initial exhalations will assist in moving the abdominal contents to the side.) As you compress further, keep your fingerpads stationary and direct your fingers downward toward the table.
3) Check that you are palpating the psoas, not the surrounding tissues, by asking your partner to flex his hip ever so slightly. If your fingers are accessing the psoas, you will feel a definite, solid contraction (6.125).

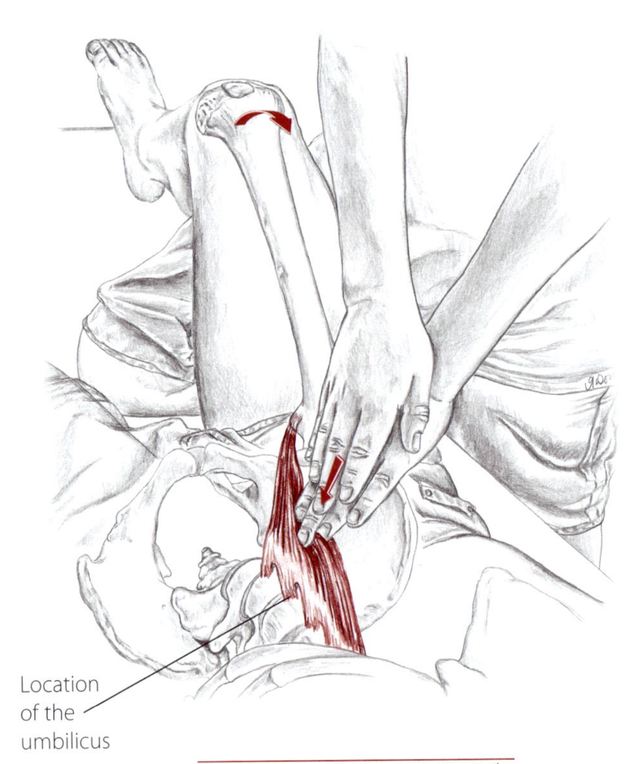

(6.125) Partner supine, accessing the psoas while your partner flexes his hip

Are you between the ASIS and navel? Is the direction of your fingers at a slight angle toward the spine? Have you compressed slowly, allowing the overlying tissue to relax? If you did not feel the muscle contract, try again with the fingers repositioned further inferiorly.

Sidelying position allows the abdominal contents to shift away from the psoas and, oftentimes, offers a less invasive position for your partner.

1) With the hips in a flexed position, place a bolster between your partner's knees. Locate the navel and ASIS, placing your fingerpads hand-on-hand between these points (6.126).
2) Following your partner's breath, curl your fingers into the abdomen and onto the surface of the psoas. Ask your partner to flex his hip slightly so you can feel for the psoas' contraction.

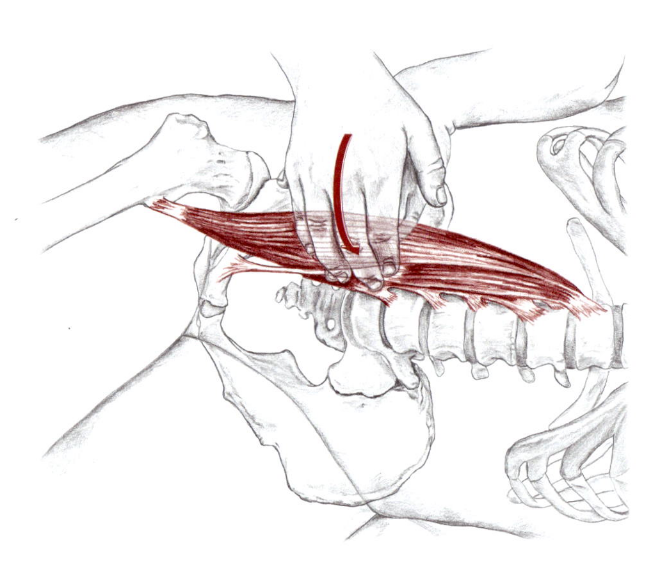
(6.126) Partner sidelying with hips flexed, curling your fingers into the abdomen

Psoas major primarily flexes the hip. But when the femur is stabilized, the psoas, in conjunction with iliacus, can increase the lordotic curvature in the lumbar spine and create anterior tilt (downward rotation) of the pelvis. It has also been proposed that only the superficial fibers of the psoas increase the lordotic curve, whereas the deeper fibers may decrease it.

Iliacus

1) Partner supine, with the hip slightly flexed and laterally rotated. Support your partner's thigh by placing your thigh underneath it.
2) Locate the anterior portion of the iliac crest and place your fingerpads hand-on-hand an inch medial to its ridge. (Beginning medially will allow you to penetrate more easily through the abdominal muscles.)
3) Slowly curl your fingers into the iliac fossa, moving only when your partner exhales (6.127). Your fingers may sink only a short distance into the tissue. Here's a hint: The intention of your touch needs to go beyond the superficial abdominal muscles and be directed toward the anterior surface of the ilium.
4) Ask your partner to flex his hip slightly, with your fingers in place. You will feel the strong iliacus contract.

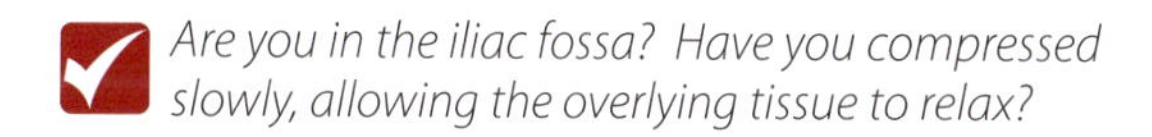

Are you in the iliac fossa? Have you compressed slowly, allowing the overlying tissue to relax?

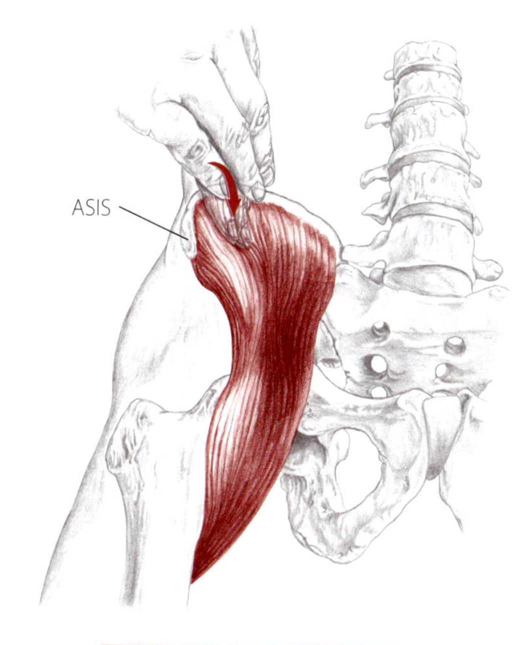

(6.127) Anterior/inferior view, partner supine

As with the psoas, sidelying position allows the abdominal contents to fall away from the iliacus and can be a more comfortable position for your partner. With the hips in a flexed position, place a bolster between your partner's knees and follow the instructions above (6.128).

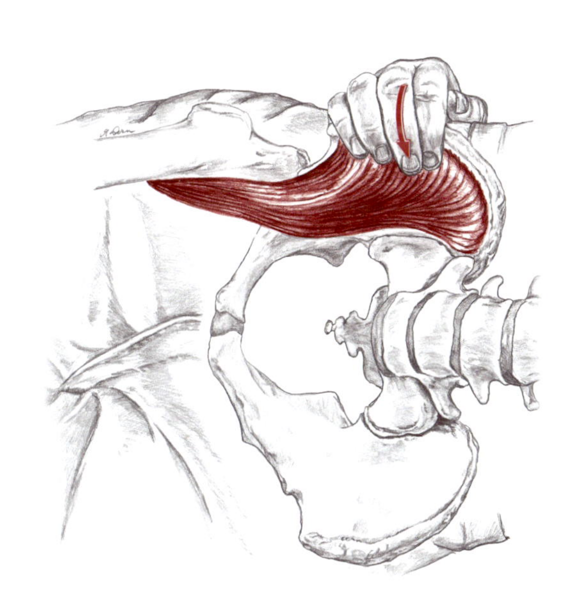

(6.128) Partner sidelying with hips flexed, curling your fingers into the abdomen

Following the above procedure, access the iliacus from the opposite side of the table. Try curling into the iliac fossa with your thumbs (6.129).

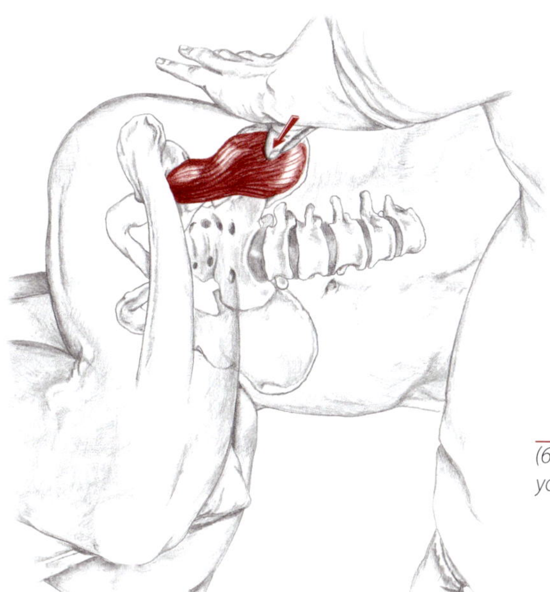

(6.129) Palpating with your partner sidelying

Other Structures of the Pelvis and Thigh

The femoral triangle is located on the anterior, medial surface of the thigh (6.130). It is formed by the inguinal ligament, adductor longus and sartorius. Several important vessels, including the femoral artery, nerve and vein, pass superficially through the femoral triangle.

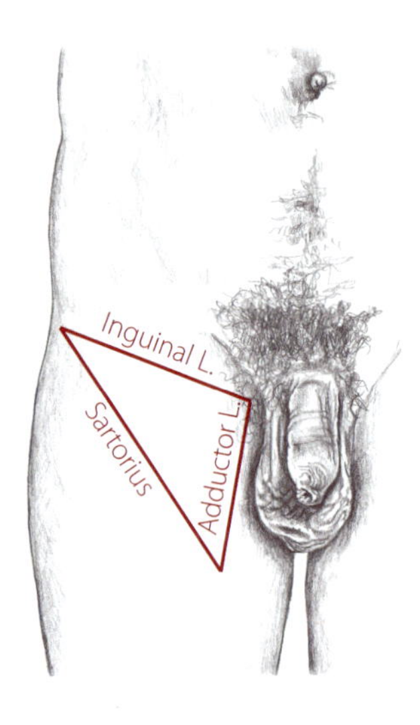

(6.130) The three borders of the femoral triangle

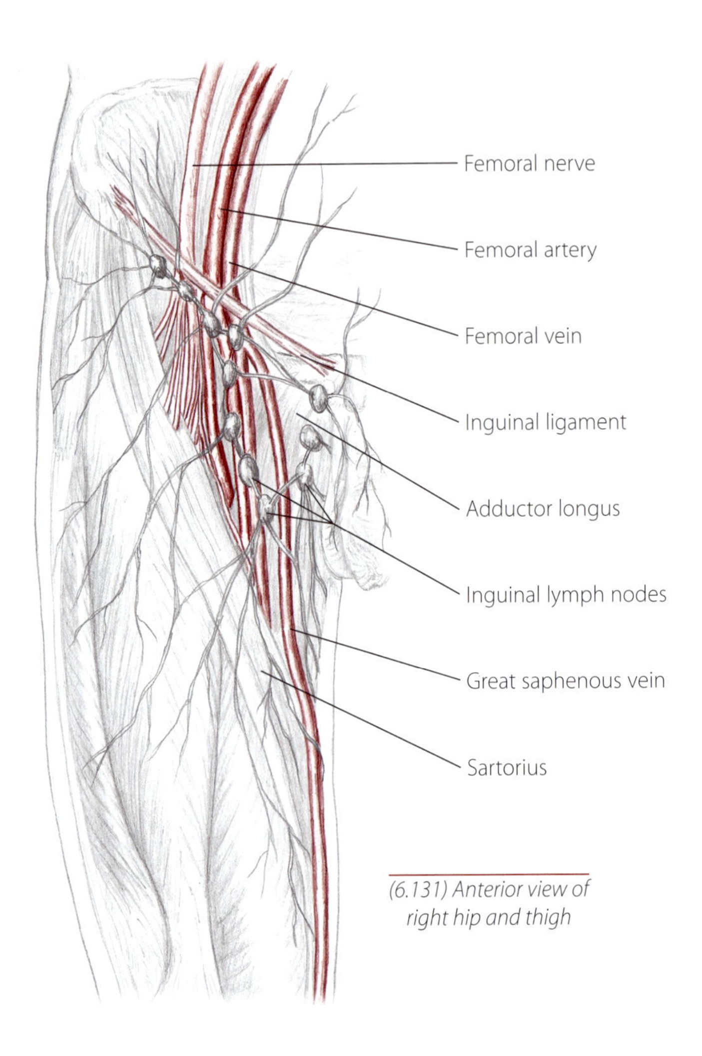

(6.131) Anterior view of right hip and thigh

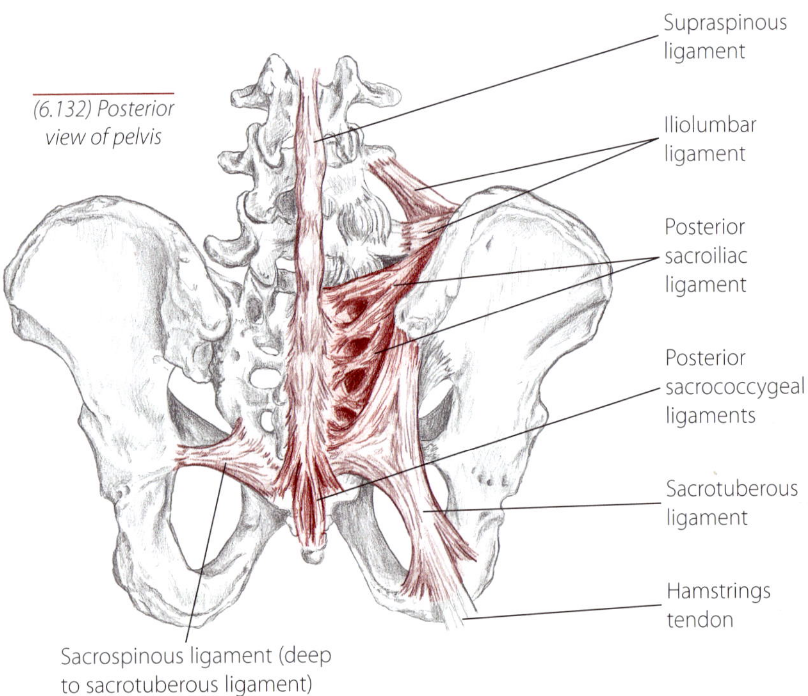

(6.132) Posterior view of pelvis

The great saphenous vein is a superficial vessel traveling the length of the lower extremity. Often visible, it begins near the ankle, passes along the medial aspect of the tibia and follows the sartorius up the thigh to empty into the femoral vein at the femoral triangle. Since it is long and easily accessible, the saphenous vein is often used for grafts in coronary bypass surgery.

saphenous **sa**-fe-nus origin unclear, perhaps Arabic *saphin*, standing; or Greek *saphen*, clearly visible

Ligaments of the Pelvis

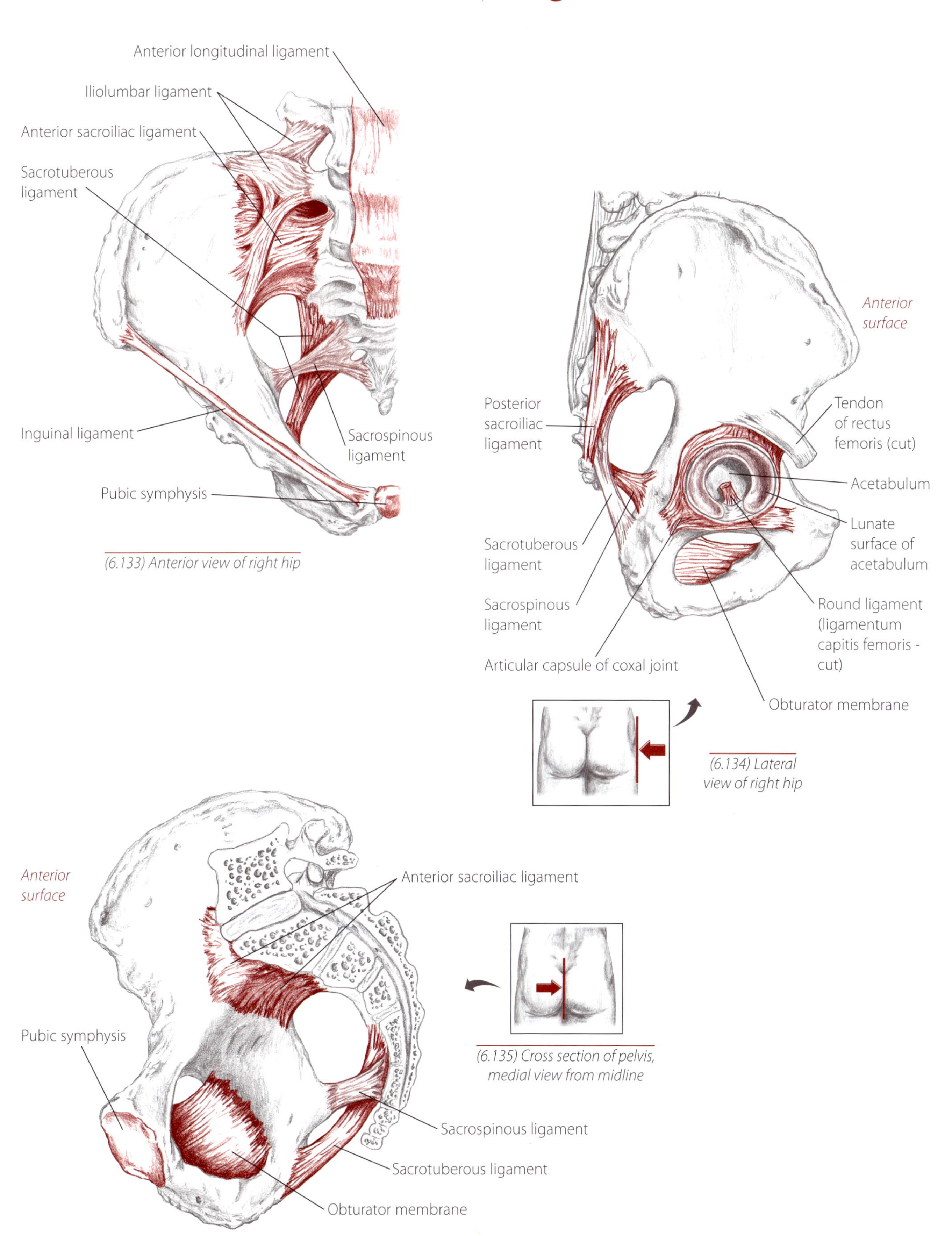

(6.133) Anterior view of right hip

(6.134) Lateral view of right hip

(6.135) Cross section of pelvis, medial view from midline

sacrococcygeal **sa**-kro-kok-**sij**-e-al
sacrotuberous **sa**-kro-**tu**-ber-us

Ligaments of the Coxal Joint

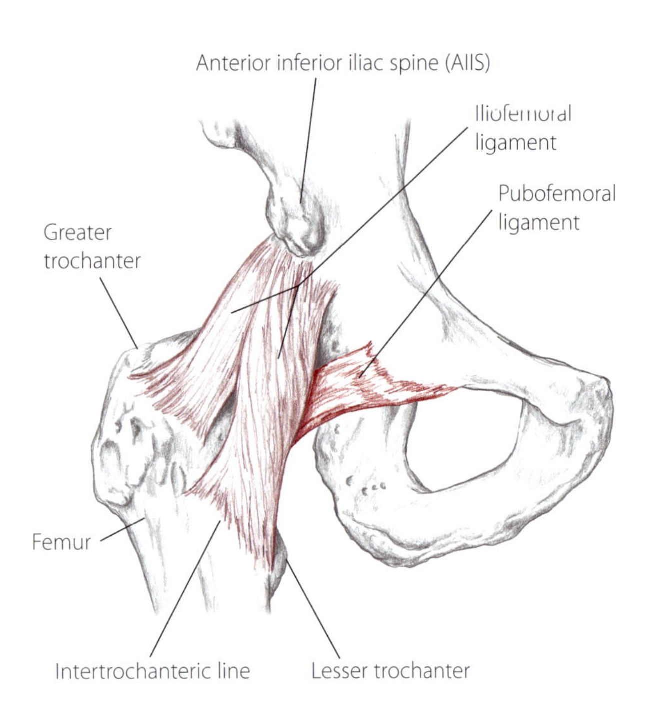

(6.136) Anterior view of right coxal joint

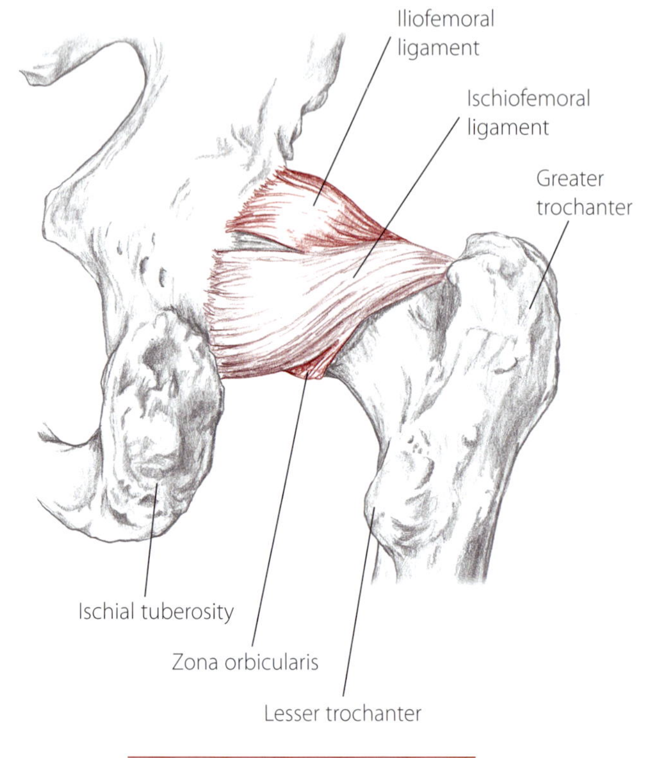

(6.137) Posterior view of right coxal joint

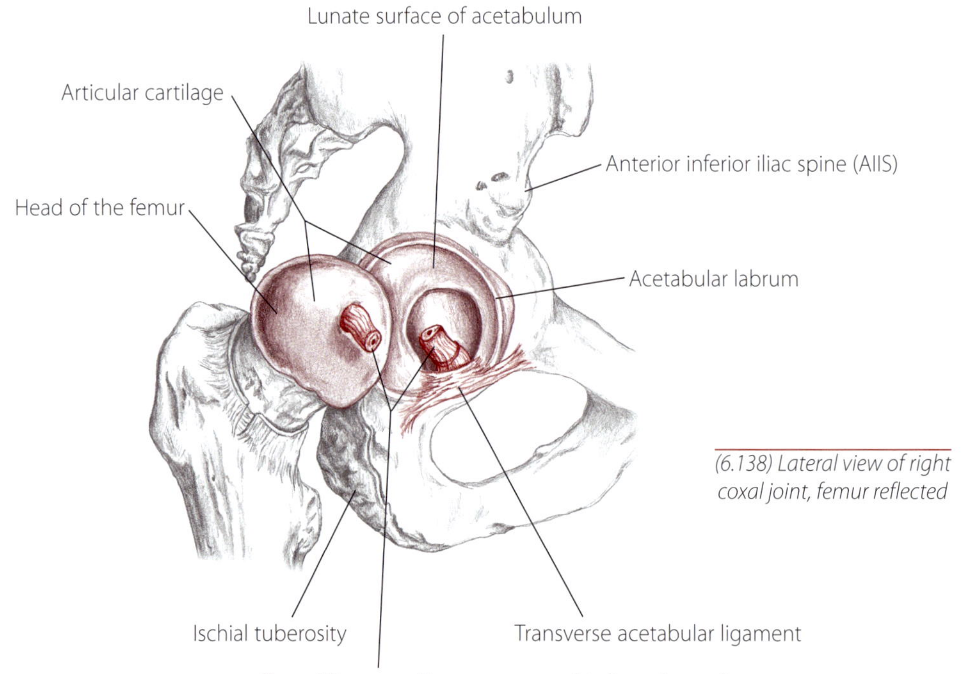

(6.138) Lateral view of right coxal joint, femur reflected

zona orbicularis **zo**-na or-**bik**-u-lar-is L. girdle + little circle

Inguinal Ligament

The inguinal ligament is a superficial band stretching between the ASIS and the pubic tubercle. It forms the superior border of the femoral triangle and the lower edge of the abdominal aponeurosis. It is an attachment site for the lower portion of the external oblique muscle.

1) Partner supine. Soften the surrounding tissue of the ligament by bolstering your partner's knee.
2) Locate the ASIS and slide diagonally in the direction of the pubic tubercle (6.139).
3) Strum gently across the slender ligament, feeling its cordlike quality.

Can you feel a thin, superficial band just beneath the skin? Does the band stretch from the ASIS and extend to the pubic tubercle?

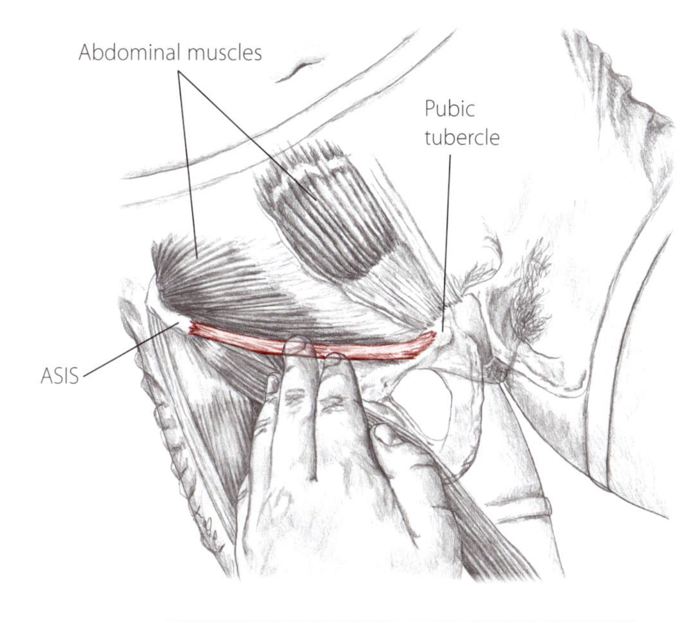

(6.139) Partner supine, locating the inguinal ligament

Femoral Artery, Nerve and Vein

The femoral artery, nerve and vein form a neurovascular bundle that courses through the femoral triangle. These vessels lie beneath the inguinal ligament and extend distally into the tissue of the thigh. The bundle is relatively superficial; the pulse of the femoral artery can be easily felt.

The penis contains no muscle tissue. During sexual arousal, the arteries of the penis dilate and a small muscle (ischiocavernosus) at the base of the penis helps to maintain an erection. The testicles are enwrapped by the cremaster muscle. It protects the sperm inside by lowering the testes when they become too warm and pulling them up closer to the body when they become too cold.

Pulse of the femoral artery

1) Partner supine. Slide your flexed knee behind your partner's knee. This position will flex and laterally rotate the hip, allowing for easier palpation.
2) Place the flat of your fingerpads halfway between the ASIS and the pubic tubercle just distal to the inguinal ligament. Feel for the strong pulse of the femoral artery (6.140).

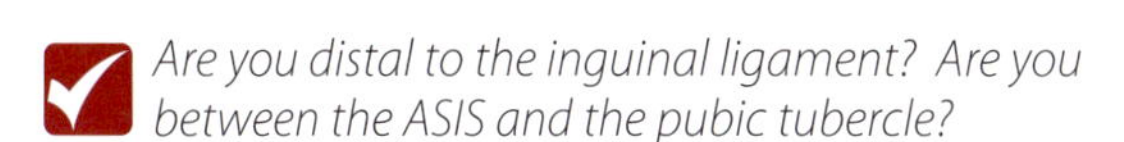

Are you distal to the inguinal ligament? Are you between the ASIS and the pubic tubercle?

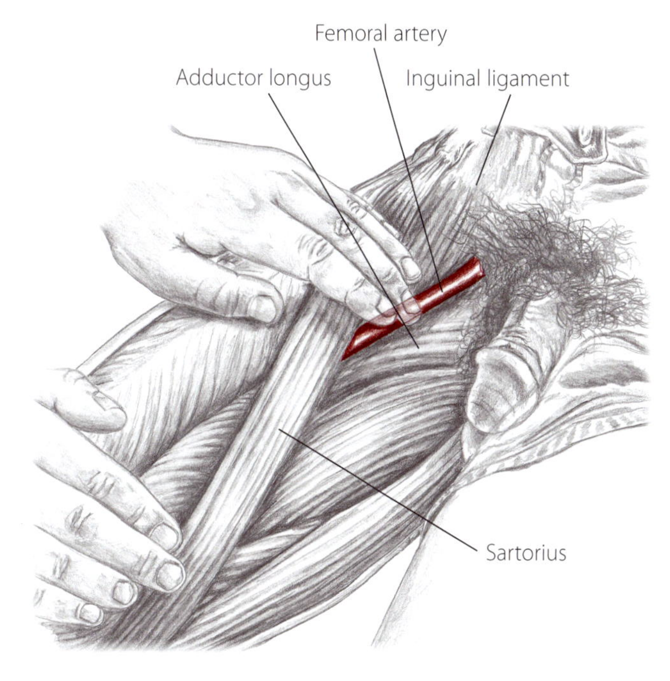

(6.140) Partner supine, locating the femoral artery

inguinal	**ing**-gwi-nal	L. of the groin
penis		L. tail

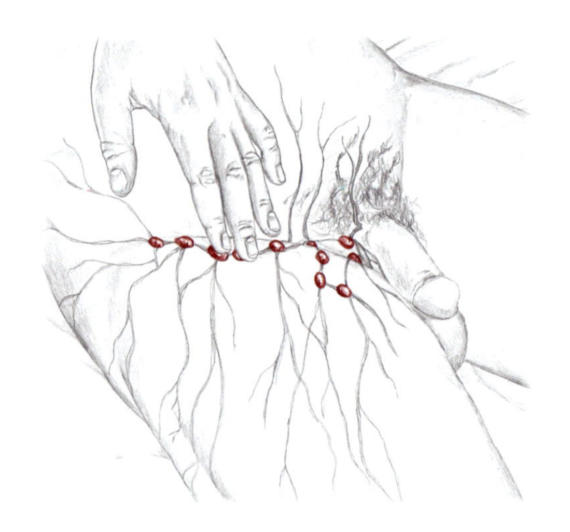

(6.141) Partner supine, palpating the inguinal lymph nodes

Inguinal Lymph Nodes

Distal to the inguinal ligament, the superficial inguinal lymph nodes are easily palpable. They number between eight and ten and vary in size from a small lentil to a raisin.

1) Partner supine, with the knee bolstered. This position will flex and laterally rotate the hip, allowing for easier palpation.
2) Locate the inguinal ligament. Slide inferiorly and explore for the superficial nodes (6.141).

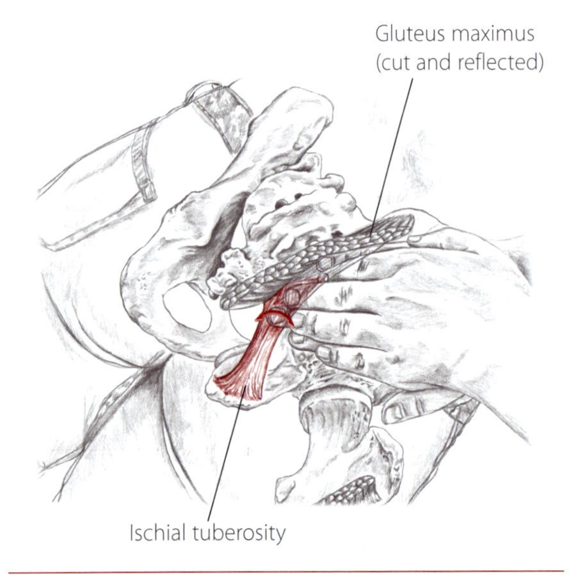

(6.142) Partner prone, rolling across the sacrotuberous ligament

Sacrotuberous Ligament

This broad, solid ligament stretches between the ischial tuberosity and the edge of the sacrum. Although it is deep to the gluteus maximus muscle, it is distinctly palpable and may feel like a span of bone.

1) Partner prone. Locate the ischial tuberosity. Locate the edge of the sacrum (p. 288).
2) Slide your fingertips off the tuberosity toward the edge of the sacrum. Using firm pressure, palpate through the gluteus maximus belly and strum broadly across the ligament (6.142).

Are you deep to the gluteus maximus fibers? Is the structure you are rolling over roughly an inch wide and inflexible? Does it stretch from the ischial tuberosity toward the sacrum?

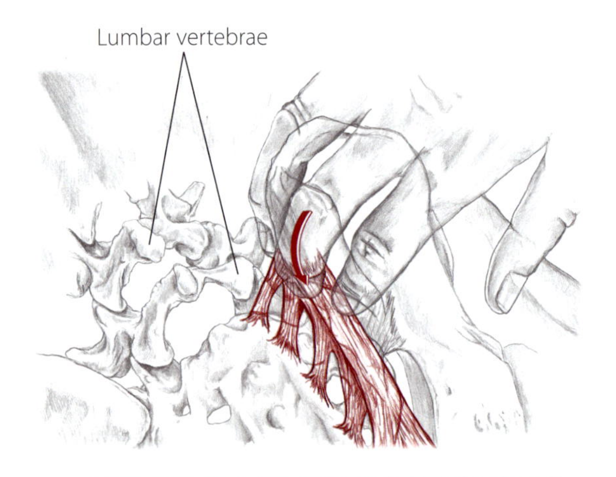

(6.143) Partner prone, locating the sacroiliac ligaments

Posterior Sacroiliac Ligaments

Superficial to the sacroiliac joint, the dense sacroiliac ligaments support the union of the posterior sacrum and the ilium. It has several segments which attach from the sacrum to the area around the PSIS. The ligaments are deep to the thoracolumbar aponeurosis, and their oblique fibers may be difficult to distinguish.

1) Partner prone. Locate the surface of the sacrum.
2) Using firm pressure, strum your fingertips across the dense fibers of the sacroiliac ligaments (6.143).

Are you medial to the PSIS, on top of the sacroiliac joint space (p. 289)?

Iliolumbar Ligament

The iliolumbar ligament is located between the transverse processes of the fourth and fifth lumbar vertebrae and the posterior iliac crest. The strong, horizontal fibers of the ligament is important in stabilizing L-4 and L-5. Deep to the thoracolumbar aponeurosis, the thick multifidi (p. 206) and the quadratus lumborum (p. 213), the ligament is difficult to palpate; however, its location and density can be determined.

1) Partner prone. Locate the PSIS.
2) Slide your thumb straight superior from the PSIS to the level of L-4 and L-5. Your thumb should be between the iliac crest and the transverse processes of the lumbar vertebrae.
3) Using firm pressure, sink into the dense muscles of the low back and attempt to roll vertically across the ligament's taut fibers (6.144).

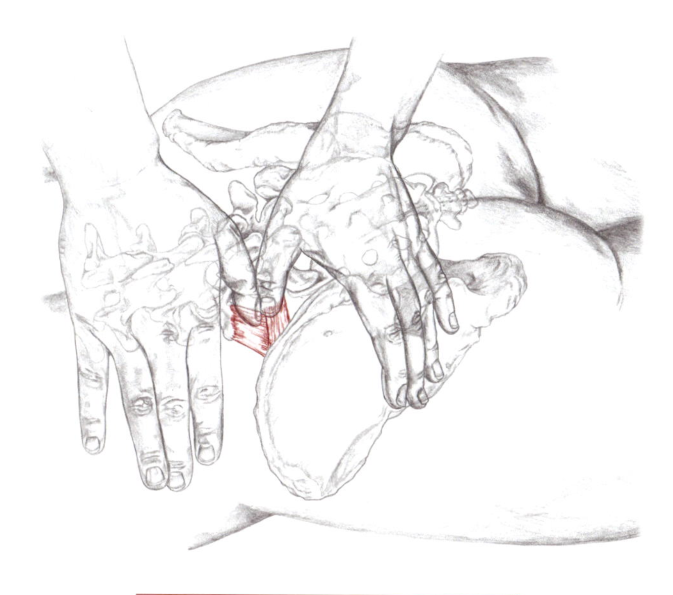

(6.144) Posterior/lateral view, partner prone, sinking your thumb onto the iliolumbar ligament

Can you palpate its dense, horizontal fibers?

Sciatic Nerve

The sciatic nerve is the largest nerve in the body - sometimes measuring three-quarters of an inch in diameter. It is formed by the spinal nerves of L-4 through S-3. The nerve passes through the greater sciatic notch, between the ischial tuberosity and greater trochanter, and extends down the posterior thigh. Distally, it branches into the tibial and peroneal nerves.

Because the sciatic nerve runs deep to the piriformis (p. 322), there is considerable potential for the piriformis to compress or entrap the nerve. In general, the sciatic nerve is difficult to isolate in the gluteal region and, of course, is best avoided.

1) First, *outline* the placement of the sciatic nerve. With your partner prone, locate the edge of the sacrum.
2) Draw a line down the buttock between the ischial tuberosity and greater trochanter. Continue down the middle of the posterior thigh. This is the location of the nerve.
3) To *access* the sciatic nerve, turn your partner sidelying and flex the hip. Locate the ischial tuberosity and greater trochanter.
4) Palpate between these landmarks for the pathway of the sciatic nerve (6.145). You can avoid pinching the nerve by palpating with the pad of your thumb.

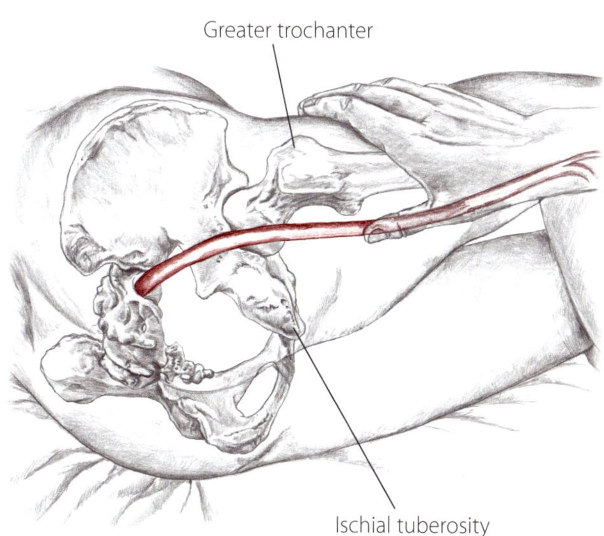

(6.145) Partner sidelying

sciatic   si-**at**-ik   Grk. *ischion*, hip joint

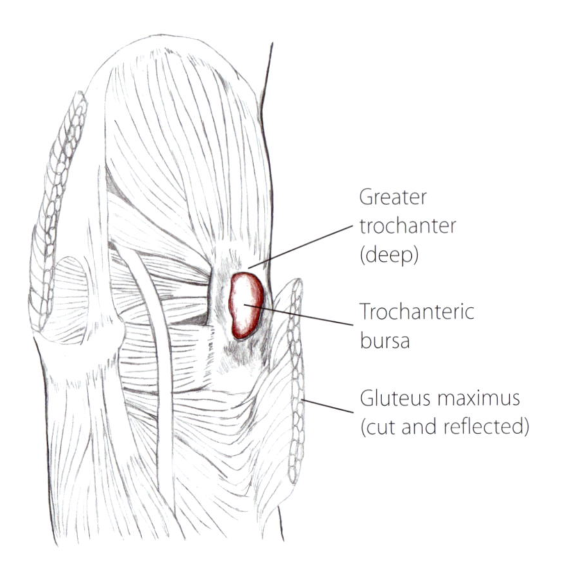

(6.146) Posterior view of right hip

Trochanteric Bursa

Positioned along the posterior/lateral aspect of the greater trochanter, this large bursa reduces friction between the trochanter and the gluteus maximus (6.146). Other bursae are located along the lateral and anterior sides of the trochanter and separate it from the gluteus medius and minimus. Unless they are inflamed or distended, the bursae will be impalpable.

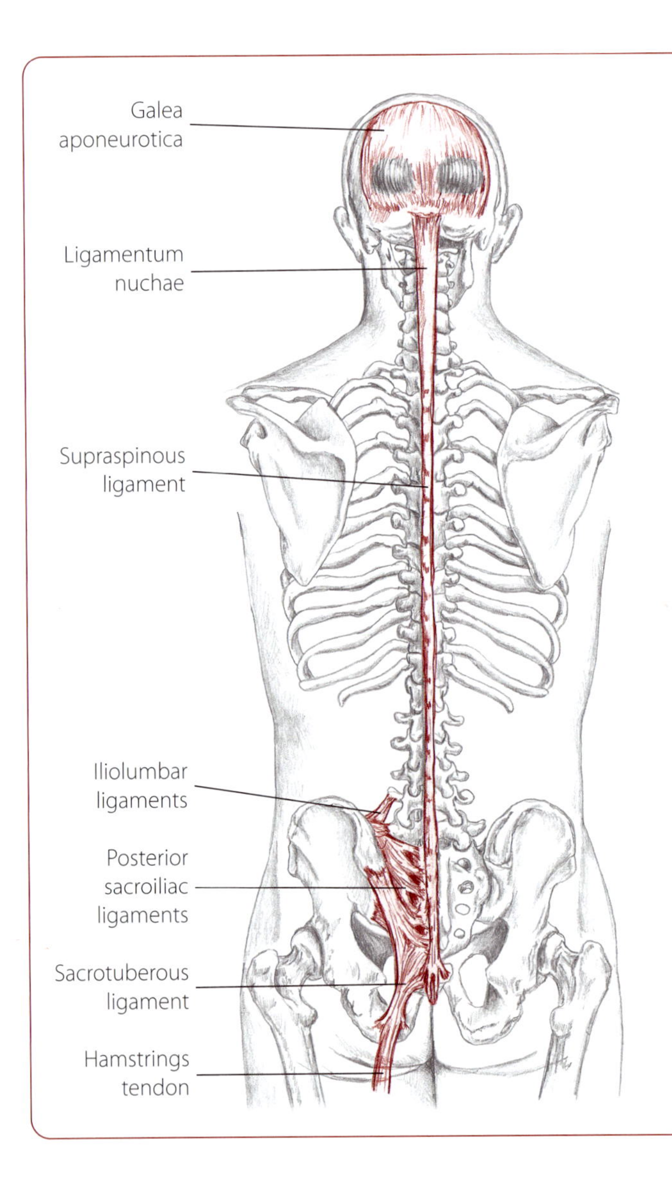

Ligaments, tendons, fasciae and retinacula are all forms of connective tissue. They are composed of virtually the same ingredients (collagen, elastin and ground substance) differing only in the proportions of these materials. For anatomical purposes, these bands and sheaths have been categorized individually although they are not separate structures. Collectively they form an incredible matrix that spreads throughout the entire body and supports it.

Now that you have explored the location of several connective tissue structures, here is an exercise to help you get a feel for how some of them connect together.

1) Partner prone. Locate the proximal hamstrings tendon (p. 307) as it attaches to the ischial tuberosity.
2) Follow the tendon superiorly as it melds into the sacrotuberous ligament (p. 334) and then to the posterior sacroiliac and iliolumbar ligaments (p. 334, 335) on the sacrum.
3) Continue superiorly as the sacroiliac ligaments blend into the thoracolumbar aponeurosis (p. 226) and the supraspinous ligament (p. 225) between the spinous processes of the vertebrae.
4) Ultimately, follow the supraspinous ligament all the way up the spine to the ligamentum nuchae (p. 224) and, finally, to the galea aponeurotica (p. 263) surrounding the cranium.

7
Leg & Foot

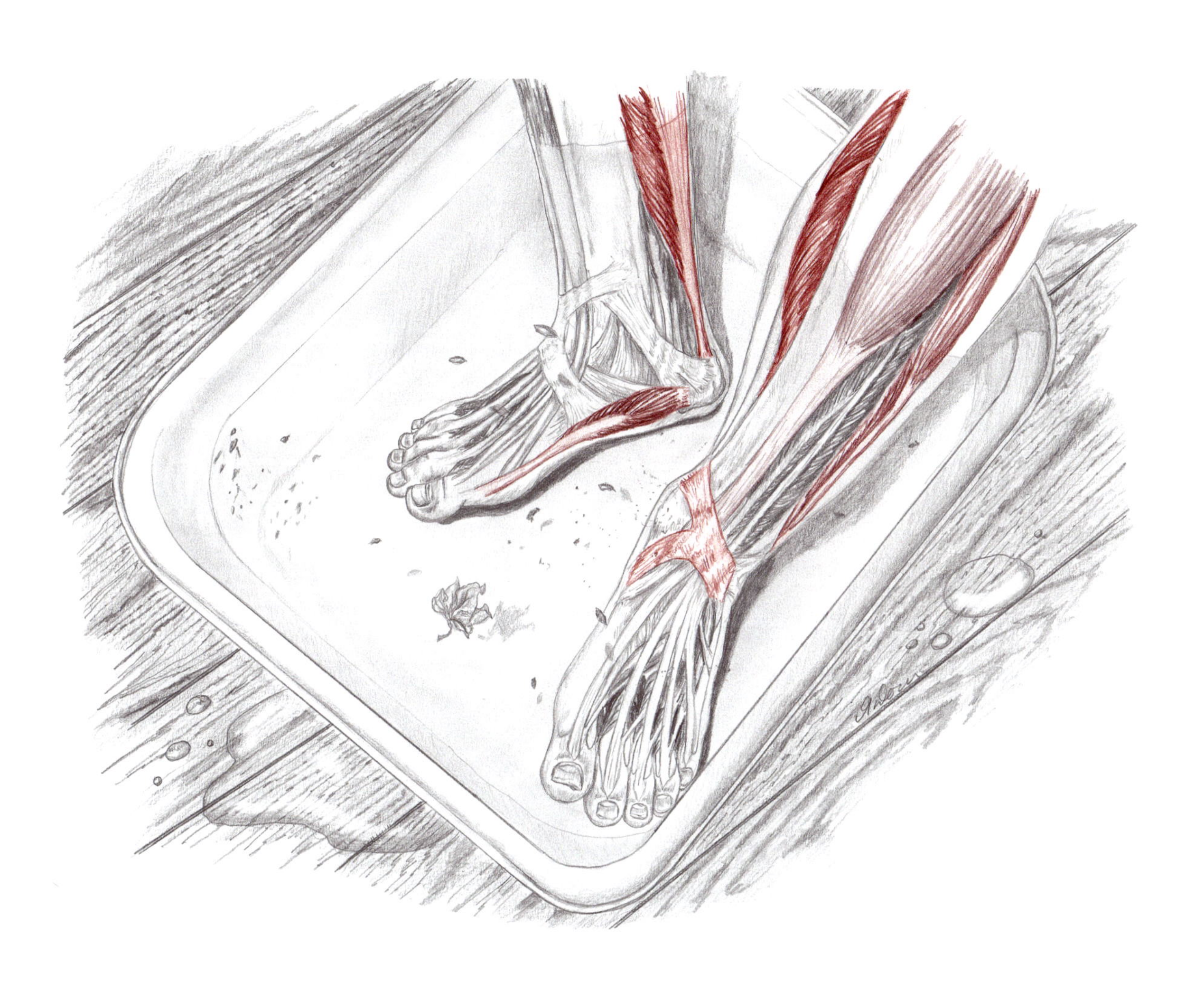

Topographical Views

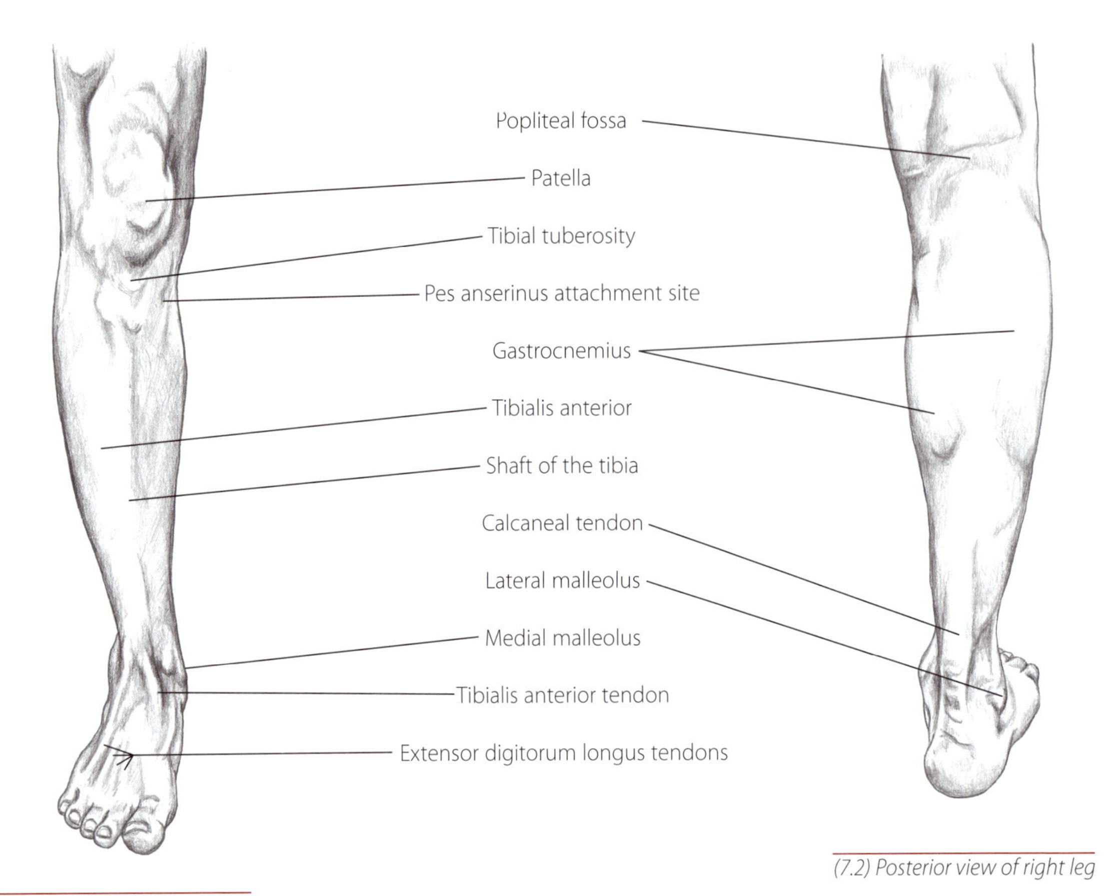

(7.1) Anterior view of right leg

(7.2) Posterior view of right leg

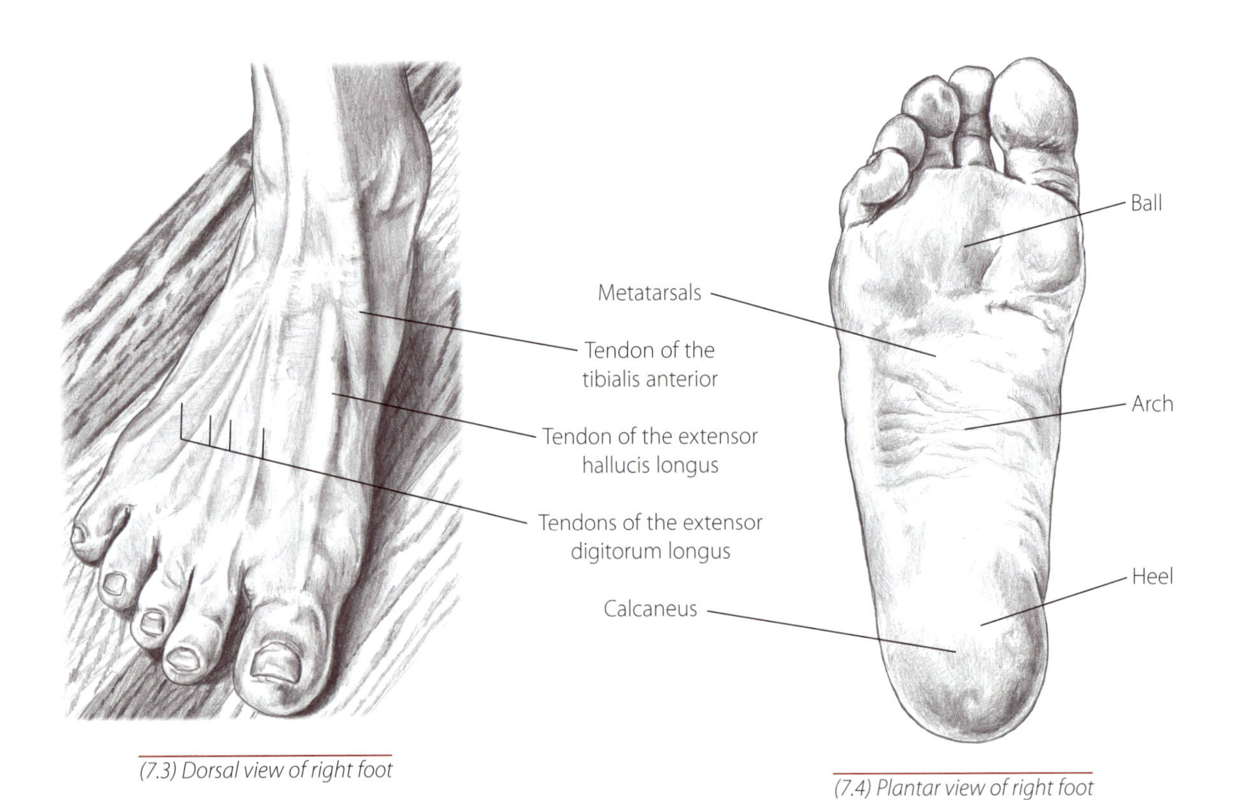

(7.3) Dorsal view of right foot

(7.4) Plantar view of right foot

Exploring the Skin and Fascia

1) Partner seated. Using one hand to stabilize the leg, use your other hand to gently twist the skin and fascia around the leg's shaft (7.5). Note the tissue's mobility or resistance as you pull it in opposite directions.
2) Now try to tug the skin superiorly and inferiorly. Oftentimes the tissue has less mobility when moving in the vertical direction than in the horizontal direction.

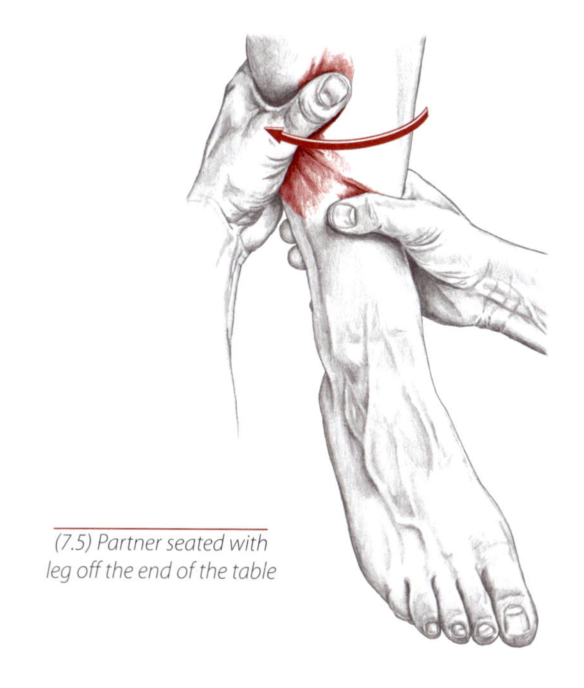

(7.5) Partner seated with leg off the end of the table

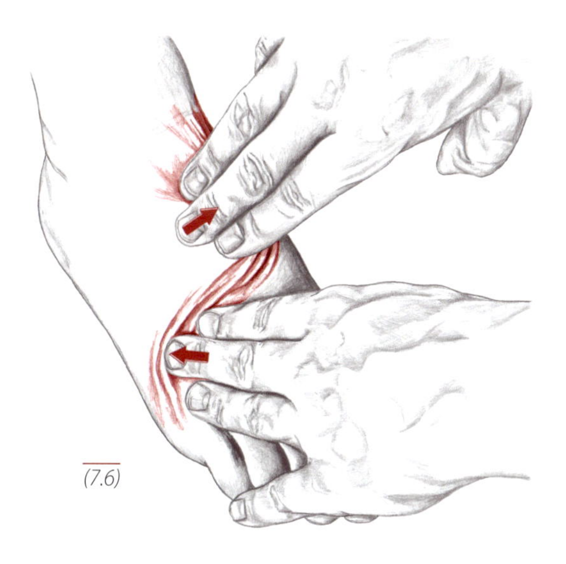

(7.6)

1) Partner seated. Place your hands on the foot's dorsal and plantar surfaces and sense their respective temperatures. Is one side warmer than the other?
2) Explore the dorsal surface by shifting the skin from side to side (7.6). Note the thin, delicate quality of its skin and fascia. Now palpate the plantar surface and note the tissue's thick, tough quality.

1) Partner prone. Here is an opportunity to feel the skin and fascia stretch upon passive movement. Grasp the tissue of the posterior ankle and passively plantar flex and dorsiflex the ankle (7.7). Feel how supple the tissue is when the ankle is plantar flexed. When you dorsiflex the ankle, however, the skin may be pulled out from between your fingers.
2) Continue to move the ankle while grasping the tissue on all sides of the leg. Now ask your partner to actively move his ankle and toes while you grasp the skin and fascia. Encourage him to move slowly. Play with isolating specific actions - such as plantar flexion of the ankle versus flexion of the toes - to feel how the tissue shifts upon different movements.

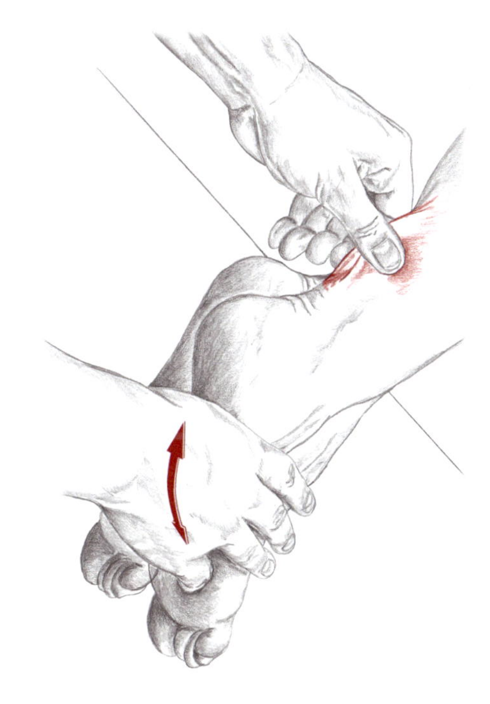

(7.7) Partner prone, with foot off the end of the table

Bones of the Knee, Leg and Foot

The knee is formed by the articulation of the distal femur and proximal tibia (7.8). The tibiofemoral (knee) joint, the largest synovial joint in the body, is a modified hinge joint. It is capable of flexion and extension, and when the knee is in a flexed position, it can medially and laterally rotate the knee (p. 342).

The region of the knee also includes the small **patella** ("kneecap") and the proximal fibula. The bony surfaces of the knee are superficial and easily accessible.

The **tibia** and **fibula** are the bones of the leg. The tibia ("shinbone") runs superficially from the knee to the ankle just as the ulna runs superficially from the elbow to the wrist. The fibula's relationship to the tibia is also similar to the radius' relationship to the ulna: It is lateral to the tibia and virtually deep to the surrounding muscles. The fibula bears only 10% of the body's weight and rightfully so: It is the thinnest bone in the body in proportion to its length.

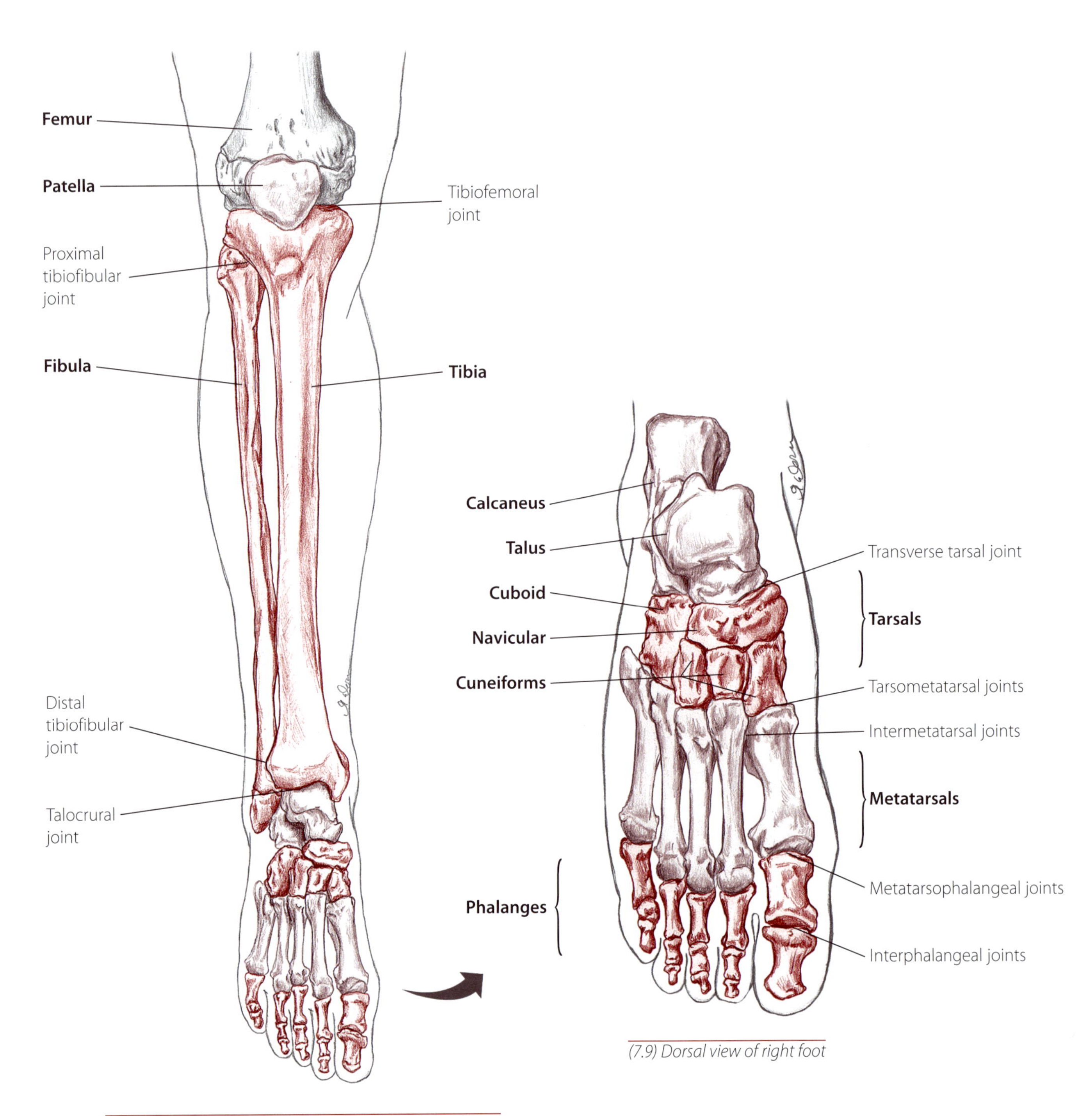

(7.8) Anterior view of right leg and foot, foot plantar flexed

(7.9) Dorsal view of right foot

fibula	**fib**-u-la	L. pin or buckle
patella	pa-**tel**-a	L. small pan

Bony Landmarks of the Knee and Leg

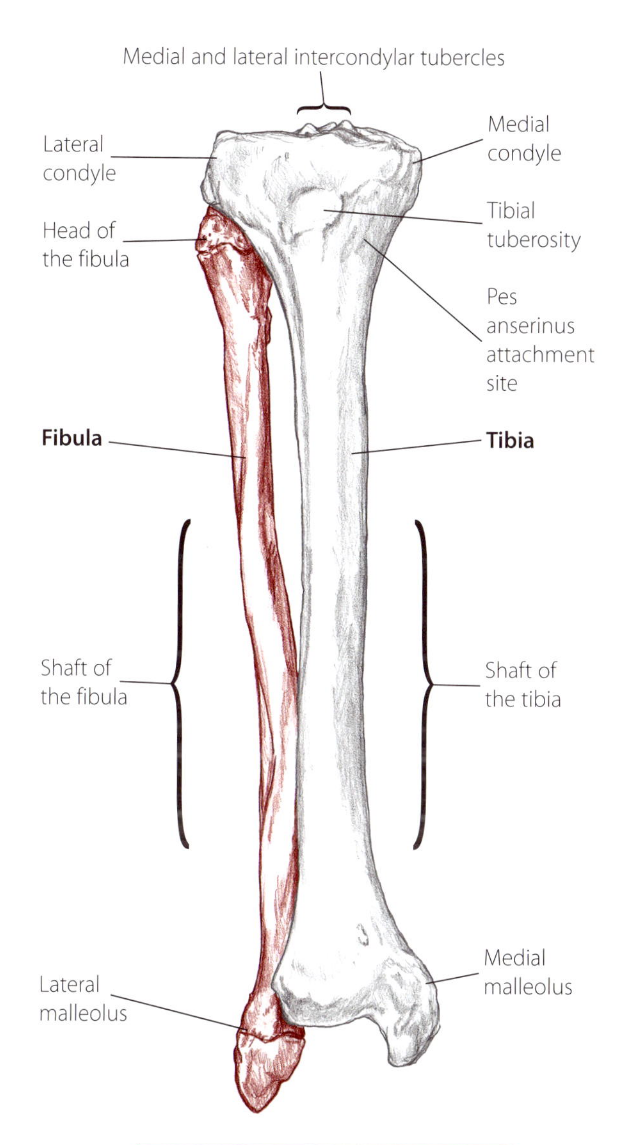

(7.10) Anterior view of right tibia and fibula

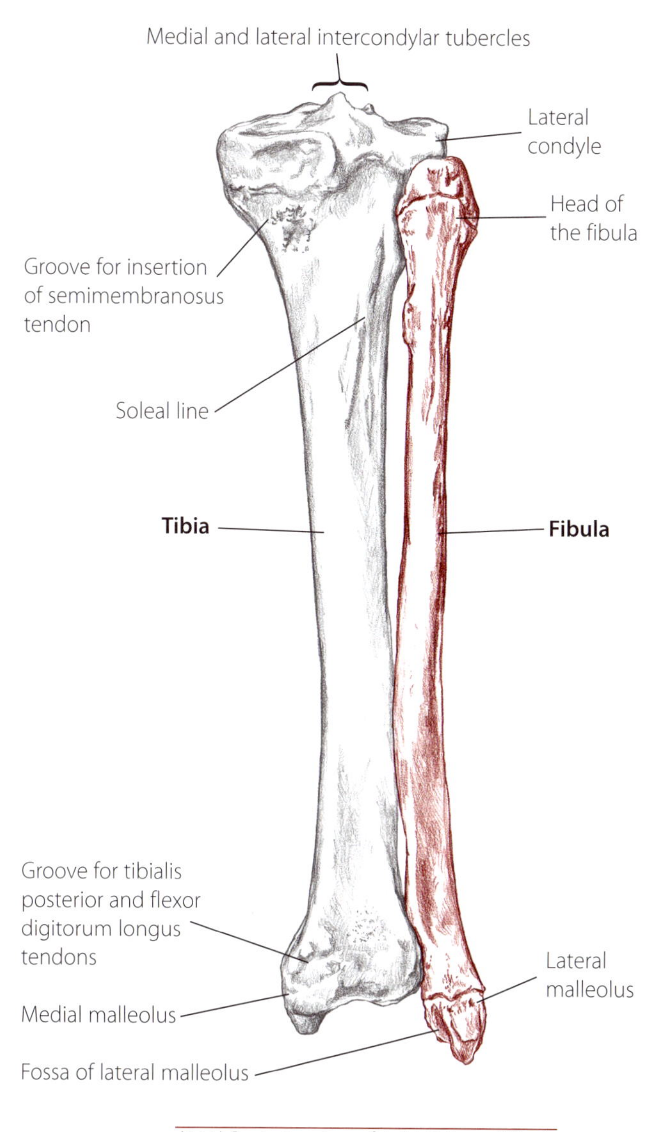

(7.11) Posterior view of right tibia and fibula

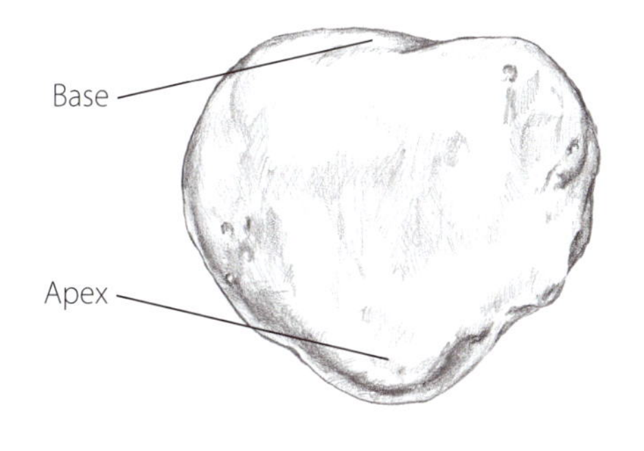

(7.12) Anterior view of right patella

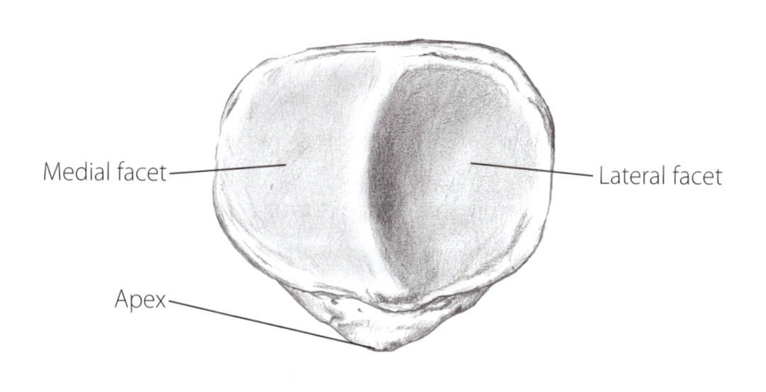

(7.13) Posterior view of right patella

Bony Landmarks of the Knee and Leg

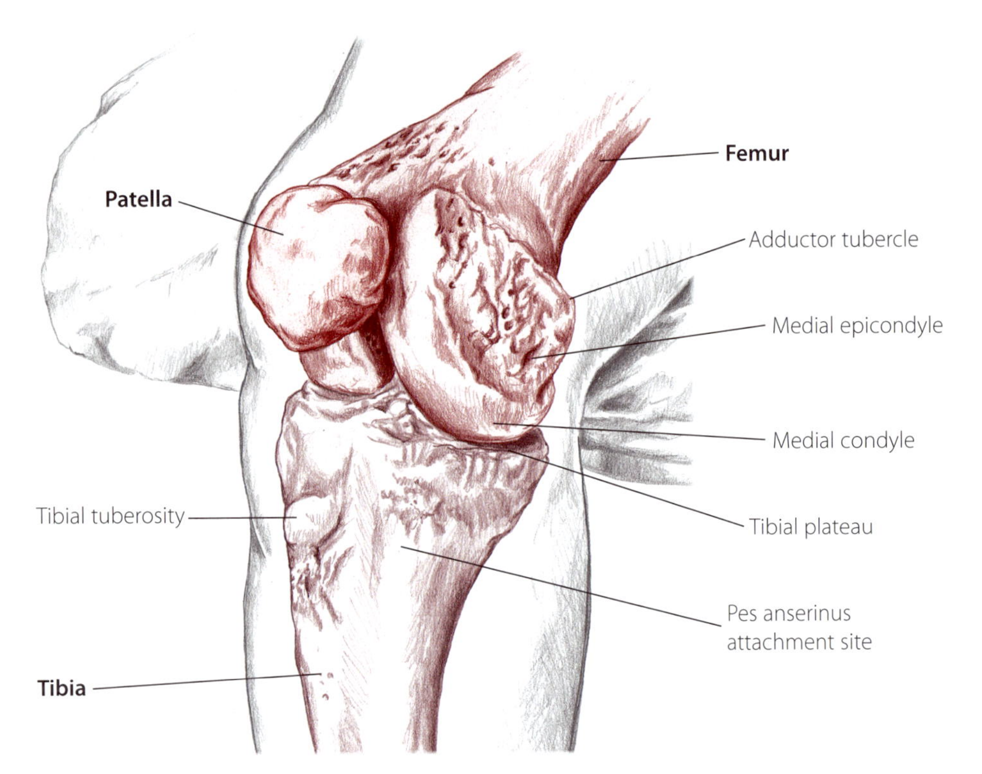

(7.14) Anterior/medial view of right knee showing tibiofemoral joint

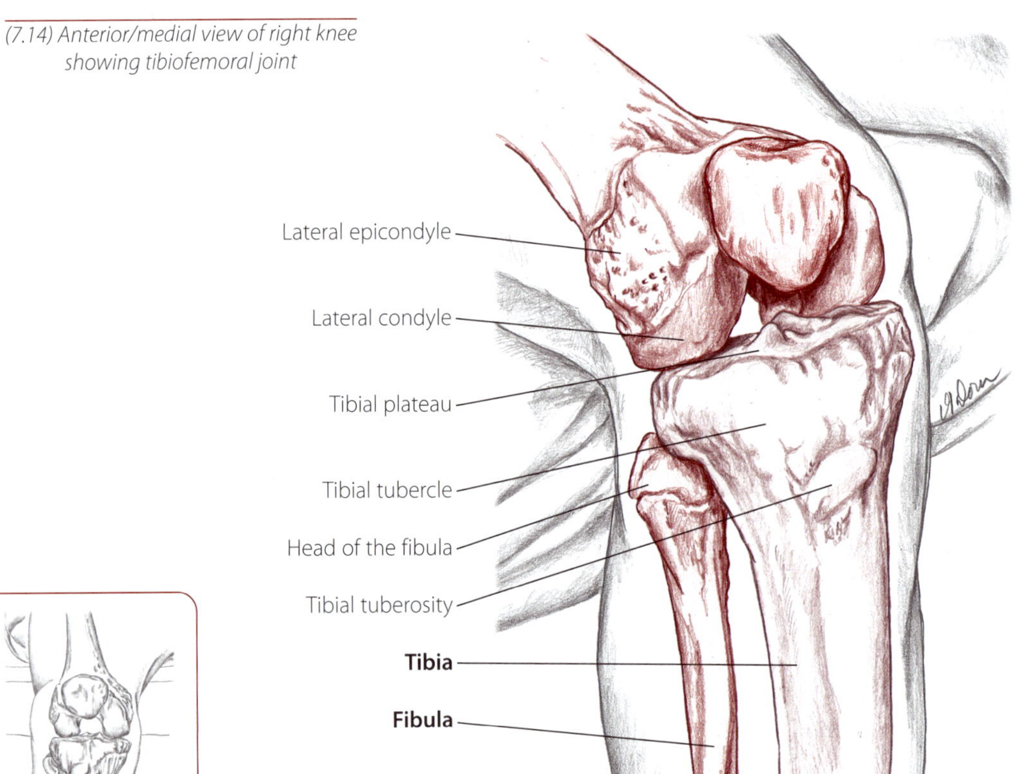

(7.15) Anterior/lateral view of right knee showing tibiofemoral joint

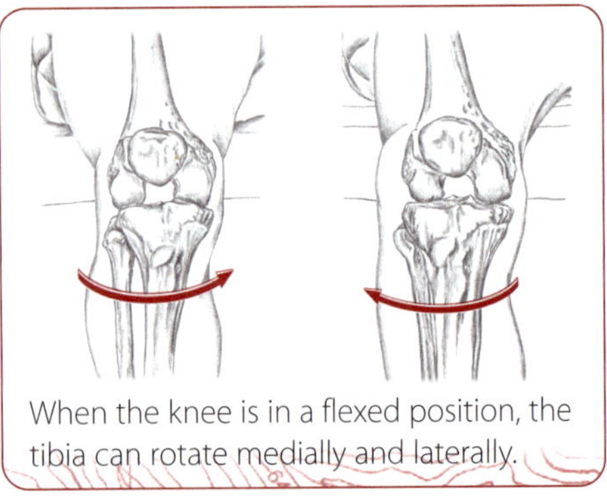

When the knee is in a flexed position, the tibia can rotate medially and laterally.

tibia	**tib**-e-a	L. shinbone
talocrural	**ta**-lo-**kroo**-ral	L. ankle + *crus*, leg

Bony Landmark Trails of the Knee

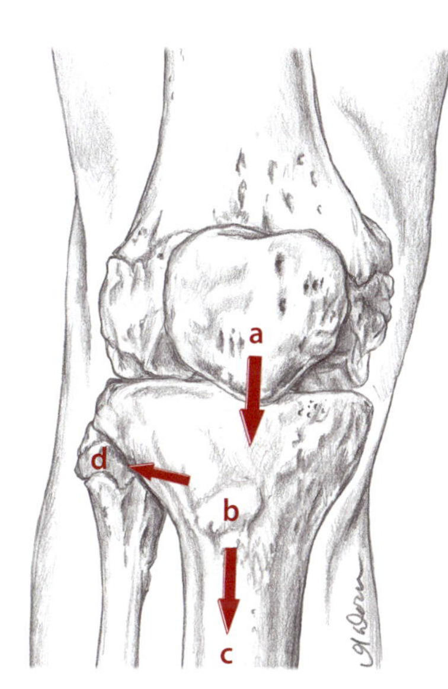

Trail 1 "Landmark Trail" links together the most prominent landmarks of the knee.

a Patella
b Tibial tuberosity
c Shaft of the tibia
d Head of the fibula

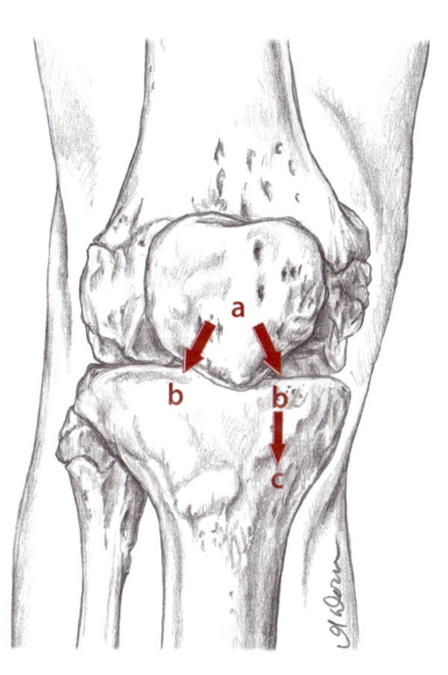

Trail 2 "Waddle Walk" has two paths exploring the medial and lateral aspects of the proximal tibia. It ends at the pes anserinus ("goose foot" in Latin) attachment site.

a Patella
b Medial and lateral tibial plateaus
c Pes anserinus attachment site

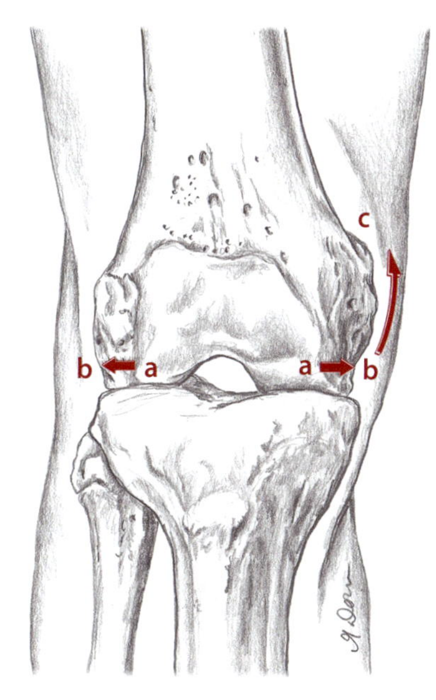

Patella removed

Trail 3 "Hills on Both Sides" explores the bumps of the distal end of the femur.

a Edges of the medial and lateral femoral condyles
b Medial and lateral epicondyles of the femur
c Adductor tubercle

The thickest layering of cartilage in the body can be found on the posterior surface of the patella. This eighth-of-an-inch-thick coating protects the patella from the incredible pressure applied by the quadriceps when the knee is flexed. Simply walking up or down stairs can place as much as six hundred pounds of pressure on the patella.

tuberosity **tu**-ber-**os**-i-tee L. a swelling

Trail 1 "Landmark Trail"

Patella

The patella is located on the anterior surface of the knee. It is superficial and round with an apex that points distally. The largest sesamoid bone in the body, the patella is an attachment site for the rectus femoris tendon (p. 300). When the knee is flexed, the patella seems to disappear, sinking into the space between the proximal tibia and femoral condyles.

1) Partner supine with the knee extended. In this position the quadriceps tendon is shortened and the patella is more mobile and easier to access.
2) Locate the patella on the anterior knee and palpate its round surface and edges. Try gently shifting it from side to side (7.16). Note any bumps or crevices along its edges.
3) Have your partner sit with his legs hanging off the table. Passively flex and extend the knee as you explore the patella's movements and its relationship to the rectus femoris tendon (7.17).

Tibial Tuberosity and Shaft of the Tibia

The **tibial tuberosity** is a superficial knob located distal to the patella on the shaft of the tibia. It is roughly half an inch in diameter and serves as an attachment site for the patellar ligament. It sometimes protrudes visibly. The **shaft of the tibia** runs superficially along the anterior leg. From the tibial tuberosity to the medial malleolus (p. 351), its edges and flat surface are easily palpable.

1) Partner seated with the knee flexed. Locate the patella. Slide your fingers three or four inches inferior from the patella and, using your thumbpad, explore for the tuberosity (7.18).
2) Continue to palpate inferiorly along the shaft of the tibia. Determine the width of the shaft by palpating along its edges. Follow it down to the medial malleolus.

With your fingers at the tibial tuberosity, ask your partner to extend his knee slightly. With this action, the patellar ligament will tighten, and you will be able to feel where it attaches to the tibial tuberosity. When palpating the tibial shaft, can you feel its distinct edges leading toward the medial ankle?

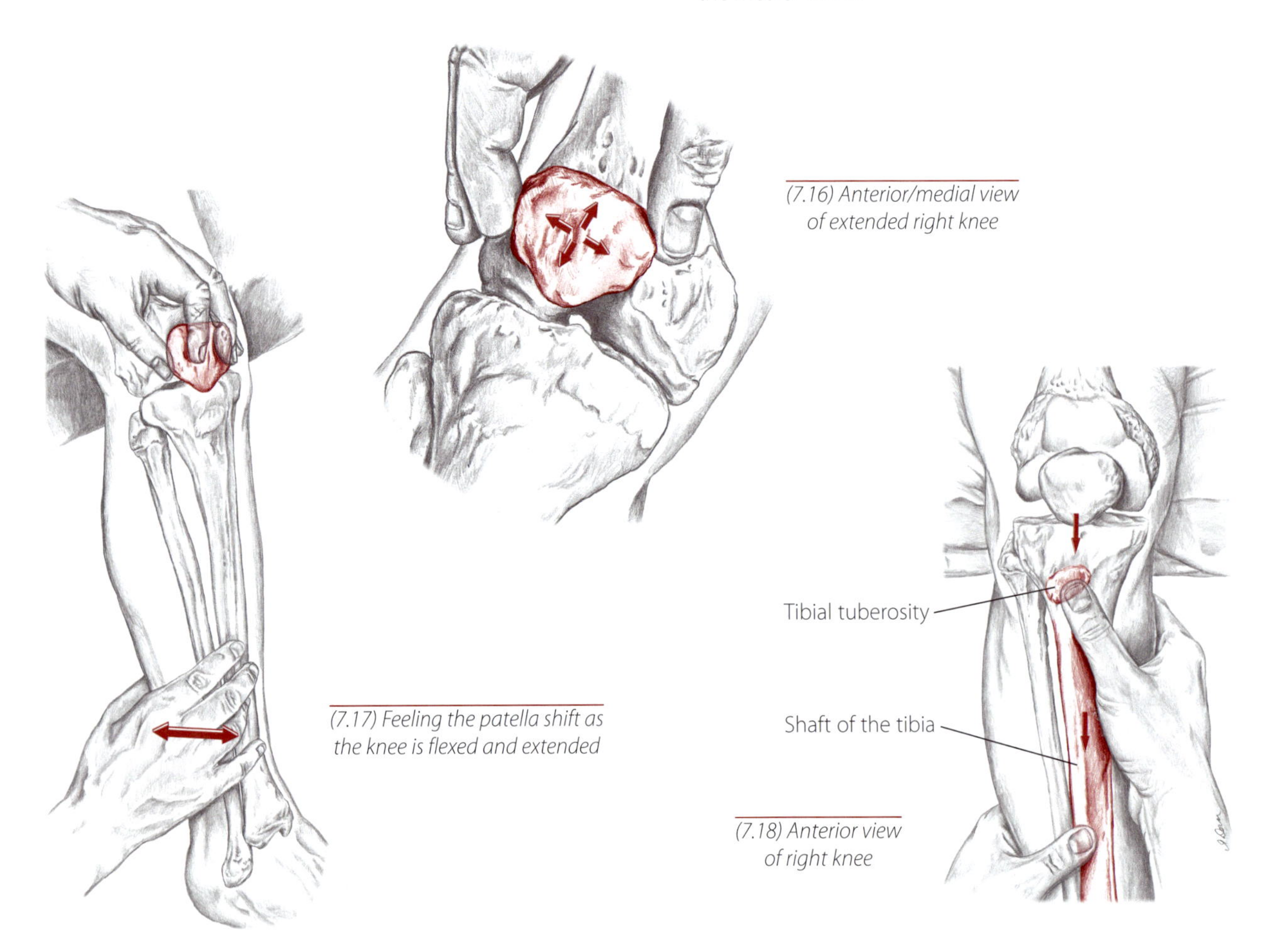

(7.16) Anterior/medial view of extended right knee

(7.17) Feeling the patella shift as the knee is flexed and extended

(7.18) Anterior view of right knee

sesamoid **ses**-a-moyd a bone or fibrocartilage in a tendon playing over a bony surface, L. resembling a sesame seed

Head of the Fibula

The head of the fibula is located on the lateral side of the leg and sometimes protrudes visibly. It is the attachment site for the biceps femoris muscle and a portion of the soleus muscle as well as the fibular collateral ligament.

1) Partner seated with the knee flexed. Locate the tibial tuberosity.
2) Slide your fingers laterally three to four inches toward the outside of the leg. Palpate for the head of the fibula (7.19). Explore its inch-wide tip.

Is the knob you are palpating lateral to the tibial tuberosity? Can you sculpt a circle around it outlining its shape? Does the biceps femoris tendon lead to the head of the fibula?

With your partner prone, bend the knee to 90° and follow the biceps femoris tendon (p. 305) distally to where it inserts at the head of the fibula.

Be aware of the common peroneal nerve (p. 387) which lies along the posterior aspect of the head of the fibula.

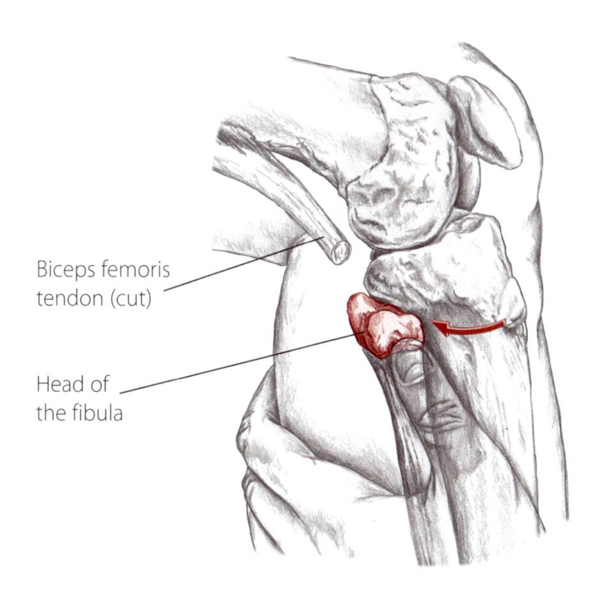

(7.19) Lateral view of flexed right knee

Trail 2 "Waddle Walk"

Tibial Plateaus

The medial and lateral plateaus are located on the proximal end of the tibia. Situated inside the knee joint, the plateaus cannot be palpated, but their edges, located superficially on either side of the patellar ligament, are easily accessible.

1) Partner seated with the knee flexed. Place your thumbs on either side of the patella.
2) Slide inferiorly, compressing into the tissue. You will feel a softening in the knee as your thumbs sink into the joint space between the femur and tibia.
3) Continue inferiorly until you feel the plateau edges (7.20). Palpate both edges and follow them in either direction.

Can you follow the edges of both plateaus horizontally to the sides of the knee? Can you feel the soft joint space superior to them? If you passively extend the knee with one hand, while palpating the edges with the other, can you feel the edges move closer to the patella?

Distal to the lateral plateau is a swelling of bone called the tibial tubercle (7.15) which is the attachment site for the iliotibial tract (p. 318). Slide distally off the lateral plateau and explore the tubercle's girth. When the knee is extended, the tubercle usually lies between the patella and the head of the fibula.

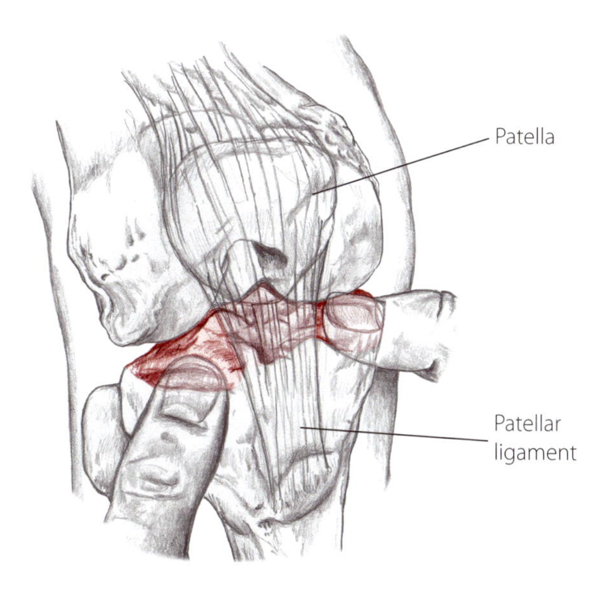

(7.20) Anterior view of flexed right knee

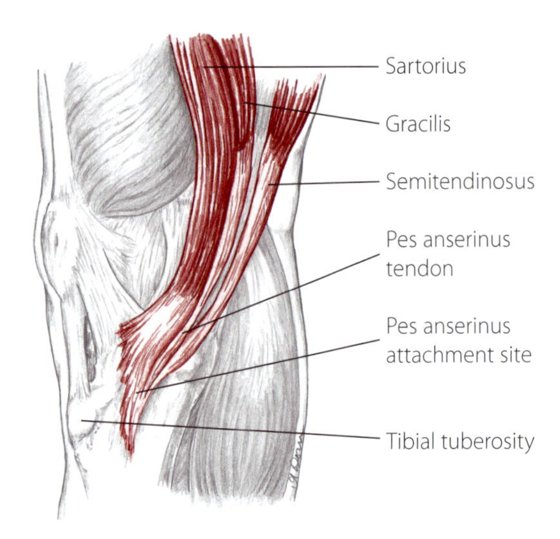

(7.21) Medial view of right knee

Pes Anserinus Attachment Site

Three tendons of the thigh - the sartorius, gracilis and semitendinosus - converge at the medial knee to form the larger pes anserinus tendon which attaches to the proximal, medial shaft of the tibia (7.21). More generally, the pes anserinus is the flat area medial to the tibial tuberosity.

1) Partner seated with the knee flexed. Locate the tibial tuberosity.
2) Slide medially one inch and explore its flat surface and any palpable tendons (7.22).

Is the region you are isolating medial to the tibial tuberosity? Is it on the anterior/medial shaft of the tibia?

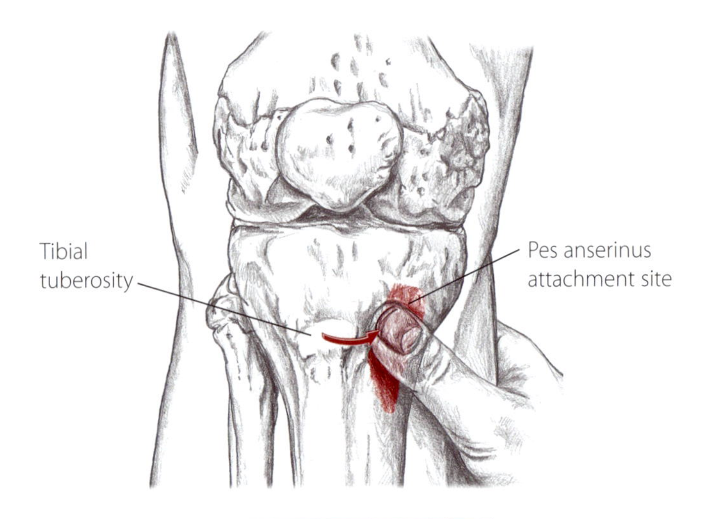

(7.22) Anterior view of right knee

Trail 3 "Hills on Both Sides"

Edges of Femoral Condyles

The two large, round femoral condyles are mostly inaccessible; however, their edges, located on either side of the patella, are easily accessible. The edges of the femoral condyles play an important role in the tracking of the patella when the knee is flexed and extended.

1) Partner supine with the knee fully extended. Locate the sides of the patella.
2) Shift the patella medially and slide off of it onto the lateral condyle. Explore the condyle's distinct edge (7.23) and follow it distally as it continues toward the joint space.
3) Palpate the edge of the medial condyle in the same manner. Compare the size and height of the two edges and the relationship of both to the patella.

Are the edges slightly underneath the patella? Can you follow them distally toward the joint space of the knee?

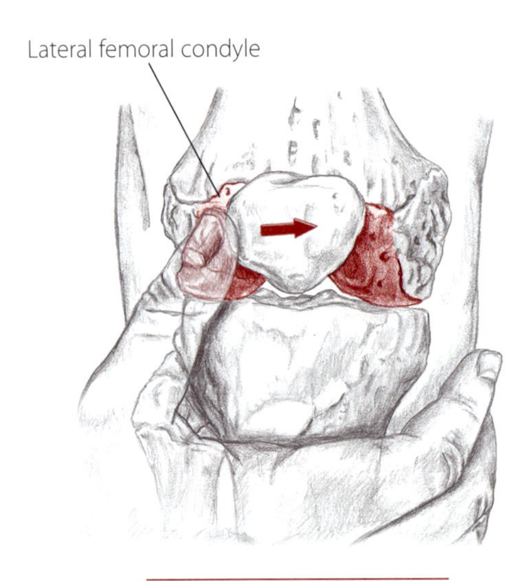

(7.23) Anterior view of right knee

pes anserinus pes **an**-ser-**i**-nus L. *pedis*, foot; *anserinus*, gooselike

Epicondyles of the Femur

The lateral epicondyle is a bald, knobby area located on the lateral surface of the knee. It serves as an attachment site for the fibular collateral ligament. It is deep to the iliotibial tract and anterior to the biceps femoris tendon.

The medial epicondyle is deep to the tendon of the sartorius, distal to the vastus medialis muscle and serves as an attachment site for the tibial collateral ligament.

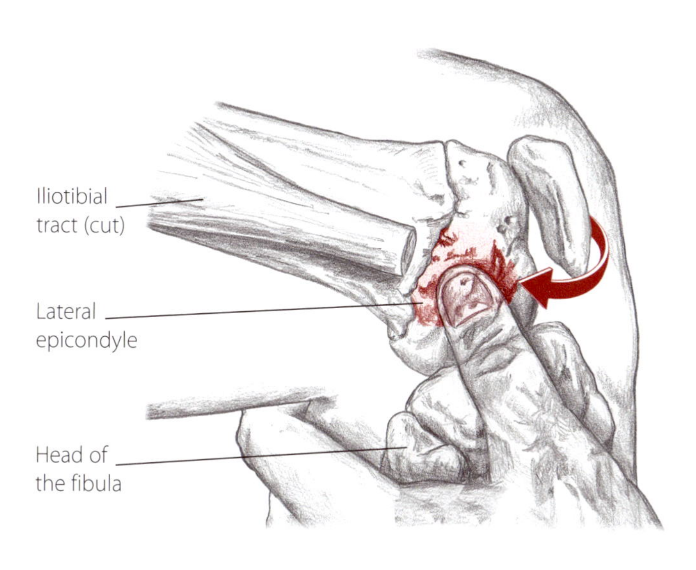

(7.24) Lateral view of right knee

1) Partner supine with the knee flexed. Locate the patella.

2) Slide directly lateral from the patella to the outside of the knee. Explore this region, noting the lateral epicondyle's location proximal to the head of the fibula (7.24).

3) Return to the patella and slide to the medial epicondyle on the inside of the knee. Note the epicondyle's superficial quality and round surface, as well as its location superior to the tibiofemoral joint.

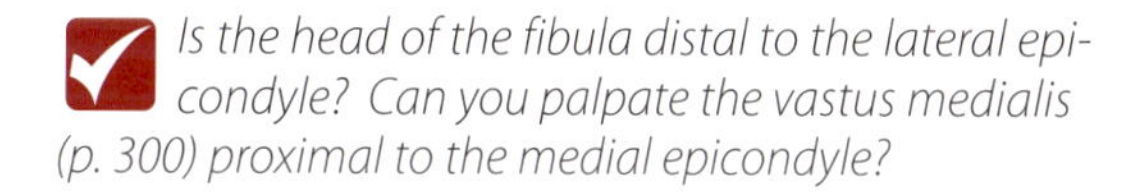

Is the head of the fibula distal to the lateral epicondyle? Can you palpate the vastus medialis (p. 300) proximal to the medial epicondyle?

Adductor Tubercle

The adductor tubercle is located proximal to the medial epicondyle, between the belly of the vastus medialis and the hamstring tendons. Its small tip sticks out from the top of the medial epicondyle and is an attachment site for the adductor magnus tendon (p. 313). It is often tender to the touch.

1) Partner supine with the knee flexed. Locate the medial epicondyle of the femur.

2) Slide superiorly along the medial side of the femur. As the outline of the femur drops off into the soft tissue, explore for the small point of the tubercle (7.25).

3) Strum across the adductor magnus tendon by rubbing your thumbpad anteriorly and posteriorly.

Are you directly proximal to the medial epicondyle? With your thumb on the proximal aspect of the tubercle (on the adductor magnus tendon), have your partner gently adduct his hip. Does the tendon of the magnus become taut and press into your finger?

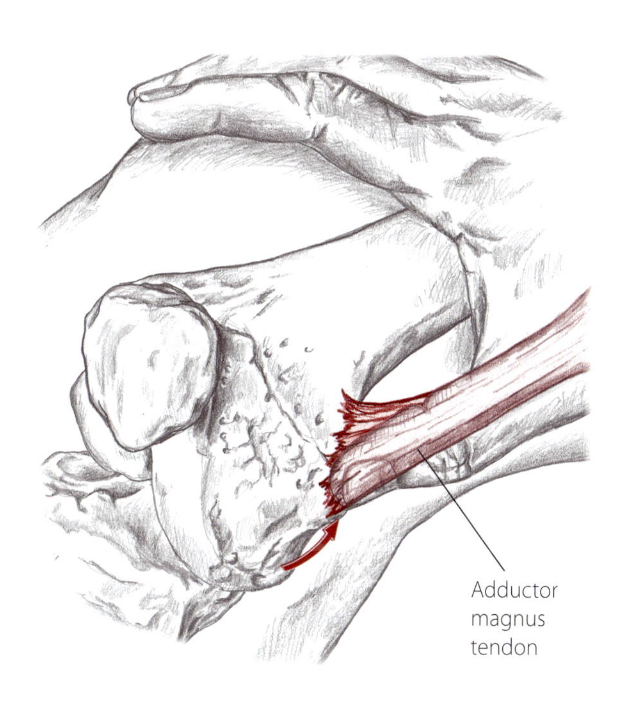

(7.25) Anterior/medial view of right knee

Bones and Bony Landmarks of the Ankle and Foot

The foot contains twenty-six bones (7.9, 7.26 - 7.30). The hind foot is the union of the talus and calcaneus. The **talus** articulates with the tibia and fibula to form the talocrural, or ankle, joint. The large, chunky **calcaneus** is the bone at the heel of the foot.

The mid foot is composed of five tarsals. Small and uniquely shaped like the carpal bones of the wrist, the **tarsals** are tightly wedged together. They are most accessible along the dorsal surface of the foot.

The forefoot is formed by the long, superficial metatarsals and phalanges. Similar to a metacarpal, each **metatarsal** consists of a proximal base, a shaft and a distal head. The first toe is formed by two sizable **phalanges**; the remaining toes have three phalanges each. The phalanges are accessible on all sides.

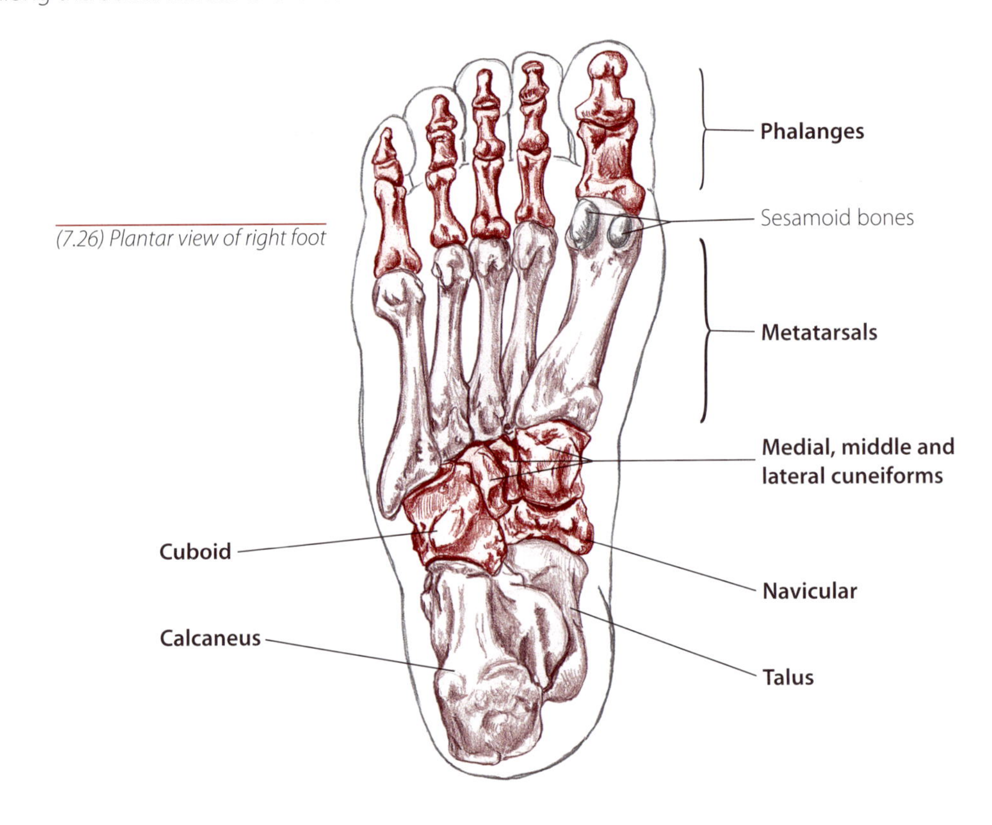

(7.26) Plantar view of right foot

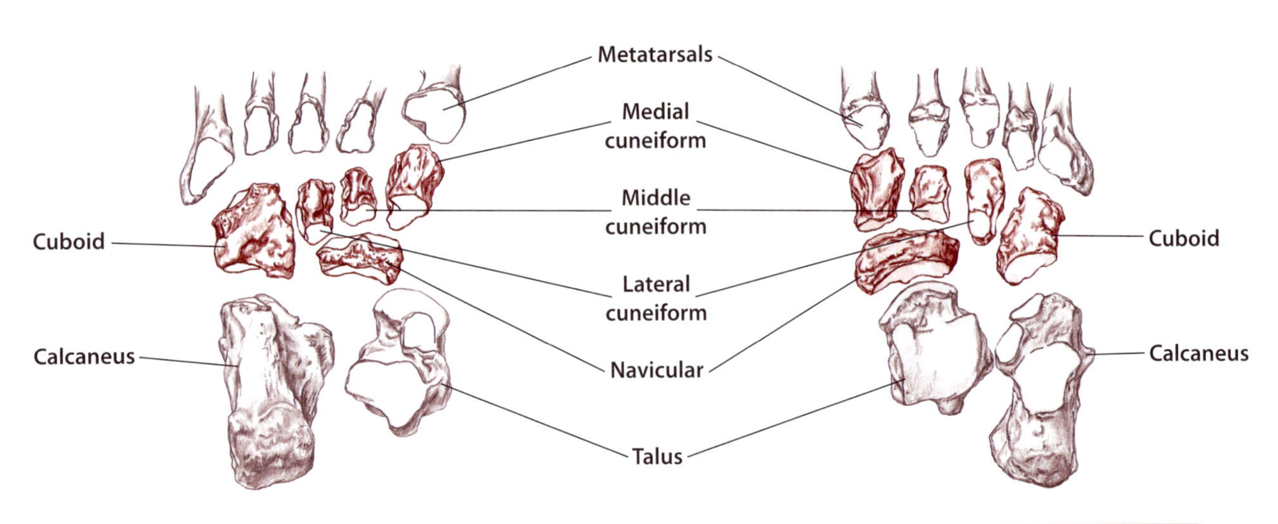

(7.27) Plantar view of right foot, bones separated

(7.28) Dorsal view of right foot, bones separated

metatarsophalangeal **met**-a-**tar**-so-**fa**-lan-**jee**-al

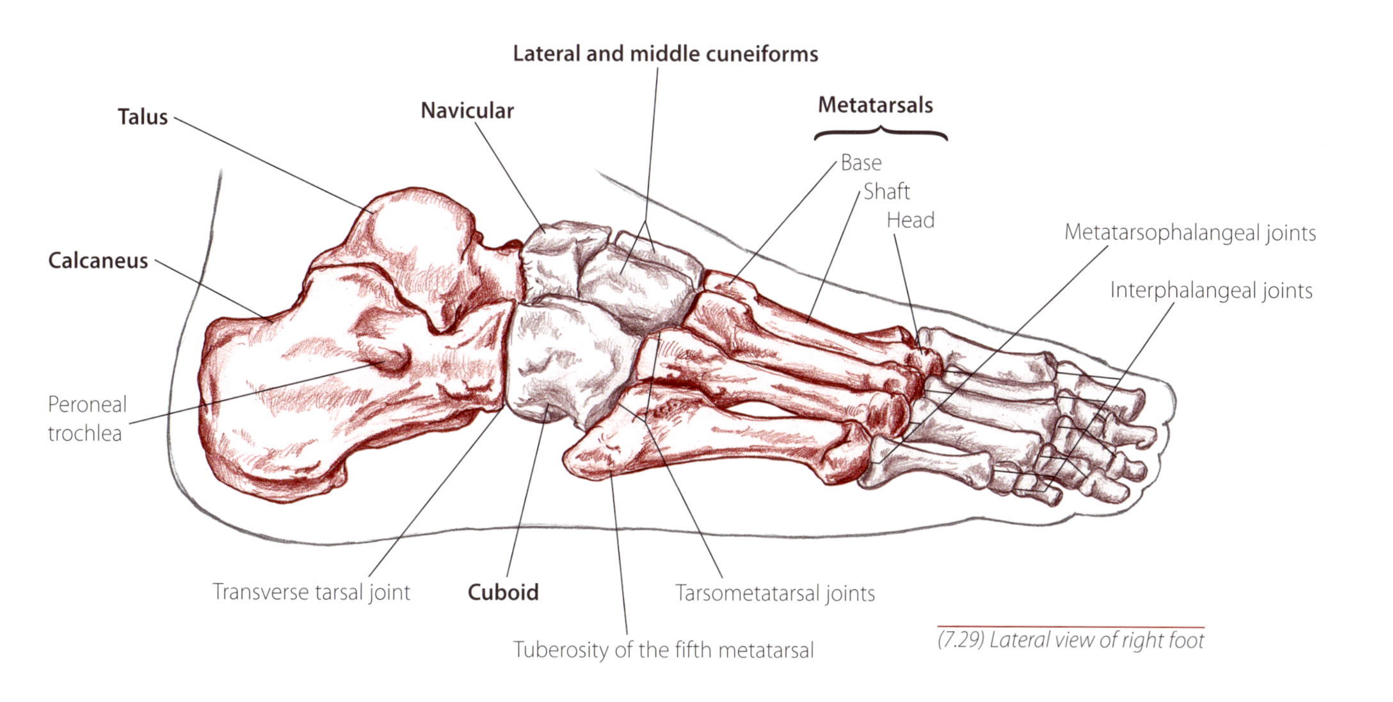

(7.29) Lateral view of right foot

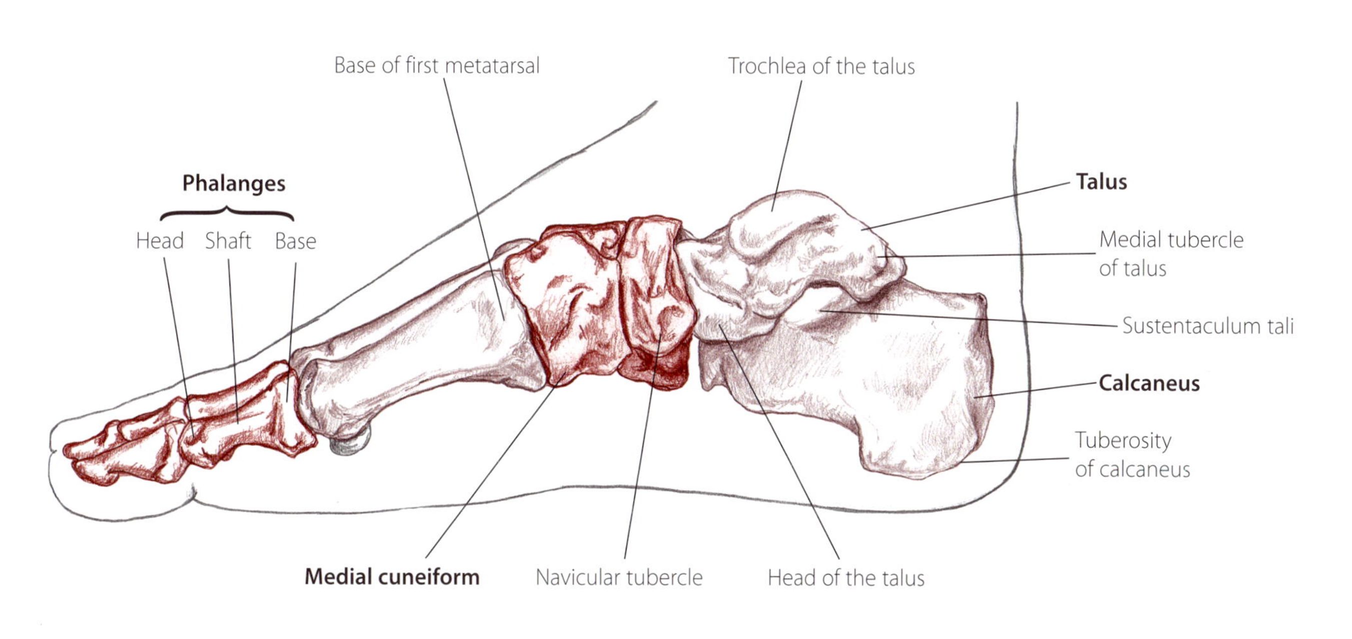

(7.30) Medial view of right foot

Bony Landmark Trails of the Ankle and Foot

The Bony Landmark Trails of the Foot present the hind and forefoot first, followed by the more challenging structures of the mid foot.

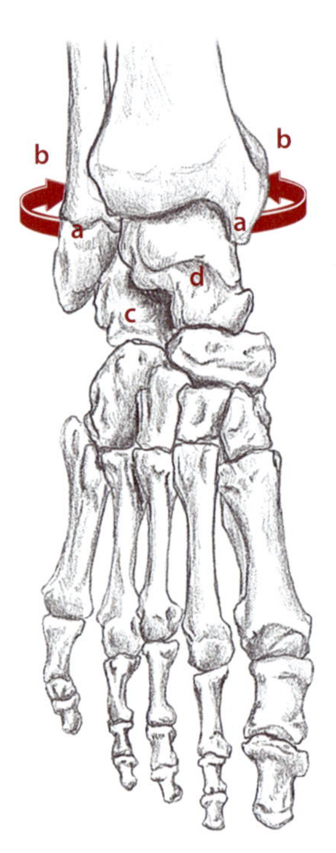

Trail 1 "The Back Road" locates the bones and landmarks of the hind foot and ankle.

a Lateral and medial malleoli
b Malleolar grooves
c Calcaneus
  Tuberosity of calcaneus
  Sustentaculum tali
  Peroneal trochlea
d Talus
  Head
  Trochlea
  Medial tubercle

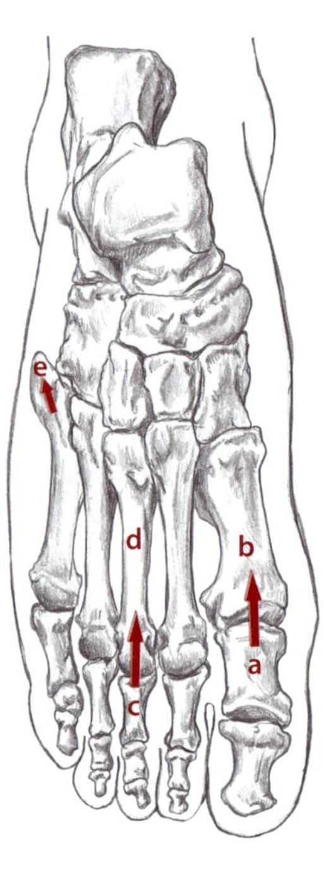

Trail 2 This route, "Little Piggies," palpates the bones and joints of the toes and forefoot.

a Hallucis
b First metatarsal
c Second through fifth phalanges
d Second through fifth metatarsals
e Tuberosity of fifth metatarsal

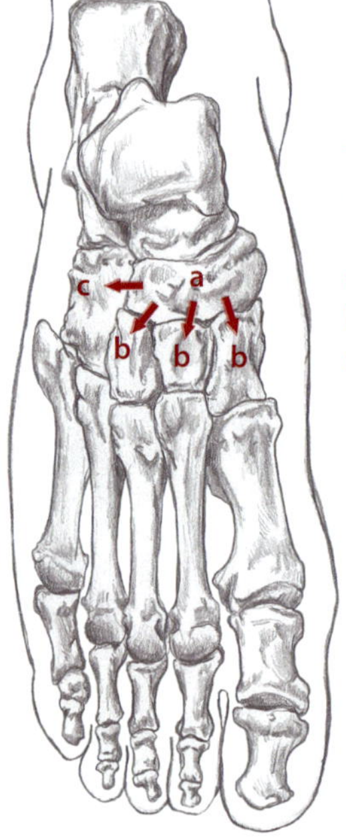

Trail 3 "The Archway" explores the bones of the mid foot located at the arch of the foot.

a Navicular and navicular tuberosity
b Medial, middle and lateral cuneiforms
c Cuboid

As with the carpals of the wrist, the names of the tarsals speak for themselves:

cuboid	L. cube-shaped
cuneiform	L. wedge-shaped
navicular	L. boat-shaped

metatarsal	**met**-a-**tar**-sal	Grk. *meta*, after or beyond
phalange	fa-**lan**-jee	Grk. closely knit row, line of battle
tarsal	**tar**-sul	Grk. wicker basket

Trail 1 "The Back Road"

Lateral and Medial Malleoli

The lateral and medial malleoli are the large conspicuous knobs on either side of the ankle. The broader medial malleolus is located at the distal end of the tibia while the more slender lateral malleolus protrudes from the distal fibula.

1) Partner seated or supine. Explore and compare the shapes and sizes of the two malleoli. Palpating all sides of their surfaces, note how the lateral malleolus extends further distally than its medial counterpart (7.31).

Sliding proximally, can you connect the medial malleolus to the shaft of the tibia and then to the tibial tuberosity? Is the medial malleolus more proximal than the lateral?

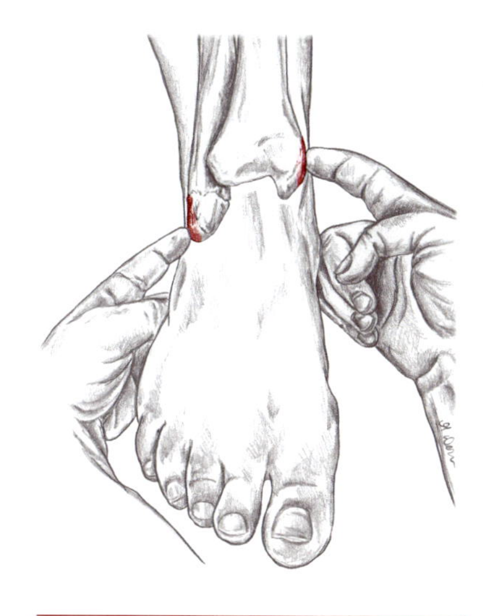

(7.31) Palpating the level of each malleolus

Malleolar Grooves

Both the medial and lateral malleoli have small vertical grooves carved into their posterior surfaces. These grooves are designed to offer stability and leverage to tendons that bend around the ankle. Because these tendons lie either inside the groove or beside it, it can be difficult to feel the actual depression of the groove.

1) Supine or seated. Locate the medial malleolus.
2) Slide roughly half an inch posteriorly to palpate the posterior aspect of the malleolus for a slender, vertical groove (7.32).
3) Shorten the surrounding tissue by passively inverting the foot and explore the length of the medial groove and the superficial tendons.
4) Try this same method along the lateral malleolus. Only now passively evert the foot to shorten the surrounding tissue and locate the lateral groove (7.33).

Since each groove runs vertically, can you roll your finger horizontally across each vertical groove to determine its location and shape?

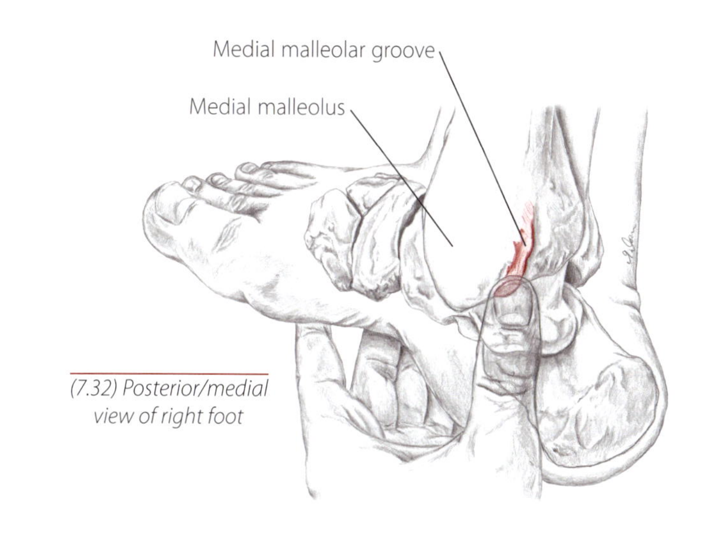

(7.32) Posterior/medial view of right foot

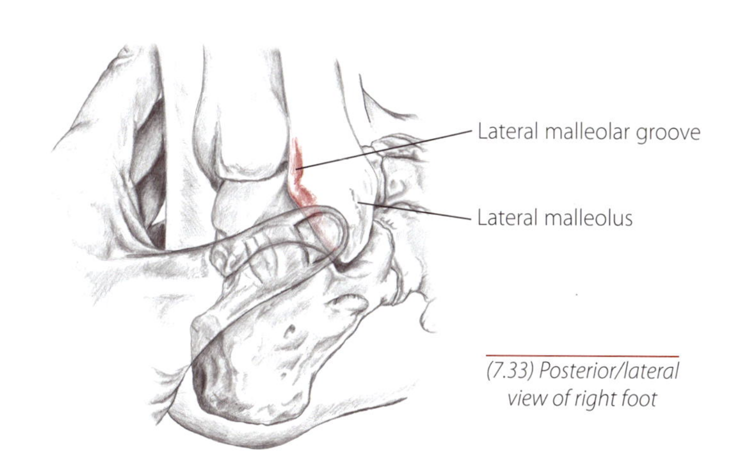

(7.33) Posterior/lateral view of right foot

calcaneus	kal-**kay**-nee-us	L. heel
malleolus	**mal**-e-**o**-lus	L. little hammer

Calcaneus

Tuberosity, Sustentaculum Tali and Peroneal Trochlea

The large, solid **calcaneus** forms the heel of the foot. It is situated beneath the talus and projects two inches posteriorly from the malleoli. The medial and lateral sides of the calcaneus are deep to tendons, yet easily palpable. The **tuberosity** of the calcaneus is a rounded region located along its posterior surface. The calcaneal tendon attaches to the superior aspect of the tuberosity.

The **sustentaculum tali** is located on the medial side of the calcaneus, roughly one inch distal to the medial malleolus (7.35). Shaped like a plank, the sustentaculum supports the talus on the calcaneus. It is also an attachment site for the deltoid ligament (p. 391) and is deep to the flexor tendons. Only its small tip is accessible.

The **peroneal trochlea** is located on the lateral side of the foot (7.34). Roughly an inch distal to the lateral malleolus, the trochlea is a small, superficial prominence that protrudes from the calcaneal surface to help stabilize the peroneal muscles (p. 369).

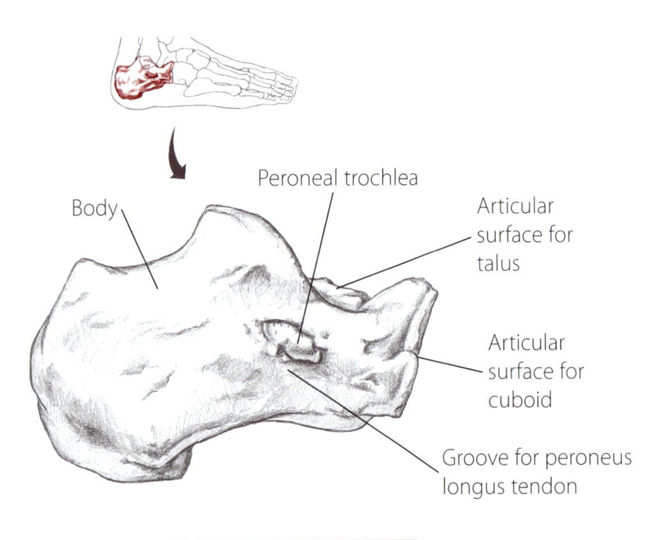

(7.34) Lateral view of right calcaneus

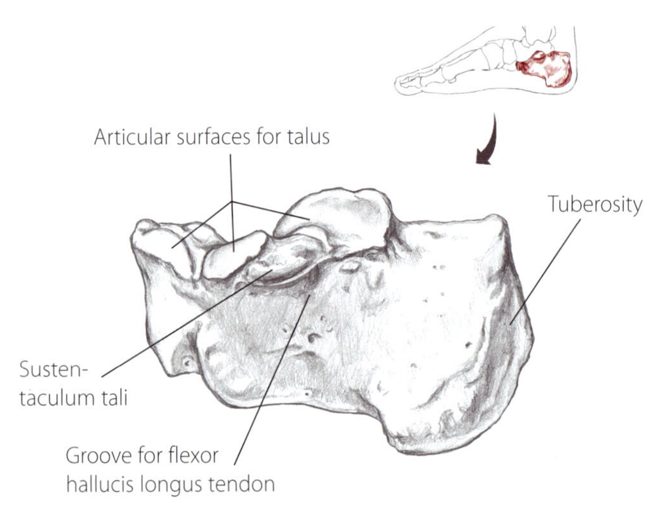

(7.35) Medial view of right calcaneus

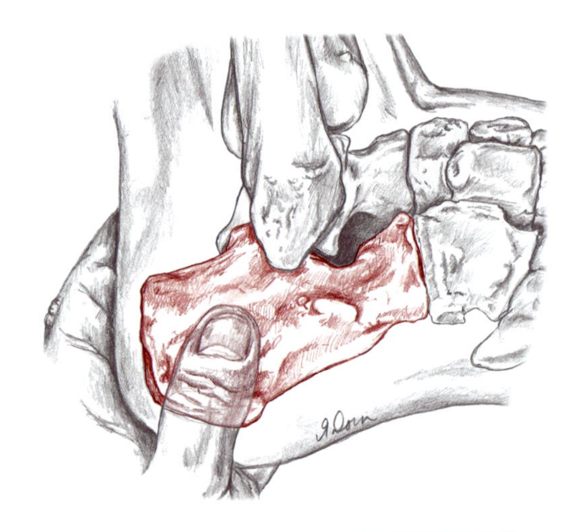

(7.36) Lateral view, exploring the calcaneus

Calcaneus

1) Partner supine or seated. Walk your fingers distally from the malleoli down to the heel. Palpate and explore the shape and girth of the posterior calcaneus.
2) Move to the plantar surface to isolate the tuberosity at the base of the heel (7.36). The tuberosity will feel more like a flat region than a distinct bump.

Place one hand at the malleoli and the other at the tuberosity. Note how far the calcaneus extends posteriorly.

sustentaculum **sus**-ten-**tak**-u-lum L. support

Sustentaculum tali

1) Supine or seated. Place the ankle in a neutral position and locate the medial malleolus.
2) Slide approximately one inch distal to the small tip of the sustentaculum (7.37). Passively inverting the foot will soften the surrounding tissues.
3) Sculpt around its sides noting the soft tissues just distal to it.

Are you distal to the medial malleolus? If you slide distally off the sustentaculum tali, do you feel the thick tissues at the sole of the foot?

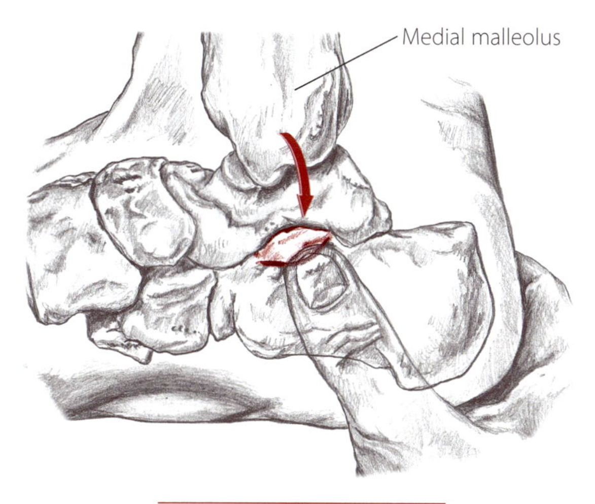

(7.37) Medial view of right ankle, locating the sustentaculum tali

Peroneal trochlea

1) Supine or seated. With the ankle in a dorsiflexed position, locate the lateral malleolus.
2) Slide roughly an inch inferiorly and explore for the small, superficial trochlea. It may feel like a short ridge on the surface of the calcaneus (7.38). Passively everting the foot will soften the surrounding tissues.
3) Sculpt around its edges, noting the soft tissues just distal to the trochlea.

Are you distal to the lateral malleolus? If you slide off the trochlea distally, do you feel the thick tissues of the foot? Ask your partner to alternately evert and relax her foot. Do the peroneal tendons pass along either side of the trochlea?

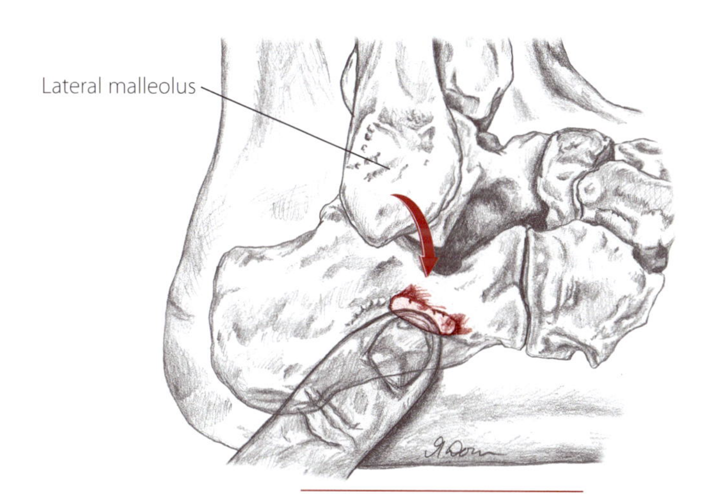

(7.38) Lateral view of right ankle, isolating the peroneal trochlea

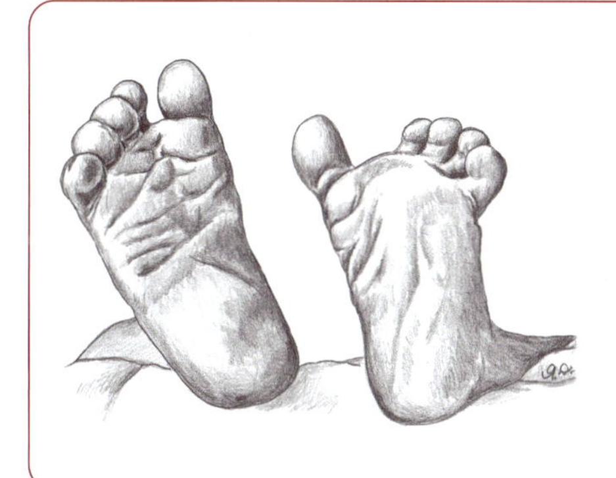

Of the two hundred different kinds of primates in the world, humans are the only ones with a nongrasping first toe. Since we are no longer tree climbers, our foot has lost its handlike capabilities in order to become a platform for an upright body.

This does not mean, however, that the toes were designed to be inactive. An infant's foot has twenty times the toe-grasping capacity of a shoe-wearing adult. And in shoeless cultures, people retain the prehensile abilities of their feet throughout adulthood, using them for sewing and even threading needles.

Talus

Head, Trochlea and Medial Tubercle

The **talus** has three accessible landmarks. The **head** is the round, anterior portion that articulates with the navicular (7.39). The medial aspect of the head is accessible posterior to the navicular tubercle (p. 359). The **trochlea**, the large, superior prominence of the body of the talus, is wedged between the distal ends of the fibula and tibia (7.40). The anterior part of the trochlea is located between the malleoli. Finally, the small **medial tubercle** of the talus (7.40) is posterior to the medial malleolus and serves as an attachment site for the deltoid ligament (p. 391).

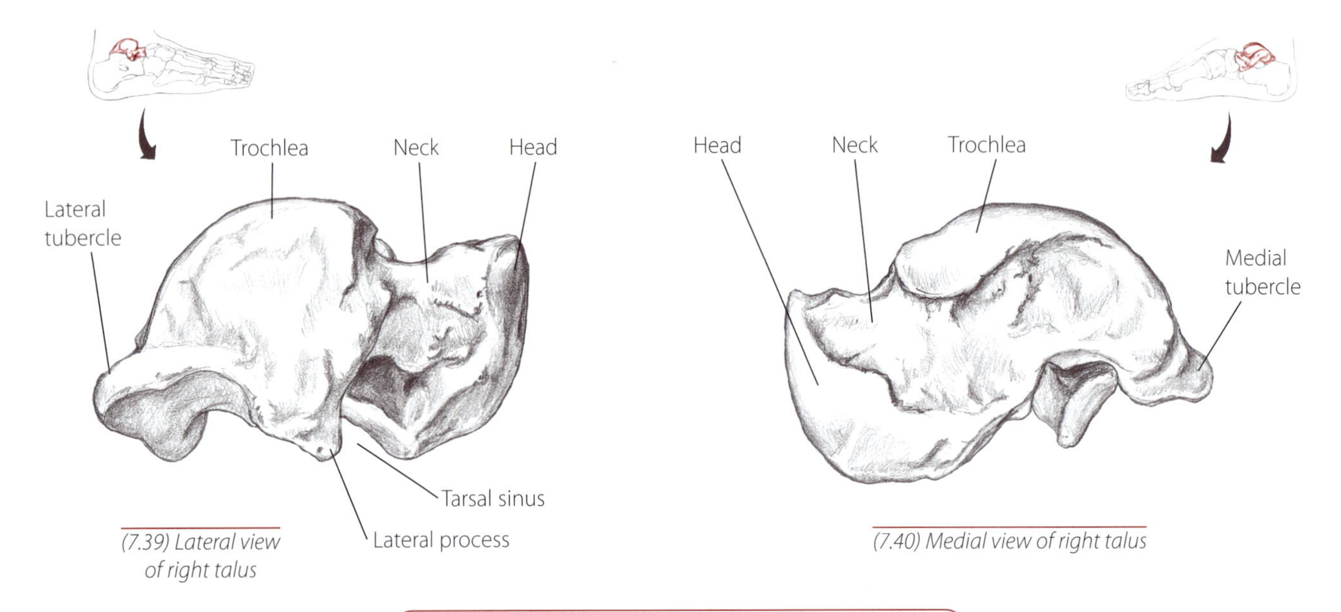

(7.39) Lateral view of right talus

(7.40) Medial view of right talus

The calcaneus, talus and cuboid bones are all roughly cube-shaped. The soldiers of ancient Rome used these bones (probably from horses) to carve out playing dice. For this reason, the talus is sometimes called the astragalus, which in Latin means *die*, the singular form of the plural *dice*.

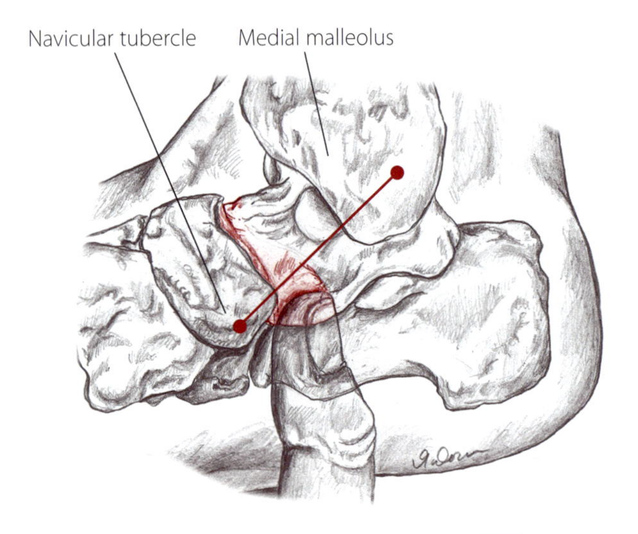

(7.41) Medial view of right ankle, accessing the medial portion of the head of the talus

Head of the talus

1) Partner supine or seated, with the ankle in a neutral position. Locate the navicular tubercle (p. 359).
2) Slide proximally off the tubercle to the head of the talus. The head may feel like a depression in comparison to the tubercle.
3) Passively invert and evert the foot in order to distinguish clearly between these two landmarks. When the foot is inverted, the navicular tubercle will become more prominent; when the foot is everted, the talar head will be more pronounced.

If you draw a line between the medial malleolus and navicular tubercle, the head of the talus will be located along that line (7.41).

talus	**ta**-lus	L. ankle
trochlea	**trok**-lee-ah	Grk. pulley

Trochlea of the talus

1) Partner supine. Passively invert and plantar flex the foot.
2) Draw a horizontal line connecting the malleoli and drop inferiorly off the center of the line, looking for a bony prominence. The trochlea will be deep to the overlying tendons and more prominent near the lateral malleolus (7.42).

Is the tissue you are palpating hard and immovable like bone, or firm and mobile like tendon? If you passively move the foot back to neutral, does the bony mound you are palpating seem to disappear into the ankle?

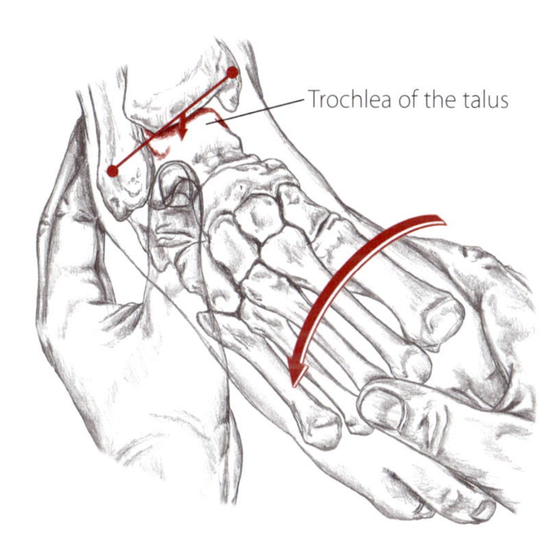

(7.42) Dorsal/lateral view of right foot, with foot inverted and plantar flexed

Medial tubercle

1) Partner supine. Locate the medial malleolus. Instead of sliding straight distally for the sustentaculum tali (7.37), slide posteriorly just off the malleolus at a 45° angle in order to locate the medial tubercle (7.43).
2) Passively dorsiflex and plantar flex the ankle, noting how the tubercle seems to slide around the malleolus.

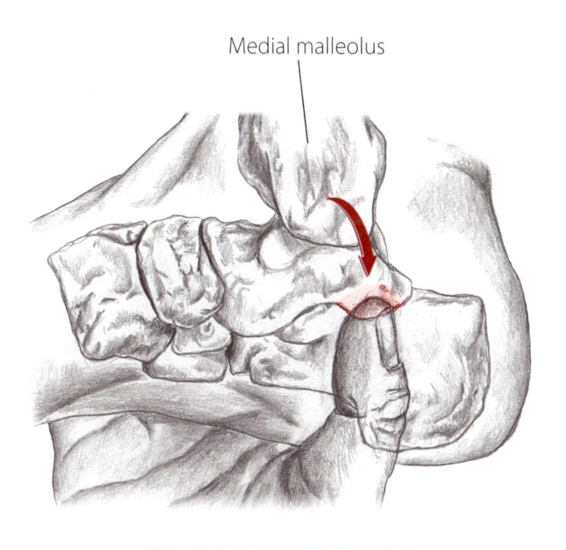

(7.43) Medial view of right foot, contacting the medial tubercle

Trail 2 "Little Piggies"

Hallucis

The hallucis is composed of two phalanges. The joint between the phalanges, the interphalangeal, is a hinge joint wrapped in supportive ligaments. The first metatarsophalangeal joint is located at the ball of the foot. It is an ellipsoid joint with a large, bulbous shape.

1) Partner seated or supine. Palpate the entire surface of the first toe, noting the differences in thickness and texture between its dorsal and plantar surfaces (7.44).
2) Explore the surface of each joint by passively moving it through its range of motion.

Is the proximal phalange nearly twice as long as its distal counterpart?

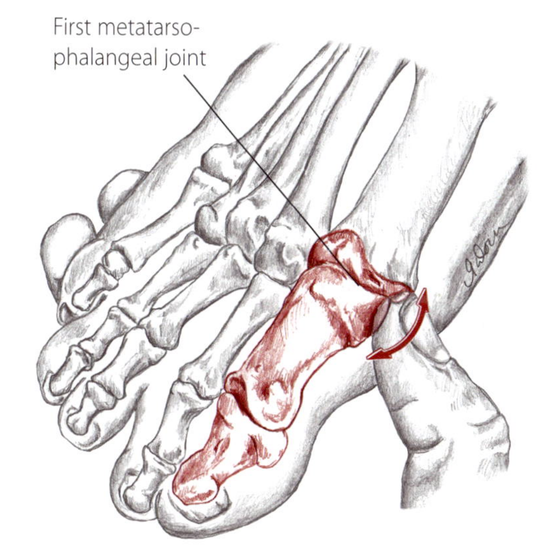

(7.44) Dorsal/medial view of right foot

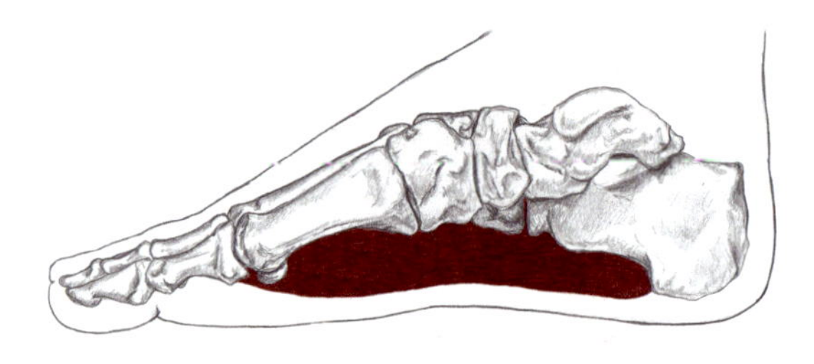

(7.45) Medial view of right foot, shaded area indicating muscle mass along the foot's plantar surface

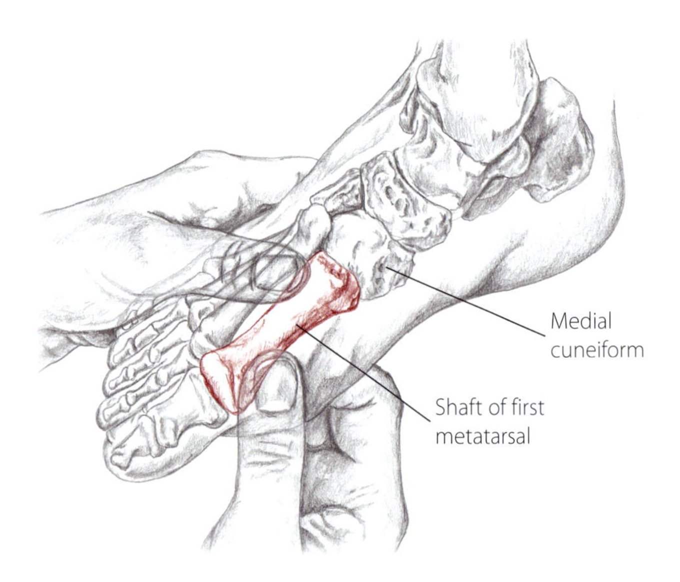

(7.46) Dorsal/medial view of right foot

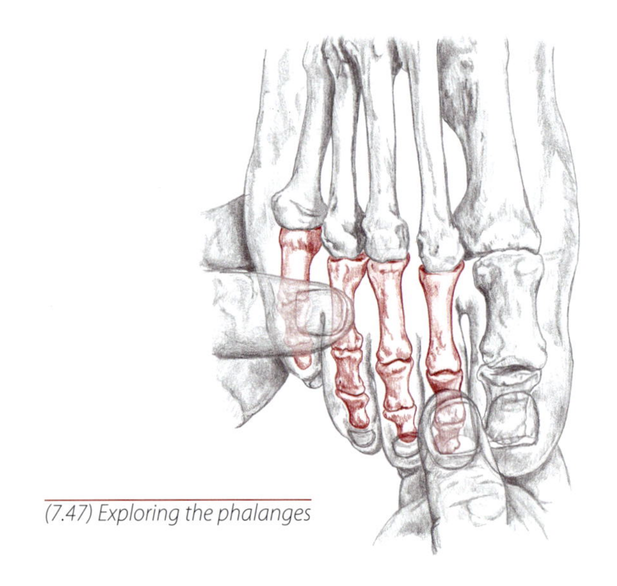

(7.47) Exploring the phalanges

First Metatarsal

Unlike the long, slender metatarsals of toes two through five, the first metatarsal is short and stocky. Its dorsal and medial sides are superficial and easily accessible; its plantar surface is deep to several thick muscles (7.45). The proximal end of the first metatarsal flares to articulate with the medial cuneiform. This articulation often forms a visible crest on the top of the foot that can be irritated by wearing tight shoes.

1) Partner seated or supine. Locate the metatarsal shaft along the medial side of the foot.

2) Explore the shaft's size and length by sliding across its entire surface. Palpate the junction and crest at the metatarsal head and medial cuneiform (7.46).

✔ *Are the head and base broader than the shaft of the metatarsal? Can you feel the cylindrical shape of the shaft?*

Phalanges

Unlike the hallucis, the second through fifth toes contain three phalanges each. In each toe there are two articulations, the **p**roximal **i**nter**p**halangeal (or "**pip**" joint) and the **d**istal **i**nter**p**halangeal (or "**dip**" joint).

1) Seated or supine. Palpate along all surfaces of the toes, noting the thin tissue along their inner sides. Explore one toe at a time, slowly moving each one through its range of motion (7.47).

hallux	**hal**-uks	L. first toe
hallucis	**hal**-ah-sis	

Second through Fifth Metatarsals

Each of the long, slender bones of the second through fifth metatarsals has an enlarged base and head. The bases are set close together in articulation with the tarsals. The spaces between the metatarsals are filled with the small intrinsic muscles of the toes and are easily palpated on the dorsal surface of the foot.

The tuberosity of the fifth metatarsal is a superficial knob that extends laterally off the base of the metatarsal and is the attachment site for the peroneus brevis (p. 369).

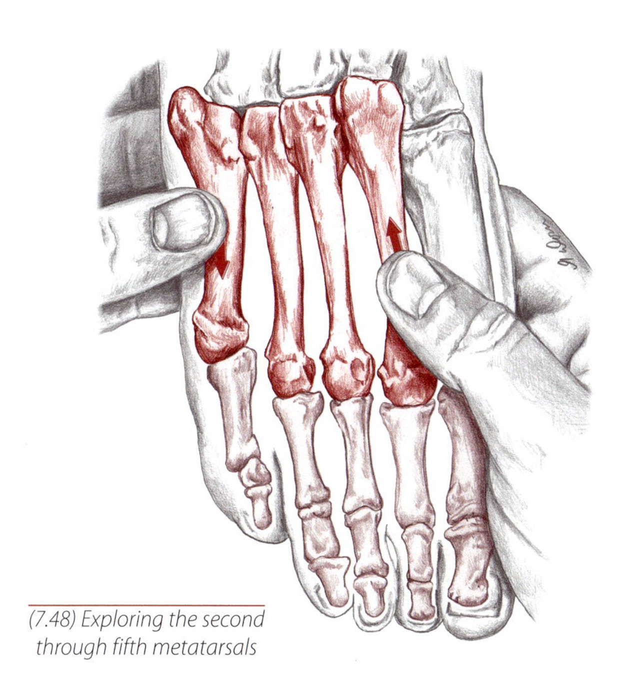

(7.48) Exploring the second through fifth metatarsals

Metatarsals

1) Partner seated or supine. Grasp the foot with both hands and palpate the head of each metatarsal on the dorsal side of the foot.
2) Use both thumb tips to explore the length of each bone and its surrounding spaces. Follow the shaft of each metatarsal proximally (7.48). Note how it widens to form the base of the metatarsal.

Tuberosity of fifth metatarsal

1) Partner seated or supine. Locate the shaft of the fifth metatarsal.
2) Follow the shaft proximally to where the base bulges laterally (7.49). Explore the superficial shape of the tuberosity and its surrounding landmarks as it projects from the side of the foot.

When the ankle is dorsiflexed, are you roughly two inches distal (anterior) to the lateral malleolus? Is the tip you are palpating connected to the fifth metatarsal?

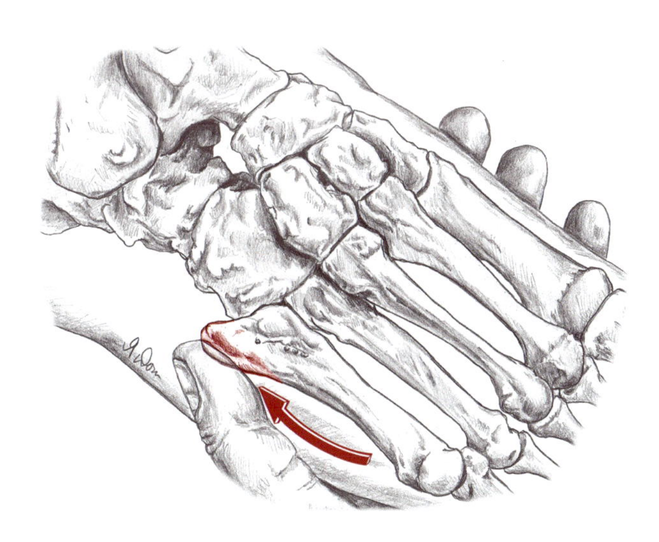

(7.49) Dorsal/lateral view of right foot, accessing the tuberosity of the fifth metatarsal

While strapping on shoes has certainly protected our feet and reduced the number of sprained ankles, it has also wreaked havoc on our arches. With the external support of shoes, our arches no longer need to adapt to varying terrain and so the normally supportive musculature weakens. Eventually, the arch on the medial side of the foot collapses, resulting in a condition commonly known as "flat foot."

Trail 3 "The Archway"

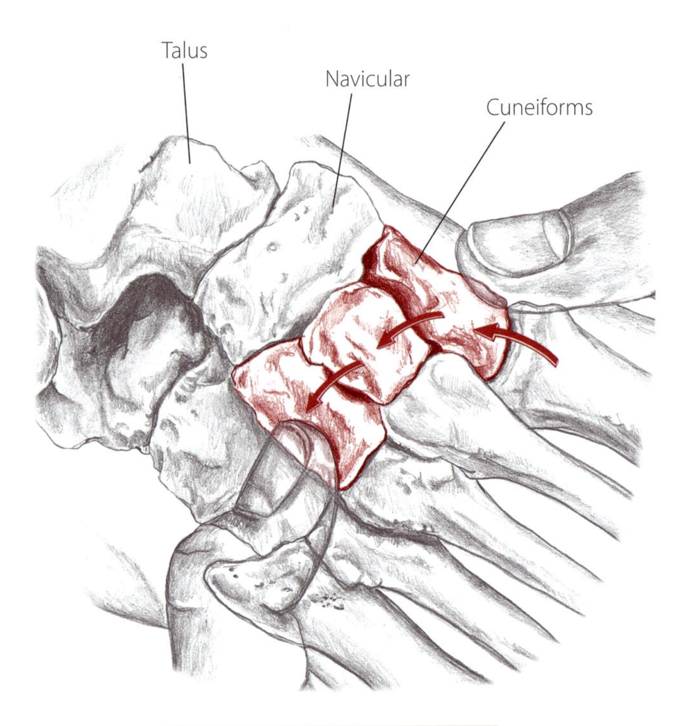

(7.50) Dorsal/lateral view of right foot

Medial, Middle and Lateral Cuneiforms

The three cuneiforms lie in a row between the navicular, talus and metatarsals. The medial cuneiform serves as an attachment for the tibialis anterior and tibialis posterior muscles. It can be isolated along its dorsal and medial surfaces. The middle and lateral cuneiforms, sandwiched between the medial cuneiform and the cuboid, are accessible on their dorsal surfaces.

1) Partner seated or supine. Locate the base of the first metatarsal.
2) Glide proximally to the skinny ditch of the first tarsometatarsal joint. Continue proximally onto the surface of the medial cuneiform.
3) Slide laterally from the medial cuneiform along the dorsal surface of the foot and explore the surfaces of the middle and lateral cuneiforms (7.50).

Are you proximal to the base of the first metatarsal and can you isolate the joint between these two bones? If you follow the tibialis anterior tendon, does it lead to the same location at which you were palpating the medial cuneiform?

The tibialis anterior tendon (p. 371) runs superficially down the dorsal surface of the ankle and leads directly to the medial side of the medial cuneiform. Have your partner dorsiflex his foot and follow the tendon distally as it blends into the medial cuneiform.

Mammals such as cats and dogs are called digitigrades, meaning they walk on their toes (digits). When digitigrades are standing, their tarsals and metatarsals are off the ground, forming what appears to be the leg. For this reason, the ankle of a dog or cat is often mistaken for the knee, while the actual knee appears to be hyperextended.

Digitigrades are raised up on their toes in such a way for additional height. The extra height enhances sensory perception and increases stride length.

Hoofed animals, call unguligrades, go a step further than digitigrades in lifting themselves up on all their phalanges except the distal one. With this wide, four-point stance, these animals literally walk on the tips of their toes all the time.

Walking "tippy-toe" will quickly tell you that neither of these designs work for humans. We are plantigrades, meaning we walk on the soles of our feet. As we are also bipeds, we must spread our feet out, pressing all of our foot bones firmly on the ground in order to keep our balance.

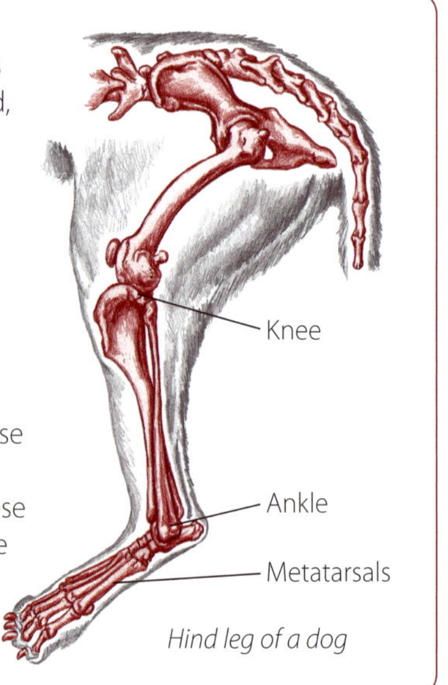

Hind leg of a dog

cuneiform ku-**ne**-i-form L. *cuneus*, wedge-shaped

Navicular

The bean-shaped navicular is sandwiched between the medial and middle cuneiforms and the talus. Its dorsal and medial surfaces are superficial and palpable. The superficial tuberosity bulges out of the medial side of the foot and is an attachment site for the tibialis posterior muscle (p. 374) and the spring ligament (p. 391).

1) Partner seated or supine. Locate the base of the first metatarsal.
2) Sliding along the foot's medial side, move proximally across the surface of the medial cuneiform and the slender joint between the medial cuneiform and the navicular.
3) As you move onto the surface of the navicular, explore the shape and size of the navicular tuberosity (7.51). The tuberosity will lie approximately one to two inches distal to the medial malleolus.

Does the bone you are palpating project more medially than the surfaces of the other bones on the medial foot? If you place a finger on the tuberosity of the fifth metatarsal and the navicular tuberosity simultaneously, does the metatarsal tuberosity lie slightly distal to the navicular tuberosity? (see box to the right)

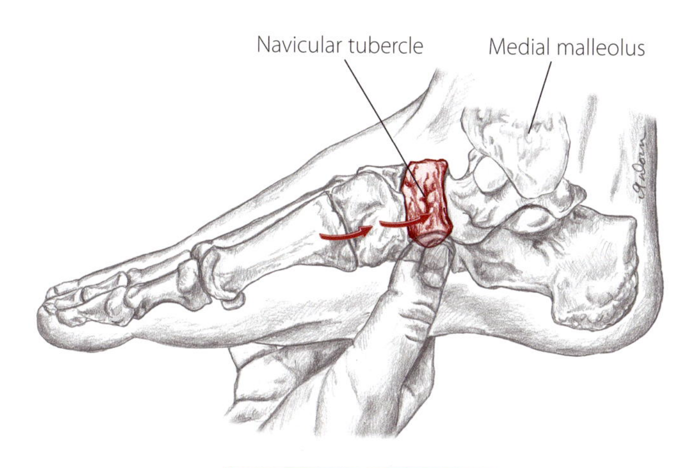

(7.51) Medial view of right foot

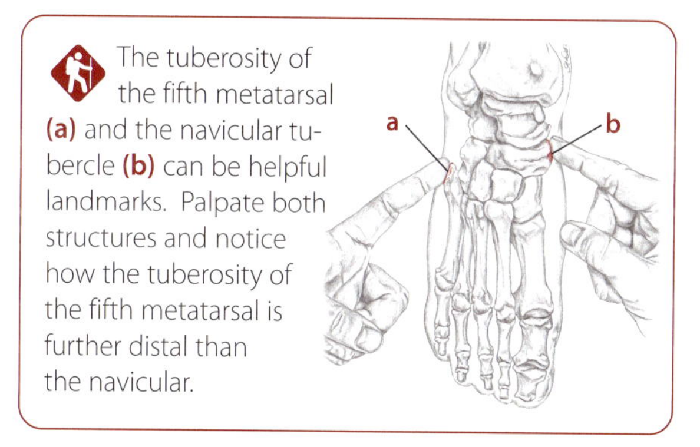

The tuberosity of the fifth metatarsal **(a)** and the navicular tubercle **(b)** can be helpful landmarks. Palpate both structures and notice how the tuberosity of the fifth metatarsal is further distal than the navicular.

Cuboid

As the translation of its name suggests, the cuboid is cube-shaped. It is surrounded on three of its four sides by the fourth and fifth metatarsals, the lateral cuneiform and the calcaneus. The cuboid's dorsal surface is partially covered by the belly of the extensor digitorum brevis (p. 377). Because of its cramped location and the covering of the brevis, the cuboid is only partially accessible.

1) Partner seated or supine. Draw an imaginary line from the tuberosity of the fifth metatarsal to the lateral malleolus.
2) Following this line, at roughly half an inch from the tuberosity, is the cuboid (7.52).

Are you proximal to the tuberosity of the fifth metatarsal? With the foot dorsiflexed, are you roughly an inch dorsal/distal to the lateral malleolus?

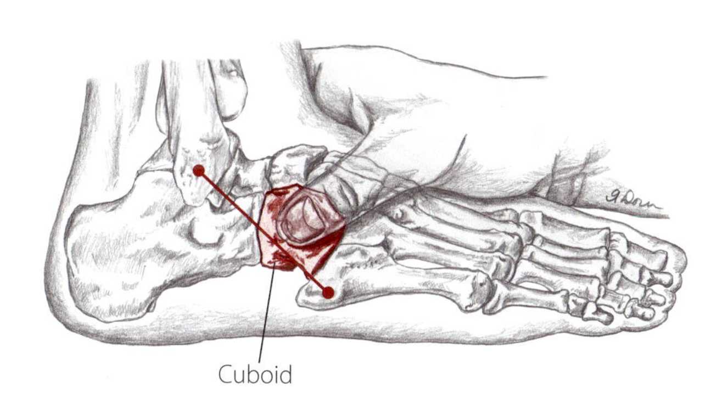

(7.52) Lateral view of right foot

navicular	na-**vik**-u-lar	L. boat-shaped
cuboid	**ku**-boyd	Grk. cube-shaped

Muscles of the Leg and Foot

Like the forearm and hand, the leg and foot feature numerous muscles. Most are directly or partially accessible, and their names reveal a great deal about their actions. Most of the muscles of the leg can be divided into four groups:

a) The large **gastrocnemius** and **soleus** form the "calf muscles" of the posterior leg.

b) The **peroneus longus and brevis** are slender muscles located along the lateral side of the leg.

c) The **extensors** of the ankle and toes (tibialis anterior, extensor digitorum longus and extensor hallucis longus) are layered together on the anterior leg and the dorsum of the foot.

d) The small **flexors** of the ankle and toes include tibialis posterior, flexor digitorum longus and flexor hallucis longus. They are deep to the gastrocnemius and soleus on the posterior leg.

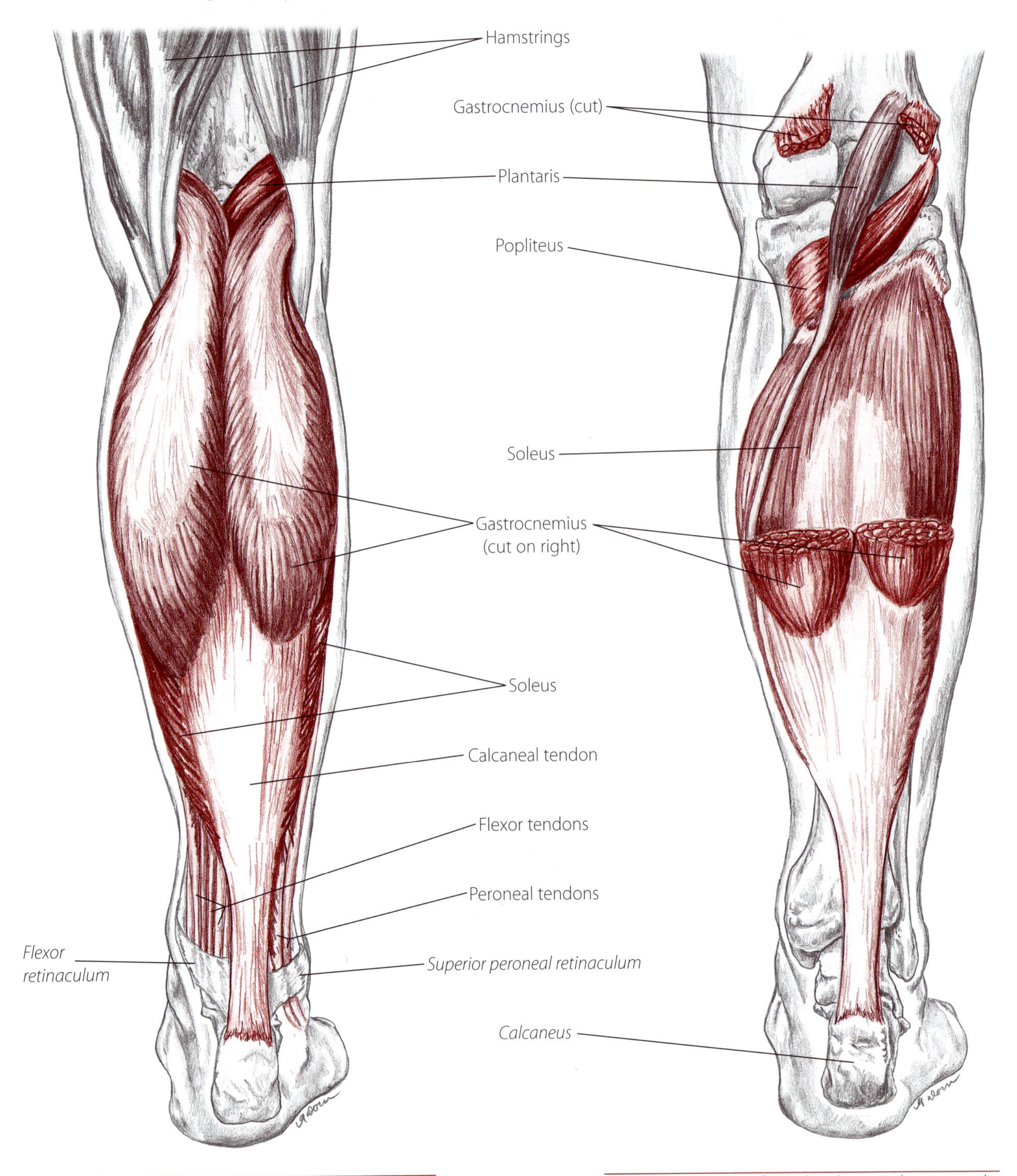

(7.53) Posterior view of right leg showing superficial muscles

(7.54) Posterior view of right leg showing deeper muscles

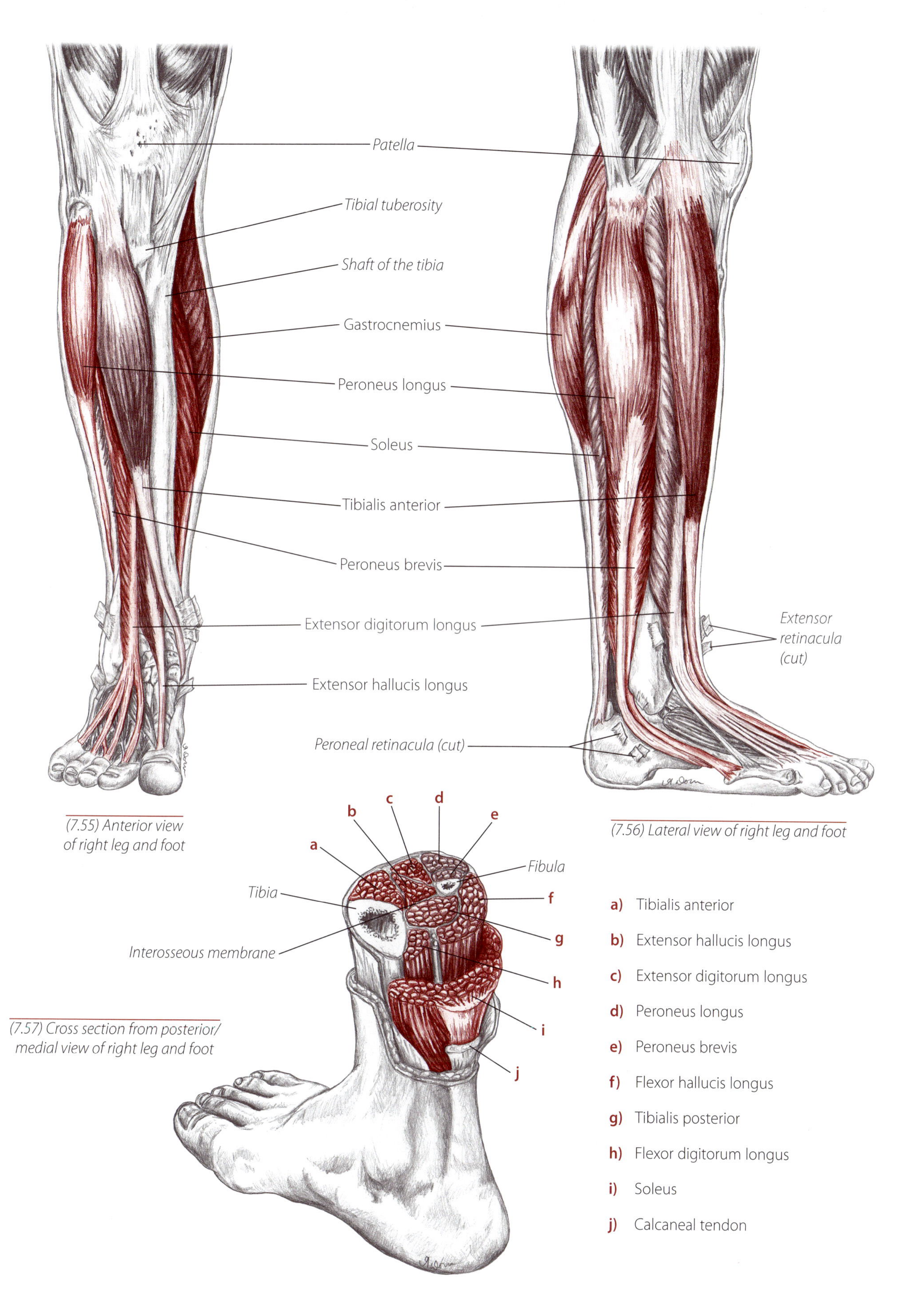

(7.55) Anterior view of right leg and foot

(7.56) Lateral view of right leg and foot

(7.57) Cross section from posterior/ medial view of right leg and foot

 Synergists - Muscles Working Together

*muscles not shown

Ankle

(talocrural joint)

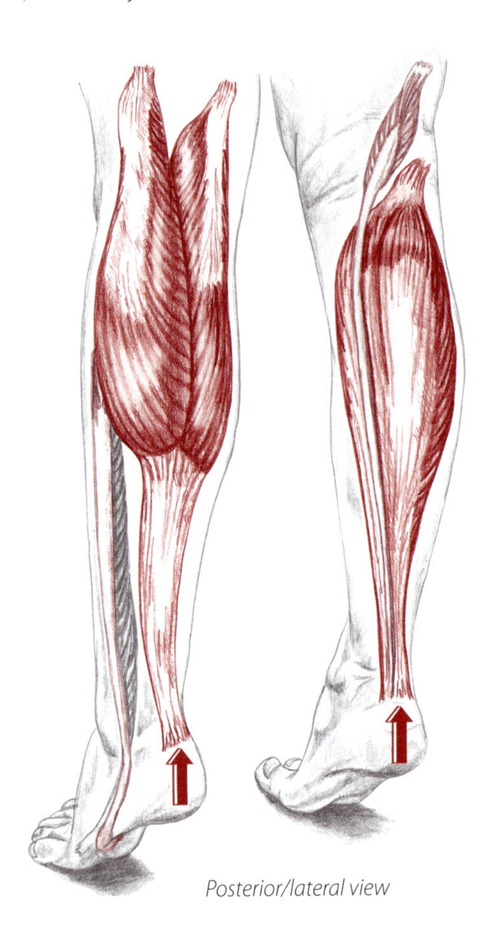

Posterior/lateral view

Plantar Flexion

Gastrocnemius
Soleus
Tibialis posterior
Peroneus longus (assists)
Peroneus brevis (assists)
Flexor digitorum longus (weak)
Flexor hallucis longus (weak)
Plantaris (weak)

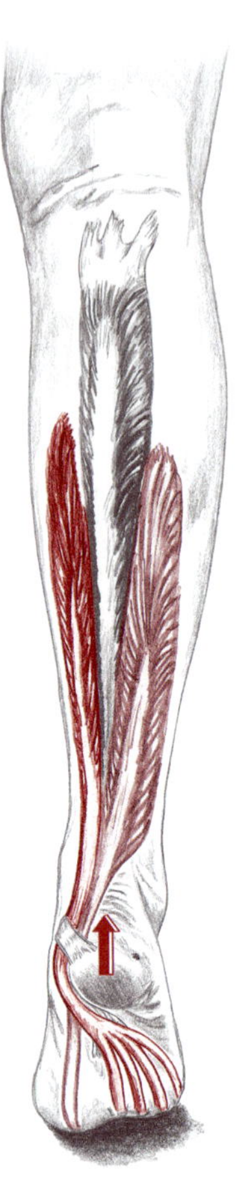

Posterior view

Dorsiflexion

Tibialis anterior
Extensor digitorum longus
Extensor hallucis longus

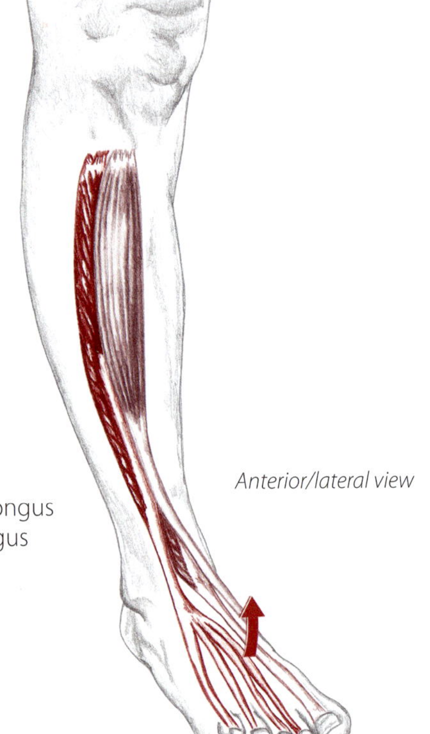

Anterior/lateral view

Foot and Toes

(talotarsal, midtarsal, tarsometatarsal, metatarsophalangeal, proximal and distal interphalangeal joints)

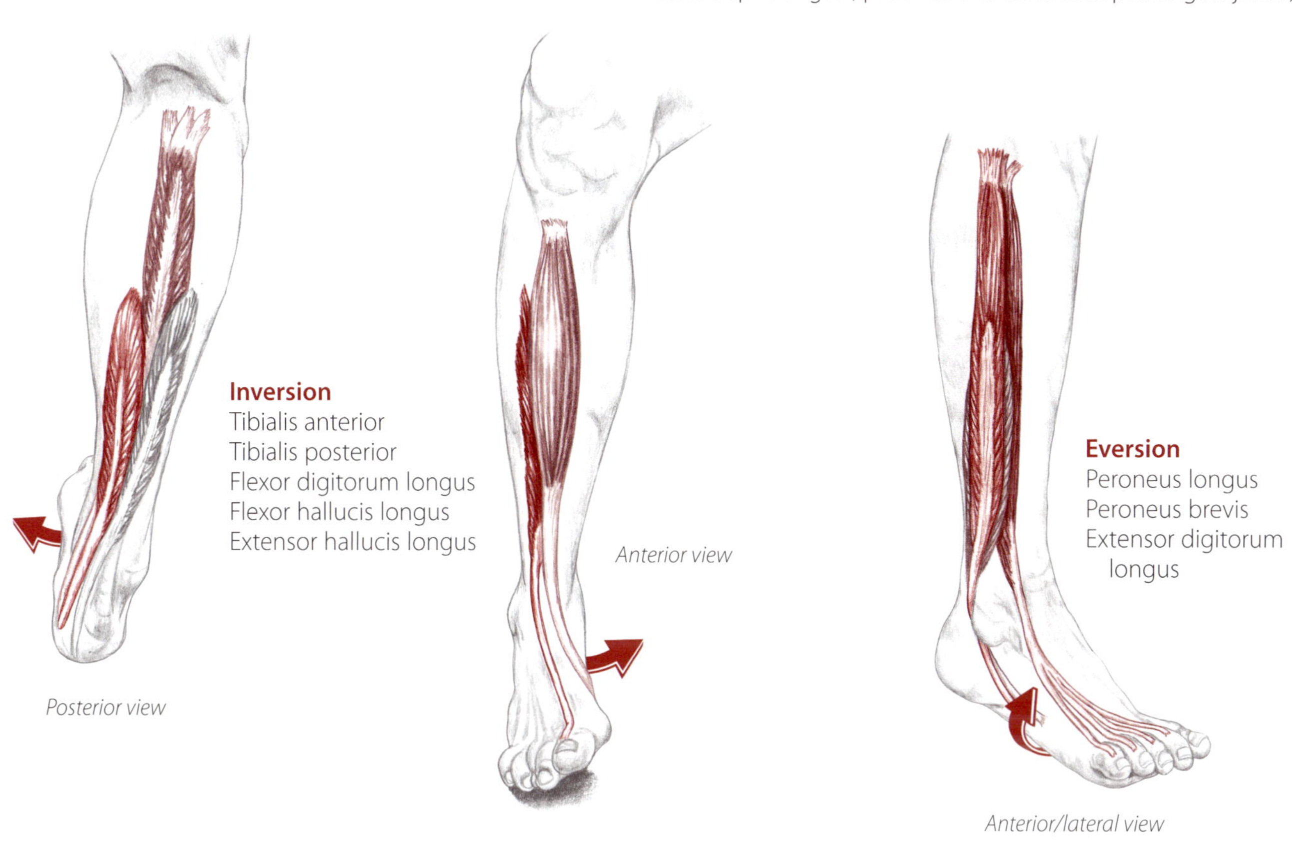

Inversion
Tibialis anterior
Tibialis posterior
Flexor digitorum longus
Flexor hallucis longus
Extensor hallucis longus

Posterior view

Anterior view

Eversion
Peroneus longus
Peroneus brevis
Extensor digitorum longus

Anterior/lateral view

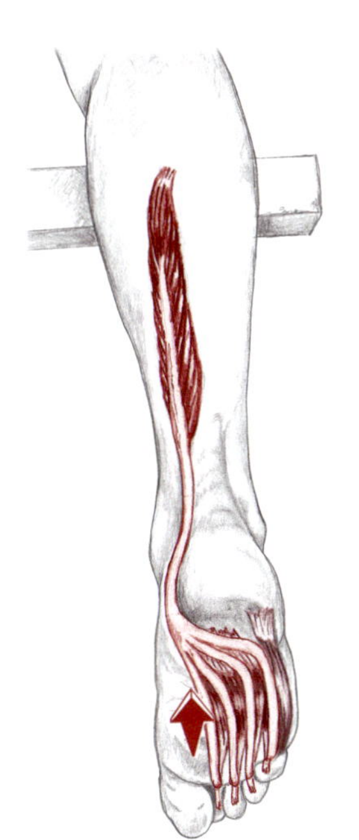

Flexion of Second through Fifth Toes
Flexor digitorum longus
Flexor digitorum brevis
Lumbricals*
Quadratus plantae (assists)*
Dorsal interossei (2nd - 4th toes)*
Plantar interossei (3rd - 5th)*
Abductor digiti minimi (5th)
Flexor digiti minimi (5th)*

Posterior/plantar view

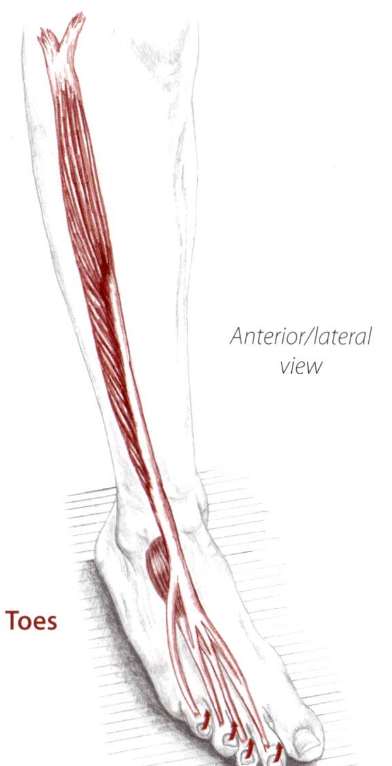

Anterior/lateral view

Extension of Second through Fifth Toes
Extensor digitorum longus
Extensor digitorum brevis (2nd - 4th)
Lumbricals*

See p. 399 for a list of muscles performing flexion, extension, abduction and adduction of the toes

Gastrocnemius and Soleus

The large muscle mass of the posterior leg is composed of the gastrocnemius and the soleus muscles. Together they form what is known as the "triceps surae" that attaches to the strong calcaneal (Achilles) tendon. Both the gastrocnemius and soleus are easily accessible.

The superficial **gastrocnemius** has two heads and crosses two joints - the knee and ankle (7.58). Emerging from between the hamstring tendons, the short gastrocnemius heads extend halfway down the leg before blending into the calcaneal tendon. Although its name (Greek for "belly of the leg") suggests that the gastrocnemius is rotund, it is actually quite thin when compared to the thick soleus.

The **soleus** is deep to the gastrocnemius, yet its medial and lateral fibers bulge from the sides of the leg and extend further distal than the gastrocnemius heads (7.59). The soleus is sometimes called the "second heart" because of the important role its strong contractions play in returning blood from the leg to the heart.

Gastrocnemius

A Flex the knee (tibiofemoral joint)
Plantar flex the ankle (talocrural joint)

O Condyles of the femur, posterior surfaces

I Calcaneus via calcaneal tendon

N Tibial

Soleus

A Plantar flex the ankle (talocrural joint)

O Soleal line, posterior surface of tibia and proximal, posterior surface of fibula

I Calcaneus via calcaneal tendon

N Tibial

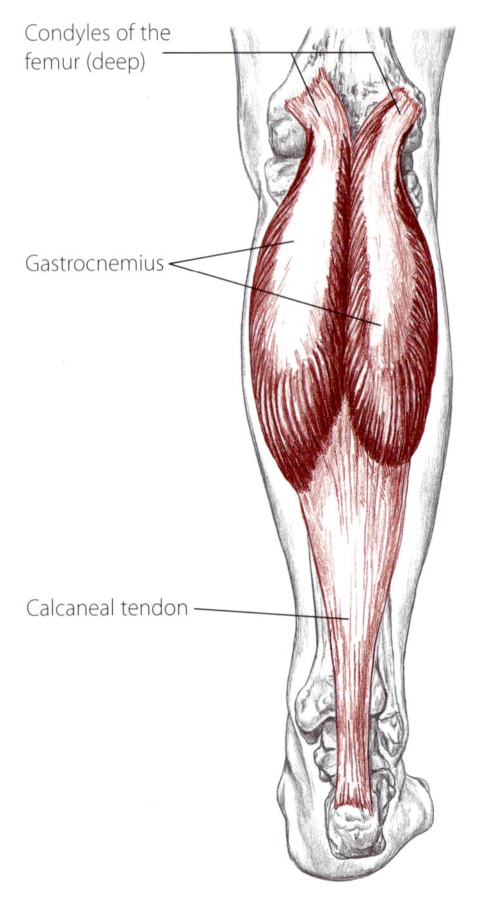

(7.58) Posterior view of right leg

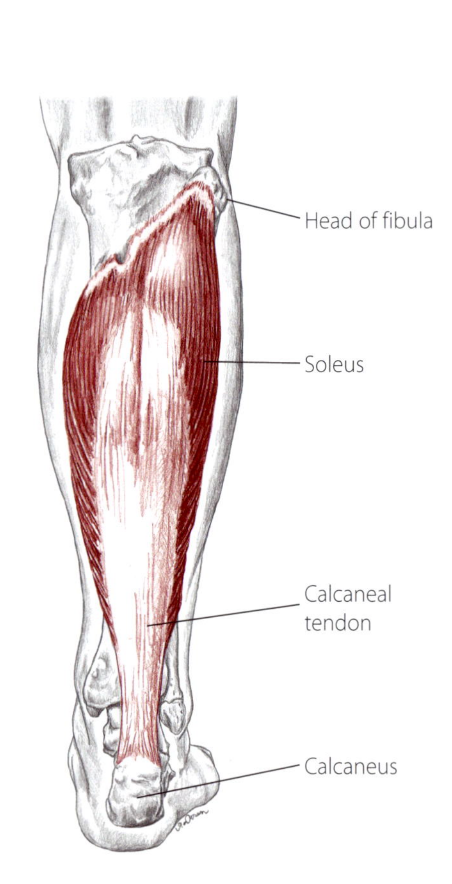

(7.59) Posterior view of right leg, with gastrocnemius removed

(7.60) Posterior view of right leg with foot plantar flexed, showing origins and insertion

gastrocnemius — **gas**-trok-**ne**-me-us — Grk. *gaster*, stomach + *kneme*, leg
soleus — so-**lay**-us — L. *solea*, as in a sole fish (right)

Gastrocnemius and soleus - standing

1) Ask your partner, supported by a chair, to stand on her toes.
2) Palpate the posterior leg, sculpting out the gastrocnemius' oval heads. Follow both heads proximally to the back of the knee. Then follow them distally, noting how the medial head extends further distal than the lateral head (7.61).
3) Move distal to the gastrocnemius and palpate the distal portion of the soleus (7.62). Also explore the medial and lateral sides of the soleus that bulge out from the gastrocnemius.
4) Follow both muscles distally as they blend into the calcaneal tendon.

Can you follow the gastrocnemius heads proximally between the hamstring tendons? Is the medial gastrocnemius head slightly longer than the lateral? Can you feel the difference in texture between the fleshy muscle bellies and the tough, dense calcaneal tendon?

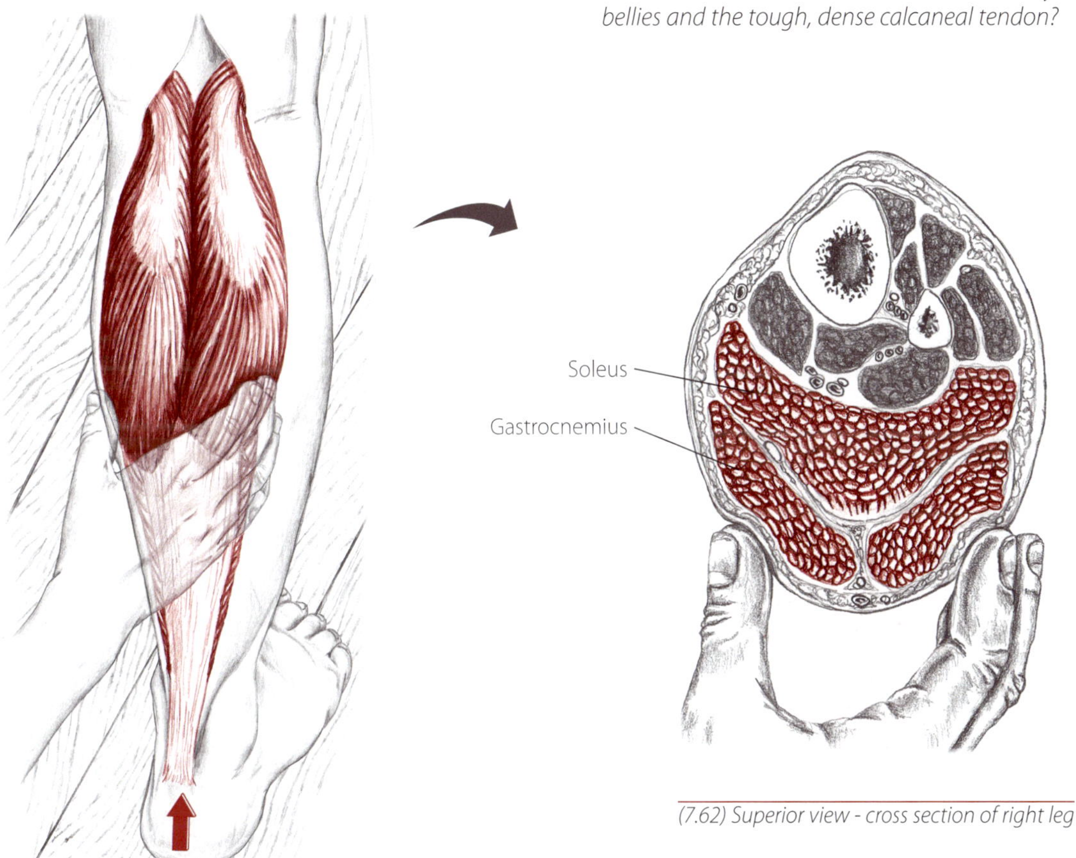

(7.61) Partner standing on her toes

(7.62) Superior view - cross section of right leg

Why was the calcaneal tendon originally called the Achilles' tendon? As a baby, the Greek mythological warrior Achilles was dipped in the River Styx by his mother to make him invulnerable. He was completely immersed except for the ankle by which she held him. After fighting in the Trojan War, Achilles was mortally wounded when an arrow penetrated his heel. Hence, "Achilles' heel" refers to a small but fatal weakness.

Gastrocnemius and soleus - standing

1) Although these muscles are located on the posterior leg, they are accessible from an *anterior* direction. With your partner standing, locate the tibial shaft.
2) Slide medially off the shaft of the tibia and feel the wad of muscle that bulges along the medial side of the leg (7.63). This tissue is the triceps surae.
3) Ask your partner to lie supine and, with the tissue relaxed, note how your thumb can sink around the medial edge of the tibial shaft to specifically locate the soleus.

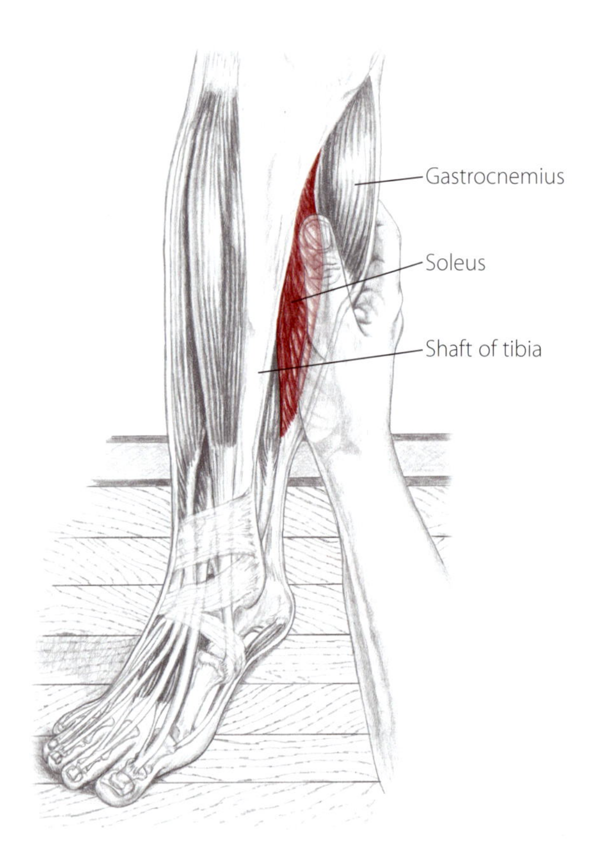

(7.63) Anterior/medial view of right leg with partner standing

Gastrocnemius and soleus - prone

1) Partner prone. Bend the knee to 90° and investigate the soft, massive bellies of the gastrocnemius and soleus and the thick calcaneal tendon.
2) When the knee is flexed, the gastrocnemius muscle is shortened and ineffectual as a plantar flexor. Isolate the soleus by asking your partner to gently plantar flex against your resistance. Notice how the thick soleus contracts while the thin, superficial bellies of the gastrocnemius remain flaccid (7.64).

Can you feel the difference in texture between the fleshy muscle bellies and the tough, dense calcaneal tendon (7.65)?

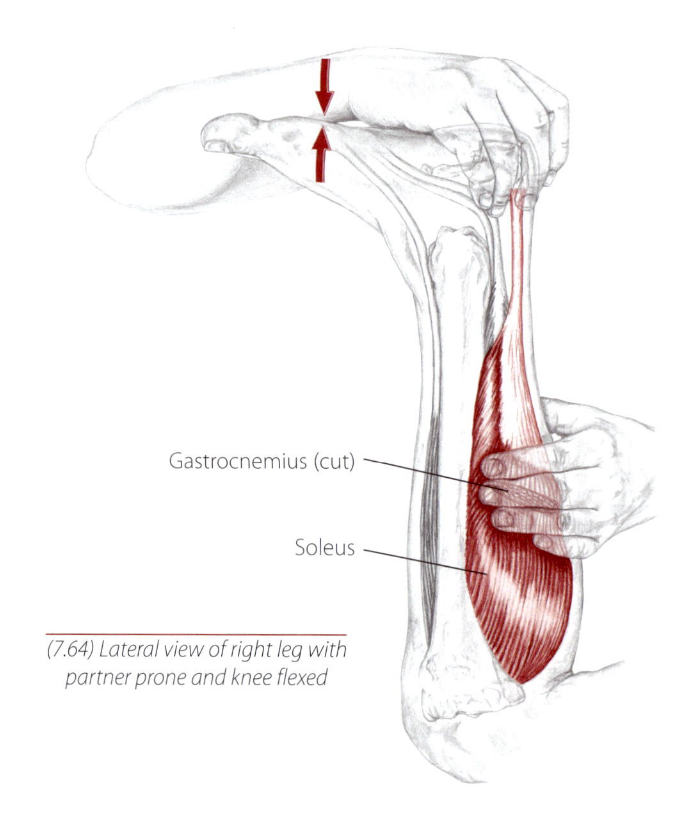

(7.64) Lateral view of right leg with partner prone and knee flexed

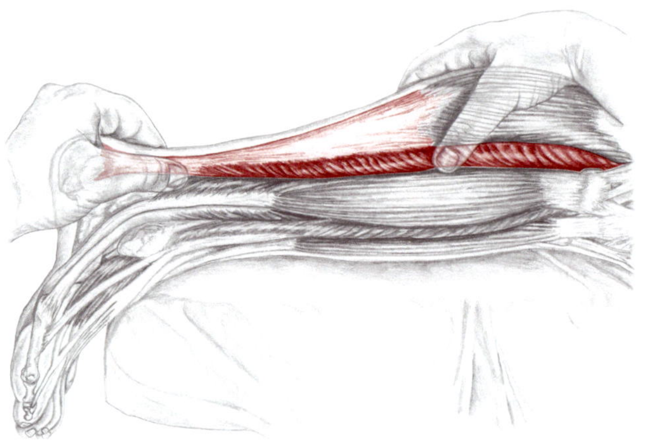

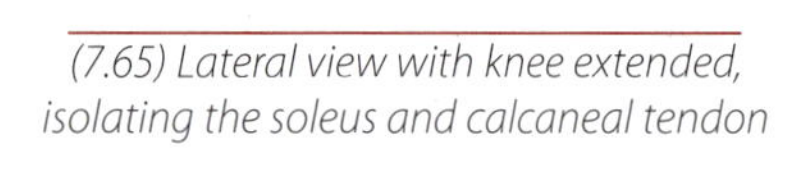

(7.65) Lateral view with knee extended, isolating the soleus and calcaneal tendon

triceps surae **tri**-seps **sir**-eye L. three-headed muscle of the calf

Plantaris

The plantaris has a short muscle belly but the longest tendon in the body. Its belly lies at an oblique angle along the popliteal space of the posterior knee between the gastrocnemius heads; its tendon extends down the length of the leg and attaches to the calcaneus (7.66). Although the plantaris belly is situated in a small, cramped area, it can be readily accessed.

From the standpoint of evolution, the plantaris is thought to be what remains of a larger plantar flexor of the foot. The plantaris of reptiles, which serves as an important muscle of propulsion, still retains much of the basic character of this older, larger flexor.

A Weak plantar flexion of the ankle (talocrural joint)

Weak flexion of the knee (tibiofemoral joint)

O Lateral condyle of the femur

I Calcaneus via calcaneal tendon

N Tibial

1) Partner prone with the knee flexed. Locate the head of the fibula.
2) Move your thumb medial into the popliteal space between the gastrocnemius heads. (Sliding your thumb a little more proximally in the popliteal space will position it off the gastrocnemius' heads.)
3) With your thumb between the gastrocnemius heads, slowly sink into the tissue of the posterior knee (7.67). Explore for an inch-wide belly that runs at an oblique angle from lateral to medial. When you believe you have located the plantaris, outline its shape by strumming your thumb across its belly.

Are you medial and proximal to the head of the fibula? Are you accessing between the gastrocnemius heads? Is the belly you are palpating one to two fingers wide with oblique fibers?

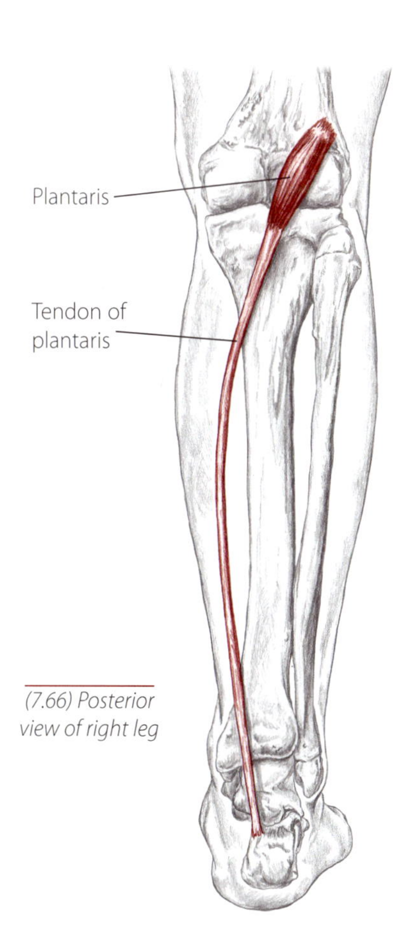

(7.66) Posterior view of right leg

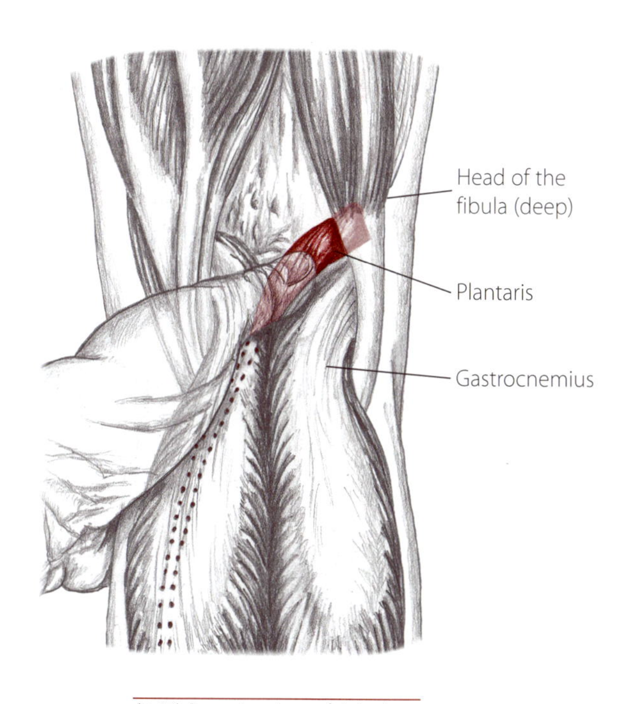

(7.67) Posterior view of right knee

It is not mere coincidence that the plantaris of the leg bears a marked resemblance to the palmaris longus (p. 149) in the forearm: The two muscles have short bellies followed by long tendons, limited capabilities and are absent in nearly 10% of the population.

plantaris   plan-**tar**-is   Fr. sole of the foot

Popliteus

As its name suggests, the popliteus is located in the popliteal space. This muscle has a small, short belly with diagonal fibers (7.68). Lying beneath the upper fibers of the gastrocnemius and plantaris, it is the deepest muscle of the posterior knee. Because of its depth, the popliteus is inaccessible; nevertheless, its tendinous insertion on the posterior tibia can be palpated. Although the popliteus is a weak flexor of the knee, it is vital in "unlocking" the joint from an extended position; hence its nickname, "the key which unlocks the knee."

1) Prone with the knee flexed. Access a portion of the popliteus by locating the tibial tuberosity and sliding medially around the tibia to the posterior surface of its shaft.
2) Explore the posterior surface of the tibia for the popliteus tendon by pushing the overlying edge of the soleus and gastrocnemius muscles to the side (7.70).
3) Although the popliteus will not readily present itself as a palpable structure, if you are accessing the posterior region of the tibial shaft, you will be on its tendinous attachment.

A Medially rotate the flexed knee (tibiofemoral joint)
Flex the knee (t/f joint)

O Lateral epicondyle of the femur

I Proximal, posterior aspect of tibia

N Tibial

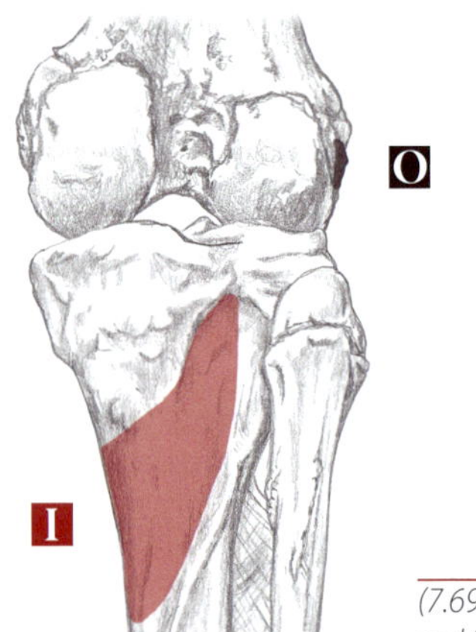

(7.69) Origin and insertion

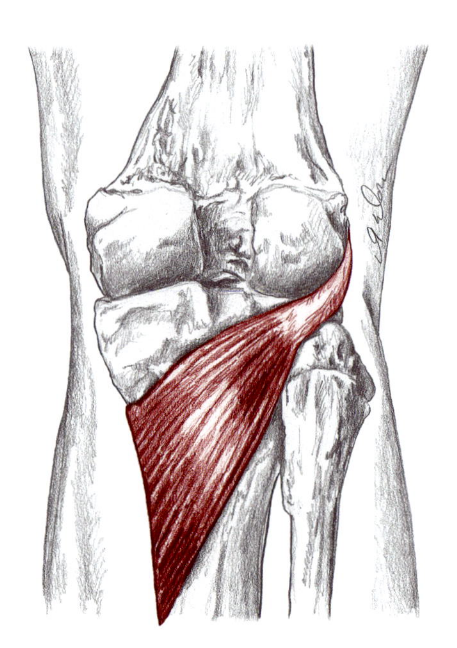

(7.68) Posterior view of right knee showing the popliteus

MMT: start w/ tibia ER'd. (seated)
Have clt. IR tibia w/ foot on floor.

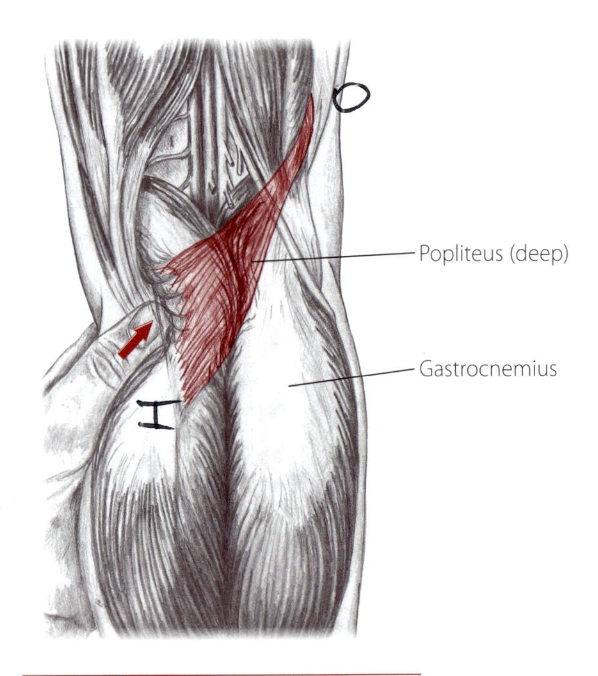

(7.70) Posterior view of right knee, sliding your thumb deep to the gastrocnemius and soleus

popliteus pop-**lit**-e-us L. *poples*, ham of the knee

Peroneus Longus and Brevis

The slender peroneal muscles are located on the lateral side of the fibula (7.71). More specifically, they lie between the extensor digitorum longus and the soleus. A portion of the peroneus brevis lies deep to the peroneus longus, yet both are accessible. Their distal tendons are superficial and palpable behind the lateral malleolus and along the side of the heel (7.72).

Peroneus Longus

A Evert the foot

Assist to plantar flex the ankle (talocrural joint)

O Proximal two-thirds of lateral fibula

I Base of the first metatarsal and medial cuneiform

N Superior peroneal

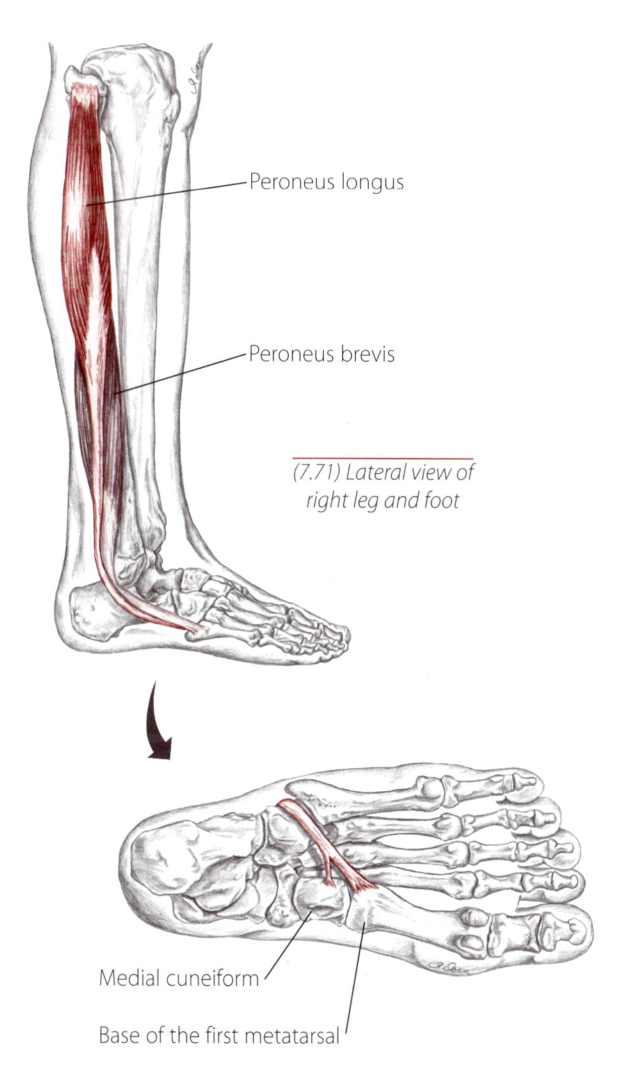

(7.71) Lateral view of right leg and foot

(7.73) Plantar view of right foot showing peroneus longus tendon

Peroneus Brevis

A Evert the foot

Assist to plantar flex the ankle (talocrural joint)

O Distal two-thirds of lateral fibula

I Tuberosity of fifth metatarsal

N Superior peroneal

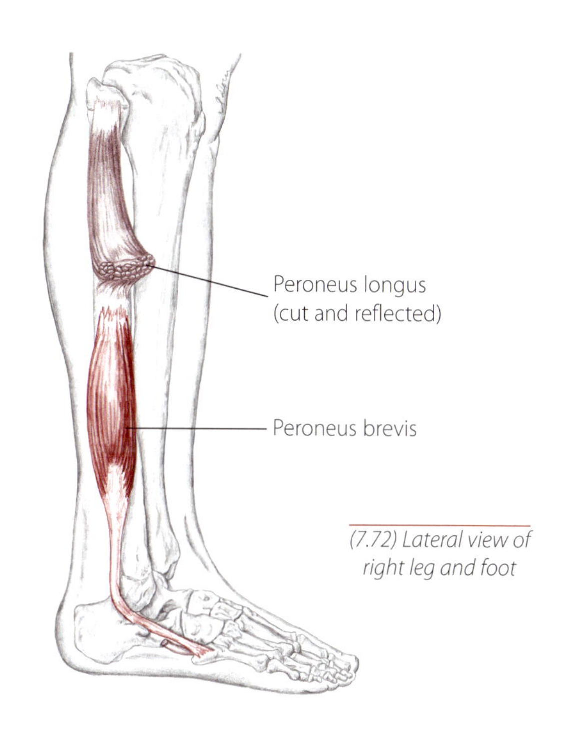

(7.72) Lateral view of right leg and foot

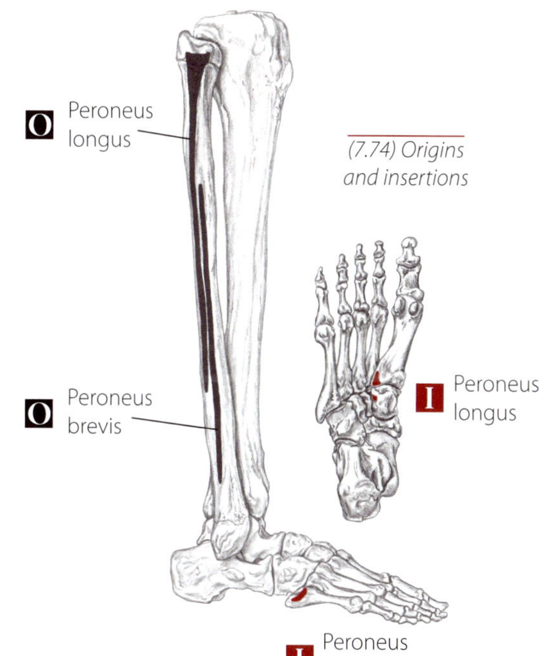

(7.74) Origins and insertions

peroneus per-**o**-ne-us Grk. pin, buckle

Peroneals

1) Partner supine, prone or sidelying. Place a finger at the head of the fibula and the lateral malleolus. The peroneal bellies are located between these two landmarks (7.75).
2) Lay your fingers between these landmarks and ask your partner to alternately evert and relax her foot. Feel the peroneals tighten upon eversion. This action will sometimes create a visible dimple or depression along the side of the leg (7.76).
3) As your partner continues to evert and relax her foot, follow the peroneus longus proximally toward the head of the fibula. Now follow both muscles distally to where their tendons wrap around the back of the lateral malleolus.
4) Follow the peroneus brevis tendon to the base of the fifth metatarsal (7.77).

Are you on the lateral side of the leg between the head of the fibula and the lateral malleolus? Can you differentiate the slender peroneals from the lateral edge of the larger gastrocnemius and soleus? Can you feel the tendon of the peroneus brevis attach to the base of the fifth metatarsal?

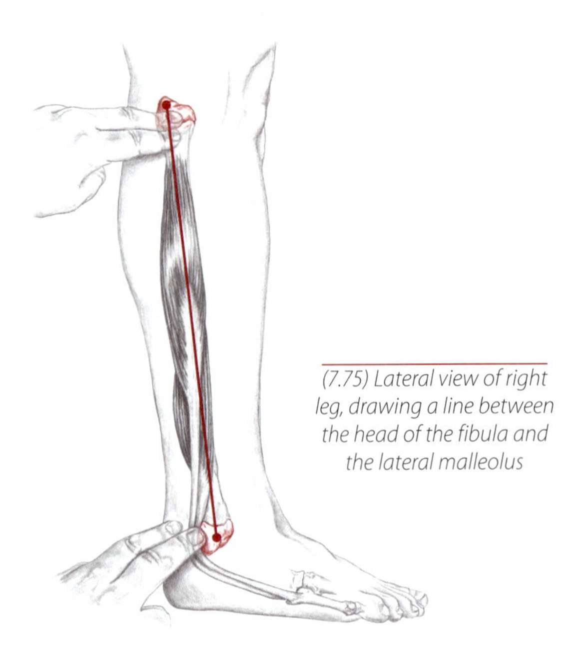

(7.75) Lateral view of right leg, drawing a line between the head of the fibula and the lateral malleolus

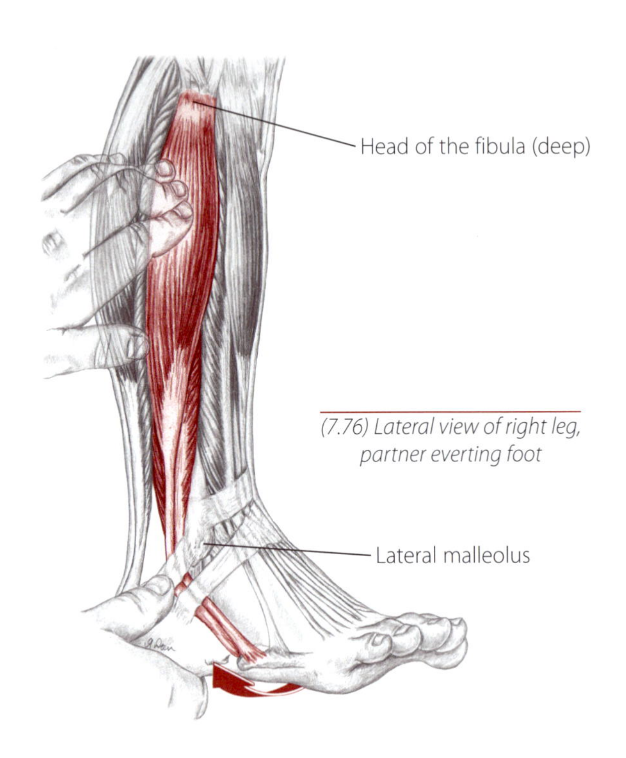

(7.76) Lateral view of right leg, partner everting foot

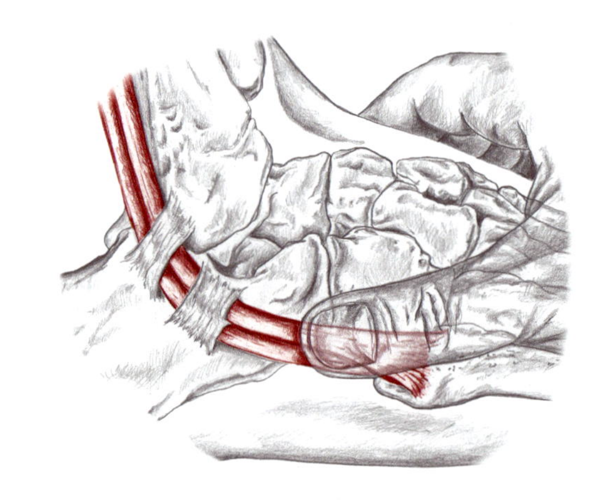

(7.77) Lateral view of right ankle, accessing the peroneus tendons

It is not uncommon for there to be a third peroneal. If present, the peroneus tertius **(a)** will be found anterior to the lateral malleolus on the front of the ankle. Oddly enough, the tertius is actually a short branch of the extensor digitorum longus **(b)** that attaches, along with peroneus brevis, at the tuberosity of the fifth metatarsal.

tertius **ter**-she-us L. third

Extensors of the Ankle and Toes

Tibialis Anterior
Extensor Digitorum Longus
Extensor Hallucis Longus

These extrinsic muscles are located on the anterior aspect of the leg between the shaft of the tibia and the peroneal muscles. The tendons of all three muscles cross beneath the extensor retinaculum at the ankle (p. 392). The **tibialis anterior** is large, superficial and the most clearly isolated of the group. It lies directly lateral to the tibial shaft (7.78).

Squeezed between the tibialis anterior and the peroneal muscles, the **extensor digitorum longus** is partially superficial. Its four tendons are clearly palpable on the dorsal surface of the foot (7.79, 7.80).

The muscle belly of the **extensor hallucis longus** lies deep to the other two muscles and can be accessed only indirectly (7.81); however, like the extensor digitorum, its distal tendon is easily found on the dorsal surface of the foot as it leads toward the first toe.

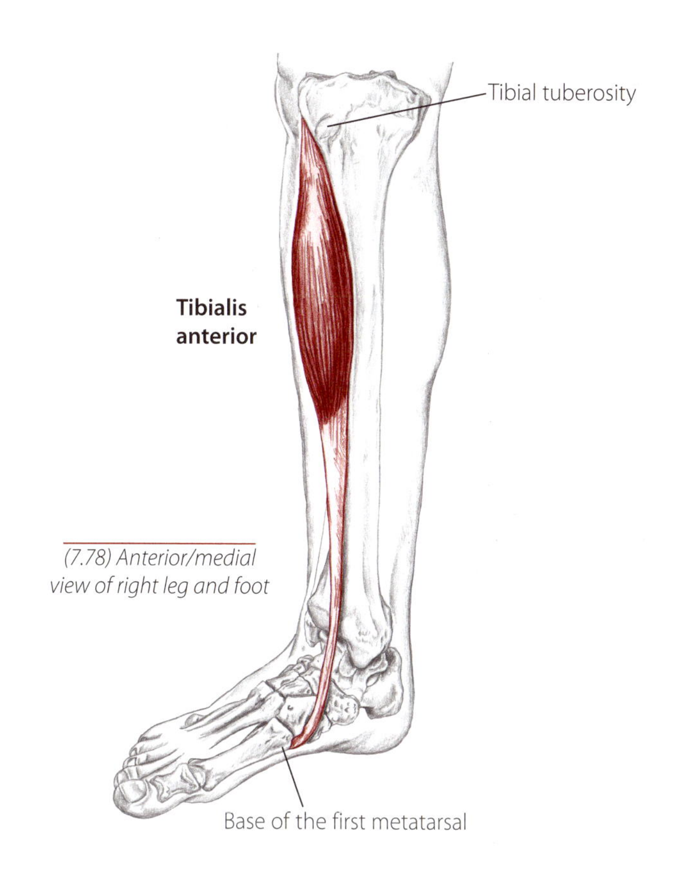

(7.78) Anterior/medial view of right leg and foot

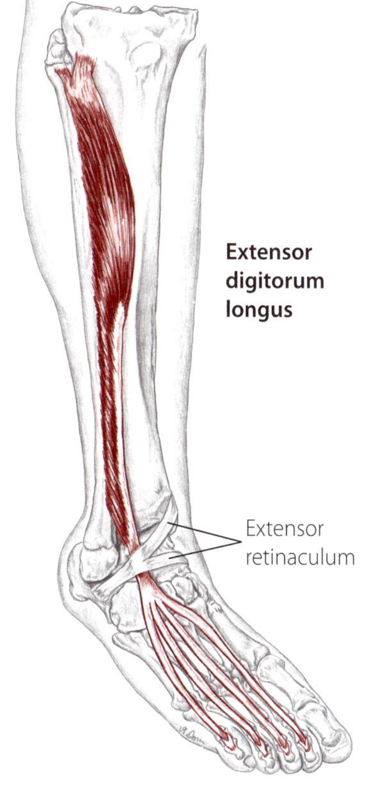

(7.79) Anterior/lateral view of right leg and foot

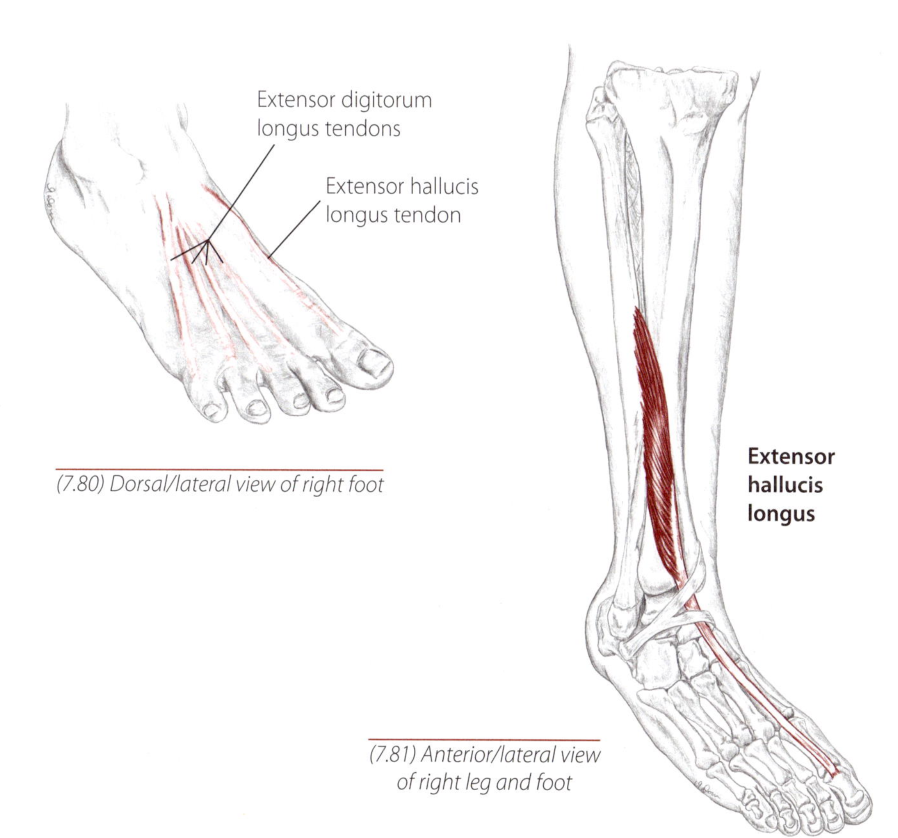

(7.80) Dorsal/lateral view of right foot

(7.81) Anterior/lateral view of right leg and foot

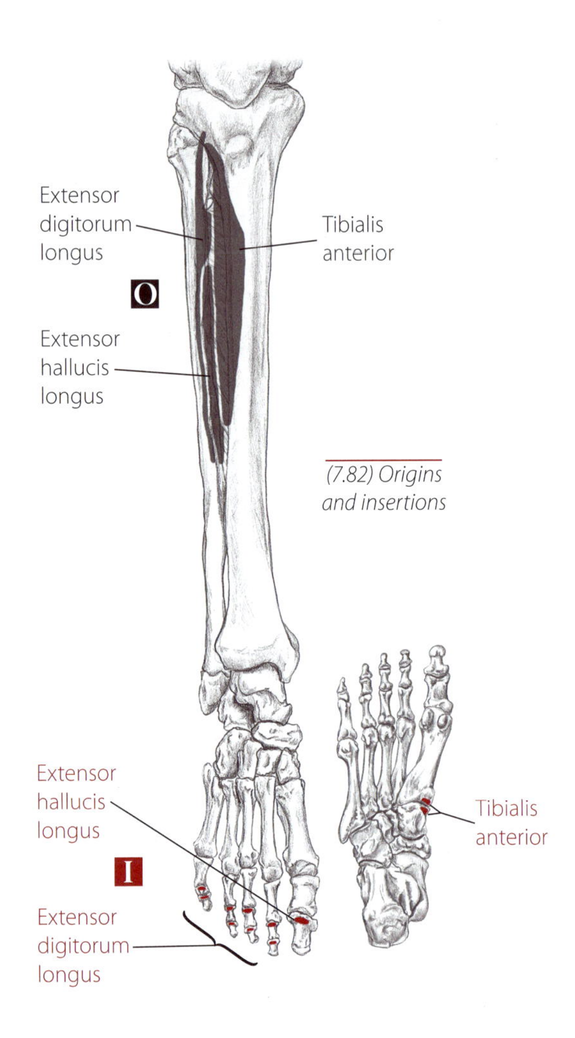

(7.82) Origins and insertions

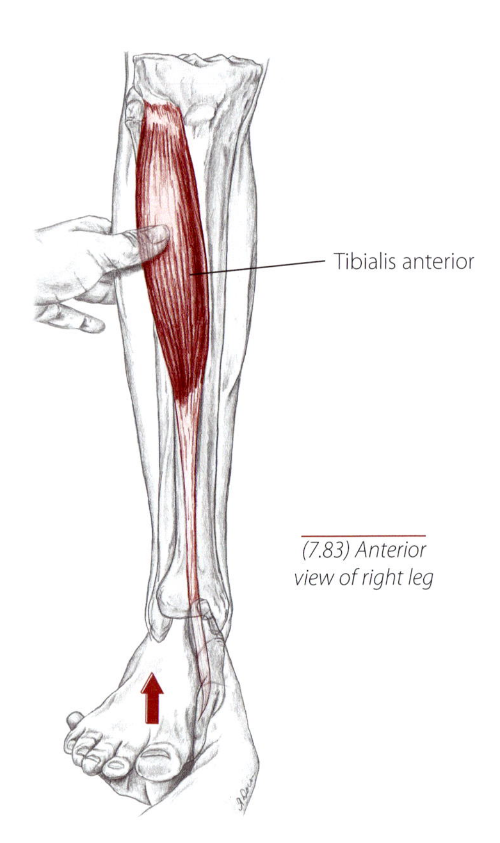

(7.83) Anterior view of right leg

Tibialis Anterior

A Invert the foot

Dorsiflex the ankle (talocrural joint)

O Proximal lateral surface of tibia and interosseous membrane

I Medial cuneiform and base of the first metatarsal

N Deep peroneal

Extensor Digitorum Longus

A Extend the second through fifth toes (metatarsophalangeal and interphalangeal joints)

Dorsiflex the ankle (talocrural joint)

Evert the foot

O Proximal anterior shaft of fibula and interosseous membrane

I Middle and distal phalanges of second through fifth toes

N Deep peroneal

Extensor Hallucis Longus

A Extend the first toe (metatarsophalangeal and interphalangeal joints)

Dorsiflex the ankle (talocrural joint)

Invert the foot

O Middle anterior surface of fibula and interosseous membrane

I Distal phalange of first toe

N Deep peroneal

Tibialis anterior

1) Partner supine. Locate the shaft of the tibia and slide off it laterally onto the tibialis anterior.
2) Ask your partner to dorsiflex his ankle and palpate its long, inch-wide belly (7.83).
3) With the ankle dorsiflexed, palpate the muscle distally as it becomes a thick, tendinous cord. Follow it to the medial side of the foot as it disappears at the medial cuneiform.

As your partner alternately dorsiflexes and relaxes his ankle, can you feel and see the tendon that crosses the top of the ankle? Ask your partner to invert his foot and note whether the tibialis anterior is involved. Can you feel where the tendon passes under the extensor retinaculum?

Extensor digitorum longus

1) Supine. The easiest way to begin is by palpating the digitorum's distal tendons. Ask your partner to extend his toes. Visibly identify and palpate the four tendons of the digitorum on the top of the foot.
2) With the toes still extended, follow the tendons toward the ankle. Notice how they converge into a single tendinous bundle that loops underneath the extensor retinacula (7.84).
3) Follow this tendon proximally as it merges into its muscle belly. Explore the slender belly of the digitorum as it squeezes between the tibialis anterior and the peroneal muscles.

Locate the digitorum and tibialis anterior tendons on the top of the ankle. With the ankle dorsiflexed, ask your partner to slowly invert and evert his foot. Do you feel the tibialis tighten upon inversion and the digitorum upon eversion?

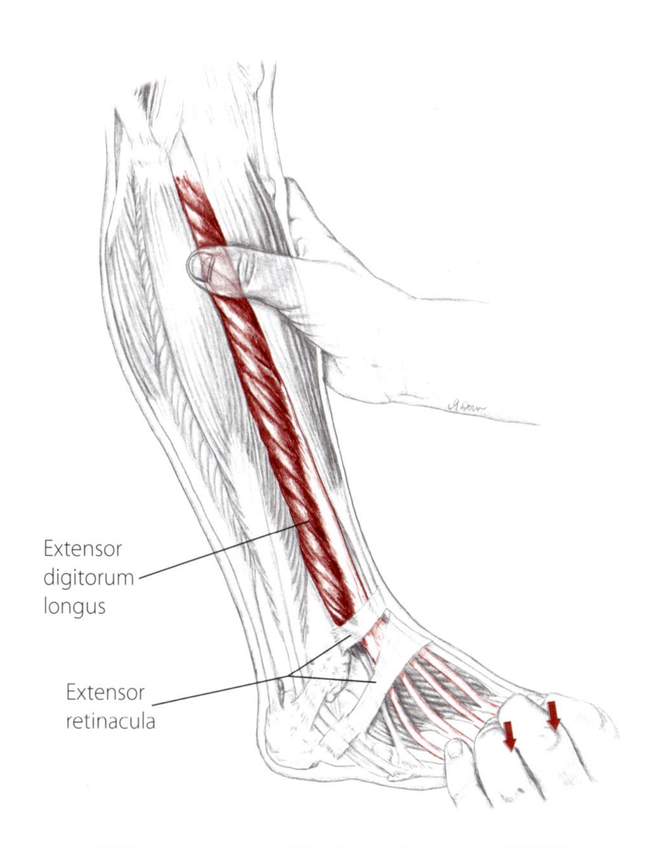

(7.84) Lateral view of right leg, resisting extension of the toes

Extensor hallucis longus

1) Supine. Ask your partner to extend his first toe. Visibly identify and palpate the solid tendon running along the dorsal surface of the foot to the first toe (7.85).
2) With the toe still extended, follow the tendon toward the ankle. Note how it snuggles between and underneath the extensor digitorum and tibialis anterior tendons.

Can you follow the tendon from the first toe to the dorsal surface of the ankle? Can you distinguish the three separate tendons of the extensors (hallucis, digitorum and tibialis anterior) along the dorsal surface of the ankle?

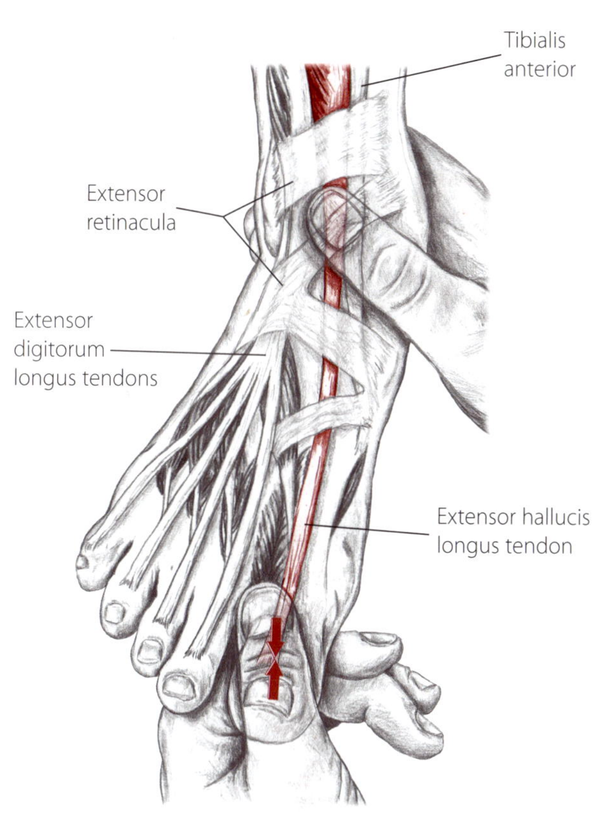

(7.85) Dorsal view of right foot, resisting extension of the first toe

Flexors of the Ankle and Toes

Tibialis Posterior
Flexor Digitorum Longus
Flexor Hallucis Longus

Buried deep to the gastrocnemius and soleus on the posterior leg are three slender muscles primarily responsible for inverting the foot and flexing the toes. All three muscles are virtually inaccessible, except at the small region on the medial side of the leg. This small gap between the tibial shaft and the edge of the calcaneal tendon is where the most distal fibers and tendons of the flexors can be palpated directly (7.89). The tendons of these three muscles curve around the medial malleolus and pass deep to the flexor retinaculum. The tibial artery and tibial nerve are situated between the tendons at the medial ankle.

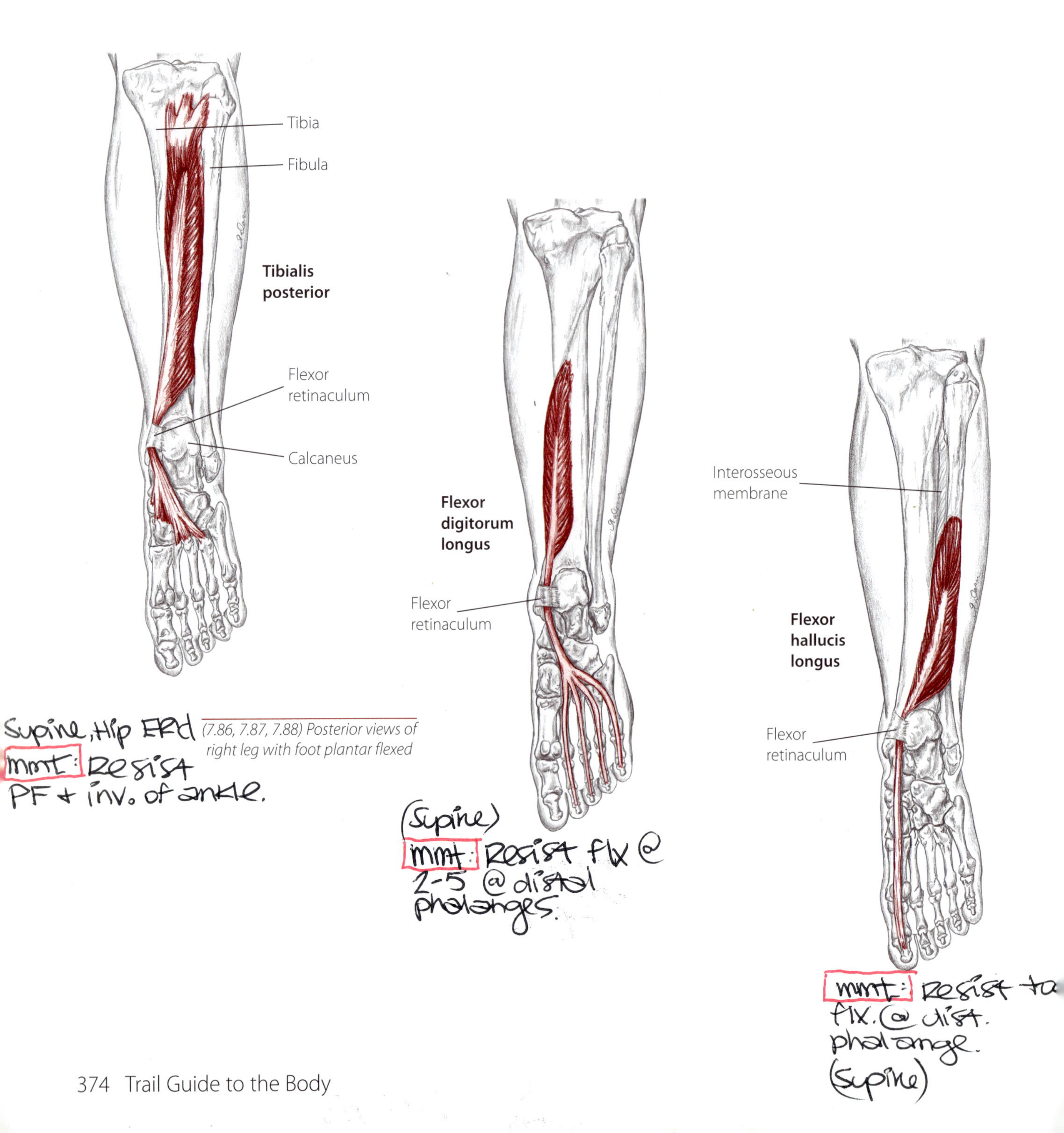

(7.86, 7.87, 7.88) Posterior views of right leg with foot plantar flexed

Tibialis Posterior

A Invert the foot

Plantar flex the ankle (talocrural joint)

O Proximal posterior shaft of tibia, proximal fibula and interosseous membrane

I Navicular, cuneiforms, cuboid and bases of second through fourth metatarsals

N Tibial

Flexor Digitorum Longus

A Flex the second through fifth toes (metatarsophalangeal and interphalangeal joints)

Weak plantar flexion of ankle (talocrural joint)

Invert the foot

O Middle posterior surface of tibia

I Distal phalanges of second through fifth toes

N Tibial

Flexor Hallucis Longus

A Flex the first toe (metatarsophalangeal and interphalangeal joints)

Weak plantar flexion of ankle (talocrural joint)

Invert the foot

O Middle half of posterior fibula

I Distal phalange of first toe

N Tibial

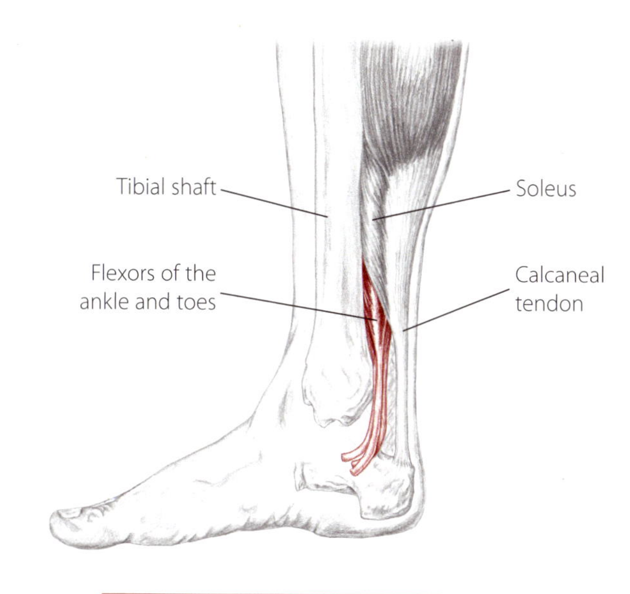

(7.89) Medial view of right leg and foot

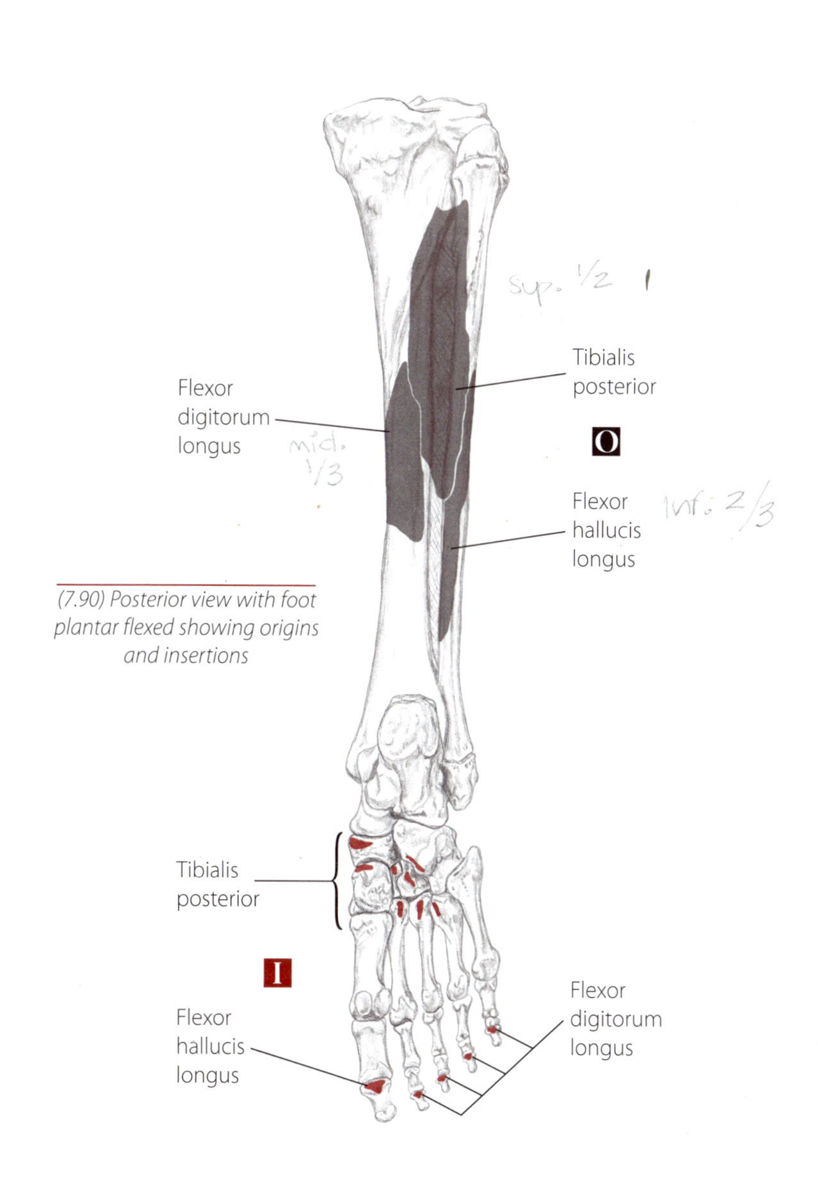

(7.90) Posterior view with foot plantar flexed showing origins and insertions

All flexors

1) Partner supine, prone or sidelying. Locate the medial malleolus. Slide off the malleolus posteriorly and proximally into the space between the posterior shaft of the tibia and the calcaneal tendon.
2) Explore this region for the distal bellies and tendons of these muscles (7.91). Follow the tendons distally around the back of the medial malleolus.
3) It is difficult to isolate specific tendons; however, tibialis posterior will be the most anterior. Have your partner invert his foot as you follow this tendon around the ankle to the underside of the foot.

Place your fingers on the distal bellies and ask your partner to slowly wiggle all his toes. Can you feel the muscles or tendons shift? Can you locate the medial malleolar groove (p. 351) and feel the tendons in and posterior to it? Can you locate the pulse of the tibial artery?

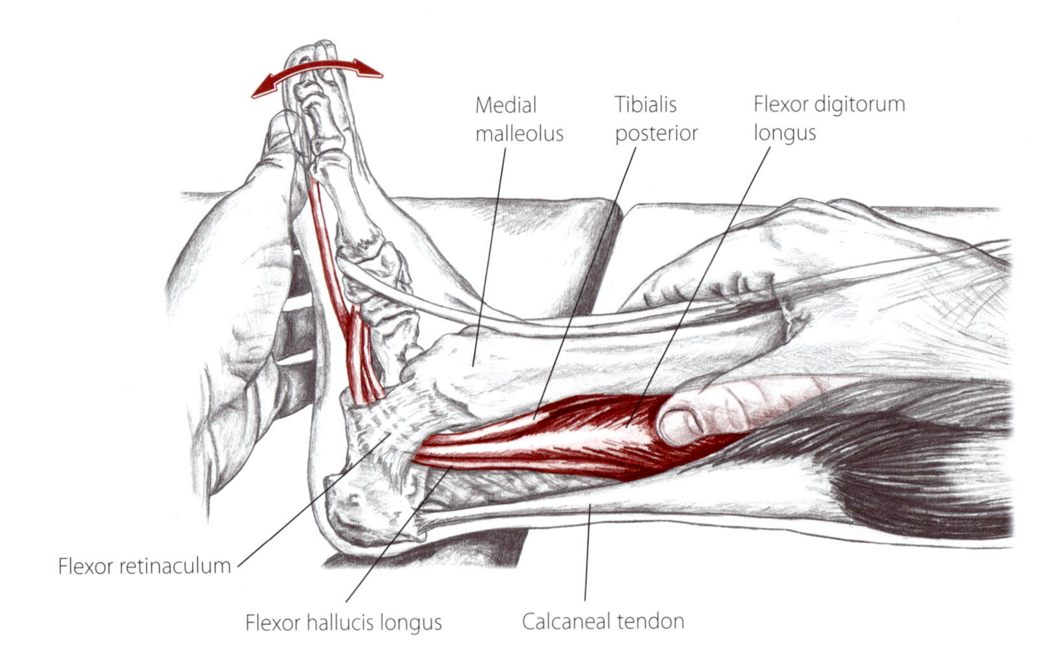

(7.91) Medial view of right leg and foot, partner wiggling his toes

"Tom, Dick AN' Harry" is a mnemonic device that corresponds to the initials of the tendons and vessels in the order that they pass by the medial malleolus. **T**ibialis posterior is the most anterior, followed by flexor **D**igitorum, the tibial **A**rtery, tibial **N**erve and then flexor **H**allucis.

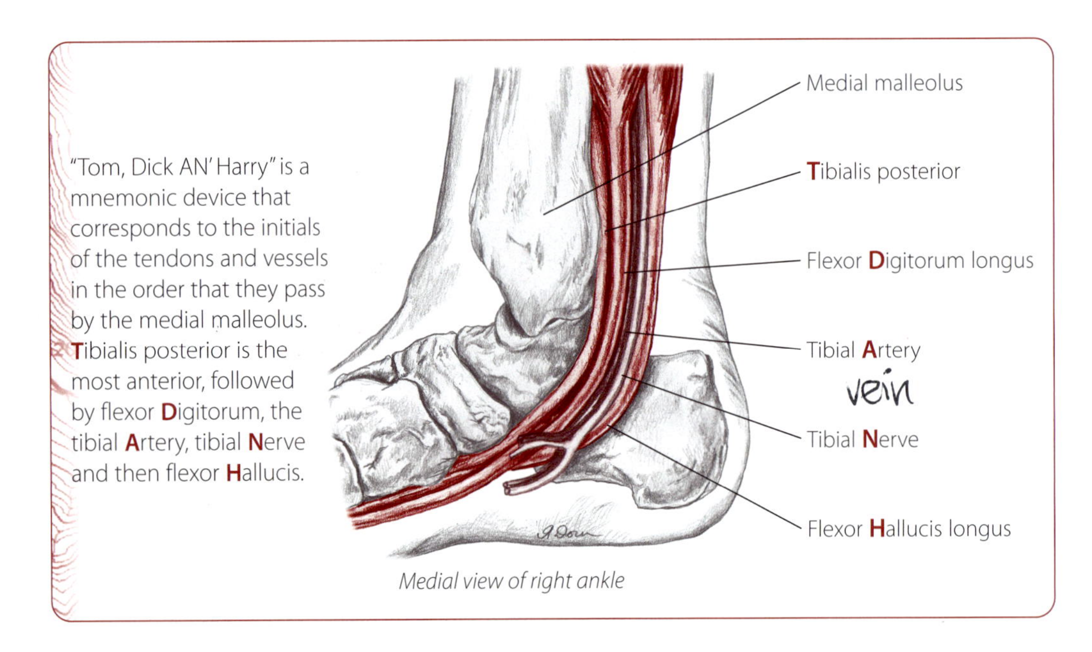

Medial view of right ankle

Muscles of the Foot

Extensor Digitorum Brevis
Flexor Digitorum Brevis
Abductor Hallucis
Abductor Digiti Minimi

The dorsal surface of the foot is home to the **extensor digitorum brevis**. Its small belly lies deep to the extensor digitorum longus tendons, but is still palpable (7.92).

Unlike the minimally muscled dorsal surface, the foot's plantar surface is made up of several layers of muscle. The first layer, located deep to the plantar aponeurosis, (p. 393), is formed by three muscles that lie side by side. The center muscle is the **flexor digitorum brevis** (7.93). It extends down the center of the foot from the calcaneus to the phalanges. Medial to the flexor digitorum brevis is the thick, superficial **abductor hallucis**; lateral to it is the superficial **abductor digiti minimi** (7.94). Both abductors are easily accessible and often visible along the sides of the foot. Although deep to the plantar aponeurosis, all three muscles are relatively superficial on the sole of the foot and are thus palpable.

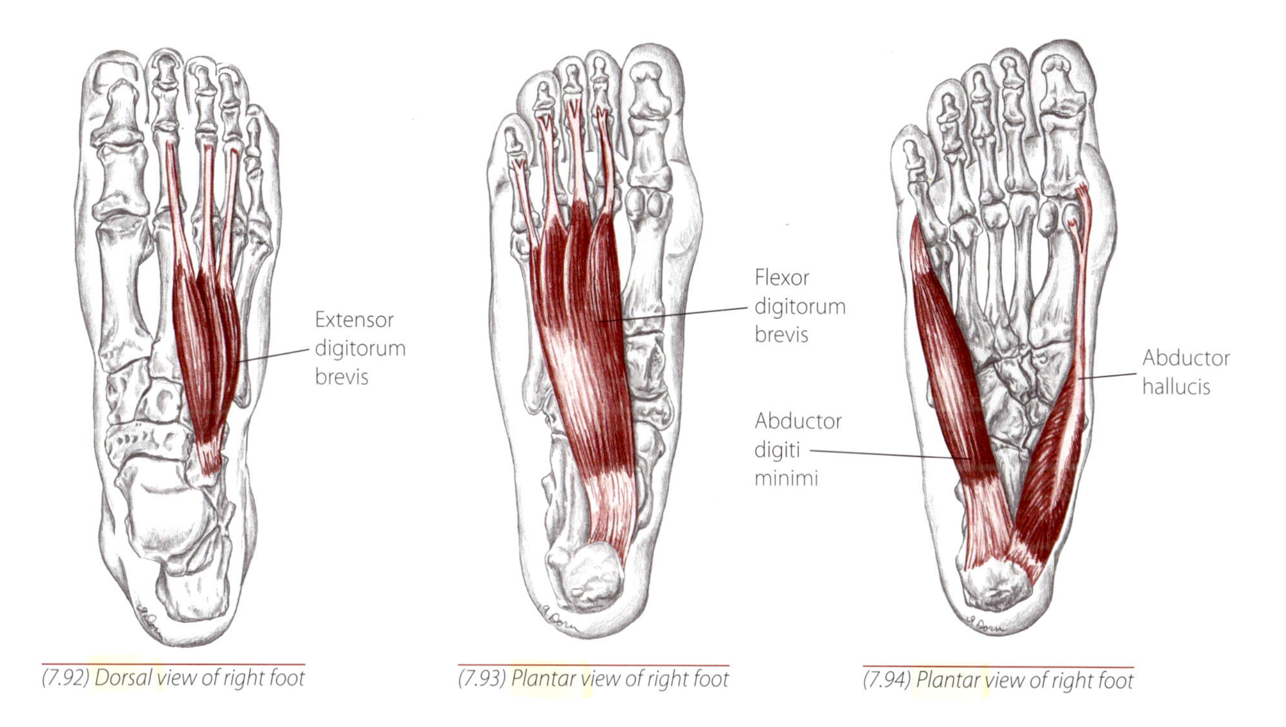

(7.92) Dorsal view of right foot

(7.93) Plantar view of right foot

(7.94) Plantar view of right foot

Extensor Digitorum Brevis

A Extend the second through fourth toes (metatarsophalangeal and interphalangeal joints)

O Calcaneus (dorsal surface)

I Second through fourth toes via the extensor digitorum longus tendons

N Deep peroneal

Flexor Digitorum Brevis

A Flex middle phalanges of the second through fifth toes (proximal interphalangeal joints)

O Calcaneus (plantar surface)

I Middle phalanges of second through fifth toes

N Medial plantar

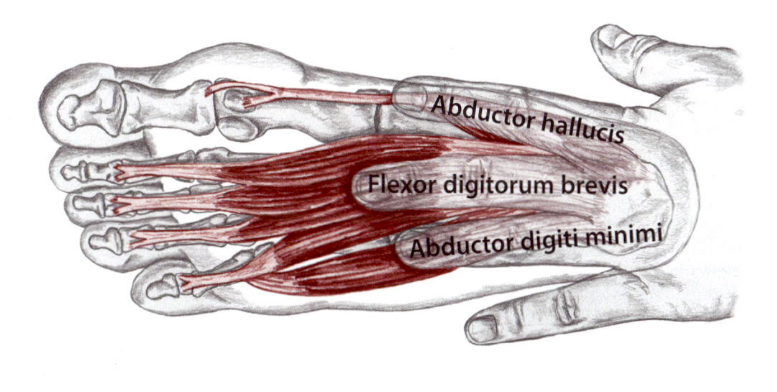

(7.95) Fingers on the plantar surface of the foot showing order of muscles

minimi **min**-i-mee L. smallest

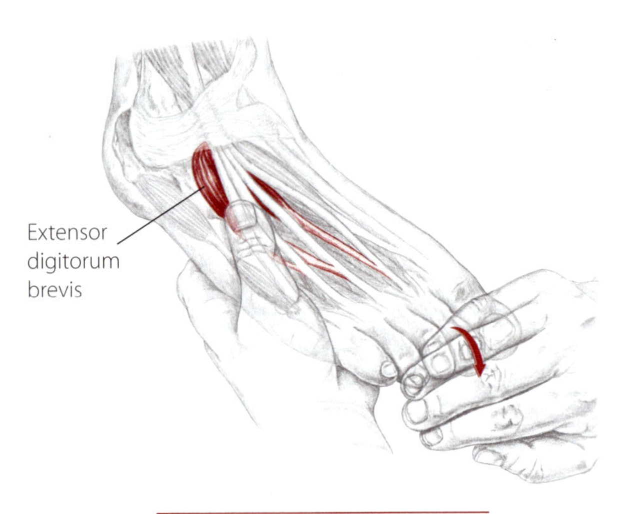

(7.96) Dorsal/lateral view of right foot, partner extends toes against your resistance

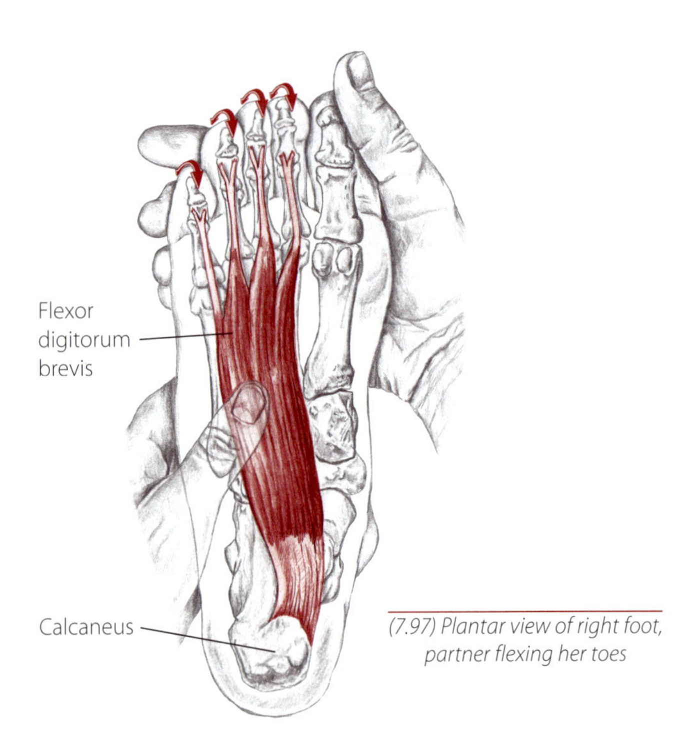

(7.97) Plantar view of right foot, partner flexing her toes

If the discomfort of wearing high-heel shoes were not enough, try this on for size: The point of a spike heel worn by the average-sized woman is subjected to nearly 2,000 pounds of pressure per square inch with every step she takes. This force is shot into the heel and reverberates up the entire body. When air travel was in its infancy, women wearing high heels were actually prohibited from boarding airplanes because the heels of their shoes might pierce the thin metal floors.

Abductor Hallucis

A Abduct the first toe (metatarsophalangeal joint)

Assist to flex the first toe (metatarsophalangeal joint)

O Calcaneus (plantar surface)

I Proximal phalange of first toe (medial side) and medial sesamoid bone

N Medial plantar

Abductor Digiti Minimi

A Flex the fifth toe

Assist to abduct the fifth toe (metatarsophalangeal joint)

O Calcaneus (plantar surface)

I Proximal phalange of fifth toe (lateral side)

N Lateral plantar

Extensor digitorum brevis

1) Partner supine, with the feet off the end of the table. Locate the lateral malleolus. Slide two inches off the malleolus toward the fifth toe. Palpate beneath and lateral to the extensor digitorum longus tendons to locate the small belly of extensor digitorum brevis.
2) Ask your partner to extend her toes against your resistance to feel the muscles contract (7.96). Note how the belly forms a dense mound over the cuboid and lateral cuneiform upon contraction.

Flexor digitorum brevis

1) Supine, with the feet off the end of the table. Locate the plantar surface of the heel and the second through fifth toes. Visualize this muscle's location by drawing imaginary lines between these points.
2) Palpating along the arch of the foot, sink your thumbs along these lines and roll across the muscle fibers (7.97). Ask your partner to alternately flex and relax her toes. It may be challenging to isolate the flexor digitorum brevis belly, but have faith that you are in the correct location.

Abductor hallucis

1) Partner supine, with the feet off the end of the table. Locate the medial surface of the heel and the medial side of the first toe.
2) Palpate between these points and note the thick, superficial tissue running alongside the medial/plantar surface of the foot (7.98).
3) Ask your partner to flex his first toe against your resistance and note the strength and density of the abductor hallucis belly.

Abductor digiti minimi

1) Supine, with the feet off the end of the table. Locate the plantar surface of the heel and the lateral surface of the fifth toe.
2) Palpate between these points for the thick, superficial tissue running alongside the lateral/plantar surface of the foot (7.99).
3) Ask your partner to abduct or flex his fifth toe against your resistance in order to feel the fibers contract.

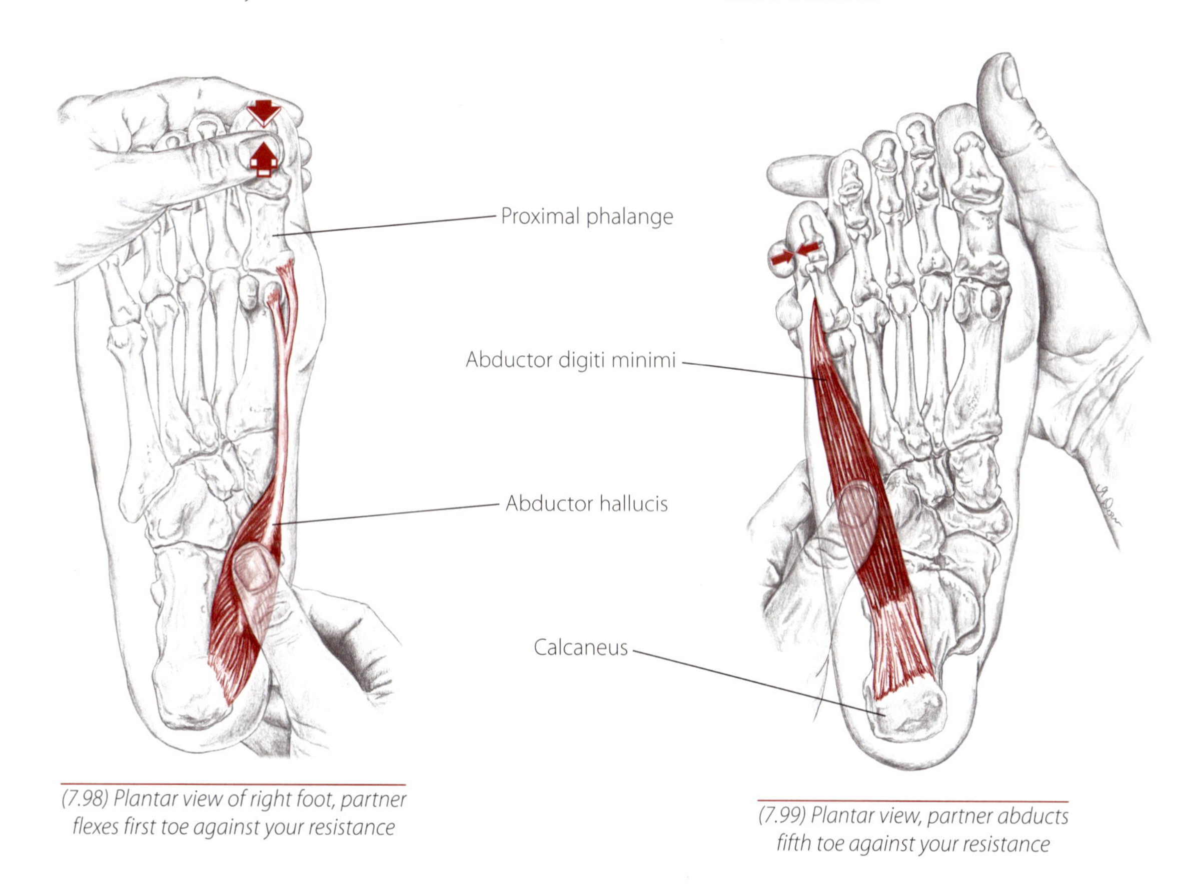

(7.98) Plantar view of right foot, partner flexes first toe against your resistance

(7.99) Plantar view, partner abducts fifth toe against your resistance

Because they bear the weight of the body when standing, walking and running, the feet are sometimes known as the "little soldiers." In comparison to standing, walking increases the pressure on the feet twofold while running increases it fourfold.

These stresses demand that the foot be designed for more than lying flat and idle on the ground. Thus the bones and ligaments of the foot are arranged to form three arches - the medial longitudinal, lateral longitudinal and transverse. These arches connect with three points of contact - the calcaneus and the heads of the first and fifth metatarsals.

The three arches together raise the center of the foot, creating a structure that is ideally shaped to distribute and absorb the weight of the body. The arches also help the plantar surface of the foot adapt to uneven terrain while hiking or climbing.

Other Muscles of the Foot

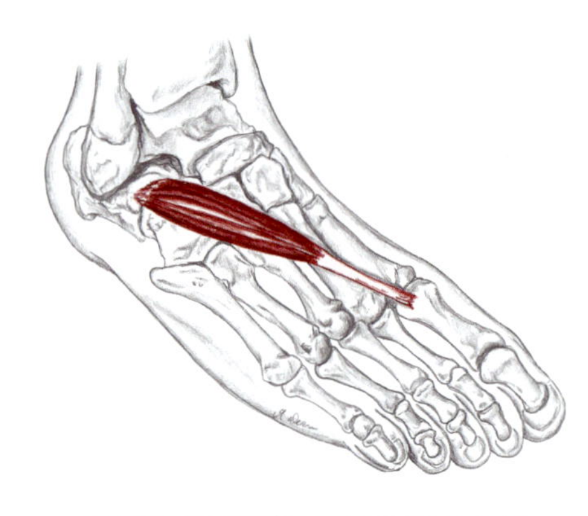

(7.100) Dorsal/lateral view of right foot

Extensor Hallucis Brevis (left)

A Extend the first toe (metatarsophalangeal joint)

O Calcaneus (dorsal surface)

I Proximal phalange of first toe

N Peroneal

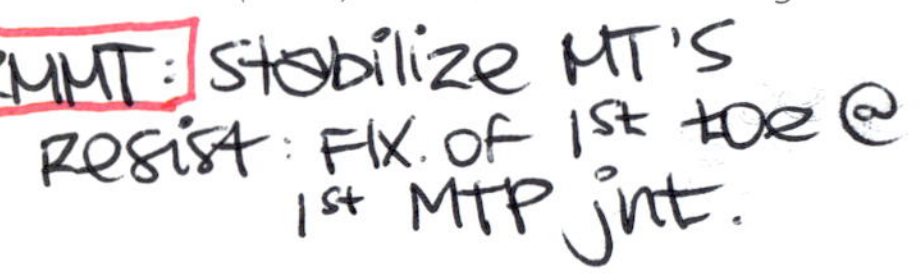

Flexor Hallucis Brevis (right)

A Flex the first toe (metatarsophalangeal joint)

O Plantar surfaces of cuboid and lateral cuneiform

I Medial and lateral sides of base of proximal phalange of first toe

N Medial plantar

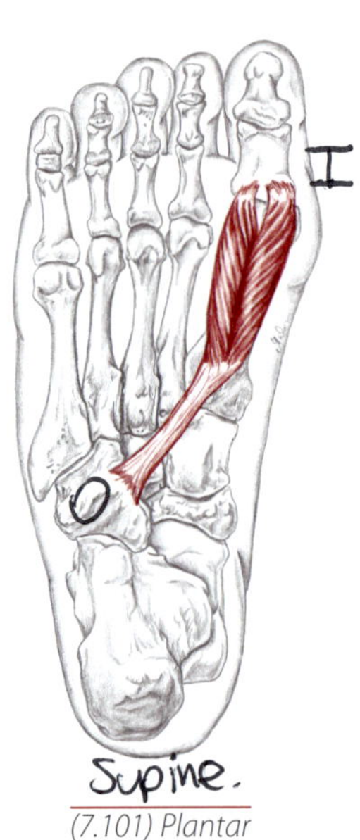

(7.101) Plantar view of right foot

Adductor Hallucis (left)

A Adduct the first toe

Assist to maintain transverse arch of foot

O *Oblique head:*
Bases of second through fourth metatarsals
Transverse head:
Plantar ligament of third through fifth metatarsophalangeal joints

I Lateral surface of base of proximal phalange of first toe and lateral sesamoid bone

N Lateral plantar

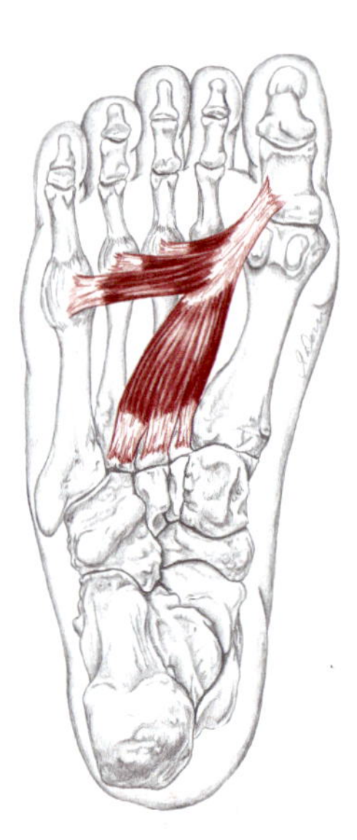

(7.102) Plantar view of right foot

Flexor Digiti Minimi (right)

A Flex the fifth toe (metatarsophalangeal joint)

O Base of fifth metatarsal

I Base of proximal phalange of fifth toe

N Lateral plantar

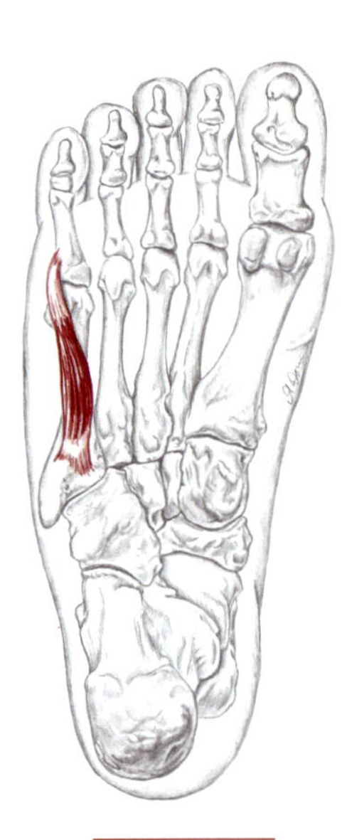

(7.103) Plantar view of right foot

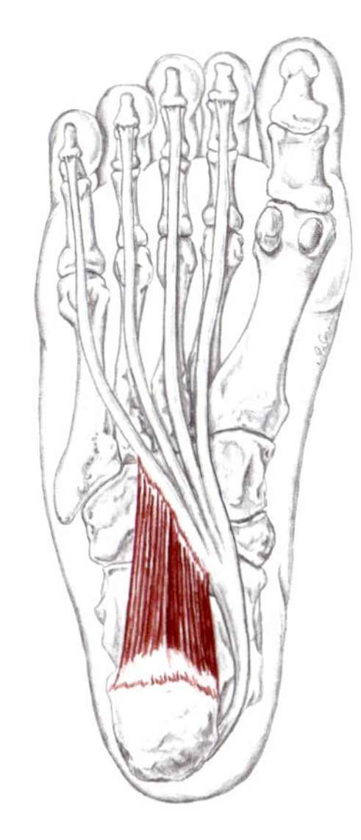

(7.104) Plantar view of right foot

Quadratus Plantae (left)

A Assist flexor digitorum longus to flex second through fifth toes

O Plantar surface of calcaneus

I Posterior, lateral aspect of flexor digitorum longus tendon

N Lateral plantar

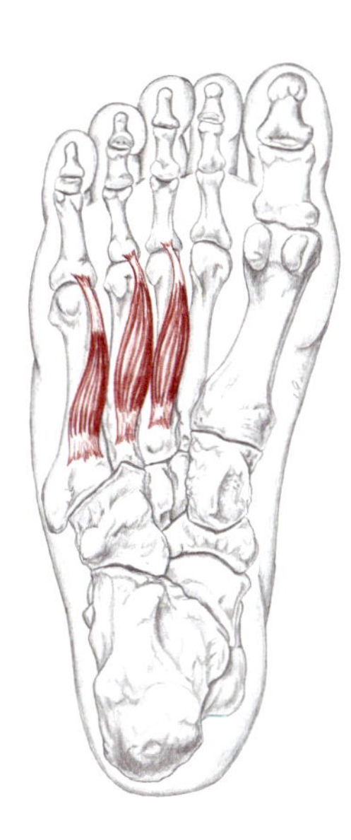

(7.105) Plantar view of right foot

Plantar Interossei (right)

A Adduct third through fifth toes (metatarsophalangeal joints)

Flex third through fifth toes (metatarsophalangeal joints)

O Bases of third through fifth metatarsals

I Medial surfaces of proximal phalanges of third through fifth toes

N Lateral plantar

MMT: w/Lumbricals

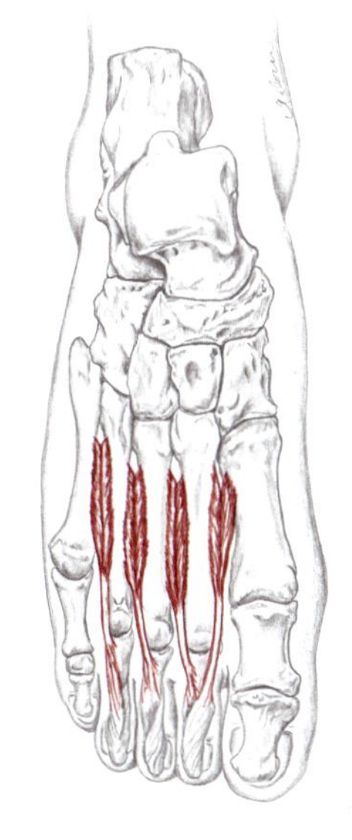

(7.106) Dorsal view of right foot

Dorsal Interossei (left)

A Abduct second through fourth toes (metatarsophalangeal joints)

Flex second through fourth toes (metatarsophalangeal joints)

O Shafts of first through fifth metatarsals

I *First:*
Medial surface of proximal phalange of second toe
Second through fourth:
Lateral surfaces of proximal phalange of second through fourth toes

N Lateral plantar

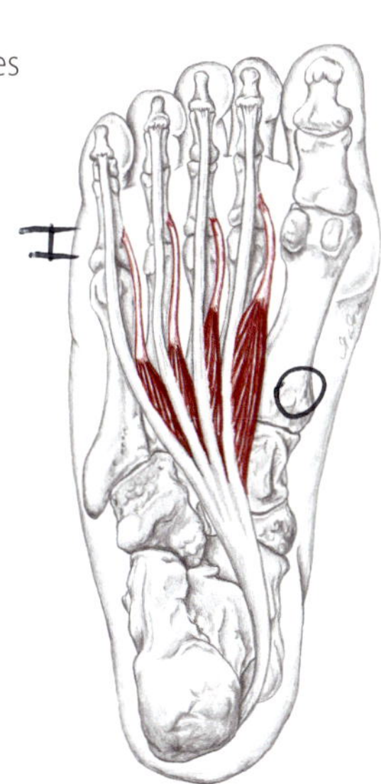

(7.107) Plantar view of right foot

Lumbricals of the Foot (right)

A Flex the proximal phalanges of the second through fifth toes at the metatarsophalangeal joints

Extend the middle and distal phalanges of the second through fifth toes at the interphalangeal joints

O Tendons of flexor digitorum longus

I Bases of proximal phalanges of second through fifth toes and expansions of extensor digitorum longus tendons (on dorsal surface of the toes)

N Medial and lateral plantar

RMMT: Stabilize MT's.
→Resist: Flx of 2-5 @ MTP jnts. +
→Resist ext. of IP jnts @ distal phalanges.

Other Structures of the Knee and Leg

Tibiofemoral Joint

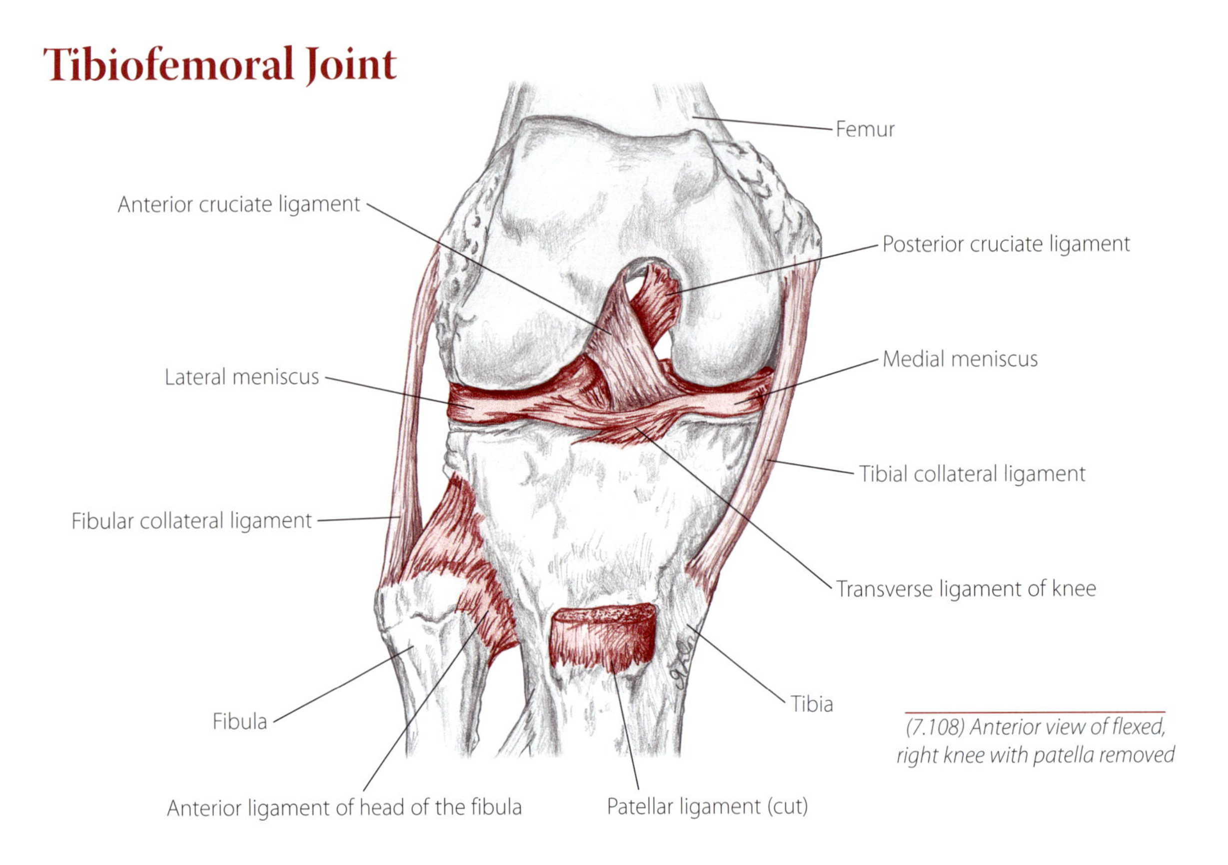

(7.108) Anterior view of flexed, right knee with patella removed

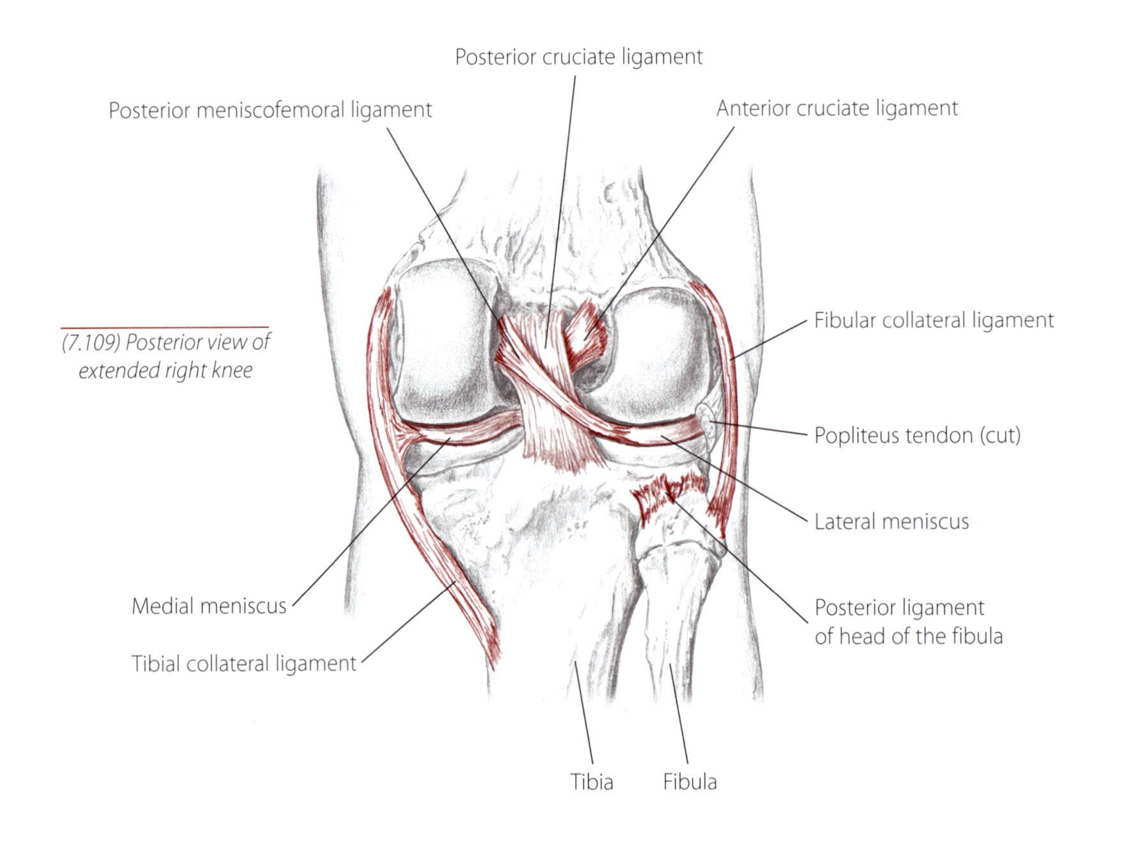

(7.109) Posterior view of extended right knee

cruciate	**kroo**-she-at	L. cross-shaped
meniscofemoral	men-**is**-ko-fem-**or**-al	

Tibiofemoral and Tibiofibular Joints

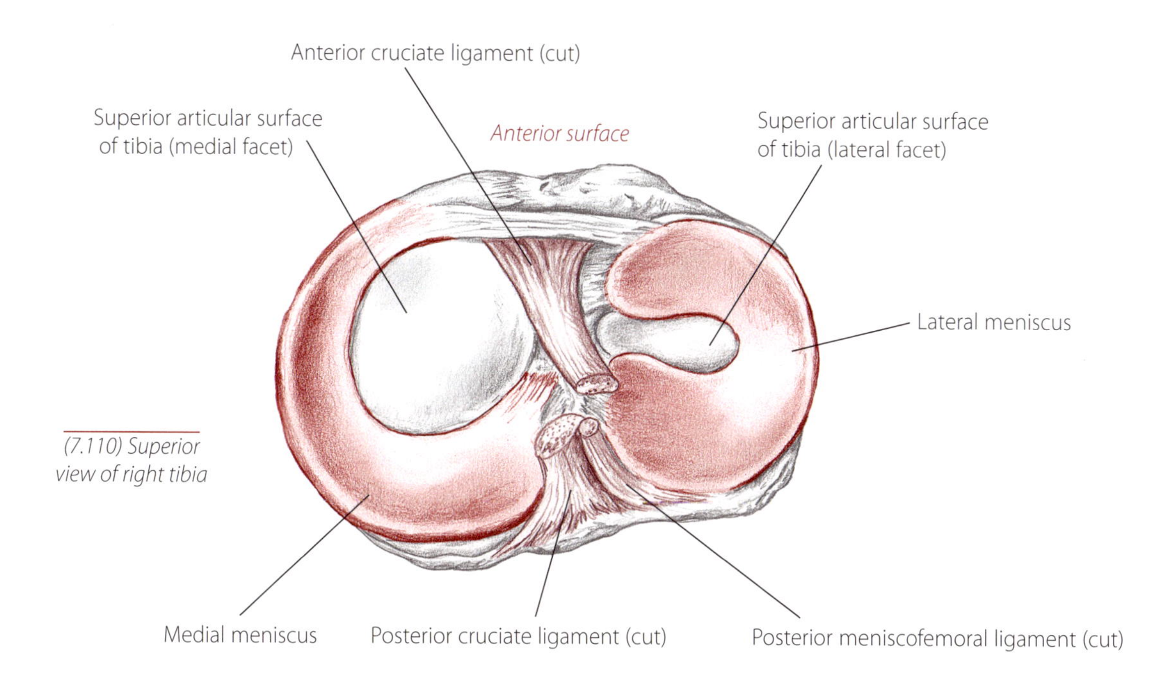

(7.110) Superior view of right tibia

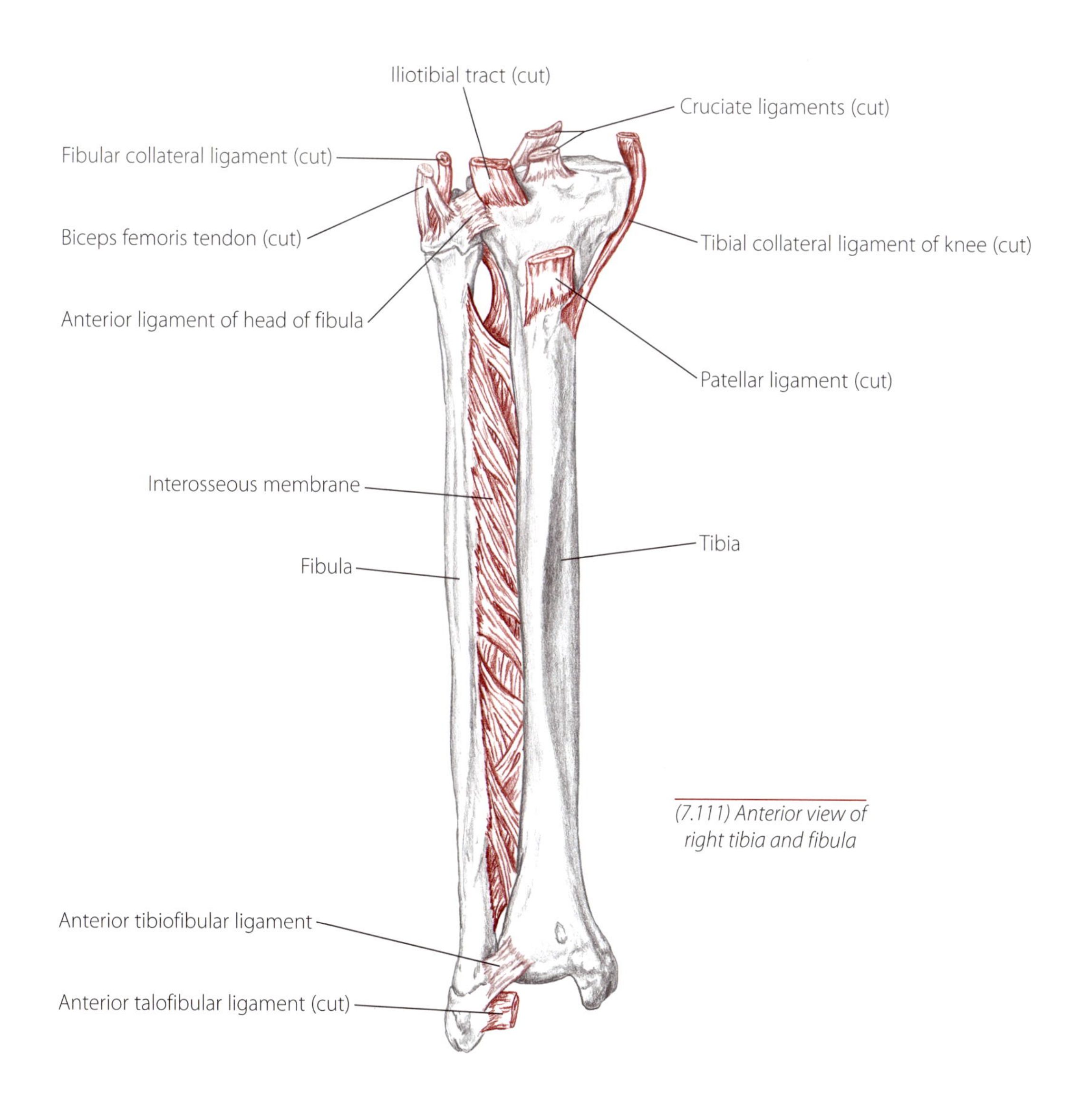

(7.111) Anterior view of right tibia and fibula

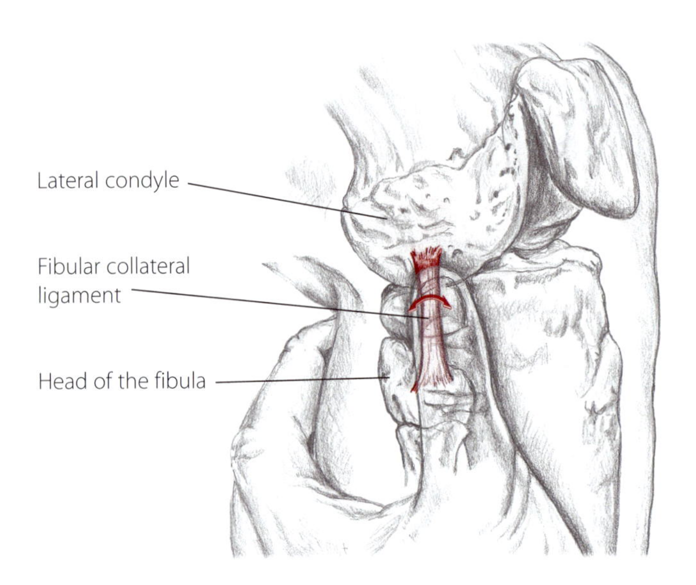

(7.112) Lateral view of right knee

Fibular and Tibial Collateral Ligaments

The **fibular collateral ligament** is a strong, thin strap that crosses the knee joint running from the lateral epicondyle of the femur to the head of the fibula (7.108). It is superficial and located between the biceps femoris tendon and the iliotibial tract.

The broad **tibial collateral ligament** lies superficial to the joint capsule of the knee, but may not be as easy to isolate as its lateral counterpart. Stretching nearly two inches distal to the knee joint, it is deep to the pes anserinus tendon (7.108).

Both collateral ligaments resist medial rotation of the knee. The fibular collateral also stabilizes the knee against genu varum stresses (often seen in bowlegged cowboys) while the tibial collateral protects against genu valgum (knock-knee) stresses. An example would be a blow from a football helmet to the lateral side of the knee joint.

Fibular collateral ligament

1) Partner seated with knee flexed. Locate the head of the fibula and the lateral epicondyle (7.112).
2) Slide your finger between these points and gently strum in a horizontal direction across this superficial ligament.

Ask your partner to cross his leg so the ankle is resting on top of the opposite knee. This position will allow the fibular collateral ligament to be easily accessed. Roll your finger between the epicondyle and the head of the fibula and palpate the ligament (7.113).

Is the band of tissue you feel the width of a pencil? Does it run from the epicondyle to the fibular head? Is it anterior to the biceps femoris tendon?

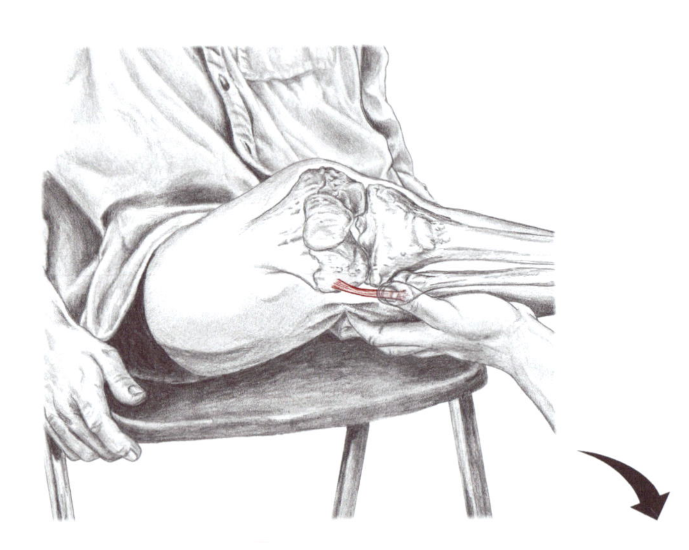

(7.113) An alternate method to locating the fibular collateral ligament

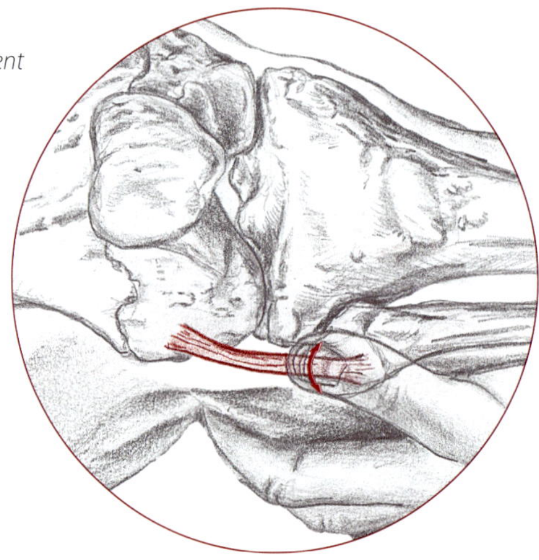

collateral	ko-**lat**-er-al	L. of both sides
genu valgum	**je**-noo **val**-gum	
genu varum	**je**-noo **va**-rum	

Tibial collateral ligament

1) Partner seated with the knee flexed. Locate the medial epicondyle of the femur. Slide distally to the joint space, the thin crevice between the tibia and femur.
2) Strum your fingertip horizontally across this space, exploring for the broad fibers of the ligament (7.114).

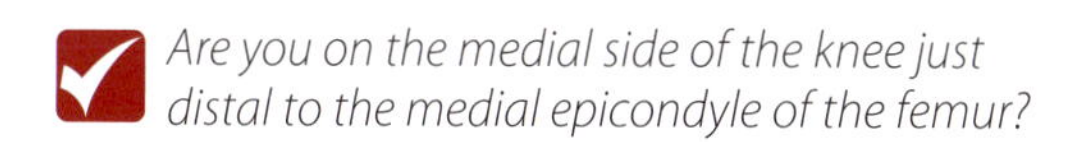

Are you on the medial side of the knee just distal to the medial epicondyle of the femur?

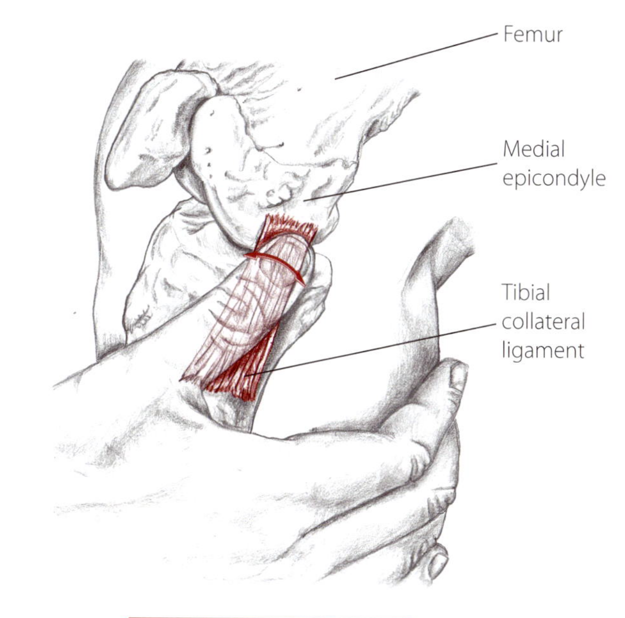

(7.114) Medial view of right knee

Menisci of the Knee

The menisci are fibrocartilaginous discs attached to the tibial condyles (7.110). They are not only important for weight distribution and friction reduction, but also help the round femoral condyles sit comfortably upon the flat tibial plateaus. The edge of the medial meniscus can be palpated just above the edge of the medial tibial plateau. The smaller, more mobile lateral meniscus is difficult to access.

Medial meniscus

1) Partner seated with the knee flexed. Place your thumb superior to the medial tibial plateau in the joint space between the femur and tibia.
2) Grasp the leg with your other hand and slowly rotate the knee medially (7.115).
3) As the medial side of the tibia rotates posteriorly, the edge of the medial meniscus will be pushed anteriorly into your thumb. The sensation may be quite subtle - a gentle pressure against your thumbpad.

Is your thumb in the knee joint space? If you slowly switch from lateral to medial rotation of the knee, do you feel a difference under your thumb?

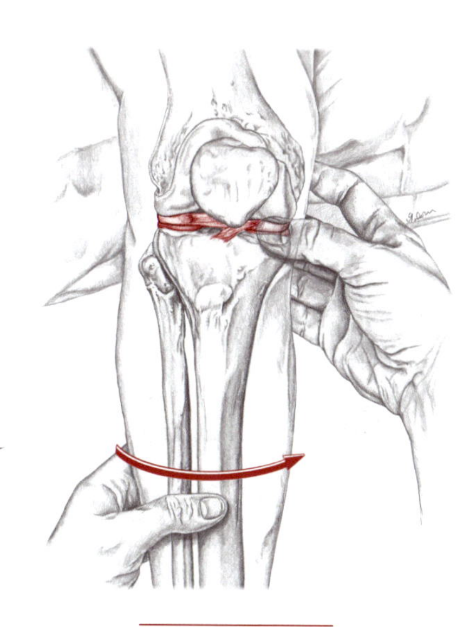

(7.115) Anterior view of flexed right knee

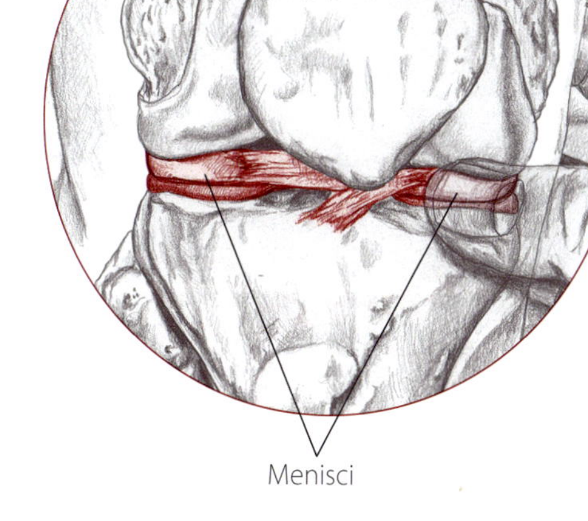

meniscus	men-**is**-kus	Grk. crescent-shaped
menisci	men-**is**-ki	plural for meniscus

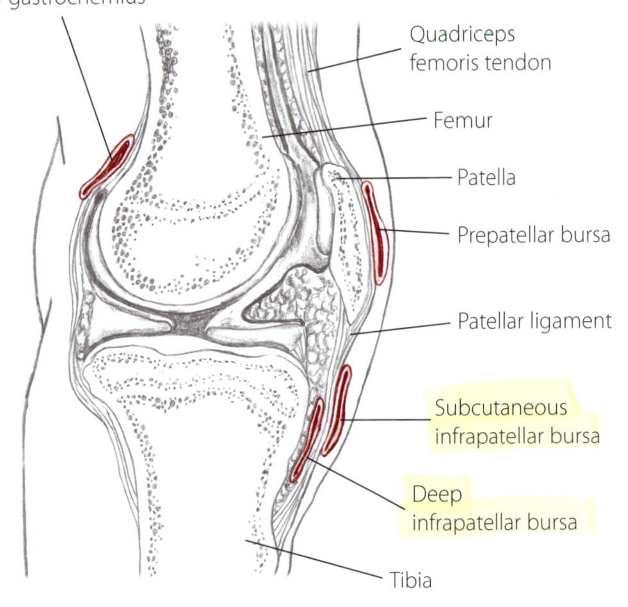

(7.116) Lateral cross section of the knee, pes anserine bursa not shown

Bursae of the Knee

Tremendous pressures, repetitive movements and chafing tendons constantly challenge the knee joint. Helping to protect the joint and its surrounding structures are nearly a dozen bursae. The primary bursae are included here (7.116).

Lying superficial to the patella, the **prepatellar bursa** helps the skin to move freely over the patella, even with the knee flexed. Excessive kneeling and squatting can inflame the vulnerable bursa, a condition called "housemaid's knee" (or "carpenter's knee," depending on your vocation).

The **subcutaneous infrapatellar bursa** and the **deep infrapatellar bursa** are located on either side of the patellar ligament. The deep bursa reduces friction between the tendon and the surface of the tibia. The superficial bursa can become irritated with prolonged kneeling, sometimes called "parson's knee" (or "Buddhist's knee," depending on your faith).

The **pes anserine bursa** serves as a buffer between the pes anserius tendons (those of the gracilis, sartorius and semitendinosus) and the tibia.

When inflamed, the superficial bursae of the knee are palpable and sometimes visible; under normal conditions, however, they are impalpable.

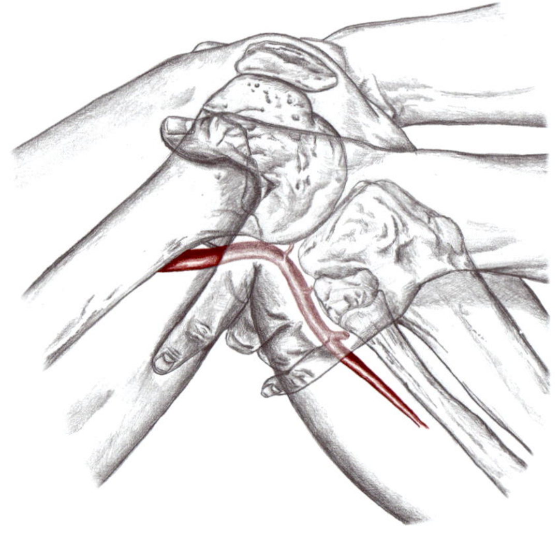

(7.117) Lateral view of right knee, feeling the pulse of the popliteal artery

Popliteal Artery

The popliteal artery branches from the femoral artery to pass through the popliteal fossa at the back of the knee. It is situated deep in the fossa and, for this reason, its pulse can be difficult to detect.

1) Partner supine. Flex your partner's knee in order to soften the overlying tissues. Hold the knee so that the fingertips of both hands are at the midline of the posterior knee.
2) Sink your fingerpads deep into the popliteal fossa and explore for the subtle pulse (7.117).

If the pulse is undetectable, follow the same instructions with your partner prone.

Common Peroneal Nerve

Branching off from the sciatic nerve, the peroneal nerve courses superficially along the posterior/lateral side of the knee. Roughly the diameter of a thick piece of spaghetti, it lies medial to the biceps femoris tendon and lateral to the gastrocnemius belly. It becomes particularly accessible (and vulnerable) along the posterior surface of the head of the fibula (7.118).

Common peroneal nerve

1) Partner prone. Passively flex the knee and locate the biceps femoris tendon and head of the fibula.
2) Gently roll your thumb from side to side, exploring the region just distal to the biceps tendon, on the posterior surface of the fibular head.
3) Distinguish between the slender, slightly mobile nerve and the gastrocnemius fibers by asking your partner to gently flex her knee against your resistance. The nerve, of course, will remain soft and mobile, while the muscle fibers will become taut (7.119).

Locate the biceps femoris tendon by asking your partner to flex her knee against your resistance. Follow the tendon to the head of the fibula, noting the nerve pathway that runs alongside it. If you follow the nerve past the head of the fibula, does it continue down the lateral side of the leg?

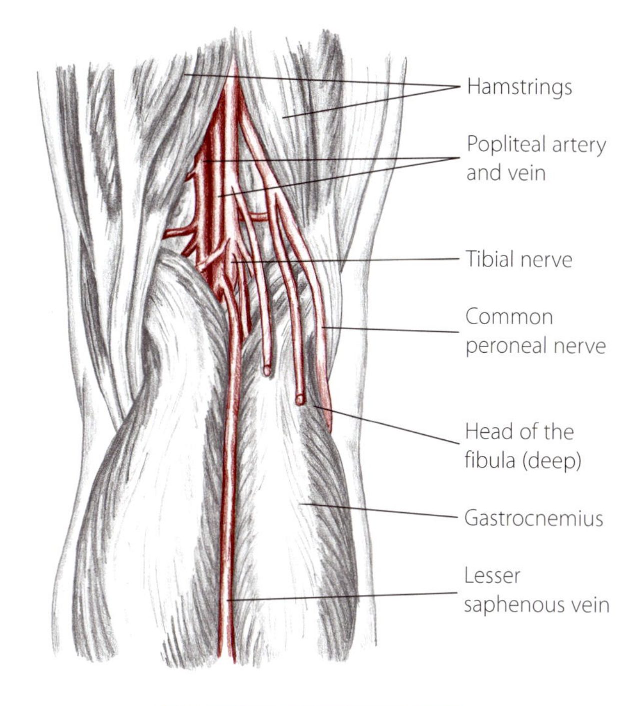

(7.118) Posterior view of right knee

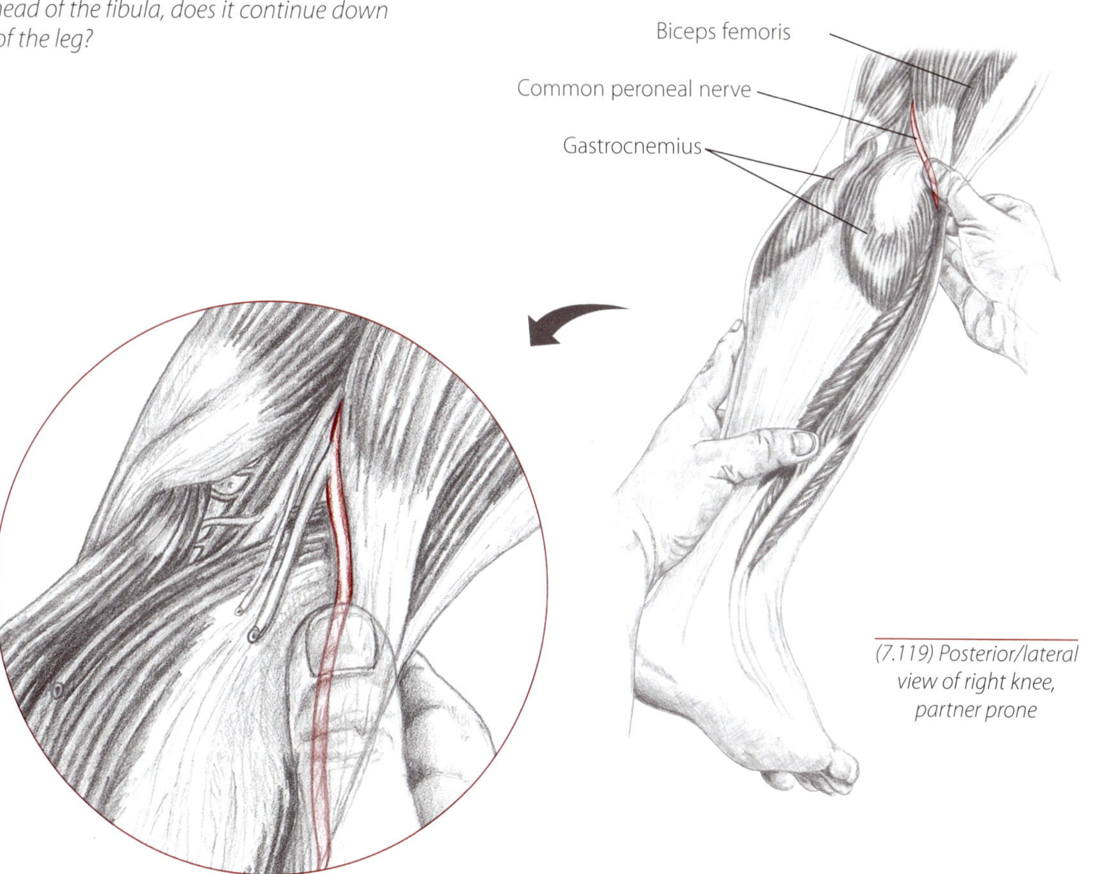

(7.119) Posterior/lateral view of right knee, partner prone

Other Structures of the Ankle and Foot

Talocrural Joint

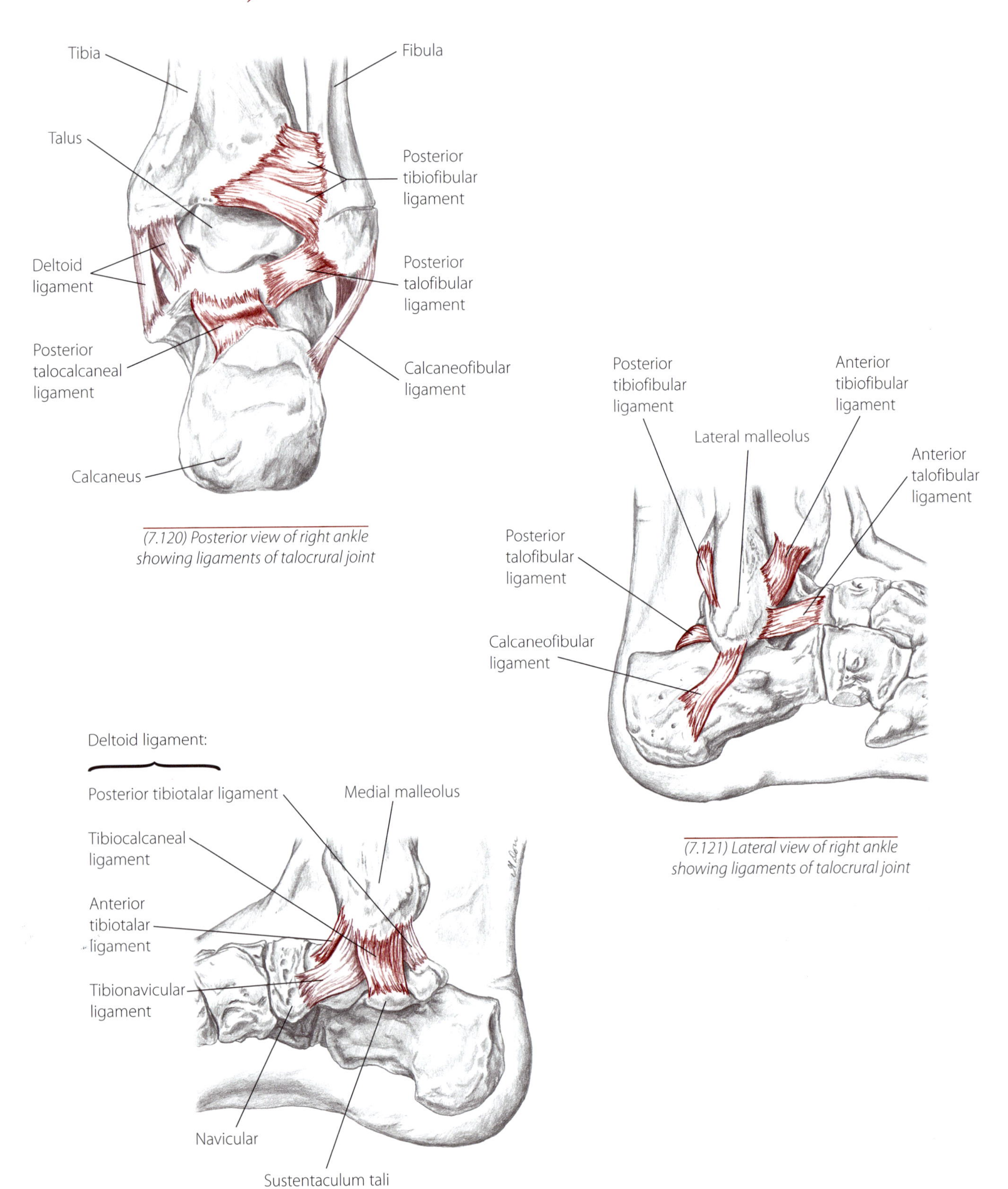

(7.120) Posterior view of right ankle showing ligaments of talocrural joint

(7.121) Lateral view of right ankle showing ligaments of talocrural joint

(7.122) Medial view of right ankle showing ligaments of talocrural joint

calcaneofibular	kal-**ka**-ne-o-**fib**-u-lar
talocalcaneal	**ta**-lo-kal-**ka**-ne-al
talofibular	**ta**-lo-**fib**-u-lar

Talotarsal Joints and Ligaments of the Foot

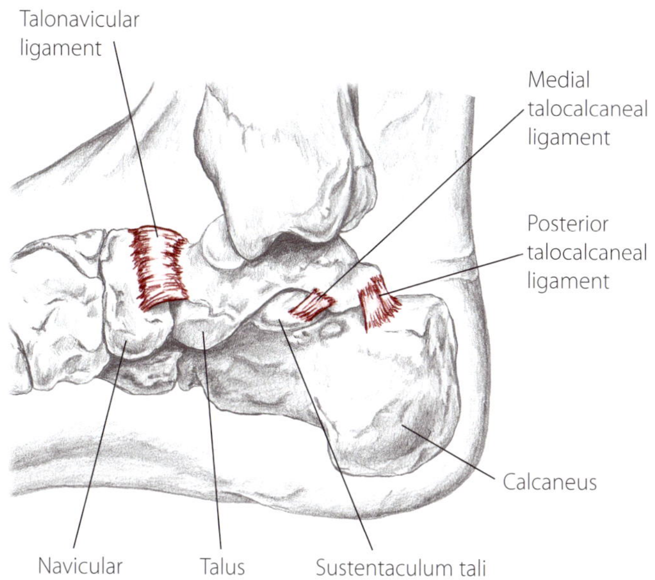

(7.123) Medial view of right ankle showing ligaments of talotarsal joints

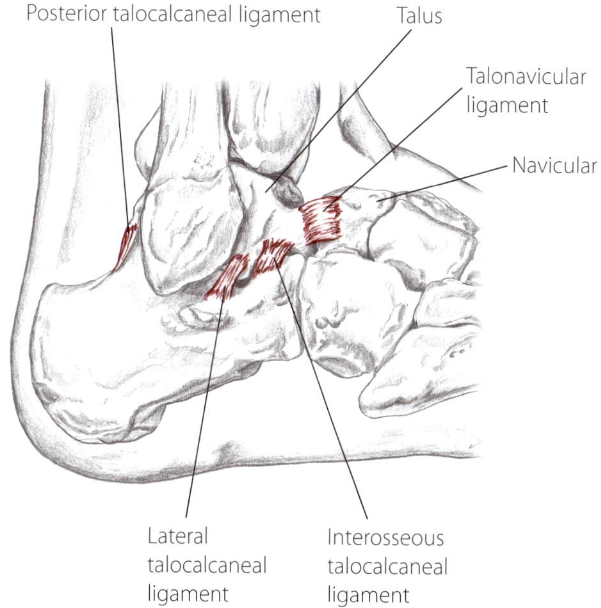

(7.124) Lateral view of right ankle showing ligaments of talotarsal joints

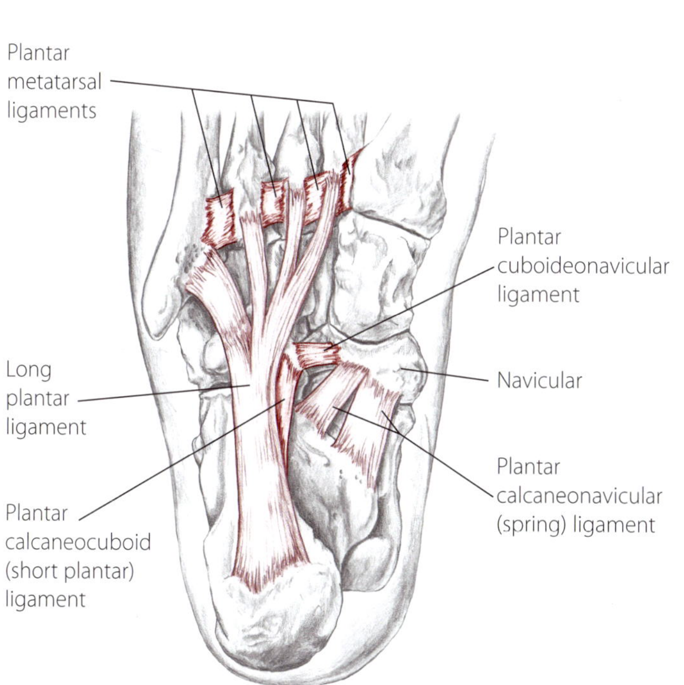

(7.125) Plantar view of right foot showing ligaments of foot

calcaneocuboid	kal-**ka**-ne-o-**ku**-boyd
cuboideonavicular	**ku**-boyd-e-o-na-**vik**-u-lar
talonavicular	**ta**-lo-na-**vik**-u-lar

Ligaments of the Foot

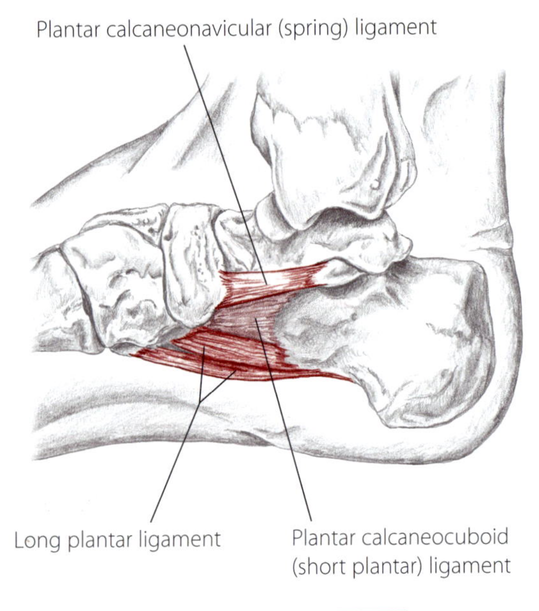

(7.126) Medial view of right ankle showing ligaments of subtalar joints

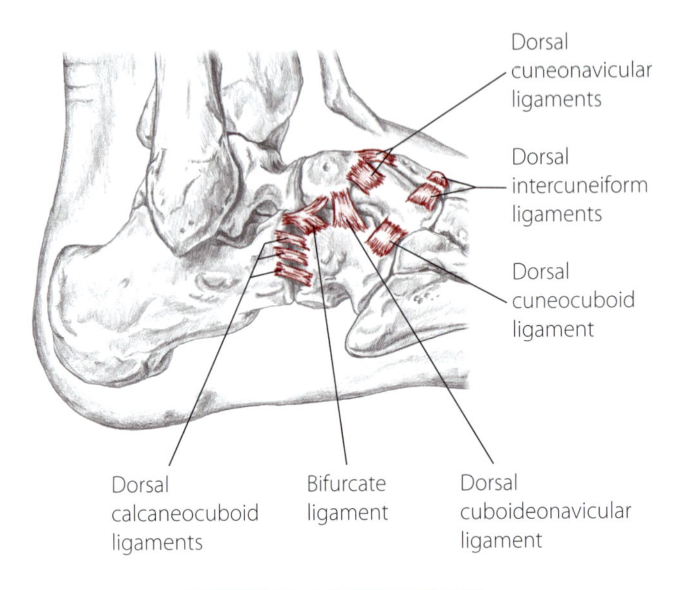

(7.127) Lateral view of right ankle showing ligaments of intertarsal joints

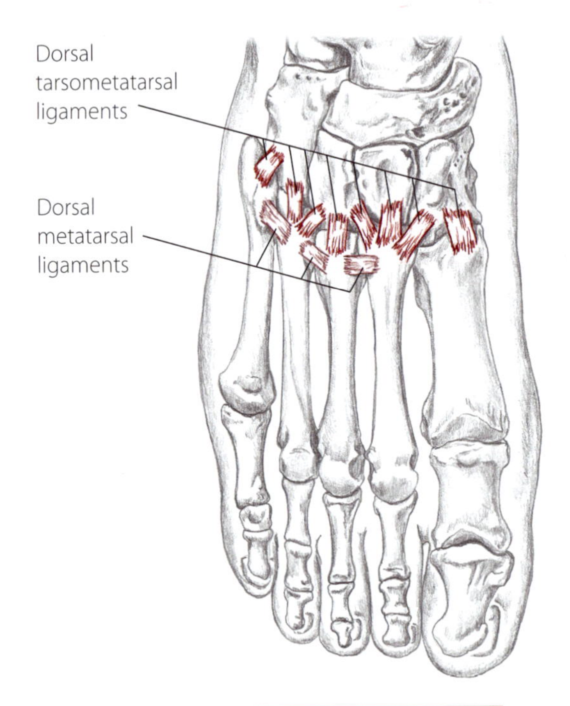

(7.128) Dorsal view of right foot showing ligaments of tarsometatarsal and intermetatarsal joints

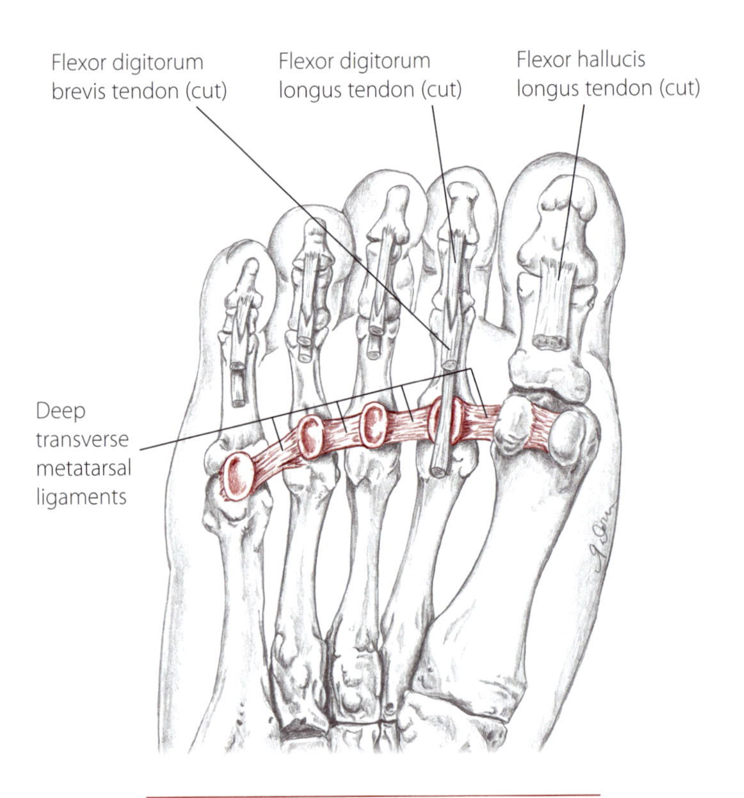

(7.129) Plantar view of right foot showing ligaments of metatarsophalangeal and interphalangeal joints

Deltoid Ligament

The deltoid ligament is composed of several ligaments that originate at the medial malleolus and fan distally to attach at the talus, sustentaculum tali and navicular (7.130). The ligament is designed to protect against medial stress of the talocrural joint. The deltoid ligament is deep to the flexor retinaculum and flexor tendons (p. 374), yet is palpable.

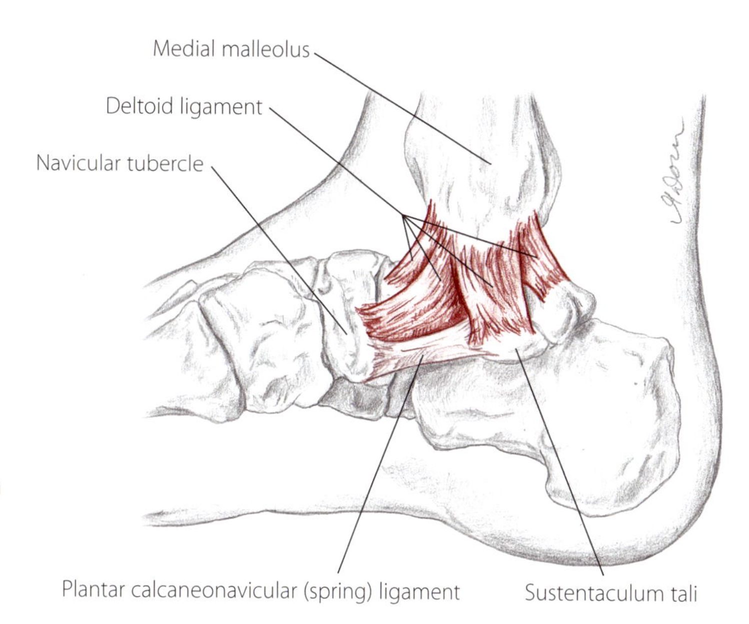

(7.130) Medial view of right ankle

1) Partner supine or seated. Locate the medial malleolus and sustentaculum tali.
2) Place your finger between these points and strum horizontally to isolate the fibers of the ligaments.
3) Slide distally from the medial malleolus at a 45° angle and palpate its angled fibers to define the anterior and posterior aspects of the deltoid ligament (7.131).

Are you palpating in the space between the medial malleolus and sustentaculum tali? Do the fibers you feel fan out from the medial malleolus and have a firm, dense texture?

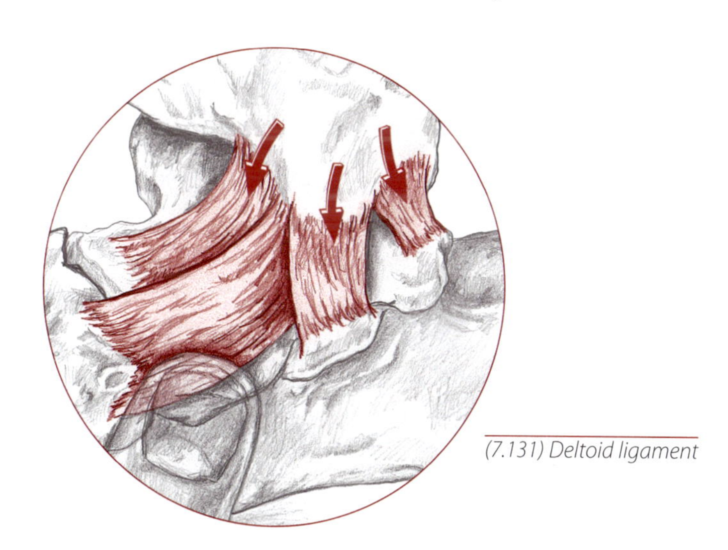

(7.131) Deltoid ligament

Plantar Calcaneonavicular (Spring) Ligament

The spring ligament is a small, tough band of tissue that plays an important role in stabilizing the medial longitudinal arch of the foot (7.130). Located along the medial side of the foot, the ligament stretches from the sustentaculum tali to the navicular tubercle and may be positioned deep to the tibialis posterior tendon. The spring ligament may be extremely tender and should be accessed slowly. Be sure to communicate with your partner.

1) Supine or seated. Passively invert the foot to soften any surrounding tissue and locate the sustentaculum tali and navicular tubercle.
2) Palpating between these bony landmarks, use a fingertip to slowly explore the taut surface of the spring ligament (7.132).

Are you between the sustentaculum tali and navicular tubercle? Can you roll your fingertip slowly across the surface of the ligament?

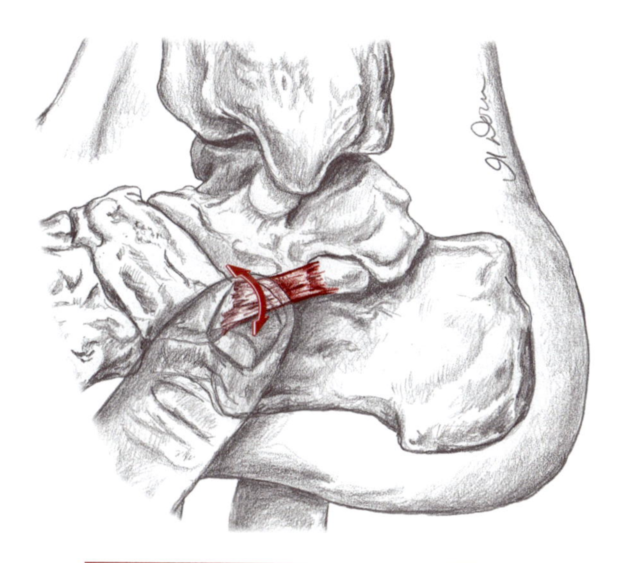

(7.132) Medial view, palpating the spring ligament

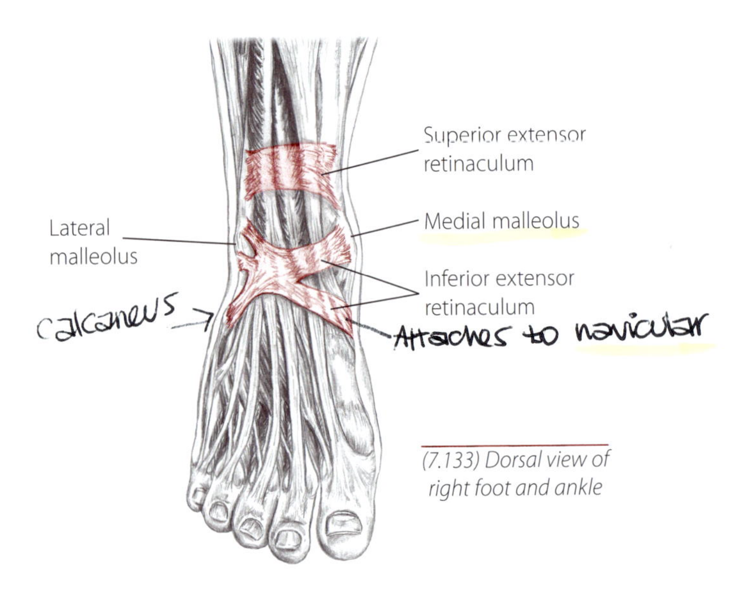

(7.133) Dorsal view of right foot and ankle

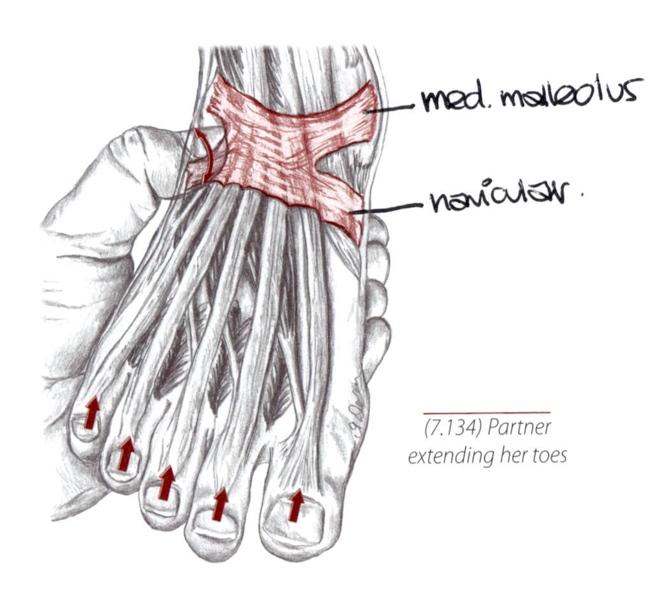

(7.134) Partner extending her toes

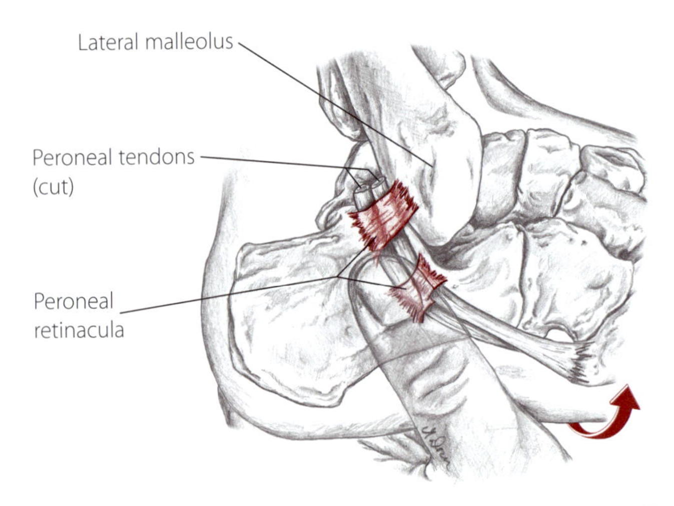

(7.135) Lateral view of right ankle, partner everting foot

Retinacula of the Ankle

The tendons of the extensor muscles (p. 371) are supported by the superior and inferior extensor retinacula. The **superior extensor retinaculum** is broad and crosses the front of the ankle just proximal to the malleoli. The **inferior extensor retinaculum** is Y-shaped and begins distal to the lateral malleolus on the calcaneus. It spans the ankle and then divides, with one fork attaching at the medial malleolus and the other connecting to the navicular (7.133).

The peroneal muscles are stabilized by the superior and inferior peroneal retinacula. The **superior peroneal retinaculum** stretches from the lateral malleolus to the calcaneus, and the **inferior peroneal retinaculum** pins the peroneal tendons down at the peroneal tubercle.

The **flexor retinaculum** is a broad strap extending from the medial calcaneus to the medial malleolus. It is designed to hold in place the tendons of the flexor muscles and the tibial artery and nerve (7.136).

Extensor retinacula

1) Partner supine. Ask your partner to dorsiflex her ankle and extend her toes. The pressure from the bulging tendons will make the retinacula more pronounced.
2) Palpate an inch proximal to the medial malleolus for the broad fibers of the superior extensor retinaculum.
3) Locate the inferior extensor retinaculum by moving distally to the level of the malleolus (7.134). Explore either side of the large tibialis anterior tendon for easy access to this retinaculum.

Are the fibers you are accessing superficial and perpendicular to the extensor tendons? Ask your partner to relax her ankle and notice how the retinacula soften.

Peroneal retinacula

1) Ask your partner to evert her foot. The tension from the peroneal tendons will make the retinacula more pronounced.
2) Locate the peroneal tendons between the lateral malleolus and lateral calcaneus (7.135). Roll your fingers along either side of the peroneal tendons to feel the small, short fibers of the retinacula.

For the superior retinaculum, does the tissue you feel strap across the peroneal tendons from the lateral calcaneus to the lateral malleolus? For the inferior retinaculum, do you feel a short band crossing over the peroneal tubercle?

retinaculum	**ret**-i-**nak**-u-lum	L. halter, band, rope
retinacula	**ret**-i-**nak**-u-la	plural for retinaculum

Flexor retinaculum

1) Ask your partner to dorsiflex and invert her foot. The tension from the flexor tendons will elevate the retinaculum closer to the surface.
2) Locate the medial malleolus and the medial side of the calcaneus.
3) Palpate between these landmarks, strumming across the broad, superficial fibers of the flexor retinaculum (7.136).

Are you between the medial calcaneus and medial malleolus? Continue to explore the retinaculum with the foot relaxed.

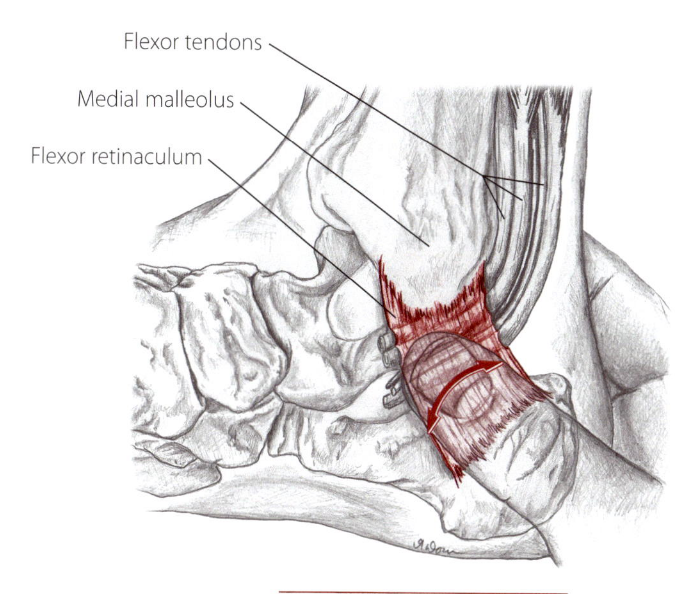

(7.136) Medial view of right ankle

Plantar Aponeurosis

The plantar aponeurosis is a thick, superficial band of fascia stretching from the heel to the ball of the foot (7.137). Originating from the tuberosity of the calcaneus and expanding toward the metatarsal heads, it is important for supporting the longitudinal arch of the foot. Because the aponeurosis is located between the skin and muscles of the foot, it can be difficult to isolate from the surrounding tissues.

1) Partner seated or supine. Crossing the ball of the foot, draw an imaginary triangle extending down to the heel.
2) Within this triangle explore the superficial layers of tissue along the sole of the foot. Passively flex and extend the toes, noting how this movement affects the tension of the plantar aponeurosis.

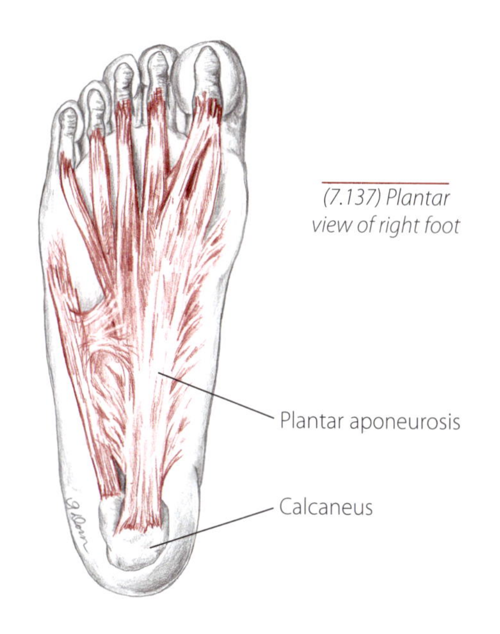

(7.137) Plantar view of right foot

Posterior Tibial Artery

The posterior tibial artery extends from the popliteal artery. It is superficial and its pulse can be felt just inferior and posterior to the medial malleolus.

1) Partner supine. Locate the medial malleolus. Using two fingerpads, slide posterior to the malleolus and feel for the pulse of the artery (7.138).

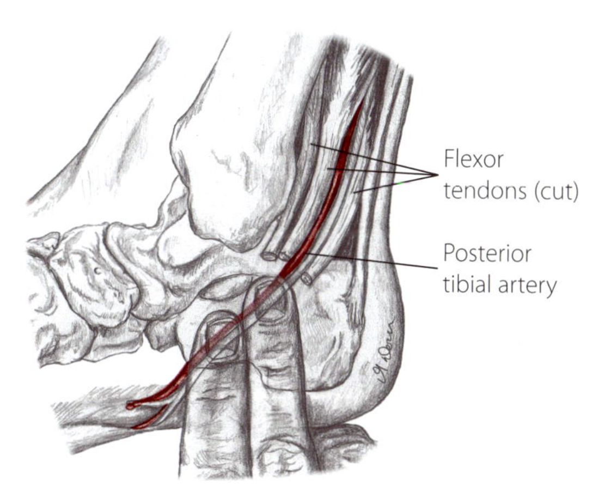

(7.138) Medial view of right ankle

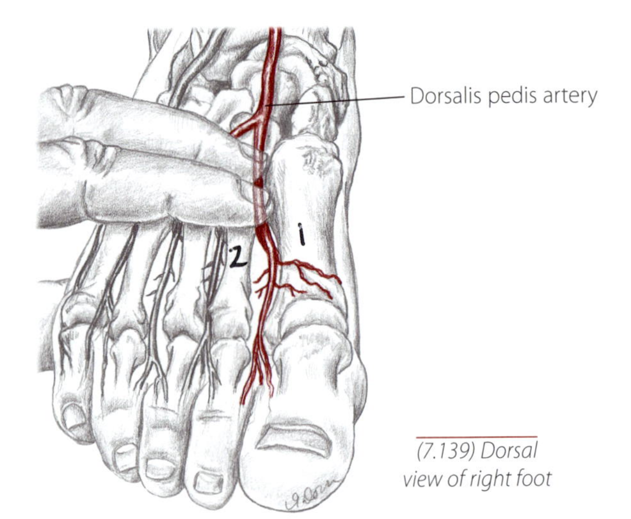

(7.139) Dorsal view of right foot

Dorsalis Pedis Artery

Located between the first and second metatarsal bones, the dorsalis pedis artery lies superficial along the dorsal side of the foot.

1) Partner supine. Locate the first and second metatarsals. Place two fingerpads between the two bones and, using gentle pressure, explore for the pulse of the dorsalis pedis artery (7.139).

Are you lateral to the extensor hallucis longus tendon? If the pulse is undetectable, move slightly lateral.

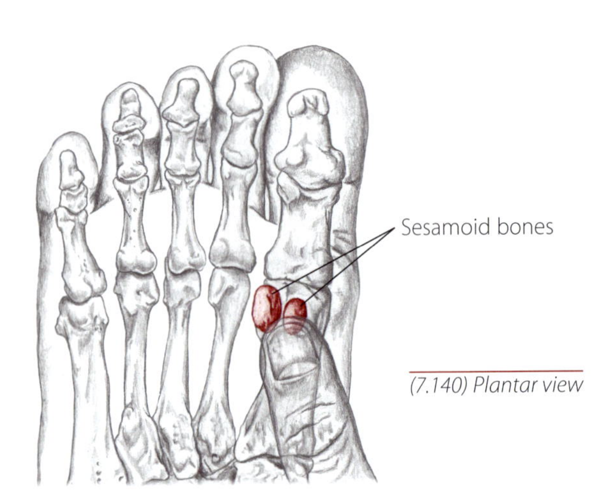

(7.140) Plantar view

Sesamoid Bones of First Metatarsal

The sesamoid bones of the first metatarsal are located along the plantar surface of the first metatarsal head. Usually there are two of these bones, but sometimes more are present. The sesamoids are spherical and imbedded in the tendon of the flexor hallucis brevis. Often only their location and density, not their specific shapes, are palpable.

1) Partner seated or supine. Locate the head of the first metatarsal. Slide around to its plantar surface at the ball of the foot.
2) Using your thumbpad, explore this surface for the small sesamoid bones. Passively flex and extend the first toe to soften the surrounding tissues (7.140).

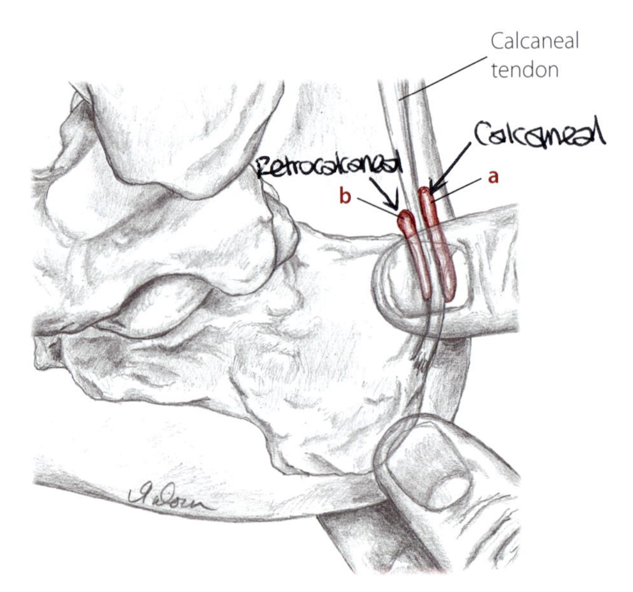

(7.141) Medial view of right foot

Calcaneal Bursae

The calcaneal bursa **(a)** is located between the attachment of the calcaneal tendon and the overlying skin. The retrocalcaneal bursa **(b)** is positioned on the opposite side of the calcaneal tendon. It serves to cushion the tendon against the calcaneus.

Both bursae are impalpable; however, they can become thick or inflamed from tight shoes and, especially, from high heels.

1) Partner seated or supine. With the ankle in a neutral position, locate the calcaneal tendon.
2) Follow it distally until it merges into the calcaneus. Gently squeeze the skin posterior to the tendon. This is where the calcaneal bursa is located (7.141).
3) Shift your fingers between the tendon and calcaneus. If the bursae are healthy, they will not be palpable.

dorsalis pedis — **dor**-sal-is **peh**-dis — L. *dorsum*, back; *pedis*, foot

aponeurosis — **ap**-o-nu-**ro**-sis — Grk. *apo*, from + *neuron*, nerve or tendon

NOTES

NOTES

Now back to the beginning...

Synergists - Muscles Working Together

Shoulder (p. 71-73)

(glenohumeral joint)

Flexion
Deltoid (anterior fibers)
Pectoralis major (upper fibers)
Biceps brachii
Coracobrachialis

Extension
Deltoid (posterior fibers)
Latissimus dorsi
Teres major
Infraspinatus
Teres minor
Pectoralis major (lower fibers)
Triceps brachii (long head)

Horizontal Abduction
Deltoid (posterior fibers)
Infraspinatus
Teres minor

Horizontal Adduction
Deltoid (anterior fibers)
Pectoralis major (upper fibers)

Abduction
Deltoid (all fibers)
Supraspinatus

Adduction
Latissimus dorsi
Teres major
Infraspinatus
Teres minor
Pectoralis major (all fibers)
Triceps brachii (long head)
Coracobrachialis

Lateral Rotation (external rotation)
Deltoid (posterior fibers)
Infraspinatus
Teres minor

Medial Rotation (internal rotation)
Deltoid (anterior fibers)
Latissimus dorsi
Teres major
Subscapularis
Pectoralis major (all fibers)

Scapula (p. 73-74)

(scapulothoracic joint)

Elevation
Trapezius (upper fibers)
Rhomboid major
Rhomboid minor
Levator scapula

Depression
Trapezius (lower fibers)
Serratus anterior (with the origin fixed)
Pectoralis minor

Adduction (retraction)
Trapezius (middle fibers)
Rhomboid major
Rhomboid minor

Abduction (protraction)
Serratus anterior (with the origin fixed)
Pectoralis minor

Upward Rotation
Trapezius (upper and lower fibers)

Downward Rotation
Rhomboid major
Rhomboid minor
Levator scapula

Elbow (p. 138)

(humeroulnar and humeroradial joints)

Flexion
Biceps brachii
Brachialis
Brachioradialis
Flexor carpi radialis
Flexor carpi ulnaris (assists)
Palmaris longus
Pronator teres (assists)
Extensor carpi radialis longus (assists)
Extensor carpi radialis brevis (assists)

Extension
Triceps brachii (all heads)
Anconeus

Forearm (p. 138)

(proximal and distal radioulnar joints)

Supination
Biceps brachii
Supinator
Brachioradialis (assists)

Pronation
Pronator teres
Pronator quadratus
Brachioradialis (assists)

Wrist (p. 138-139)

(radiocarpal joint)

Extension
Extensor carpi radialis longus
Extensor carpi radialis brevis
Extensor carpi ulnaris
Extensor digitorum (assists)

Flexion
Flexor carpi radialis
Flexor carpi ulnaris
Palmaris longus
Flexor digitorum superficialis
Flexor digitorum profundus (assists)

Abduction (radial deviation)
Extensor carpi radialis longus
Extensor carpi radialis brevis
Flexor carpi radialis

Adduction (ulnar deviation)
Extensor carpi ulnaris
Flexor carpi ulnaris

Hand and Fingers (p. 139)

(metacarpophalangeal, proximal and distal interphalangeal joints)

Flexion of the second through fifth fingers
Flexor digitorum superficialis
Flexor digitorum profundus
Flexor digiti minimi brevis (5th)
Lumbricals
Dorsal interossei (2nd - 4th) (assists)
Palmar interossei (2nd, 4th, 5th) (assists)

Extension of the second through fifth fingers
Extensor digitorum
Extensor indicis (2nd)
Lumbricals
Dorsal interossei (2nd - 4th) (assists)
Palmar interossei (2nd, 4th, 5th) (assists)

Abduction of the second through fifth fingers
Dorsal interossei (2nd - 4th)
Abductor digiti minimi (5th)

Adduction of the second through fifth fingers
Palmar interossei (2nd, 4th, 5th)
Extensor indicis (2nd) (assists)

Opposition of the fifth finger
Opponens digiti minimi
Abductor digiti minimi (assists)
Flexor digiti minimi brevis (assists)

Thumb (p. 139)

(first carpometacarpal and metacarpophalangeal joints)

Flexion
Flexor pollicis longus
Flexor pollicis brevis
Adductor pollicis (assists)
Palmar interossei (1st) (assists)

Extension
Extensor pollicis longus
Extensor pollicis brevis
Abductor pollicis longus
Palmar interossei (1st) (assists)

Abduction
Abductor pollicis longus
Abductor pollicis brevis

Adduction
Adductor pollicis
Palmar interossei (1st)

Opposition
Opponens pollicis
Flexor pollicis brevis (assists)
Abductor pollicis brevis (assists)

Vertebral Column (p. 200-201)

Flexion
Rectus abdominis
External oblique (bilaterally)
Internal oblique (bilaterally)

Extension
Spinalis (bilaterally)
Longissimus (bilaterally)
Iliocostalis (bilaterally)
Multifidi (bilaterally)
Rotatores (bilaterally)
Semispinalis capitis
Quadratus lumborum (assists)
Intertransversarii (bilaterally)
Interspinalis
Latissimus dorsi (when arm is fixed, p. 79)

Rotation (all unilaterally)
Multifidi (to the opposite side)
Rotatores (to the opposite side)
External oblique (to the opposite side)
Internal oblique (to the same side)

Lateral Flexion
(unilaterally to the **same** side)
Spinalis
Longissimus
Iliocostalis
Quadratus lumborum
External oblique
Internal oblique
Intertransversarii
Latissimus dorsi (p. 79)

Ribs/Thorax (p. 201)

Elevation/Expansion
Anterior scalene (bilaterally)
Middle scalene (bilaterally)
Posterior scalene (bilaterally)
Sternocleidomastoid (assists)
External intercostals (assists)
Serratus posterior superior
Pectoralis major
(may assist if arm is fixed)
Pectoralis minor (if scapula is fixed)
Serratus anterior (if scapula is fixed)
Subclavius (first rib)

Depression/Collapse
Internal intercostals (assists)
Serratus posterior inferior

Muscles of Inhalation
Diaphragm
Anterior scalene (bilaterally)
Middle scalene (bilaterally)
Posterior scalene (bilaterally)
Sternocleidomastoid (assist)
External intercostals (assists)
Serratus posterior superior
Quadratus lumborum
Pectoralis major (assist if arm is fixed)
Pectoralis minor (if scapula is fixed)
Serratus anterior (if scapula is fixed)
Subclavius (first rib)

Muscles of Exhalation
Internal intercostals (assists)
Serratus posterior inferior
External oblique
(by compressing abdominal contents)
Internal oblique
(by compressing abdominal contents)
Transverse abdominis
(by compressing abdominal contents)
Quadratus lumborum

Cervical Spine (p. 248-249)

Flexion
Sternocleidomastoid (bilaterally)
Anterior scalene (bilaterally)
Longus capitis (bilaterally)
Longus colli (bilaterally)

Extension
Trapezius - upper fibers (bilaterally)
Levator scapula (bilaterally)
Splenius capitis (bilaterally)
Splenius cervicis (bilaterally)
Rectus capitis posterior major
Rectus capitis posterior minor
Oblique capitis superior
Semispinalis capitis
Longissimus capitis (assists, p. 202)
Longissimus cervicis (assists, p. 202)
Iliocostalis cervicis (assists, p. 202)

Rotation (unilaterally to the **same** side)
Levator scapula
Splenius capitis
Splenius cervicis
Rectus capitis posterior major
Oblique capitis inferior
Longus colli
Longus capitis
Longissimus capitis (assists, p. 202)
Longissimus cervicis (assists, p. 202)
Iliocostalis cervicis (assists, p. 202)

Rotation (unilaterally to the **opposite** side)
Trapezius (upper fibers)
Sternocleidomastoid
Anterior scalene
Middle scalene
Posterior scalene

Lateral Flexion
(unilaterally to the **same** side)
Trapezius (upper fibers)
Levator scapula
Splenius capitis
Splenius cervicis
Sternocleidomastoid
Longus capitis
Longus colli
Anterior scalene (with ribs fixed)
Middle scalene (with ribs fixed)
Posterior scalene (with ribs fixed)
Longissimus capitis (assists, p. 202)
Longissimus cervicis (assists, p. 202)
Iliocostalis cervicis (assists, p. 202)

Mandible (p. 249)

(temporomandibular joint)

Elevation
Masseter
Temporalis
Medial pterygoid

Depression
Geniohyoid
Mylohyoid
Stylohyoid
Digastric (with hyoid bone fixed)
Platysma (assists)

Protraction
Lateral pterygoid (bilaterally)
Medial pterygoid (bilaterally)

Retraction
Temporalis
Digastric

Lateral Deviation (unilaterally)
Lateral pterygoid (to the opposite side)
Medial pterygoid (to the opposite side)

Pelvis

Anterior Tilt (downward rotation)
Latissimus dorsi (assists, p. 79)

Posterior Tilt (upward rotation)
Biceps femoris
Semitendinosus
Semimembranosus
Psoas minor

Lateral Tilt (elevation)
Quadratus lumborum
Latissimus dorsi (assists, p. 79)

Coxal (p. 296-298)
(hip joint)

Flexion
Rectus femoris
Gluteus medius (anterior fibers)
Gluteus minimus
Adductor magnus (assists)
Adductor longus (assists)
Adductor brevis (assists)
Pectineus (assists)
Tensor fasciae latae
Sartorius
Psoas major
Iliacus

Extension
Biceps femoris
Semitendinosus
Semimembranosus
Gluteus maximus (all fibers)
Gluteus medius (posterior fibers)
Adductor magnus (posterior fibers)

Medial Rotation (internal rotation)
Semitendinosus
Semimembranosus
Gluteus medius (anterior fibers)
Gluteus minimus
Adductor magnus
Adductor longus
Adductor brevis
Gracilis
Pectineus
Tensor fasciae latae

Lateral Rotation (external rotation)
Biceps femoris
Gluteus maximus (all fibers)
Gluteus medius (posterior fibers)
Sartorius
Piriformis
Quadratus femoris
Obturator internus
Obturator externus
Gemellus superior
Gemellus inferior
Psoas major
Iliacus

Abduction
Gluteus maximus (all fibers)
Gluteus medius (all fibers)
Gluteus minimus
Tensor fasciae latae
Sartorius
Piriformis (when the hip is flexed)

Adduction
Adductor magnus
Adductor longus
Adductor brevis
Pectineus
Gracilis
Psoas major
Iliacus
Gluteus maximus (lower fibers)

Knee (p. 299)
(tibiofemoral joint)

Flexion
Biceps femoris
Semitendinosus
Semimembranosus
Gracilis
Sartorius
Gastrocnemius
Popliteus
Plantaris (weak)

Extension
Rectus femoris
Vastus lateralis
Vastus medialis
Vastus intermedius

Medial Rotation of Flexed Knee
Semitendinosus
Semimembranosus
Gracilis
Sartorius
Popliteus

Lateral Rotation of Flexed Knee
Biceps femoris

Ankle (p. 362)
(talocrural joint)

Plantar Flexion
Gastrocnemius
Soleus
Tibialis posterior
Peroneus longus (assists)
Peroneus brevis (assists)
Flexor digitorum longus (weak)
Flexor hallucis longus (weak)
Plantaris (weak)

Dorsiflexion
Tibialis anterior
Extensor digitorum longus
Extensor hallucis longus

Foot and Toes (p. 363)
(talotarsal,
midtarsal,
tarsometatarsal,
metatarsophalangeal,
proximal and distal
interphalangeal joints)

Inversion
Tibialis anterior
Tibialis posterior
Flexor digitorum longus
Flexor hallucis longus
Extensor hallucis longus

Eversion
Peroneus longus
Peroneus brevis
Extensor digitorum longus

Flexion of Second through Fifth Toes
Flexor digitorum longus
Flexor digitorum brevis
Lumbricals
Quadratus plantae (assists)
Dorsal interossei (2nd - 4th)
Plantar interossei (3rd - 5th)
Abductor digiti minimi (5th)
Flexor digiti minimi (5th)

Extension of Second through Fifth Toes
Extensor digitorum longus
Extensor digitorum brevis (2nd - 4th)
Lumbricals

Adduction of Second through Fifth Toes
Plantar interossei (3rd - 5th)

Abduction of Second through Fifth Toes
Dorsal interossei (2nd - 4th)
Abductor digiti minimi (5th) (assists)

Flexion of First Toe
Flexor hallucis longus
Flexor hallucis brevis
Abductor hallucis (assists)

Extension of First Toe
Extensor hallucis longus
Extensor hallucis brevis

Adduction of First Toe
Adductor hallucis

Abduction of First Toe
Abductor hallucis

Glossary of Terms

abdomen - the region between the diaphragm and the pelvis

acetabulum - the rounded cavity on the external surface of the coxal bone; the head of the femur articulates with the acetabulum to form the coxal joint

adhesion - abnormal adherence of collagen fibers to surrounding structures during immobilization, following trauma or as a complication of surgery, which restricts normal elasticity of the structures involved

anatomical position - erect posture with face forward, arms at sides, forearms supinated (so that palms of the hands face forward) and fingers and thumbs in extension

antagonist - a muscle that performs the opposite action of the prime mover and synergist muscles

antecubital - the anterior side of the elbow

anterior - toward the front or ventral surface

anterior tilt of pelvis - tilt in which the vertical plane through the anterior superior iliac spines (ASISes) are anterior to the vertical plane through the symphysis pubis

appendage - a structure attached to the body such as the upper and lower extremities

arm - the portion of the upper limb between the shoulder and elbow joints

arthrology - the study of joints

articular facet - a small articular surface of a bone, especially a vertebra

articular process - a small flat projection found on the surfaces of the arches of the vertebrae on either side incorporating the articular surface

articulation - a joint or connection of bones

atlas first cervical vertebra, articulating with the occipital bone and rotating around the odontoid process of the axis

axis -the second cervical vertebra

bilateral - pertaining to two sides

bursa - a small, fluid-filled sack that reduces friction between two structures

cartilaginous joint - a joint in which two bony surfaces are united by cartilage; the two types of cartilaginous joints are **synchondroses** and **symphyses**

caudal - downward, away from the head (toward the tail)

cephalic - toward the head

collagen - the protein of connective tissue fibers

concentric contraction - a shortening of the muscle during a contraction; a type of isotonic exercise

condyle - a rounded articular surface at the extremity of a bone

connective tissue - the supportive tissues of the body, made of ground substance and fibrous tissues, taking a wide variety of forms

contraction - an increase in muscle tension, with or without change in overall length

coronal - a vertical plane perpendicular to the sagittal plane dividing the body into anterior and posterior portions, also called the frontal plane

coronal axis - a horizontal line extending from side to side, around which the movements of flexion and extension take place

cramp - a spasmodic contraction of one or many muscles

cranial - upward, toward the head

crepitation - an audible and/or palpable crunching during movement of tendons or ligaments over bone

cutaneous - referring to the skin

deep - away from the surface of the body; the opposite of superficial

distal - farther from the center or median line or from the thorax

dorsal - relating to the back; posterior

eccentric muscle contraction - an overall lengthening of the muscle while it is contracting or resisting a workload

edema - a local or generalized condition in which body tissues contain an excessive amount of fluid

facet - a small plane or concave surface

fascia - a general term for a layer or layers of loose or dense fibrous connective tissue

fascicle - a bundle of muscle fibers

fibrous joint - a joint in which the components are connected by fibrous tissue

flexibility - the ability to readily adapt to changes in position or alignment; may be expressed as normal, limited, or excessive

forearm - the portion of the upper limb between the elbow and wrist joints

frontal plane - a vertical plane perpendicular to the sagittal plane dividing the body into anterior and posterior portions, also called the coronal plane

genu valgum - "knock-knees," defined as a lateral displacement of the distal end of the distal bone in the joint

genu varum - "bowlegs," defined as a medial displacement of the distal end of the distal bone in the joint

impingement - an encroachment on the space occupied by soft tissue, such as nerve or muscle

inferior - away from the head

insertion - the more mobile attachment site of a muscle to a bone; the opposite end is the origin

interstitial - the space within an organ or tissue

interstitial fluid - the fluid that surrounds cells

isometric - increase in tension without change in muscle length

isotonic - increase in tension with change in muscle length (in the direction of shortening); concentric contraction

isotonic contraction (dynamic) - a concentric or eccentric contraction of a muscle; a muscle contraction performed with movement

kinesiology - the study of movement

kyphosis - a condition characterized by an abnormally increased convexity in the curvature of the thoracic spine as viewed from the side

lateral - away from the midline

lateral tilt - pelvic tilt in which the crest of the ilium is higher on one side than on the other

leg - the portion of the lower extremity between the knee and ankle joints

ligament - a fibrous connective tissue that connects bone to bone

longitudinal axis - a vertical line extending in a cranial/caudal direction about which movements of rotation take place

lordosis - an abnormally increased concavity in the curvature of the lumbar spine as viewed from the side

lymph node - a small oval structure located along lymphatic vessels

lymphatic - pertains to the system of vessels involved with drainage of bodily fluids (lymph)

medial - toward the midline

muscle - an organ composed of one of three types of muscle tissue (skeletal, cardiac or visceral), specialized for contraction

muscle contracture - an increase of tension in the muscle caused by activation of the contractile mechanism of the muscle

myofascial - pertains to skeletal muscles ensheathed by fibrous connective tissue

occipital condyles - elongated oval facets on the undersurface of the occipital bone on either side of the foramen magnum, which articulate with the atlas vertebra

odontoid process (or dens) - a process projecting upward from the body of the axis vertebra around which the atlas rotates

origin - the more stationary attachment site of a muscle to a bone; the opposite end is the insertion

palmar - toward the palm

palpable - touchable, accessible

palpate - to examine or explore by touching (an organ or area of the body), usually as a diagnostic aid

paravertebrals - alongside or near the vertebral column

pelvic girdle - the two hip bones

pelvic tilt - an anterior (forward), a posterior (backward) or a lateral (vertical) tilt of the pelvis from neutral position

pelvis - composed of the two hip bones, sacrum and coccyx

periosteum - the fibrous connective tissue which surrounds the surface of bones

posterior - toward the back or dorsal surface

posterior tilt of pelvis - tilt in which the vertical plane through the anterior superior iliac spines (ASISes) are posterior to the vertical plane through the symphysis pubis

prime mover - a muscle that carries out an action

process - a projection or outgrowth from a bone

proximal - nearer to the center or midline of the body

range of motion - the range, usually expressed in degrees, through which a joint can move or be moved

range of motion, active - the free movement across any joint of moving levers that is produced by contracting muscles

range of motion, passive - the free movement that is produced by external forces across any joint or moving levers

retinaculum - a network, usually pertaining to a band of connective tissue

sagittal axis - a horizontal line extending from front to back, about which movements of abduction and adduction take place

sagittal plane - a plane that divides the body into left and right portions

soft tissue - usually referring to myofascial tissues, or any tissues which do not contain minerals (such as bone)

superficial - nearer to the surface of the body; the opposite of deep

superior - toward the head

surface anatomy - the study of structures that can be identified from the outside of the body

symphysis - a union between two bones formed by fibrocartilage

synchondrosis - a union between two bones formed either by hyaline cartilage or fibrocartilage

synergist - a muscle that supports the prime mover

synovial joint - a joint containing a lubricating substance (synovial fluid) and lined with a synovial membrane or capsule

tactile - pertaining to touch

tendon - a fibrous tissue connecting skeletal muscle to bone

thigh - the portion of the lower extremity between the coxal and knee joints

thorax - the region between the neck and abdomen

tightness - shortness; denotes a slight to moderate decrease in muscle length; movement in the direction of lengthening the muscle is limited

transverse plane - a plane that divides the body into superior and inferior (or proximal and distal) portions

trunk - the part of the body to which the upper and lower extremities attach

unilateral - pertaining to one side

ventral - a synonym for anterior, usually applied to the torso

Pronunciation and Etymology

etymology **et**-i-**mol**-o-gee the science of the origin and development of a word

Term	Pronunciation	Etymology
ab- (as in abduct)		L. away from
abdomen	**ab**-do-men	L. belly
abdominis	ab-**dah**-min-is	
abduct	**ab**-duct	L. to lead away, bring apart
acetabulum	**as**-e-**tab**-u-lum	L. a little saucer for vinegar
acromioclavicular	a-**kro**-me-o-kla-**vik**-u-lar	
acromion	a-**cro**-me-on	Grk. *akron*, top + *omos*, shoulder
ad- (as in adduct)		L. toward
adduct	**ad**-duct	L. to bring together
adipose	**a**-di-**pose**	L. fat, copious
alar	**ay**-lar	
anconeus	an-**ko**-nee-us	Grk. elbow
annular	**an**-u-ler	L. ringlike
annulus	**an**-u-lus	L. ring
aponeurosis	**ap**-o-nu-**ro**-sis	Grk. *apo*, from + *neuron*, nerve or tendon
appendicular	**ap**-en-**dik**-u-lar	L. to hang to
arrector pili	**a**-rek-tor **pee**-li	L. *arrector*, lifter; *pilus*, hair
artery	**ar**-ter-e	Grk. windpipe
atlantoaxial	at-**lan**-to-**ak**-se-al	
atlantooccipital	at-**lan**-to-ok-**si**-pi-tal	
axial	**ak**-see-al	L. axle
axilla	**ak**-sil-a	L. armpit
axillary	**ak**-si-**lar**-ee	
basilic	bah-**sil**-ic	Arabic *basilik*, inner
biceps	**bi**-seps	L. *bis*, twice + *caput*, head
biceps brachii	**bi**-seps **bray**-key-i	L. two-headed muscle of the arm
biceps femoris	**bi**-seps fe-**mor**-is	Grk. the two-headed (muscle) of the thigh
brachial	**bray**-key-al	L. relating to the arm
brachialis	**bray**-key-**al**-is	
brachii	**bray**-key-i	L. of the arm
brachioradialis	**bray**-key-o-**ra**-de-**a**-lis	
brevis	**breh**-vis	L. short
bursa	**bur**-sah	L. a purse
calcaneocuboid	kal-**ka**-ne-o-**ku**-boyd	
calcaneofibular	kal-**ka**-ne-o-**fib**-u-lar	
calcaneus	kal-**kay**-nee-us	L. heel
capillary	**kap**-i-**lar**-ee	L. hairlike
capitate	**kap**-i-tate	L. head-shaped
capitis	**kap**-i-tis	L. of the head
capitulum	ka-**pit**-u-lum	L. small head
carotid	ka-**rot**-id	Grk. causing deep sleep
carpal	**kar**-pul	Grk. pertaining to the wrist
carpi	**kar**-pi	L. of the wrist
cartilage	**kar**-ti-lij	L. gristle
cauda equina	**kaw**-da eh-**kwy**-na	L. horse's tail
cephalic	se-**fa**-lic	Grk. pertaining to the head
cervical	**ser**-vi-kal	L. referring to the neck
cervicis	**ser**-vi-sis	L. neck
chest		AS. box
cisterna chyli	sis-**turn**-a **ki**-lee	
clavicle	**klav**-i-k'l	L. little key
coccyx	**kok**-siks	Grk. cuckoo
collateral	ko-**lat**-er-al	L. of both sides
condyle	**kon**-dial	Grk. knuckle
conoid	**ko**-noid	Grk. cone-shaped
coracoacromial	**kor**-a-**ko**-a-**cro**-mi-ul	
coracobrachialis	**kor**-a-ko-**bra**-kee-**al**-is	
coracoclavicular	**kor**-a-**ko**-cla-**vic**-u-lar	
coracoid	**kor**-a-koyd	Grk. raven's beak
coronal	ko-**ro**-nal	L. crownlike
coronoid	**kor**-a-noyd	Grk. crown-shaped
costal	**kos**-tal	L. rib
coxal	**kox**-sal	L. hip
cranio-	**cra**-nee-o	Grk. skull
cranium	**cra**-nium	Grk. skull
cremaster	kre-**mas**-ter	L. to suspend
cricoid	**kri**-koyd	Grk. ring-shaped
cruciate	**kroo**-she-at	L. cross-shaped
cuboid	**ku**-boyd	Grk. cube-shaped
cuboideonavicular	**ku**-boyd-e-o-na-**vik**-u-lar	
cuneiform	ku-**ne**-i-form	L. wedge-shaped
deltoid	**del**-toid	Grk. *delta*, capital letter D (Δ) in the Greek alphabet
diaphragm	**di**-a-**fram**	Grk. a partition, wall
digastric	di-**gas**-trik	Grk. double-bellied
digit	**di**-jit	L. finger
digitigrade	**di**-ji-tah-grade	L. toe-walking
dorsalis pedis	**dor**-sal-is **peh**-dis	L. *dorsum*, back; *pedis*, foot

Grk.	Greek	Fr.	French
L.	Latin	ME.	Middle English
AS.	Anglo-Saxon		

dorsi	**dor**-si	L. of the back
dura mater	**dyoo**-ra **ma**-ter	L. tough mother
epi-	**eh**-pee	Grk. above, upon
facet	**fac**-et	Fr. small face
facial	**fa**-shal	L. pertaining to the face
fascia	**fash**-ah	L. a band, bandage
fasciae	**fash**-ay	plural for fascia
fascicle	**fas**-i-kl	L. little bundle
femur	**fee**-mur	L. thigh
fibula	**fib**-u-la	L. pin or buckle
flavum	**flay**-vum	
flex		L. to bend
foot		AS. *fot*
foramen	for-**ay**-men	L. a passage or opening
fossa	**fos**-a	L. a shallow depression
furcula	**fur**-ku-la	L. a little fork
gastrocnemius	**gas**-trok-**ne**-me-us	Grk. *gaster*, stomach + *kneme*, leg
gemellus	jem-**el**-us	L. twins
geniohyoid	**je**-ne-o-**hi**-oyd	Grk. *genion*, chin
genu valgum	**je**-noo **val**-gum	
genu varum	**je**-noo **va**-rum	
gland		L. acorn
glenoid	**glen**-oid	Grk. eyeball
glossus	**glah**-sis	Grk. tongue
gluteus	**gloo**-te-us	Grk. *gloutos*, buttocks
gracilis	gra-**cil**-is	L. slender, graceful
hallucis	**hal**-ah-sis	
hallux	**hal**-uks	L. first toe
ham		AS. haunch
hamate	**ham**-ate	L. hooked
hamulus	**ham**-u-lus	L. a small hook
humerus	**hu**-mer-us	L. upper arm
hyoid	**hi**-oyd	Grk. U- shaped
hypothenar	**hi**-po-**thee**-nar	Grk. *hypo*, under or below
iliacus	i-**lee**-a-cus	L. pertaining to the loin
iliocostalis	**il**-ee-o-kos-**ta**-lis	L. from hip to rib
ilium	**il**-ee-um	L. groin, flank
indicis	**in**-di-kis	
inferior	in-**fe**-ree-or	L. below
infraspinatus	**in**-fra-spi-**na**-tus	
inguinal	**ing**-gwi-nal	L. of the groin
interdigitate	**in**-ter-**dij**-i-tate	L. to interlock, as fingers of clasped hands
interroseus	**in**-ter-**ah**-see-us	L. between bones

interspinalis	**in**-ter-spi-**na**-lis	
interstitial	**in**-ter-**stish**-al	L. placed between
intertransverserii	**in**-ter-trans-**verse**-er-**i**	
intertubercular	**in**-tur-tu-**ber**-ku-lar	
ischiocavernosus	**ish**-she-o-**ka**-ver-**no**-sus	
ischium	**ish**-ee-um	Grk. hip
jaw		ME. iawe
joint		L. to join
jugular	**jug**-u-lar	L. throat
kyphosis	ki-**fo**-sis	Grk. bent, curved, or stooped
labrum	**lay**-brum	L. lip
lamina	**lam**-i-na	L. thin plate, leaf
latae	**la**-ta	L. broad
lateral	**lat**-er-al	L. to the side
latissimus dorsi	la-**tis**-i-mus **dor**-si	L. widest of the back
levator	leh-**va**-tor	L. lifter
levator scapula	leh-**va**-tor **skap**-u-la	
ligament	**lig**-a-ment	L. a band
linea aspera	**lin**-e-a **as**-per-a	L. rough line
longissimus	lon-**jis**-i-mus	L. longest
longus colli	**long**-us **ko**-li	L. long (muscle) of the neck
lordosis	lor-**doh**-sis	Grk. bent backward
lumbar	**lum**-bar	L. loin
lumborum	lum-**bor**-um	
lumbrical	**lum**-bri-kal	L. earthworm
lunate	**lu**-nate	L. crescent-shaped
lymph	limf	L. pure spring water
magnus	**mag**-nus	L. large
malleolus	**mal**-e-**o**-lus	L. little hammer
mandible	**man**-di-ble	L. lower jawbone
manubrium	ma-**nu**-bree-um	L. handle
masseter	**mas**-se-ter	Grk. chewer
mastoid	**mas**-toyd	Grk. breast-shaped
maxilla	**max**-il-a	L. jawbone
medial	**me**-dee-ul	L. middle
menisci	men-**is**-ki	plural for meniscus
meniscofemoral	men-**is**-ko-fem-**or**-al	
meniscus	men-**is**-kus	Grk. crescent-shaped
mentalis	men-**tal**-is	L. chin
meta-	**met**-a	Grk. after or beyond
metacarpal	**met**-a-**kar**-pul	
metacarpophalangeal	**met**-a-**kar**-po-**fa**-lan-**jee**-al	
metatarsal	**met**-a-**tar**-sal	

Term	Pronunciation	Origin
metatarsophalangeal	**met**-a-**tar**-so-**fa**-lan-**jee**-al	
minimi	**min**-i-mee	L. smallest
multifidi	mul-**tif**-i-di	L. *fidi*, to split
muscle	**mus**-el	L. *musculus*, a little mouse
mylohyoid	**my**-lo-**hi**-oyd	Grk. *myle*, mill
myo-		Grk. muscle
nape		ME. back of the neck
nasal	**na**-zl	L. nose
navicular	na-**vik**-u-lar	L. boat-shaped
neck		AS. nape
nerve		L. sinew
nuchae	**nu**-kay	L. nape of neck
nuchal	**nu**-kal	L. back of the neck
oblique	o-**bleek**	L. diagonal, slanted
obturator	**ob**-tu-**ra**-tor	L. obstructor
occipitofrontalis	ok-**sip**-i-to-fron-**ta**-lis	
occiput	**ok**-si-put	L. the back of skull
odontoid	o-**don**-toyd	Grk. toothlike
olecranon	o-**lek**-ran-on	Grk. elbow
omohyoid	**o**-mo-**hi**-oyd	Grk. *omos*, shoulder
opponens	o-**po**-nens	L. opposing
palpate	**pal**-pate	L. *palpare*, to touch
panniculus carnosus	pan-**ik**-u-lus car-**no**-sis	L. small, fleshy garment
parietal	puh-**ri**'e-tul	L. wall
parotid	pa-**rot**-id	Grk. beside the ear
patella	pa-**tel**-a	L. small pan
pectineus	pek-**tin**-e-us	L. comblike
pectoralis	**pek**-to-**ra**-lis	L. chest
pedicle	**ped**-i-k'l	L. a little foot
pelvis	**pel**-vis	L. basin
penis		L. tail
peroneus	per-**o**-ne-us	Grk. pin, buckle
pes anserinus	pes **an**-ser-**i**-nus	L. *pedis*, foot; L. *anserinus*, gooselike
phalange	fa-**lan**-jee	Grk. closely knit row, line of battle
phalanx	**fal**-anks	singular for phalange
piriformis	**pir**-i-**form**-is	L. pear-shaped
pisiform	**pi**-si-form	L. pea-shaped
plantar	**plan**-tar	L. the sole of the foot
plantaris	plan-**tar**-is	Fr. pertaining to the sole of the foot
plantigrade	**plant**-i-grad	L. sole-walking
platysma	pla-**tiz**-ma	Grk. plate
plexus	**plek**-sus	L. interwoven
pollex	**pol**-eks	L. thumb
pollicis	**pol**-li-sis	L. thumb
popliteus	pop-**lit**-e-us	L. ham of the knee
process	**pros**-es	L. going forth
profundus	pro-**fun**-dus	L. deep
pronate	**pro**-nate	L. bent forward
psoas	**so**-as	Grk. muscle of the loin
pterygoid	ter-i-**goyd**	Grk. wing-shaped
pubis	**pu**-bis	NL. bone of the groin
quadratus	**kwod**-rait-us	L. squared, four-sided
quadratus lumborum	**kwod**-rait-us lum-**bor**-um	L. four-sided muscle of the lumbar region
quadriceps	**kwod**-ri-seps	L. four-headed
quadruped		Grk. four-footed
radiocapitate	**ray**-dee-o-**kap**-i-tate	
radioscapholunate	**ray**-dee-o-**skaf**-o-**loo**-nate	
radiotriquetrum	**ray**-dee-o-tri-**kwe**-trum	
radius	**ray**-dee-us	L. staff, spoke of a wheel
ramus	**ray**-mus	L. branch
rectus	**rek**-tus	L. straight
retinacula	**ret**-i-**nak**-u-la	plural for retinaculum
retinaculum	**ret**-i-**nak**-u-lum	L. halter, band, rope
retinacula	**ret**-i-**nak**-u-la	plural for retinaculum
rhomboid	**rom**-boyd	Grk. geometry, a parallelogram with oblique angles and only the opposite sides equal
rotatores	**ro**-ta-**tor**-ays	L. plural for rotators
sacrococcygeal	**sa**-kro-kok-**sij**-e-al	
sacrotuberous	**sa**-kro-**tu**-ber-us	
sacrum	**sa**-krum	L. sacred or holy thing, from the use of the sacrum in Roman animal sacrifice
sagittal	**saj**-i-tal	L. arrowlike
saphenous	**sa**-fe-nus	Grk. *saphen*, clearly visible
sartorius	sar-**tor**-ee-us	L. tailor
scalene	**skay**-leen	Grk. uneven
scaphoid	**skaf**-oyd	L. boat-shaped
scapula	**skap**-u-la	L. shoulder, blade
scapulae	**skap**-u-lay	plural for scapula
sciatic	si-**at**-ik	Grk. *ischion*, hip joint
sciatica	si-**at**-ika	L. suffering in the hip
semimembranosus	**sem**-eye-**mem**-bra-**no**-sus	L. half membranous

semispinalis	**sem**-eye-spi-**na**-lis	L. half spinal
semitendinosus	**sem**-eye-**ten**-di-**no**-sus	L. half tendinous
septa	**sep**-ta	plural for septum
septum	**sep**-tum	L. enclosure
serratus	ser-**a**-tus	L. notched
sesamoid	**ses**-a-moyd	L. resembling a sesame seed
skeleton	**skel**-et-on	Grk. dried up
skull		ME. bow
soleus	so-**lay**-us	L. *solea*, as in a sole fish
sphenoid	**sfe**-noyd	Grk. wedge-shaped
spinalis capitis	**spi**-na-lis **kap**-i-tis	
spinalis cervicis	**spi**-na-lis **ser**-vi-sis	
spine		L. thorn
splenius	**sple**-nee-us	Grk. bandage
splenius capitis	**sple**-nee-us **kap**-i-tis	L. bandage-like (muscle) of the head
splenius cervicis	**sple**-nee-us **ser**-vi-sis	
stapedius	sta-**pe**-de-us	L. stirrup
sternoclavicular	**ster**-no-kla-**vik**-u-lar	
sternocleidomastoid	**ster**-no-**kli**-do-**mas**-toyd	
sternohyoid	**ster**-no-**hi**-oyd	
sternothyroid	**ster**-no-**thi**-royd	
sternum	**ster**-num	Grk. chest
stylohyoid	**sti**-lo-**hi**-oyd	
styloid	**sti**-loyd	Grk. a pillar
subclavius	sub-**klay**-vee-us	
subscapularis	sub-**skap**-u-**lar**-is	
superficialis	**soo**-per-**fish**-ee-**a**-lis	L. on the surface
supinate	**su**-pi-nate	L. bent backward
supraspinatus	**soo**-pra-spi-**na**-tus	
sustentaculum	**sus**-ten-**tak**-u-lum	L. support
suture	**su**-chur	L. a seam
symphysis	**sim**-fi-sis	Grk. growing together
synchondrosis	**sin**-con-**dro**-sis	
synovial	sin-**o**-ve-al	L. *synovia*, joint fluid
talocalcaneal	**ta**-lo-kal-**ka**-ne-al	
talocrural	**ta**-lo-**kroo**-ral	L. ankle + *crus*, leg
talofibular	**ta**-lo-**fib**-u-lar	
talonavicular	**ta**-lo-na-**vik**-u-lar	
talus	**ta**-lus	L. ankle
tarsal	**tar**-sul	Grk. wicker basket
temporalis	**tem**-po-**ra**-lis	L. time, seen by the graying of hairs in this region
tendon	**ten**-dun	L. to stretch
tensor	**ten**-sor	L. a stretcher
teres	**teh**-reez	L. rounded, finely shaped
tertius	**ter**-she-us	L. third
thenar	**thee**-nar	Grk. palm, flat of the hand
thoracic	tho-**ras**-ik	Grk. chest
thoracolumbar	tho-**rak**-o-**lum**-bar	
thorax	**tho**-raks	Grk. chest
thyrohyoid	**thi**-ro-**hi**-oyd	
thyroid	**thi**-royd	Grk. shield
tibia	**tib**-e-a	L. shinbone
trachea	**tray**-ke-a	Grk. rough
tract		L. extent, drawn out
transverse	**trans**-verse	L. across, turned across
trapezium	tra-**pee**-ze-um	Grk. little table
trapezius	tra-**pee**-ze-us	Grk. a little table or trapezoid shape
trapezoid	**trap**-e-zoid	Grk. table-shaped
triceps brachii	**tri**-seps **bray**-key-i	L. three-headed muscle of the arm
triceps surae	**tri**-seps **sir**-eye	L. three-headed muscle of the calf
triquetrum	tri-**kwe**-trum	L. three-cornered
trochanter	tro-**kan**-ter	Grk. to run
trochlea	**trok**-lee-ah	Grk. pulley
tubercle	**tu**-ber-kl	L. a little swelling
tuberosity	tu-ber-**os**-i-tee	L. a swelling
ulna	**ul**-na	L. elbow, arm
ulnolunate	**ul**-no-**lu**-nate	
ulnotriquetrum	**ul**-no-tri-**kwe**-trum	
umbilicus	um-**bil**-i-kus	L. navel, center
uvula	**uv**-u-la	L. a little grape
vastus	**vas**-tus	L. vast
vein		L. vessel
vertebra	**ver**-ta-bra	L. joint
xiphoid	**zif**-oyd	Grk. sword-shaped
zona orbicularis	**zo**-na or-**bik**-u-lar-is	L. girdle + little circle
zygomatic	**zy**-go-**mat**-ik	Grk. cheekbone
zygapophyseal	**zy**-gah-**pof**-i-se-al	

Alexander, R. McNeill, *The Human Machine*, Columbia University Press, New York, 1992

Anson, Barry, *An Atlas of Human Anatomy*, W.B. Saunders, Philadelphia, 1963

Asimov, Isaac, *The Human Body*, Houghton Mifflin Co., Boston, 1963

Backhouse, Kenneth and Hutchings, Ralph, *Color Atlas of Surface Anatomy*, Williams & Wilkins, Baltimore, 1986

Bates, Barbara, *A Guide to Physical Examination and History Taking*, 4th ed., J. B. Lippincott, Philadelphia, 1987

Bergman, Ronald; Thompson, Sue Ann and Afifi, Adel K., *Catalog of Human Variation*, Urban and Schwarzenberg, Baltimore, 1984

Bodanis, David, *The Body Book*, Little, Brown and Company, Boston, 1984

Calais-Germain, Blandine, *Anatomy of Movement*, Eastland Press, Seattle, 1993

Cartmill, Hylander and Shafland, *Human Structure*, Harvard University Press, Cambridge, 1987

Chaitow, Leon, *Palpatory Literacy*, Thorsons, London, 1991

Chaitow, Leon, *Palpatory Skills*, Churchill Livingstone, New York, 1997

Clemente, Carmine, *Anatomy: A Regional Atlas of the Human Body*, 3rd edition, Urban & Schwarzenberg, Baltimore, 1987

Clemente, Carmine, *Gray's Anatomy*, 30th edition, Lea & Febiger, Philadelphia, 1985

Craig, Marjorie, *Miss Craig's Face Saving Exercises*, Random House, New York, 1970

Cyriax, J.H. and Cyriax, P.J., *Cyriax's Illustrated Manual of Orthopaedic Medicine*, 2nd ed., Butterworth/Heinemann Ltd., Oxford, 1992

Dorland's Illustrated Medical Dictionary, 24th edition, W.B. Saunders, Philadelphia, 1965

Eaton, Theodore Jr., *Comparative Anatomy of the Vertebrates*, 2nd edition, Harper and Brothers, 1971

Feher, Gyorgy and Szunyoghy, Andras, *Cyclopedia Anatomicae*, Black Dog & Leventhal Publishers, New York, 1996

Field, E. J., *Anatomical Terms: Their Origin and Derivation*, W. Heffer & Sons, Cambridge, UK, 1947

Gebo D., *Plantigrady and foot adaptation in African apes: implications for hominid origins*, American Journal of Physical Anthropology 89: 29-58, 1992

Gehin, Alain, *Atlas of Manipulative Techniques for the Cranium and Face*, Eastland Press, Seattle, 1985

Greene, Lauriann, *Save Your Hands! Injury Prevention for Massage Therapists*, Infinity Press, Seattle, 1995

Gross, Fetto, and Rosen, *Musculoskeletal Examination*, Blackwell Sciences, Malden, 1996

Guillen, Michael, *Five Equations That Changed the World*, Hyperion, New York, 1995

Hamrick, M.W. and Inouye, S.E., *"Thumbs, tools, and early humans,"* Science, p. 586-7, April 1994

Handy, Chester, *A History of Cranial Osteopathy*, Journal of American Osteopathic Association, vol. 47, pp. 269-272, January 1948

Hertling, Darlene and Kessler, Randolph M., *Management of Common Musculoskeletal Disorders*, 3rd ed., JB Lippincott, Philadephia, 1996

Hildebrand, Milton, *Analysis of Vertebrate Structure*, 4th ed., John Wiley & Sons, New York, 1995

Hole, John, *Essentials of Human Anatomy and Physiology*, 4th edition, Wm. C. Brown, Dubuque, 1992

Hoppenfeld, Stanley, *Physical Examination of the Spine and Extremities*, Appleton & Lange, Norwalk, 1976

Jamieson, E. B., *Illustrations of Regional Anatomy, Sections I - VII*, E.S. Livingstone, Edinburgh, 1946

Jenkins, David, *Hollinshead's Functional Anatomy of the Limbs and Back*, 6th ed., W.B. Saunders, Philadelphia, 1991

Juhan, Deane, *Job's Body: A Handbook for Bodywork*, Station Hill, Barrytown, New York, 1987

Kapandji, I. A., *The Physiology of the Joints, Volumes 1, 2 & 3*, 5th ed., Churchill Livingstone, New York, 1982

Kapit, Wynn and Elson, Lawrence, *The Anatomy Coloring Book*, 2nd edition, HarperCollins College Publishers, 1993

Kendall, F.P., McCreary E.K., Provance P.G., *Muscles: Testing and Function*, 4th edition, Williams & Wilkins, Baltimore, 1993

Kent, George, *Comparative Anatomy of the Vertebrates*, 6th edition, Mosby, St. Louis, 1987

Koch, Tankred, *Anatomy of the Chicken and Domestic Birds*, Iowa State University Press, Ames, Iowa, 1973

Lumley, John, *Surface Anatomy*, Churchill Livingstone, Edinburgh, 1990

Luttgens, Kathryn and Wells, Katharine, *Kinesiology: Scientific Basis of Human Motion*, Saunders College Publishing, Philadelphia, 1982

MacClintock, Dorcas, *A Natural History of Giraffes*, Charles Scribner's Sons, New York, 1973

Magee, David, *Orthopedic Physical Assessment*, 2nd edition, W.B. Saunders, Philadelphia, 1992

Marzke, MW, *Evolutionary development of the human thumb*, Hand Clinics, p. 1-9, Feb 1992

McAleer, Neil, *The Body Almanac*, 1st ed., Doubleday & Co., New York, 1985

McMinn, R.M.H., Hutchings, R.T., *Color Atlas of Human Anatomy*, Year Book Medical Publishers, Chicago, 1985

Melloni, John, *Melloni's Illustrated Dictionary of the Musculoskeletal System*, Parthenon Publishing, New York, 1998

Moore, Keith, *Clinically Oriented Anatomy*, 3rd ed., Williams and Wilkins, Baltimore, 1992

Montagna, William, *Comparative Anatomy*, John Wiley and Sons, 1970

Myers, Thomas, *Anatomy Trains: Myofascial Meridians for Manual and Movement Therapists*, Churchill Livingstone, 2001

Napier, John, *Hands*, Princeton Science Library, Princeton, 1993

Netter, Frank, *Atlas of Human Anatomy*, CIBA-GEIGY, Summit, New Jersey, 1989

Neumann, Donald, Kinesiology of the Musculoskeletal System, Mosby, St. Louis, 2002

Norkin, Cynthia and Levangie, Pamela, *Joint Structure and Function*, 2nd ed., F.A. Davis, Philadelphia, 1992

Olsen, Andrea, *Bodystories: A Guide to Experiential Anatomy*, Station Hill Press, Barrytown, New York, 1991

Olsen, Todd, *A.D.A.M: Student Atlas of Anatomy*, Williams and Wilkins, Baltimore, 1996

Parker, Steve, *Natural World*, Dorling Kindersley, London, 1994

Peck, Stephen Rogers, *Atlas of Human Anatomy*, Oxford University Press, Oxford, 1982

Platzer, Werner, *Color Atlas and Textbook of Human Anatomy, Volume 1: Locomotor System*, Thieme Inc., New York, 3rd ed., 1986

Rohen, Johannes and Yokochi, Chihiro, *Color Atlas of Anatomy*, 3rd ed., Igaku-Shoin Publishers, New York, 1993

Rolf, Ida, *Rolfing and Physical Reality*, Healing Arts Press, Rochester, Vermont, 1990

Rolf, Ida, *Rolfing: Integration of Human Structures*, Harper Row, New York, 1977

Rossi, William, *Shoes and the "Normal" Foot*, Podiatry Management, February, 1997

Schider, Fritz, *An Atlas of Anatomy for Artists*, 3rd ed., Dover Publishing, New York, 1957

Schultz, R. Louis and Feitis, Rosemary, *The Endless Web - Fascial Anatomy and Physical Reality*, North Atlantic Books, Berkeley, 1996

Searfoss, Glenn, *Skulls and Bones*, Stackpole Books, Mechanicsburg, Pennsylvania, 1995

Seig, Kay and Adams, Sandra, *Illustrated Essentials of Musculoskeletal Anatomy*, 2nd ed., Megabooks, Gainesville, 1993

Stern, Jack, *Core Concepts in Anatomy*, Little, Brown and Company, Boston, 1997

Stern, Jack, *Essentials of Gross Anatomy*, F.A. Davis, Philadelphia, 1988

Stone, Robert and Stone, Judith, *Atlas of the Skeletal Muscles*, Wm. C. Brown, Dubuque, 1990

Sutcliffe, Jenny and Duin, Nancy, *A History of Medicine*, Barnes and Noble, New York, 1992

Taber's Cyclopedic Medical Dictionary, 17th ed., F.A. Davis, Philadelphia, 1993

Thompson, Clem, *Manual of Structual Kinesiology*, 11th edition, Times Mirror/Mosby College, St. Louis, 1989

Thompson, Diana, *Hands Heal: Documentation for Massage Therapy, 2nd ed.*, Lippincott Williams & Wilkins, 2000

Todd, Mabel Elsworth, *The Thinking Body*, Dance Horizons, Brooklyn, 1979

Tortora, Gerald, *Principles of Human Anatomy*, 5th edition, Harper & Row, New York, 1989

Traupman, John, *New College Latin and English Dictionary*, Bantam Books, New York, 1995

Travell, Janet and Simons, David, *Myofascial Pain and Dysfunction: Trigger Point Manual, Volume 1*, Williams and Wilkins, Baltimore, 1983

Travell, Janet and Simons, David, *Myofascial Pain and Dysfunction: Trigger Point Manual, Volume 2*, Williams and Wilkins, Baltimore, 1992

Upledger, John and Vredevoogd, Jon, *Craniosacral Therapy*, Eastland Press, Seattle, 1983

Walker, Judith, *NeuroMuscular Therapy I - IV*, International Academy of NMT, St. Petersburg, 1994

Walker, Warren, *A Study of the Cat in Reference to the Human*, 5th ed., Saunders College Publishers, Fort Worth, 1993

Walker, Warren, *Functional Anatomy of the Vertebrates: An Evolutionary Perspective*, Saunders College Publishers, Fort Worth, 1987

Way, Robert, *Dog Anatomy - Illustrated*, Dreenan Press, Ltd., New York, 1974

Zihlman, Adrienne, *Human Evolution Coloring Book*, Harper & Row, New York, 1982

Index

Structures are grouped together by type. See the following headings to find individual structures. (For example, to find palmar aponeurosis look under **Aponeurosis**.*)*

Aponeurosis	**Movements of the Body**
Artery	**Muscle (terminology)**
Bones	**Muscles (listing of muscles)**
Bursa	**Nerve**
Comparative Anatomy	**Palpation**
Fascia	**Retinaculum**
Joint	**Synergists**
Ligament	**Systems of the Body**

A

B

C

D

E

F

M

S

T

U

V

W

Z

meow!

NOTES

NOTES

Andrew Biel is a licensed massage therapist. He has served on the faculties of Boulder College of Massage Therapy and Ashmead College and has taught Cadaver Studies for Bodyworkers at Bastyr Naturopathic University. He lives outside of Lyons, Colorado with his wife, Lyn Gregory.

Robin Dorn is an artist, illustrator and licensed massage practitioner. She specializes in bodywork illustration and exhibits on the West Coast and in France.

Bones

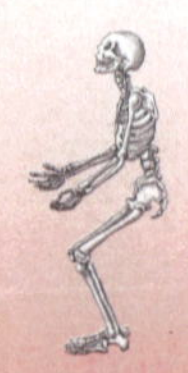